Sedimentation and Tectonics of Western North America

Volume 2

Accreted Terranes of the North Cascades Range, Washington

Geologic Evolution of the Northernmost Coast Ranges and Western Klamath Mountains, California

The San Andreas Transform Belt

Tectonic Evolution of Northern California

Early Mesozoic Tectonics of the Western Great Basin, Nevada

Tectonics of the Eastern Part of the Cordilleran Orogenic Belt, Chihuahua, New Mexico and Arizona

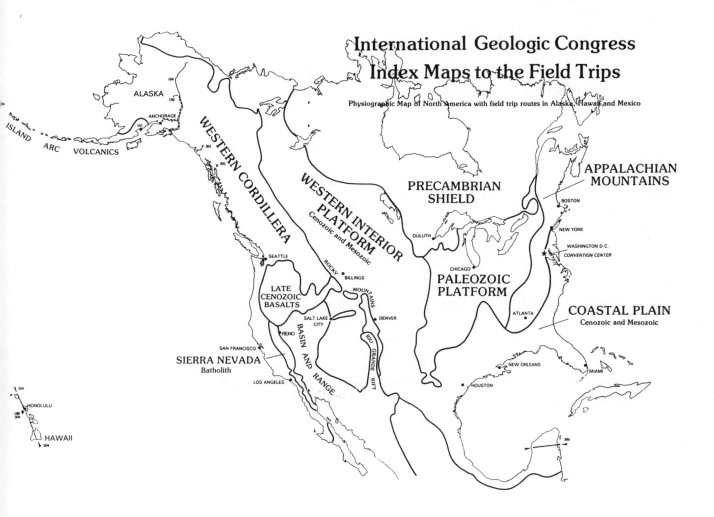

International Geologic Congress
Index Maps to the Field Trips

Physiographic Map of North America with field trip routes in Alaska, Hawaii and Mexico

ISLAND ARC VOLCANICS

ALASKA

ANCHORAGE

WESTERN CORDILLERA

WESTERN INTERIOR PLATFORM
Cenozoic and Mesozoic

PRECAMBRIAN SHIELD

APPALACHIAN MOUNTAINS

BOSTON

NEW YORK

WASHINGTON D.C.
CONVENTION CENTER

SEATTLE

DULUTH

CHICAGO

PALEOZOIC PLATFORM

LATE CENOZOIC BASALTS

ROCKY

BILLINGS

MOUNTAINS

SALT LAKE CITY

DENVER

ATLANTA

COASTAL PLAIN
Cenozoic and Mesozoic

RENO

BASIN AND RANGE

RIO GRANDE RIFT

NEW ORLEANS

MIAMI

SAN FRANCISCO

SIERRA NEVADA
Batholith

LOS ANGELES

HOUSTON

HONOLULU

HAWAII

Legend

Field Trip Maps

Cities

175 329   Field trip numbers (refer to text)

Specific routes; line type for differentiating
field trip routes; limited to area of
geologic interest

Field trip routes which extend over
a broad geographic area

Field trip routes which focus around
a single area

PENELOPE M. HANSHAW, *Field Trip Series Editor*

Sedimentation and Tectonics of Western North America

Volume 2

American Geophysical Union, Washington, D.C.

Library of Congress Cataloging-in-Publication Data

Sedimentation and tectonics of western North America.
    p. cm.
    Publications for the 28th International Geological Congress. IGC,
    held in Washington, D.C., July 1989.
    ISBN 0-87590-675-3
    1. Geology, Structural—Guide-books.   2. Geology—West (U.S.)—
    -Guide-books.   3. Sedimentation and deposition—West (U.S.)—Guide
    -books.   4. West (U.S.)—Description and travel—1981- —Guide-
    books.   I. International Geological Congress (28th : 1989 :
    Washington, D.C.)
    QE601.S335   1989   89-7001
    557.8—dc20   CIP

TABLE OF CONTENTS

PREFACE

Virtually every geologic province in the United States is encountered in these combined volumes of the field trip guides from the 28th International Geological Congress. Some guides extend into adjoining parts of Canada, Mexico, the Bahamas and the Caribbean. One provides an extensive view of the Scotia Arc, Antarctica. The length of trips and their complexity reflect the great variety of terranes of the North American continent.

These volumes combine trips with similar geography and geology. The main divisions are geographical. Within them an attempt has been made thematically to group various geologic topics.

Field trips for the 28th International Geological Congress (IGC) were conceived and developed with the cooperation of all segments of the Earth Science community in the United States. The Field Trip Committee solicited proposals for trips from all academic institutions, from State and Federal geological surveys, and through a number of widely circulated professional journals. About 250 proposals were received, of which the Committee selected 188 for listing in the IGC First Circular; 100,000 copies of this Circular were distributed in January 1987. Responses indicated that about 140 of these trips were of sufficient interest to go ahead with planning and preparation of guidebooks. Of these, 126 have been published in these volumes and as separate guidebooks.

We feel these guidebooks represent a valuable compendium of the current status of knowledge of the regional geology of the United States. We hope that you, the reader, will use and enjoy these volumes. We believe that publication of the guidebooks is a fitting tribute to the leaders who gave so generously of their time and talents in planning trips and preparing guidebooks. It has been our pleasure to work with them.

For the 28th International Geological Congress

John C. Reed, Jr., *Vice President, Field Trips*
Juergen Reinhardt, *Co-Chairman, Field Trip Committee*
Penelope M. Hanshaw, *Co-Chairman, Field Trip Committee*
Robert M. Mixon, *Chairman, Local Field Trips*
Tom Freeman
Robert N. Ginsburg
A. R. Palmer
J. Keith Rigby
John Rodgers
P. K. Sims

Accreted Terranes of the North Cascades Range, Washington

Spokane to Seattle, Washington
July 21–29, 1989

Field Trip Guidebook T307

Leaders:
Rowland W. Tabor   Ralph H. Haugerud   Edwin H. Brown
R. Scott Babcock   Robert B. Miller

American Geophysical Union, Washington D.C.

COVER Skagit orthogneiss exposed in Northern Picket Range. View
SW from Luna Peak.

IGC FIELD TRIP T307:
ACCRETED TERRANES OF THE NORTH CASCADES RANGE, WASHINGTON

R. W. Tabor[1], R. A. Haugerud[1], R. B. Miller[2],
E. H. Brown[3], and R. S. Babcock[3]

TABLE OF CONTENTS

[1]U.S. Geological Survey, Menlo Park, California
[2]Department of Geology, San Jose State University, San Jose, California
[3]Department of Geology, Western Washington University, Bellingham, Washington

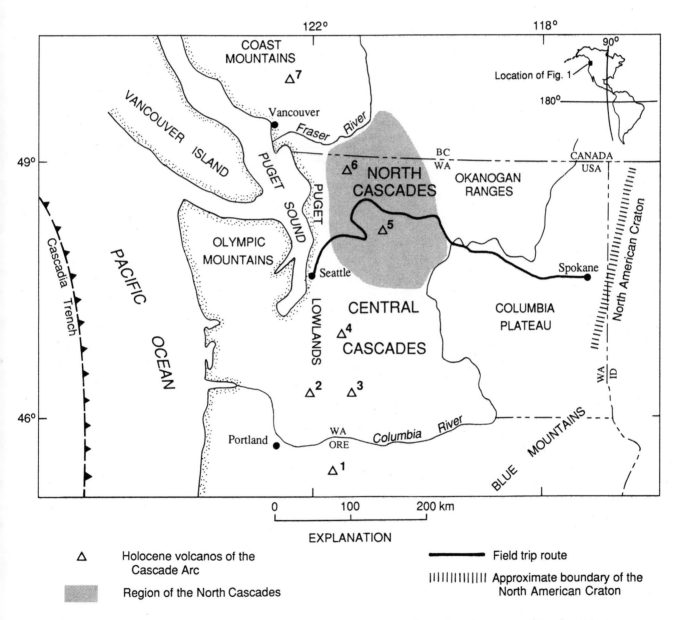

FIGURE 1. General location of the North Cascades Range and route of the field trip. Volcanoes of the Cascade Arc include: 1-Mt Hood, 2-Mt St Helens, 3-Mt Adams, 4-Mt Rainier, 5-Glacier Peak, 6-Mt Baker, and 7-Mt Garibaldi.

Leaders:

Rowland W. Tabor and Ralph Haugerud
U.S. Geological Survey, MS 975
345 Middlefield Road
Menlo Park, CA 94025

Edwin H. Brown and R. Scott Babcock
Department of Geology
Western Washington University
Bellingham, WA 98225

Robert B. Miller
Department of Geology
San Jose State University
San Jose, CA 95192-0102

OVERVIEW OF THE GEOLOGY OF THE NORTH CASCADES

R. W. Tabor, R. A. Haugerud, and R. B. Miller

INTRODUCTION

The Cascade Range is an active north-trending volcanic arc at the western edge of North America (Figure 1). At the northern end of the range, between 47°N and 49°N, the average elevation increases, peaks become sharper, numerous small glaciers survive on the higher slopes, and volcanic rocks of the Cascade arc are scarce. This region is the North Cascades Range. The North Cascades are bounded on the west by the fore-arc basin of the Puget Lowland, on the south by the arc volcanic rocks of the Central Cascades, and on the southeast by the back-arc flood basalts of the Columbia Plateau. The geologic identity of the range is not so clearly defined to the north, but it is geographically bounded on the northeast by the Okanogan Ranges and on the northwest by the Fraser River, which separates the Cascades from the Coast Mountains.

The geology of the North Cascades strongly reflects processes of terrane accretion and dispersion at the western edge of the Cordilleran orogen. The range is remarkable for the number of apparently distinct tectonostratigraphic terranes exposed in a small region and the extent to which rocks of these terranes are involved in Cordilleran orogeny. This field trip traverses the range from E to W, viewing as many terranes as possible in the allotted time, and examining the evidence for terrane accretion and dispersion and the deformation, plutonism, and metamorphism that here constitute Cordilleran orogeny.

We intend this guide to be an introduction to the pre-Oligocene geology of the range, as well as a handbook to the stops of the field trip itself. Other works on the geology of the North Cascades and surroundings are cited below as appropriate. For brief geologic overviews of the range we especially recommend Misch [1988], Babcock and Misch [1988], Brown [1988], and the short synthesis of north-Cordilleran accretion by Monger and others [1982]. More detailed accounts of North Cascades geology by Misch [1966] and McTaggart [1970] predate the realization that parts of the range may be exotic; Davis and others [1978] and Hamilton [1978] briefly discuss the range from more mobilistic perspectives.

Our interpretations span much (though not all) of the range of opinion amongst workers in the range. In this guide we try to preserve some of this invigorating diversity. Many of the interpretations presented below stem from ongoing mapping projects and will undoubtedly be modified in the course of further work.

Acknowledgments

Many others have helped us in our studies, discussed Cascades geology with us, and kept us informed of their work. We are especially grateful to those geochronologists whose numbers grace the pages below: R. L. Armstrong, S. A. Bowring, J. S. Stacey, Peter van der Heyden, and J. A. Vance. Miller's work has been supported by NSF grant EAR-8707956. The text of this guide has been improved with the aid of careful reviews by F. K. Miller and R. E. Wells. Workers in North Cascades geology owe a special debt to the late Peter Misch, whose work forms the foundation of our present knowledge of the range. All of us were Peter's students.

THE GEOLOGY OF THE NORTH CASCADES RANGE

The complex geology of the North Cascades results from superposition of at least four sets of phenomena. Bedrock largely consists of (I) several *pre-Late Cretaceous tectonostratigraphic terranes*. The time at which the terranes were accreted is not well defined, though most were involved in (II) *Late Cretaceous to Eocene(?) orogeny* that reflects significant crustal thickening within the North American continental margin. A poorly-understood (III) *Eocene event* has fragmented the Late Cretaceous orogen, and superimposed on all is (IV) the *Oligocene to Holocene Cascade magmatic arc*. The geology outlined below and described in the trip log is organized around these four headings.

The N-trending Straight Creek fault and several NW-trending fault zones divide the North Cascades Range into several tectonic blocks (Figure 2) within which strata, tectonic styles, and (or) facies of metamorphism are relatively coherent. Between most blocks there are substantial differences. We share the prejudice of most workers in the region that Late Cretaceous plutonism, metamorphism, and deformation tie the pre-Late Cretaceous terranes of the North Cascades together and that differences between the various blocks of Figure 2 reflect kilometer-scale to perhaps 100-kilometer-scale late-orogenic to post-orogenic displacement *within* this part of the Cordilleran orogen. However, we cannot satisfactorily restore most of the block-bounding faults. Displacements on the faults may be significantly greater and parts of the range could have been widely separated prior to the earliest Oligocene. Paleomagnetic studies of Late Cretaceous plutons in the North Cascades and the Coast Mountains to the north suggest that the western Cordilleran orogen as a whole is substantially allochthonous [Beck and others, 1981a, b; Irving and others, 1985].

Following the early work of Misch [1952, 1966], many workers have considered the North Cascades to be a complete two-sided orogen that could be understood in tectonic isolation. This tendency persists—indeed we are guilty of it in this guide—despite advances in our understanding of regional geology and orogenic dynamics. The reader should keep in mind that: (1) Many pre-Late Cretaceous terranes of the North Cascades may be parts of much larger units which can be recognized in northern California, in NE Oregon and W-central Idaho, and in western British Columbia and SE Alaska. (2) Similarities in timing of deformation, as well as the size of modern orogens, suggest that the North Cascades were but a small part of a much larger Cretaceous orogen that included most of the northern Cordillera, with its eastern limit in the Rocky Mountain foothills. (3) The Eocene event that modified the North Cascades was part of a regime of crustal extension in the core of the Cordilleran orogen, northward translation along the continental margin, and widespread magmatism and basin development that affected a region extending from the Pacific coast to the Rocky Mountains and from central Oregon to northern British Columbia.

Present Topography and Glaciation

Most summit elevations in the North Cascades range from 1800 to 2700 m, with intervening valleys not far above sea level. Most ridges and valleys trend NW-SE and reflect the structural grain of the underlying rock. Superimposed on this topography are Mt Baker and Glacier Peak, late Quaternary stratovolcanoes of the Cascade Arc (see Figure 8).

Uplift of the North Cascades is young. Near Wenatchee, Miocene flows of the Yakima Basalt Subgroup (of the Columbia River Basalt Group) are draped over the range (for example, Russell [1900]; Mackin and Cary [1965]; Tabor and others [1982a]). Westward thinning of Yakima flows along the east flank of the range, interbedded west-derived fluvial and lacustrine sediments, and deltas of Yakima basalt pillows produced by flow of lava into lakes ponded against the margin of the basalt field all point to some uplift prior to and during the middle Miocene. Uplift probably continues today, given the relief that characterizes much of the range.

Willis [1903] suggested that the concordant summits of the North Cascades outlined a dissected warped peneplain and that peneplanation postdated eruption of the Yakima Basalt Subgroup. The existence of a unique high surface has since been questioned [Waters, 1939]. A surface did develop during and after eruption of the Columbia River Basalt Group as streams responded to the temporary raising of local base level by the flood basalts [Waitt *in* Tabor and others, 1987a]. Remnants of this surface are readily identified adjacent to the lavas and in the Okanogan ranges to the east of the North Cascades, but in the high mountains subsequent glaciation has been so intense that relics of this surface have yet to be found.

Much of the mountainous scene traversed by the field trip owes its form to Pleistocene and, in the higher regions, Holocene glacial erosion. At the time of maximum growth of the Cordilleran ice sheet (about 15 ka) this region was all but covered with ice flowing south from Canada. In the mountains farther south, separate alpine glaciers had retreated prior to withdrawal of the Cordilleran ice in Puget Sound. West-side drainages were dammed by the Cordilleran ice, forming lake deposits, deltas, and moraine in valleys near the range front. For a summary and references, see Booth [1987].

The mountains strip moisture from the wet Pacific winds with remarkable efficiency. To the west, in the Puget Lowland, annual rainfall is a meter or less. As the winds rise over the western foothills, annual precipitation locally in excess of three meters waters one of the world's great forests. Old-growth douglas fir (*Pseudotsuga menziesii*), western hemlock (*Tsuga heterophylla*), and western red cedar (*Thuja plicata*) reached heights of 60 m, with trunks 2-4 m in diameter near the base. Logging, now waning because of overcutting and competition with cheaper lumber from lands of lower labor costs and less environmental protection, has long been the base of the regional economy. The forest hides much: natural outcrop is poor, cross-country travel is difficult, and most of the geologist's day is spent examining roadcuts along the net of logging roads that ensnares the western foothills.

Heavy winter snows in the heart of the range support the largest collection of glaciers in the United States outside of Alaska. Despite relatively warm average temperatures, snow-free ground can only be expected between July and October, and even during these months minor snowfall at higher elevations is likely. Outcrop varies from nonexistent to spectacular where recent deglaciation has exposed hectares of bedrock. Most of the center of the range is now dedicated wilderness managed by the National Park and National Forest Services: hikers and climbers are the major users. Within the past generation this region was summer home to the prospector, the sheepherder, and before them the aboriginal hunter. There has never been significant permanent settlement. Even now most access requires much walking (or a helicopter).

On the east slope annual precipitation drops to a half meter or less, ponderosa pine (*Pinus ponderosa*) becomes a dominant species in the forest, cross-country travel is easier, and outcrops are more abundant. The style of logging is different, reflecting smaller timber and a more open forest than on the west side. Run-off from the mountains has shaped more than the landscape: cheap hydroelectric power from the Columbia River (see figure 8) and river-irrigated apple, peach, and pear orchards are important to the local economy.

(I) Pre-Late Cretaceous terranes

Going west across northern Washington, from the fringing miogeocline of stable North America, the traveler encounters a succession of tectonostratigraphic terranes (Figures 1, 2, 3).

The closest exposures of unremobilized **North American craton** lie far to the east, in the Laramide uplifts of the central Rockies. North of the 49th parallel, continental basement west of the Rocky Mountain trench has been recognized on seismic records. Overlying this basement are three sedimentary sequences: the thick (perhaps ~20 km) Middle Proterozoic Belt (Purcell in Canada) Supergroup, the Middle(?) and Late Proterozoic Windermere Group, thought to be rift-related, and the Cambrian to Jurassic Cordilleran miogeoclinal sequence. At the latitude of this trip the Phanerozoic miogeocline is not well preserved. Scattered pre-Tertiary outcrops near Spokane include metamorphosed Belt Supergroup, metamorphosed pre-Belt plutons, and abundant Mesozoic granitic rocks.

Westward from Spokane lies the Quesnel terrane, or **Quesnellia**, consisting largely of upper Paleozoic and Mesozoic arc and oceanic rocks which were accreted to North America in the Jurassic(?) [Monger and others, 1982]. Our route lies south of most exposures of these rocks. Though the western extent of Quesnellia is a matter for debate, we tentatively place its boundary at the Chewack-Pasayten fault (Figure 2). Plutons of the loosely-defined *Okanogan arc* intrude Upper Triassic arc strata at the western edge of Quesnellia. Correlative plutons north of 49°N are largely Late Jurassic and the complex seems to have cooled in the Early Cretaceous [Greig, 1988; Todd, 1987], unlike the Late Cretaceous and younger crystalline rocks of the North Cascades (cf. Davis and others [1978]).

Pre-Late Cretaceous terranes east of the Straight Creek fault. Mesozoic, mostly marine, sedimentary and minor volcanic strata constitute a coherent stratigraphic sequence which we include here in the **Methow terrane**. The rocks of this terrane form a NW-trending belt along the east edge of the North Cascades, between the Hozameen fault (including its probable southern extensions) and the Chewack-Pasayten fault. The margins of the depositional basin are not preserved: the Hozameen and Chewack-Pasayten faults appear to be younger and unrelated to Mesozoic sedimentation [Trexler and Bourgeois, 1985].

To the northwest, the belt is offset ~110 km to the north by the Straight Creek fault [Kleinspehn, 1985]. To the southeast, strata of the belt are in fault contact with the Leecher Metamorphics and Methow Gneiss of Barksdale [1948]. We include these metamorphic rocks in the Methow terrane, although their actual tectonic affinity is uncertain; they could be part of Quesnellia (but see **Stop 1-**

1).

Aggregate thickness of the Mesozoic Methow section is on the order of 12 km or more [Barksdale, 1975; Coates, 1974] (Figures 3 and 4). The terrane is intruded by plutons of Late Jurassic, Cretaceous, and Eocene age [Tabor and others, 1968; Barksdale, 1975; Todd, 1987]. Many workers [e.g. Tennyson and Cole, 1978] have considered the marine sedimentary rocks to comprise a fore-arc or perhaps interarc sequence analogous to coeval parts of the Great Valley sequence of California; others [e.g. Kleinspehn, 1985] have characterized them as successor-basin deposits. Trexler and Bourgeois [1985] inferred an active strike-slip fault on the west margin of the basin during the mid-Cretaceous, based on facies patterns in sedimentary rocks of the Pasayten Group of Coates (1974) (see **Stops 2-1, 2-2**, and Figure 4).

A late Early Cretaceous change from a volcanic to a plutonic source for sandstone in the Jackass Mountain and Pasayten Groups (Figure 4) led Tennyson and Cole [1978] to infer that these strata record unroofing of the Okanogan arc (see Quesnellia, above) to the east. There is, at present, no other good evidence for a pre-Eocene link between the Methow terrane and Quesnellia. Volcanic rocks of the Upper Cretaceous Midnight Peak Formation of Barksdale [1948] (Kmm, map 1*) and coeval(?) plutons are similar in age to volcanic rocks which lie on Quesnellia to the northeast [Thorkelson, 1985]. This age similarity would suggest Late Cretaceous proximity, were it not that similar-age magmatic rocks are present throughout large parts of the Cordillera.

Poorly preserved Mesozoic radiolarians in chert pebbles from conglomerates of the Albian and Cenomanian Virginian Ridge Formation of Barksdale [1948] (Kms, map 1; equivalent to part of the Pasayten Group), as well as sparse greenstone pebbles [Trexler, 1985], suggest that the Hozameen Group (described below) or some other oceanic assemblage cropped out to the west during deposition of Virginian Ridge strata [Tennyson and Cole, 1978] (**Stop 2-1**).

The late Paleozoic to middle Mesozoic *Hozameen Group* of Cairnes [1944] constitutes the **Hozameen terrane** (MzPzh, map 4). It is composed of basaltic greenstone (metamorphosed pillow basalt, tuff, breccia, massive lava), chert, and argillite with minor amounts of limestone, gabbro, sandstone, and dacite(?) [McTaggart and Thompson, 1967; Haugerud, 1985; Ray, 1986].

On the northeast side, the Hozameen terrane is bounded by the Hozameen fault. On the southwest the Hozameen terrane is separated from the Skagit Gneiss of Misch (1966), in the Chelan Mountains terrane, by the Ross Lake fault zone and the intervening Little Jack terrane.

Permian radiolarians have been found in the Hozameen Group southeast of Jack Mountain (see Figure 9, map 4) [Tennyson and others, 1982], where the unit is mostly basalt with minor gabbro, chert, and limestone. In British

---

*Unit symbols and map numbers in parentheses refer to Figure 9.

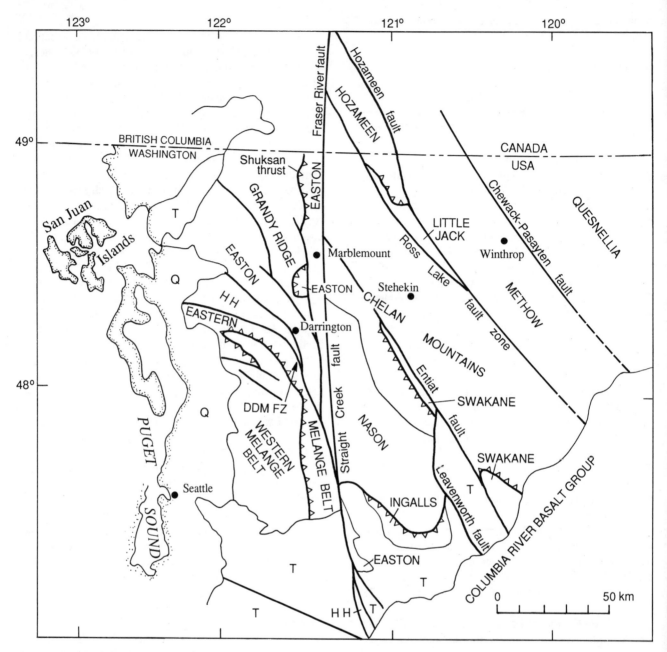

FIGURE 2. Sketch showing major terranes and faults. Q = unconsolidated deposits; T = Tertiary sedimentary and volcanic rocks; HH = Helena-Haystack melange.

Columbia, Middle and Late Triassic radiolarians have been identified in the Hozameen [Haugerud, 1985].

Ages, lithologies, and tectonic position demonstrate that the Hozameen Group is correlative with the Bridge River Complex of southern British Columbia. Correlation with other oceanic assemblages such as the Cache Creek Group of southern and central British Columbia, the Deadman Bay terrane of the San Juan Islands [Brandon and others, 1988] and parts of the Baker terrane of the Blue Mountains of northeast Oregon [Silberling and others, 1987] is attractive, though the Tethyan fusulinids distinctive of these terranes have not been found in the Hozameen.

The Twisp Valley Schist of Adams [1964] (Rct, map 1),

here included in the Chelan Mountains terrane because of its similarity to other oceanic assemblages in that terrane, lies along strike to the SE of the Hozameen Group and may be its metamorphosed equivalent [Misch, 1966; Miller, 1987].

We lump together schists, phyllites, slates, and argillites and metavolcanic rocks of Misch's [1966] *Elija Ridge Schist*, *Jack Mountain Phyllite* and *North Creek Volcanics* as the **Little Jack terrane** (Mzlj, maps 1, 2, 4), named for exposures on Little Jack Mountain. All observed contacts of this provisional terrane with other strata are faults. To the southwest, the Ross Lake fault and intrusive rocks of the Ruby Creek area separate the Little Jack unit from the Skagit Gneiss and associated schist of the Chelan Mountains

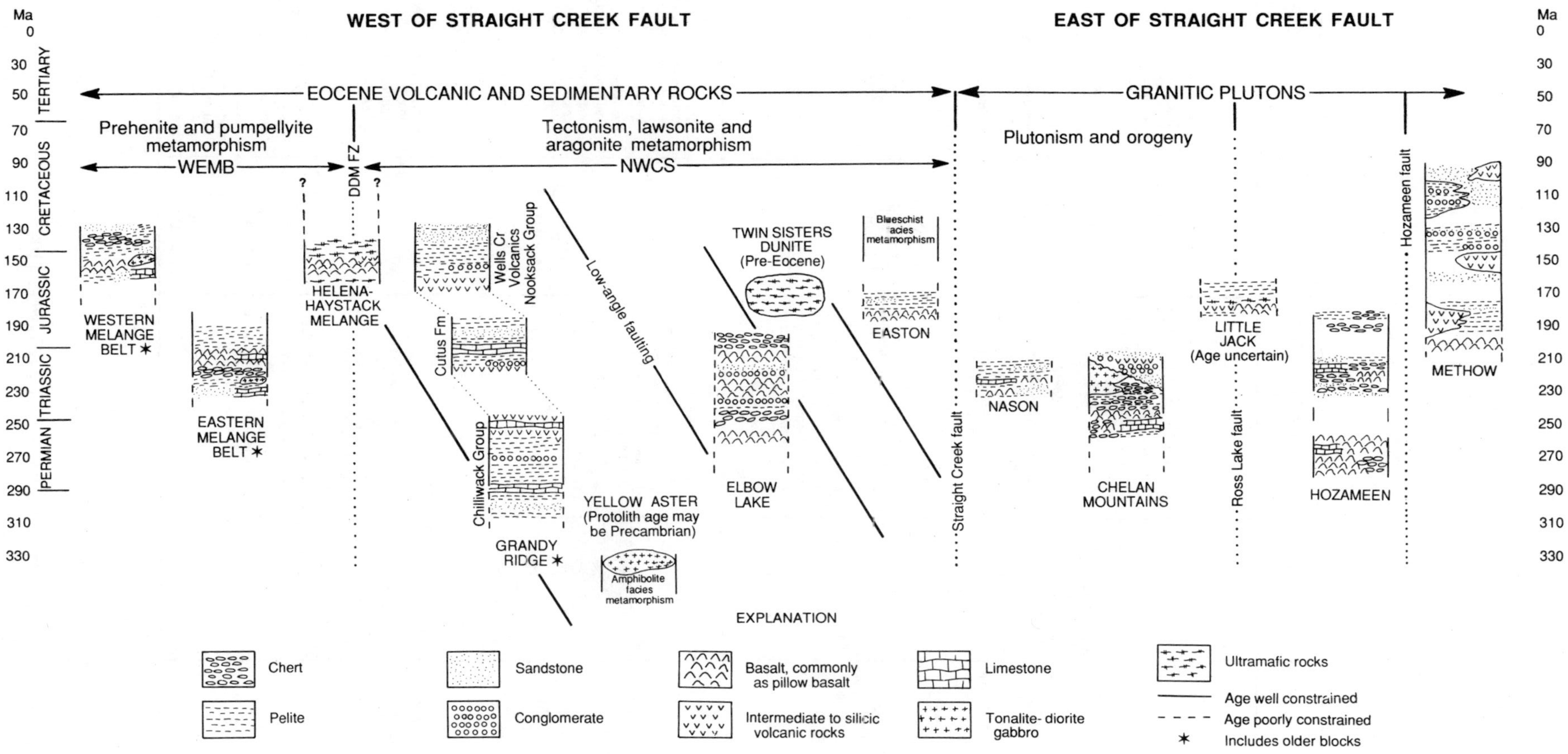

FIGURE 3. Summary of terrane protolith lithologies, approximate age ranges, and structural relationships. For sources of lithologic data see text. WEMB = western and eastern melange belts; DDMFZ = Darrington-Devils Mountain fault zone; NWCS = Northwest Cascades System.

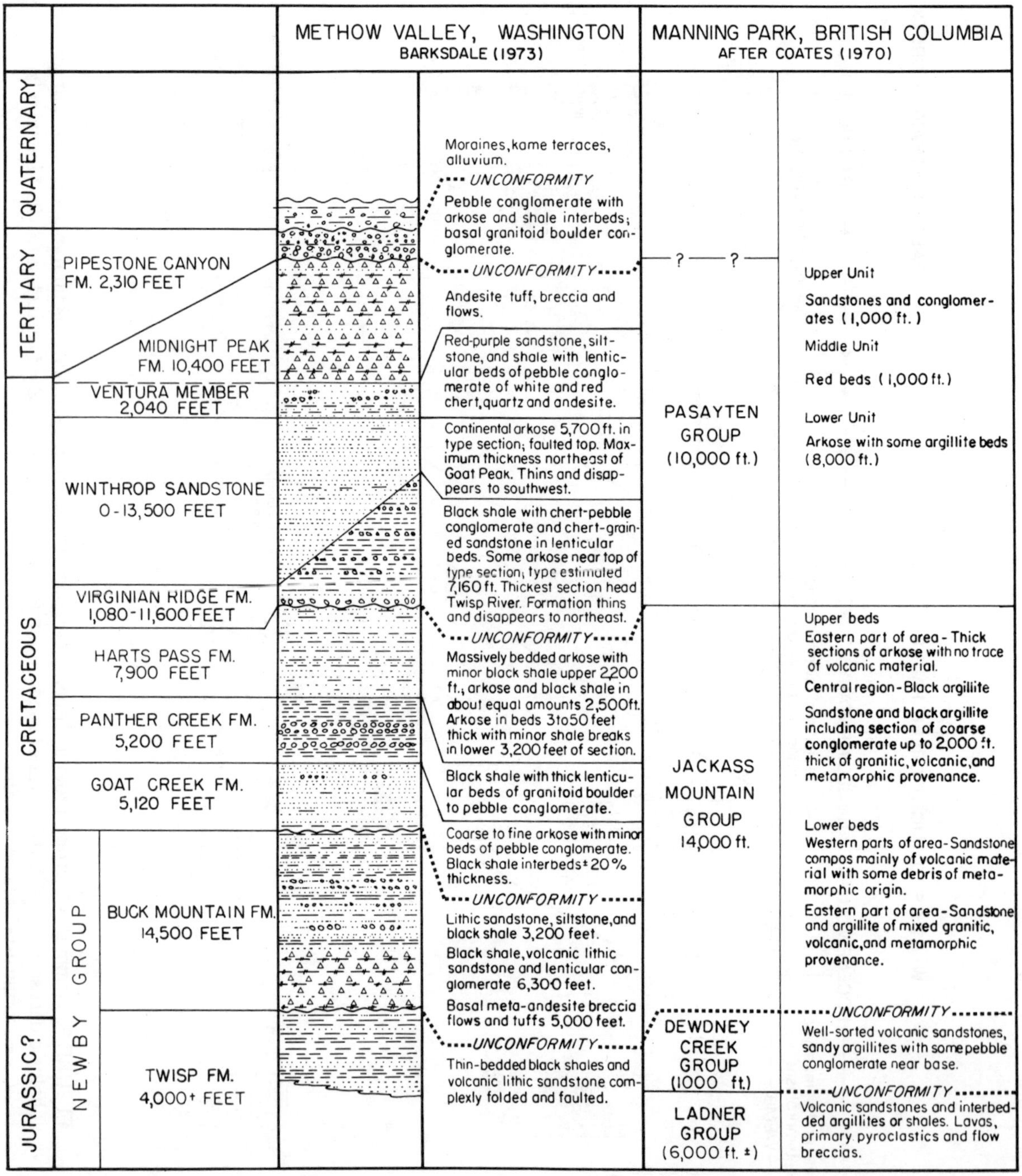

METHOW VALLEY, WASHINGTON
BARKSDALE (1973)

MANNING PARK, BRITISH COLUMBIA
AFTER COATES (1970)

Period	Unit (Methow Valley)	Description (Methow Valley)	Unit/Description (Manning Park)
QUATERNARY		Moraines, kame terraces, alluvium.	
		···· *UNCONFORMITY* ···	
TERTIARY	PIPESTONE CANYON FM. 2,310 FEET	Pebble conglomerate with arkose and shale interbeds; basal granitoid boulder conglomerate.	
		···· *UNCONFORMITY* ·····	? — ?
		Andesite tuff, breccia and flows.	Upper Unit — Sandstones and conglomerates (1,000 ft.)
	MIDNIGHT PEAK FM. 10,400 FEET	Red-purple sandstone, siltstone, and shale with lenticular beds of pebble conglomerate of white and red chert, quartz and andesite.	Middle Unit — Red beds (1,000 ft.)
	VENTURA MEMBER 2,040 FEET		PASAYTEN GROUP (10,000 ft.)
CRETACEOUS	WINTHROP SANDSTONE 0-13,500 FEET	Continental arkose 5,700 ft. in type section; faulted top. Maximum thickness northeast of Goat Peak. Thins and disappears to southwest.	Lower Unit — Arkose with some argillite beds (8,000 ft.)
		Black shale with chert-pebble conglomerate and chert-grained sandstone in lenticular beds. Some arkose near top of type section; type estimated 7,160 ft. Thickest section head Twisp River. Formation thins and disappears to northeast.	
	VIRGINIAN RIDGE FM. 1,080-11,600 FEET		Upper beds — Eastern part of area - Thick sections of arkose with no trace of volcanic material. Central region - Black argillite
		···· *UNCONFORMITY* ·····	
	HARTS PASS FM. 7,900 FEET	Massively bedded arkose with minor black shale upper 2,200 ft.; arkose and black shale in about equal amounts 2,500 ft. Arkose in beds 3 to 50 feet thick with minor shale breaks in lower 3,200 feet of section.	Sandstone and black argillite including section of coarse conglomerate up to 2,000 ft. thick of granitic, volcanic, and metamorphic provenance.
	PANTHER CREEK FM. 5,200 FEET		JACKASS MOUNTAIN GROUP 14,000 ft.
	GOAT CREEK FM. 5,120 FEET	Black shale with thick lenticular beds of granitoid boulder to pebble conglomerate.	
	NEWBY GROUP — BUCK MOUNTAIN FM. 14,500 FEET	Coarse to fine arkose with minor beds of pebble conglomerate. Black shale interbeds ± 20% thickness.	Lower beds — Western parts of area - Sandstone composed mainly of volcanic material with some debris of metamorphic origin.
		···· *UNCONFORMITY* ·····	Eastern part of area - Sandstone and argillite of mixed granitic, volcanic, and metamorphic provenance.
		Lithic sandstone, siltstone, and black shale 3,200 feet. Black shale, volcanic lithic sandstone and lenticular conglomerate 6,300 feet. Basal meta-andesite breccia flows and tuffs 5,000 feet.	
		···· *UNCONFORMITY* ·····	DEWDNEY CREEK GROUP (1000 ft.) — Well-sorted volcanic sandstones, sandy argillites with some pebble conglomerate near base.
JURASSIC?	NEWBY GROUP — TWISP FM. 4,000+ FEET	Thin-bedded black shales and volcanic lithic sandstone complexly folded and faulted.	····· *UNCONFORMITY* ·····
			LADNER GROUP (6,000 ft. ±) — Volcanic sandstones and interbedded argillites or shales. Lavas, primary pyroclastics and flow breccias.

FIGURE 4. Correlation chart for units in the Methow terrane. After Barksdale [1975].

terrane. The Hozameen fault separates the Little Jack terrane from the Methow terrane to the northeast, although Misch [1966, 1977b] considered the Jack Mountain Phyllite to grade eastward into unmetamorphosed Lower Cretaceous rocks of the Methow terrane. On Jack Mountain and Crater Mountain, the low-angle Jack Mountain fault separates the Little Jack terrane from the structurally overlying Hozameen terrane.

In the area of Little Jack Mountain and Ross Lake, most of the Jack Mountain Phyllite is indeed phyllite and semischist with minor amounts of intercalated ribbon chert and greenish tuffaceous schist as well as scattered ultramafic bodies, some up to several hundred meters in length. Metamorphic grade within the the Jack Mountain Phyllite increases rapidly across strike from northeast to southwest, from sub-biotite grade phyllite to garnet- and andalusite-bearing (rare sillimanite) biotite schist. Numerous sills and dikes of plagioclase-quartz porphyry intrude the Jack Mountain Phyllite and most are visibly deformed.

We know little of the age or stratigraphic affinity of the rocks in the Little Jack terrane, but except for the ultramafic rocks, the premetamorphic lithologies are common to many of the Mesozoic strata in the Methow terrane.

The **Chelan Mountains terrane** includes: (1) the Napeequa unit (equivalent to the "rocks of the Napeequa River area" of Cater and Crowder [1967]), including correlative rocks mapped by Tabor and others [1987a, 1988] and the Twisp Valley Schist of Adams [1964]; (2) the Marblemount Meta Quartz Diorite of Misch [1966] and correlative plutons [Misch, 1966; Mattinson, 1972]; and (3) part of the Cascade River Schist of Misch [1966], which we refer to herein as the Cascade River unit. Migmatized correlatives of these units occur within the Skagit Gneiss of Misch [1966]. Tabor and others [1987a,b] include in this terrane the Chelan Complex of Hopson and Mattinson [1971] which crops out south of the field trip route.

The present eastern boundary of the Chelan Mountains terrane is probably the Ross Lake fault zone. To the southwest, the Chelan Mountains terrane adjoins the Nason terrane along a zone (a metamorphosed fault?) marked by numerous ultramafic bodies; in early Late Cretaceous time this zone was intruded by tonalitic plutons.

The *Napeequa unit* (Rcn, maps 2, 3, 4, 5; **Stop 6-7**) is mostly quartz-rich mica schists and quartzite and fine-grained amphibolite, strongly reflecting its origin as a chert- and basalt-rich oceanic unit. It also includes greenschist, quartz-rich phyllite, very rare Al-rich schist, amphibole-bearing mica schist, metagabbro, and rare marble [Cater and Crowder, 1967; Misch, 1966; Tabor, 1961; Fugro NW, 1979; Tabor and others, 1987a and b]. Ultramafic pods and lenses, probably metaperidotite, are scattered throughout.

Much of our Napeequa unit was formerly included in the Mad River terrane [Tabor and others, 1987b]; its inclusion here in the Chelan Mountains terrane follows the revised interpretation of Tabor and others [1988]. Several large, as yet undated, granodiorite to granite plutons (Figure 9-3) within the Napeequa unit are possibly pre-Triassic and are included in the Chelan Mountains terrane here. Arguments for this age assignment and further description of the plutons are in Tabor and others [1988].

The *Twisp Valley Schist* (Rct, map 1; **Stop 3-1**) consists mainly of interlayered biotite schist, siliceous schist and impure quartzite (metachert?). Widespread amphibolite, greenschist, calc-silicate rock and marble, and less common ultramafic rock (metaperidotite?) generally form lenses which are <50 m in thickness [Adams, 1961; Miller, 1987]. The interleaving of metaperidotite with supracrustal rocks in both the Napeequa unit and the Twisp Valley Schist probably records tectonic mixing that predated metamorphism and polyphase folding of the units [Miller, 1987]. Dynamothermal metamorphism of the Twisp Valley Schist ranges from the biotite zone to the sillimanite zone and apparently occurred at least in part during the Paleocene and (or) early Eocene.

The Cascade River Schist as described by Misch [1966, 1968, 1979] includes oceanic rocks that we assign to the Napeequa unit, as well as distinctive granitoid-clast metaconglomerate and associated metasandstone, rare metapelites, and a variety of metavolcanic rocks. Following the usage of Tabor and others [1988], we intend to restrict the Cascade River Schist to this coarse-clastic-rich package and its associated metavolcanic rocks, but for now refer to it as the *Cascade River unit* (Rcc, maps 2 & 3; **Stop 5-8**). Some metavolcanic rocks mapped as Cascade River unit, especially those in the southeastern part of the belt (Figure 9-2), may belong to the Napeequa unit.

The *Marblemount Meta Quartz Diorite* of Misch [1966] (Rcm, map 3; **Stop 6-8**) is a chlorite-grade gneissic tonalite where it crops out along WA 20. Metamorphic grade and textures increase along strike from northwest to southeast where the unit becomes hornblende tonalite gneiss. U-Pb zircon dating by Mattinson [1972] demonstrated a 220 Ma age for this body and the correlative plutons in the Holden area to the southeast [Crowder, 1959; Cater and Crowder, 1967; Cater, 1982].

Metaconglomerate in the Cascade River unit contains boulders of tonalite derived from the Marblemount. This unconformable relationship has long been recognized [Tabor 1961; Misch 1966, 1977a] but there has been controversy because in some areas Marblemount plutons or their correlatives appear to intrude the (unrestricted) Cascade River Schist [Cater, 1982]. The recognition that parts of the (unrestricted) Cascade River Schist are actually parts of the older Napeequa unit reconciles these enigmatic observations and suggests that the protolith of the Cascade River unit unconformably overlay the protolith of the Napeequa unit as well as the Marblemount [Tabor and others, 1988]. Rocks invaded by plutons of the Marblemount belt in the Holden area [Cater, 1982] have been dated as Permian [Mattinson, 1972] and farther south granodioritic orthogneiss apparently intrusive into the Napeequa unit is Paleozoic or older [Tabor and others, 1982].

The Cascade River unit appears to have an age of about 220 Ma, based on a recent U-Pb analysis of zircon (J. S. Stacey, written communication, 1987) obtained from a metadacite tuff collected by Jeff Cary (M.S. thesis in progress, Western Washington University). Brown [in Fugro NW, 1979] suggested that metavolcanic rocks in the Cascade River unit were comagmatic with Marblemount, and to reconcile the apparently coeval ages of the basement Marblemount and overlying Cascade River units, Tabor and others [1988] suggest the two assemblages formed in an arc environment where intrusion of the plutonic rocks, uplift, erosion, and deposition could take place in rapid succession.

The Napeequa unit of the Chelan Mountains terrane and the correlative Twisp Valley Schist may be the metamorphosed equivalents of the older part of the Hozameen Group (which would also make them possible correlatives of the Bridge River Complex, etc.) Prior to Cretaceous orogeny the two units may have been continuous; however, different metamorphic grades indicate that the present-day Hozameen and Chelan Mountains terranes must have been at different structural levels following Cretaceous tectonism. The Chelan Mountains terrane in general has some similarity to the region studied by Rusmore [1987], at the east edge of the Coast Mountains at latitude 51°N, where the Upper Triassic Cadwallader Group contains conglomerates with granitic cobbles and is faulted against the oceanic Bridge River Complex [Rusmore, 1987].

The *Skagit Gneiss* of Misch [1966] is a complex of metamorphosed plutons (TKso, maps 1, 2, 3, 4) with intimately-related paragneisses (TKsp, maps 2, 3, 4). Extensive lit-par-lit intrusion of orthogneiss into paragneiss and development of pegmatitic leucosomes in both orthogneiss and paragneiss contribute to an overall migmatitic aspect. Although even on close inspection the distinction between orthogneiss and paragneiss is commonly difficult, in the Skagit gorge type section [Misch, 1987] the unit appears to be more than half orthogneiss. Orthogneisses are even more abundant in the Lake Chelan area (see Figure 9-1, 2). The remainder of the Skagit is banded gneiss composed of orthogneiss interlayered with biotite and hornblende-bearing schist and paragneiss, amphibolite, marble, calc-silicate rock, and ultramafic rock.

On the high ridge south of Newhalem (maps 3 & 4), Skagit paragneiss grades into the Napeequa unit of the Chelan Mountains terrane; similar schists reappear to the east, on the upper slopes of Ruby Mountain and in the Twisp Valley Schist of Adams [1964]. Overall the rocks appear to define a complex N-trending antiform with Skagit migmatites in its core (see Figure 15).

The **Nason terrane** includes the *Chiwaukum Schist* (Rnc, map 5) and an unnamed unit of banded gneiss (Kng, map 5) [Tabor and others, 1987]. To the the terrane is bounded by the Straight Creek fault, to the south it is structurally overlain by the Ingalls terrane along the Windy Pass thrust, and on the east it is in premetamorphic fault (?) contact with the Chelan Mountains terrane. The southeast

extent of the Nason terrane is limited by the Leavenworth fault, which places the metamorphic rocks against Eocene fluvial sedimentary rocks of the Chumstick Formation.

In the context of the North Cascades (and the western portion of the northern Cordillera) the Chiwaukum Schist is unusual: it is mostly a regionally-metamorphosed pelite. The unit is described by Evans and Berti [1986] and Plummer [1980], among others. Besides garnet-, staurolite-, and Al_2SiO_5-bearing mica schists, the Chiwaukum includes amphibolite, rare marble, and ultramafic rocks. Graded bedding and basalt pillows are locally preserved. Similar composition, metamorphic history, and associated plutons suggest that the Chiwaukum Schist is the lateral equivalent of the Settler Schist of southwest British Columbia, offset by the Straight Creek fault [Misch, 1966, also see below]. Rb and Sr isotope analyses of both units outline similar Late Triassic "scatterchrons" interpreted to record the depositional age [Magloughlin, 1986].

The Chiwaukum Schist locally grades into banded gneiss consisting of interlayered biotite gneiss, gneissic tonalite, and relicts of mica schist and amphibolite. The transition reflects addition of material by igneous intrusion and (perhaps) metasomatism [Tabor and others, 1987b].

Additional terranes making up the North Cascades east of the Straight Creek fault are the **Ingalls terrane** (Figure 2 only) composed of the ophiolitic Ingalls Complex of Miller [1985], a Late Jurassic ophiolite which probably was modified in an intraoceanic transform fault [Miller, 1985; Miller and Mogk, 1987] and the **Swakane terrane** (Figure 2 and pЄs, map 5) [Tabor and others, 1987b], which is composed of the Swakane Biotite Gneiss, inferred with some controversy to have a Proterozoic protolith [Mattinson, 1972; see discussions in Tabor and others, 1987b].

Pre-Late Cretaceous terranes west of the Straight Creek fault. Rocks west of the Straight Creek fault are separated into two blocks by the NW- to WNW-trending *Darrington-Devils Mountain fault zone* (DDMFZ) (Figure 2). To the northeast lies the Northwest Cascades System (NWCS) of Misch [1966] and Brown [1987]. To the southwest are the western and eastern melange belts (WEMB) of Tabor and others, [1988] (see also Frizzell and others [1987]). Within the DDMFZ lies the Helena-Haystack melange [Tabor and others, 1988], a mixture of rocks of the NWCS, the WEMB, and rock types found in neither of the adjoining blocks. High-angle faults of the DDMFZ were active after the middle Eocene, a point discussed in more detail under heading (III) below.

The *Northwest Cascades System* is a tectonically imbricated stack consisting mostly of, from structurally lowest to highest, the Jurassic Wells Creek Volcanics of Misch [1966] and stratigraphically overlying Jurassic and Cretaceous Nooksack Group of Misch [1966], the Paleozoic Chilliwack Group of Cairnes [1944], with the overlying(?) Cultus Formation of Monger [1970], the probably mostly Mesozoic oceanic Elbow Lake Formation of Brown and others [1987], and the blueschist-facies Easton Metamorphic Suite. A minor but significant component of the stack is the crystalline, Paleozoic and older(?) Yellow Aster Complex of Misch [1966]. Also present, but not seen on this trip and not discussed here are Permian and Triassic blueschist and amphibolite of the Vedder Complex [Armstrong and others, 1983], and the Twin Sisters Dunite of Misch [1952; see also Ragan, 1963; Onyeagocha, 1978]. Each of these packets may be part of a unique terrane. Most units have participated to some extent in high-pressure facies metamorphism, though differences in metamorphism indicate substantial post-metamorphic tectonism [Brown and others, 1981]. The scale and complexity of imbrication has led Brown [1987] to liken the NWCS to a regional melange.

The Chilliwack Group, Cultus Formation, Wells Creek Volcanics, and Nooksack Group are here provisionally included in the **Grandy Ridge terrane**, though depositional contacts above and below the Cultus Formation have not been observed. The *Chilliwack Group* (Pzc, maps 3, 5, 6) and *Cultus Formation* are composed of arc volcanic rocks, mostly basalt and andesite but ranging to rhyolite; limestone; and associated greywacke and argillite of probable deep-water, marine-fan origin [Danner, 1966; Monger, 1966, 1970; Franklin, 1974; Christenson, 1981; Blackwell, 1983; Brown, 1987]. Chilliwack limestones yield fossils ranging from Silurian(?) and Devonian to Permian. Late Triassic and Jurassic faunas are recognized in the Cultus. Chilliwack and Cultus rocks range from slightly to moderately recrystallized and characteristically have a low-angle planar fabric. Aragonite and lawsonite occur locally [Vance, 1968; Monger, 1970; Brown and others, 1981]. Permian fusulinid faunas in the Chilliwack Group resemble those of the Stikine terrane in British Columbia and also the Quesnel terrane [Price and others, 1985]; intermediate to felsic arc volcanism in both the Permian and Triassic of the Grandy Ridge terrane invites correlation with the Seven Devils Group of NE Oregon and W-central Idaho [Vallier, 1977], considered by many to be part of Wrangellia [Jones and others, 1977].

The Jurassic, intermediate-composition *Wells Creek Volcanics* and overlying clastic, Late Jurassic to Early Cretaceous *Nooksack Group* were interpreted by Sondergaard [1979] to represent a volcanic arc and associated turbiditic submarine fan. The units are less recrystallized and deformed than most other rocks of the NWCS. Misch [1966] believed these units to be autochthonous or parautochthonous beneath the remainder of the NWCS, but Brown and others [1987] interpret poorly exposed outcrops of Nooksack Group south of Mt Baker to be slices at least locally imbricated within the NWCS. We will not see these units on this trip.

The *Easton Metamorphic Suite* [Tabor and others, in press] or **Easton terrane**, commonly referred to as the Shuksan Metamorphic Suite of Misch [1966], consists of the *Darrington Phyllite* and *Shuksan Greenschist* (which is in part blueschist) (Jed and Jes, maps 3, 5, 6, 7), and

associated higher-grade blueschist, amphibolite, and eclogite in the Iron Mountain - Gee Point area (see Figure 9-6). Protoliths of the Darrington and Shuksan were marine shales and minor sandstone and MORB-type submarine basalt which probably mostly underlay the shale [Haugerud and others, 1981; Dungan and others, 1983; Brown, 1986]. Gallagher and others [1988] describe more-felsic volcanic and plutonic rocks in the Easton and propose that the terrane includes an arc component. Based on isotopic studies, Brown and others [1982], Brown [1986], Armstrong and Misch [1987], and Gallagher and others [1988] suggest that the protolith age of the Easton Metamorphic Suite is Jurassic and blueschist-facies metamorphism is Early Cretaceous (pre-120 Ma). Well-recrystallized blue amphibole-bearing assemblages testify to metamorphism at 7-9 kb and 300-400°C [Brown, 1986]. Blueschists with similar metamorphic history and lithologies occur in northern California, where such rocks are considered part of the Franciscan Complex [Brown and Blake, 1987].

Tonalitic to gabbroic igneous rocks and older quartz-rich clinopyroxene gneisses which occur as slabs and slivers in the Chilliwack Group or as fragments along faults of the NWCS are generally lumped together as the *Yellow Aster Complex* of Misch [1966] (ya, maps 3, 5, 6), here called the **Yellow Aster terrane**. Rocks of the complex are generally mylonitic and (or) strongly cataclastically deformed. Some of the more mafic igneous rocks of the Yellow Aster could be late Paleozoic intrusive rocks that invaded both the Chilliwack Group and the older gneisses.

On the basis of a concordant U-Pb sphene age, Mattinson [1972] assigned metamorphism of the older clinopyroxene gneisses to about 415 Ma; a strongly discordant U-Pb zircon age suggests a Precambrian component. The presence of marble and calc-silicate rocks amongst the clinopyroxene gneisses suggests that the older gneisses are dominantly metasedimentary. Discordant ~370 Ma zircon ages from a younger tonalitic orthogneiss were interpreted by Mattinson [1972] to represent a primary intrusive age of ~460 Ma and metamorphism at ~415 and ~90 Ma.

The *western and eastern melange belts* differ significantly from the NWCS in structure, metamorphic facies, and overall lithologic mix. Whereas bedding and foliation in the NWCS commonly have low to moderate dips, cleavage and bedding in the WEMB are steep. Phyllite of the western melange belt locally resembles the Darrington Phyllite of the NWCS, but nowhere in the WEMB have mafic rocks developed blue amphibole. Lawsonite and aragonite, which are common in the NWCS, may be present in the WEMB in the vicinity of Whitehorse Mountain, but most rocks of the WEMB only attained the metamorphic conditions which produce prehnite and pumpellyite.

An exception to the generally steep dips of the WEMB occurs southwest of Darrington, where rocks of the eastern melange belt overlie rocks of the western melange belt in a broad faulted antiform (see Figure 21). The contact between the units is no more disrupted than the units themselves, but the configuration suggests that the eastern melange belt has been thrust over the western melange belt or slid into it as an olistostromal mass [Tabor and Booth, 1986].

Although some of the disruption of beds in both melange belts may be of olistostromal origin (**Stop 9-7**), the last episode of mixing in both belts is probably tectonic, as judged by (a) overall penetrative deformation, (b) the local synkinematic metamorphism in the western belt, and (c) strong cataclasis in crystalline megaclasts.

The **eastern melange belt** (MzPzem, maps 6 & 7) consists predominantly of mafic volcanic rocks and chert with minor argillite and greywacke. Marble is conspicuous locally and large and small pods of ultramafic rock occur throughout the belt. Phacoids of metadiabase and mafic migmatitic gneiss complex are also present.

We include in the eastern melange belt the *Trafton sequence* of Danner [1966], which Dethier and others [1980] and Whetten and others [1988] referred to as the Trafton melange or terrane. The distinctive Tethyan fusulinid fauna of Permian limestones in the Trafton was one of the first clues that this part of the Cordillera contained the exotic tectonic elements [e.g. Danner, 1966; Monger and Ross, 1971] later to become known as "terranes".

The eastern melange belt, including the Trafton melange, yields diverse ages ranging from Mississippian to Jurassic. The oldest ages are generally from marble blocks or radiometric ages from tonalite blocks. The youngest ages are from chert and argillite [Tabor and others, 1988].

Southwest of Darrington the eastern part of the eastern melange belt is lithologically and structurally more coherent than the rest of the belt. From west to east, exposed on and in the vicinity of Whitehorse Mountain, is a sequence of internally disrupted units consisting of greenstone, greywacke, argillite, rare marble, and metatonalite (MzPzem) overlain(?) on the east by argillite and greenstone which contain abundant Upper Triassic chert (ℝec, maps 6 & 7). These are overlain by Upper Triassic massive greenstone—the volcanic rocks of Whitehorse Mountain (ℝew, map 6,7)—and this in turn is overlain by Middle and Upper Jurassic massive argillite (Jea, maps 6 & 7). Many of the ages in this sequence are from radiolarians and conodonts recently identified by Charles Blome and Anita Harris, respectively. Most of the bedding in these rocks is vertical (**Stop 8-7**; see Figure 21).

The **western melange belt** (JKws and JKwp, maps 6 & 7) is predominantly argillite and greywacke (subquartzose sandstone) containing generally lesser amounts of mafic volcanic rocks, conglomerate, chert, and marble. Ultramafic rocks, mostly serpentinite, are present but very scarce. Outcrop- to mountain-sized phacoids of metagabbro, metadiabase, resistant sandstone, chert, and marble generally stand out boldly in a matrix of poorly foliated argillite or thin-bedded argillite and disrupted sandstone beds. Commonly the matrix is not well exposed,

but disruptions of beds, crude foliation, or pervasive cataclasis is apparent in most outcrops. The eastern part of the western belt is more thoroughly metamorphosed than the rest of the belt; the disrupted rocks grade eastward to slate, phyllite, and semischist, with minor greenschist and chert.

The ages of most components of the western melange belt appear to be Jurassic and Early Cretaceous, based on fossils from several locations throughout the belt (many of them radiolarians identified by Charles Blome) and concordant and slightly discordant U-Th-Pb zircon ages in the range of 170-150 Ma from four metatonalite-metagabbro masses [Frizzell and others, 1987; Tabor and others, 1988].

Rocks similar to the western melange belt comprise the Russell Ranch Complex of Miller [in press] some 90 km to the south. The Constitution and Lummi Formations of the San Juan Islands (Figure 2) are less disrupted rocks of similar general age and lithologies [Vance, 1975; 1977; Brandon and others, 1988]. Also similar is the Pacific Rim Complex [Brandon, 1985] on the southwest coast of Vancouver Island (Figure 1).

The **Helena-Haystack melange** consists of a wide variety of blocks in a serpentinite matrix (Mzh, maps 6 & 7) and is mostly bounded by high-angle faults, many of which cut middle Eocene sandstone and volcanic rocks [Vance and others, 1980]. Many of the blocks are of rock types found in the eastern melange belt (such as metagabbro, chert, and greenstone) and rocks of the Easton terrane, but other blocks are of lithologies occurring only in the Helena-Haystack melange. Blocks range from a few centimeters to greater than 8 km across. Metagabbro on Jumbo Mountain is a single block at least 10 km long. The serpentinite matrix is very poorly exposed and locally the extent of the melange may only be surmised from subdued topography interrupted by steep-sided hillocks of resistant rock [Tabor and others, 1988].

Rock types unique to the Helena-Haystack melange are actinolitic greenstone [Brown and others, 1981; Reller, 1987], foliated silicic metaporphyry, and micaceous quartz-feldspar schist (metarhyolite) (Jhmv, maps 6 & 7). Very distinctive are fine-grained amphibolite and hornblende tonalite (Jhat, maps 6 & 7). Some of the unique components yield Jurassic ages comparable to ages of units in the NWCS and WEMB (**Stop 9-2**), but the metarhyolite yields radiometric ages suggesting Late Cretaceous metamorphism (about 90 Ma; **Stop 9-4**). Assembly of the Helena-Haystack melange could not have taken place until after this metamorphism, that is, in Late Cretaceous or early Tertiary time.

The northern extent of the Helena-Haystack melange is not known but, based on the distribution of the contiguous Haystack terrane mapped by Whetten and others [1988], it appears to widen to the northwest. This could reflect a change in dip. To the south the Helena-Haystack melange appears to intersect the Straight Creek fault and reappear on the east side of the fault about 100 km farther south (see Figure 6; Tabor [1987]).

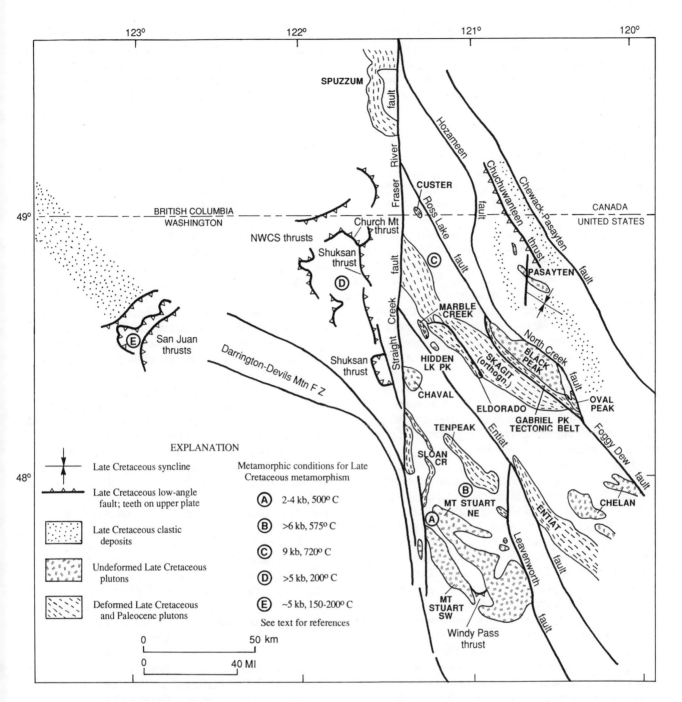

FIGURE 5. Sketch map of North Cascades and vicinity showing Late Cretaceous to Paleocene features described in text (Section II).

(II) Late Cretaceous to Eocene(?) orogeny

Late Cretaceous orogeny in the North Cascades Range was a crustal thickening event with overthrusting, metamorphism at relatively high pressures, and syn- to late-orogenic calc-alkaline plutonism. Tectonism, metamorphism, and associated plutonism (Figure 5) are overlap events, that is, events which affected all the pre-Late Cretaceous terranes and thus establish their proximity at the time of orogeny. Locally, Late Cretaceous orogeny

may have continued into the Eocene, at which time it was associated with strike-slip faulting. Alternately, Eocene phenomena may be temporally and genetically distinct.

Two hypotheses for Late Cretaceous North Cascades orogeny are current in the literature: W-vergent thrusting [Misch, 1966, 1977a; Brandon and Cowan, 1985; McGroder, 1988] and dextral wrench-faulting [Pacht, 1984; Trexler and Bourgeois, 1985; Brown, 1987]. The evidence at hand is insufficient to favor one hypothesis over the other, or to rule out other explanations.

Deformation. Cretaceous deformation can be demonstrated in most of the terranes. Albian and older sedimentary rocks in the Methow terrane are east-derived marine greywackes with no hint of a landmass to the west. In the late Albian to Cenomanian, chert clast-rich strata of the shallow-marine and deltaic Virginian Ridge Formation record uplift of a western source (probably the Hozameen Group) and rise of the basin surface above sea level [Barksdale, 1975; Tennyson and Cole, 1978; Trexler, 1985] (Figure 5). Deformation continued and migrated into the basin, as shown by a significant angular unconformity within the overlying Midnight Peak Formation (M. McGroder, written communication, 1987). The steeply W-dipping Chuchuwanteen thrust places Jurassic Ladner Group over Albian or younger rocks of the Pasayten Group [Coates, 1974] (M. McGroder, written communication, 1988). Dominant structures in the Methow block are large-wavelength NNW-trending folds which involve the volcanic rocks of the Upper Cretaceous Midnight Peak Formation. If these volcanic rocks are coeval with the 89 Ma Pasayten pluton, the upper part of the Methow section may be as young as Turonian and folding is, at least in part, younger still.

Early Late Cretaceous deformation in metamorphic rocks of the Chelan Mountains and Nason terranes is well documented by ~90 Ma ages of synkinematic plutons (see below) and constrained also by widely distributed post-90 Ma K-Ar cooling ages (**Stop 7-3**).

Late Cretaceous deformation in the Skagit Gneiss is less easily proven, though it is widely assumed (cf. Mattinson [1972]). An older NW-trending planar fabric in the gneiss which is cross-cut by ~46 Ma late lineated dikes (see Section III, below, and **Stop 6-1**) is at least in part younger than 65 Ma (see **Stop 6-6**). Some Skagit deformation probably is pre-Tertiary, as the older fabric is continuous with that in the Marblemount Meta Quartz Diorite to the southwest, which contains metamorphic muscovite yielding a 94 Ma age [Tabor and others, 1988].

Within the NWCS, orogeny postdates deposition of the Valanginian and older Nooksack Group. Misch [1966, 1977b; see also Brandon and Cowan, 1985] ascribed juxtaposition of the various units to the west-verging Cretaceous Shuksan and Church Mountain thrusts. However, structures in the fault zones [Brown, 1987; Smith, 1988] indicate NNW-SSE transport. Brown [1987] and Brown and others [1987] interpret the overall structure to be a regional melange formed by S-verging dextral transpression, with simultaneous S-verging overthrusts and dextral strike-slip faults. Brown [1987] cites early Late Cretaceous K-Ar ages on mylonites from fault zones as evidence that northwest Cascades fault fabrics and geometry are fundamentally Late Cretaceous. However, some of these structures may also have had significant post-middle Eocene motion, judging from their similarity with structures in the DDMFZ (described below). Relative importance of Cretaceous and Tertiary motion on these structures should be a rich source of discussion on the trip.

Brandon and others [1988] have documented early Late Cretaceous deformation in the San Juan Islands to the west of the North Cascades. Significant thrust faulting and consequent high P and low T metamorphism (prehnite, aragonite, lawsonite) occurred between 99 and 83 Ma. At the southern edge of the North Cascades, the Windy Pass thrust placed the Ingalls terrane above Chiwaukum Schist at about 94 Ma [Miller, 1985].

Metamorphism. In most of the North Cascades metamorphism occurred in the Late Cretaceous, but in the Skagit Gneiss and perhaps in nearby rocks it extended into the Eocene. Areas to the SW of the Skagit were also deformed and recrystallized locally in the Eocene, as described below. Separating these events is convenient for didactic purposes, but we realize the separation probably masks a continuous process.

Late Cretaceous to Eocene recrystallization of rocks in the North Cascades is genetically related to mid-Cretaceous to Eocene tectonism and magmatism. This metamorphism (Figure 5) includes: (1) greenschist to upper amphibolite facies metamorphism in the core of the range, with locally extensive migmatization; (2) metamorphism of the Little Jack terrane with grade rapidly increasing to the southwest; (3) low-grade metamorphism west of the Straight Creek fault; and (4) loss of porosity and pervasive remagnetization in strata of the Methow terrane [Granirer and others, 1986]. Blueschist-facies metamorphic events recorded in the Easton terrane (Shuksan metamorphism of Misch [1966], Brown [1986], and Brown and Blake [1987], among others) and the Vedder Complex [Armstrong and others, 1983] pre-date the incorporation of these units into the North Cascades.

Some of the most detailed studies of metamorphic history in the North Cascades have focused on the pelitic rocks of the Nason terrane south of the field trip route (see **Stop 7-3**). Peak metamorphism occurred at a variety of P-T conditions, with P varying from ~2 kb on the southwest to >6 kb on the northeast [Frost, 1976; Evans and Berti, 1986; Magloughlin, 1986]. Mineral parageneses and isotopic ages indicate a complex change in P-T conditions in the Late Cretaceous, with amphibolite-facies metamorphism peaking slightly before 90 Ma [Haugerud, 1987; Magloughlin and Brasaemle, 1988; Tabor and others, in press].

In the Skagit Gneiss, local sillimanite, the absence of epidote in calcic granitoid gneisses, and the assemblage enstatite + forsterite in metamorphosed ultramafic rocks attest to high metamorphic temperatures. Except for these assemblages, the predominantly tonalitic gneisses of the Skagit preserve little paragenetic evidence for the conditions of metamorphism [Misch, 1968; Yardley, 1978; Haugerud, 1985, Whitney and Evans, 1988]. Mineral chemistry in rare pelitic assemblages suggests that peak metamorphic conditions in the heart of the Skagit Gneiss were on the order of 720°C and 9 kb [Whitney and Evans, 1988]. See **Stops 6-3** and **6-4**.

West of the Straight Creek fault, widespread partial

recrystallization to aragonite and lawsonite-bearing assemblages in the Church Mountain terrane, parts of the Nooksack Group, and other rocks of the NWCS probably occurred in the Late Cretaceous [Misch, 1966; Vance, 1968; Brown and others, 1981].

South of the DDMFZ, metamorphism of the WEMB terrane was mostly at low P and low T, as evidenced by the local development of prehnite and pumpellyite. Locally, near the Helena-Haystack melange, rocks of the eastern melange belt contain lawsonite(?), possibly indicating higher P/T conditions. The age of this metamorphism is not well constrained.

Syn- to late-orogenic plutons. Tonalitic to granodioritic plutons intrude most of the terranes east of the Straight Creek fault. These plutons—the Eldorado, Chaval, Black Peak, Sloan Creek, Tenpeak, Mt Stuart, and others (Figure 5; **Stops 5-1,2,3** and **7-2**)—are overlap units that tie the various terranes together. Most plutons have both metamorphic and igneous characteristics, generally lack thermal aureoles, and yield U-Pb zircon ages of about 90 Ma. They appear to have been more or less synkinematic. A few plutons—such as the southwestern pluton of the Mt Stuart batholith and the Pasayten pluton in the Methow terrane—are higher-level bodies of the same age and general calc-alkaline aspect. Plutons of the same general age and lithology make up much of the Coast Plutonic Complex to the northwest, and especially in this respect the North Cascades appear to be a southern extension of the Coast Mountains of British Columbia. Some workers [e.g., Snyder and others, 1976] have interpreted this plutonic assemblage to represent a mid-Cretaceous subduction-related arc, though it is possible that they originated by partial melting of tectonically-thickened crust [see Zen, 1985, 1988].

Recent U-Pb dating has produced a number of 65-75 Ma zircon ages from tonalitic to granodioritic plutons within the Skagit Gneiss and adjacent schist (**Stops 2-4, 6-6**). Analytical work on some of these bodies has not been detailed enough to confirm that they are not yielding composite ages, the result of analyzing mixed populations of older (90 Ma?) magmatic zircon and younger metamorphic zircon. However, detailed work on the Custer orthogneiss unit north of 49°N suggests its zircons are essentially of a single 60-65 Ma igneous age (P. van der Heyden, written communication, 1987). By analogy the zircons of the other plutons give magmatic ages as well. We do not know why plutons of this age are not present outside of the higher-grade region.

(III) The Eocene event

The geology of the North Cascades Range was strongly modified by an Eocene tectonic event whose nature is not yet fully understood. Significant Eocene motion has been demonstrated for all the faults of Figure 2, though many may have had pre-Eocene motion as well. Preceding, accompanying, and following Eocene faulting were the filling and folding of sedimentary basins, magmatism, and local ductile deformation (Figure 6). K-Ar ages of some crystalline rocks record rapid unroofing at this time.

Eocene faulting. First recognized by Vance [1957a] and interpreted as a dextral strike-slip fault by Misch [1966], the **Straight Creek fault** (Fraser River fault in Canada) is arguably the most significant geologic boundary within the North Cascades. Post-mid-Cretaceous offset on the fault has been estimated at 80 to 190 km (Figure 7 and **Stop 7-1**). The fault is plugged by undeformed early Oligocene plutons of the composite Chilliwack batholith.

Faults in the **Darrington-Devils Mountain fault zone (DDMFZ)** cut strongly deformed middle Eocene (~42-44 Ma) volcanic and sedimentary rocks and are cut by the unfaulted 35 Ma Squire Creek stock and its satellitic plutons. The western end of the DDMFZ may offset rocks as young as early Oligocene (i.e. ~36 Ma) [Lovseth, 1975; Marcus, 1981; Whetten and others, 1988]. See **Stop 9-2**.

To the south the DDMFZ appears to intersect the Straight Creek fault at an acute angle, although the intersection is obscured by younger rocks (Figure 9-5). Pieces of the Helena-Haystack melange appear again about 100 km to the south of the intersection suggesting offset on the Straight Creek fault (Figure 6). However, because of the acute angle of intersection between the DDMFZ and the Straight Creek fault, and because of the intruding younger batholiths, the faults south of the zone of intersection could be the DDMFZ, not the Straight Creek. If so, the DDMFZ cuts the Straight Creek fault or the faults are in part contemporaneous [Tabor, 1987].

The development of mylonites in crystalline rocks along the southern extent of the **Entiat fault** and its straightness suggest that it was a strike-slip feature of considerable displacement [Laravie, 1976; Whetten and Laravie, 1976]. Maximum dextral offset as measured by the displacement of a major fold axis in metamorphic units to the south is about 30-40 km [Tabor and others, 1987a]. However, the fault is not a well-defined single strand in the Cascade-Skagit River area and significant strike slip is not readily demonstrated. There was probably strike slip on the Entiat fault during middle Eocene filling of the Chiwaukum graben [Johnson, 1985]. Dip slip during the middle Eocene was demonstrated by Gresens and others [1981].

The **Ross Lake fault zone (RLFZ)** is one of the most complex of the early Cenozoic tectonic elements in the North Cascades and apparently records both dip slip and dextral strike slip. The RLFZ forms the northeastern border of the crystalline core and is the site of numerous plutons ranging from Late Cretaceous to Eocene in age (Figures 6 and 9-1,2,4). These intrusions have obliterated significant parts of the fault zone, but also commonly have been deformed by subsequent movements in the zone. The width of the belt of high strain and the evidence of continued activity suggest significant displacement, but how much is difficult to document. Some workers [Kriens and Wernicke,

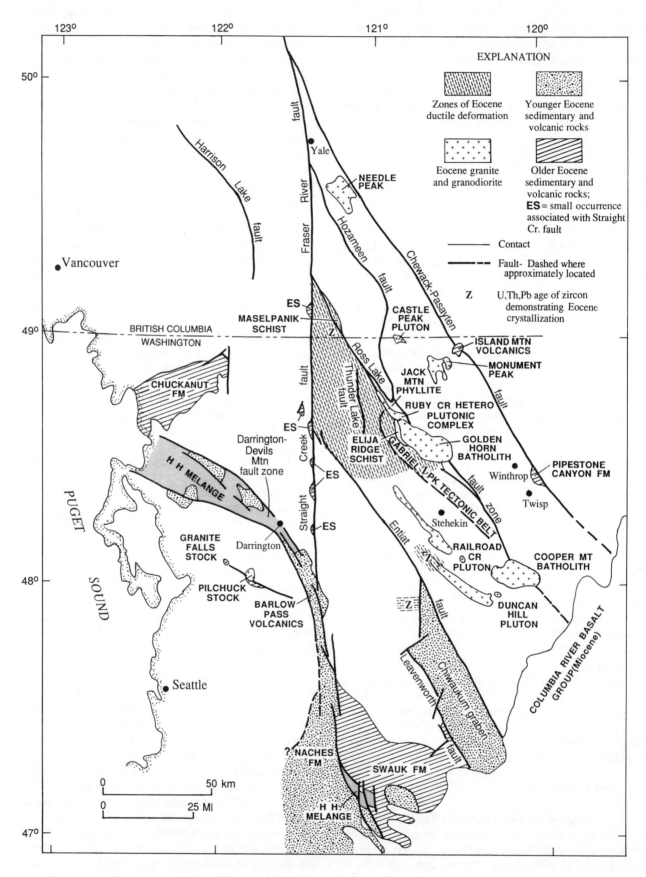

FIGURE 6. Principle features of the Eocene event (Section III).

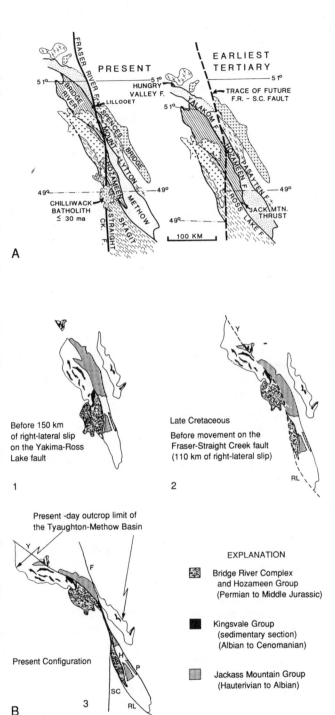

A

Before 150 km
of right-lateral slip
on the Yakima-Ross
Lake fault

1

Late Cretaceous

Before movement on the
Fraser-Straight Creek fault
(110 km of right-lateral slip)

2

Present -day outcrop limit of
the Tyaughton-Methow Basin

EXPLANATION

Bridge River Complex
and Hozameen Group
(Permian to Middle Jurassic)

Kingsvale Group
(sedimentary section)
(Albian to Cenomanian)

Jackass Mountain Group
(Hauterivian to Albian)

Present Configuration

B

3

FIGURE 7. A. Determination of 110 km of offset on the Straight Creek fault based on correlation of the Hozameen Group with the Bridge River Complex. After Monger [*in* Price and others, 1985]. B. The same based on correlations of the Late Cretaceous sedimentary rocks in the Methow terrane, after Kleinspehn [1985].

1986; Kriens, 1987] do not consider the Ross Lake fault zone to have significant displacement.

Misch [1966] defined the RLFZ as consisting of a series of major faults and associated fault zones. It includes the Ross Lake fault *sensu stricto*, the Gabriel Peak tectonic belt, the Hozameen-North Creek fault and its probable extension the Foggy Dew fault, and the Twisp River fault.

The *Ross Lake fault sensu stricto* places the Skagit Gneiss against the Hozameen terrane and the Little Jack terrane [Misch, 1966; Haugerud, 1985], within which metamorphic grade decreases sharply to the northeast (see **Stop 6-2**). The fault is plugged by an undeformed 30 Ma phase of the Chilliwack batholith [Richards and McTaggart, 1976]. North of 49°N, a well-developed horizontal stretching lineation and scattered kinematic indicators indicate dextral strike slip [Haugerud, 1985].

The continuation of the fault south of its intersection with the Black Peak batholith has not been resolved. It may step to the west side of the batholith to the *Gabriel Peak tectonic belt*, a high strain zone that extends NW-SE for about 60 km along the NE border of the Skagit Gneiss [Miller, 1987] (see **Stops 3-1** and **3-2**). The major component of this zone is the Gabriel Peak Orthogneiss of Misch [1966], a partly-mylonitic, epidote-amphibolite facies, 65-68 Ma tonalitic gneiss [Hoppe, 1982], which in part represents a deformed phase of the Black Peak batholith [Misch, 1977b; Hoppe, 1982]. To the south the deformed southwestern margin of the Paleocene Oval Peak pluton best defines the tectonic belt. Overall, the belt probably is kinematically complex, as dextral(?) strike slip may have dominated to the north, whereas dip slip clearly occurred at the south end.

The *Hozameen-North Creek fault* and *Foggy Dew fault*, which may be the same structure, form the east margin of the RLFZ (Figures 6 and 9-1; Misch [1966]; Staatz and others [1971]; Tennyson [1974]; M. F. McGroder, written communication, 1988). Mylonites in the Foggy Dew fault have a prominent SE-plunging lineation recording oblique slip. Abundant kinematic indicators indicate motion was dextral and normal, that is, NE side down and to the SE (see **Stop 1-4**).

South of the Black Peak batholith, between the Gabriel Peak tectonic belt and the North Creek fault, lies the *Twisp River fault* (see Figure 9-1,3 and **Stop 2-3**). It separates the Twisp Valley Schist and Oval Peak pluton from less metamorphosed strata of the North Creek Volcanics of Misch [1966]. Steeply dipping mylonites with a strong subhorizontal lineation are well-developed near the fault and suggest strike-slip motion. Barksdale [1975] proposed sinistral slip, but if the fault is a splay of the dextral Foggy Dew fault dextral motion is more likely [Miller, 1987].

Evidence for the timing of motion on the various strands of the RLFZ is inconsistent. At its south end the RLFZ is plugged by the 48 Ma Cooper Mountain batholith [Barksdale, 1975; Wade and others, 1987]. Paleocene and (or) early Eocene motion is probable in the Foggy Dew fault zone and the south end of the Gabriel Peak tectonic belt, given the radiometric age constraints (**Stops 1-4, 3-1**). Latest motion in the northern part of the Gabriel Peak tectonic belt postdates 50 Ma orthogneiss in Bridge Creek (Figure 6) [Hoppe, 1982]. The Hozameen fault is plugged by the 50 Ma Golden Horn batholith, yet north of 49°N it seems to have been active after 46 Ma [Ray, 1986]. A few km north of 49°N, the Ross Lake fault was active after ~46 Ma (P. van der Heyden, written communication, 1988). These ages may be imprecise, and if so we can only surmise that the age of latest tectonism in the Ross Lake zone is middle Eocene; alternatively, latest penetrative deformation in the RLFZ may have been diachronous, younging across strike to the SW and along strike to the NW.

Consistently horizontal lineations and a narrower, better-defined shear zone north of 49°N suggest a northward increase in the amount of Eocene strike-slip displacement [Haugerud, 1985]. Along-strike variations in displacement on both the Entiat and Ross Lake faults can be reconciled if, from northwest to southeast, displacement on the Ross Lake fault had been progressively transferred to the Entiat fault. Ductile extension within the Skagit Gneiss may have been part of the transfer mechanism.

Several north-south to northwest-southeast brittle faults within the Skagit Gneiss are late Eocene or early Oligocene, as they crosscut rocks with ductile Middle Eocene fabric (below) and predate intrusion of the Chilliwack batholith. Noteworthy is a zone of faulting along Stetattle and Thunder Creeks, the most obvious strand of which is the **Thunder Lake fault** (**Stop 6-5**).

Motion on the **Chewack-Pasayten fault** is at least in part Eocene; east of Twisp it cuts the Eocene(?) (an old paleobotanical determination indicates the unit may be Paleocene) Pipestone Canyon Formation of Barksdale [1975]. Farther to the north, Eocene feldspathic sandstone and felsic to intermediate volcanic rocks of Island Mountain were deposited during active faulting [White, 1986]. North of 49°N faults presumed related to the Chewack-Pasayten fault cut middle Eocene clastic rocks before intrusion of the 46 Ma Needle Peak pluton [Greig, 1988].

Eocene deposition. Thick sections of Eocene continental strata crop out at several locations in and near the North Cascades. Dominantly fluvial sandstone of the *Chuckanut Formation* [Johnson, 1984b, c] crops out northwest of Darrington (Figure 6). Johnson [1984c] estimates an average subsidence rate of greater than 0.5 mm/yr for the lower half of the 6 km thick section. Most sands are feldspar-rich, though basal strata commonly contain detritus from underlying rocks. Johnson [1984a, 1985] suggested that Chuckanut sedimentation occurred in a pull-apart basin within an active strike-slip regime, though this basin is larger than better-preserved and documented pull-apart basins elsewhere. At the south end of the North Cascades, similar fluvial strata of the *Swauk Formation* also accumulated rapidly in an active strike-slip regime [Tabor and others, 1984].

Particularly conspicuous in the vicinity of the Helena-Haystack melange are middle Eocene feldspathic sandstone and mafic to silicic volcanic rocks. These constitute the *Barlow Pass Volcanics* of Vance [1957b] (Tbv and Tbs, maps 5, 6, 7) and were deposited on the already-assembled NWCS and WEMB, probably in fault-controlled basins. Further faulting along the DDMFZ in post-Barlow Pass time locally imbricated sandstone, volcanic rocks, and the underlying Helena-Haystack melange.

Barlow Pass volcanic rocks are somewhat bimodal with basalt and rhyolite predominating. Interbeds of fluvial sandstone increase towards the bottom of the pile and to the northwest and southeast away from what we presume to be the center of volcanism near Barlow Pass. The unit appears to have an age of about 40 Ma (latest middle Eocene) based on numerous fission-track ages [Tabor and others, 1988] and correlates with similar bimodal volcanic rocks of the Naches Formation, located some 65 km to the south, and the Puget Group, about 75 km to the southwest [Tabor and others, 1984].

Eocene high-level deformation. Deformation of the sedimentary basins accompanied their filling or followed shortly thereafter. The Chuckanut and Swauk Formations and similar Eocene sandstones elsewhere in the range are strongly folded, typically on N- to NW-trending folds with wavelengths of a kilometer or more. Deformed, scaly shales suggest decollement at the basal contact (cf. Clayton and Miller [1977], Ashleman [1979]). On Hagan Mountain, near the Straight Creek fault (Figure 9-3), fold wavelengths are shorter and intraformational faulting becomes important. This deformation preceded intrusion of the Chilliwack batholith, seems to be related to the Straight Creek stress regime, and provisionally may be considered Eocene as well.

Eocene magmatism. Eocene plutons in the North Cascades are distinctively more potassic than younger and older plutons in the range. Best-known is the 50 Ma *Golden Horn batholith* (Tgg, map 4) [Misch, 1965; Stull, 1969; Boggs, 1984]. Others are the roughly contemporaneous Needle Peak, Castle Peak, Monument Peak, Cooper Mountain, Duncan Hill, Railroad Creek, and Mt Pilchuck plutons (Figure 6). The Duncan Hill is distinctive for its shape, compositional zoning and and structural style, which reflect up-to-the-northwest tilting since intrusion: it resembles a southeast-swimming tadpole, with a discordant, miarolitic, epizonal head and a concordant, gneissose, deep-seated tail [Hopson and others, 1970; Cater, 1982; Dellinger and Hopson, 1986]. Most other Eocene plutons are relatively shallow. Many of the Eocene plutons bristle with swarms of porphyritic, commonly granophyric dikes that generally have N and NE strikes, perhaps reflecting the dextral shear regime which induced motion on the Straight Creek fault.

Distinctive late-metamorphic granitic dikes, sills, and small plugs within the Skagit Gneiss are also Eocene. These bodies typically crosscut foliation, yet share the lineation of the enclosing gneiss. Informally we call them the *late lineated dikes*. Whole-rock Rb-Sr analyses of late lineated dikes from along WA 20 indicate an age of ~45 Ma [Babcock and others, 1985]. U-Pb analyses of zircon, xenotime, and monazite from a small plug of lineated garnet-biotite-muscovite granite 3 km north of 49°N indicate crystallization at 46 Ma (P. Van der Heyden, written communication, 1988).

Eocene ductile deformation. Much of the northern Skagit block is characterized by a NW-trending sub-horizontal stretching lineation (e.g. Haugerud, 1985; **Stop 6-1**). The lineation at least partly postdates intrusion of late lineated dikes, yet predates early Oligocene plutons of the Chilliwack batholith. Stretching was heterogeneous, with intense, mylonitic, deformation in some quartz-rich rocks and little deformation in quartz-poor rocks.

Penetrative Eocene deformation also affected rocks along the Ross Lake zone at the east margin of the Skagit block. Extensive Paleocene to early Eocene metamorphism is recorded by orthogneisses along the northeastern margin of the crystalline core east of Lake Chelan. Dynamothermal greenschist- and amphibolite-facies metamorphism of the Twisp Valley Schist, which may be the southern continuation of the Hozameen Group, also occurred during this time interval [Miller and others, 1988]. South of the field trip route, a tectonic zone bounding the Tenpeak pluton (NW of the Chiwaukum graben) contains zircons with a strong ~45 Ma overprint [Tabor and others, 1987a].

Eocene K-Ar cooling ages. Eocene K-Ar ages from some crystalline rocks in the Skagit Gneiss appear not to reflect reheating by Eocene magmas, but probably record cooling by unroofing of rocks that were hot before the middle Eocene. In the Tenpeak area, Late Cretaceous and earliest Tertiary K-Ar ages reflect slow cooling at depth following early Late Cretaceous orogeny, terminated by Eocene differential unroofing [Haugerud, 1987].

West of the Straight Creek fault, phyllites of the western melange belt give a 48 Ma K-Ar sericite age [Frizzell and others, 1987]. Farther northwest along the regional strike, poorly defined early Tertiary fission track ages from the SE San Juan Islands [Johnson and others, 1986] also suggest cooling after an Eocene thermal pulse and (or) cooling by Eocene unroofing.

(IV) Cascade magmatic arc

South of the North Cascades the Cascade range is dominated by eruptive rocks of the Cascade arc. Within the North Cascades the arc is represented by abundant plutons as well as small remnants of Neogene tuff, volcanic breccia, flows, and associated sedimentary rocks. Youngest of all are the Quaternary volcanic centers of Mt Baker and Glacier Peak.

All these rocks seem to be derived from typical calc-alkaline arc magmas. Pb isotope ratios suggest mantle sources with little crustal contamination [Church and Tilton, 1973]. Plutons are high-level and isotropic, with discordant contacts and local andalusite-bearing contact aureoles. Some are associated with explosion breccias suggesting venting to the surface [Fiske and others; 1963, Tabor, 1963; Tabor and Crowder, 1969; Cater, 1969].

The oldest suite of plutons, at about 34 Ma, includes the Index batholith and associated stocks and part of the Chilliwack composite batholith [Vance, and others, 1986; Richards and McTaggart, 1976; Engels and others, 1976; Tabor and others, in press]. We refer to it here as the *Index family* (Tin, maps 3, 4, 6, 7). A voluminous 25-30 Ma suite includes much of the Snoqualmie and Grotto batholiths and part of the Chilliwack batholith: this is the *Snoqualmie family* (Tsn, maps 4, 5). Other plutons are younger still: the *Cascade Pass family* (Tcp, maps 2, 3, 4, 5) ranges from about 20 to 3 Ma, the latter age determined from the Lake Ann stock, east of Mount Baker and associated with eruptive rocks of about the same age [Engels and others, 1976; James, 1980].

Shapes of some individual plutons suggest strong structural control. Plutons in the oldest two families tend to be aligned along major N- or NW-trending faults whereas plutons of the youngest family, such as the Cascade Pass Dike (18 Ma) (Tcp, map 3) and the Silver Lake pluton (10 Ma) (Tcp, map 4) of the Chilliwack composite batholith, tend to be aligned along NE-trending structures. The Cloudy Pass batholith (20-22 Ma) (Tcp, map 5) is aligned both ways, but may be composite.

Volcanic rocks in the vicinity of Glacier Peak (Tvn, map 5) began erupting about 2 Ma [Tabor and others, 1988] and similar-age volcanic rocks crop out east of and under(?) Mount Baker [Engels and others, 1976; James, 1980]. The volcanic cones themselves (Qgv, map 5) are probably less than a million years old.

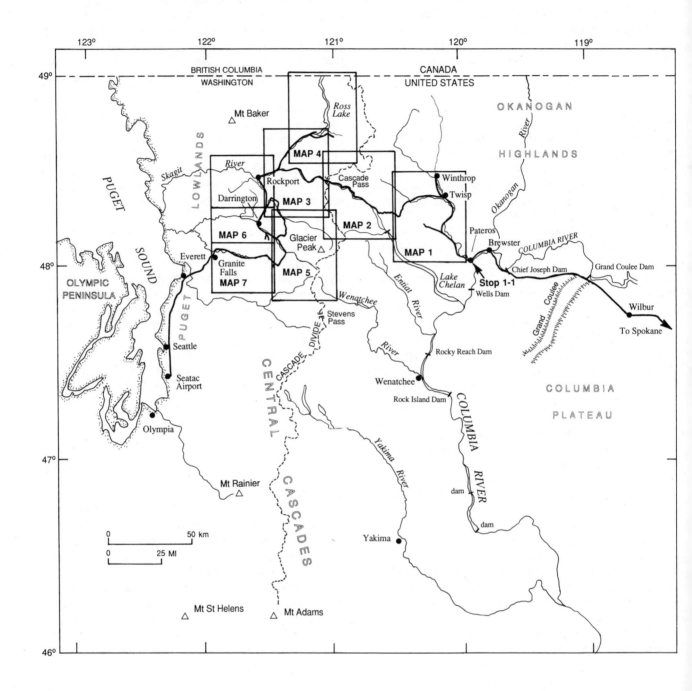

FIGURE 8. General location of North Cascades and field trip route. Numbered rectangles refer to geologic maps of Figure 9.

FIGURE 9. (On following pages) Geologic sketch map of North Cascades and field guide stops. See Figure 8 for map locations. In explanations, metamorphic and melange units are arranged by probable ages of their protoliths. Very heavy lines show route and roads except where dashed (route and trails). Heavy lines are faults; thin lines are geologic contacts (dotted under alluvium). Solid dots are field trip stops. Maps compiled from the following sources: (numbers in parentheses refer to map numbers) Barksdale [1975] (1); Brown and others [1987] (3,6); Cater and Crowder [1967] (2); Cater and Wright [1967] (2); Ford and others [in press] (2); M.F. McGroder (written communication 1988) (1,2); Miller [1987] (1,2); R.B. Miller, unpublished field maps (2,4); Misch [1977b] (2,4); Staatz and others [1972] (4); Tabor [1961] (2); Tabor and others [1988] (2,3,5,6,7); Tabor and others [in press] (5,7); R.W. Tabor and R.A. Haugerud, unpublished USGS field maps (2,3,4); Tipper and others [1981] (4).

EXPLANATION
(Figure 9, map 1)

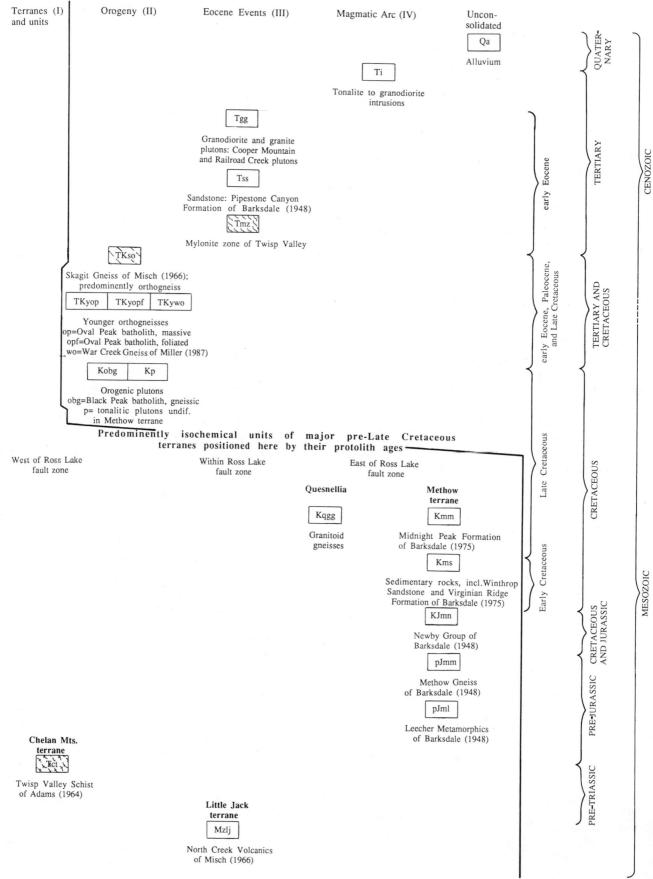

Terranes (I) and units

Orogeny (II)

Eocene Events (III)

Magmatic Arc (IV)

Unconsolidated

Qa

Alluvium

Ti

Tonalite to granodiorite intrusions

Tgg

Granodiorite and granite plutons: Cooper Mountain and Railroad Creek plutons

Tss

Sandstone: Pipestone Canyon Formation of Barksdale (1948)

Tmz

Mylonite zone of Twisp Valley

TKso

Skagit Gneiss of Misch (1966); predominantly orthogneiss

TKyop	TKyopf	TKywo

Younger orthogneisses
op=Oval Peak batholith, massive
opf=Oval Peak batholith, foliated
wo=War Creek Gneiss of Miller (1987)

Kobg	Kp

Orogenic plutons
obg=Black Peak batholith, gneissic
p= tonalitic plutons undif.
in Methow terrane

Predominantly isochemical units of major pre-Late Cretaceous terranes positioned here by their protolith ages

West of Ross Lake fault zone

Within Ross Lake fault zone

East of Ross Lake fault zone

Quesnellia

Kqgg

Granitoid gneisses

Methow terrane

Kmm

Midnight Peak Formation of Barksdale (1975)

Kms

Sedimentary rocks, incl. Winthrop Sandstone and Virginian Ridge Formation of Barksdale (1975)

KJmn

Newby Group of Barksdale (1948)

pJmm

Methow Gneiss of Barksdale (1948)

pJml

Leecher Metamorphics of Barksdale (1948)

Chelan Mts. terrane

Rct

Twisp Valley Schist of Adams (1964)

Little Jack terrane

Mzlj

North Creek Volcanics of Misch (1966)

early Eocene

early Eocene, Paleocene, and Late Cretaceous

Late Cretaceous

Early Cretaceous

QUATERNARY

TERTIARY

TERTIARY AND CRETACEOUS

CRETACEOUS

CRETACEOUS AND JURASSIC

PRE-JURASSIC

PRE-TRIASSIC

CENOZOIC

MESOZOIC

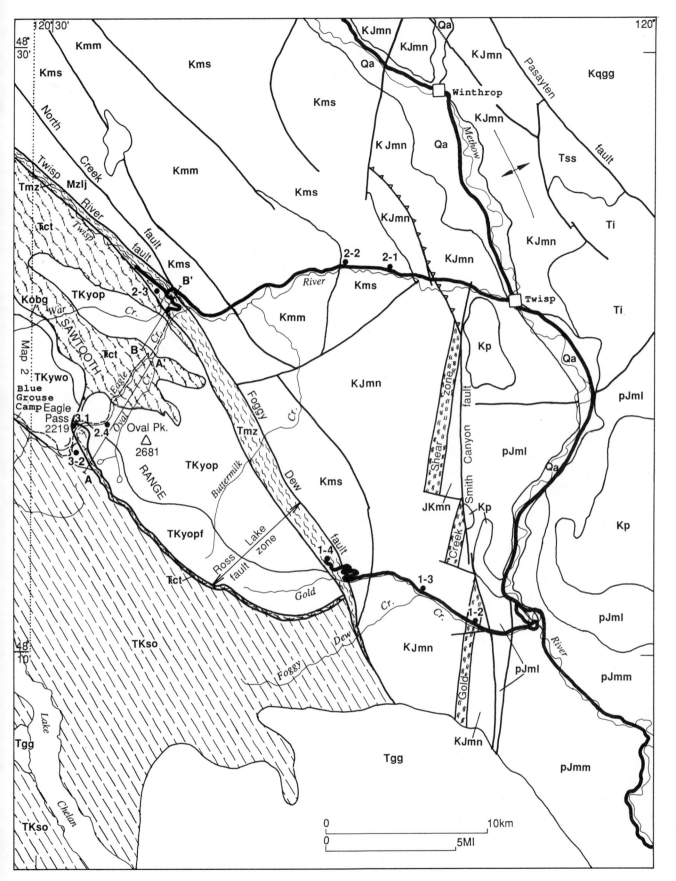

Figure 9, Map 1

EXPLANATION
(Figure 9, map 2)

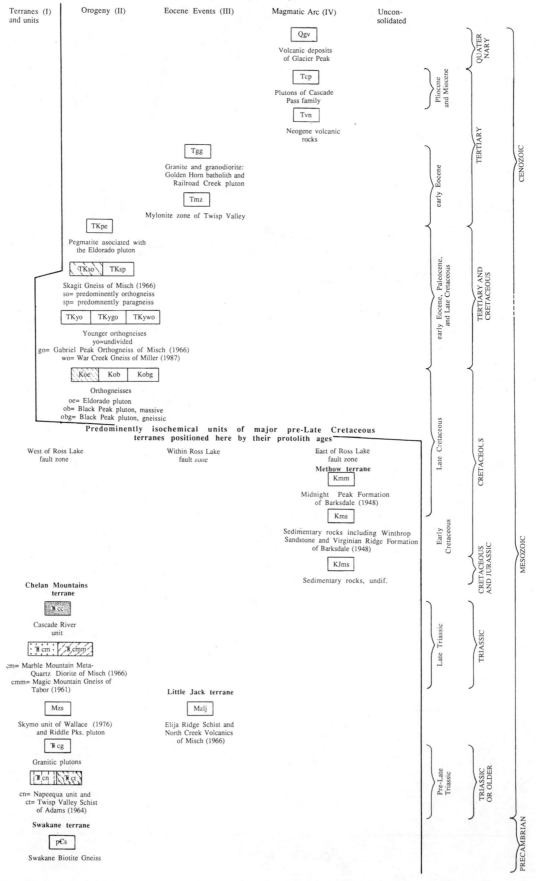

Terranes (I) and units

Orogeny (II)

Eocene Events (III)

Magmatic Arc (IV)

Uncon- solidated

Qgv

Volcanic deposits of Glacier Peak

Tcp

Plutons of Cascade Pass family

Tvn

Neogene volcanic rocks

Tgg

Granite and granodiorite: Golden Horn batholith and Railroad Creek pluton

Tmz

Mylonite zone of Twisp Valley

TKpe

Pegmatite asociated with the Eldorado pluton

TKso | TKsp

Skagit Gneiss of Misch (1966)
so= predominently orthogneiss
sp= predomnently paragneiss

TKyo | TKygo | TKywo

Younger orthogneises
yo=undivided
go= Gabriel Peak Orthogneiss of Misch (1966)
wo= War Creek Gneiss of Miller (1987)

Koe | Kob | Kobg

Orthogneisses
oe= Eldorado pluton
ob= Black Peak pluton, massive
obg= Black Peak pluton, gneissic

Predominently isochemical units of major pre-Late Cretaceous terranes positioned here by their protolith ages

West of Ross Lake fault zone

Within Ross Lake fault zone

East of Ross Lake fault zone

Methow terrane

Kmm

Midnight Peak Formation of Barksdale (1948)

Kms

Sedimentary rocks including Winthrop Sandstone and Virginian Ridge Formation of Barksdale (1948)

KJms

Sedimentary rocks, undif.

Chelan Mountains terrane

℞ cc

Cascade River unit

℞ cm | ℞ cmm

cm= Marble Mountain Meta- Quartz Diorite of Misch (1966)
cmm= Magic Mountain Gneiss of Tabor (1961)

Mzs

Skymo unit of Wallace (1976) and Riddle Pks. pluton

℞ cg

Granitic plutons

℞ cn | ℞ ct

cn= Napeequa unit and
ct= Twisp Valley Schist of Adams (1964)

Little Jack terrane

Mzlj

Elija Ridge Schist and North Creek Volcanics of Misch (1966)

Swakane terrane

p€s

Swakane Biotite Gneiss

QUATER- NARY

Pliocene and Miocene

early Eocene

TERTIARY

CENOZOIC

early Eocene, Paleocene, and Late Cretaceous

TERTIARY AND CRETACEOUS

Late Cretaceous

CRETACEOUS

Early Cretaceous

CRETACEOUS AND JURASSIC

Late Triassic

TRIASSIC

MESOZOIC

Pre-Late Triassic

TRIASSIC OR OLDER

PRECAMBRIAN

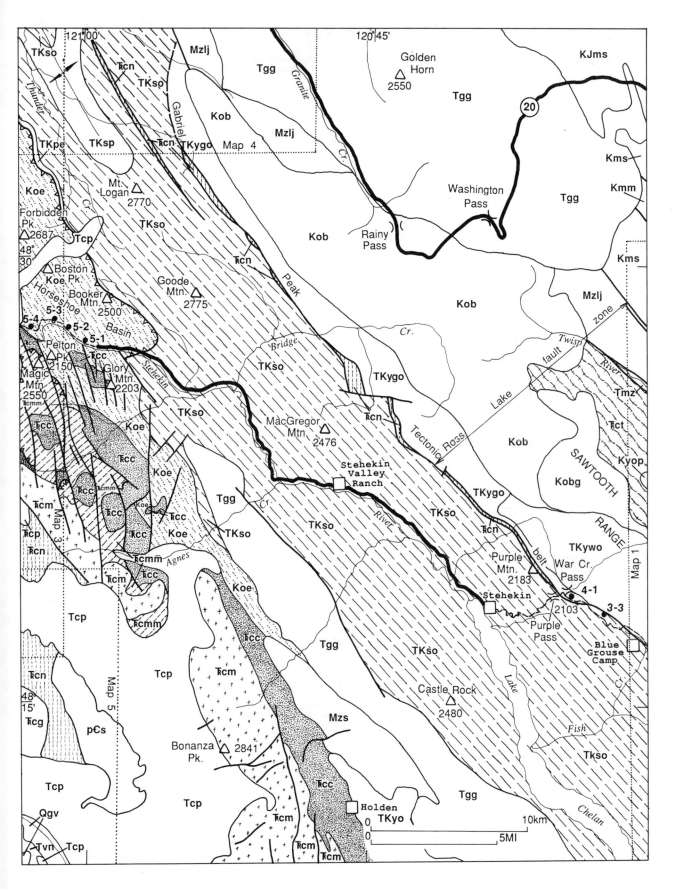

Figure 9, Map 2

EXPLANATION
(Figure 9, map 3)

Terrances (I) and units | Orogeny (II) | Eocene Events (III) | Magmatic Arc (IV) Orogeny (II) | Unconsolidated

Qa
Alluvial and glacial deposits

Tcp
Plutons of Cascade Pass family

Tsn
Plutons of Snoqualmie family

Tin
Plutons of Index family

Tss
Sandstone

TKpe
Pegmatite asociated with the El Dorado pluton

TKso **TKsp**
Skagit Gneiss of Misch (1966)
so= predominently orthogneiss
sp= predominently paragneiss

TKyo
Younger orthogneisses
includes Alma Cr. and Marble Cr. plutons and Hidden Lake Peak stock

Koe **Koc** **Kobl** **Kos**
Orthogneisses
oc= Eldorado pluton
oc= Chaval pluton
bl= Bench Lake pluton
os= Sloan Cr. plutons

Right-side time columns:
- QUATERNARY
- Pliocene and Miocene / Miocene and Oligocene / early Oligocene / TERTIARY
- early Eocene, Paleocene, and Late Cretaceous / TERTIARY AND CRETACEOUS
- Late Cretaceous / CRETACEOUS
- CENOZOIC

Predominently isochemical units of major pre-Late Cretaceous terranes positioned here by their protolith ages

West of Straight Creek fault

East of Straight Creek fault

Northwest Cascades System (NWCS)

Jes **Jed**
Easton Metamorphic Suite
es= Shuksan Greenschist
ed= Darrington Phyllite

Nason terrane

Ћ nc
Chiwaukum Schist

Chelan Mountains terrane

Ћ cc
Cascade River unit

Ћ cm **Ћ cmm**
cm= Marble Mountain Meta-Quartz Diorite of Misch (1966)
cmm= Magic Mountain Gneiss of Tabor (1961)

Mzeb
Elbow Lake Formation of Brown and others (1987)

Ћ cg
Granitic plutons

Ћ cn
Napeequa unit

Grandy Ridge terrane

Pzc
Chilliwack Group of Cairnes (1944)

Yellow Aster terrane

ya
Yellow Aster Complex of Misch (1966)

Right-side time columns (lower):
- JURASSIC
- Late Triassic / TRIASSIC
- Pre-Late Triassic / SILURIAN TO PERMIAN OR OLDER
- TRIASSIC
- PALEOZOIC AND PRECAMBRIAN
- MESOZOIC
- PALEOZOIC

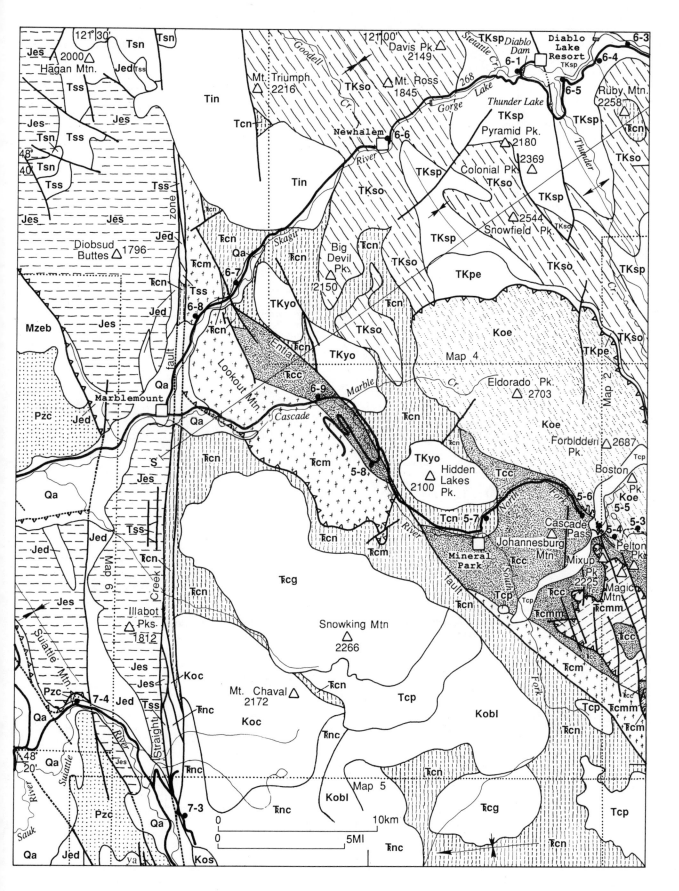

Figure 9, Map 3

EXPLANATION
(Figure 9, map 4)

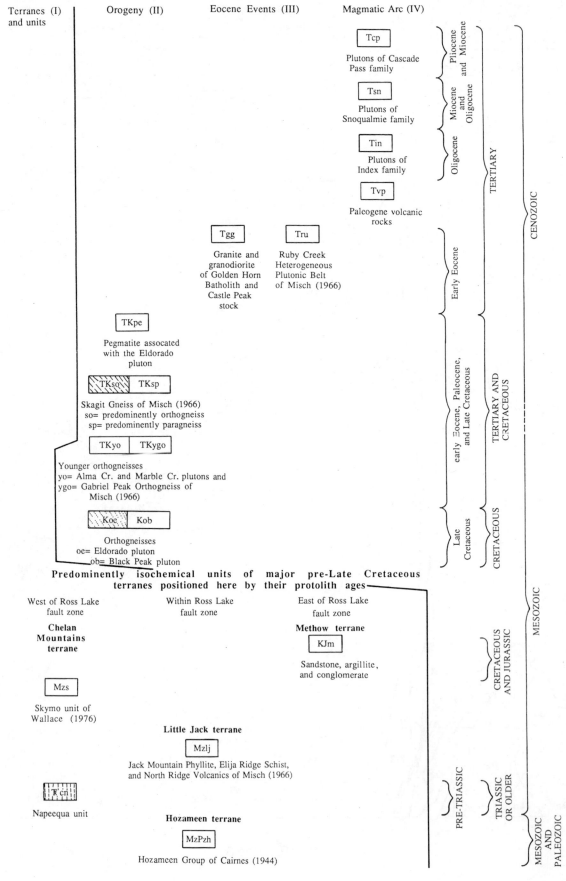

Terranes (I) and units

Orogeny (II)

Eocene Events (III)

Magmatic Arc (IV)

Tcp
Plutons of Cascade Pass family

Tsn
Plutons of Snoqualmie family

Tin
Plutons of Index family

Tvp
Paleogene volcanic rocks

Pliocene and Miocene

Miocene and Oligocene

Oligocene

TERTIARY

Tgg
Granite and granodiorite of Golden Horn Batholith and Castle Peak stock

Tru
Ruby Creek Heterogeneous Plutonic Belt of Misch (1966)

Early Eocene

TKpe
Pegmatite associated with the Eldorado pluton

TKso TKsp
Skagit Gneiss of Misch (1966)
so= predominently orthogneiss
sp= predominently paragneiss

early Eocene, Paleocene, and Late Cretaceous

TERTIARY AND CRETACEOUS

CENOZOIC

TKyo TKygo
Younger orthogneisses
yo= Alma Cr. and Marble Cr. plutons and
ygo= Gabriel Peak Orthogneiss of Misch (1966)

Koe Kob
Orthogneisses
oe= Eldorado pluton
ob= Black Peak pluton

Late Cretaceous

CRETACEOUS

Predominently isochemical units of major pre-Late Cretaceous terranes positioned here by their protolith ages

West of Ross Lake fault zone

Within Ross Lake fault zone

East of Ross Lake fault zone

Chelan Mountains terrane

Methow terrane

KJm
Sandstone, argillite, and conglomerate

CRETACEOUS AND JURASSIC

MESOZOIC

Mzs
Skymo unit of Wallace (1976)

Little Jack terrane

Mzlj
Jack Mountain Phyllite, Elija Ridge Schist, and North Ridge Volcanics of Misch (1966)

Ŧcn
Napeequa unit

PRE-TRIASSIC

TRIASSIC OR OLDER

Hozameen terrane

MzPzh
Hozameen Group of Cairnes (1944)

MESOZOIC AND PALEOZOIC

T307: 26

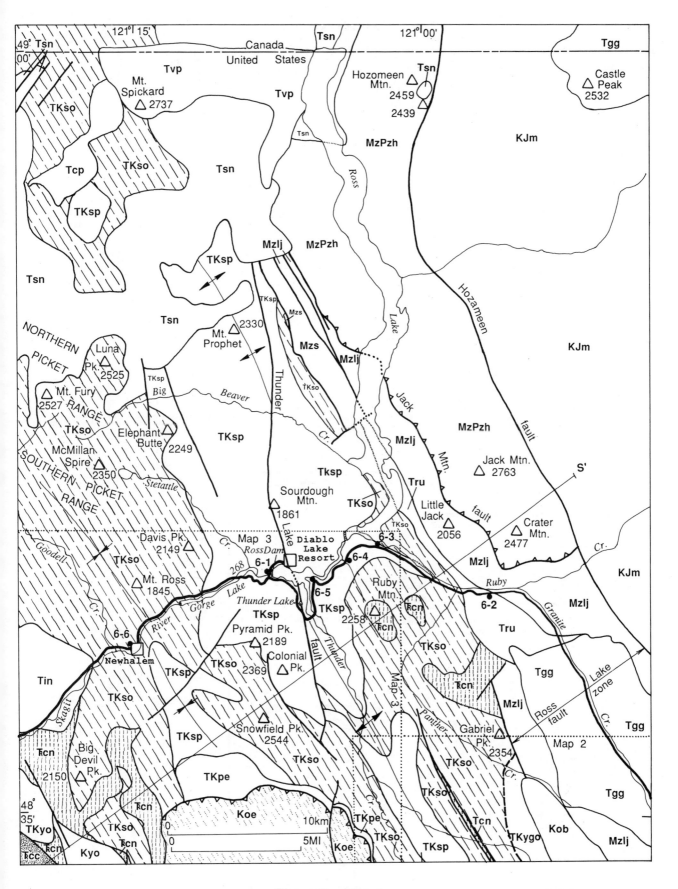

Figure 9, Map 4

EXPLANATION

(Figure 9, map 5)

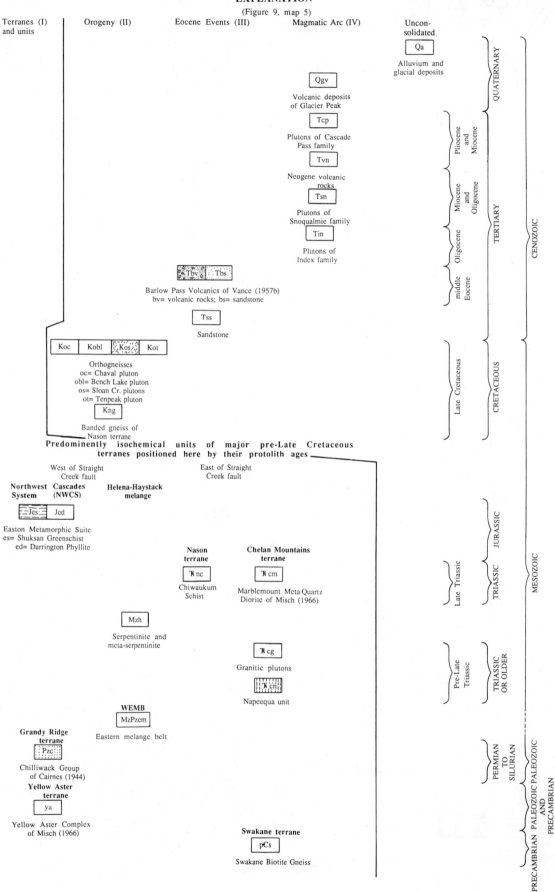

Terranes (I) and units	Orogeny (II)	Eocene Events (III)	Magmatic Arc (IV)	Unconsolidated

Qa

Alluvium and glacial deposits

Qgv

Volcanic deposits of Glacier Peak

Tcp

Plutons of Cascade Pass family

Tvn

Neogene volcanic rocks

Tsn

Plutons of Snoqualmie family

Tin

Plutons of Index family

Tbv Tbs

Barlow Pass Volcanics of Vance (1957b)
bv= volcanic rocks; bs= sandstone

Tss

Sandstone

Koc Kobl Kos Kot

Orthogneisses
oc= Chaval pluton
obl= Bench Lake pluton
os= Sloan Cr. plutons
ot= Tenpeak pluton

Kng

Banded gneiss of Nason terrane

Predominently isochemical units of major pre-Late Cretaceous terranes positioned here by their protolith ages

West of Straight Creek fault

East of Straight Creek fault

Northwest Cascades System (NWCS)

Helena-Haystack melange

Jes Jed

Easton Metamorphic Suite
es= Shuksan Greenschist
ed= Darrington Phyllite

Nason terrane

Ʀ nc

Chiwaukum Schist

Chelan Mountains terrane

Ʀ cm

Marblemount Meta Quartz Diorite of Misch (1966)

Mzh

Serpentinite and meta-serpentinite

Ʀ cg

Granitic plutons

Ʀ cn

Napeequa unit

WEMB

MzPzem

Eastern melange belt

Grandy Ridge terrane

Pzc

Chilliwack Group of Cairnes (1944)

Yellow Aster terrane

ya

Yellow Aster Complex of Misch (1966)

Swakane terrane

pЄs

Swakane Biotite Gneiss

QUATERNARY
Pliocene and Miocene
Miocene and Oligocene
Oligocene
middle Eocene
TERTIARY
CENOZOIC
Late Cretaceous
CRETACEOUS
JURASSIC
Late Triassic
TRIASSIC
Pre-Late Triassic
TRIASSIC OR OLDER
MESOZOIC
PERMIAN TO SILURIAN
PALEOZOIC AND PRECAMBRIAN
PRECAMBRIAN

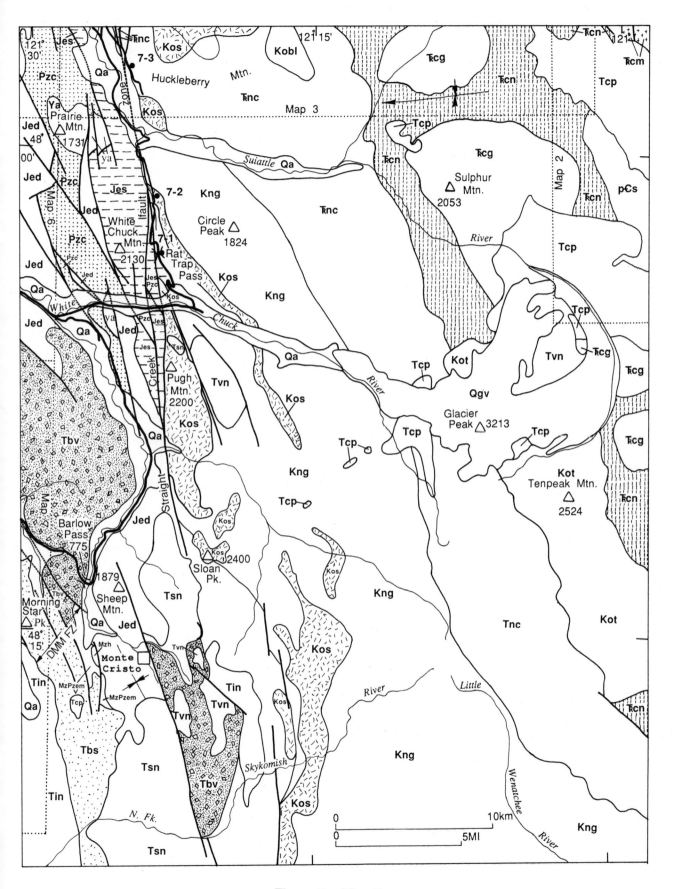

Figure 9, Map 5

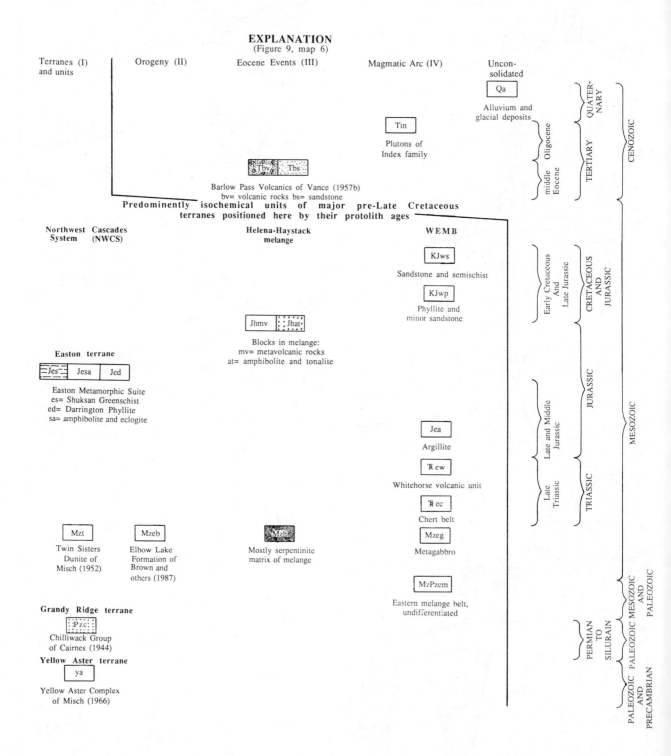

EXPLANATION
(Figure 9, map 6)

Terranes (I)
and units

Orogeny (II)

Eocene Events (III)

Magmatic Arc (IV)

Uncon-
solidated

Qa

Alluvium and
glacial deposits

Tin

Plutons of
Index family

Tbv   Tbs

Barlow Pass Volcanics of Vance (1957b)
bv= volcanic rocks bs= sandstone

**Predominently isochemical units of major pre-Late Cretaceous
terranes positioned here by their protolith ages**

**Northwest Cascades
System (NWCS)**

**Helena-Haystack
melange**

WEMB

KJws

Sandstone and semischist

KJwp

Phyllite and
minor sandstone

Jhmv   Jhat

Blocks in melange:
mv= metavolcanic rocks
at= amphibolite and tonalite

Easton terrane

Jes   Jesa   Jed

Easton Metamorphic Suite
es= Shuksan Greenschist
ed= Darrington Phyllite
sa= amphibolite and eclogite

Jea

Argillite

ℝew

Whitehorse volcanic unit

ℝec

Chert belt

Mzt

Twin Sisters
Dunite of
Misch (1952)

Mzeb

Elbow Lake
Formation of
Brown and
others (1987)

Mzh

Mostly serpentinite
matrix of melange

Mzeg

Metagabbro

MzPzem

Eastern melange belt,
undifferentiated

Grandy Ridge terrane

Pzc

Chilliwack Group
of Cairnes (1944)

Yellow Aster terrane

ya

Yellow Aster Complex
of Misch (1966)

QUATER-
NARY

middle Oligocene
Eocene

TERTIARY

CENOZOIC

Early Cretaceous
And
Late Jurassic

CRETACEOUS
AND
JURASSIC

Late and Middle
Jurassic

JURASSIC

Late
Triassic

TRIASSIC

MESOZOIC

PERMIAN
TO
SILURAIN

PALEOZOIC

PALEOZOIC
MESOZOIC
AND
PALEOZOIC

PRECAMBRIAN

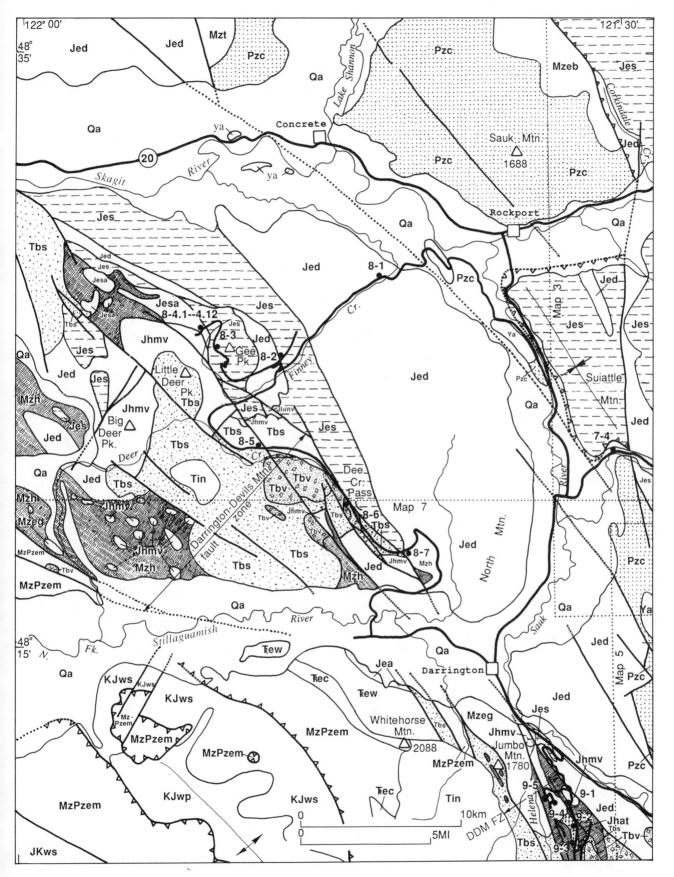

Figure 9, Map 6

EXPLANATION
(Figure 9, map 7)

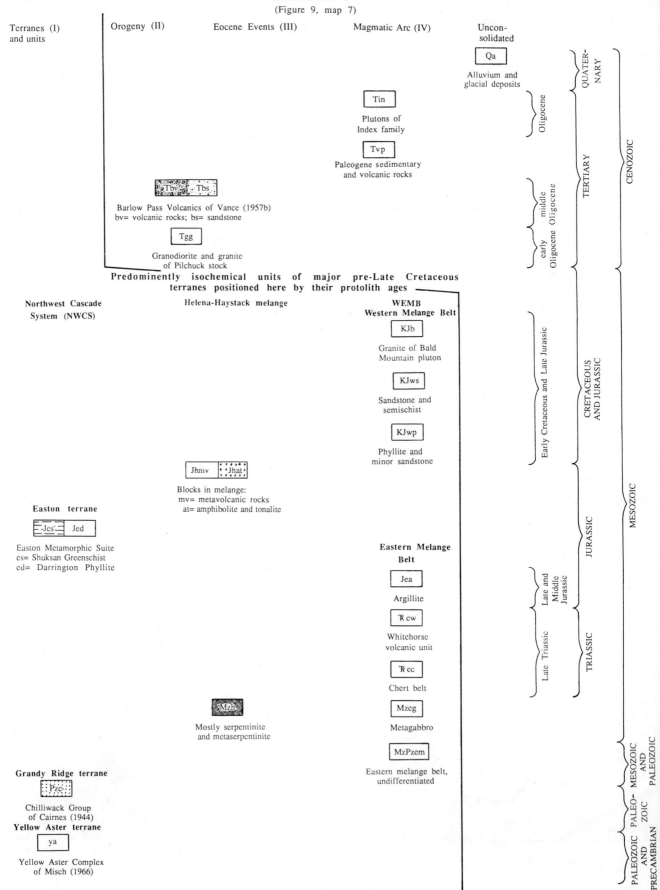

Terranes (I)
and units

Orogeny (II)

Eocene Events (III)

Magmatic Arc (IV)

Uncon-
solidated

Qa

Alluvium and
glacial deposits

Tin

Plutons of
Index family

Tvp

Paleogene sedimentary
and volcanic rocks

Tbv Tbs

Barlow Pass Volcanics of Vance (1957b)
bv= volcanic rocks; bs= sandstone

Tgg

Granodiorite and granite
of Pilchuck stock

**Predominently isochemical units of major pre-Late Cretaceous
terranes positioned here by their protolith ages**

**Northwest Cascade
System (NWCS)**

Helena-Haystack melange

**WEMB
Western Melange Belt**

KJb

Granite of Bald
Mountain pluton

KJws

Sandstone and
semischist

KJwp

Phyllite and
minor sandstone

Jhmv Jhat

Blocks in melange:
mv= metavolcanic rocks
at= amphibolite and tonalite

Easton terrane

Jes Jed

Easton Metamorphic Suite
es= Shuksan Greenschist
cd= Darrington Phyllite

**Eastern Melange
Belt**

Jea

Argillite

℞cw

Whitehorse
volcanic unit

℞cc

Chert belt

Mzeg

Metagabbro

Mzh

Mostly serpentinite
and metaserpentinite

MzPzem

Eastern melange belt,
undifferentiated

Grandy Ridge terrane

Pzc

Chilliwack Group
of Cairnes (1944)

Yellow Aster terrane

ya

Yellow Aster Complex
of Misch (1966)

QUATER-
NARY

Oligocene

TERTIARY

CENOZOIC

early   middle
Oligocene Oligocene

Early Cretaceous and Late Jurassic

CRETACEOUS
AND JURASSIC

MESOZOIC

Late and
Middle
Jurassic

JURASSIC

Late Triassic

TRIASSIC

MESOZOIC
AND
PALEOZOIC

PALEO-
ZOIC

PALEOZOIC
AND
PRECAMBRIAN

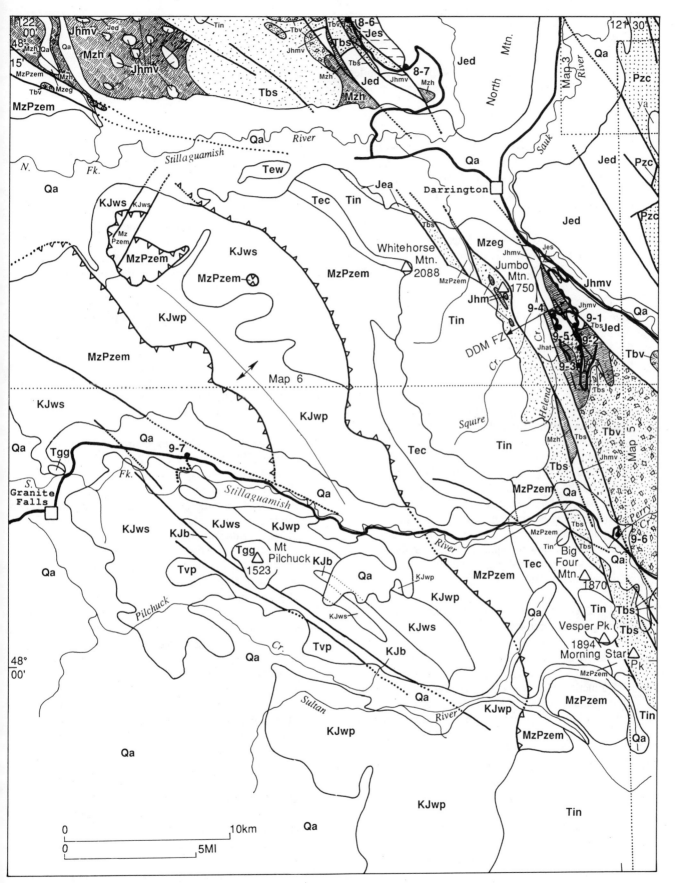

Figure 9, Map 7

Distances are in miles (kilometers in parentheses)
References to Sections refer to the Overview

Day 1. *Spokane to Twisp and Winthrop.* We cross the
Columbia River plateau with views of plateau basalts
and glacial features. In the late afternoon we make a
few stops in rocks of the Methow terrane and look at the
Ross Lake fault zone.

0.0 Intersection of I-90 and Maple Street, Spokane. Take I-
90 west 2 miles (3 km) to US 2, follow US 2 west
through Reardon (22 miles/35 km) and Davenport (35
miles/56 km).

For the next two hours we drive across the north
edge of the Columbia River plateau, a moderate-sized
flood basalt province. The bulk of the lava erupted in
the Miocene between 16.5 and 6 Ma. The volume of
tholeiitic basalt has been estimated to be 2×10^5 km^3
[Swanson and Wright, 1978].

During the late Pleistocene the surface of the
plateau was modified by catastrophic floods
(jokulhlaup) issuing from glacial Lake Missoula in
western Montana. The high-velocity flood waters
(locally up to 30 m/sec, V. Baker, *in* Baker and
Nummedal [1978]) stripped the eolian cover off the
basalt, carved large channels (coulees), built mile-long
gravel bars, and laid down ripple marks that are small
hills [Baker and Nummedal, 1978]. The resulting
topography is known as channeled scabland. Some 40
jokulhlaups are now recognized [Waitt, 1980], though
the effects were originally attributed to a single great
"Spokane flood" [Bretz, 1925].

52(84) US 2 leaves the scattered ponderosa pine forest and
enters grass-covered loess and channeled scabland.

63(101) Town of Wilbur. Continue west on US 2.

64(103) Junction with WA 174. Turn right, and proceed
west (actually north at this point) on WA 174. In 0.4
mile (0.6 km) pass junction with WA 21; continue on
WA 174.

75(121) Leave wheat-fields (loess dunes) and begin descent
into Grand Coulee, a major channel during the late
Pleistocene jokulhlaups.

83(134) Junction with WA 155.

85.3(137.3) Turnoff to right (north) goes to Crown Point
Vista and view of Grand Coulee Dam. Our route
continues ahead.

Hills to the north are the crystalline rocks of the
Okanogan gneiss dome, a middle Eocene metamorphic
core complex emplaced into rocks of the Quesnel
terrane.

104.6(168.3) Junction. Turn right onto WA 17.

120.4(193.8) Junction with WA 173 to Brewster. Continue
on WA 17, crossing bridge over the Columbia River,
here ponded behind Wells Dam.

128.3(206.5) Junction with US 97. Follow US 97 towards
Brewster.

132(212) Enter town of Brewster.

140.3(225.8) Junction with WA 153 and town of Pateros.
Turn right onto WA 153.

140.8(226.6) ***Stop 1-1*** Good exposures of the migmatite
complex of Alta Lake. Park 0.2 miles (0.3 km) beyond
and walk back (see Figure 8).

The Alta Lake unit consists mainly of variably
migmatized amphibolite and trondhjemitic orthogneiss,
with lesser biotite schist and rare quartzite [Raviola,
1988]. Migmatization resulted from metamorphic
differentiation and injection, according to Raviola.
These rocks resemble migmatites in the Chelan
Complex of Hopson and Mattinson [1971] (see Section
I) but may be separated from them by the Ross Lake
fault [Raviola, 1988]. Hornblende from an amphibolite
in the Alta Lake unit yields a K-Ar age of 104 ± 5 Ma
[Raviola, 1988]. This is much older than most
hornblende ages from the Chelan Complex (maximum
84 Ma [Tabor and others, 1987]) and, at the least,
reflects a different cooling history.

The Alta Lake unit occupies a critical position
between the southern extensions of the Ross Lake fault
zone and the Chewack-Pasayten fault, but the actual
relationship of the unit to gneisses of the Chelan
Mountains terrane of the North Cascades to the west
and gneisses of the Okanogan Ranges to the east
(Quesnel terrane) is yet to be determined.

Along the Methow River Valley upstream are
numerous glacial outwash terraces and occasional
outcrops of the pre-Late Cretaceous Methow Gneiss.

144.1(231.9) Bridge over Methow River. A large raft of
schist in the Methow Gneiss is exposed along the river.

151.3(243.3) Community of Methow.

157.9(254.1) Junction of Gold Creek Road (USFS 3201) and WA 153. Turn left onto Gold Creek Road.

158.0(254.3) Outcrop of Methow Gneiss.

158.8(255.6) Junction with USFS road that returns to WA 153. Stay left on USFS 3201. We return to this junction after Stop 1-4.

159.9(257.3) Junction with South Fork Gold Creek Road (USFS 3107). Keep straight.

161.3(259.6) *Stop 1-2* Strongly foliated greenschist and minor phyllite (east end of outcrop) of the Gold Creek shear zone (Figure 9, map 1) This 1-km-wide shear zone strikes NNE, at a significant angle to the regional NW trend of the North Cascades. The gently N-plunging stretching lineation and steep foliation suggest that this is a strike-slip zone, and DiLeonardo [1987] has postulated dextral slip on the basis of sparse kinematic indicators. This high strain zone deforms the pre-Lower Cretaceous Newby Group of Barksdale [1948]—perhaps correlative with the Ladner Group (Figure 4)—of the Methow terrane and is truncated to the south by the middle Eocene Cooper Mountain batholith.

161.6(260.1) Road junction, keep straight.

161.8(260.4) Begin large outcrops of the Newby Group.

163.9(263.8) *Stop 1-3* Outcrops of well-bedded volcaniclastic sedimentary rocks of the Newby Group, which forms the base of the southern Methow terrane. Hopkins [1987] describes graded and cross-laminated beds in these rocks which he interprets as turbidites apparently derived from a volcanic arc to the east [Tennyson and Cole, 1978]. These rocks probably are the protolith for some of the strongly foliated greenschist in the Gold Creek shear zone (**Stop 1-2**).

164.0(263.9) Turnoff to Foggy Dew Campground. Pavement ends. Keep straight.

165.5(266.3) Junction with Crater Creek Road. Turn left, toward the Crater Creek trail.

168.2(270.7) Amphibolite of the Foggy Dew fault zone on right and on hill above.

168.5(271.2) Mylonitic gneiss and amphibolite of the Foggy dew fault zone in roadcut on right. Gneiss from this outcrop yielded a discordant U-Pb zircon age of ~50 Ma (S. A. Bowring, written communication, 1988).

Park and walk up trail about 0.3 miles (0.5 km) to a long series of outcrops of mylonitic gneiss, amphibolite, and rare schist of the Foggy Dew fault zone (Figure 9-1). This ~1-km-wide mylonite zone marks the southernmost segment of the Ross Lake fault zone (Section III). The Foggy Dew fault separates the amphibolite-facies mylonites from weakly metamorphosed rocks of the Methow terrane to the northeast.

NW-striking mylonitic foliation in the Foggy Dew fault zone dips moderately steeply to steeply to the NE and the stretching lineation typically plunges about 25-30° to the SE. Abundant type-1 and type-2 S-C fabrics [Lister and Snoke, 1984] and lesser shear bands, asymmetric porphyroclasts, and foliation fish provide evidence for oblique dextral slip. Asymmetric folds of foliation are also common in the amphibolites. The structures record oblique slip with components of dextral strike slip and normal dip slip, down to the NE [Miller, 1987].

Some of the mylonitic gneiss in these outcrops is derived from the ~64 Ma Oval Peak pluton (U-Pb zircon, Miller and others [1988]). Intrusive contacts are locally preserved between the Oval Peak-derived gneiss and the amphibolites. The latter have yielded a K-Ar hornblende age of 55.8 ± 3.6 Ma (J. K. Nakata, written communication, 1987) and may be derived from metadiabase and metagabbro which occur locally in the fault zone. These ages date ductile deformation as Paleocene and (or) early to possible middle Eocene. The fault is plugged by the middle Eocene Cooper Mountain batholith.

Rare metasedimentary rocks in these outcrops include sillimanite-bearing schists that may be derived from the Twisp Valley Schist of Adams [1964].

Turn around and retrace route ~10 miles (16 km) to road junction (mile 158.8, above).

0.0 Junction of USFS Road 3201 and shortcut to WA 153. Reset odometer and turn left (north) towards Twisp.

~0.4(0.6) Large outcrops of Methow Gneiss.

1.6(2.6) Junction with WA 153. Turn left (north) towards Twisp.

4.1(6.6) Junction with Libby Creek Road. Stay on WA 153.

~4.2(6.8) Large outcrops of the Leecher Metamorphics of Barksdale [1975].

5.1(8.2) Cross the Methow River.

5.2(8.4) Enter Carlton.

13.9(22.4) Junction with WA 20. Keep straight on WA 20.

15.9(25.6) Cross Methow River and enter Twisp.

16.0(25.7) Downtown Twisp. Junction with Twisp River Road. Keep straight (north) on WA 20 toward Winthrop.

17.0(27.4) Views across Methow River to eroded hoodoos of Eocene(?) Pipestone Canyon Formation of Barksdale [1948] (Section III). These terrestrial sedimentary rocks comprise fanglomerate, sandstone, and lacustrine silstsone [Barksdale, 1975]. Their deposition may have been controlled by the Chewack-Pasayten fault.

Continue ~7 miles (11 km) north to Winthrop, lodging, and dinner.

Day 2. *Winthrop to upper Eagle Creek.* We examine Cretaceous sedimentary rocks in the Methow terrane and mylonites in the Twisp River fault zone, then hike to upper Eagle Creek Camp, passing outcrops of the Twisp Valley Schist and the Oval Creek pluton.

0.0 Retrace route to Twisp and junction with the Twisp River Road. Reset odometer and turn right (west).

5.1(8.2) ***Stop 2-1*** Steeply-dipping sandstone, shale, and chert-pebble conglomerate of the Virginian Ridge Formation of Barksdale [1948]. The shales show weak pencil structure.

Albian and older sub-Virginian Ridge strata are marine and east-derived [Barksdale, 1975; Tennyson and Cole, 1978]. Albian and Cenomanian strata of the Virginian Ridge Formation were derived from the west and deposited in fluvial, deltaic, and shallow-marine settings [Trexler, 1985]. Virginian Ridge strata interfinger to the east with non-marine arkose of the Winthrop Sandstone (seen at **Stop 2-2**). Both units grade up into the subaerial Midnight Peak Formation of Barksdale [1948]; collectively, the three units are laterally equivalent to the Pasayten Group recognized in Manning Park to the north [Coates, 1974; Tennyson and Cole, 1978].

The western chert-rich source may have been the Hozameen Group, uplifted by crustal thickening in the early stages of mid-Cretaceous orogeny (Section II). This thickening seems to have affected the Methow basin as well, as no post-Cenomanian marine strata are present. By Midnight Peak time deformation had migrated into the Methow basin, as an angular unconformity is present at least locally [McGroder, 1988].

Cretaceous Methow strata are preserved in the cores of large NNW-trending synclines. McGroder [1988] interprets the related thrusts, which verge both east and west, to be related to a west-verging mid-Cretaceous Cascade thrust system. However, facies patterns within the Virginian Ridge Formation suggested to Trexler [1985] that its fan-delta system moved north through

time, and thus that dextral wrench faulting was the dominant tectonic theme.

6.8(10.9) ***Stop 2-2*** Roadcuts display feldspathic sandstone, siltstone, and shale of the Albian to Cenomanian Winthrop Sandstone of Barksdale [1948] which are cut by normal faults.

7.0(11.3) Roadcuts in the Winthrop Sandstone. More carbonaceous shale is present than at Stop 2-2. The sequence is cut by small faults and is weakly folded (drag?).

14.7(23.7) Intersection with War Creek road (USFS 4430). Turn left and cross Twisp River.

14.9(24.0) Intersection with USFS road 4420. Stay right to reach:

15.0(24.1) ***Stop 2-3*** Mylonites exposed in a series of roadcuts are associated with the Twisp River fault, here concealed by alluvium in the valley of the Twisp River. This major strand of the Ross Lake fault zone separates the polydeformed Twisp Valley Schist of Adams [1964] and other rocks of the North Cascade crystalline core from the lower-grade North Creek Volcanics of Misch [1966]. The latter consist of altered intermediate volcanic rocks and feldspathic sandstone of uncertain age which may be part of the lower section of the Methow terrane or may be a separate terrane (M. F. McGroder, written communication, 1987) such as the Little Jack terrane (Figure 9-1).

Rocks in the Twisp River fault zone are mainly mylonitic gneiss, garnet-biotite schist, and metaporphyry. The latter may have been intruded while the fault was active and forms the lineated mylonites exposed at this locality. They contain relict plagioclase phenocrysts which can be recognized in thin section, but are difficult to see with a hand lens.

The subhorizontal stretching lineation and the steep foliation suggest strike-slip motion in the mylonite zone. Kinematic indicators are rare, but dextral motion is tentatively favored as the mylonite zone is continuous with, and is probably a splay of, the Foggy Dew fault zone on which there is ample evidence of dextral slip (see Section III and **Stop 1-4**).

A few 10s of meters southwest (uphill) of these roadcuts is a narrow belt of mylonitic tonalitic gneiss which contains the assemblage quartz + plagioclase + biotite + muscovite + epidote, and displays plagioclase porphyroclasts and prominent quartz ribbons. The gneiss may be derived from the Oval Peak pluton, but is separated from it by garnet-biotite schist of the Twisp Valley Schist.

15.7(25.3) Turn around and retrace route to intersection with USFS 4420. Just down the road, turn right and

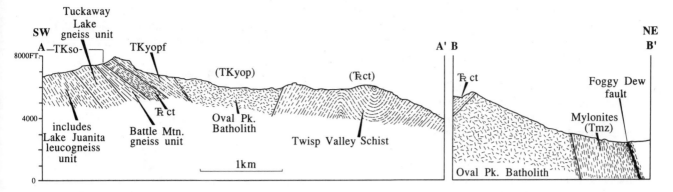

FIGURE 10. Diagrammatic cross sections of Oval Peak batholith and associated structures. See Figure 9, map 1 for location. After Miller [1987].

continue southeast along Twisp River to:

16.1(25.9) Intersection with Eagle Creek Road (USFS 080) climbing up hill to west. Turn right and drive up glacial outwash to:

17.6(28.3) **Eagle Creek trailhead** (~920 m/3010') where our hiking excursion begins. Distances given along the trail are approximate. The 1:24,000 geologic maps of the Oval Peak and Sun Mountain quadrangles [Miller, 1987] may be useful companions. The trail starts on glacial deposits, but a short distance to the east below the level of the trail Eagle Creek has cut down into the Twisp Valley Schist.

Exposures are absent in the first stretch of trail which stays on the west side of Eagle Creek. Large boulders from northern "finger" of the Paleocene Oval Peak pluton occur along the trail and are from outcrops on Snowshoe Ridge to the north. In the boulders note magmatic foliation defined by biotite books and rounded aggregates of weakly-strained quartz.

1.6(2.6) Boundary of the Lake Chelan-Sawtooth Wilderness and junction with the Oval Lake trail. Go right on the Eagle Creek trail.

2.2(3.5) Outcrops of Twisp Valley Schist occur a short distance above the trail and a pegmatite related to the Oval Peak pluton is exposed next to the trail. Some of the schist has relict clastic texture and may be metamorphosed siltstone, a relatively uncommon lithology within the Twisp Valley Schist.

4.7(7.6) (~1750 m/5750') Good view to the SE of Oval Peak (2681 m/8795'), the highest peak in the area. View to the WNW of the contact between Twisp Valley Schist and the foliated margin of the Oval Peak pluton. Note that the pluton overlies the schist along this steep intrusive contact (Figure 10, NE portion of section A-A').

5.3(8.5) Short trail to left goes to a good campsite, a lunch

stop, and then on to Silver Lake. After lunch our route continues up main Eagle Creek trail.

5.6(9.0) *Stop 2-4* (~1830 m/6000') Biotite leucotonalite at this stop is part of the isotropic to weakly foliated core of the Oval Peak pluton. This compositionally rather homogeneous pluton contains probable magmatic epidote and thus may have crystallized at relatively high pressures, although there is andalusite in the Twisp Valley Schist. A tonalite collected near here has yielded a preliminary U-Pb zircon age of ~63 Ma (S. A. Bowring, written communication, 1987), which likely is the crystallization age.

7.1(11.4) Well-used campsite (~2035 m/6680') in basin where we stop for the night.

Day 3. *Upper Eagle Creek to Lake Juanita.* We traverse Sawtooth Ridge with a chance to examine more of the Twisp Valley Schist, the Oval Peak pluton, and other orthogneisses associated with the RLFZ.

Stop 3-1 From the campsite traverse northward to the ridge which extends NE from Battle Mountain. A second traverse (**Stop 3-2**) begins at Eagle Pass, where we scramble up the ridge south of the pass to the ridge above Tuckaway Lake. Both traverses examine contact relations between the foliated margin of the Oval Peak pluton and structurally underlying rocks.

The foliated margin of the pluton sharply overlies a moderately steep, NE-dipping thin envelope of the Twisp Valley Schist which in turn structurally overlies, in descending order, the Battle Mountain gneiss, Tuckaway Lake gneiss, and Lake Juanita leucogneiss units of Miller [1987] (Figure 10). These orthogneisses are metamorphosed intrusive bodies within the Skagit Gneiss, and are probably among the youngest orthogneisses in the Skagit. The deformed SW margin of the Oval Peak pluton, underlying Twisp Valley Schist, Battle Mountain and Tuckaway Lake units, and highly foliated part of the Lake Juanita unit together constitute the Gabriel Peak tectonic belt, a high strain zone which extends northwestward for ~60 km (Section III; Figures 5, 6)).

The Twisp Valley Schist is an oceanic assemblage which we assign to the Chelan Mountains terrane. It may be correlative with part of the Permian to Jurassic Hozameen Group. In this area, the Twisp Valley Schist consists of siliceous schist, impure quartzite (metachert), amphibolite and calc-silicate rocks. A slice of metaperidotite occurs on the ridge NE of Battle Mountain (Stop 3-2). The intense deformation of the Twisp Valley Schist is best recorded by calc-silicate rocks. Isoclinal refolding of isoclinal folds has formed type-2 and type-3 interference patterns [Ramsay, 1967]. Gently (5-25°) NW-plunging sheath folds occur locally. Hornblende in an amphibolite in the Twisp Valley Schist, about 2 km to the SE of Eagle Pass, gives a K-Ar age of 54.1 ± 1.6 Ma (J. K. Nakata, written communication, 1987).

The tonalitic Battle Mountain unit is the oldest of the orthogneisses in this area and is intruded by sills of the structurally lower Tuckaway Lake unit. The Battle Mountain unit is distinguished from the latter unit by its coarser grain size and the presence of hornblende and large sphene.

The Tuckaway Lake unit is a collection of metamorphosed intrusive sheets which form a narrow belt between the Battle Mountain and Lake Juanita leucogneiss units. This biotite tonalite gneiss is distinguished from the structurally lower Lake Juanita unit by its finer grain size (medium-grained), stronger foliation, and higher color index.

The heterogeneous Lake Juanita leucogneiss unit consists mainly of leucogranodioritic and trondhjemitic biotite gneisses. Most of the unit is lineated, but only weakly to moderately foliated, except near its intrusive contact with the older Tuckaway Lake unit where it is well foliated and locally mylonitic. Leucogneiss from the Lake Juanita unit 1.2 km south of Eagle Pass has yielded a concordant U-Pb zircon age of about 59 Ma (S. A. Bowring, written communication, 1988). We interpret this as a crystallization age, and to indicate that dynamothermal metamorphism of the gneiss is Cenozoic (Paleocene-early Eocene).

The intensity of foliation and a down-dip stretching lineation in the underlying orthogneisses increases toward the envelope of Twisp Valley Schist. Foliation intensity in the Oval Peak also increases towards the schist envelope, but the pluton shows only a weak lineation. Much of this deformation may be related to forceful emplacement of the Oval Peak into a pre-existing shear zone.

0.0 Campsite, upper Eagle Creek.

1.8(2.9) Eagle Pass (~2220 m/7280'); scramble south up ridge to *Stop 3-2* (see discussion above). After finishing Stop 3-2, go west down hill to :

2.8(4.5) Junction with Summit trail (~2040 m/6680'). Continue to right (north) on Summit trail.

4.3(6.9) Junction with Fish Creek trail (~1720 m/5640'); continue straight on Summit trail.

6.0(9.7) Trail passes campsite ("Camp Comfort") on left.

7.1(11.4) Ridge crest (~2220 m/7280'). Excellent views of peaks to the west of Lake Chelan (Figure 11). Enter Lake Chelan National Recreation Area. Collecting rock samples is not permitted within the North Cascades National Park and Recreation Complex. Please put your hammers away to avoid temptation.

7.2(11.6) *Stop 3-3* As the trail begins to descend to the west, continue straight (north) along the ridge (Figure 9-2). Here, in the Gabriel Peak tectonic belt, the Lake Juanita leucogneiss unit is in contact with the probable southern continuation of the Gabriel Peak Orthogneiss of Misch [1966].

The Lake Juanita unit has developed a strong fabric and displays S-C relations near the contact with the Gabriel Peak. Foliation is subparallel to the contact which, in this area, typically dips 65-70° to the northeast. Strong down-dip streaking lineation suggests that this is a zone of dip-slip deformation. Approximately 15 km to the northwest, the lineation in this zone plunges gently, indicating a complex kinematic history for the Gabriel Peak tectonic belt.

At this point the Gabriel Peak Orthogneiss is a strongly foliated hornblende tonalite gneiss. Zircons from the Gabriel Peak farther to the northwest have been dated at 65-68 Ma [Hoppe, 1982].

Retrace route to the trail and continue northwest toward Lake Juanita. The trail stays below the ridge and passes scattered outcrops of Lake Juanita leucogneiss.

9.6(15.4) Junction with the War Creek trail. Continue about 0.1 mile (0.2 km) to campsites near Lake Juanita (2031 m/6665').

Day 4. Lake Juanita to Stehekin and Stehekin Valley Ranch. We look at schist along the Gabriel Peak tectonic zone and more orthogneisses of the Skagit Gneiss.

0.2(0.3) *Stop 4-1* War Creek Pass (~2070 m/6780'). At the pass, climb up the ridge to the southeast for a short distance (< 100 m) to again examine the Lake Juanita leucogneiss unit-Gabriel Peak Orthogneiss contact.

At this locality, a small lens of biotite schist and amphibolite of the Rainbow Lake Schist of Adams [1961, 1962] intervenes between the gneiss units. The Rainbow Lake Schist extends discontinuously

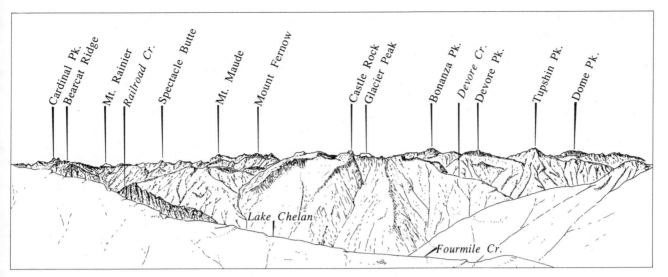

FIGURE 11. View southwest from pass south of Stop 3-3. Most of the mountainside west of Lake Chelan is in orthogneisses of the Skagit Gneiss, including Cardinal Peak, part of Bearcat Ridge, Castle Rock and Tupshin Peak. Spectacle Butte, Mt. Maude, Mt. Fernow, and Bonanza Peak are eroded from Late Triassic tonalite gneiss of the Marblemount belt. Devore Peak is in the granodiorite of the middle Eocene Railroad Creek pluton (figure 6). Dome Peak is in the Oligocene and Miocene tonalite of the Cloudy Pass batholith. Mount Rainier and Glacier Peak are Holocene volcanoes. Mount Rainier (4392 m/14410') is the highest peak in Washington and Bonanza Peak (2899 m/9511') is the highest non-volcanic peak.

northwestward for ~45 km and forms a narrow screen along the southwest side of the Gabriel Peak Orthogneiss. The unit consists of siliceous biotite schist and quartzite (metachert?), less common amphibolite, marble and calc-silicate rock, and rare metaperidotite [Adams, 1961; Miller, 1987]. Lithologies are similar to those of the Twisp Valley Schist and the oceanic Napeequa unit of the Chelan Mountains terrane, but the relationship between these units is unknown. The Rainbow Lake Schist may well be a relict of the Napeequa unit all but engulfed in orthogneiss and forming the eastern limb of the Skagit anticlinorium (see Section II and Figure 14).

Return to War Creek Pass. Turn right (north) on the War Creek trail for a short distance, examining good outcrops of the Gabriel Peak Orthogneiss. The junction with the Boulder Creek trail is reached after ~100 m. Continue north for 50 m, past large outcrops, to a 15 m wide zone of spectacular folded mylonites and ultramylonites of the Gabriel Peak Orthogneiss. Fold axes are parallel to the stretching lineation and we find sheath folds in the float. The most intense deformation occurred in this zone some 50 m from the contact with the Lake Juanita leucogneiss unit. Foliation intensity diminishes to the NE.

Return to War Creek Pass, descend to Lake Juanita, and continue west to Purple Pass.

1.1(1.8) Purple Pass (2098 m/6884'). Begin switchback descent to Stehekin. From Lake Juanita to Stehekin the trail passes through leucogneiss of the Lake Juanita unit and other, little studied, weakly metamorphosed to non-

metamorphosed intrusive rocks. All are labeled Skagit orthogneiss on Figure 9-2.

6.5(10.5) Cross Purple Creek (~790 m/2600'). Do not drink the water unless chemically treated or filtered.

8.4(13.5) Stehekin (335 m/1100'). The small settlement of Stehekin can be reached only by float plane, boat along Lake Chelan which stretches 55 miles to the southeast, or by trail on foot or horseback. The community has only a few permanent year-around residents, but in the summer boatloads of visitors arrive and depart daily from Stehekin Landing. The first settlers in Stehekin arrived in the late 1800s, and the community gradually grew as prospectors, trappers, and early tourists passed through in larger numbers. The town is now the hub of Lake Chelan National Recreation Area, a part of the North Cascades National Park complex which was established in 1968. The weary hiker (or geologist) can find meals, lodging, and a few supplies here.

0.0 Begin road miles at Stehekin.

0.6(1.0) A cliff along the east side of the road is hornblende-biotite tonalite orthogneiss of the Skagit Gneiss. The rock displays an unusual texture of small light-colored spots in a more uniform gneiss. Adams [1961] called these rocks "spotted granofels" and noted that the light-colored spots were equant in undeformed rock and elongate spindles where deformed, with length:width ratios as high as 4:1. Orientations of the spindles are the same as other stretching lineations in the Skagit Gneiss (see Section III). We have noted similar rocks throughout the Skagit as far north as Elephant Butte (Figure 9-4).

The spots are largely quartz and plagioclase, locally with minor chlorite, commonly with a grain of yellow sphene in the center (may not be visible at this outcrop), and are slightly bounded by aligned tangential subidioblastic plagioclase crystals. Are the spots recrystallized deformed phenocrysts? Pseudomorphs after an earlier mineral? Relicts of igneous glomerocrysts? Metamorphic differentiates?

1.4(2.3) The road leaves Lake Chelan and enters second-growth forest.

3.2(5.1) On the right is the old one-room Stehekin school house which was built in 1921 and which most Stehekin children attended prior to high school. It has recently been replaced by a newer and larger building.

4.4(7.1) Field of large talus blocks of heterogeneous biotite-hornblende gneiss derived from cliffs to northeast.

8.5(13.7) Stehekin Valley Ranch (~430 m/1420').

Day 5. Stehekin Valley to Skagit Valley. We hike over Cascade Pass examining rocks in the Chelan Mountains terrane and the Eldorado Orthogneiss of Misch [1966].

8.5(13.7) Stehekin Valley Ranch.

10.6(17.1) High Bridge over Stehekin River. The road enters a rocky canyon cut into the lower end of the hanging valley of upper Stehekin River. At this point the glacier flowing down the Stehekin valley was augmented by ice from Agnes Creek on the west and the increased eroding power left the upper Stehekin River and other tributaries hanging. The inner gorges here probably have been cut since the alpine glaciers retreated about 10 to 15 ka.

Exposures along the road are of orthogneiss typical of the Skagit Gneiss.

14.8(23.8) Bridge Creek Campground.

15.2(24.5) Bridge Creek.

21.7(34.9) Cottonwood Camp (~835 m/2750'). End of road. Beginning of trail to Cascade Pass.

0.0 Cottonwood Camp. Begin trail mileage, which is approximate.

0.3(0.5) *Stop 5-1* Small stream on side of Booker Mountain with considerable debris of the Late Cretaceous Eldorado Orthogneiss of Misch [1966]. The

main phase of the Eldorado pluton is a hornblende-biotite tonalite to granodiorite gneiss. It commonly has conspicuous, well-aligned hornblende prisms. In thin section the rock can be seen to have been an igneous rock, but is also clearly metamorphosed with zones of mylonitic to crystalloblastic texture and considerable replacement of plagioclase by epidote minerals. Some epidote may be igneous. On the margins of the pluton, especially to the southwest, a strong flaser texture has developed and locally the rock looks porphyroclastic.

The Eldorado contains zircons that yield concordant U-Pb ages of about 90 Ma. It is one of many 90 Ma synkinematic plutons exposed south of this area (see Section II). To the north of this stop, on the northern and eastern sides (Figure 9, map 3), the pluton and adjoining country rocks are intruded by numerous K-feldspar-rich pegmatite dikes and sills. Quartz in most of the pegmatites has a strong mylonitic fabric aligned parallel to the regional northwest strike. A large mass of pegmatite obscures the northern end of the pluton, but the map pattern suggests that the gross lithologic layering defined by alternating orthogneiss sills and paragneiss layers in the Skagit Gneiss has been truncated by the Eldorado.

Some of these orthogneiss bodies truncated by the 90 Ma Eldorado pluton (see **Stop 6-6**) yield 65 to 74 Ma concordant zircon ages. To explain this apparent enigma, we propose that the Eldorado pluton has been faulted against these younger, deep-seated rocks, either dropped down (see Figure 14) from a shallower level in the Paleogene orogen or thrust from the southwest into and over the Skagit Gneiss in a nappe-like fold during the latest Paleocene or early Eocene. Misch [1966] suggested that the Eldorado was thrust over the Skagit, and that pegmatite at the contact was derived from the Skagit and intruded along and near the thrust [Babcock and Misch, 1988].

Views to the southwest along here are of Glory and Trapper Mountains, eroded from the Cascade River unit interlayered with strongly flaseroid Eldorado pluton. Interlayering may reflect tectonic imbrication or lit-par-lit intrusion.

0.8(1.3) *Stop 5-2* A large block of schistose dike rock with folded leucosomes of pegmatite and leucotonalite. Schistose mafic dikes in the Eldorado, many with deformed phenocrysts, probably record the same deformation as the late lineated dikes in the Skagit (Section III; **Stop 6-1**), although we do not know if they intruded at the same time. Such dikes in the Eldorado are commonly strongly discordant but clearly bear the regional foliation.

1.2(1.9) Meadow with rusted iron relicts of early mining activity.

2.1(3.4) *Stop 5-3* (~1105 m/3630') Junction with trail to

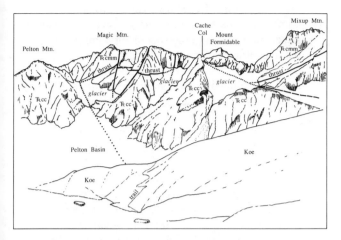

FIGURE 12. View from Doubtful Lake south across Pelton Basin to Magic Mountain Gneiss (ℸRcm) above Cascade River unit (ℸRcc). Eldorado Orthogneiss (Koe) and trail to Cascade Pass in foreground.

Horseshoe Basin (Figure 9-2). Just to the southeast are excellent exposures of Eldorado Orthogneiss. Much of the mountain above us to the north looks like this rock. The Horseshoe basin trail follows the remnants of an old wagon road to the mines in the basin some 350 m above this point.

The trail climbs seriously now and, crossing water-smoothed outcrops of Eldorado Orthogneiss, winds its way into the hanging valley of Pelton Basin.

3.8(6.1) *Stop 5-4* (~1490 m/4900') Views to the south of Cascade River unit and Magic Mountain Gneiss of Tabor [1961, 1962] on Pelton Mountain and the upper ramparts of Magic Mountain, respectively (Figure 9-3). In the Cascade River unit are abundant fine-grained chlorite-biotite-quartz plagioclase schists, many with considerable epidote and/or carbonate, quartz-rich amphibolites with blue-green hornblende, rare garnet-two mica schist and fine-grained garnet biotite gneiss with relict clastic grains (metasandstone). These rocks appear to be metavolcanic and metaclastic rocks.

Above the Cascade River unit on Magic Mountain and continuing around to Mixup Mountain on the west (Figure 12) is the Magic Mountain Gneiss. This well-banded gneiss is retrogressive or diapthoritic; it is derived from the Marblemount Meta Quartz Diorite of Misch [1966] exposed about 6 km to the southwest. The Late Triassic Marblemount pluton forms the basement for the Cascade River unit (see Section I). The Marblemount was thrust over the Cascade River unit, presumably during Late Cretaceous metamorphism [Tabor, 1961]. The relatively homogeneous original pluton was metamorphically segregated in the epidote-amphibolite facies prior to and during(?) thrusting. Layers of hornblende-biotite-rich gneiss formed, interlayered with quartzofeldspathic gneiss. Garnets were formed locally. As the gneiss was thrust up(?) and over greenschist facies rocks, it was recrystallized in the greenschist facies. The Magic Mountain Gneiss is now mostly interlayered greenschist and chlorite-epidote-albite-quartz gneiss. Relict garnets and rare hornblende occur in some rocks. To the south, metamorphic grade in the gneiss increases and it eventually merges with more typical Marblemount which is mostly tonalitic gneiss.

Discordant U-Pb zircon ages of 195-209 Ma (J. S. Stacey, written communication, 1988) from a sample collected just south of Cascade Pass suggest a Late Triassic protolith and support the petrogenesis of the gneiss outlined here.

The trail continues in talus of Eldorado Orthogneiss. The western part of the body becomes more and more flaseroid and some debris here may show the high strain.

5.0(8.0) Cascade Pass (~1650 m/5400'). Climbing up the last heather-covered slopes to the pass, we pass white outcrops of hornblende-biotite tonalite of the large, Miocene, Cascade Pass dike.

5.3(8.5) *Stop 5-5* Views to south of (east to west) the Triplets, Cascade Peak, and Johannesburg Mountain (~2505 m/8220'). Underfoot is the 18 Ma Cascade Pass dike. Look across at the dark cliffs of the ridge east of The Triplets and on Johannesburg which are made of the Cascade River unit, mostly mica and hornblende-mica schists, but with some metaconglomerate along the summit ridge. The large Cascade Pass dike, which is about 15 km long and up to 1.5 km wide, can be seen on the slopes below the glacial bench at the foot of and east of the The Triplets (Figure 13). A coarse lit-par-lit contact breccia is exposed at the base of The Triplets. It grades upwards into a breccia of hornblende schist fragments cemented with vuggy quartz and calcite. Tabor [1963] suggests that the breccia formed at the top of the dike below where it vented to the surface. The dike widens in the valleys and pinches down on the

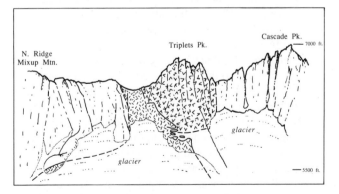

FIGURE 13. View to south from near Stop 5-5 showing 18 Ma Cascade Pass dike (random dashes) and explosion breccia on Triplets Peak (chevrons) Lit-par-lit contact breccia shown diagrammatically. From Tabor [1963].

ridge-tops indicating that it has the shape of a blunt blade. It shares its northeast orientation with a few other young plutons of the Cascade Pass family (Section IV) and many smaller dikes throughout the North Cascades.

K-Ar ages of hornblende and biotite pairs are concordant at about 18 Ma, that is, early Miocene [Engels and others, 1976]. Surficial volcanic rocks of this age do not occur this far north in the uplifted North Cascades, but are voluminous 120 km to the south (Section IV).

The trail switchbacks interminably down the steep hillside, mostly in forest, and passes poor outcrops of hornblende-biotite tonalite of the Cascade Pass dike and, in the lower parts, hornfelsic Eldorado Orthogneiss.

8.0(12.9) *Stop 5-6* (~1110 m/3650') Parking lot at end of the Cascade River Road. Views across the north Fork of the Cascade River to Mount Johannesburg, patched with hanging glaciers. Most years, visitors here will hear and see ice falling from the glaciers. We board vans for remainder of field trip and:

0.0 Begin road miles. The steep mountain road descends the slopes of the lower cirque, crossing tumultuous , bouldery streams, mostly laden with flaser gneiss debris of the Eldorado pluton. Good view to the northeast of brushy avalanche chutes, swept clear of trees by snow slides.

2.1(3.4) On left are the remains of Gilbert Cabin, an old prospector's cabin. Near here was a cable tramway to mining propects on the shoulder of Mount Johannesburg.

2.5(4.0) Bridge across the North Fork of the Cascade River.

5.9(9.5) *Stop 5-7* Outcrop of fine-grained amphibolite in the Napeequa unit. The schistose mafic amphibolite is coarse-grained but otherwise is typical of the oceanic facies of the Chelan Mountains terrane. The amphibolite here has two lineations: green hornblende prisms plunge steeply, but a crinkle lineation is subhorizontal. Metaclastic rocks of the Cascade River unit crop out to the southeast on Johannesburg Mtn.

The relation between the oceanic Napeequa unit and the metaclastic rocks of the Cascade River unit has not been worked out. Northeast of Hidden Lakes Peak (Figure 9-3), the two units are interlayered on a large scale, suggesting folding or fault imbrication prior to or during metamorphism (see Section I).

To the northwest view light-colored cliffs of the Hidden Lake Peak stock, a ~75 Ma (latest Cretaceous) biotite metagranodiorite. This small, slightly lensoid pluton intrudes the Napeequa unit with strong discordance. The pluton is partially recrystallized but

has no foliation. The Marble Creek pluton, a hornblende-biotite metatonalite gneiss of essentially the same age (~76 Ma), crops out about 4 km to the north but in contrast is strongly deformed, especially at its northern end. The contrast in deformation between these contemporaneous plutons demonstrates the rapid transition from rocks showing only the effects of the Late Cretaceous(?) metamorphism, such as those here, to rocks of the Skagit Gneiss that were further metamorphosed in the Paleocene and Eocene.

7.2(11.6) Bridge across the North Fork of the Cascade River. The junction of the north and south forks was the site of an early mining camp called Mineral Park. Prospectors stopped here on their way to the silver diggings (none could really be called mines) above Cascade Pass and beyond in Horseshoe Basin.

Beyond are scattered outcrops of quartz-rich schist (metachert) on the road to Marblemount.

13.0(20.9) Turn left (southwest) on Irene Creek Road (USFS 1550).

14.0(22.5) Bridge across Cascade River.

14.1(22.7) *Stop 5-8* Outcrops of Cascade River unit in the Chelan Mountains terrane. Plagioclase-rich muscovite schist, sericite-chlorite schist, and quartz-pebble metaconglomerate.

The Cascade River follows the Entiat fault along this straight stretch. Here the fault separates metaclastites of the Cascade River unit from oceanic rocks of the Napeequa unit. Studies by Jeff Cary and Joe Dragovich (M.S. theses in progress at Western Washington University) indicate a considerable metamorphic pressure increase from west to east across the fault.

Turn around here and retrace route to the Cascade River road. Turn left (north).

16.8(27.0) On left are outcrops of the Marble Creek pluton, here a light-colored biotite tonalite with little deformation, but across Marble Creek on the ridge to the north, the pluton is highly flaseroid (see **Stop 5-7**).

17.2(27.7) Bridge across Marble Creek.

21.0(33.8) Begin pavement on Cascade River Road.

25.5(41.0) Junction with WA 20 at town of Marblemount. Turn right, north.

39.5(63.6) Town of Newhalem (see Day 6)
Above Newhalem the river has cut a spectacular canyon into the Skagit Gneiss and the narrowness and steepness of the canyon suggest that this part of the range is still rising, or has only recently stopped.

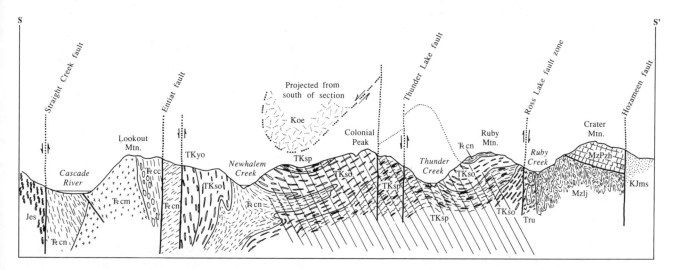

FIGURE 14. Diagrammatic cross section of antiform in Skagit Gneiss of Misch [1966] (TKso and TKsp). Eldorado Orthogneiss (Koe) shown down-faulted into orogen. Diagonal lines denote strongest migmatization. Location of section and other unit symbols shown in Figure 9, map 4.

Roadcuts provide a wonderful display of Skagit migmatites; we will examine some of them tomorrow.

45.4(73.0) Bridge across Gorge Lake (Skagit River). View of the town of Diablo, a private community for employees of Seattle City Light, which owns the dams along the Skagit River.

46.6(75.0) Turn left (northwest) to Diablo Dam.

46.9(75.5) Diablo Dam. After crossing dam, turn right.

47.8(76.9) Diablo Lake Resort. The resort is built on an active fan from Sourdough Creek, a straight drainage eroded along shattered rock of the Thunder Lake fault (see **Stop 6-5**).

Day 6. Diablo Lake to Darrington. Today's route traverses the antiformal Skagit Gneiss, flanked by the Napeequa unit (Figure 14). Much of the log for day 6 has been extracted from Misch [1977b].

0.0 Diablo Lake Resort.

0.2(0.3) Talus of mostly biotite orthogneiss of the Skagit Gneiss, derived from the large cliff above.

0.7(1.1) *Stop 6-1* Diablo Dam. The dam was built in the late '20s to supply hydroelectric power for the city of Seattle. Supplies were brought in by railroad expressly built for the dam construction. When Ross Dam was built at the upper end of Diablo Lake, a railroad car lift was used to raise the cars up the steep mountainside west of the dam so they could proceed on rails to Diablo Lake where they were loaded on barges to be towed up to the construction site. The lift now moves tourists up

and down the mountainside.

At the north end of Diablo Dam, lineated dikes of late granite crosscut migmatitic banded gneiss of the Skagit Gneiss. Good outcrops continue to the west along the road to the top of the Diablo inclined railroad. Look across river and below the dam to see snake-like dikes on canyon walls.

These dikes commonly cross-cut foliation in the host gneiss yet are themselves lineated parallel to the NW-trending low angle lineation in the gneiss. Locally the dikes are foliated. Dikes with these structural features have a wide range of compositions, but most common are cream-colored granites like those that occur here, commonly with minor biotite, locally with garnet and (or) muscovite. The granitic composition is distinctive; along the Skagit River older Skagit gneisses mostly lack potassium feldspar. Textures of the late lineated dikes are commonly mylonitic: aggregates of quartz form ribbons and elongate pools within which grain boundaries are sutured, shadowy strain lamellae are evident, and a strong preferred c-axis orientation is developed; plagioclase is broken, with fractures healed by quartz or later plagioclase; potassium feldspar has undulose extinction and sub-grains are developed without obvious fracture; and biotite is bent and shredded. Absence of visible foliation in typical late lineated dikes suggests a prolate uniaxial strain ellipsoid, that is, shortening in all directions normal to the lineation and lengthening along the lineation.

The late lineated dikes were intruded after main regional metamorphism but while the gneiss complex was still hot. They were subsequently stretched NW-SE along with the rest of the complex. Isotopic evidence for their middle Eocene age is discussed in Section III.

Retrace route across Diablo Dam to WA 20.

1.2(1.9) Turn left on WA 20 and drive east about 12.5 miles (20 km) to stop 6-2. For most of today, mileage is given both as distance from stop 6-2 and referenced to mile markers posted along WA 20.

8.1(13.0) Panther Creek bridge (milepost 138.5). 1.4 miles (2.3 km) farther to:

0.0 *Stop 6-2* (milepost 139.9) Park on left side of road and reset odometer. Exposures of agmatite and schist of the Ruby Creek Heterogeneous Plutonic Belt of Misch [1966].

This stop is in the Ross Lake fault zone, the zone of faulting, ductile deformation, and westward increase in metamorphic grade that separates the Skagit Gneiss from lower-grade rocks to the east (Section III). On days 2 and 3 we saw this zone SE of the 50 Ma Golden Horn granite, which is exposed a few kilometers SE of here. From here to the west side of Ross Lake, the fault zone has been invaded by numerous granitoid intrusions, some foliated, some not, ranging in composition from granite to hornblende diorite. These intrusions are the *Ruby Creek Heterogeneous Plutonic Belt* of Misch [1966]. The complicated intrusive relations and large amounts of included country rock are probably the result of intrusion into an active fault zone [Misch, 1977b].

Agmatites at this stop range from varieties with equant and angular inclusions to varieties with elongate, rounded, foliated fragments. The latter are Misch's "pollywog agmatite". Most inclusions are probably Jack Mountain Phyllite and (or) Elija Ridge Schist. The trondhjemite matrix of the pollywog agmatite has given a 48 Ma U-Pb zircon age [Miller and others, 1988].

Turn around and drive west on WA 20.

1.4(2.3) Panther Creek Bridge (milepost 138.5).

4.8(7.7) *Stop 6-3* Ross Lake overlook (milepost 135.1, ~660 m/2200'). Park in large turnout on right.

Good views to the north of (right to left) Little Jack, Jack, and Hozomeen Mountains, Ross Lake, and Mount Prophet, the highest point on the left. Mount Prophet (~2330 m/7650') is carved from marble, paragneiss, and orthogneiss of the Skagit Gneiss. One of the few Al$_2$SiO$_5$ localities in the Skagit—sillimanite in sulphidic gneisses—is on its southern slopes. It appears to be in the core of the antiformal Skagit Gneiss (Figure 14). Ridges to the right of Mt Prophet are underlain by the Skymo unit [Wallace, 1976]: olivine gabbro and olivine norite intruded by medium- to coarse-grained gabbro. Brittle, N- and NW-trending faults separate the Skymo unit from Skagit Gneiss but on the western edge of the unit, Skymo gabbros are intruded by tonalitic pegmatite of the Skagit Gneiss [Staatz and others, 1971]. Rocks of the Skymo unit extend southeastward to the shore of Ross Lake.

To the NE, but on the west side of the lake, are schists, semi-schists and phyllites of the Little Jack terrane, with metamorphic grade decreasing to the east. Further to the NE are greenstone and chert of the Hozameen terrane. In the distance, beyond the head of the lake in Canada, is Silvertip Mountain. Closer, the sharp horns to the right are Mount Hozomeen (2459 m/8066'). Small stocks of the Chilliwack batholith are present on both Silvertip and Hozomeen and their contact-metamorphic effects have helped produce rugged mountains out of otherwise incompetent chert, argillite, and greenstone of the Hozameen Group.

The massive peak to the right is Jack Mountain (2763 m/9066'). On its lower slopes, near timberline, the gently NE-dipping Jack Mountain fault places the Hozameen terrane above semischist, phyllite, talc schist, and metaporphyry of the Little Jack terrane. Little Jack Mountain (2056 m/6745') is the grassy ridge below and to the right of Jack Mountain.

Although the Little Jack terrane here is in brittle fault contact with the Hozameen terrane and the Skymo complex of the Chelan Mountains terrane, it displays a sharp metamorphic gradient away from the high grade Skagit core to the little-metamorphosed Hozameen. A similar gradient occurs in British Columbia where the Maselpanik schist unit of Haugerud [1985] (Figure 6) also grades abruptly from sillimanite-grade Skagit Gneiss to prehnite-pumpellyite-facies Hozameen. Brittle faults do not significantly interrupt the gradation there. Haugerud [1985] correlated the Maselpanik with the Hozameen, but the abundance of metapelite in the schist suggests it might be correlative with the Jack Mountain Phyllite (and Elija Ridge Schist) of Misch [1966].

The low peninsula extending from the east into Ross Lake, separating Ross Lake from Ruby Arm to the right, is underlain by homogeneous orthogneiss with nearly flat foliation and a well developed N- to NE-trending stretching lineation, an unusual orientation for the North Cascades. Between this orthogneiss and the Little Jack terrane on the slopes above are intrusive rocks and migmatites of the Ruby Creek Heterogeneous Plutonic Belt.

Complex migmatites of the Skagit Gneiss are exposed across the road.

6.4(10.3) *Stop 6-4* Horsetail Creek and John Pierce Falls (milepost 133.5). Park on right just before bridge. To the east of the bridge are spectacular roadcuts of banded migmatitic gneiss, locally with layers of metamorphosed ultramafic rock.

The origin of ultramafic pods in the Chelan Mountains terrane is a matter of some debate. Their emplacement is pre-metamorphic and they are present throughout much of the Napeequa unit and the Skagit Gneiss. Peridotitic compositions and relict chromite grains suggest that they ultimately came from the mantle. Misch (written communication, 1981) believed all ultramafic bodies are located along pre-metamorphic faults. Alternately, some pods may have been emplaced as submarine slide masses of mostly serpentine.

Metasomatic reaction zones ("blackwalls") at the contacts of ultramafic bodies in the Skagit Gneiss and Napeequa unit record mass transfer during metamorphism caused by the large differences in chemical potentials between the Mg-rich ultramafic pod and adjacent K-, Si-, and Al-rich schist. Misch [1977b] reports zoning at this stop that ranges from a core of tan anthophyllite + silvery talc + emerald green Cr-bearing tremolite + light-brown phlogopite outward to tremolite + anthophyllite + phlogopite, to tremolite + phlogopite, to rock with darker-green actinolitic hornblende and darker-brown phlogopitic biotite, to a margin of biotite-bearing hornblendite.

Mineral compositions in rare sillimanite- and kyanite-bearing assemblages in the Skagit Gneiss indicate equilibrium pressures of ~9 kb, corresponding to some 30 km of burial, and temperatures of ~720°C [Whitney and Evans, 1988]. This estimated temperature is consistent with the local occurrence of enstatite in some Skagit ultramafic rocks, which requires temperatures in excess of 680°C, depending on the activity of water (Figure 15). As many of the ultramafic pods are probably metaserpentinites, we assume that progressive metamorphism produced H_2O and kept its activity high. These P-T conditions are in the stability field of kyanite; sillimanite in Skagit Gneiss, as well as

common anthophyllite + forsterite in Skagit ultramafic rocks, must have formed during near-isothermal unroofing.

There is no consensus on the origin of Skagit migmatites. Misch [1968] demonstrated that plagioclase compositions and zoning patterns in leucosomes were similar to those of adjacent wallrock. On this basis he ruled out an origin by igneous injection and proposed that the migmatites formed by metasomatic replacement coupled with metamorphic segregation. Misch dismissed an anatectic origin for reasons which included inferred peak metamorphic temperatures of ~600°C and differences in composition between leucosome + wallrock and their inferred protolith.

Microprobe analyses by Yardley [1978] showed a correspondence of biotite as well as plagioclase compositions in adjacent layers; he argued that metamorphic segregation was the primary process of migmatization. Whitney and Evans [1988] noted that melting would have occurred at the ~720°C - ~9 kb conditions they estimated for the peak of Skagit metamorphism and suggested the leucosomes are anatectic melts, perhaps localized along fluid-intrusion pathways. Babcock and Misch [in press] present mass-balance calculations which they interpret to indicate that the Skagit migmatites could not have formed by any strictly closed system process, be it partial melting or metamorphic segregation.

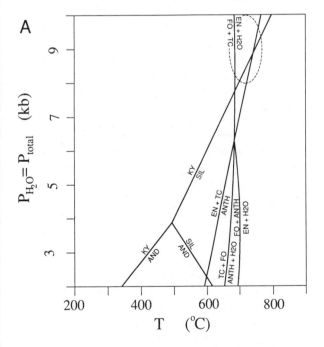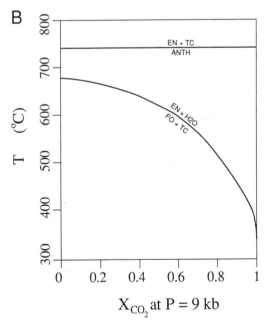

FIGURE 15. A. P-T plot for system $MgO-SiO_2-H_2O$, with Al_2SiO_5 reactions, showing peak metamorphic conditions in Skagit Gneiss as inferred by Whitney and Evans [1988] (dashed ellipse). B. $T-X(CO_2)$ plot for the system $MgO-SiO_2-H_2O$, showing how temperature of reaction varies with composition of mixed H_2O-CO_2 fluid phase. AND=andalusite, ANTH=anthophyllite, EN=enstatite, FO=forsterite, KY=kyanite, SIL=sillimanite, TC=talc.

Clearly, at peak metamorphic conditions most Skagit gneisses would have been partially molten if water saturated; however, Clemens and Vielzeuf [1987] suggest that water saturation is unlikely in upper amphibolite facies rocks and that widespread melting would only occur above the dehydration-reaction curves of muscovite, biotite, or amphibole. Melting related to muscovite dehydration can begin at temperatures as low as 650°C for pelitic compositions [Vielzeuf and Holloway, 1988; Le Breton and Thompson, 1988]. Muscovite-bearing rocks are uncommon in the lower-grade schists of the Chelan Mountains terrane and unmigmatized metasediments of the Skagit Gneiss, so widespread melting in the Skagit might have required the dehydration of biotite at 850-880°C [Vielzeuf and Holloway, 1988; Le Breton and Thompson, 1988]. It is possible that some of the migmatites formed in more permeable shear zones accessible to water or by selective melting of small amounts of muscovite-bearing rock. However, most Skagit leucosomes lack potassium feldspar and this composition is inconsistent with their origin by partial melting of a pelitic protolith.

The main problem with an anatectic or segregation model is that there appears to be no correspondence between the composition and volume of the migmatite leucosomes and the wallrock with which they are associated. The problem with an injection or metasomatic model is that the migmatites cannot be simple mixtures of an injected igneous (or hydrothermal) phase and a host rock; mineral compositions (at least plagioclase and biotite) seem to be controlled locally rather than by some external fluid reservoir. Thus the origin of the Skagit migmatites remains enigmatic and should provide a topic for spirited discussion during this excursion.

8.1(13.0) *Stop 6-5* Diablo Lake overlook. (milepost 131.8) Park in large turnout on right View SW across Thunder Arm of Diablo Lake to the Colonial Peak (2369 m/7771') group. To the N, across Diablo Lake, peaks made of Skagit Gneiss are (left to right) Davis Peak (2149 m/7051'), Elephant Butte (2249 m/7380') and Sourdough Mountain (1861 m/6106'), with a fire lookout made famous by poet Gary Snyder.

The valleys of Stetattle Creek and Thunder Creek are localized along strands of a group of N- and NNW-trending brittle faults. These faults must be of late Eocene or earliest Oligocene age, as they cut Skagit Gneiss with its ductile middle Eocene fabric and do not offset the early Oligocene phases of the Chilliwack batholith to the north. Sourdough Creek (which drains towards us on S face of Sourdough Mountain), Thunder Lake and the large notch on east ridge of Colonial Peak are located on the *Thunder Lake fault*, the most conspicuous of these structures.

Outcrop across the highway is massive biotite-hornblende quartz diorite orthogneiss, veined by

extensive light-colored dikes. From this point to beyond the turnoff to Diablo outcrops are dominantly paragneiss.

9.4(15.1) Bridge across Thunder Arm (milepost 130.5). Thunder Creek drains some of the most extensive glaciers in the North Cascades. Glacially-ground rock flour produces the milky green color of Thunder Arm.

10.4(16.7) Thunder Lake (milepost 129.5). Lake is in a glacier-carved depression along the Thunder Lake fault (see **Stop 6-5**).

12.5(20.1) Junction with road to Diablo Dam and Diablo Lake resort (milepost 127.4) Roadcuts at this junction display both older, pegmatitic, main Skagit leucosomes and younger, fine-grained, cross-cutting late lineated dikes.

13.7(22.0) Bridge across Gorge Lake (milepost 126.2) followed by turnoff to town of Diablo.

The mountain ahead is Davis Peak. Roadcuts on the right after crossing Gorge Lake are in banded gneiss, including garnet biotite schist, amphibolite, and thin layers of (mostly) biotite tonalite orthogneiss, all threaded with coarse-grained leucotonalite (trondhjemite). Massive orthogneiss becomes more abundant to the SW and beyond Gorge Lake is the predominant rock of the canyon.

16.5(26.6) Gorge Creek bridge (milepost 123.5)

19.0(30.6) Enter town of Newhalem (milepost 120.9) (160 m/525'). Newhalem was built by Seattle City Light in the late 1920s. On the east side of the Skagit is the power house for Gorge Dam. Above on the west are the orthogneiss buttresses of Mount Ross (1845 m/6052').

19.6(31.5) Turn left near milepost 120.3, onto parallel gravel road, and park by prominent buttress of rock.

Stop 6-6 Bluffs of gray rock immediately north of the road are "incipiently migmatized" [Misch, 1977a, p. 35] gneissic hornblende-biotite tonalite orthogneiss. The igneous protolith is surmised mostly on the basis of lithologic uniformity and very rare relict euhedral oscillatory zoning in plagioclase as seen in thin section (Figure 16). Textures are predominantly crystalloblastic. Orthogneiss from this outcrop has yielded roughly concordant U-Pb zircon ages of ~65 Ma, but similar-looking rock a few kilometers across strike to the south yields U-Pb zircon ages of ~74 Ma (J. S. Stacey, written communications, 1987, 1988).

Magmatic epidote is abundant in many of the Late Cretaceous plutons of the North Cascades, including those exposed on the ridges to the south and west of this point. Because of different P-T slopes for the solidus and epidote-breakdown reactions, the two cross, and at

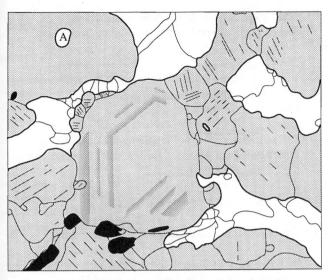

FIGURE 16. Relict oscillatory zoning in orthogneiss of the Skagit Gneiss. Quartz is clear, plagioclase is gray, A = apatite, dark grains are biotite, and there is a small grain of zircon to the right of center. View is 3 mm wide. Sketch of thin section of sample RWT 35-86, from Stop 6-6.

higher pressures epidote is stable with melt [Zen and Hammarstrom, 1984]; that is, melting and solidification can occur in the epidote-amphibolite facies. In the epidote-bearing plutons of the North Cascades hornblende and significant potassium feldspar seem antipathetic, suggesting a reaction relationship such as

> *low P*   *high P*
> hornblende + potassium feldspar = biotite + epidote
> (± quartz, plagioclase, muscovite, sphene, etc.)

At higher P the reaction proceeds until either hornblende or potassium feldspar is exhausted, producing an epidote-biotite granodiorite or an epidote-biotite-hornblende tonalite. This may be a reason for the low potassium feldspar content of many of these plutons: K_2O is tied up in the assemblage biotite + epidote, rather than in the more typical granitoid assemblage hornblende + potassium feldspar.

19.8(31.9) As we cross the bridge across Goodell Creek, look to the right (north) for a view of the Southern Picket range, eroded out of orthogneiss. This cluster of peaks is supposedly named for a 19th century U.S. Army officer, not their uncanny resemblance to a fence.

21.7(34.9) First outcrops of the Chilliwack composite batholith.

25.0(40.2) (mileposts 114.9 - 114.6) Light-colored waterlaid volcanic ash is exposed in roadcuts. This deposit must be post-glacial, yet the present-day Skagit

River is an unlikely agent to deposit such fine-grained material. Probably a landslide farther downstream dammed the Skagit and these ash beds were laid down in the resulting lake.

25.3(40.7) *Stop 6-7* (milepost 114.6) Park on river side of road. Schists of the Napeequa unit, with sills of late-metamorphic orthogneiss. At the upstream end of the outcrop are intricately folded quartz-rich biotite schists—with thicker quartzite and thinner biotite-rich laminae—typical of metacherts in the Napeequa unit. The intricate fold style is common in much lower-grade ribbon chert, and by analogy it is likely that these folds largely predate amphibolite-facies metamorphism.

For the next 2.1 miles (3.4 km), outcrops are in the toes of large landslides. Rapids in this stretch of the Skagit River are formed by large blocks of landslide debris. Rock types in the roadcuts include biotite schist, two-mica schist, quartzose schist, amphibole-bearing schist, and chlorite-rich mafic schist. Less abundant are pods and layers of talc and talc-carbonate schists locally with forsterite, tremolite, and anthophyllite. All are metamorphosed in sodic andesine-epidote to oligoclase-epidote amphibolite facies [Misch, 1977b]. Vein quartz is locally extensive enough to have been mined for glass-making.

27.5(44.3) *Optional Stop* (milepost 112.4) Abandoned talc quarry in Napeequa unit (unrestricted Cascade River Schist of Misch [1966]) (Figure 9-3). Park on river side of road. The best grades of talc are used for carving, and in order not to shatter the rock the miner's tool of choice has been a chainsaw.

Talc schists here carry porphyroblasts of magnesite and, locally, dolomite. A well developed metasomatic "blackwall" is present. Misch [1977b] reports outward progressing zones of essentially mono-mineralic or bi-mineralic tremolite schist (emerald green from Cr-bearing tremolite), chlorite schist, chlorite-biotite schist, biotite schist, and biotite-muscovite schist.

Massive light-colored outcrops across the river are weakly gneissose Alma Creek Leucotrondhjemite of Misch [1966]. Textures suggest the Alma Creek pluton is late-metamorphic [Misch, 1966, 1977b]; it may belong to the group of younger orthogneiss plutons (Figure 9), but here is outside of the region of intense Eocene deformation.

29.2(47.0) Bridge over Bacon Creek (milepost 110.7) Bacon Creek flows along the Straight Creek fault zone a few kilometers to the north. Eocene(?) siltstone, feldspathic sandstone, and pebble to small-cobble conglomerate occur within the Straight Creek fault zone along the bottom and east side of the Bacon Creek valley (see section III) Most of these strata are contact-metamorphosed by the nearby Chilliwack batholith. Near the forks of Bacon Creek, penetrative deformation and subsequent contact metamorphism have produced a "pseudo-gneiss" with flattened pebbles and well-oriented contact-metamorphic biotite in Eocene conglomerate (see discussion at **Stop 9-2**).

29.5(47.5) *Stop 6-8* Marblemount Meta Quartz Diorite of Misch [1966] (milepost 110.4). Pull off onto shoulder. Here the Marblemount is predominantly a muscovite-chlorite-epidote-albite-quartz rock. Misch [1977b] reports local green-brown biotite and rare relict hornblende. From these outcrops Mattinson [1972] obtained a 220 Ma (Late Triassic) zircon U-Pb age. The Marblemount and its equivalent Late Triassic plutons form a belt extending from the Straight Creek fault, only about 1 km to the west of this stop, southeastward for 150 km before disappearing under the cover of Miocene Columbia River Basalt Group.

33.8(54.4) Town of Marblemount (milepost 106.1). WA 20 turns right at restaurant. We turn left onto the Cascade River Road, across bridge over Skagit River.

The broad valley junction here is eroded out along two strands of the Straight Creek fault. Lookout Mountain looming up to the east is underlain by Marblemount Meta Quartz Diorite. Forested bluffs to the west are Shuksan Greenschist of the NWCS.

40.7(65.5) *Stop 6-9* Metaconglomerate. Roadcut on north (left) side of road exposes metaconglomerate of the Cascade River unit. At first glance this rock might appear to be an orthogneiss, but it is mostly large clasts of metatonalite identical in appearance to the Marblemount, stretched out so that they are difficult to recognize as clasts except on a fresh surface normal to the lineation.

Look upvalley from the bend in the road for a view up Marble Creek to the glacial cirque at its head and Eldorado Peak (2703 m/8868'), type area of the Eldorado Orthogneiss.

Turn around and retrace route along the Cascade River road to Marblemount.

47.7(76.8) Stop sign in Marblemount. Continue west (straight ahead) on WA 20, after having:

0.0 Reset odometer at stop sign.

Marblemount has long been a way-station for travelers in the North Cascades. The inn here was built in 1898 and has operated as a tavern, hotel, and restaurant ever since.

3.1(5.0) Corkindale Creek. To the north, east of Corkindale Creek, steep wooded hillsides are underlain by the Shuksan Greenschist of the Easton Metamorphic Suite. West of Corkindale Creek are greenstones and low-grade metasedimentary rocks of the upper Paleozoic Chilliwack Group of Cairnes [1944], and high on the

ridge thick chert beds and greenstone of the Mesozoic(?) Elbow Lake Formation of Brown and others [1987].

8.4(13.5) Immediately before the turnoff to Darrington, cliffs on the right are basaltic greenstone of the Chilliwack Group.

8.5(13.7) Rockport and the junction with WA 530, the Darrington-Rockport road (milepost 97.7). Turn left on the Darrington-Rockport road.

8.6(13.8) Riverhouse Restaurant. Dinner.

0.0 Reset odometer. Continue south to Darrington.

0.1(0.2) Bridge over Skagit River. Look upriver (east) for good views of Marble Creek cirque and Mount Eldorado.

6.3(10.1) Junction with south Skagit road. Stay left, south to Darrington.

11.0(17.7) Junction with Suiattle River Road. Stay right, west to Darrington.

17.7(28.5) Sawmill.

18.4(29.6) Darrington (167 m/549'). A large proportion of Darrington's population originally came from the state of North Carolina on the east coast of the U.S. When the Great Smokies National Park was established in the 1930s, many people in the logging industry, fearing loss of jobs as the government land was closed to logging, moved to Darrington. The town has a strong North Carolina tradition, including the production of bluegrass music. Many residents still refer to themselves as tarheels (residents of North Carolina). The accents are those of the American South, even in the younger generations.

Turn right (west) to reach:

18.5(29.8) Stagecoach Inn.

Day 7. Darrington to Rat Trap Pass and Suiattle Mountain and return. We look at the rocks along the Straight Creek fault, the Nason terrane, and the Shuksan thrust of Misch (1966) between Grandy Ridge terrane rocks and the Easton terrane.

0.0 Stagecoach Inn. Proceed south through Darrington to the main street,

0.4(0.6) Turn left (east) and follow the Mountain Loop Highway.

3.4(5.5) Outcrops along a new section of the Mountain

Loop Highway are in greenschist and phyllite of the Easton Metamorphic Suite.

8.9(14.3) Bridge across Sauk River.

9.2(14.8) Junction of White Chuck River Road (USFS 23) and Mountain Loop Highway. Turn left (east).

10.0(11.1) Begin views to northeast of White Chuck Mountain composed of Shuksan Greenschist.

14.8(23.8) Bridge across White Chuck River.

15.1 (24.3) Turn left (north) on Rat Trap Pass Road (USFS 27). The road climbs up glacial debris and finally reaches:

16.8(27.0) Outcrops of banded biotite gneiss. The road follows a steep-walled side canyon eroded along an eastern strand of the Straight Creek fault. Rocks are highly broken and fractured. There are numerous light-colored tonalite and pegmatite leucosomes. To the south are views of Mount Pugh (2202 m/7224') carved from the resistant gneiss of the Sloan Creek plutons (see **Stop 7-2**).

19.1(30.7) *Stop 7-1* Rat Trap Pass (~975 m/3200'). Here we are on the main strand of the Straight Creek fault (see Section III). On the west, White Chuck Mountain (2130 m/6989') rises abruptly above the pass. It is carved from Shuksan Greenschist of the Easton Metamorphic Suite, a part of the Northwest Cascades System (NWCS) exposed west of the Straight Creek fault. Slopes east of the pass are eroded from biotite gneiss of the Nason terrane, part of the higher grade core of the North Cascades.

Offset on the Straight Creek fault has been estimated by several means (Figure 7). Correlation of the Mt Stuart batholith and Chiwaukum Schist near Stevens Pass with the Spuzzum pluton and Settler Schist near Yale, B.C. led Misch [1977a] to propose 190 km of offset. More detailed mapping in the U.S. [Tabor and others, in press] suggests this is a maximum and that offset based on the Chiwaukum-Settler correlation might be as little as 130 km. Correlation of the Hozameen Group with the Bridge River Complex and the Ross Lake fault near Hope, B.C. with the Yalakom fault north of Lillooet, B.C. suggests about 110 km of offset [Monger *in* Price and others 1985; Kleinspehn, 1985].

The above estimates are suspect because they either (a) correlate non-unique entities—schists and plutons very similar to the Settler and Spuzzum are present 60 km farther north in British Columbia [Hollister, 1969], (b) correlate bits and pieces of a Hozameen-Bridge River-Cache Creek terrane that must have been regionally extensive, or (c) assume that NW-trending faults pre-date the Straight Creek fault and are offset by it; some may be in part coeval. The best-founded estimate of Straight Creek displacement is provided by Kleinspehn's [1985] observation that undoing 110 km of strike-slip allows restoration of Albian depositional system preserved in Jackass Mountain Group strata of the Methow terrane. However, no suggested restoration of the Straight Creek fault is entirely satisfactory. Substantial vertical offset must contribute to the misfit, as must Eocene NW-SE stretching of the Skagit Gneiss.

North of the Skagit River, latest motion on the Straight Creek fault is constrained by: (1) the fault cuts undated, probable Eocene, fluvial strata [Misch, 1979; Monger, 1970]; (2) the subsidiary Yale fault crosscuts Skagit Gneiss with a ~46 Ma ductile fabric [Haugerud, 1985]; and (3) the fault is plugged by 30-35 Ma plutons of the Chilliwack batholith [Richards and McTaggart, 1976; Vance and others, 1986]. Southeast of Darrington there appears to have been little strike-slip motion since ~40 Ma (latest middle Eocene), to judge from the pattern of the Barlow Pass Volcanics of Vance [1957b] cropping out on both sides of the fault (Figure 6; Tabor and others [in press]). However, in this area the exact position of the fault is obscured by younger batholithic rocks (Figure 9-5).

20.8(33.5) Outcrops of feldspathic sandstone. This small sliver of probable Eocene sandstone is one of many distributed along the Straight Creek fault. No ages have been obtained from these rocks, but they probably are remnants of a wrench basin deposit that formed along the Straight Creek fault in the early Tertiary, similar to the Swauk Formation exposed 64 km to the south (see Section III).

22.0(35.4) *Stop 7-2* Biotite-hornblende tonalite gneiss of the Sloan Creek plutons is well exposed in water-washed outcrops a few tens of meters up the bouldery creek on east. Good examples of the gneiss can be seen in the creek boulders as well. The Sloan Creek plutons are synkinematic 90 Ma intrusions, members of the class of stitching plutons that characterize the Late Cretaceous metamorphic event (see Section II). They commonly show deformed and partially recrystallized igneous textures in thin section. The exposure here is in a large sill-like body, the characteristic structure of the Sloan Creek plutons, although some of them are of mountain-size scale. These rocks, resistant to erosion, hold up several prominent peaks to the south including Mount Pugh and Sloan Peak (Figure 9-5).

The road continues down Straight Creek, passing outcrops of banded gneiss well dusted by logging trucks. BE CAREFUL!

25.5(41.0) Junction of Straight Creek Road and Suiattle South Side Road (USFS 25). Turn left (west).

28.7(46.2) Bridge across Suiattle River (239 m/784') and junction with main Suiattle River road (USFS 26). Turn left (west).

30.4(48.9) Junction with Tenas Creek Road (USFS 2660). Turn right (north) onto Tenas Creek Road.

31.9(51.3) Views of Prairie Mountain to west and White Chuck Mountain to southwest. We are looking across the Straight Creek fault to the NWCS. Prairie Mountain is eroded from the Chilliwack Group and, on the summit, large slivers of the Yellow Aster Complex of Misch (1966).

32.5(52.3) Intersection with logging spur (USFS 1825). Turn right (southeast).

33.6(54.1) *Stop 7-3* Outcrops of Chiwaukum Schist. This is typical staurolite-garnet-biotite schist of the Chiwaukum Schist and characteristic of the Nason terrane. Kyanite is well developed regionally beyond the kyanite isograd about 2 km to the east.

Farther south, Al_2SiO_5 minerals show a complex history related to the intrusion of Late Cretaceous, roughly syn-metamorphic plutons (of the Mt Stuart batholith) (Figure 5) From southwest to northeast across the Nason terrane, peak-metamorphic pressures vary: contact-metamorphic assemblages suggest 2-4 kb near Mt Stuart on the southwest [Frost, 1976]; andalusite pseudomorphed by kyanite near the Mt Stuart pluton indicates pressures changed with time, but were in the vicinity of the Al_2SiO_5 triple point (4 kb) [Evans and Berti, 1986]; and thermobarometry on kyanite-bearing schists [Magloughlin, 1986] and magmatic epidote in the Tenpeak pluton indicate pressures of 6 kb or more.

The timing of deformation has been hotly debated [Plummer, 1980; Evans and Berti, 1986; Magloughlin, 1986] but the development of kyanite appears to post-date the intrusion of a 95 Ma phase of the Mt Stuart batholith. K-Ar ages record cooling in the Late Cretaceous.

Greenschist-facies retrogression of unknown age is widespread in the Chiwaukum Schist. Mineralogy and textures commonly suggest the reaction: kyanite (and (or) staurolite) + biotite —> garnet + chlorite + muscovite.

Turn around and return to Suiattle River Road.

38.8(62.4) Junction of Tenas Creek road and Suiattle River Road. Turn right (west) and:

0.0 Reset odometer.

5.2(8.4) *Stop 7-4* Outcrop of metabasalt and (or) meta-andesite in Chilliwack Group (Figure 9-6). Exposures are in a small creek coming down to road from the

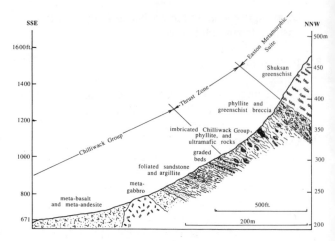

FIGURE 17. Sketch cross section of creek on south side of Suiattle Mountain (Stop 7-4).

north. We begin a hike here to view the thrust between the Chilliwack Group below and the Shuksan Greenschist (Easton Metamorphic Suite) above. The hike is not long but requires considerable scrambling over slippery rocks, logs, and waterfalls in the creek. A fixed line has been attached on one steep, loose hillside. Round-trip time is about 3 hours.

Beginning at the road the creek has exposed massive greenstone followed on up the creek by variously foliated metavolcanic rocks and some metasedimentary rocks typical of the Chilliwack (Figure 17). Structures are low-angle, as is typical of much of the NWCS. Approaching the fault note the marked increase in the amount of strain, especially noticeable in slivers of mafic plutonic rocks (Yellow Aster Complex?). Just below the upper plate metasedimentary rocks are imbricated with greenschist and phyllite of the Easton Metamorphic Suite. The base of the Easton is highly deformed, consisting of contorted Darrington Phyllite with slivers of greenschist.

This exposure is one of a few easily attained for a close-up view of the Shuksan thrust of Misch [1966]. Misch believed that the Easton Metamorphic Suite (his Shuksan Metamorphic Suite) had been thrust on a regional scale over the Chilliwack Group from the east. More recent workers [Brown 1986; Smith, 1988] have found stretching lineations suggesting that the thrust moved NNW-SSE, and infer that that the deformation is the result of Late Cretaceous oblique convergence (see section II). Unfortunately, stretching lineations in these exposures are not consistent with the regional pattern.

Return with great care to the Suiattle River Road and continue west.

6.7(10.8) Views of Mount Baker volcano to the north. The road traverses glacial deposits, mostly till of the Vashon Drift which was deposited by the Canadian ice sheet. Several large pits along the road here are kettles.

8.3(13.4) Junction of Suiattle River road and Darrington-Rockport Road. Return to Darrington.

On the last straight stretches of road coming into Darrington look up to good views of surrounding peaks. The high reddish peak straight ahead is Whitehorse Mountain (2088 m/6852'), carved mostly out of meta-andesite of the Whitehorse Volcanics of Vance [1957b] (Section I) of the eastern melange belt, here thermally metamorphosed by the 35 Ma Squire Creek stock. The smaller, needle-sharp peak immediately to the east is Buckeye Mountain, eroded from tonalite of the Squire Creek stock. Farther to the east are: Jumbo Mountain underlain by a megablock of gabbro in the the Helena-Haystack melange (within the DDMFZ); Mt Pugh, held up by Sloan Creek plutons, east of the Straight Creek fault; and White Chuck Mountain made of Shuksan Greenschist.

The wooded hill to the left is Gold Hill, so-named for gold prospects first discovered at the turn of the century, and the type locality [Vance, 1957a; Misch 1966] of the Darrington Phyllite of the Easton Metamorphic Suite (see also Figure 20).

15.8(25.4) Stagecoach Inn, Darrington.

Day 8. Darrington to Gee Point area, Deer Creek Pass and return to Darrington. On this day we examine the Easton terrane, in particular the Shuksan Greenschist and related rocks, as well as a bit of the Helena-Haystack melange and Eocene sandstone caught up in the Darrington-Devils Mountain fault zone.

0.0 Stagecoach Inn. Drive north on Darrington-Rockport Road (WA 530).

7.2(11.6) Bridge across Sauk River. Cross and continue north towards Rockport. At this point, just below the confluence with the Suiattle River, the eastern portion of the Sauk is muddier than the main stream. Both a large contribution of glacial melt to the Suiattle and its excavation of a post-glacial volcanic debris flow on the east flank of Glacier Peak color the river here.

12.1(19.5) Junction with Skagit River South Side Road. Turn left (west) and cross bridge over Sauk River.

12.3(19.8) Outcrops of Chilliwack Group, mostly highly deformed metasedimentary rocks and marble. Continuing up road to north, pass outcrops of mafic igneous rocks of the Yellow Aster Complex. Isolated outcrops of the Chilliwack forming resistant knobs here are aligned northwesterly along probable faults between Shuksan Greenschist exposed to the east and Darrington Phyllite exposed to the west. We prefer to consider the faults high angle, but another interpretation would make these Chilliwack outcrops a half-window below the folded Shuksan thrust.

20.0(32.2) Finney Creek Road (USFS 17). Turn left, southwest. This paved, snake-swallowed-the-bird road is heavily traveled by fast-moving logging trucks. BE CAREFUL!

24.2(38.9) Finney Creek bridge.

24.3(39.1) *Stop 8-1* (~300 m/980') Descend through brush on south side of road to reach gravel bars of Finney Creek. Walk downstream to examine broken formation in the Darrington Phyllite. The distinctive unit of the Easton Metamorphic Suite is the blueschist-facies Shuksan Greenschist, largely metabasalt, locally with recognizable relict pillows or layering in metatuff. Associated with the greenschist are quartzose metasediments, typically well-recrystallized quartz-chlorite-muscovite-graphite (± albite, garnet, lawsonite) schists with well-developed pressure solution cleavage (S_2) and abundant quartz veins. Here in the Finney Creek region coarser-grained, more feldspathic metasediments are associated with Shuksan Greenschist, indicating a facies change in the sedimentary protolith for the Darrington.

30.3(48.8) Intersection with Gee Point logging road (USFS 1720). Turn right (north) to USFS 1720.

31.3(50.4) Intersection with logging spur (USFS 1721). Turn right (east) and proceed 0.65 miles (1.0 km) to:

31.9(51.3) *Stop 8-2* Relict stratigraphy in metabasite, Fe-Mn formation, and pelitic schist of the Easton Metamorphic Suite. The beds are deformed into F_2 folds. Such folds are abundant in this area, where they range from microscopic crenulations to megascopic folds mappable by tracing metabasite-metasediment contacts [Morrison, 1977]. The metabasite is greenschist containing quartz + albite + epidote + chlorite + actinolite + sphene. Fe-Mn metasediment contains magnetite + hematite + quartz + spessartine garnet + crossite + stilpnomelane + aegirine. In the pelitic schist are quartz + porphyroblastic albite + muscovite + epidote + chlorite + sphene + carbonaceous material + lawsonite.

Retrace route 0.65 miles to USFS 1720. At intersection, turn right and continue for 4.4 miles up 1720, past junction with 1722, to

37.0(59.5) *Stop 8-3* From this viewpoint serpentinite and amphibolite outcrops of the next stop are visible across the valley in a clear-cut area. Structures and minerals seen here at Stop 8-3 appear to overprint rocks at Stop 8-4. Actinolite and crossite grains at 8-3 have hornblende or barroisite cores. Hornblende at 8-4 has actinolite or crossite rims and fracture fillings. S_1 at 8-3 appears to overprint amphiboles at 8-4. F_2 folds mapped in the regional schists at 8-3 are also mappable, with the same orientation, at 8-4. From these observations, Wilson [1978] and Wilson and Brown [1979] concluded that the amphibolite complex at 8-4 was incorporated into the regional schists of the Easton Metamorphic Suite prior to or during regional blueschist-facies metamorphism.

Walk down road to the end of outcrop (600 m) observing interlayered lithologies of the Easton Metamorphic Suite: greenschist, blueschist, quartz schist, carbonaceous schist, and iron formation. $^{18}O/^{16}O$ partitioning between quartz and magnetite in rock from this outcrop suggests a temperature of 350°C [Brown and O'Neil, 1982].

During Shuksan metamorphism the stability of blue amphibole depended on the bulk composition, especially the $Fe^{2+}-Fe^{3+}$ ratio. More ferric rocks bear crossite, whereas less ferric rocks bear actinolite [Brown, 1974; Dungan and others, 1983]. Actinolite-bearing greenschist and crossite-bearing blueschist are closely interlayered at some Shuksan outcrops. Some of the local differences in oxidation state may result from submarine weathering. Greenschist is more common overall, hence the name of the unit, though metamorphism was at blueschist-facies conditions.

Return to vehicles and continue up USFS 1720.

38.7(62.3) Beginning of foot traverse (~1070 m/3500'). Leave vehicles for several hours, to walk about 1.5 miles (2.4 km) and gain and lose about 800' (240 m) elevation. Walk undrivable spur road 017, branching to the right at a small saddle, approximately 10 minutes to:

Stop 8-4 Here at Gee Point is evidence for an early, high-T high-P metamorphic event which preceded regional low-T high-P metamorphism of the Easton Metamorphic Suite. The high-T event involves Easton compositions and ultramafic rocks. Similar high T-high P rocks on Catalina Island, California, were explained by Platt [1975] as developing during the early stages of subduction when ocean-crust protolith was juxtaposed with hot upper mantle peridotite.

Stop 8-4-1 (location of this and subsequent stops on Figure 18) Foliated serpentinite on the left side of the road. Foliation is defined by magnetite clusters and platy texture of antigorite. The foliation is parallel to the regional Shuksan S_1. Brucite, magnesite, chlorite, and rare diopside and andradite are accessory minerals.

Continue on road 50 m around the corner to a small creek (may be dry), then walk up in the vicinity of the creek, climbing about 65 vertical meters to a knob on the left for:

Stop 8-4-2 Garnet (40% grossularite) + salite rocks occurring as an enclave within serpentinite. Garnet veins and irregular patches. This rock is perhaps a high-grade rodingite. Garnet-pyroxene temperatures are 700-800°C for similar rocks in the region based on the Ellis and Green [1979] geothermometer, but this calibration probably incorrectly treats the effect of grossularite in the garnet and overestimates the temperature of this rock.

Continue up hill to:

Stop 8-4-3 Fine-grained, well, foliated, quartzose, carbonaceous, garnet hornblende schist.

Stop 8-4-4 Non-foliated albite + tremolite-actinolite rock. Mg-chlorite, clinozoisite, sphene, and jadeite-omphacite are accessory minerals. Some Na-pyroxene veins and albite veins. No quartz in this rock.

Stop 8-4-5 Foliated serpentine with tremolite.

Stop 8-4-6 Coarse, well-foliated, folded muscovite schist with patches and lenses of metasomatic talc-tremolite rock.

Stop 8-4-7 Talus with blocks representing lithologies from outcrops above. Note the presence of coarse talc-tremolite rock, garnet amphibolite, garnetite, coarse muscovite schist, and quartzite. The association of amphibolite, quartzite, and muscovite schist possibly represents ocean-crust protolith. The talc-tremolite rock is blackwall material produced by metasomatism in a high-temperature contact zone between ultramafic rock and amphibolite and mica schists. We interpret the garnetite to also be a high-grade blackwall metasomatic rock. This association is virtually identical to that of basic gneiss and ultramafic rock on Catalina Island, California. The strong fabric in the high grade rocks indicates dynamic metamorphism. We speculate that this metamorphism was caused by slices of hot ultramafic rock faulted into an ocean-crust protolith (Figure 19).

Stop 8-4-8 Serpentinite with coarse radial chlorite and finer tremolite.

Stop 8-4-9 Amphibolite, quartzite, and coarse muscovite schist with a strong fabric and isoclinal folds. Hornblende from this locality gave a K-Ar age of 148 ± 5 Ma; muscovite from the nearby hill gave K-Ar ages of 164 ± 6 Ma; nearby regionally metamorphosed schists gave concordant K-Ar and Rb-Sr ages of 129 ± 5 Ma [Brown and others, 1982].

Stop 8-4-10 Here garnet amphibolite containing brown hornblende occurs in a thin layer (~10 meters wide) between serpentinite and Na-amphibole schist. The Na-amphibole schist is barroisite-bearing in most places, but glaucophane is the primary amphibole in some rocks. Garnet is present in the Na-amphibole

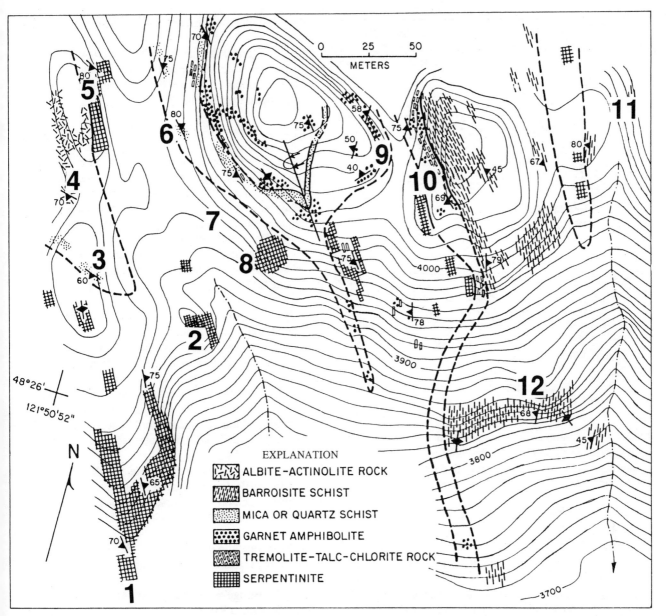

FIGURE 18. Detailed geologic map of Gee Point area showing field trip Stops 8-4-1 to 8-4-12. Modified from Brown and others [1982].

schist only near the brown hornblende amphibolite. Some rocks in the transition zone between amphibolite and Na-amphibole schist contain almandine garnet + glaucophane + albite; others contain glaucophane + omphacite + albite. Near Iron Mountain, the eclogitic assemblage garnet + omphacite + glaucophane + albite is developed, but tends to be strongly retrograded so distinction of primary and secondary assemblages is not always easy.

Pumpellyite, actinolite, Fe-chlorite, and crossite are abundant late-stage minerals developed along a secondary foliation. Note the coarse-grained tremolite lenses deformed by late foliation.

Stop 8-4-11 Talus of regional blueschist and greenschist. Crossite and actinolite in these rocks have barroisite cores.

Stop 8-4-12 Coarse barroisite schist.

Return to vehicles. Retrace route to Finney Creek Road (USFS 17). At intersection:

0.0 Reset odometer. Turn right and continue westerly on Finney Creek road.

1.0(1.6) Bridge over Finney Creek.

3.0(4.8) Junction of Finney Creek road (USFS 17) and Deer Creek (Segelson) Pass road (USFS 18). Take left-hand fork to the pass.

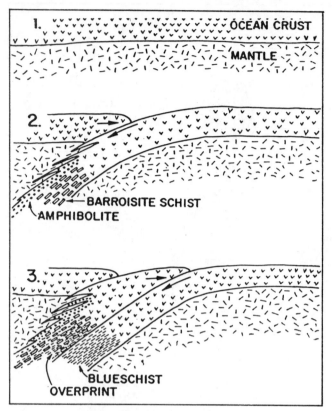

FIGURE 19. Hypothetical tectonic and metamorphic events by which the serpentinite-amphibolite-blueschist assemblage of the Easton Metamorphic Suite may have originated. From Brown and others [1982].

4.0(6.4) Pass outcrops of foliated meta-andesite and metadacite. The prominences of Big and Little Deer Peak seen to the west are metavolcanic rocks similar to these. A U-Pb age of zircon from the metadacite is 168 Ma (Middle Jurassic) [N. Walker *in* Brown and others, 1987]. Similar rocks crop out in small slices within the Helena-Haystack melange to the southeast (**Stops 8-6, 9-4**). Tabor and others [1988] interpret the Deer Peak rocks to be blocks in the Helena-Haystack melange. Reller [1987] and Brown and others [1987] cite evidence for thrust emplacement of the Deer Peak rocks over the Easton Metamorphic Suite.

6.3(10.1) *Stop 8-5* Fluvial sandstone and argillite of the Barlow Pass Volcanics of Vance [1957b]. Note leaf fossils. These are typical beds of the non-volcanic part of this unit, although in this place the sedimentary rocks could as well be called Chuckanut Formation, a correlative unit exposed to the NW (see Section III). We correlate these rocks with the Barlow Pass Volcanics because rhyolitic volcanic rocks are interbedded with the sedimentary rocks just across Deer Creek to the southeast. The coincidence of these Tertiary deposits with the Helena-Haystack melange (Figure 6) may indicate deposition in fault-controlled

basins along the earlier melange structure (see Section III).

The road winds up Deer Creek along many exposures of sandstone and argillite. Rapid changes in dip and topping direction indicate complex folding and faulting in the section.

12.3(19.8) *Stop 8-6* (~1025 m/3360') The narrow gap of the pass is eroded from contorted phyllite in a fault zone separating Eocene sandstone and a sliver of foliated metatuff on the west from Shuksan Greenschist on the east. The metatuff is locally blastomylonitic, but relict plagioclase microporphyroclasts in a schistose chlorite-epidote-sericite-quartz-plagioclase matrix reveal the probable volcanic origin.

Continuing south on the road are outcrops of sheared middle Eocene sandstone, argillite, and welded dacite tuff.

15.7(25.3) *Stop 8-7* Views of the North Cascades. West of the Straight Creek fault, rocks of the Chilliwack Group and the Yellow Aster complex are exposed on Prairie Mountain to the east and Shuksan Greenschist on White Chuck Mountain to the southeast. Farther to the east, behind the NWCS, are rocks of the North Cascades crystalline core, mostly schist and gneiss of the Nason terrane and Late Cretaceous orthogneiss bodies. Glacier Peak (3213 m/10541'), a Quaternary stratovolcano, crowns them all.

To the south we view the Helena-Haystack melange on Helena Ridge and Jumbo Mountain, the eastern melange belt on Whitehorse Mountain and the western melange belt west of Whitehorse (Figure 20).

25.4(40.9) Junction of Deer Creek (Segelson) Pass road (USFS 18) and Swede Heaven road. Turn left, east.

26.1(42.0) Bridge across the North Fork of the Stillaguamish River.

27.2(43.8) Junction of Swede Heaven road and WA 530 (~140 m/460'). Whitehorse Store. Turn left, east.

28.7(46.2) Low drainage divide just east of Squire Creek bridge. Volcanic rock clasts from Glacier Peak volcano exposed here show that the underfit North Fork of the Stillaguamish River once carried drainage from Glacier Peak that today goes north to the Skagit River [Vance, 1957a]. The Skagit River drained this way when its valley was dammed by stagnant ice of the Canadian icesheet near the town of Concrete.

28.9(46.5) Look back to the northwest for views of Mount Higgins (1567 m/5142') held up by a faulted syncline in thick beds of sandstone of the Barlow Pass Volcanics.

32.2(51.8) Stagecoach Inn.

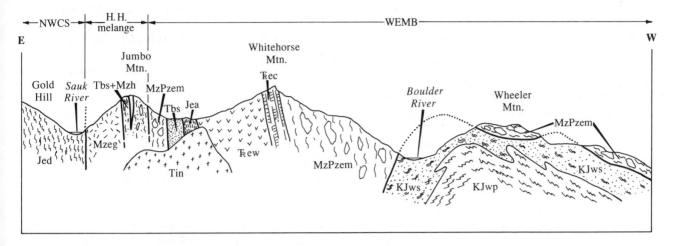

FIGURE 20. Sketch of geologic relations in view from Deer Creek Pass road (Stop 8-7). For map-unit symbols refer to Figure 9, map 6.

Day 9. Darrington to Helena Ridge, Barlow Pass, and Seattle. We look at the Helena-Haystack melange, views of the DDMFZ, and the eastern and western melange belts.

0.0 Stagecoach Inn. Before leaving Darrington look up to the south to see Jumbo Mountain (~1780 m/5840'), a megablock of gabbro in the serpentinite matrix of the Helena-Haystack melange. Prominent reddish rock fins on the east side of the summit are contact-metamorphosed fault slivers of serpentinite imbricated in Eocene sandstone (see **Stop 9-5**).

Proceed south from Darrington on the Mountain Loop Highway as described for Day 7.

4.8(7.7) Junction with Helena Ridge road (USFS 2070) (~225 m/740'). Turn right (west) and begin ascent of Helena Ridge.

6.0(9.0) Pass outcrops of Shuksan Greenschist. The road winds up along Barns Creek, which is eroded out along a fault separating the Helena-Haystack melange from the Easton Metamorphic Suite.

9.1(114.6 Intersection with logging spur (USFS 2075).

9.5(15.3) ***Stop 9-1*** (~850 m/2790) Brecciated chert and greenstone. Subdued slopes with scarce outcrops all around this knocker are the serpentinite matrix of the Helena-Haystack melange (and landslides from it). The age of this knocker is not known, but it probably is a block of the eastern melange belt. The melange here on Helena Ridge was first described by Vance [1957] and Vance and others [1980], who suggested that it was a klippe of sheared rocks of the Shuksan thrust zone.

10.8(17.4) ***Stop 9-2*** (~875 m/2870') Eocene fluvial conglomerate. This vertical bed with tops toward the road (east) rests unconformably on fine-grained amphibolite exposed just up the road. The conglomerate and amphibolite are a fault-bounded block within the Helena-Haystack melange, where latest faulting is of Eocene (or earliest Oligocene) age, as the DDMFZ is plugged by the 35 Ma Squire Creek stock. Tertiary deformation was not restricted to brittle faulting: cobbles in Eocene conglomerate in the DDMFZ to the west of this stop are stretched along N and NW trends (also see Day 6, mile 29.2).

At issue here is just how much of the deformation in the DDMFZ—and, by extrapolation, much of the NWCS and WEMB—is Eocene. Pre-Eocene deformation is difficult to demonstrate directly, though deformation which affects Tertiary strata does not seem extensive enough to juxtapose the diverse pre-Tertiary rocks now adjacent to each other in the DDMFZ. N- and NNW-trending stretching lineations in the NWCS are interpreted to have formed during Late Cretaceous orogeny [Brown, 1987; Smith, 1988], yet similar fabrics in the DDMFZ were developed in the Eocene. Was the DDMFZ formed in the Late Cretaceous and reactivated in the Eocene? Did Late Cretaceous orogeny continue into the Eocene? Or is all deformation Eocene?

Walk up the road ~0.1 mile to a quarry in the amphibolite. Hornblende from this amphibolite yielded a K-Ar age of 141 ± 6 Ma [Armstrong and Misch, 1987]. Just upslope amphibolite is interlayered with gneissic hornblende tonalite on the margins of more massive tonalite (**Stop 9-3**) which yields concordant U-Pb zircon ages of ~150 Ma (written communication from N. Walker to J. A. Vance, *in* Tabor and others [1988]). Similar amphibolite and associated tonalite are not found elsewhere west of the Straight Creek fault at this latitude, but similar rocks with the same ages crop out across the fault in the Manastash Ridge area about 100 km to the south. See Figure 6 and Section III for discussion.

11.1(17.9) *Stop 9-3* Hornblende tonalite. This is the megablock in the serpentinite matrix which is associated with the amphibolite. Turn around and return to logging spur USFS 2075.

13.1(21.1) Intersection with logging spur (USFS 2075). Turn left, west. Outcrops along the first part of this road are serpentinite and blocks of greenstone.

14.2(22.9) *Stop 9-4* Fine-grained mica schist, probably meta-rhyolite, a coarser-grained variety of metavolcanic rock seen at Deer Creek Pass (Stop 8-6). A K-Ar age on muscovite from this rock is about 90 Ma, roughly the same age as that derived from a two point Rb-Sr isochron (R. L. Armstrong, written communication to J. A. Vance, *in* Tabor and others [1988]). If this age is correct the Helena-Haystack melange formed in the latest Cretaceous or early Tertiary (see Section III).

15.0(24.1) *Stop 9-5* (~900 m/2950') Metagabbro. This rock is crushed and totally altered to pumpellyite, actinolite, albite and prehnite(?); it resembles altered gabbro on Jumbo Mountain (visible to the west) and metagabbro knockers in the eastern melange belt west of Jumbo. In the summit area of Jumbo Mountain, locally strongly tectonized feldspathic sandstone of Tertiary age is faulted against elongate slivers of metamorphic dunite which then resemble dikes. Vance and Dungan [1977] show clearly that these ultramafic "dikes" are metaserpentinite. They recrystallized in the contact aureole of the Oligocene Squire Creek stock seen on the south ridge of Jumbo (but for an alternate hypothesis see Vance and Dungan [1977]).

Turn around here and return to the Mountain Loop Highway.

0.0 Junction of Helena Ridge Road (USFS 2070) and Mountain Loop Highway. Reset odometer, turn right, and continue south on the Mountain Loop Highway.

4.4(7.1) Junction of White Chuck River Road (USFS 23) and Mountain Loop Highway (USFS 20). Take right (south) fork.

16.2(26.1) U.S. Forest Service Guard Station. During the summer months, a Forest Service employee lives here to look after back roads, trails, and camping grounds and to help and inform forest visitors.

16.5(26.6) Bridge over the North Fork of the Sauk River.

20.0(32.2) Outcrops along road are dacite porphyry of the Sauk "ring dike", a large crescent-shaped intrusive body exposed in the lower part of the valley (J. A. Vance, written communication, 1988). Swarms of porphyry dikes extend into the adjacent middle Eocene Barlow Pass Volcanics. The ring dike appears to be about the

same age as or a little younger than the extrusive rocks.

20.3(32.7) The road winds around a large landslide from Sheep Mountain (1879 m/6166') on the east, which dammed the south fork of the Sauk. Marshes and beaver ponds to the west outline the former lake.

23.1(37.2) Barlow Pass (720 m/2361'). This low pass was the route of a railroad to the mining town of Monte Cristo about 13 km to the southeast. The railroad was busy in the early 1900s. West of the pass sections of railroad grade, cut into basalt of the Barlow Pass Volcanics, are visible above the highway.

27.6(44.4) Junction of Mountain Loop Highway and Coal Creek logging road (USFS 4060). Turn right (north).

28.1(45.2) Junction with short logging spur (USFS 4061). Turn right (east).

28.3(45.5) *Stop 9-6* End of road on logging platform. View of Big Four Mountain (1870 m/6135'). Across the South Fork of the Stillaguamish River, high-angle faults on the north face of Big Four separate Eocene sandstone at the base (and eroded out of the main valley) and greenstone, metachert, marble, and metasandstone of the eastern melange belt which form the summit cliffs. The Oligocene Vesper Peak stock, mostly hidden behind Big Four, has thermally metamorphosed the rocks, making them cliff-formers.

To the south, view sharply folded and faulted sandstone in Morning Star Peak (1835 m/6020'). To the east, above and behind us, are basalt and rhyolite of the Barlow Pass Volcanics. These bimodal volcanic rocks with interbeds of fluvial sandstone suggest deposition in an extensional tectonic setting (see Section III). Mild deformation in the volcanic rocks to the east in contrast to the strong deformation in the predominantly sedimentary part of the section to the south and west suggests deformation along the DDMFZ was strongly focused here. The disappearance of the Helena-Haystack melange under the Tertiary rocks at this point suggests the area is a major structural sag.

Turn around and return to Mountain Loop Highway.

29.0(46.7) Junction of the Coal Creek road and Mountain Loop Highway. Turn right and proceed west on the highway.

32.7(52.6) Town of Silverton was once a mining camp for sulphide prospects associated with the Vesper Peak and Squire Creek stocks and associated plugs.

36.8(59.2) Bridge over South Fork of the Stillaguamish. Outcrops along the road to the west are highly deformed greywacke, argillite, and chert of the western melange belt.

43.8(70.5) Verlot Ranger Station.

47.7(76.8) **Stop 9-7** (~295 m/970'). Beginning of Washington State Department of Natural Resources trail to lower Stillaguamish Gorge. We will walk about 3.5 miles (5.6 km) round trip.

0.0 (Trail mileage is approximate) Follow the trail a few hundred meters to edge of glacial terrace and descend to lower river terrace.

0.5(0.8) Trail reaches old road; turn right (north) and follow switchback down to river level.

1.0(1.6) Trail emerges at river's edge. Follow abandoned railroad grade downstream through first tunnel (Tunnel 6). The old railroad tunnels which allow access to good outcrops in the lower Stillaguamish gorge were constructed in 1892 for the Everett and Monte Cristo Railroad. The tunnels and gorge were the downfall of the railroad, and probably the mining endeavors in Monte Cristo, because the river when in flood rises up into the tunnels.

1.8(2.9) Tunnel 5. Go through tunnel and continue along railroad grade. Just before a steepwalled cut partly filled with landslide debris, scramble down loose rocks to:

1.9(3.1) Riverside outcrops of melange. The disruption of beds here is typical of the non-slatey western melange belt except in large blocks of structurally resistant sandstone and chert. The deformation here with its lenticular and locally necked sandstone blocks, greenstone, limestone, and rare metatonalite resembles Cowan's [1985] type-II melange evolved from submarine landslides of a once coherent stratigraphic sequence. However, the regional scale of mixing suggests that much of it is tectonic (Section I).

Return to vehicles on Mountain Loop Highway.

47.7(76.8) Resume road miles and continue west to:

53.0(85.3) Granite Falls. End of road log. The drive to downtown Seattle is about 50 miles (80 km), about 1 hour driving time. Follow WA 92 west to WA 9 and turn left (south) on WA 9. At the Frontier Village stop light, turn right on WA 204. Follow signs to US 2 and, just beyond, Interstate 5. Take Interstate 5 south to Seattle.

REFERENCES

Adams, J.B., Petrology and structure of the Stehekin-Twisp Pass area, northern Cascades, Washington, Ph.D. thesis, University of Washington, Seattle, 172 p., 1961.

Adams, J.B., Petrology and structure of the Stehekin-Twisp Pass area, northern Cascades, Washington, *Dissertation Abstracts*, 22, p. 3981, 1962.

Adams, J.B., Origin of the Black Peak Quartz Diorite, Northern Cascades, Washington, *American Journal of Science*, 262, p. 290-306, 1964.

Ashleman, J.C., The geology of the western part of the Kachess Lake quadrangle, Washington, M.S. thesis, University of Washington, Seattle, 88 p., 1979.

Armstrong, R.L., J.E. Harakal, E.H. Brown, M.L. Bernardi, and P.M. Rady, Late Paleozoic high-pressure metamorphic rocks in northwestern Washington and southwestern British Columbia: The Vedder Complex, *Geological Society of America Bulletin*, 94, p. 451-458, 1983.

Armstrong, R.L., and P. Misch, Rb-Sr and K-Ar dating of mid-Mesozoic blueschist and late Paleozoic albite-epidote-amphibolite and blueschist facies metamorphism in the North Cascades, Washington and British Columbia, and fingerprinting of eugeosynclinal rock assemblages, in Selected papers on the geology of Washington, edited by J.E. Schuster, *Washington Division of Geology and Earth Resources Bulletin 77*, p. 85-105, 1987.

Babcock, R.S., Geochemistry of the main-stage migmatitic gneisses in the Skagit Gneiss Complex, Ph.D. thesis, University of Washington, Seattle, 147 p., 1970.

Babcock, R.S., R.L. Armstrong, and P. Misch, Isotopic constraints on the age and origin of the Skagit Metamorphic Suite and related rocks, *Geological Society of America Abstracts with Programs*, 17, p. 339, 1985.

Babcock, R.S., and P. Misch, Evolution of the crystalline core of the North Cascades Range, in *Metamorphism and crustal evolution of the western United States*, Rubey Volume VII, edited by W.G. Ernst, Prentice Hall, Englewood Cliffs, New Jersey, p. 214-232, 1988.

Babcock, R.S., and P. Misch, Origin of the Skagit migmatites, North Cascades Range, Washington State, *Contributions to Mineralogy and Petrology*, in press.

Baker, V.R., and Dag Nummedal, editors, *The channeled scabland*, Washington, D.C., National Aeronautics and Space Administration, 186 p., 1978.

Barksdale, J.D., Stratigraphy in the Methow quadrangle, Washington, *Northwest Science*, 22, p. 164-176, 1948.

Barksdale, J.D., Geology of the Methow Valley, Okanogan County, Washington, *Washington Division of Geology and Earth Resources Bulletin 68*, 72 p., 1975.

Beck, M.E., Jr., R.F. Burmester, and R. Schoonover, Paleomagnetism and tectonics of the Cretaceous Mt. Stuart Batholith of Washington: translation or tilt?, *Earth and Planetary Science Letters*, 56, p. 336-342, 1981a.

Beck, M.E., Jr., R.F. Burmester, D.C. Engebretson, and R. Schoonover, Northward translation of Mesozoic batholiths, western North America: paleomagnetic evidence and tectonic significance: *Geofisica Internacional*, 20, p. 144-162, 1981b.

Blackwell, D.L., Geology of the Park Butte-Loomis Mountain area, Washington, M.S. thesis, Western Washington University, Bellingham, 253 p., 1983.

Boggs, R.C., Mineralogy and geochemistry of the Golden Horn batholith, northern Cascades, Washington, Ph.D. thesis, University of California, Santa Barbara, 187 p., 1984.

Booth, D.B., Timing and processes of deglaciation along the southern margin of the Cordilleran ice sheet, in *North America and adjacent oceans during the last glaciation*, The Geology of North America, K-3, edited by W.F. Ruddiman and H.E. Wright, Jr., Geological Society of America, Boulder, Colorado, p. 71-90, 1987.

Brandon, M.T., Pacific Rim Complex of western Vancouver Island: tectonic evolution of a late Mesozoic active margin, *Geological Society of America Abstracts with Programs, 17*, p. 343, 1985.

Brandon, M.T., and D.S. Cowan, The Late Cretaceous San Juan Islands - northwestern Washington thrust system, *Geological Society of America Abstracts with Programs, 17*, p. 343, 1985.

Brandon, M.T., D.S. Cowan, and J.A. Vance, The late Cretaceous San Juan thrust system, San Juan Islands, Washington, *Geological Society of America Special Paper 221*, 81 p., 1988.

Bretz, J H., The Spokane flood beyond the Channeled Scabland, *Journal of Geology, 33*, p. 97-115 and 236-259, 1925.

Brown, E.H., Comparison of the mineralogy and phase relations of blueschists from the North Cascades, Washington, and greenschists from Otago, New Zealand, *Geological Society of America Bulletin, 85*, p. 333-344, 1974.

Brown, E.H., Geology of the Shuksan Suite, North Cascades, Washington, in Blueschists and related eclogites, edited by B.W. Evans and E.H. Brown, *Geological Society of America Memoir 164*, p. 143-154, 1986.

Brown, E.H., Structural geology and accretionary history of the Northwest Cascades system, Washington and British Columbia, *Geological Society of America Bulletin, 99*, p. 201-214, 1987.

Brown, E.H., Metamorphic and structural history of the Northwest Cascades, Washington and British Columbia, in *Metamorphism and crustal evolution of the western United States*, Rubey Volume VII, edited by W.G. Ernst, Prentice Hall, Englewood Cliffs, New Jersey, p. 196-213, 1988.

Brown, E.H., and J.R. O'Neil, Oxygen isotope geothermometry and stability of lawsonite and pumpellyite in the Shuksan Suite, North Cascades, Washington, *Contributions to Mineralogy and Petrology, 80*, p. 240-244, 1982.

Brown, E.H., M.L. Bernardi, B.W. Christenson, J.A. Cruver, R.A. Haugerud, P.M. Rady, and J.N. Sondergaard, Metamorphic facies and tectonics in part of the Cascade Range and Puget Lowland of northwest Washington, *Geological Society of America Bulletin, 92*, p. 170-178, 1981.

Brown, E.H., D.L. Wilson, R.L. Armstrong, and J.E. Harakal, Petrologic, structural, and age relations of serpentine, amphibolite, and blueschist in the Shuksan Suite of the Iron Mountain-Gee Point area, North Cascades, *Geological Society of America Bulletin, 92*, p. 1087-1098, 1982.

Brown, E.H., D.L. Blackwell, B.W. Christenson, F.I. Frasse, R.A. Haugerud, J.T. Jones, P.A. Leiggi, M.L. Morrison, P.M. Rady, G.J. Reller, J.H. Sevigny, D.S. Silverberg, M.T. Smith, J.N. Sondergaard, and C.B. Ziegler, Geologic map of the northwest Cascades, Washington, *Geological Society of America Map and Chart Series MC-61*, 1987.

Brown, E.H., and M.C. Blake, Jr., Correlation of Early Cretaceous blueschists in Washington, Oregon, and northern California, *Tectonics, 6*, p. 795-806, 1987.

Cairnes, C.E., Hope Sheet, British Columbia, *Geological Survey Canada Map 737A*, 1944.

Cater, F.W., The Cloudy Pass epizonal batholith and associated subvolcanic rocks, *Geological Society of America Special Paper 116*, 53 p., 1969.

Cater, F.W., Intrusive rocks of the Holden and Lucerne quadrangles, Washington—the relation of depth zones, composition, textures, and emplacement of plutons, *U.S. Geological Survey Professional Paper 1220*, 108 p., 1982.

Cater, F.W., and D.F. Crowder, Geologic map of the Holden quadrangle, Snohomish and Chelan counties, Washington, *U.S. Geological Survey Geological Quadrangle Map GQ-646*, scale 1:62,500, 1967.

Cater, F.W., and T.L. Wright, Geologic map of the Lucerne quadrangle, Chelan County, Washington, *U.S. Geological Survey Geological Quadrangle Map GQ-647*, scale 1:62,500, 1967.

Christenson, B.W., Structure, petrology and geochemistry of the Chilliwack Group near Sauk Mountain, Washington, M.S. thesis, Bellingham, Western Washington University, 181 p., 1981.

Church, S.E., and G.R. Tilton, Lead and strontium isotopic studies in the Cascade Mountains: Bearing on andesite genesis, *Geological Society of America Bulletin, 84*, p. 431-454, 1973.

Clayton, D.N., and R.B. Miller, Geologic studies of the southern continuation of the Straight Creek Fault, Snoqualmie area, Washington, in Washington Public Power Supply System PSAR for WPSS Nuclear Projects 1 and 4, Amendment 23, Shannon and Wilson, Inc., 31 p., 1977.

Clemens, J.D., and D. Vielzeuf, Constraints on melting and magma production in the crust, *Earth and Planetary Science Letters, 86*, p. 287-306, 1987.

Coates, J.A., Stratigraphy and structure of Manning Park area, Cascade Mountains, British Columbia, *Geological Association of Canada Special Paper 6*, p. 149-154, 1970.

Coates, J.A., Geology of the Manning Park area, British Columbia, *Geological Survey of Canada Bulletin 238*, 177 p., 1974.

Cowan, D.S., Structural styles in Mesozoic and Cenozoic melanges in the western Cordillera of North America, *Geological Society of America Bulletin, 96*, p. 451-462, 1985.

Crowder, D.F., Granitization, migmatization, and fusion in the northern Entiat Mountains, Washington, *Geological Society of America Bulletin, 70*, p. 827-878, 1959.

Danner, W.R., Limestone resources of western Washington, Washington Division of Mines and Geology, Bulletin 52, 474 p., 1966.

Davis, G.A., J.W.H. Monger, and B.C. Burchfiel, Mesozoic construction of the Cordilleran "Collage", central British Columbia to central California, in *Mesozoic paleogeography of the western United States*, Pacific Coast Paleogeography Symposium 2, edited by D.G. Howell and K.A. McDougall, Pacific Section, Society of Economic Paleontologists and Mineralogists, p. 1-32, 1978.

Dellinger, D.A., and C.A. Hopson, Age-depth compositional spectrum through the diapiric Duncan Hill pluton, North Cascades, Washington, *Geological Society of America Abstracts with Programs, 18*, p. 100-101, 1986.

Dethier, D.P., J.T. Whetten, and P.R. Carroll, Preliminary geologic map of the Clear Lake SE quadrangle, Skagit County, Washington, U.S. Geological Survey Open-File Report 80-303, 2 plates, scale 1:24,000, 11 p., 1980.

DiLeonardo, C.G., Structural evolution of the Smith Canyon fault, northeastern Cascades, Washington, M.S. thesis, San Jose State University, San Jose, California, 85 p., 1987.

Dungan, M.A., J.A. Vance, and D.P. Blanchard, Geochemistry of the Shuksan greenschists and blueschists, North Cascades, Washington: Variably fractionated and altered metabasalts of oceanic affinity, *Contributions to Mineralogy and Petrology, 82*, p. 131-146, 1983.

Ellis, D.J., and D.H. Green, An experimental study of the effect of Ca upon garnet-clinopyroxene Fe-Mg exchange equilibria, *Contributions to Mineralogy and Petrology, 71*, p. 13-22, 1979.

Engels, J.C., R.W. Tabor, F.K. Miller, and J.D. Obradovich, Summary of K-Ar, Rb-Sr, U-Pb, Pb-alpha, and fission-track ages for rocks from Washington State prior to 1975 (exclusive of Columbia Plateau basalts), *U.S. Geological Survey Map MF-710*, 1976.

Evans, B.W., and J.W. Berti, A revised metamorphic history for the Chiwaukum Schist, North Cascades, Washington, *Geology, 14*, p. 695-698, 1986.

Fiske, R.S., C.A. Hopson, and A.C. Waters, Geology of Mount Rainier National Park, Washington, *U.S. Geological Survey Professional Paper 444*, 93 p., 1963.

Ford, A.B., W.H. Nelson, R.A. Sonnevil, R.A. Loney, Carl Huie, R.A. Haugerud, and S.L. Garwin, Geologic map of the Glacier Peak Wilderness and adjacent areas, Chelan, Skagit, and Snohomish Counties, Washington, *U.S. Geological Survey Miscellaneous Field Studies Map MF-1652*, in press.

Franklin, W.E., Structural significance of meta-igneous fragments in the Prairie Mountain area, North Cascade Range, Snohomish County, Washington, M.S. thesis, Oregon State University, Corvallis, 109 p., 1974.

Frizzell, V.A., Jr., R.W. Tabor, R.E. Zartman, and C.D. Blome, Late Mesozoic or early Tertiary melanges in the western Cascades of Washington, in Selected papers on the geology of Washington, edited by J.E. Schuster, *Washington Division of Mines and Geology Bulletin 77*, p. 129-148, 1987.

Frost, B.R., Limits to the assemblage forsterite-anorthite as inferred from peridotite hornfelses, Icicle Creek, Washington, *American Mineralogist, 61*, p. 732-750, 1976.

Fugro NW, Interim report on geologic feasibility studies for Copper Creek dam, report to Seattle City Light, 145 p., 1979.

Gallagher, M.P., E.H. Brown, and N.W. Walker, A new structural and tectonic interpretation of the western part of the Shuksan blueschist terrane, *Geological Society of America Bulletin, 100*, p. 1415-1422, 1988.

Granirer, J.L., R.F. Burmester, and M.E. Beck, Jr., Cretaceous paleomagnetism of the Methow-Pasayten belt, Washington, *Geophysical Research Letters, 13*, p. 733-736, 1986.

Greig, C.J., Geology and geochronometry of the Eagle plutonic complex, Hope map area, southwestern British Columbia, in Current Research, Part E, Cordillera and Pacific margin, *Geological Survey of Canada Paper 88-1E*, p. 177-184, 1988.

Gresens, R.L., C.W. Naeser, J.T. and Whetten, Stratigraphy and age of the Chumstick and Wenatchee Formations: Tertiary fluvial and lacustrine rocks, Chiwaukum graben, Washington: Summary, *Geological Society of America Bulletin, Pt I, 92*, p. 233-236, 1981.

Hamilton, W., Mesozoic tectonics of the western United States, in *Mesozoic paleogeography of the western United States*, Pacific Coast Paleogeography Symposium 2, edited by D.G. Howell and K.A. McDougall, Pacific Section, Society of Economic Paleontologists and Mineralogists, p. 33-70, 1978.

Haugerud, R.A., Geology of the Hozameen Group and the Ross Lake shear zone, Maselpanik area, North Cascades, southwest British Columbia, Ph.D. dissertation, University of Washington, Seattle, 263 p., 1985.

Haugerud, R.A., Argon geochronology of the Tenpeak pluton and untilting of the Wenatchee block, North Cascades Range, Washington (abstract), *EOS, 68*, p. 1814, 1987.

Haugerud, R.A., M.L. Morrison, and E.H. Brown, Structural and metamorphic history of the Shuksan Metamorphic Suite in the Mount Watson and Gee Point areas, North Cascades, Washington, *Geological Society of America Bulletin, 92*, p. 374-383, 1981.

Hollister, L.S., Metastable paragenetic sequence of andalusite, kyanite, and sillimanite, Kwoiek area, British Columbia, *American Journal of Science, 271*, p. 352-370, 1969.

Hopkins, W.N., Geology of the Newby Group and adjacent units in the southern Methow trough, northeast Cascades, Washington, M.S. thesis, San Jose State University, San Jose, California, 95 p., 1987.

Hoppe, W.J., Structure and geochronology of the Gabriel Peak orthogneiss and adjacent crystalline rocks, North Cascades, Washington, *Geological Society of America Abstracts with Programs, 14*, p. 173, 1982.

Hopson, C.A., F.W. Cater, and D.F. Crowder, Emplacement of plutons, Cascade Mountains, Washington, *Geological Society of America Abstracts with Programs, 2*, p. 104, 1970.

Hopson, C.A., and J.M. Mattinson, Metamorphism and plutonism, Lake Chelan region, northern Cascades, Washington, in Metamorphism in the Canadian Cordillera, Programme and Abstracts, Geological Association of Canada, Cordilleran Section, Vancouver, p. 13, 1971.

Irving, E., G.J. Woodsworth, P.J. Wynne, and A. Morrison, Paleomagnetic evidence for displacement from the south of the Coast Plutonic Complex, British Columbia, *Canadian Journal of Earth Sciences, 22*, p. 584-598, 1985.

James, E.W., Geology and petrology of the Lake Ann stock and associated rocks, M.S. thesis, Western Washington University, Bellingham, 58 p., 1980.

Johnson, S.Y., Evidence for a margin-truncating transcurrent fault (pre-Late Eocene) in western Washington, *Geology, 12*, p. 538-541, 1984a.

Johnson, S.Y., Stratigraphy, age, and paleogeography of the Eocene Chuckanut Formation, northwest Washington, *Canadian Journal of Earth Sciences, 21*, p. 92-106, 1984b.

Johnson, S.Y., Cyclic fluvial sedimentation in a rapidly subsiding basin, northwest Washington, *Sedimentary Geology, 38*, p. 361-392, 1984c.

Johnson, S.Y., Eocene strike-slip faulting and nonmarine basin formation in Washington, in *Strike-slip deformation, basin formation, and sedimentation*, edited by K.T. Biddle and N. Christie-Blick, Society of Economic Paleontologists and Mineralogists Special Publication 37, p. 283-302, 1985.

Johnson, S.Y., R.A. Zimmermann, C.W. Naeser, and J.T. Whetten, Fission-track dating of the tectonic development of the San Juan Islands, Washington, *Canadian Journal of Earth Sciences, 23*, p. 1318-1330, 1986.

Jones, D.L., N.J. Silberling, and John Hillhouse, Wrangellia—A displaced terrane in northwestern North America, *Canadian Journal of Earth Sciences, 14*, p. 2565-2577, 1977.

Kleinspehn, K.L., Cretaceous sedimentation and tectonics, Tyaughton-Methow basin, southwestern British Columbia, *Canadian Journal of Earth Sciences, 22*, p. 154-174, 1985.

Kriens, Bryan, and Wernicke, Brian, Crustal sections and arc magmatism: New findings from the Ross Lake fault zone, North Cascades, Washington (abstract), *EOS, 67*, p. 1189, 1986.

Kriens, Bryan, Cretaceous-Tertiary tectonic evolution of the North Cascades, Washington: New findings from the Ross Lake fault zone, *Geological Society of America Abstracts with Programs, 19*, p. 396, 1987.

Laravie, J.A., Geological field studies along the eastern border of the Chiwaukum graben, central Washington, M.S. thesis, University of Washington, Seattle, 56 p., 1976.

Le Breton, N., and A.B. Thompson, Fluid-absent (dehydration) melting of biotite in the early stages of crustal anatexis, *Contributions to Mineralogy and Petrology, 99*, p. 226-237, 1988.

Lister, G.S., and A.W. Snoke, S-C mylonites, *Journal of Structural Geology, 6*, p. 617-638, 1984.

Lovseth, T.P., The Devils Mountain fault zone, northwestern Washington, M.S. thesis, University of Washington, Seattle, 29 p., 1975.

Mackin, J.H., and A.S Cary, Origin of Cascade landscapes, *Washington Division of Mines and Geology Information Circular 41*, 35 p., 1965.

Magloughlin, J.F., Metamorphic petrology, structural history, geochronology, tectonics and geothermometry/geobarometry in the Wenatchee Ridge area, North Cascades, Washington, M.S. thesis, University of Washington, Seattle, 344 p., 1986.

Magloughlin, J.F., and B.L. Brasaemle, Petrology and structural implications of a suite of peraluminous trondhjemite pegmatites from the North Cascade Mountains, Washington, *Geological Society of America Abstracts with Programs, 20,* p. 177, 1988.

Marcus, K.L., Eocene-Oligocene sedimentation and deformation in the northern Puget Sound area, Washington, *Northwest Geology, 9,* p. 52-58, 1980.

Mattinson, J.M., Ages of zircons from the Northern Cascade Mountains, Washington, *Geological Society of America Bulletin, 83,* p. 3769-3784, 1972.

McGroder, M.F., Yalakom-Foggy Dew fault: a 500(+) km long Late Cretaceous-Paleogene oblique-slip fault in Washington and British Columbia, *Geological Society of America Abstracts with Programs, 19,* p. 430, 1987.

McGroder, M.F., Structural evolution of eastern Cascades foldbelt, *Geological Society of America Abstracts with Programs, 20,* p. 213, 1988.

McTaggart, K.C., Tectonic history of the northern Cascade Mountains, *Geological Society of Canada Special Paper 6,* p. 137-148, 1970.

McTaggart, K.C., and R.M. Thompson, Geology of part of the northern Cascades in southern British Columbia, *Canadian Journal of Earth Sciences, 4,* p. 1199-1228, 1967.

Miller, R.B., The ophiolitic Ingalls Complex, north-central Cascade Mountains, Washington, *Geological Society of America Bulletin, 96,* p. 27-42, 1985.

Miller, R.B., Geology of the Twisp River - Chelan Divide region, North Cascades, Washington, *Washington Division of Geology and Earth Resources Open-File Report 87-17,* 12 p., 1987.

Miller, R.B., The pre-Tertiary Rimrock Lake inlier, southern Cascades, Washington, *Geological Society of America Bulletin,* in press.

Miller, R.B., and D.W. Mogk, Ultramafic rocks of a fracture-zone ophiolite, North Cascades, Washington, *Tectonophysics, 142,* p. 261-289, 1987.

Miller, R.B., S.A Bowring, and W.J. Hoppe, New evidence for extensive Paleogene plutonism and metamorphism in the crystalline core of the North Cascades, *Geological Society of America Abstracts with Programs, 20,* p. 432-433, 1988.

Misch, P., Geology of the Northern Cascades of Washington, *The Mountaineer, 45,* p. 3-22, 1952.

Misch, P., Alkaline granite amidst the calc-alkaline intrusive suite of the Northern Cascades, Washington (abstract), *Geological Society of America Special Paper 87,* p. 216-217, 1965.

Misch, P., Tectonic evolution of the Northern Cascades of Washington State, in *Tectonic history and mineral deposits of the Western Cordillera,* edited by H.C. Gunning, Canadian Institute of Mining and Metallurgy Special Volume 8, p. 101-148, 1966.

Misch, P., Plagioclase compositions and non-anatectic origin of migmatitic gneisses in Northern Cascades of Washington State, *Contributions to Mineralogy and Petrology, 17,* p. 1-70, 1968.

Misch, P., Dextral displacements at some major strike faults in the North Cascades, Geological Association of Canada Cordilleran Section, Programme with Abstracts, Vancouver, p. 37, 1977a.

Misch, P., Bedrock geology of the North Cascades, in *Geological excursions in the Pacific Northwest,* edited by E.H. Brown and R.C. Ellis, Geological Society of America Field Guide, Annual Meeting, Seattle, p. 1-62, 1977b.

Misch, P., Geologic map of the Marblemount quadrangle, Washington, *Washington Division of Geology and Earth Resources Geologic Map GM-23,* scale 1:48,000, 1979.

Misch, P., Tectonic and metamorphic evolution of the North Cascades: an overview, in *Metamorphism and crustal evolution of the western United States,* Rubey Volume VII, edited by W.G. Ernst, Prentice Hall, Englewood Cliffs, New Jersey, p. 180-195, 1988.

Monger, J.W.H., Stratigraphy and structure of the type area of the Chilliwack Group, southwestern British Columbia, Ph.D. thesis, University of British Columbia, Vancouver, 158 p., 1966.

Monger, J.W.H., Hope map-area, west half, British Columbia, *Geological Survey of Canada Paper 69-47,* 75 p., 1970.

Monger, J.W.H., R.A. Price, and D.J. Tempelmann-Kluit, Tectonic accretion and the origin of the two major metamorphic and plutonic welts in the Canadian Cordillera, *Geology, 10,* p. 70-75, 1982.

Monger, J.W.H., and C.A. Ross, Distribution of Fusulinaceans in the western Canadian Cordillera, *Canadian Journal of Earth Sciences, 8,* p. 259-278, 1971.

Morrison, M.L., Structure and stratigraphy of the Shuksan metamorphic suite in the Gee Point-Finney Peak area, North Cascades, M.S. thesis, Bellingham, Western Washington State College, 69 p., 1977.

Onyeagocha, A.C., Twin Sisters dunite: petrology and mineral chemistry, *Geological Society of America Bulletin, 89,* p. 4159-4174, 1978.

Pacht, J.A., Petrologic evolution and paleogeography of the Late Cretaceous Nanaimo Basin, Washington and British Columbia: Implications for Cretaceous tectonics, *Geological Society of America Bulletin, 95,* p. 766-778, 1984.

Platt, J.P., The petrology, structure, and geologic history of the Catalina Schist terrane, Southern California, *University of California Publications in Geological Sciences, 112,* 111 p., 1976.

Plummer, C.C., Dynamothermal contact metamorphism superposed on regional metamorphism in the pelitic rocks of the Chiwaukum Mountains area, Washington Cascades, *Geological Society of America Bulletin, Pt II, 91,* p. 1627-1668, 1980.

Price, R.A., J.W.H. Monger, and J.A. Roddick, Cordilleran cross-section Calgary to Vancouver, in *Field Guides to Geology and Mineral Deposits in the Southern Canadian Cordillera,* edited by D. Tempelmann-Kluit, Geological Society of America, Cordilleran Section Meeting, Vancouver, British Columbia, May 1985, p. 3-1 - 3-85, 1985.

Ragan, D.M., Emplacement of the Twin Sisters dunite, Washington, *American Journal of Science, 261,* p. 549-565, 1963.

Ramsay, J.G., *Folding and fracturing of rocks,* New York, McGraw-Hill, 568 p., 1967.

Raviola, F.P., Metamorphism, plutonism and deformation in the Pateros-Alta Lake region, north-central Washington, M.S. thesis, San Jose State University, San Jose, California, 182 p., 1988.

Ray, G.E., The Hozameen fault system and related Coquihalla serpentine belt of southwestern British Columbia, *Canadian Journal of Earth Sciences, 23,* p. 1022-1041, 1986.

Reller, G.J., Structure and petrology of the Deer Peak area, North Cascades, Washington, M.S. thesis, Western Washington

University, Bellingham, 105 p., 1986.

Richards, T.A., and K.C. McTaggart, Granitic rocks of the southern Coast Plutonic Complex and northern Cascades of British Columbia, *Geological Society of America Bulletin, 87,* p. 935-953, 1976.

Rusmore, M.E., Geology of the Cadwallader Group and the Intermontane-Insular superterrane boundary, southwestern British Columbia, *Canadian Journal of Earth Sciences, 24,* p. 2279-2291, 1987.

Russell, I.C., A preliminary paper on the geology of the Cascade Mountains in northern Washington, U.S. Geological Survey, Twentieth Annual Report, v. 2, p. 83-210, 1900.

Silberling, N.J., D.L. Jones, M.C. Blake, Jr., and D.G. Howell, Lithotectonic terrane map of the western conterminous United States, *U.S. Geological Survey Miscellaneous Field Studies Map MF-1874-C,* scale 1:2,500,000, 20 p., 1987.

Smith, M., Deformational geometry and tectonic significance of a portion of the Chilliwack Group, northwestern Cascades, Washington, *Canadian Journal of Earth Sciences, 25,* p. 433-441, 1988.

Snyder, W.S., W.R. Dickinson, and M.L. Silberman, 1976, Tectonic implications of space-time patterns of Cenozoic magmatism in the western United States, *Earth and Planetary Science Letters, 32,* p. 91-106.

Sondergaard, J.N., Stratigraphy and petrology of the Nooksack Group in the Glacier Creek-Skyline Divide area, North Cascades, Washington, M.S. thesis, Western Washington University, Bellingham, 103 p., 1979.

Staatz, M.H., P.L. Weis, R.W. Tabor, J.F. Robertson, R.M. Van Noy, E.C. Pattee, and D.C. Holt, Mineral resources of the Pasayten Wilderness area, Washington, *U.S. Geological Survey Bulletin 1325,* 255 p., 1971.

Staatz, M.H., R.W. Tabor, P.L. Weis, J.F. Robertson, R.M. Van Noy, and E.C. Pattee, Geology and mineral resources of the northern part of the North Cascades National Park, Washington, *U.S. Geological Survey Bulletin 1359,* 132 p., 1972.

Stull, R.J., The geochemistry of the southeastern portion of the Golden Horn Batholith, Northern Cascades, Washington, Ph.D. thesis, University of Washington, Seattle, 127 p., 1969.

Swanson, D.A., and T.L. Wright, Bedrock geology of the northern Columbia Plateau and adjacent areas, in *The Channeled Scabland,* edited by V.R. Baker and Dag Nummedal, National Aeronautics and Space Administration, Washington D.C., p. 37-57, 1978.

Tabor, R.W., The crystalline geology of the area south of Cascade Pass, Northern Cascade Mountains, Ph.D. thesis, Washington, University of Washington, Seattle, 205 p., 1961.

Tabor, R.W., The crystalline geology of the area south of Cascade Pass, Northern Cascade Mountains, Washington, *Dissertation Abstracts,* v. 22, p. 3160, 1962.

Tabor, R.W., Large quartz diorite dike and associated explosion breccia, Northern Cascade Mountains, Washington, *Geological Society of America Bulletin, 74,* p. 1203-1208, 1963.

Tabor, R.W., The Helena-Haystack melange, a fundamental western Washington structure, and its relation to the Straight Creek fault (abstract), *EOS, 69,* p. 1814, 1987.

Tabor, R.W., and D.B. Booth, Folded thrust fault between major melange units of the western North Cascades, Washington, and its relationship to the Shuksan thrust, *Geological Society of America Abstracts with Programs, 17,* p. 412, 1985.

Tabor, R.W., and D.F. Crowder, On batholiths and volcanoes: Intrusion and eruption of late Cenozoic magmas in the Glacier Peak area, North Cascades, Washington, *U.S. Geological Survey Professional Paper 604,* 67 p., 1969.

Tabor, R.W., J.C. Engels, and M.H. Staatz, Quartz diorite-quartz

monzonite and granite plutons of the Pasayten River area, Washington: petrology, age, and emplacement, *U.S. Geological Survey Professional Paper 600-C,* p. C45-C62, 1968.

Tabor, R.W., R.B. Waitt, V.A. Frizzell, Jr., D.A. Swanson, and G.R. Byerly, Geologic map of the Wenatchee 1:100,000 quadrangle, central Washington, *U.S. Geological Survey Miscellaneous Investigations Map I-1311,* 1982a.

Tabor, R.W., V.A. Frizzell, Jr., D.B. Booth, J.T. Whetten, and R.B. Waitt, Jr., Preliminary geologic map of the Skykomish River 1:100,000 quadrangle, Washington, *U.S. Geological Survey Open-File Map 82-747,* 1982b.

Tabor, R.W., V.A. Frizzell, Jr., J.A. Vance, and C.W. Naeser, Ages and stratigraphy of lower and middle Tertiary sedimentary and volcanic rocks of the central Cascades, Washington: Application to the tectonic history of the Straight Creek fault, *Geological Society of America Bulletin, 95,* p. 26-44, 1984.

Tabor, R.W., V.A Frizzell, Jr., J.T. Whetten, R.B. Waitt, D.A. Swanson, G.R. Byerly, D.B. Booth, M.J. Hetherington, and R.E. Zartman, Geologic map of the Chelan 30-minute by 60-minute quadrangle, Washington, *U.S. Geological Survey Miscellaneous Investigations Map I-1661,* scale 1:100,000, 1 sheet and 33 p., 1987a.

Tabor, R.W., R.E. Zartman, and V.A. Frizzell, Jr., Possible tectonostratigraphic terranes in the North Cascades crystalline core, Washington, in Selected papers on the geology of Washington, edited by J.E. Schuster, *Washington Division of Mines and Geology Bulletin 77,* p. 107-127, 1987b.

Tabor, R.W., D.B. Booth, J.A. Vance, and M.H. Ort, Preliminary geologic map of the Sauk River 30 x 60 minute quadrangle, Washington, *U.S. Geological Survey Open-File Report 88-692,* 1988.

Tabor, R.W., V.A. Frizzell, Jr., D.B. Booth, R.B. Waitt, J.T. Whetten, and R.E. Zartman, Geologic map of the Skykomish 30-minute by 60-minute quadrangle, Washington, *U.S. Geological Survey Miscellaneous Investigations Map I-1963,* scale 1:100,000, in press.

Tennyson, M.E., Stratigraphy, structure, and tectonic setting of Jurassic and Cretaceous sedimentary rocks in the west-central Methow-Pasayten area, northeastern Cascade Range, Washington and British Columbia, Ph.D. dissertation, University of Washington, Seattle, 112 p., 1974.

Tennyson, M.E., and M.R. Cole, Tectonic significance of upper Mesozoic Methow-Pasayten sequence, northeastern Cascade Range, Washington and British Columbia, in *Mesozoic paleogeography of the western United States,* Pacific Coast Paleogeography Symposium 2, edited by D.G. Howell and K.A. McDougall, Pacific Section, Society of Economic Paleontologists and Mineralogists, p. 499-508, 1978.

Tennyson, M.E., D.L. Jones, and B. Murchey, Age and nature of chert and mafic rocks of the Hozameen Group, North Cascade Range, Washington, *Geological Society of America Abstracts with Programs, 14,* p. 239-240, 1982.

Thorkelson, D.J., Geology of the mid-Cretaceous volcanic units near Kingsvale, southwestern British Columbia, in Current Research, Part B, *Geological Survey of Canada Paper 85-1B,* p. 333-339, 1985.

Tipper, H.W., G.J. Woodsworth, and H. Gabrielse, H., Tectonic assemblage map of the Canadian Cordillera and adjacent parts of the United States of America, *Geological Survey of Canada Map 1505A,* scale 1:2,000,000, 1981.

Todd, V.R., Jurassic and Cretaceous plutonism in and near the Methow basin, north-central Washington: Preliminary K-Ar data, *Geological Society of America Abstracts with Programs, 19,* p. 458, 1987.

Trexler, J.H., Jr., Sedimentology and stratigraphy of the

Cretaceous Virginian Ridge Formation, Methow Basin, Washington, *Canadian Journal of Earth Sciences, 22*, p. 1274-1285, 1985.

Trexler, J.H., Jr., and J. Bourgeois, Evidence for mid-Cretaceous wrench-faulting in the Methow Basin, Washington: tectonostratigraphic setting of the Virginian Ridge Formation, *Tectonics, 4*, p. 379-394, 1985.

Vallier, T.L., The Permian and Triassic Seven Devils Group, western Idaho and northeastern Oregon, *U.S. Geological Survey Bulletin 1437*, 58 p., 1977.

Vance, J.A., The geology of the Sauk River area in the Northern Cascades of Washington, Ph.D. thesis, University of Washington, Seattle, 312 p., 1957a.

Vance, J.A., The geology of the Sauk River area in the Northern Cascades of Washington, *Dissertation Abstracts, 17*, p. 1984-1985, 1957b.

Vance, J.A., Metamorphic aragonite in the prehnite-pumpellyite facies, northwest Washington, *American Journal of Science, 266*, p. 299-315, 1968.

Vance, J.A., Bedrock geology of San Juan County, in Geology and water resources of the San Juan Islands, San Juan County, Washington, edited by R.H. Russell, *Washington Department of Ecology Water-Supply Bulletin 46*, p. 3-19, 1975.

Vance, J.A., The stratigraphy and structure of Orcas Island, San Juan Islands, in *Geological excursions in the Pacific Northwest*, edited by E.H. Brown and R.C. Ellis, Geological Society of America field guide, Annual Meeting, Seattle, p. 170-203, 1977.

Vance, J.A., and M.A. Dungan, Formation of peridotites by deserpentinization in the Darrington and Sultan areas, Cascade Mountains, Washington, *Geological Society of America Bulletin, 88*, p. 1497-1508, 1977.

Vance, J.A., M.A. Dungan, D.P. Blanchard, and J.M. Rhodes, Tectonic setting and trace element geochemistry of Mesozoic ophiolitic rocks in western Washington, *American Journal of Science, 280-A*, p. 359-388, 1980.

Vance, J.A., N.W. Walker, and J.M. Mattinson, U/Pb ages of early Cascade plutons in Washington State, *Geological Society of America Abstracts with Programs, 18*, p. 194, 1986.

Vielzeuf, D., and J.R. Holloway, Experimental determination of the fluid-absent melting relations in the pelitic system, *Contributions to Mineralogy and Petrology, 98*, p. 257-276, 1988.

Wade, W.M., Frank Raviola, and Will Hopkins, Emplacement of the Eocene Cooper Mountain batholith into the Ross Lake fault zone, NE Cascades, Washington, *Geological Society of America Abstracts with Programs, 19*, p. 460, 1987.

Wallace, W.K., Bedrock geology of the Ross Lake fault zone in the Skymo Creek area, North Cascades National Park, Washington, M.S. thesis, University of Washington, Seattle, 111 p., 1976.

Waitt, R.B., About forty last-glacial Lake Missoula jokulhlaups through southern Washington, *Journal of Geology, 88*, p. 653-679, 1980.

Waters, A.C., Resurrected erosion surface in central Washington, with discussion by Bailey Willis, *Geological Society of America Bulletin, 50*, p. 635-659, 1939.

Whetten, J.T., and J.A. Laravie, Preliminary geologic map of the Chiwaukum 4 NE quadrangle, Chiwaukum graben, Washington, *U.S. Geological Survey Miscellaneous Field Studies Map MF-794*, scale 1:24,000, 1976.

Whetten, J.T., P.I. Carroll, H.D. Gower, E.H. Brown, and F. Pessl, Jr., Bedrock geologic map of the Port Townsend quadrangle, Washington, *U.S. Geological Survey Miscellaneous Investigations Map I-1198-G*, scale 1:100,000, 1988.

White, P.J., Geology of the Island Mountain area, Okanogan County, Washington, M.S. thesis, University of Washington, Seattle, 80 p., 1986.

Whitney, D.L., and B.W. Evans, Revised metamorphic history for the Skagit Gneiss, North Cascades: Implications for the mechanism of migmatization, *Geological Society of America Abstracts with Programs, 20*, p. 242-243, 1988.

Willis, Bailey, Physiography and deformation of the Wenatchee-Chelan district, Cascade Range, in Contributions to the geology of Washington, *U.S. Geological Survey Professional Paper 19*, p. 41-97, 1903.

Wilson, D.L., The origin of serpentinites associated with the Shuksan Metamorphic Suite near Gee Point, Washington, M.S. thesis, Western Washington University, Bellingham, 53 p., 1978.

Wilson, D.L., and E.H. Brown, Garnet amphibolite and eclogite in the Shuksan blueschist terrane, North Cascades, Washington, *Geological Society of America Abstracts with Programs, 11*, p. 541, 1979.

Yardley, B.W.D., Genesis of the Skagit migmatites, Washington and the distinction between possible mechanisms of migmatization, *Geological Society of America Bulletin, 89*, p. 941-951, 1978.

Zen, E-an, Implications of magmatic epidote-bearing plutons on crustal evolution in the accreted terranes of northwestern North America, *Geology, 13*, p. 266-269, 1985.

Zen, E-an, Tectonic significance of high-pressure plutonic rocks in the western Cordillera of North America in *Metamorphism and crustal evolution of the western United States*, Rubey Volume VII, edited by W.G. Ernst, Prentice Hall, Englewood Cliffs, New Jersey, p. 42-67, 1988.

Zen, E-an, and J.M. Hammarstrom, Magmatic epidote and its petrologic significance, *Geology, 12*, p. 515-518, 1984.

Geologic Evolution of the Northernmost Coast Ranges and Western Klamath Mountains, California

Galice, Oregon to Eureka, California
July 20–28, 1989

Field Trip Guidebook T308

Leaders:
K. R. Aalto   G. D. Harper

Associate Leaders:
G. A. Carver   S. M. Cashman
W. C. Miller III   H. M. Kelsey

American Geophysical Union, Washington, D.C.

TABLE OF CONTENTS

Leaders:

K. R. Aalto
Department of Geology
Humboldt State University
Arcata, CA 95521

G. D. Harper
Department of Geological Sciences
State University of New York
Albany, NY 12222

Associate Leaders:

G. A. Carver, S. M. Cashman,
and W. C. Miller III
Department of Geology
Humboldt State University
Arcata, CA 95521

H. M. Kelsey
Department of Geology
Western Washington University
Bellingham, WA 98225

IGC FIELD TRIP T308:
GEOLOGIC EVOLUTION OF THE NORTHERNMOST COAST RANGES AND
WESTERN KLAMATH MOUNTAINS, CALIFORNIA

PREFACE

This guidebook describes the geology of areas we will visit in the western Klamath Mountains and Coast Ranges of northernmost California. Articles are arranged in the order of our travel, from Medford to Crescent City on highway 199, and south to Eureka on coastal highway 101 (Figure 1). Authors summarize the regional geologic setting and results of their research and earlier studies of the areas we will visit in review papers. Thus the geologic background necessary for interpretation of field stops is provided without the need of additional papers.

The Josephine ophiolite and overlying Galice Formation are part of the western Jurassic belt of the Klamath Mountains geologic province (Harper, 1984). These units, discussed herein by G.D. Harper, are the subject of our first two days of field study. Western Klamath rocks have overthrust the Franciscan Complex some tens of kilometers along an east-dipping, low angle fault that is well exposed on ridges north and south of highway 199, our route from Medford to Crescent City (Aalto and Harper, 1982). Adjacent to the fault, Franciscan turbidites of the eastern Yolla Bolly subterrane have suffered a small amount of textural reconstitution and folding. Scattered tectonic blocks, chiefly composed of metachert or greenstone, are concentrated in the gouge developed beneath the fault and may have been derived from Franciscan melange that presently underlies the western Jurassic belt (Aalto, 1982).

In the Crescent City area, discussed herein by K.R. Aalto and W.C. Miller, Late Mesozoic Franciscan Complex rocks of the eastern and western Yolla Bolly belts are well exposed along the coast and will be the focus of our second two days of field study. Both belts are part of the Jurassic-Cretaceous Eastern Franciscan belt. In the Crescent City area, eastern subterrane rocks consist exclusively of tectonized turbidites, and western subterrane rocks of intercalated melange and turbidite units. The Franciscan has traditionally been interpreted as a subduction complex, constructed of dismembered fragments of oceanic crust and forearc sediments fronting the Mesozoic Sierran-Klamath magmatic arc.

Should time permit, we may also visit outcrops of the Late Jurassic Otter Point Formation, which is faulted against Franciscan rocks in southwestern Oregon (Figure 1). Although coeval with the Franciscan, this unit is more volcaniclastic and is interpreted as having formed adjacent to an andesitic oceanic island arc (Dott, 1971).

Progressing south from Crescent City, we will spend a day visiting Redwood National Park, and outcrops of the Eastern Franciscan belt Pickett Peak terrane, discussed herein by S.M. Cashman. These rocks are exposed within the fault-bounded *Redwood Mountain outlier* (Irwin, 1960), which contains greenschist to blueschist facies metasedimentary and metavolcanic rocks that are similar to those exposed adjacent to the Klamath Mountains to the east (Cashman *et al.*, 1986). The outlier, however, has been translated northwards some 100 km by strike-slip faulting during the Tertiary (Kelsey and Hagans, 1982).

Articles by G.A. Carver, R. Burke and H.M. Kelsey review the Late Cenozoic tectonic development of northernmost California, the subject of our last days of field study. Kelsey describes evidence for recent strike-slip faulting in Pleistocene terraces in Redwood National Park. Carver and Burke describe thrust faults and large fault-bend folds that have deformed Neogene and Quaternary rocks in the vicinity of Eureka and Trinidad. Kelsey and Carver (1988) relate both styles of deformation to the northward migration of the San Andreas transform boundary and the Mendocino triple junction.

The authors have spent a combined total of some fifty field seasons studying the geology of the western Klamaths and Coast Ranges, a region which had previously been mapped only in reconnaissance. We hope that our familiarity with the North Coast, its natural beauty and the wide range of geologic features to be seen will assure that our trip be a success.

K. R. Aalto
Humboldt State University
Arcata, California

REFERENCES CITED

Aalto, K.R., The Franciscan Complex of northernmost California: sedimentation and tectonics, in *Trench-Forearc Geology*, edited by J.K. Leggett, *Geological Society of London Special Publication* 10, 419-432, 1982.

Aalto, K.R., and G.D. Harper, Geology of the Coast Ranges in the Klamath and Ship Mountain quadrangles, Del Norte County, California, *California Division of Mines and Geology Open File Map* QFR 82-16-SF, scale 1:62,500, 1982.

Blake, M.C., Jr., D.G. Howell, and D.L. Jones, Tectonostratigraphic terrane map of California, *U.S. Geological Survey Open-File Report* 82-593, scale 1:750,000, 1982.

Blake, M.C., Jr., D.C. Engebretson, A.S. Jayko, and D.L. Jones, Tectonostratigraphic terranes in southwestern Oregon, in *Tectonostratigraphic terranes of the Circum-Pacific region*, edited by D.G. Howell, *Circum-Pacific Council for Energy and Mineral Resources, Earth Science Series*, 1, 147-157, 1985.

Cashman, S.M., P.H. Cashman, and J.D. Longshore, Deformational history and regional tectonic significance of the Redwood Creek schist, northwestern California, *Geological Society of America Bulletin*, 97, 35-47, 1986.

Dott, R.H., Jr., Geology of southwestern Oregon west of the 124th meridian, *Oregon Department of Geology and Mineral Industries Bulletin* 69, 1971.

Harper, G.D., The Josephine ophiolite, northwestern California, *Geological Society of America Bulletin*, 95, 1009-1026, 1984.

Irwin, W.P., Geological reconnaissance of the northern Coast Ranges and Klamath Mountains, California, *California Division of Mines and Geology Bulletin*, 179, 1960.

Kelsey, H.M., and G.A. Carver, Late Neogene and Quaternary tectonics associated with the northward growth of the San Andreas transform fault, northern California, *Journal of Geophysical Research*, 93, (in press), 1988.

Kelsey, H.M., and D.K. Hagans, Major right-lateral faulting in the Franciscan assemblage of northern California in Late Tertiary time, *Geology*, 10, 387-391, 1982.

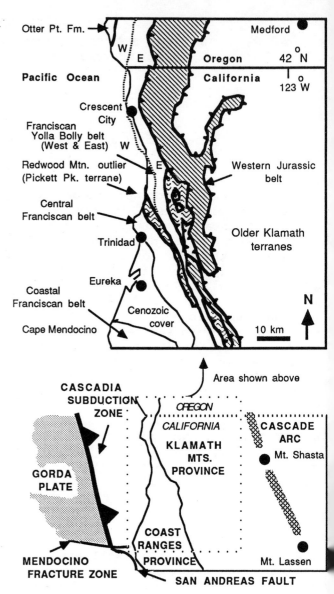

FIGURE 1 Generalized geologic map showing basement terranes in southern Oregon and northwestern California. Nomenclature and some terrane boundaries are from Irwin (1960) and Blake, *et al.* (1982, 1985), modified according to mapping by the guidebook authors. The Franciscan Coastal-Central belt contact is covered near Eureka. The east and west Yolla Bolly belt may in places be in depositional contact (Aalto, this volume). All other terranes are in fault contact. Faults east of the Otter Point Formation and *Redwood Mountain outlier* are strike-slip in nature (shown in heavy black. All others are thrust faults (shown with teeth). The lower map shows principal geologic provinces and plate boundaries.

COVER Franciscan Complex melange exposed along the coast north of Trinidad, California.

FIELD GUIDE TO THE JOSEPHINE OPHIOLITE AND COEVAL ISLAND ARC COMPLEX, OREGON-CALIFORNIA

Gregory D. Harper

Department of Geological Sciences, State University of New York at Albany

INTRODUCTION

The Klamath Mountains occupy the fore-arc region of the Cascade volcanic arc. They consist of east-dipping thrust sheets ranging from early Paleozoic to late Mesozoic. The terranes and thrust faults generally become younger from east to west [Irwin, 1981; Wright and Fahan, 1988]. There is no Precambrian basement present; the Klamath terranes consist of magmatic arcs, ophiolites, and accretionary prisms.

The western Klamath terrane (Fig. 1) consists of two east-dipping thrust sheets emplaced during the Late Jurassic Nevadan orogeny. The lower sheet (Fig. 2) includes the Rogue Formation and calc-alkaline plutonic rocks of the Chetco intrusive complex. The upper sheet includes the Josephine ophiolite which is in thrust contact with the Chetco intrusive complex along an amphibolite sole (Figs. 1, 2) [Dick, 1976; Cannat and Boudier, 1985]. Late Jurassic flysch of the Galice Formation conformably overlies both the Rogue Formation and the Josephine ophiolite (Fig. 2) [Harper, 1984; Wood, 1987]. The Rogue-Chetco complex and the Josephine ophiolite apparently represent a Late Jurassic island arc [Dick, 1976; Garcia, 1979] and back-arc basin [Saleeby et al., 1982; Harper and Wright, 1984]. This interpretation is based on similarities in radiometric and fossil ages (Fig. 2), overlying flysch, age of emplacement, and regional geology [Saleeby et al., 1982; Harper and Wright, 1984; Harper et al., 1986].

THE ROGUE AND GALICE FORMATIONS (ISLAND ARC COMPLEX)

On the first day of the trip, we will be rafting down the Rogue River to observe the Rogue and Galice Formations (Fig. 3). The Rogue Formation consists of a thick sequence of tuffs, breccias, and rare flows which range from mafic to silicic in composition [Garcia, 1979; Riley, 1987]. Riley [1987] subdivided the Rogue Formation into units consisting, from oldest to youngest, of (1) andesitic breccia, (2) volcaniclastic turbidites, (3) massive fine-grained tuff (Figs. 2, 3). The presence of pillow lavas, Bouma sequences, and cherty tuffs containing radiolaria and sponge spicules indicate submarine deposition [Riley, 1987]. Trace-element geochemistry of the volcanic rocks indicates that they are calc-alkaline [Garcia, 1979]. Saleeby

[1984] reports a concordant Pb/U zircon age of 157 ± 2 Ma on a tuff-breccia in the Rogue Formation (Fig. 2) and a 150 ± 2 Ma age on a felsic dike.

Kuroko-type massive sulfide deposits occur in the Rogue Formation [Koski and Derky, 1981; Wood, 1987]. We will visit one of these (Almeda Mine, Stop 3) which is exposed on the Rogue River, just below the contact between the Rogue Formation and overlying Galice Formation.

The Galice Formation consists of graywacke, slaty shale, and rare conglomerate. Bouma sequences are common and sole marks are locally present, indicating deposition by turbidity currents.

Bedding and foliation in the Rogue Formation generally dip steeply to the southeast, and the Galice Formation has a steeply dipping slaty cleavage. Tight to isoclinal folds occur in both the Rogue and Galice Formations (Fig. 3) [Kays, 1968; Riley, 1987; Park-Jones, 1988]. Folds are overturned to the west and axial planes generally strike northeast (Fig. 3). Fold hinges in the Galice plunge from 0-90° [Kays, 1968; Harper and Park, 1986; Park-Jones, 1988]; it is uncertain whether this variation is primary or the result of post-Nevadan cross folding. Volcanic rocks east of the village of Galice (Fig. 3) were originally mapped as volcanic members in the Galice Formation [Wells and Walker, 1953]; however, at least some of these rocks are overturned and may in fact be the Rogue Formation [Park-Jones, 1988].

We will not be visiting other parts of the island-arc complex. The probable plutonic roots of the arc are represented by the gabbroic Chetco intrusive complex which has yielded Late Jurassic K/Ar hornblende ages (Fig. 2). The southern part of the complex is predominantly gabbro, whereas the northern part is largely quartz diorite.

Other rocks associated with the Rogue Formation and Chetco intrusive complex include the Briggs Creek amphibolite, Rum Creek metagabbro (gneissic amphibolite), and minor peridotite (Figs. 2, 3) [Garcia, 1982]. The Briggs Creek amphibolite contains minor quartzites and has been interpreted to be metamorphosed pillow lava and chert. The association of metagabbro, peridotite, metabasalt, and metachert suggests that these rocks may represent a dismembered and regionally metamorphosed ophiolite [Coleman et al., 1976]. This interpretation is tentative, however, because of the possibility of large fault displacements, lack of structural work, and limited geochronology.

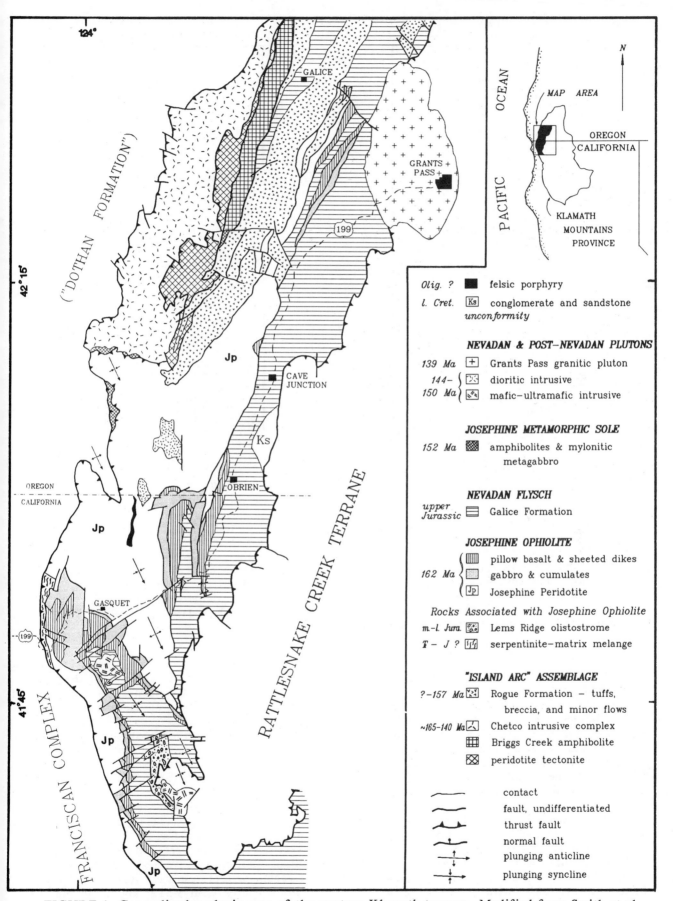

FIGURE 1 Generalized geologic map of the western Klamath terrane. Modified from Smith et al. [1982], Harper [1984], and Wagner and Saucedo [1987].

T308: 3

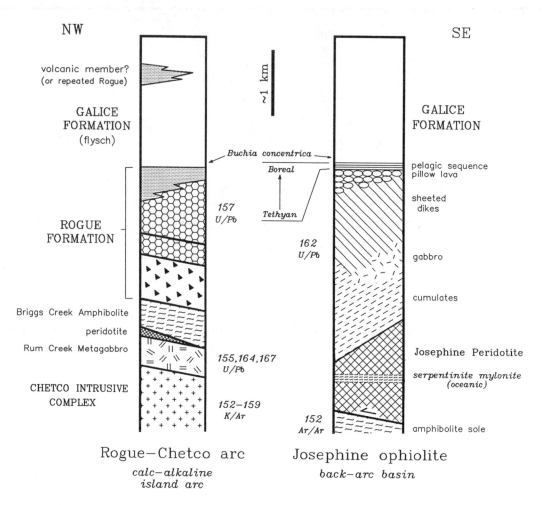

NW

SE

volcanic member?
(or repeated Rogue)

GALICE
FORMATION
(flysch)

GALICE
FORMATION

Buchia concentrica

Boreal

ROGUE
FORMATION

157
U/Pb

Tethyan

pelagic sequence
pillow lava

sheeted
dikes

162
U/Pb

gabbro

cumulates

Briggs Creek Amphibolite
peridotite
Rum Creek Metagabbro

Josephine Peridotite

serpentinite mylonite
(oceanic)

155,164,167
U/Pb

CHETCO INTRUSIVE
COMPLEX

152–159
K/Ar

152
Ar/Ar

amphibolite sole

Rogue–Chetco arc
calc–alkaline
island arc

Josephine ophiolite
back–arc basin

FIGURE 2 Late Jurassic island-arc complex and the Josephine ophiolite exposed along the Rogue River, Oregon, and the Smith River, California, repectively (Figs. 3 and 4) [Garcia, 1979, 1982; Riley, 1987; Park, 1987]. Symbols for the Rogue Formation are the same as those in Figure 3. The Josephine ophiolite was thrust over the island arc and thrust beneath western North America at ~150 Ma, at which time both sequences were intruded by calc-alkline dikes and sills [Saleeby, 1984; Harper and Wright, 1984; Harper et al., 1987]. K/Ar hornblende (recalculated using standardized decay constants), Ar/Ar hornblende, and U/Pb zircon ages are from Hotz [1971] and Dick [1976], Saleeby, [1984, 1987] and Harper [unpublished data], respectively.

THE JOSEPHINE OPHIOLITE AND GALICE FORMATION

The Josephine ophiolite is one of the largest and most complete ophiolites in the world. The ophiolite and overlying metasedimentary rocks compose a >10-km-thick, east dipping thrust sheet (Figs. 2, 4). The ophiolite consists of (1) depleted peridotite (mostly harzburgite) covering >800 km^2 (Fig. 1), (2) gabbroic and ultramafic cumulates (Fig. 5), (3) high-level gabbro (Fig. 6), (4) sheeted dikes (Figs. 7, 8), and (4) pillow lava (Fig. 9), massive lava, and pillow breccia.

The ophiolite was originally dated at 157±2 Ma by Pb/U on zircon from two plagiogranites [Saleeby et al., 1982]. With the use of newer techniques, it was discovered that the ages were

slightly discordant. After abrasion of the zircon, one of the samples has yielded a concordant age of 162±1 [Saleeby, 1987]. In addition, a small fragment of the ophiolite exposed in the southwestern Klamath Mountains has yielded a zircon age of 164±1 Ma [Wright and Wyld, 1986]. In both areas, the dated plagiogranites are clearly part of the ophiolite because they are cut by mafic dikes.

Pelagic/hemipelagic rocks conformably overlie the Josephine ophiolite (Fig. 10) and grade upwards into the Galice flysch (Fig. 2). These rocks consist of chert, tuffaceous chert, and radiolarian argillite. Many of these rocks have high contents of terrigenous or tuffaceous detritus. Metalliferous horizons locally occur 8-23 m above the ophiolite and formed from low-T off-axis hydrothermal

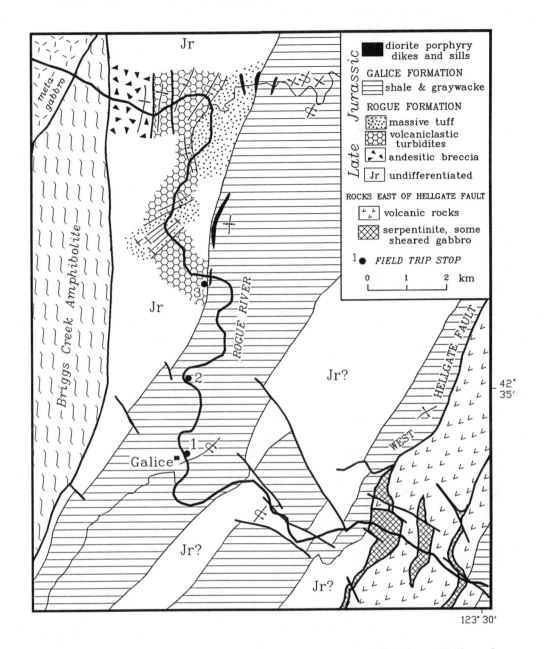

FIGURE 3 Geologic map of the Rogue River area modified from Wells and Walker [1953], Garcia [1982], Riley [1987], and Park-Jones [1988]. Fold hinges show locations of Nevadan folds in the Rogue Formation [Riley, 1987] and Galice Formation [Park-Jones, 1988].

springs [Pinto-Auso and Harper, 1985]. Well-preserved radiolaria preserved in limestone nodules at Stop 2 indicate an age of middle-upper Oxfordian [Pessagno and Mizutani, 1988]. <u>Buchia concentrica</u> (a bivalve) occurs in the lower part of the flysch (Fig. 2), and has a range of middle Oxfordian to upper Kimmeridgian [Imlay, 1980].

The pelagic/hemipelagic rocks grade upwards into flysch with the appearance of thick turbidites interbedded with radiolarian argillites (see Stop 2). Several hundred meters of massive and thick-bedded graywacke and rare conglomerate overlie this transition, whereas most of the remaining Galice consists of slate with some packets of medium bedded graywacke.

Although the Josephine ophiolite and overlying Galice Formation were regionally metamorphosed to low grade, the ophiolite shows little penetrative deformation except where the metamorphic grade reaches lower greenschist facies in the southern part of the area shown in Figure 4. A strong, typically bedding-parallel cleavage is present in the Galice Formation. Isoclinal folds are locally well developed in the Galice. Calc-alkaline dikes, sills, and stocks were intruded during metamorphism; they are regionally metamorphosed and commonly deformed (Fig. 11). U/Pb zircon and Ar/Ar ages on these intrusives range from 148-151 Ma [Saleeby

FIGURE 4 (Caption and key on next page.)

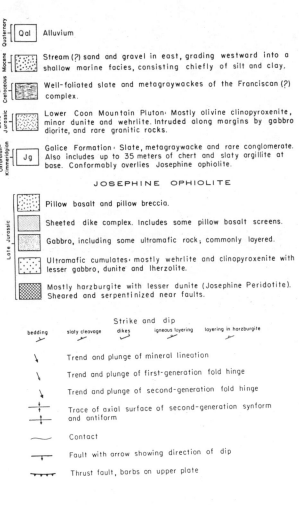

Stream (?) sand and gravel in east, grading westward into a shallow marine facies, consisting chiefly of silt and clay.

Well-foliated slate and metagraywackes of the Franciscan (?) complex.

Lower Coon Mountain Pluton: Mostly olivine clinopyroxenite, minor dunite and wehrlite. Intruded along margins by gabbro diorite, and rare granitic rocks.

Jg — Galice Formation: Slate, metagraywacke and rare conglomerate. Also includes up to 35 meters of chert and slaty argillite at base. Conformably overlies Josephine ophiolite.

JOSEPHINE OPHIOLITE

Pillow basalt and pillow breccia.

Sheeted dike complex. Includes some pillow basalt screens.

Gabbro, including some ultramafic rock; commonly layered.

Ultramafic cumulates: mostly wehrlite and clinopyroxenite with lesser gabbro, dunite and lherzolite.

Mostly harzburgite with lesser dunite (Josephine Peridotite). Sheared and serpentinized near faults.

Strike and dip

bedding | slaty cleavage | dikes | igneous layering | layering in harzburgite

Trend and plunge of mineral lineation

Trend and plunge of first-generation fold hinge

Trend and plunge of second-generation fold hinge

Trace of axial surface of second-generation synform and antiform

Contact

Fault with arrow showing direction of dip

Thrust fault, barbs on upper plate

FIGURE 4 Geologic map of the Josephine ophiolite and overlying Galice Formation [Harper, 1984]. The ophiolite is broadly folded into an F_2 anticline and syncline, and the upper part of the ophiolite is repeated by three reverse faults in the northeastern part of the area. Numbers refer to field trip stops.

FIGURE 5 Layered gabbroic and ultramafic cumulates from near the base of the cumulate sequence, Josephine ophiolite.

FIGURE 6 Complex intrusive breccia in high-level gabbro cut by a mafic dike and a thin plagiogranite dike (near Stop 5), Josephine ophiolite.

et al., 1982; Harper et al., 1986]. Dikes both intrude and are cut by small Nevadan faults; northwestward thrusting is indicated by fiberous slickensides and offset dikes and beds [G. Harper, in prep.]. Thrust directions inferred from the basal amphibolite sole of the Josephine ophiolite (Fig. 1) are north-northeast [Cannat and Boudier, 1985; K. Grady and G. Harper, in prep.].

Continued deformation after peak metamorphism is indicated by extensive folding of slaty cleavage, numerous reverse faults, and the local presence of a crenulation cleavage and lineation. The entire ophiolite was also folded during this later phase of deformation (Figs. 1, 4).

Igneous Geochemistry

Dikes and lavas of the Josephine ophiolite show a wide range in composition because of extreme fractionation and multiple types of parental magmas. The great majority of dikes and lavas have "immobile" trace element concentrations that indicate magmatic affinities transitional between island-arc tholeiites and mid-ocean ridge basalt (Figs. 12 and 13) [Harper, 1984, 1988]. Highly

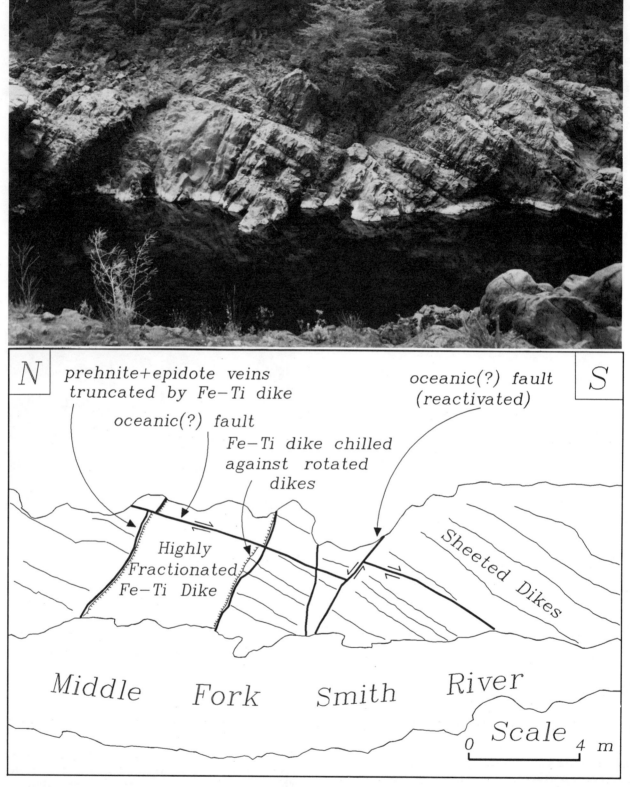

FIGURE 7 Sheeted dikes at Stop 5 dipping towards the south, viewed from Highway U.S. 199. This outcrop is in the hinge of a gently plunging syncline and thus the sheeted dikes are orientated essentially the same as when they were part of the oceanic crust (prior to ophiolite emplacement). The dip of the sheeted dikes reflects tilting at the spreading axis [Harper, 1982], probably by rotational faulting. The sketch shows the location of several probable oceanic faults, some of which are the locus of abundant prehnite veins. A late highly fractionated dike was intruded along one of these faults and much of its margins were subsequently sheared and veined with prehnite.

FIGURE 8 Sheeted dikes and gabbro screens (gb) at Stop 5. The center dike is ~0.4 m wide. In the right foreground is a probable oceanic normal fault; the lowermost gabbro screen is displaced 30 cm as shown by arrows. Prehnite veins (white), some with small offsets, occur on either side of the fault.

fractionated Fe-Ti basalts occur in the uppermost pillow lavas (Stop 2), and a Fe-Ti late dike (Fig. 7) and plagiogranite dikes and screens occur at the sheeted-dike/gabbro transition (Stop 4).

Very primitive lavas and dikes are widespread in Josephine ophiolite, and are highly variable in composition, ranging from primitive island-arc tholeiites to boninites (Fig. 12) We will look at a primitive pillow lava at Stop 1; black porphyritic primitive dikes also occur within high-level gabbro at Stop 4). The pillow lavas have thick chilled rims, distinctive variolitic textures, and abundant vesicles. Some of the lavas and dikes plot near melting curves, suggesting that they are primary mantle melts.

The geochemistry of the dikes and lavas clearly indicates a suprasubduction zone setting. Most of the lavas and dikes appear to have been generated by melting of a MORB-like source, but apparently involved higher degrees of partial melting than MORB (Fig. 12; or less melting of a depleted source). The high Th contents of the Josephine rocks compared to MORB (Fig. 13) reflects a "subduction-zone component" added to the source. The depleted primitive lavas (low Y, high Cr samples) apparently formed by hydrous partial melting of a depleted mantle source (Fig. 12).

These inferred variations in conditions and source rocks during partial melting is consistent with studies of the residual Josephine Peridotite which suggest two stages of partial melting [Dick and Bullen, 1984].

Subseafloor Hydrothermal Metamorphism

A study of the subseafloor metamorphism of the Josephine ophiolite has recently been published by Harper et al. [1988]. The ophiolite shows an overall downward increase in metamorphic grade and decrease in ^{18}O, similar to other ophiolites and resulting from metamorphism under a steep thermal gradient. Most rocks in the extrusives and sheeted dike complex have lost Ca and gained Mg and Na (Fig. 14), but relict igneous textures are well preserved. These chemical chemical changes and pervasive metamorphism have been interpreted as the alteration by downwelling seawater (recharge, Fig. 15).

Alteration during upwelling (discharge) is much more localized at an outcrop scale. Mineral zonation resulting from discharging fluids is shown in Figure 15. One of the most significant recent discoveries in ophiolites is that the path of discharging fluids is represented by granoblastic

FIGURE 9 Pillow lavas ~50 m below the contact between the Josephine ophiolite and overlying sediments shown in Figure 10.

FIGURE 10 Depositional contact (15 cm scale rests on contact) between the uppermost pillow lavas of the Josephine ophiolite (left) and overlying thinly bedded chert and siliceous argillite (Stop 2). The contact dips 60° toward the left (east).

FIGURE 11 Boudinage in a sill within the pelagic/hemipelagic sequence overlying the Josephine ophiolite, near the mouth of Little Jones Creek (near Stop 1). Such calc-alkaline dikes and sills were intruded, deformed and regionally metamorphosed during the Nevadan orogeny at 145-150 Ma.

epidote + quartz ± chlorite rocks called epidosites (Fig. 16) [Harper et al., 1988; Schiffman and Smith, 1988; Seyfried et al., 1988]. The epidosites are metasomatized and strongly enriched in Ca and depleted in Na and Mg (Fig. 14), as well as Cu and Zn. Chemical, isotopic, and experimental work indicates that the epidosites represent pathways for huge volumes of fluids discharging at high velocities (meters/sec); the discharging fluids were probably similar to 350°C fluids exiting from "black smokers" at modern mid-ocean ridges [Seyfried et al., 1988].

We will see abundant dike-parallel epidosites in the sheeted dike complex at Stop 1, repsectively. The lowermost pillow lavas at Stop 1 are largely silicified and rich in sulfides. The silicification is almost certainly due to cooling of discharging fluids as they flow upwards into the more permeable pillow lavas. Above the lowermost pillows throughout the Josephine extrusives, bulbous "albite" epidosites and abundant hematite are evident in outcrop, and muscovite and K-

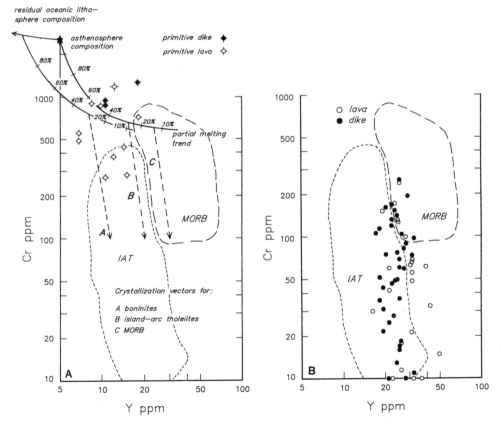

FIGURE 12 A. Y vs. Cr plot for primitive dikes and lavas from the Josephine ophiolite [Harper, 1988]. Note that several of the Josephine rocks plot near partial melting trends, indicating that they are essentially primary mantle partial melts. It is likely that the more depleted samples (lower Y) were derived by melting of a depleted mantle source because unrealistically high degrees of partial melting would be required for an undepleted source. B. Y vs. Cr plot for typical dikes and lavas of the Josephine ophiolite. This plot and other trace element data [Harper, 1984] indicate magmatic affinities transitional between IAT and MORB. Extrapolation upwards along fractionation trends suggests that these rocks were derived from partial melting of a relatively undepleted mantle source.

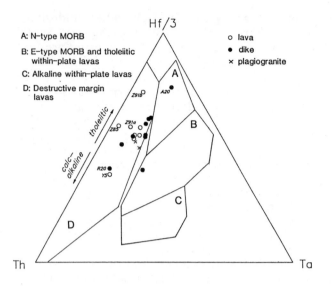

FIGURE 13 Hf/3-Ta-Th diagram showing affinities of Josephine ophiolite dikes and lavas to island-arc tholeiites [Harper, unpublished data]. The linear trend extending away from the Th apex may be the result of a variable "subduction component" present during melting or mixing of MORB magmas with depleted primitive lavas. Numbered samples are very primitive and show a wide variation in composition.

feldspar commonly occur in amygdules and in interpillow matrix (Fig. 15).

Locally in the Josephine ophiolite, hot fluids discharged directly onto the seafloor to form massive sulfide deposits. The Turner-Albrigt deposit, which occurs southwest of Obrien, Oregon (Fig. 15) is gold-rich [Koski and Derkey, 1981].

Spreading Rate and Magmatic/Tectonic Cycles

Modern mid-ocean ridges vary dramatically in morphology, intensity of faulting, and average depth and size of earthquakes. One of the most important differences appears to be the presence of long-lived magma chambers at fast spreading ridges, although they are much smaller (<4 km wide) than previously proposed, whereas magma chambers at slow spreading centers (<5 cm/yr) are almost certainly episodic as indicated, for example, by the presence of deep microearthquakes beneath the Mid-Atlantic Ridge (MAR) [Harper, 1985]. Although high-T springs (black smokers) have recently been discovered on the Mid-Atlantic Ridge, hydrothermal activity at slow spreading centers appears to be episodic and multistage [Sempere and Macdonald, 1987].

The widespread occurrence of very primitive Cr-spinel bearing lavas in the basal pillow lavas of the Josephine ophiolite requires that magma chambers periodically solidified because these mantle-derived melts would have otherwise mixed into an existing magma chamber [Harper, 1988]. Solidifying magma chambers would also produce

highly fractionated magmas such as Fe-Ti basalts that occur at Stops 1 and the late Fe-Ti dike and plagiogranite that occurs at Stop 4 (Fig. 7). The small size and episodic nature of magma chambers would also account for features of the cumulate sequence including a generally fine-grain size, poorly developed and discontinuous igneous layering (Stop 5), and the occurrence of ultramafic cumulates throughout the sequence [Harper, 1984].

During periods when no magma chamber is present, extension will be taken up entirely by structural processes such as fissuring and faulting [Harper, 1985, 1988]. Normal faulting often takes place by rotation of fault blocks. Rotated fault blocks have been observed at the MAR, but the amount of rotation is controversial [Sempere and Macdonald, 1987] and much be obscured by eruption of lava flows during rotation (growth faulting).

In the Josephine ophiolite the sheeted dikes and cumulate sequence are inclined with repect to conformably overlying sediments (Fig. 2) [Harper, 1982]. Assuming the dikes were intruded vertically and igneous layering was subhorizontal, >50° tilting prior to deposition of overlying pelagic rocks (i.e. at the spreading axis) is required. This tilting is especially evident at Stop 4 where the basal sheeted-dike complex occurs in the hinge of a large gently plunging sycline (Fig. 7). In this outcrop, the dikes have a dip of ~40° S and commonly show evidence of faulting in two directions: (1) along dike margins (Fig. 7), (2) along steep normal faults (Figs. 7, 8) which are similar in strike to the dikes. Both of these sets of

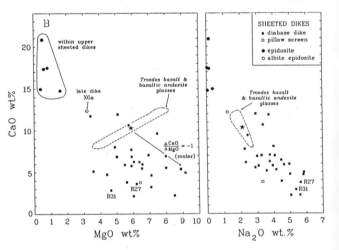

FIGURE 14 Plots of CaO versus MgO and Na$_2$O for sheeted dikes from the Josephine ophiolite [Harper et al., 1988]. The fields for Troodos glasses illustrate variation due to igneous fractionation. Most of the dikes show the effects of seawater alteration during discharge which results in Ca loss and a gain in Mg and Na; these rocks typically have well-preserved relict textures. In contrast, epidosites altered during discharge are strongly enriched in Ca and depleted in Mg and Na..

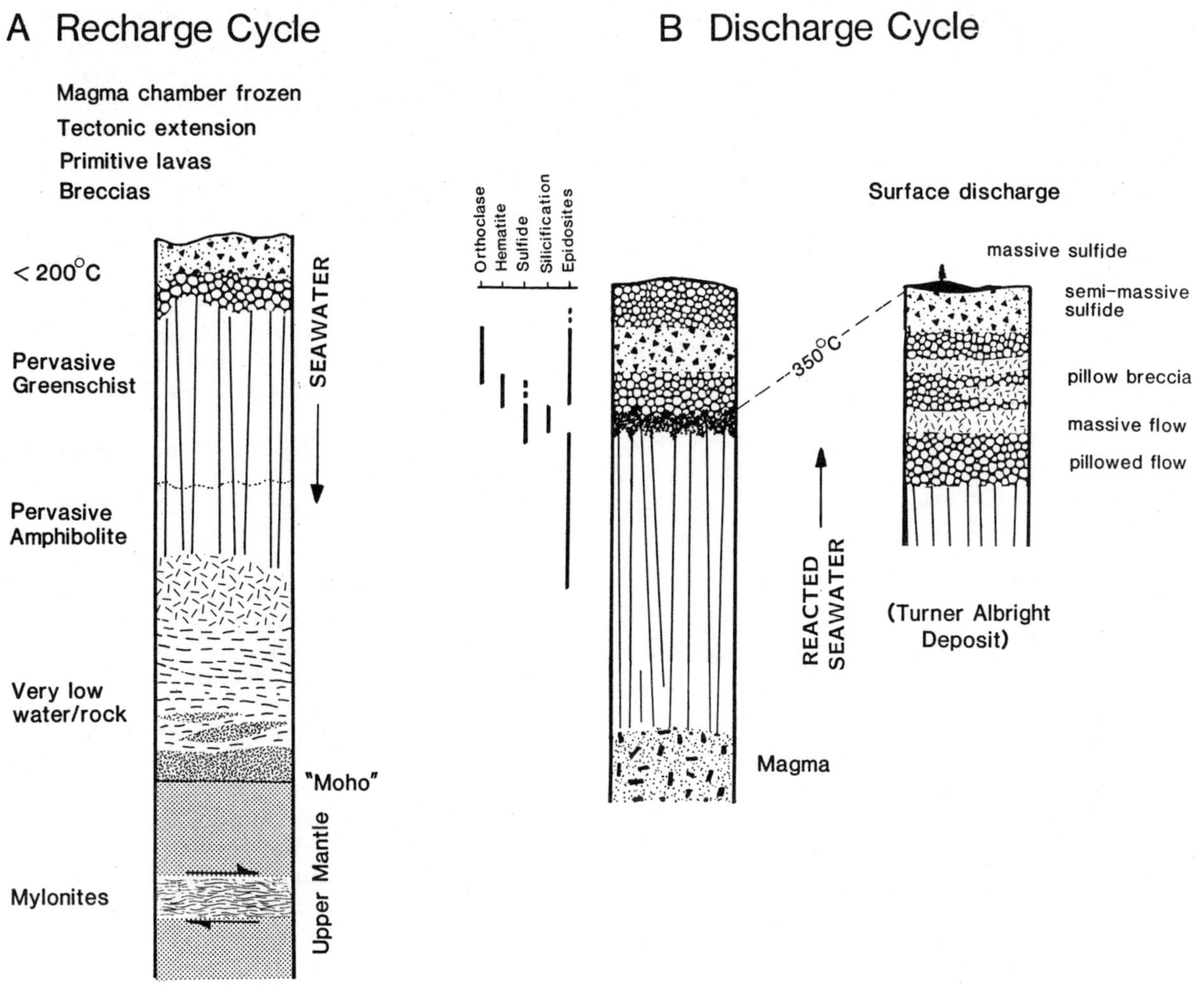

A Recharge Cycle

Magma chamber frozen
Tectonic extension
Primitive lavas
Breccias

B Discharge Cycle

FIGURE 15 Summary diagram showing subseafloor metamorphism during recharge and discharge cycles [Harper et al., 1988]. Discharging fluids typically cooled as they flowed upward through the extrusives producing a distinctive mineral zonation. At the Turner-Albright deposit, however, high-T fluids vented onto the seafloor to form massive sulfide deposits [Zierenberg et al, 1988].

faults are probably oceanic because they have subseafloor hydrothermal alteration along including prehnite veins (Fig. 8), epidosite veins and epidosites. In addition, the late Fe-Ti at Stop 4 (Fig. 7) was intruded parallel to the steep faults, and much of its margins are sheared and contain abundant prehnite veins; this indicates that this dike was intruded during the faulting.

A likely mechanism for tilting is rotational normal faulting. In this case, both the fault blocks and the faults rotate. As the faults rotate, they become unfavorably oriented, and new steeper faults map form; the steep faults at Stop 4 (Fig. 7) may represent such new faults.

Because the entire crustal sequence of the Josephine ophiolite is tilted, oceanic faults would be expected to have extended into the upper mantle. Such faults could have died out into a zone of uniform ductile flow, or perhaps into a discrete subhorizontal horizon [Harper, 1985].

A regionally extensive flat shear zone in the upper mantle peridotite of the Josephine ophiolite has recently been mapped and consists of unusual antigorite mylonites and peridotite mylonites [Norrell et al., 1988]. This shear zone is interpreted to represent an oceanic detachment above which the crustal sequence was rotated by block faulting [Norrell and Harper, 1988]. If we assume that the >50° tilting above this fault took place by rotational faulting, then the ophiolite must have been stretched by ~100%. This would indicate the the crustal sequence, which is now ~3 km thick (Fig. 2), was originally ~6 km thick, similar to modern oceanic and back-arc basin crust.

These aspects of the Josephine ophiolite imply that it formed at a spreading center where the magma supply was low relative to the amount of extension. Slow spreading mid-ocean ridges appear to have a low magma supply as indicated by the absence of magma chambers along much of the

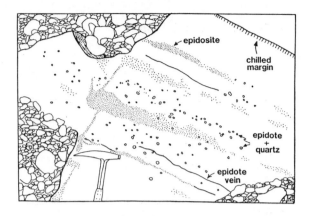

FIGURE 16 Sketch of epidosites in the upper-sheeted dike complex at Stop 1. These probably represent pathways for discharging, ~350°C fluids similar to those at modern "hot smokers" at mid-ocean ridges [Harper et al., 1988].

Mid-Atlantic Ridge [Sempere and Macdonald, 1987]. By analogy, the Josephine probably formed at a slow spreading center to account for the apparent episodic nature of magma chambers [Harper, 1988] and the extreme tilting of the crustal sequence [Harper, 1985; Norrell and Harper, 1988]. It is possible, however, that the spreading rate was intermediate or even fast if the spreading segments were short and bounded by transforms (i.e., cold edge effect). In any case, the Josephine is different from the Oman or Bay of Islands ophiolites which have thick, well layered cumulate sequence, little evidence for large-scale tilting, and which probably formed by fast spreading with steady-state magma chambers.

TECTONIC EVOLUTION

The similarities in age and overlying flysch of the Galice Formation (Fig. 2) suggests that the Rogue island-arc complex and Josephine ophiolite suggest that they were formed in a single magmatic arc. Because the Rogue arc is thrust beneath the Josephine ophiolite, which is in turn thrust beneath older rocks of the Klamath Mountains, it is likely that the arc was west-facing with the Josephine ophiolite forming in a back-arc basin between the arc and western North America [Harper and Wright, 1984]. The remains of an extensive Middle Jurassic magmatic arc occur east and structurally above the Josephine ophiolite, and is in the appropriate position to represent a remnant arc left behind as the Josephine back-arc basin opened. If this model is correct, the active arc (Rogue Formation and Chetco intrusive complex) migrated away from North America as spreading occurred in the back-arc basin.

Although it is difficult to directly tie the Rogue-Chetco arc and Josephine ophiolite to the western Klamath Mountains (i.e. rule out that they are exotic to North America), there are several strong arguments. Recently, it has been recognized that several areas within the western Klamath terrane contain fragments of what appear to be the Late Triassic and Early Jurassic Rattlesnake Creek terrane. This terrane consists of a suite of disrupted ophiolitic and metasedimentary rocks, much of which forms a distinctive serpentinite melange, and which currently is thrust over the Josephine ophiolite and Galice Formation (Fig. 1). These fragments may have been rifted off western North America during Late Jurassic extension [Wyld and Wright, 1988]. Possible Rattlesnake Creek fragments include the Lems Ridge olistostrome (Fig. 4), serpentinite melange along the western edge of the ophiolite in the area of Figure 4 and in the southern Klamath Mountains, and ophiolitic rocks containing Triassic chert [Roure and DeWever, 1983] west of Cave Junction, Oregon (Fig. 1) which appear to lie beneath the Rogue Formation (these workers mistakenly assigned these cherts to the Josephine ophiolite). The presence of older Klamath basement is required by the occurrence of xenocrystic Precambrian zircon in a southern remnant of the Josephine ophiolite in the southern Klamath Mountains [Wright and Wyld, 1986]; the dated plagiogranites are clearly part of the ophiolite and the xenocrystic zircon is probably from epiclastic rocks in Klamath basement. The lower part of this ophiolite remnant is not exposed, but it must have been built on the edge of the Josephine basin during extension but just prior to sea-floor spreading.

New data suggests that the back-arc basin model needs to be modified or replaced by a more complex tectonic scenario. Harper et al. [1985] noted that the orientation of spreading centers inferred from the strike of sheeted dikes, however, are approximately east-west [Harper et al., 1985; Norrell and Harper, 1988], suggesting that the back-arc basin may have had a geometry like the Gulf of California where long transforms separate short spreading segments. This geometry may result from intra-arc strike-slip faulting in response to oblique subduction as in the modern modern Anadaman Sea [Harper et al., 1985; Wyld and Wright, 1988]. In addition, preliminary dating suggested that the Middle Jurassic arc shut off just as the Josephine ophiolite formed, consistent with rifting to form a back-arc basin and remnant arc [Harper and Wright, 1984]. Recent Pb/U dating, however, has shown that the ophiolite is slightly older (164-162 Ma), and that the "Middle Jurassic" arc was active until 160 Ma [Saleeby, 1987; Wright and Fahan, 1988]. Thus, if the Rogue arc and Josephine ophiolite formed by rifting of the western Klamath Mountains, then the ophiolite must have been an intra-arc basin with active volcanism on both sides.

Radiolaria extracted from the pelagic\hemipelagic sequence overlying the Josephine ophiolite at Stop 2 indicates northward migration of the ophiolite prior to deposition of the overlying flysch. Radiolaria indicative of the Central to Northern Tethyan Province occur in the lower 45 m, above which the first graywackes appear and radiolaria from interbedded argillites are indicative of the Southern Boreal Province [Pessagno and Blome, 1988]. Buchia concentrica and plant fossils in the Galice flysch are also indicative of Boreal Realm [E. Pessagno, personal communication, 1988]. The amount of relative motion with respect to the North American plate is uncertain, however, because paleomagnetic studies indicate that the North American plate was also moving northward during this time [Debiche et al., 1987].

The Galice flysch may represent the earliest sign of the Nevadan orogeny which culminated with thrusting and regional metamorphism of the Josephine and Rogue-Chetco terranes. The Galice sandstones contain clasts of volcanic rock fragments, plagioclase, chert, shale, and minor monocrystalline quartz and metamorphic rock fragments. The basal graywackes that we will see at Stop 2 are much richer in volcanic rock fragments and plagioclase than those higher in the sequence; they also have abundant clinopyroxene and hornblende, but pebbles at the base are mostly radiolarian chert. Other minerals in Galice graywackes include zircon, tourmaline, apatite, chromian spinel, and muscovite; less common minerals include biotite, garnet, and glaucophane. The detrital zircons are highly variable in color and morphology and have yielded Pb/U zircon isotopic data that indicate sources which are ~1500-1600 Ma and early Mesozoic [Miller and Saleeby, 1987]. The graywacke petrography, zircon ages, and sparse paleocurrent indicators [Harper, 1980; Park-Jones, 1988], indicate derivation of the flysch from older rocks of the Klamath Mountains to the east. The volcanic component of the wackes is probably derived from a Middle Jurassic arc complex built on older rocks of the Klamath Mountains [Harper and Wright, 1984; Wright and Fahan, 1988]. The absolute age range for flysch deposition is constrained by the age of basement rocks (157-162 $\pm$ 1 Ma, Fig. 2) and by the age of cross-cutting sills, dikes, and plutons (139-150 $\pm$ 1 Ma) [Saleeby et al., 1982; Harper et al., 1986; Saleeby, 1987]. Thus the Galice flysch was deposited and thrust beneath western North America within approximately 7 Ma. The cause of uplift and erosion to the east of the Josephine ophiolite may have been from underthrusting which eventually involved the Josephine and Rogue-Chetco terranes. If this is true, then the Galice is syntectonic and heralds the change from extensional to compressional tectonics.

The Rogue arc and Josephine back-arc basin underwent compression during the latest Jurassic Nevadan orogeny and were thrust beneath the Klamath Mountains. During and following thrusting, the Josephine ophiolite and overlying Galice Formation became the locus of arc magmatism as indicated by the presence of abundant calc-alkaline dikes, sills (Fig. 11), and plutons (Fig. 1) ranging in age from 139-151 Ma [Saleeby et al., 1982; Saleeby, 1984; Harper et al., 1986]. The amount of underthrusting on the roof thrust overlying the Josephine ophiolite and Galice Formation (Fig. 1) is approximately 110 km [Jachens et al., 1986]. This thrust is cut by plutons as old as 150$\pm$1 Ma [Harper et al., 1986]. Uplift to the surface occurred by ~130 Ma because the Galice Formation is overlain with angular unconformity by a small remnant of Early Cretaceous (Hauterivian-Barremian) sedimentary rocks near the Oregon-California border (Fig. 1) [Nilsen, 1984].

The change from back-arc extension to compression (Nevadan orogeny) correlates with an abrupt change in the polar wander path of cratonic North America and with plate reorganization in the Atlantic recorded by marine magnetic anomalies [Steiner, 1983; May and Butler, 1986]. Thus the Nevadan orogeny in the Klamath Mountains may have resulted from compression due to a major change in plate motions.

DAY 1

Drive from Grants Pass, Oregon, northwest on Interstate 5; the light-colored road exposures consist of the 139 Ma Grants Pass pluton (Fig. 1). Take the exit to Merlin, a few km north of Grants Pass. Follow the road west, through Merlin, and on to the small village of Galice. Exposures of sheared serpentinite, greenstones, and gabbro will be evident as we drive into the area shown on Figure 3; these may be the northward continuation of the Josephine ophiolite (Fig. 1). As we are waiting for rafts to be loaded, notice the picture on the wall of the Galice store which shows the water level during the great 1964 flood.

We will board the rafts and float downstream for approximately 13 km, with three stops (Fig. 3). We will go through numerous rapids, and you should see large birds called Great Blue Herrons.

Stop 1

Paddle across the Rogue River, directly opposite where the rafts are launched. This outcrop shows excellent turbidites with flute casts (Late Jurassic Galice Formation). The beds are overturned and the shales have a slaty cleavage formed during the Nevadan orogeny. Cleavage cuts bedding at a moderate angle. Faulting and associated veining is also present and may be related to the Nevadan folding.

Stop 2

Thin-bedded graywacke and slate are evident in

this exposure. The bedding is folded and there is an axial planar slaty cleavage. The folds and cleavage are Nevadan age. As we continue downstream, we will be approaching the Rogue Formation which is volcaniclastic and underlies the Galice Formation.

Stop 3

We will stop on the right side, just above the rapids and impressive iron staining. The staining is from weathering of a Kuroko-type massive sulfide (Almeda Mine) which is located just below the contact with the Galice Formation. The ore body was mined for gold, silver, copper, and lead. We will hike up to the mine shaft on the north side of the river. At the entrance to the mine shaft a contact between a massive sulfide and fine-grained bedded tuff is exposed. On the left side of the shaft, the massive sulfide locally contains barite. The tuff is exposed on the roof of the shaft, and on the right side of the shaft is a diorite dike [Wood, 1987].

Walk on up the road (east) to the black slate outcrops. This is the basal Galice Formation and the bivalve Buchia concentrica can be found. This is also an ammonite locality [Imlay, 1980]. The contact between the Rogue Formation and Galice Formation is not exposed here, but appears to be depositional as reported by [Wood, 1987].

Stop 3 to End of River Trip

As we continue down the river, we will enter a narrow, quiet part of the canyon with high walls. Notice the steeply dipping tuffs. Most of the tuffs are green and andesitic, but a few white silicic tuffs are evident. Upon careful observation, you may see overturned graded bedding and isoclinal fold hinges along the banks of the river. Many of the tuffs are turbidites.

Travel

We will drive back through Grants Pass and south on Highway U.S. 199 (Fig. 1), where we will stay overnight. Tomorrow we will visit the Josephine ophiolite.

DAY 2

We will leave Cave Junction and drive south on Highway U.S. 199. We are now situated on flysch of the Galice Formation which overlies the Josephine ophiolite. As we drive south, you will see the Josephine peridotite which forms the poorly vegetated mountains to the west; in this area, the entire crustal sequence of the ophiolite has been cut out by a large north-striking normal fault. As we cross into California, we will drive through a

long tunnel. As we exit the tunnel, the topography will be much steeper. We have entered the drainage area of the Smith River which is rapidly down-cutting. The erosion is the result of late Cenozoic uplift, which is probably the result of subduction of young, hot oceanic lithosphere of the Juan de Fuca plate. Relicts of an extensive Miocene erosion surface (Klamath peneplain) can be seen forming relatively flat mountain tops.

Stop 1

Continue southwest on highway U.S. 199 to a highway maintenence station (Idlewild) on the right side of the road. Walk several hundred meters further down the highway to where the trees begin on the left side of the road. Walk down the embankment and cross the stream to the other side. The ophiolite is exposed as a homocline in this area which dips approximately 60° east.

We are now in the upper part of the sheeted dike complex and will walk up-stream to the contact with pillow lavas. Notice chilled margins on sheeted dikes, epidosites (Fig. 16), and disseminated sulfides. A screen (septa of country rock) of pillow lava will be pointed out. Notice the peculiar texture (variolitic) and thick chilled margin. This is a very primitive, depleted basalt (~1000 ppm Cr, 10 ppm Y) which contains Cr-spinel [sample R20, Harper et al., 1988].

Continue walking upsection to the contact of the sheeted-dike complex with the overlying pillow lavas (where trees begin on left). Notice the presence of a quartz-rich and sulfide-rich breccia near the contact; it was formed by discharging fluids and is very similar to "stockwork-like" rocks drilled from the sheeted-dike/pillow basalt transition zone in Deep-Sea Drilling Project Hole 504b, south of the Galapagos spreading center [J. Alt, personal communication, 1987].

The basal pillow lavas are sulfide-rich and strongly silicified, and the interpillow matrix is locally completely replaced by pyrite and chalcopyrite. This type of mineralization is typical along this contact in the Josephine ophiolite, although the intensity is highly variable. As noted above, this contact is apparently where discharging fluids cooled and possibly mixed with seawater.

Continue walking up-stream and observe red hematite-bearing lavas and light-green "albite epidosites" within the red lavas. Muscovite and/or K-feldspar typically occurs in the hematitic lavas, particularly in amygdules and interpillow matrix.

Travel Between Stops 1 and 2

Return to vehicles and continue to drive southwest along U.S. 199 for ~10 km. We will be driving through sheeted dikes, gabbros, and cumulate ultramafics. This sequence is faulted at the base (Fig. 4) and as a result we will drive back into the Galice Formation overlying the ophiolite.

We will cross several more N-S reverse faults that repeat the upper part of the ophiolite (Fig. 4); these faults were probably formed during late-stage Nevadan thrusting.

Stop at large pull-out on left, just before the place where the road narrows abruptly and passes through very tight curves.

Stop 2

A depositional contact is clearly exposed on the Smith River between the uppermost pillow lavas of the ophiolite and the basal pelagic/hemipelagic sediments (Fig. 10). The pillows below the contact are large and best observed on the lower, downstream part of the outcrop. They are highly fractionated Fe-Ti basalts containing abundant microphenocrysts of plagioclase and clinopyroxene. 400 m of pillow lava (Fig. 9), massive lava, and pillow breccias are well exposed along the river below the contact (see Fig. 10 of Harper [1984] for stratigraphic section), but we will not be able to observe them because rafts and much time are needed to travel the deep canyon.

The basal 45 meters of sediments overlying the ophiolite consist of chert, tuffaceous chert, siliceous argillite (black), and rare nodules and layers of limestone (gray), and several sills. The siliceous argillites are slaty and have high Al contents indicative of a large component of terrigenous sediment [Pinto-Auso and Harper, 1985]. Suprisingly, the limestones have yielded abundant perfectly preserved radiolarians which indicate an Oxfordian age; furthermore the radiolarians indicate that the ophiolite moved northward from the Central Tethyan to Southern Boreal Realms during deposition of the basal 45 meters of the pelagic/hemipelagic sequence [Pessagno and Blome, 1984; Pessagno and Mizutani, 1988].

At 45 meters stratigraphically above the contact, two thick bedded graywackes are present and mark the beginning of flysch deposition. These are overlain by silty radiolarian argillites and a few graywacke beds exposed continuously to a sharp bend in the river (100 m above the ophiolite). Slates with abundant limestone nodules occur at the bend. Massive graywackes and some pebble conglomerate is exposed further upsection (upstream).

Many sills and dikes are present at this locality. They are regionally metamorphosed, and the sills have been extended to form boudinage (Fig. 11). The sills are mafic, calc-alkaline, and usually contain hornblende. Recent $^{40}Ar/^{39}Ar$ step heating ages on two of these sills have yielded ages of 147 ± 1 and 150 ± 1 Ma (the ophiolite is 162 ± 1 Ma) and thus tightly constrain the timing of sediment deposition and subsequent deformation and associated prehnite-pumpellyite facies metamorphism.

Nevadan structures evident in this section include a bedding-parallel slaty cleavage, boudinage, extension veins, flattened pebbles, and bedding-parallel faults. The latter are preferentially eroded, have thin sheared calcite, and are especially evident ~25-30 m above the contact. A highly disrupted zone is evident for several meters below the first graywacke; small thrusts and ramps are evident on close examination, and some thrust surfaces have extension veins beneath them with fibers up to 25 cm long. Deformation in this zone is apparently due to bedding-parallel thrusting during the Nevadan deformation.

Travel Between Stops 3 and 4

Drive 1.5 km further south and west on U.S. 199 and take a right at the road just before the bridge (Patrick Creek Road). Drive 2.5 kilometers north and turn right. We are now at the base of the cumulates and will be driving all the way through the cumulates, sheeted dike complex and into the lower pillow lavas (Fig. 4). There are only weathered outcrops until we reach the pillow lavas, so this is a good time to enjoy the scenery.

Stop 3

Pillow lavas are exposed in a quarry on the east side of the road. Pillow lavas are best viewed on huge blocks on the quarry floor. Hydrothermal metamorphism during discharge is evident from epidote $\pm$ hematite between pillows. Hematitic veins are also present in some of the outcrop. Volcanic breccias are exposed on the far end (south) of the quarry.

Travel Between Stops 3 and 4

Drive back to U.S. 199, turn right and drive 12 km toward the village of Gasquet (pronounced Gas-Key). We will be driving through the Josephine Peridotite; most of what is exposed is sheared and serpentinized along a northeast-striking fault zone (Fig. 4). Many landslides are evident, some of which are active.

Continue driving on U.S. 199 for ~20 km to locality 4 on Figure 4. Stop ~1 km past the second bridge (over Hardscrabble Creek) where the road becomes very wide, and rock outcrops are clearly evident along the river.

Stop 4

Park on the left side of the highway and walk down to the river. Exposed on the opposite side of the river and upstream on the highway side are sheeted dikes (Figs. 7, 8) that grade down-stream into high-level gabbro. Subparallel dikes with chilled margins are clearly evident, and gabbro screens are locally present (Fig. 8). A plagiogranite screen intruded by dikes which has been dated by

Pb/U occurs on the side opposite the highway. As the contact with the gabbro is approached (downstream), dikes become less abundant, and intrusive breccias such as those in Figure 6 become abundant. A late, highly fractionated Fe-Ti dike intrudes the sheeted dikes at a high angle (Fig. 7). A few very primitive dikes intrude the high-level gabbro and are recognized by their blue-black color and porphyritic texture.

Tilting of the crustal sequence is clearly apparent in this outcrop. Because it is situated in the core of a gently plunging syncline, the ophiolite is essentially horizontal and the ~40° dip of the dikes is due to tilting at the spreading axis. Probable oceanic faults are present in these outcrops (Fig. 7); an oceanic origin is indicated by the presence of hydrothermal veins (Fig. 8) or by the presence of an ophiolitic dike along the fault (Fig. 7).

The sheeted dikes and gabbros have amphibolite-facies assemblages resulting from high-temperature subseafloor hydrothermal metamorphism, but many are partially retrograded to greenschist facies. Epidote veins and abundant white prehnite veins occur throughout this exposure (Fig. 8), and quartz veins or quartz-matrix breccias are locally present. The prehnite veins are clearly oceanic in origin because some are cut by dikes.

Travel Between Stops 4 and 5

Return to vehicles and continue southwest on U.S. 199 to a large pullout on the left side of the road. Walk down the road and descend to the river before reaching guardrail.

Stop 5

The exposures along the river consist of gabbroic and ultramafic cumulates intruded by mafic dikes. We are still in the hinge area of a syncline so that the structures are essentially in their oceanic orientation. Igneous layering is present and dips steeply northwest. Most dikes are similar in orientation to the last stop, but many sills parallel to igneous layering are also present.

The igneous layering is discontinuous, often faint, and irregular. High-T hydrothermal alteration has resulted in mm-wide black hornblende veins and amphibolite-facies assemblages. Coarse-grained pegmatites are also common and may also be the result of high-T interaction with seawater.

Travel from Stop 5 to Crescent City, California

As we drive west on U.S. 199 toward Crescent City we will enter the small village of Hiouchi which is situated on the basal thrust which separates the Josephine ophiolite from the Franciscan Complex. This change in rock type is reflected in a dramatic change in vegetation from Douglas Fir to Redwoods. Many of the Redwoods are more than 1000 years old and some are over 100 meters high. We will drive through several groves of giant Redwoods starting just west of Hiouchi, and we will make a stop for taking photographs. The redwoods and associated ferns once covered much of North America in the Cretaceous as indicated by fossils in coals of the western U.S. and Canada. Thus, the Redwoods are "living fossils". You may be able to imagine dinosaurs roaming through the Redwoods as often depicted in museums.

As we continue to drive west, we will drive down onto a large marine terrace which was uplifted in the Pleistocene. Crescent City is situated along the coast, at edge of the terrace. This town was largely destroyed by a tsunami resulting from the great 1964 Alaskan earthquake, and a huge "tetrapod" washed in from the breakwater sits along the side of U.S. 101. In a rare case of thoughtful urban planning, the coastal area of the town was turned into a park. Later in 1964 there was a gigantic flood that further damaged the town as well as much of coastal northern California and southwestern Oregon. This was the same flood that inundated the Galice resort on the Rogue River.

Acknowledgments

Field work and preparation was funded by NSF EAR-8518974. I thank G. Norrell and R. Alexander for their assistance.

REFERENCES

Cannat, M., and R. Boudier, Structural study of intra-oceanic thrusting in the Klamath Mountains, northern California: Implications on accretion geometry, Tectonics, 4, 435-452, 1985.

Coleman, R.G., M.O. Garcia, and C. Anglin, The amphibolite of Briggs Creek: A tectonic slice of metamorphosed oceanic crust in southwestern Oregon?, Geol. Soc. Am. Abs. Prog., 9, 363, 1976.

Debiche, M.G., A. Cox, and D. Engebretson, The motion of allochthonous terranes across the North Pacific basin, Spec. Pap., Geol. Soc. Am., 207, 49pp., 1987.

Dick, H.J.B., The origin and emplacement of the Josephine Peridotite of southwestern Oregon, Ph.D. thesis, 409 pp., Yale Univ., New Haven, Conn., 1976.

Dick, H.J.B., and T. Bullen, Chromian spinel as a petrogenetic indicator in abyssal and alpine-type peridotites and spatially associated lavas, Contributions to Mineralogy and Petrology, 86, 54-76, 1984.

Garcia, M.O., Petrology of the Rogue and Galice Formations, Klamath Mountains, Oregon: Identification of a Jurassic island arc sequence,

J. Geol., 86, 29-41, 1979.

Garcia, M.O., Petology of the Rogue River island-arc complex, southwest Oregon, Am. J. Sci., 282, 783-807, 1982.

Harper, G.D., Evidence for large-scale rotations at spreading centers from the Josephine ophiolite, Tectonophysics, 82, 25-44, 1982.

Harper, G.D., The Josephine ophiolite, Geol. Soc. Am. Bull., 95, 1009-1026, 1984.

Harper, G.D., Tectonics of slow-spreading mid-ocean ridges and consequences of a variable depth to the brittle/ductile transition, Tectonics, 4, 395-409, 1985.

Harper, G.D., Freezing magma chambers and amagmatic extension in the Josephine ophiolite, Geology, 16, 831-834, 1988.

Harper, G.D., and J.E. Wright, Middle to Late Jurassic tectonic evolution of the Klamath Mountains, California-Oregon, Tectonics, 3, 759-772, 1984.

Harper, G.D., J.B. Bowman, and R. Kuhns, A field, chemical, and stable isotope study of subseafloor metamorphism of the Josephine ophiolite, California-Oregon, J. Geophys. Res., 93, 4625-4656, 1988.

Harper, G.D., and R. Park, Comment on "Paleomagnetism of the Upper Jurassic Galice Formation, southwestern Oregon: Evidence for differential rotation of the eastern and western Klamath Mountains", Geology, 14, 1049-1050, 1986.

Harper, G.D., J.B. Saleeby, and E. Norman, Geometry and tectonic setting of sea-floor spreading for the Josephine ophiolite, and implications for Jurassic accretionary events along the California margin, in Tectonostratigraphic Terranes of the Circum-Pacific Region, Earth Sci. Ser. vol. 1, edited by D. Howell, pp. 239-257, Circum-Pacific Council for Energy and Mineral Resources, Houston, Tex., 1985.

Harper, G.D., J.B. Saleeby, S. Cashman, and E. Norman, Isotopic age of the Nevadan orogeny in the western Klamath Mountains, California-Oregon, Geol. Soc. Am. Abs. Prog., 18, 114, 1986.

Hotz, P.E., Plutonic rocks of the Klamath Mountains, California and Oregon, U.S. Geol. Surv. Bull., 1290, 91 pp., 1971.

Imlay, R.W., Jurassic paleobiogeography of conterminous United States in its continental setting, U.S. Geol. Surv. Prof. Pap. 1062, 134 pp, 1980.

Irwin, W.P., Tectonic accretion of the Klamath Mountains, The Geotectonic Development of California, edited by W.G. Ernst, pp. 29-49, Rubey Series 1, Prentice-Hall, Englewood Cliffs, N.J., 1981.

Jachens, R.C., C.G. Barnes, and M.M. Donato, Subsurface configuration of the Orleans fault: Implications for deformation in the western Klamath Mountains, California, Geol. Soc. Am. Bull., 97, 388-395, 1986.

Kays, A., Zones of alpine tectonism and metamorphism Klamath Mountains, southwestern Oregon, J. Geol., 76, 17-36, 1968.

Koski, R.A., and R.E. Derkey, Massive sulfide deposits in oceanic-crust and island-arc terranes of southwestern Oregon, Oregon Geology, 43, 119-125, 1981.

May, S.R., and R.R. Butler, North American Jurassic apparent polar wander: Implications for plate motion, paleogeography and Cordilleran tectonics, J. Geophys. Res., 91, 1986.

Miller, M. M., and J.B. Saleeby, Detrital zircon studies of the Galice formation: common provenance of strata overlying the Josephine ophiolite and Rogue island arc -- western Klamath Mountains (abstract), Geol. Soc. Am. Abs. Prog., 19, 772-773, 1987.

Nilsen, T.H., Stratigraphy, sedimentology, and tectonic framework of the upper Cretaceous Hornbrook Formation, Oregon and California, Geology of the Upper Cretaceous Hornbrook Formation, Oregon and California, edited by T.H. Nilsen, Pacific Section S.E.P.M. 42, 51-88, 1984.

Norrell, G.T., and G.D. Harper, Detachment faulting and amagmatic extension at mid-ocean ridges: The Josephine ophiolite as an example, Geology, 16, 827-830, 1988.

Norrell, G.T., A. Teixell, and G.D. Harper, Microstructure of serpentinite mylonites from the Josephine ophiolite and serpentinization in retrogressive shear zones, Geol. Soc. Am. Bull., in press, 1988.

Park-Jones, R., Sedimentology, structure, and geochemistry of the Galice Formation: Sediment fill of a back-arc basin and island arc in the western Klamath Mountains, M.S. thesis, 165 pp., State Univ. New York, 1988.

Pessagno, E.A., Jr., and C.D. Blome, Biostratigraphic, chronostratigraphic, and U/Pb geochronometric data from the Rogue and Galice Formations, western Klamath Terrane (Oregon and California): -- Their bearing on the age of the Oxfordian-Kimmeridgian boundary and the Mirifusus first occurrence event, Proc. 2nd International Symposium on Jurassic Stratigraphy, Lisbon, Portugal, 14 pp., 1988.

Pessagno, E.A., Jr., and S. Mizutani, Correlation of radiolarian biozones of the eastern and western Pacific (North America and Japan), in Jurassic of the Circum-Pacific Region, edited by G.E.G. Westermann, in press, 1988.

Pinto-Auso, M., and G.D. Harper, Sedimentation, metallogenesis, and tectonic origin of the basal Galice Formation overlying the Josephine ophiolite, northwestern California, J. Geol., 93, 713-725, 1985.

Riley, T.A., The petrogenetic evolution of a Late Jurassic island arc: the Rogue Formation, Klamath Mountains, Oregon, M.S. thesis, 40 pp., Stanford Univ., 1987.

Roure, F., and P. DeWever, Triassic cherts discovered in the western Jurassic belt of the Klamath Mountains, southwestern Oregon, U.S.A.: Implications for the age of the Josephine ophiolite, C. R. Acad. Sci. Paris, 297, 161-164, 1983.

Saleeby, J.B., Pb/U zircon ages from the Rogue River area, western Jurassic belt, Klamath Mountains, Oregon, Geol. Soc. Am. Abs. Prog., 16, 331, 1984.

Saleeby, J.B., Discordance patterns in Pb/U zircon ages of the Sierra Nevada and Klamath Mountains (abstract), Eos Trans. AGU, 68, 1514-1515, 1987.

Saleeby, J.B., G.D. Harper, A.W. Snoke, and W. Sharp, Time relations and structural-stratigraphic patterns in ophiolite accretion, west-central Klamath Mountains, California, J. Geophys. Res., 87, 3831-3848, 1982.

Schiffman, P., and B.M. Smith, Petrology and oxygen-isotope geochemistry of a fossil seawater hydrothermal system within the Solea graben, northern Troodos Ophiolite, Cyprus, J. Geophys. Res., 93, 4612-4624, 1988.

Sempere, J.-C., and Macdonald, K.C., Marine tectonics: Processes at mid-ocean ridges, Rev. Geophys., 25, 1313-1347, 1987.

Seyfried, W.E., Jr., M.E. Berndt, and J.S. Seewald, Hydrothermal alteration processes at mid-ocean ridges: Constraints from diabase alteration experiments, hot spring fluids and composition of the oceanic crust, Can. Mineral., in press, 1988.

Smith, J.G., N.J. Page, M.G. Johnson, B.C. Moring, and F. Gray, Preliminary geologic map of the Medford 1°x2° quadrangle, Oregon and California, U.S. Geol. Surv. Open File Rep., 82-955, 1982.

Wagner, D., and G.J. Saucedo, Geologic map of the Weed quadrangle, scale 1:250,000, California, Calif. Div. of Mines Geol., Sacramento, 1987.

Wood, R.A., Geology and geochemistry of the Almeda mine, Josephine County, Oregon, M.S. thesis, California State University, Los Angeles, 237 pp., 1987.

Wright, J.E., and S.J. Wyld, Significance of xenocrystic Precambrian zircon contained within the southern continuation of the Josephine ophiolite: Devils Elbow ophiolite remnant, Klamath Mountains, northern California, Geology, 14, 671-674, 1986.

Wright, J.E., and M.R. Fahan, An expanded view of Jurassic orogenesis in the western United States Cordillera: Middle Jurassic (pre-Nevadan) regional metamorphism and thrust faulting within an active arc environment, Klamath Mountains, California, Geol. Soc. Am. Bull., 100, 859-876, 1988.

Wyld, S.J., and J.E. Wright, The Devils Elbow ophiolite remnant and overlying Galice Formation: New constraints on the Middle to Late Jurassic evolution of the Klamath Mountains, California, Geol. Soc. Am. Bull., 100, 29-44, 1988.

FRANCISCAN COMPLEX GEOLOGY OF THE CRESCENT CITY AREA, NORTHERN CALIFORNIA

K. R. Aalto

Department of Geology, Humboldt State University, Arcata, California

INTRODUCTION

The origin of melange terranes in accretionary prism settings is a controversial topic that commands the attention of researchers of both modern and ancient subduction complexes (Raymond, 1984; Moore et al., 1985). With the advent of plate tectonic theory, the late Mesozoic Franciscan Complex of California came to be considered a type example of a subduction complex, deformed in a subduction zone fronting the Great Valley forearc basin and Sierran-Klamath magmatic arc (Hsü, 1968; Bailey and Blake, 1969; Hamilton, 1969; Bailey, 1970; Dickinson, 1970b; Ernst, 1970).

Franciscan rocks consist chiefly of variably metamorphosed, texturally and compositionally immature turbidites, with lesser amounts of radiolarian chert, pelagic limestone, greenstone, ultramafic and mafic plutonic rocks and blueschist-facies metamorphic rock (Bailey et al., 1964; Blake and Jones, 1974). These minor constituents exist chiefly as blocks in melange units. The turbidites exist both in fairly coherent stratal sequences that are intercalated with melanges and as blocks in melanges. Accretion-related deformation has resulted in pervasive fracturing of Franciscan rocks and localized development of large contractional faults and folds.

Subsequent to the formation of the subduction complex, Franciscan rocks have been translated to the north along regional Cenozoic strike-slip faults (Kelsey and Hagans, 1982; Aalto, 1983; Blake et al., 1984, 1985a, 1985b; McLaughlin and Ohlin, 1984; Cashman et al., 1986; Jayko and Blake, 1987). In northern California, late Cenozoic crustal shortening has resulted in thrust faulting that has further disrupted Franciscan rocks (Carver, 1987; Kelsey and Carver, 1988).

In attempts to understand the early history of Franciscan rocks, in particular the sedimentary and tectonic processes that have interacted to form melange zones within the Franciscan Complex, detailed study has been valuable despite post-accretion, fault-related disruption (eg. Suppe, 1973; Maxwell, 1974; Cowan, 1974, 1978, 1982; Page, 1978; Aalto, 1981; Bachman, 1982; Cloos, 1982). In the Crescent City area, excellent recently quarried coastal exposures of Franciscan rocks offer an opportunity to study primary textures and structures. Franciscan units include a shale-matrix melange, interpreted as an olistostrome, bounded by coherent turbidites of various facies (Aalto, 1986b; Aalto and Murphy, 1984). Structures preserved within these rocks record early pervasive layer-parallel extension and later crustal shortening, which may be related to the growth and deformation of the Mesozoic Franciscan accretionary prism.

GEOLOGIC SETTING

The Franciscan Complex in the Crescent City area consists of intercalated melange and broken formation units (cf. Hsü, 1974). I tentatively assign these rocks to the Yolla Bolly terrane of the Eastern Franciscan belt (cf. Blake et al., 1982, 1985a, 1985b; see map in guidebook preface). This correlation is based upon their location relative to major regional faults, lack of exotic blueschist blocks in melange units, and general sandstone composition (Aalto, 1987b). The Yolla Bolly terrane of northernmost California and southwestern Oregon ranges in age from Tithonian to Hauterivian (Blake et al., 1985b; Aalto, 1987a). In the Coast Ranges to the south, Yolla Bolly rocks are schistose and contain incipient blueschist-facies minerals (Blake, Irwin and Coleman, 1967; Blake and Jayko, 1983). To the north there is a gradational decline in metamorphic grade within this terrane (Young, 1978; Monsen and Aalto, 1980). Franciscan graywackes in the Crescent City area are texturally unreconstituted and contain, in their matrix, the metamorphic mineral assemblage: quartz + albite + white mica + chlorite ± pumpellyite ± calcite (Aalto and Murphy, 1984).

The Franciscan Complex in the Crescent City area is bounded on the east by the Coast Range fault ("South Fork fault" of Irwin *et al.*, 1974), along which older Jurassic rocks of the Klamath Mountains have overthrust Franciscan rocks by some tens of kilometers (Irwin, 1960; Blake and Jones, 1977; Aalto, 1982; Aalto and Harper, 1982). This major regional structure is overlain by the upper Miocene Wimer Formation northeast of Crescent City (Dott, 1971).

The Yolla Bolly terrane in southwestern Oregon has been divided into eastern and western subterranes based upon lithologic differences (Blake *et al.*, 1985b). Western subterrane rocks consist of coarse-grained sandstone, mudstone, volcanic rock and radiolarian chert, listed in order of decreasing abundance. Eastern subterrane rocks consist chiefly of finer-grained sandstone and mudstone, with subordinate chert. The western and eastern subterranes correlate to the Macklyn and Winchuck Members of the Dothan Formation (Oregon equivalent of the Franciscan), as defined by Widmier (1962). Aalto (1982) subdivided Franciscan rocks of northernmost California into a western belt of intercalated melange and turbidite units and an eastern belt consisting solely of turbidites. These belts are correlative to the two Yolla Bolly subterranes. The olistostrome discussed herein is proposed to mark the contact between the subterranes, a contact which is therefore depositional in nature.

The Otter Point Formation of southwestern Oregon is coeval with the Franciscan Complex and consists chiefly of volcanic-rich turbidites, with minor andesitic volcanic rocks (Koch, 1966; Aalto and Dott, 1970; Dott, 1971). The contact between the two units is a strike-slip fault that may have been active during the latest Cretaceous to early Tertiary (Bourgeois and Dott, 1985). Paleomagnetic evidence indicates that Otter Point rocks may have been displaced northward a minimum of 1,200 km and, despite age overlap, originated separately from the adjacent Yolla Bolly

Franciscan rocks (Blake *et al.*, 1985b).

The Franciscan Complex is overlain unconformably by the Plio-Pleistocene St. George Formation (Back, 1957) and younger marine terraces in the vicinity of Crescent City (Figure 1 and 2). The St. George Formation consists of fossiliferous marginal marine sandstone and sandy mudstone. Both the terraces and St. George Formation tilt slightly to the northeast.

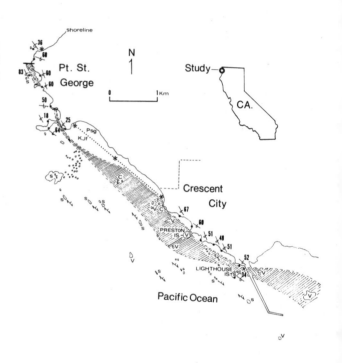

FIGURE 1 Geologic map of the Crescent City area. Only Franciscan Complex (KJf) geology is shown. Uplifted marine terrace deposits and the Plio-Pleistocene St. George Formation (Psg) underlie inland areas. The Franciscan-Neogene unconformity is situated along the line with asterisks. The main olistostrome outcrop is cross-hatched. Attitudes are given for locations indicated with dots. Composition of larger olistoliths is given as follows: C-chert, V-greenstone and T-tonalite. Sandstone stacks visited by boat are labeled "S". The barbed strike-and-dip symbols depict the approximate orientation of beds offshore, as observed from a boat. Facing direction of these beds is unknown.

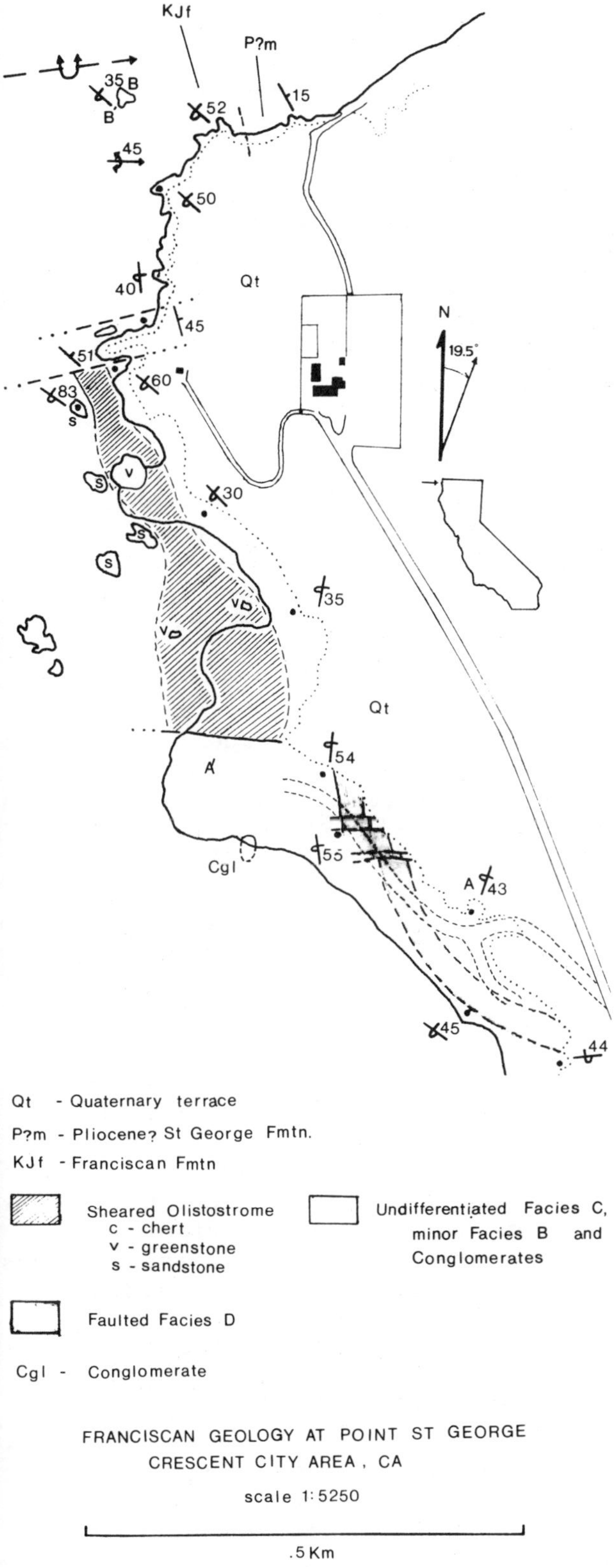

Qt - Quaternary terrace

P?m - Pliocene? St George Fmtn.

KJf - Franciscan Fmtn

Sheared Olistostrome
  c - chert
  v - greenstone
  s - sandstone

Undifferentiated Facies C,
minor Facies B and
Conglomerates

Faulted Facies D

Cgl - Conglomerate

FRANCISCAN GEOLOGY AT POINT ST GEORGE
CRESCENT CITY AREA, CA

scale 1:5250

.5 Km

FIGURE 2 Geologic map of northern Point St. George (modified from Murphy, 1982). The "Road Section" is situated between A and A' near the bottom.

SEDIMENTOLOGY

Introduction

Franciscan Complex sedimentary rocks of the Crescent City area are pre-dominantly submarine sediment gravity flow deposits that may be described using standard facies terminology of Mutti and Ricci Lucchi (1978). In areal distri-bution (Figures 1 and 2), chaotic de-posits of facies F and massive to graded sandstone turbidites of facies B and C are most abundant. Facies A con-glomeratic sandstone and facies D and E shales with sandstone interbeds are subordinate in abundance. No significant compositional differences exist among shales and sandstones of the different facies (Aalto and Murphy, 1984).

Bedding features among facies A and B deposits are described using nomenclature of Aalto (1976), who recognized a common vertical sequence of layers within beds (Table 1). These layer divisions are similar to those proposed in more recent models of Lowe (1983) and Massari (1984). See these three articles for a discussion of depositional mechanics.

Facies A

Pebbly sandstones and conglomerates crop out at Lighthouse Island (Figures 1 and 3; Aalto, 1982) and in fault-bounded patches and the "Road Section" at Point St. George (Figures 2 and 4). Beds commonly range in thickness from 1-2 m, although some are as thick as 10 m, and contain partial to complete layer division sequences: II-III-IV-V or IV-V-VI. Many of the coarser beds have basal scour structures. Most clasts are subrounded, poorly sorted and range from granule to boulder size, with a maximum observed long axis length of 0.75 m. Three field counts of approximately 100 medium pebble and coarser clasts each give an average composition of: 37% clastic sedimentary, 8% chert, 24% por-phyritic volcanic, 18% non-porphyritic volcanic and 13% tonalitic plutonic clasts (Murphy, 1982). Volcanic clasts include a variety of tuffs and micro-litic, felsitic and lathwork textural types of Dickinson (1970). Two 100 clast counts of granules in thin section give an average composition of: 11% quartz, 36% chert, 28% mudstone, 23% volcanic and

TABLE 1 Internal layer divisions of facies A and B beds, with a comparison of models of Aalto (1976), Lowe (1983) and Massari(1984)

Aalto	Lowe	Massari	Layer description
I	R2	R1	Massive sandstone, inversely graded sandstone or inversely graded pebbly sandstone
II	R2-3	R	Conglomerate, with or without imbrication
III	R3	R2	Normally graded conglomerate or pebbly sandstone
IV	S1-2	S1	Diffusely laminated and/or cross-stratified pebbly sandstone
V	$S3=T_a$	S2	Massive sandstone (if at base of bed, can be underlain by layer I type inverse grading)
VI	$S3-T_t$	S3	Laminated sandstone, dish & pillar structures

2% tonalitic plutonic grains (Aalto and Murphy, 1984).

Facies B and C

Fine to coarse, massive or normally graded, gray sandstones are exposed throughout much of Point St. George, along seacliffs below Crescent City and at Enderts Beach 6 km south of Crescent City (Figures 4 and 5). Facies B beds are commonly massive to laminated, normally graded in their uppermost portion and overlain by thin shale layers. They commonly contain layer sequences $V-T_{ae}$ or $V-VI-T_{ae}$. Bed thickness ranges from 1-6 m, averaging 1-2 m. Some beds show inverse grading at the base (sequence I-V). Others have thin pebbly layers at the base but are assigned to facies B because their massive sand divisions predominate. Shale ripup clasts, dish and pillar structures and convoluted lamination are common in the laminated upper portions of beds. Flute, groove and load casts have largely not survived deformation, but exist at several locations.

Facies C beds are commonly less than a meter thick and exhibit classic Bouma sequences: T_{ae}, T_{abe} and T_{abce}, with T_e divisions up to 50 cm, but commonly several centimeters thick.

Facies D and E

Thinly bedded sandstones and shales crop out at Point St. George (Figures 4 and 6), Lighthouse Island (Figure 7) and at Enderts Beach (Figure 5). Rocks of both facies are in places highly folded and faulted. Facies D sandstone beds range in thickness from several to 50 cm and exhibit base-cut-out Bouma sequences and, uncommonly, flute, groove and load casts and trace fossils (Miller, 1986a, 1986b). Pelitic T_e divisions up to 45 cm thick occur and the overall sandstone-shale ratio is on the order of 1:5. Facies E beds are quite discontinuous, have sandstone-shale ratios on the order of 1:1 and exhibit base-cut-out Bouma sequences and extensive ripple drift. Sandstone-shale contacts are sharp. Beds average several centimeters in thickness.

Facies F: Sandy Diamictites

Diamictites interpreted to be olistostromes and associated slump-folded turbidites crop out extensively along the coast and offshore in the Crescent City area (Figure 1). Two types of deposits exist: sandy diamictites and shaly diamictites with scaly matrix foliation.

Sandy diamictites that lack matrix foliation are exposed in the northern part of Point St. George, in the second cove north of the fault truncating the main olistostrome exposure (Figure 2), and in seacliffs due east of Preston

Island (Figure 1). At both localities diamictites are approximately 2 m thick and are composed chiefly of angular to rounded clasts of graywacke and minor chert and greenstone, dispersed in a sand-silt matrix.

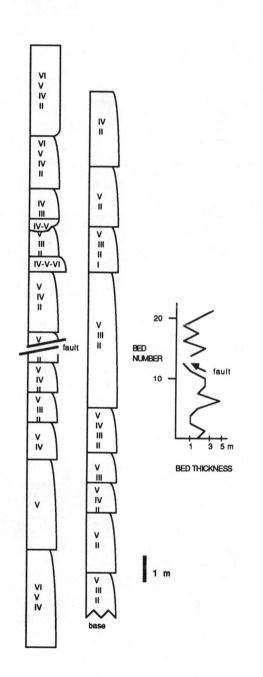

FIGURE 3 Stratigraphic section at Lighthouse Island (Figure 1). Bedding features are shown by Roman numerals following the Aalto (1976) model (Table 1). The graph portrays thicknesses of successive beds (*bed number*) from base to top of section.

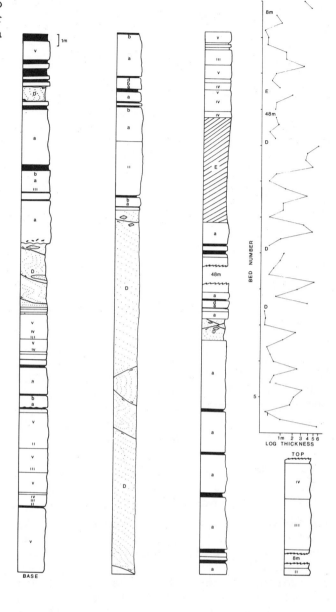

FIGURE 4 Stratigraphy of the "Road Section" at Point St. George. Bouma divisions are shown with lower-case letters. Roman numerals follow the Aalto (1976) model (Table 1). Facies D and E are labeled with upper-case letters. Black depicts thick shale beds and shale ripup clasts. Facies D beds are shown faulted, folded and containing sandstone phacoids. The graph portrays thicknesses of successive beds from base to top of section. Breaks occur where facies D beds are contorted, where thin facies E beds occur, or where the section is covered. From Murphy (1982).

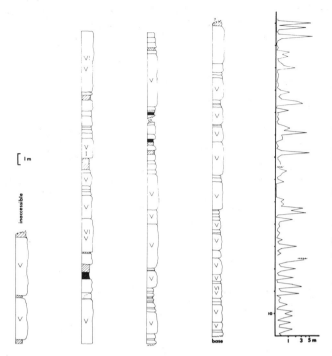

FIGURE 5 Stratigraphic section at Enderts Beach. Roman numerals follow the Aalto (1976) model (Table 1). Beds lacking symbols have chiefly T_{ae} or T_{abe} Bouma sequences. Black depicts thick shale and the diagonal rule facies E deposits. The graph portrays thicknesses of successive beds from base to top of section.

At the northern exposure (Figure 8), progressing up within one diamictite unit, thin-bedded T_{abe} and T_{abcde} turbidites become slump-folded, folds become irregular and detached, and basal portions of beds appear as isolated sandstone clasts. Clast roundness increases upwards within the bed. These changes are ascribed to downslope mass flow, soft-sediment deformation, and the progressive homogenization of a thin-bedded turbidite sequence. The appearance of chert and greenstone pebbles in the upper portion of the diamictite unit suggests that disruption of the turbidites was initiated by the overriding of a bedded turbidite sequence by a submarine debris flow. The debris flow, which contained the chert and greenstone clasts, amalgamated with the slumping turbidite sediments.

Similar relationships exist east of Preston Island, but the rocks are less well exposed. The presence of the major olistostrome unit immediately to the west of slump-folded, sand-matrix diamictites

FIGURE 6 Facies B, C and D beds within the "Road Section". Beds beneath the half-meter scale are upright, but are overturned above the scale. Folds are developed below faults A and B, which dip into the cliff and contain calcite veins. The uppermost sandstone beds are overturned.

FIGURE 7 Folded facies E beds at Lighthouse Island (Figure 1). These beds are in fault contact with those portrayed in Figure 3. Beds are overturned at the upper right. The scale is 50 cm.

suggests that a genetic relationship exists between the two. The diamictites attest to unstable submarine slopes, which resulted in slumping of turbidites immediately before the emplacement of the large olistostrome discussed below.

FIGURE 8 Submarine slump deposit with progressive development of a chaotic fabric. From left to right, upwards within the bed, slump folds become indistinct, the fabric becomes random, sandstone clast roundness increases, and chert and greenstone clasts appear. Scale is 50 cm. Outcrop located in the second cove north of the northernmost olistostrome outcrop at Point St. George.

Facies F: Shaly Diamictites

Introduction. Based upon contact relations described below, the main Facies F diamictite unit portrayed in Figures 1 and 2 is interpreted to be an olistostrome. This interpretation is further supported by its close association with sandy diamictites described above, and by the outcrop pattern of the unit. Block-in-matrix fabric is well exposed on recently quarried surfaces at the northern end of Point St. George and Preston Island. The description presented below is largely based upon observations made at these outcrops.

Contact relations. The coastal exposure of the olistostrome unit is well constrained by onshore and offshore mapping. It is up to 600 m thick and

FIGURE 9 Northernmost Point St. George exposure of the Crescent City olistostrome. The olistostrome unit is 25 m thick and contains ellipsoidal blocks of sandstone (labeled S), limestone, chert and greenstone dispersed in argillite. Overturned facies B turbidites (top to left) bound the olistostrome on both sides.

exposed for some 12 km along strike. At its northernmost outcrop, the unit is approximately 25 m thick and in depositional contact with facies B turbidites both to the east and west (Figures 2 and 9; Murphy and Aalto, 1983). To the east, the chaotic unit rests upon a 6m-thick facies B sandstone bed containing the layer sequence V-T_{bcde}. The T_e division of this bed is in some places separated from the scaly matrix of the olistostrome by a calcite-filled fracture, but in other places grades into this matrix (Figure 10). The matrix foliation passes at a shallow angle (with respect to bedding) into the T_e division, is refracted to a shallower angle in the T_{cd} divisions and fades out within the T_b division of the bed.

Similar, but less well exposed, contacts with facies B or C turbidites exist at the western margin of the olistostrome at this location, at southern Point St. George exposures and east of Lighthouse Island (Figure 1). Bedding orientation in offshore islands west of the unit appears to be similar to that east of the unit. Thus, although the olistostrome is offset by faults along strike, it is bounded by and in apparent conformable contact with turbidites on both sides, where outcrop exists.

Olistostrome matrix. The argillaceous matrix of the olistostrome has a penetrative anastomosing cleavage that causes argillite to break into lenticular chips, whose origin is discussed below (Figure 11). The matrix contains abundant clay and varying proportions of chiefly quartzofeldspathic sand and silt (Aalto, 1986b).

There is a compositional similarity between this matrix and shales in the turbidite sequences that bound the unit. X-ray diffraction study of 19 samples of the matrix and 21 samples of the shale (Aalto and Murphy, 1984; Aalto, 1986b) indicates that both contain chiefly iron-rich chlorite, white mica, quartz and sodic plagioclase, with minor kaolinite and no mixed-layer clays. There are no statistically significant differences in their illite crystallinity index (Kubler, 1969) or the b_o lattice spacing of the white mica, which is related to an increasing phengitic component (Sassi and Scolari, 1974). Thus, both the matrix and shale have been subjected to similar pressures during their deformation.

Olistoliths are surrounded by matrix. Their abundance expressed as percent per unit area was determined semi-quantitatively by estimating the percent of clasts exceeding 1 cm in length along a meter rule positioned subnormal to foliation at 174 sites (Aalto, 1986b). Thirty four sites have 5% or less clasts, 66 sites 6-14%, 51 sites 15-25% and 23 sites over 25% (maximum: 35%).

Olistolith composition and size. Olistolith composition was determined by four counts of 100 or more clasts made along lines oriented subnormal to foliation (Aalto, 1986b). Based upon a total of 1,250 clasts counted, sandstone comprises an average of 88%, greenstone 8%, chert 4% and limestone, pebble conglomerate and phyllite trace amounts

FIGURE 10 Depositional contact between the olistostrome unit shown in Figure 9 and an underlying, overturned turbidite bed. Note how the olistostrome matrix foliation (parallel to the 50 cm scale) passes into the T_{cde} divisions of the turbidite. Greenstone (labeled V) and sandstone (S) clasts are stretched parallel to foliation.

FIGURE 11 Preston Island exposure of the Crescent City olistostrome. Irregular blocks of sandstone (labeled S) and greenstone (V) are dispersed in scaly argillite.

of the population. No significant compositional differences exist between Preston Island and Point St. George exposures.

Sandstone olistoliths are chiefly feldspathic litharenites (*cf.* Folk, 1974), similar in composition to the bounding turbidites. Red, green or light tan ribbon chert commonly contains radiolarians. Greenstone includes metabasalt, dacite porphyry, quartz keratophyre and tuff that contains variable proportions of crystals, shards and lithic fragments.

Some tuffaceous olistoliths contain highly indurated black argillite that is complexly intermixed with tuff and, in some cases, graywacke and basaltic rock. This association is common in Franciscan and other late Mesozoic melanges throughout the North American Cordillera (Hsü, 1968; Aalto, 1981; Cowan, 1985). The origin of these olistoliths is attributed to the structural dismemberment of originally stratified sequences of argillite, diamictite and volcanic rocks (Aalto, 1986a). These accumulated

on an oceanic plate marginal to North America and were later incorporated into the Franciscan accretionary prism by subduction accretion. It is noteworthy that argillite within such blocks contains little white mica.

The large tonalite olistolith shown in Figure 1 is composed of abundant, subequal amounts of granophyric quartz and albite, with trace amounts of apatite, sphene and chloritized amphibole. Scattered keratophyre dikes intrude the tonalite, and both are thoroughly brecciated at thin section scale. J. Mattinson (pers. comm., 1987) obtained a U-Pb age date for this block of 162.3±1.0 Ma from sphene, feldspar and apatite fractions.

Maximum dimensions of olistoliths that are clearly surrounded by matrix are 12 m for sandstone, 37 m for greenstone, 1.8 m for chert, 2.3 m for phyllite and 27 cm for limestone. Blocks inferred to be olistoliths are over 200 m across, but contacts with matrix cannot be observed around blocks of this size.

Many olistoliths have experienced

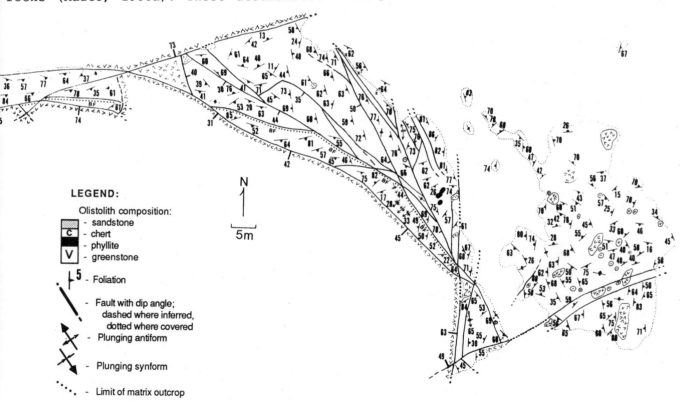

LEGEND:

Olistolith composition:
- sandstone
- C - chert
- phyllite
- V - greenstone

5 - Foliation

- Fault with dip angle; dashed where inferred, dotted where covered

- Plunging antiform

- Plunging synform

- Limit of matrix outcrop

FIGURE 12 Geologic map of the Crescent City olistostrome at Preston Island. V's along faults show where the huge Preston Island keratophyre olistolith is faulted against olistostrome matrix. Light dotted lines delineate covered areas. An area with solely sandstone olistoliths (labeled BF) exists in the western region.

obvious post-depositional form modification (discussed below). However, some appear to have suffered little structural modification. For these, there appear to be two sources: (1) a primary, first-cycle source for abundant angular, tabular blocks of sandstone, greenstone and chert; and (2) a recycled-sedimentary source for subrounded to well rounded clasts of these and other lithologies (Figure 11). The latter are commonly smaller than the former, although size overlap between the two populations is considerable. Where blocks have experienced structural form modification, it is impossible to assign them to either category.

Many of the tabular, angular sandstone olistoliths were pieces of turbidite beds. Concentrations of these blocks exist along the western margins of the Preston Island and northern Point St. George exposures (Figures 9 and 12). Here blocks are arranged in trains, some of which define folds that appear unrelated to faults that truncate them (Figure 12). Thus a ghost stratigraphy exists within the olistostrome. It is possible the olistolith trains developed by slump folding and progressive homogenization of turbidite beds which occurred during olistostrome emplacement. A 12 m-wide olistolith that is clearly encased in matrix (Figure 9, labeled "S") contains a likely slump fold, a large overturned syncline developed in massive sandstone beds. The axial surface of this fold is oriented subnormal to both the olistostrome matrix foliation and bedding of the turbidites that bound the unit.

The presence of slump folds, a ghost stratigraphy within the unit and abundant rounded clasts, presumably recycled from a conglomerate, provide support for the olistostromal origin of this deposit.

Sandstone Composition

Crescent City area. Sandstones both within and bounding the main olistostrome unit are commonly lith-arenites or feldspathic litharenites (*cf.* Folk, 1974). Based upon six counts of 300 grains per thin section, the following detrital modes (*cf.* Dickinson, 1984) have been determined: Qt=44%, Qm=21%, Qp=23%, F=P=11%, Lt=69%, L=46%, Ls+Lm=11%, and Lv=35% (Aalto and Murphy,

1984).

Most monocrystalline quartz grains exhibit undulose extinction, and some have deformation bands and lamellae. Some monocrystalline grains have the clarity, idiomorphism, fractures and embayments that indicate a volcanic

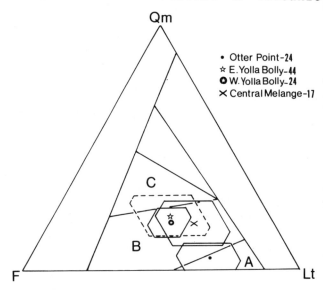

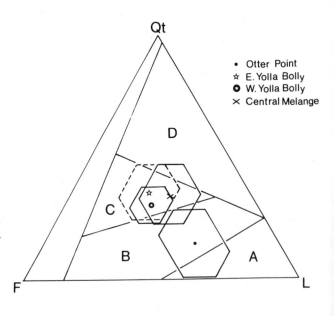

FIGURE 13 Compositional plots of Franciscan terrane and Otter Point Formation sandstones from the northern-most California and southwest Oregon Coast Ranges (see map in guidebook preface). Fields surrounding means depict calculated standard deviations. Detrital modes and provenance fields are from Dickinson (1984): A-undissected arc, B-transitional arc, C-dissected arc, and D-recycled orogenic provenance.

provenance. All varieties of poly-crystalline quartz that are described by Young (1976) were observed. Chert is commonly partially recrystallized. Plagioclase consists chiefly of un-twinned, albitized grains and twinned, relatively unaltered sodic oligoclase. Pumpellyite is present in some grains.

All four major textural types of volcanic rock fragments of Dickinson (1970a) occur, but microlitic and felsitic grains are most common. Lithic sedimentary clasts include mudstone and very fine sandstone. Matrix constitutes an average of 20% of samples counted. Chloritic orthomatrix and pseudomatrix (derived from mudstone grains and counted as Ls where recognizable), are common. Common accessory detrital minerals include pigeonite, augite, zoisite, musco-vite and sphene.

Regional trends and provenance. Modal data for Franciscan terranes and the coeval Otter Point Complex of southwestern Oregon are plotted on standard ternary provenance diagrams (Figure 13; Dickinson, 1984). Crescent City area Franciscan rocks are tentatively assigned to the western part of the Yolly Bolly terrane (Aalto, 1987b). Central Franciscan belt sandstones are somewhat richer in lithics than those of the Yolla Bolly terrane, although differences are slight.

The outlined provenance fields suggest that the Yolla Bolly sandstones were derived from a partially to deeply dissected magmatic arc. The Gazzi-Dickinson point counting technique (Ingersoll et al., 1984) was utilized to minimize the effect of grain size on composition and allow comparison with other published data. The modal data are very similar to that of Franciscan sandstones of the northern Coast Ranges to the south in California and in southern Oregon, which suggest a Klamath Mountains provenance (Dickinson et al, 1982; Blake et al., 1985a). Limited paleocurrent data from several sets of sole marks and parting lineation are compatible with a sediment source to the east.

Depositional Setting

Turbidites intercalated with melange units of the Franciscan Complex of northernmost California are predominantly inner- to mid-fan deposits (Aalto, 1982). Pervasive deformation has obscured much of the evidence needed for detailed facies reconstruction in the Crescent City area. However, the prevalence of facies F, B, C and A deposits, listed in order of decreasing abundance, and the dearth of facies D and G deposits suggest an inner- to mid-fan depositional setting for these rocks. Facies E deposits may represent over-levee spills from high-energy sediment gravity flows on the inner region of a submarine fan (Aalto and Murphy, 1984).

Progressing from east to west, upwards within the stratigraphic sections preserved at both Point St. George and Crescent City coastal exposures, the facies succession generally passes from mixed facies B, C and D, to facies F, to mixed facies A, B and E. This suggests prograding of an inner fan over a mid fan, accompanied by the emplacement of a large olistostrome. Melange units that exist further offshore (Figure 1) may have been emplaced by similar mass flow events.

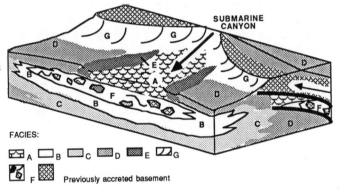

FIGURE 14 Sedimentary tectonic model for the origin of the Crescent City area Franciscan Complex. The block diagram depicts an actively deforming inner trench slope. Facies are labeled according to Mutti and Ricci Lucchi (1978). Olistoliths were derived by slumping of upfaulted basement material that occurred in the vicinity of a submarine canyon. The arrow drawn in the canyon axis depicts transport direction for sediment gravity flows (including olistostromes) into either a trench or trench slope basin. Thrust faults are shown schematically on the front of the block. The Crescent City olistostrome is represented on the front of the block.

Franciscan turbidites were deposited in a forearc that fronted the late Mesozoic Sierran-Klamath magmatic arc (Hamilton, 1969; Dickinson *et al.*, 1972, 1979, 1982; Ingersoll, 1978, 1979, 1982). This arc is inferred to have been the source for turbidites of the Crescent City area (Aalto, 1982; Aalto and Murphy, 1984). Results of this study suggest that olistoliths were formed by recycling of arc-derived sediments and previously accreted oceanic-plate igneous and sedimentary rocks. These were exposed on the inner trench slope, upslope from the depositional basin in which the olistostrome was emplaced. Sedimentation may have occurred near the mouth of a submarine canyon that exhumed previously accreted basement (Figure 14). Abundant slope mud contributed to the olistostrome matrix.

STRUCTURAL GEOLOGY

Introduction

Northwestern California has undergone a complex structural evolution that reflects deformation associated with the initial development of the Franciscan subduction complex, as well as several episodes of post-accretion crustal shortening and strike-slip faulting (Blake, 1984, *et al.*, 1985a, 1985b; Kelsey and Carver, 1988). At Crescent City it is possible to distinguish between the earliest-formed Franciscan structures and those that reflect later deformations. Because of this, it is possible to construct a structural model for the evolution of Franciscan rocks that may better constrain the depositional setting of the olistostrome.

Late Cenozoic Structures

Sandstone and sandy mudstone of the Plio-Pleistocene St. George Formation (Back, 1957) unconformably overlie Franciscan rocks along the coast between Crescent City and Point St. George (Figures 1 and 2). This unit has been tilted northeast and contains numerous northeast-striking, steeply dipping joints and normal faults (Figure 15). Apparent dip-slip on these faults is less than a meter and striations on fault surfaces plunge moderately southeast.

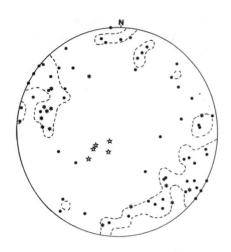

FIGURE 15 St. George Formation structural data, Crescent City area. Lower hemisphere, equal area plot depicting poles to bedding (open stars), normal faults (asterisks) and joints (dots). Contours depict concentrations of 3% per unit area for combined fracture data (n=70).

FIGURE 16 Scissor fault north of Klamath. Encircled meter scale rests on a fault that dips 70° to the left. The surface on the left is the base of a single overturned bed, while on the right overturned facies C turbidites dip at 56° into the cliff. The fault has normal drag and accounts for a major change in bed orientation along a 400 m roadcut along highway 101. Note the stepping down of layers across a smaller fault to the right.

Less prominent are northwest-striking, steeply dipping joints, and reverse faults with northeast dips. A few northeast-trending near-vertical faults have subhorizontal striations that indicate strike-slip displacement. Holocene marine terrace deposits that unconformably overlie the St. George Formation and Franciscan Complex are tilted gently to the northeast, but lack extensive fractures.

Steep, east-west to northeast-striking faults that truncate the Crescent City olistostrome at Point St. George (Figure 2) may be Quaternary in age and have appreciable dip-slip displacement. Faults with similar orientations are common among Franciscan rocks in northernmost California and account for juxtaposition of turbidite sequences with markedly different orientations (Figures 16 and 17). Although it is not possible to prove that many of these faults are young in age, it is a likely possibility. Nearly all major regional faults that are demonstrably pre-Pliocene in age strike northwest and northeast-trending faults are present in Plio-Pleistocene deposits.

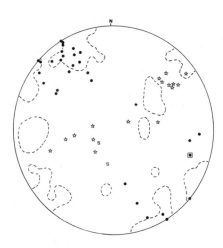

FIGURE 17 Structural data for the Figure 16 exposure. Lower hemisphere, equal area plot depicting poles to bedding for upright (solid star) and overturned (open stars) turbidites, poles to major normal faults (dots), and plunge of striations on fault surfaces (S). Countours depict concentrations of 3% per unit area of poles to minor fractures (n=71). The boxed point is the Figure 16 scissor fault.

The fracturing and tilting of the St. George Formation and overlying terrace deposits very likely resulted from Neogene through Quaternary northeast-southwest convergence of the northern Gorda and North American plates, which resulted in the uplift of the Coast Ranges and Klamath Mountains in northernmost California (Carver, 1987).

Early Structures in Turbidites

Introduction. Pre-Quaternary structures in turbidites that postdate emplacement of the Crescent City olistostrome can be divided into those that are pervasive, affecting all areas at all scales, and those that are localized. The latter include a shear zone and large folds that are described separately.

Mesoscopic structures. Pinch-and-swell and neck structures are ubiquitous among Franciscan turbidites. Bed thinning occurs along a multitude of well defined small faults, commonly oriented at moderate to high angles to bedding. These fractures may be calcite-filled or intruded by shale. The more prominent faults commonly terminate in a fan of smaller fractures that become parallel to bedding in shale layers bounding the thinned sandstone bed. Surrounding strata, in some cases, can remain unaffected and thus stratigraphic sections can thin at the expense of some layers while others survive intact.

Mesoscopic faults disrupt bedding pervasively, so that it is difficult to trace a single bed over more than several meters (Figure 18). Faults are commonly oriented at 30-60° to bedding and consistently show normal offset. Many are calcite-filled. Motion on these faults results in the formation of lozenges that can become isolated by intrusion of mudstone, which appears to have flowed ductilely into fractures and around ends of sandstone beds. Lozenges defined by faults are approximately equidimensional, viewed normal to bedding, and bed attenuation by necking and pinch-and-swell occurs in all directions parallel to bedding (Aalto, 1986b). These observations suggest that extension was largely axially symmetric in the plane of stratification.

Microstructures. Microfaults with

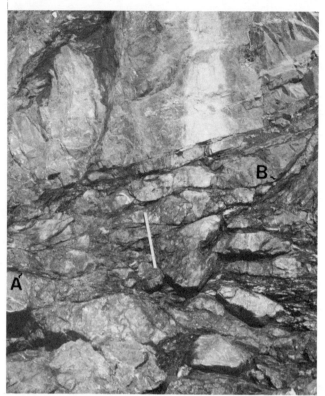

FIGURE 18 Two orders of extensional
faults at Enderts Beach. Beds are
upright but vary markedly in orientation
across the larger faults (A and B) that
crosscut the outcrop. Note the necking
and boudinage of beds along the faults
and the development of smaller-scale
faults between undeformed beds (beneath
the 50 cm scale) that has resulted in the
rotation of sandstone lozenges.

normal offset are common in the T_{bcd}
divisions of some turbidite beds. In
thin section, they appear as narrow, iron
oxide-rich surfaces that truncate sand
grains and produce normal drag in
laminae. The faults may fade into
unlaminated basal portions of turbidite
beds, whose soles exhibit no dip-slip
offset but whose massive T_a divisions
contain web structures (Figure 19; Aalto,
1986b).

In thin section, webs appear to be
cataclastic shear zones, They commonly
have distinct margins and contain broken
grains and concentrations of clay and
oxides (Figure 20), similar to webs
described by Lucas and Moore (1986) from
other locations. Webs may merge with
microfaults upwards within a bed, or may
be truncated by faults that pass through
a bed. Some webs thin and fade upwards
within a bed. Both microfaults and web

structures crosscut dish and pillar
structures that formed by dewatering
during and immediately after sediment
deposition (Figure 19).

Primary quartz and secondary calcite
veins commonly follow fractures, and both
follow and crosscut webs, dishes and
pillars. Stereoplots of 520 poles to
minor fractures in Franciscan rocks of
the Crescent City-Point St. George area
show complete scatter for both filled
and unfilled fractures and combined data.

**Spatial variations and state of
sediment during deformation.** The
style of deformation varies markedly
over short distances both between
different turbidite facies and within the

FIGURE 19 Sandy diamictite with slump
folds and angular to subrounded clasts of
sandstone and greenstone, overlain by two
graded beds with convoluted lamination,
flame structures and fluid escape
structures. Soft sediment deformation
structures are crosscut by webs (near
asterisk) and both unfilled and calcite-
filled fractures. Note how some
fractures cross-cut all beds while others
are confined to upper portions of beds.
The middle bed is approximately 40 cm
thick. Located in the second cove north
of the northernmost olistostrome outcrop
at Point St. George.

FIGURE 20 Web structure in thin section (field is 2 x 3 mm). Photomicrograph of a dark web in plane light. Stratigraphic top towards top of photo. Note the finer grain size of the web fill and the calcite vein that crosscuts the web at the X. Sample from the Figure 19 site.

same facies. Where marked contrasts in degree of stratal disruption exist, contacts between zones with contrasting deformation are either distinct faults, or shear zones oriented at low angles to bedding (Figure 21). Shear zones are planar to undulating and commonly are developed along shale interbeds. These shales have a pronounced phacoidal cleavage, in contrast to the weak fissility of the less deformed shale of undisturbed sections.

Within the "Road Section" at Point St. George, a major shallow dip-slip fault juxtaposes thickly-bedded sandstones against a predominantly shaly sequence (Figure 6). The thick sandstone beds are disrupted by numerous small extensional faults, whereas the thin sandstones in the shaly sequence are drag-folded near the fault, but are otherwise little deformed (Figure 22).

FIGURE 21 Fault contact between facies D turbidites with contrasting degrees of stratal disruption within the Point St. George "Road Section" (Figure 4). The fault, marked by 2 cm of fine crush breccia, is at the center of the 50 cm rule, and is subparallel to bedding that has similar orientation, but which is upright above and overturned below the fault. Note the bedding-parallel calcite veins among the less deformed beds and the numerous small tension fractures and extensional faults among deformed beds. Necking occurs along a myriad of small fractures in both sandstone and shale.

Sandstone dikes intrude the shaly sequence and are derived from thin sandstone beds within this sequence. In some places the dikes cut randomly and irregularly through beds and locally contain mudchips ripped from surrounding muddy strata. Where exhumed by erosion, the walls of such dikes have a fluted appearance produced by varying amounts of intrusion into bounding beds. In other places dikes follow, and in turn are cut by widely spaced fractures (Figure 23). Dikes that follow fractures have planar walls and do not contain mud chips, suggesting that the fractures were dilating during passive dike emplacement. Irregular dikes appear to have been intruded more forceably.

Folds that developed by drag along the major fault that bounds the shaly sequence of the Road Section are truncated by irregular coarse-grained sandstone dikes, suggesting that dike emplacement occurred after faulting (Figure 22). These dikes contain mudchips stoped from bounding beds. Other dikes are themselves folded and truncated by the fault. Since dike emplacement both predated and postdated faulting and fracturing of these rocks, it is likely that a single, continuous deformation event accounts for the formation of these structures, and that the sediments were unconsolidated during this deformation.

Cowan (1982) notes that Franciscan sedimentary rocks exhibit features of both cataclastic and intergranular flow, which appear mesoscopically as brittle and ductile deformation. He interprets this behavioral contrast as resulting from lateral and vertical differences in effective confining pressure within the sediment at the time of deformation. These differences existed because the

sediments were variably dewatered and consolidated. Dewatered, consolidated sediments were deformed cataclastically while fluid-rich, unconsolidated sediments flowed. All gradations in behavior existed over short distances. The Cowan model might explain the variations in style of deformation discussed above (Aalto, 1986b).

Early Structures in Olistostrome

Introduction. Pervasive structures resulting from pre-Quaternary deformation within the Crescent City olistostrome are similar to those in the turbidites discussed above, except that the

FIGURE 22 Z-fold within facies D turbidites of the "Road Section" at Point St. George (Figure 4). The fold resulted from drag on a fault immediately beneath and parallel to the 50 cm scale (note the calcite vein). A sandstone dike (B) truncates the hinge above B. Note the presence of mud chips within the dike to the right of B, where tongues of sand emanate from the dike and intrude concordantly into surrounding beds.

FIGURE 23 Sandstone dikes in thinly bedded turbidites of the Point St. George "Road Section". The dike cuts upright facies D turbidites above the arrow, then intersects and follows a higher fracture. Note the small amounts of normal offset on both the prominent fractures that dip right and less prominent antithetic fractures that dip left. The scale is 50 cm.

olistostrome matrix has a scaly foliation while turbidite mudstones away from sheared zones are weakly fissile. In addition, the spatial distribution and compositional variability of olistoliths accounts for a greater complexity in interpreting deformation history of the unit.

Scaly foliation and fracturing. The olistostrome has a scaly matrix similar to that found in melanges worldwide (Silver and Beutner, 1980; Bosworth, 1984). The scaly foliation is defined mesoscopically by a penetrative, irregular and anastomosing spaced cleavage that causes argillite to break into lenticular chips (Figure 11). Cleavage surfaces are wavy in form and spaced millimeters apart. They are most commonly dull, but may be polished or striated, especially near faults. The foliation bends around olistoliths.

In thin section, the cleavage surfaces intersect at low angles and enclose lenticles of mudstone (Figure 24). The surfaces are defined by concentrations of dark clays and oxides. Adjacent lenticles have markedly contrasting grain size distributions; one composed nearly entirely of clay can be juxtaposed against one rich in silt. The sediment within the lenticles may be laminated (with laminations defined by contrasting grain sizes), or may have massive fabric. Variations between lenticles indicate that the cleavage surfaces are microshear zones which have experienced significant slip. Lenticles are not aligned on foliation surfaces (Aalto, 1986b).

If, as discussed above, the olistostrome contains abundant turbidite debris homogenized during mass flow emplacement of the unit, likely sources for the lenticles are the finer divisions of turbidite beds. These divisions may have initially been disrupted by mass flow, and then further disrupted by the microshearing that produced the scaly foliation.

In most exposures of the olistostrome the scaly foliation is oriented subparallel to bedding of bounding turbidites. Sets of faults, similar to those discussed above, are commonly oriented at low to moderate angles to the foliation (Figure 25). At Preston Island (Figure 1), a brecciated quartz keratophyre olistolith that measures over 150 by 200 m in outcrop is faulted against

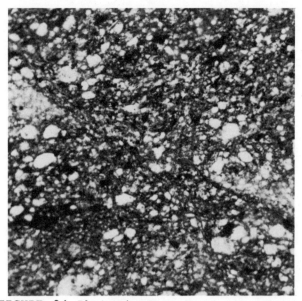

FIGURE 24 Photomicrograph of the olistostrome matrix at Preston Island (plane light; field width is 3.5 mm). Numerous dark, anastomosing, conjugate microshear zones are oriented at approximately 30° to the horizontal and juxtapose lenticles with contrasting grain size distributions.

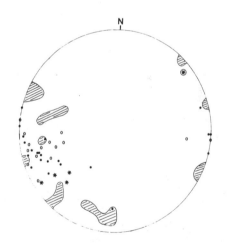

FIGURE 25 Clast orientation and matrix fabric data for the northernmost Point St. George olistostrome exposure. Lower hemisphere, equal area plot depicting poles to planes containing the X and Y axes of oblate ($R_{yz} \geq 1.5$) olistoliths (dots), poles to matrix foliation (O) and poles to overturned (asterisks) and upright (encircled asterisk) beds that bound the unit. Concentrations of $\geq 7\%$ per unit area of poles to minor fractures within the unit are shown by the pattern (n=28).

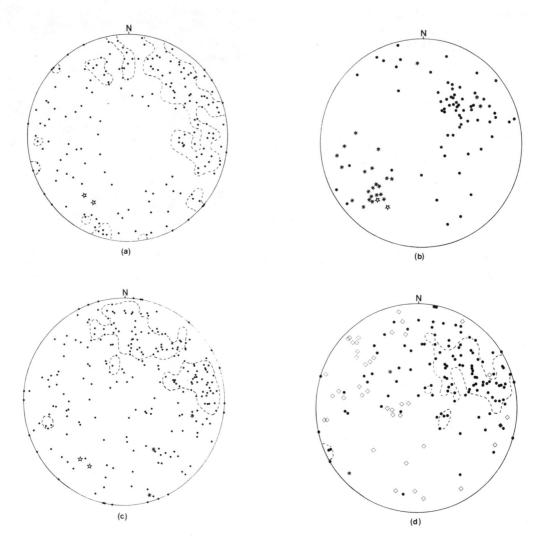

FIGURE 26 Structural data for the Crescent City olistostrome at Preston Island. The open stars show poles to bedding for overturned turbidites east of the exposure. Lower hemisphere, equal area plots depict: a) poles to fractures (n=188) with contours showing concentrations of 2% per unit area, b) poles to major faults (dots) and plunge of striations on fault surfaces (asterisks), c) poles to matrix foliation (dots, n=231) with contours showing concentrations of 2% per unit area, and d) poles to planes containing the X and Y axes of oblate ($R_{yz} \geq 1.5$) olistoliths (dots, n=115), with contours showing concentrations of 3.5% per unit area, plunge of long axes of prolate ($R_{xy} \geq 1.5$) olistoliths (diamonds) and plunge of striae on olistoliths (asterisks).

olistostrome matrix. Some of the east-trending faults at this site are probably Quaternary. The similarity in orientation of northwest-trending fractures and foliation suggests that these structures developed by progressive deformation (Figures 12 and 26).

The olistostrome scaly foliation is folded only at Preston Island. Most folds are small, irregular and have tight hinges, but some are cylindroidal and open (Aalto, 1986b). Some microscopic calcite veins are folded. However, both filled and unfilled mesoscopic fractures and faults are not folded. All folds in the scaly foliation are truncated by faults over short distances. Some folds in foliation have clearly formed by drag on faults (Figure 27). The presence of relict slump folds, discussed above, complicates fold interpretation.

Olistolith form. Olistolith form reflects several factors, including original shape, strain environment, ductility contrast between clast and matrix, orientation prior to strain, volume changes and multiple deformation events (Ramsay, 1967, 1980). In order to assess the relative significance of these factors, a total of 278 olistoliths with a minimum long axis length of 5 cm were extracted from the matrix at several exposures (Aalto, 1986b). Long (X), intermediate (Y) and short (Z) axes were measured directly to determine axial ratios, which were used to calculate olistolith ellipticity (R_{xy}, R_{yz} and R_{xz}).

Lode's parameter (ν) and the amount of

FIGURE 28 Train of sandstone blocks at the western margin of the Crescent City olistostrome at Preston Island (see Figure 13). Note the scaly matrix foliation and presence of conjugate fractures that truncate block margins and matrix. Small folds are developed in the matrix as a result of drag on the fault that parallels the pen. Bedding-parallel slip has occurred within the upper left olistolith.

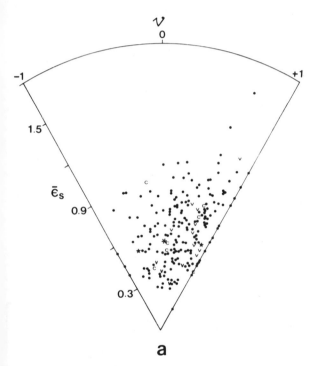

a

FIGURE 27 Graphical illustration of possible strain of olistoliths (following the method of Hossack, 1968). Amount of strain as a function of the three principal logarithmic strains ($\bar{\varepsilon}_s$) is plotted radially against Lode's parameter (ν). Composition of olistoliths is shown as follows: dots-sandstone, L-limestone, C-chert, V-greenstone, and star-phyllite. The asterisk depicts the *initial shape factor* (Hossack, 1968, p. 322) that represents the average shape of undeformed pebbles. Figure gives data for Preston Island (n=226).

strain (if any) expressed as a function of the three principal logarithmic strains ($\bar{\varepsilon}_s$) were determined according to the equations of Ramsay and Huber (1983, p. 201-203) and plotted on Hsu (1966) diagrams, following the method of Hossack (1968).

The data for Preston Island are presented in Figure 28. There is a wide scatter, but most points lie to the right in the diagram. This may in part reflect a predominantly flattening strain that is recorded in olistolith form. Oblate ($R_{yz} \geq 1.5$) olistoliths commonly lie with their XY plane parallel to matrix foliation (Figures 25 and 26). Long axes of prolate ($R_{xy} \geq 1.5$) olistoliths commonly plunge randomly within the plane of foliation. Internal attenuation of olistoliths appears both mesoscopically and in thin section to result from brecciation along fractures oriented at moderate to high angles to the XY plane.

Fracture density greatly increases in attenuated regions and veining is common. Radiolarians in chert and vesicles in basaltic clasts are not flattened.

Phyllite, ribbon chert and foliated lithic tuff clasts are more attenuated than others because of preferential slip along internal foliation or bedding planes, but overall form variation due to compositional differences is not pronounced (Figure 28). Many olistoliths have margins truncated by sets of conjugate fractures that are oriented at moderate angles to their XY planes and to foliation (Figure 27). This results in a lozenge or diamond shape. In thin section, most olistolith-matrix contacts are sharply defined, either by fractures or by matrix foliation that wraps around clasts. This suggests that they were lithified prior to deformation. However, some sandstone olistoliths have poorly defined margins and appear to grade into matrix.

As is discussed above, some clasts appear to have been recycled from preexisting conglomerates. The asterisk on the plot depicts the average shape of detrital pebbles: the *initial shape factor* of Hossack (1968). The clustering of much of the data close to this point suggests that strain recorded may not be great, and that some of the clasts retain an inherited detrital form. Other clasts are clearly deformed and it is likely that several of the other factors listed above contributed to their form modification during progressive deformation. Given the difficulty of discriminating between primary and recycled clast populations, it is impossible to assess the relative importance of these other factors. Thus the Hsu diagram provides chiefly descriptive information.

Localized Large-Scale Structures

Introduction. The style of fracturing of Franciscan rocks described above is pervasive throughout the Crescent City area. Localized large-scale deformation features in this region include large folds at Point St. George and a shear zone at Enderts Beach.

Point St. George folds. Large folds are visible on the abrasion platform at the northern extreme of Point St. George (Figures 2 and 29). They are overturned anticlines and synclines with axial surfaces that commonly dip steeply northwest to northeast and axes that

plunge moderately to steeply northeast. The folds are parallel, isoclinal to open, with rounded hinges and closures ranging from 3-30 m. Limbs are straighter among tighter folds. Folding was accommodated by flexural slip on shale interbeds. Hinge areas are characterized by brecciation and polyclinal folding of finer-grained layers. Some isoclinal folds have faulted axial surfaces.

Beds in hinge areas and on straight limbs appear to have suffered faulting prior to folding that resulted in the layer-parallel extension described above. Fractures that formed during folding are oriented subparallel to axial surfaces of folds and truncate the fractures that define earlier-formed lozenges.

Enderts Beach shear zone. At Enderts Beach a subhorizontal shear zone places northwest to southeast-dipping, thinly-bedded, facies E turbidites above thickly-bedded, facies B and C sandstones (Figures 30 and 31). The shear zone is estimated to be 10 m thick at the maximum, but its upper contact is not exposed. The thinly bedded turbidites probably are overlevee interchannel or interlobe submarine fan deposits, once in depositional contact with channel or suprafan lobe sandstones (Aalto, 1986b).

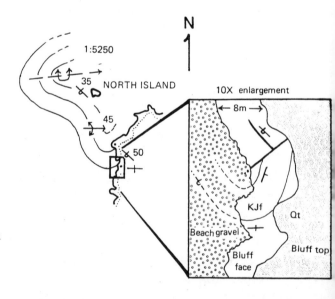

FIGURE 29 Map of folded facies B and C sandstone beds at northernmost Point St. George. The enlargement shows overturned beds, with stratigraphic tops to the west, that extend from North Island to the bluff face. From Murphy (1982).

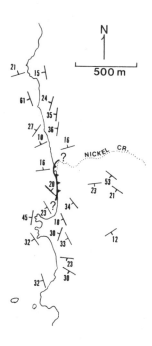

FIGURE 30 Geologic map of Franciscan rocks at Enderts Beach. The map shows bedding attitudes and the location of the Enderts Beach shear zone (thrust symbol). Minor faults, a major normal fault that follows Nickel Creek, and terrace deposits that surround the creek are not shown. The beach is located 6 km south of Crescent City at the foot of a Redwood National Park trail.

The basal contact of the shear zone is a gently dipping striated, stepped fault surface that locally may be polished, lined with fibrous calcite and/or brecciated. Absence of lineation or ductile offset in the underlying sandstone suggests that this is a brittle shear zone (cf. Ramsay, 1980). Within the shear zone are numerous closely spaced, intersecting low-angle faults that are subparallel and similar in appearance to the basal fault. Striations and steps on fault surfaces indicate upper plate transport to the northwest.

South of Nickel Creek, where bedding dips northwest, folds are present in fault-bounded packets (Figure 32). Folds are parallel to cylindroidal and tight to open, with an average mean width for fold pairs of 14 cm, height of 4 cm and height/width ratio of 0.5 cm (Aalto, 1986b). Most pairs are sinistral when viewed down plunge. The folds were analysed by the separation angle method of Hansen (1971) to estimate slip line,

slip plane and sense of relative movement that produced the folds (Figure 33). Although fold axes are scattered, the analysis indicates their rotation towards the northwest, in agreement with the slip determination from fault data. The data for large extension fissures with calcite fillings from folded intervals provide additional support for northwest movement within the shear zone (Figure 33).

North of Nickel Creek, where bedding dips southeast, sinistral shear within the zone resulted in transposition of primary layering to the horizontal along a multitude of fault surfaces spaced millimeters apart (Figure 34). Transposition was accompanied by attenuation and rotation of layers on both mesoscopic and microscopic scales. Lineation, defined by faint striations and (uncommonly) linear sandstone pods, is poorly developed on shear surfaces but supports northwest slip.

FIGURE 31 Basal contact of the Enderts Beach shear zone south of Nickel Creek. The contact is a subhorizontal fault developed on the light massive sandstone in the foreground. Bedding in the sandstone is upright and subhorizontal. Overlying turbidite beds dip moderately to the left. Folds are developed in fault-bounded packets above the massive sandstone. Extensional faults that cut folds (top of photo) dip moderately to the right. Inferred tectonic transport is to the left, approximately parallel to the outcrop face. Notebook is scale.

Early layer-parallel extension occurred among turbidites at Enderts Beach
prior to development of the shear zone.
Structures indicating this deformation
(described above) are pervasive among
turbidites bounding the zone (Figure 18).
Within the shear zone south of Nickel
Creek, layer-parallel bed attenuation
cannot logically be related to shear zone
development. Should extension have occurred in this region as a result of
sinistral shear, apparent direction of
extension on the outcrop face (Figure
32) would be subnormal to bedding, not
bedding parallel. An outcrop of folded
boudins at the southern end of the shear
zone exposure developed by the
contraction of layers that had earlier

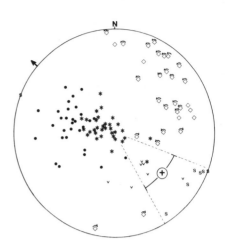

FIGURE 33 Structural data for the Enderts
Beach shear zone. Lower hemisphere,
equal area plot depicting poles to major
faults (asterisks), plunge of striations
on fault surfaces (S), poles to axial
surfaces of folds associated with these
faults (dots), poles to calcite-filled
extension fissures (>1 cm thick) in
folded areas (V), and plunge of fold axes
(diamonds). Curved arrows show sense of
fold asymmetry viewed down plunge for
fold pairs. Dashed lines encompass
striations and the separation angle (arc
with circular symbol) drawn in a plane
approximating the fold axis distribution.
Arrow on net margin shows inferred sense
of upper plate movement.

FIGURE 32 S-folds developed beneath a low
angle fault south of Nickel Creek. Note
the alignment of inclusions in the
sheared zone above the 15 cm scale.
Attenuation of beds that dip left at the
top occurred prior to development of the
shear zone.

suffered extreme attenuation.
Numerous northeast-striking, moderately to steeply dipping normal faults
truncate all other structures within the
shear zone. Most of these contain
calcite veins. Ten fault surfaces have
striations that plunge at moderate to
high angles to the northwest or
southeast, indicating northwest-southeast
extension. Where a fault juxtaposes fold
sets against attenuated turbidite beds,
it may appear to have contrary sense of
relative motion along its trace.
Northeast-striking faults that truncate
the entire shear zone and underlying
massive sandstone are very likely
Quaternary. Faults with similar
orientations that are confined to the
shear zone could be young, but could as
well have resulted from progressive
deformation and rotation of earlier
formed structures during one deformation
event.

FIGURE 34 Foliation within the Enderts Beach shear zone north of Nickel Creek. Primary sand laminae that dip to the right merge with horizontal shear surfaces. These are defined by necked regions of laminae, show sinistral offset and are parallel the base of the shear zone. Scale bar is 3 cm.

Deformational setting

The depositional basin in which the Crescent City olistostrome was emplaced may have been either a trench or trench slope basin (Aalto, 1986b). Results from this study suggest that layer-parallel coaxial extension is the dominant structural style recorded in the early deformation of these rocks. Should they have accumulated in a trench slope basin, this deformation might reflect the gravitational collapse of sediments blanketing the accretionary prism. This collapse may have resulted either from downslope movement of the slope cover (Cowan, 1982), or prism expansion by continued underplating (Aalto, 1986b; Platt, 1986). In the Cowan model, the formation of structures that postdate early extension, the large folds at Point St. George and the Enderts Beach shear zone, might have resulted from later internal shortening of the accretionary wedge. Platt (1986) has suggested that such shortening resulted from the lateral emplacement of nappes of older, more elevated Franciscan rocks westward over younger, more recently accreted Franciscan.

Alternatively, the olistostrome and bounding turbidites might have first accumulated in a trench, been subducted and then suffered layer-parallel extension due to subvertical loading beneath the decollement that underlies the toe of the inner trench slope (Byrne et al., 1987). Such extension would occur prior to underplating of the subducting sediments. In this model, the Point St. George folds and Enderts Beach shear zone might have developed when these rocks were accreted by underplating as a duplex (Silver et al., 1985). Duplex accretion is compatible with seismic observations of complex structures beneath the inner trench slope of the modern Aleutian accretionary prism (McCarthy and Scholl, 1985).

Although structural study provides little evidence to favor one of these models over the other, Franciscan rocks of the Crescent City area lack the pervasive pressure solution cleavage that researchers of other subduction complexes have related to a history of underplating (eg. Byrne, 1984; Fisher and Byrne, 1987; Moore, 1987). This alone favors a trench slope basin setting for the emplacement of the olistostrome.

Acknowledgments

Much of this study was funded by grants from the National Science Foundation (#EAR 7809957 and EAR 8418137) and the Humboldt State University Foundation. I thank Dr. Angela Jayko for reviewing the manuscript and the Pacific Section, Society of Economic Paleontologists and Mineralogists, for permission to use material from my previously published papers (Aalto, 1986b; Aalto and Murphy, 1984).

REFERENCES CITED

Aalto, K.R, Sedimentology of a melange: Franciscan of Trinidad, California, *Journal of Sedimentary Petrology*, 46, 913-929, 1976.
Aalto, K.R., Multistage melange formation in the

Franciscan Complex, northernmost California, *Geology*, 9, 602-607, 1981.

Aalto, K.R., The Franciscan Complex of northernmost California: sedimentation and tectonics, in *Trench-Forearc Geology*, edited by J.K. Leggett, *Geological Society of London Special Publication* 10, 419-432, 1982.

Aalto, K.R., Franciscan Complex geology of the Pilot Creek quadrangle, northern California, *Geological Society of America Abstracts with Programs*, 15, 275, 1983.

Aalto, K.R., Depositional sequence of argillite, diamictite, hyaloclastite, and lava flows within the Franciscan Complex, northern California, *Journal of Geology*, 94, 744-752, 1986a.

Aalto, K.R., Structural geology of the Franciscan Complex of the Crescent City area, northern California, in *Cretaceous Stratigraphy-Western North America*, edited by P. Abbott, *Pacific Section, Society of Economic Paleontologists and Mineralogists Book* 46, 197-209, 1986b.

Aalto, K.R., An accreted slab in Franciscan Complex, Redwood National Park, Del Norte County, California, *California Geology*, 40, 31-37, 1987a.

Aalto, K.R., Sandstone petrology and tectono-stratigraphic terranes of the NW California and SW Oregon Coast Ranges, *Geological Society of America Abstracts with Programs*, 19, 565, 1987b.

Aalto, K.R., and R.H. Dott, Jr., Late Mesozoic conglomeratic flysch in southwestern Oregon, and the problem of transport of coarse gravel in deep water, in *Flysch Sedimentology in North America*, edited by J. Lajoie, *Geological Association of Canada Special Paper* 7, 53-65, 1970.

Aalto, K.R., and G.D. Harper, Geology of the Coast Ranges in the Klamath and Ship Mountain quadrangles, Del Norte County, California, *California Division of Mines and Geology Open File Map* QFR 82-16-SF, scale 1:62,500, 1982.

Aalto, K.R., and J.M. Murphy, Franciscan Complex geology of the Crescent City area, northern California, in *Franciscan Geology of Northern California*, edited by M.C. Blake, Jr., *Pacific Section, Society of Economic Paleontologists and Mineralogists Book* 43, 185-201, 1984.

Bachman, S.B., The Coastal Belt of the Franciscan: youngest phase of northern California subduction, in *Trench-Forearc Geology*, edited by J.K. Leggett, *Geological Society of London Special Publication* 10, 401-418, 1982.

Back, W., Geology and ground water features of the Smith River plain, Del Norte County, California, *U.S. Geological Survey Water Supply Paper* 1254, 1957.

Bailey, E.H., On-land Mesozoic ocean crust in California Coast Ranges, *U.S. Geological Survey Professional Paper* 700C, C70-C81, 1970.

Bailey, E.H., and M.C. Blake, Jr., Tectonic development of western California during the Late Mesozoic (in Russian), *Geotektonika*, pt. 3, 17-30, pt. 4, 24-34, 1969.

Bailey, E.H., W.P. Irwin, and D.L. Jones, Franciscan and related rocks, and their significance in the geology of western California, *California Division of Mines and Geology Bulletin* 183, 1964.

Blake, M.C., Jr., W.P. Irwin, and R.G. Coleman, Upside-down metamorphic zonation, blueschist facies, along a regional thrust in California and Oregon, in *U.S. Geological Survey Research, 1967, U.S. Geological Survey Professional Paper* 575-C, C1-C9, 1967.

Blake, M.C., Jr., and D.L. Jones, Origin of Franciscan melanges in northern California, in *Modern and Ancient Geosynclinal Sedimentation*, edited by R.H. Dott, Jr. and R.H. Shaver, *Society of Economic Paleontologists and Mineralogists Special Publication* 19, 345-357, 1974.

Blake, M.C., Jr., and D.L. Jones, Plate tectonic history of the Yolla Bolly junction, northern California, *Geological Society of America Cordilleran Section Guidebook*, 1977.

Blake, M.C., Jr., A.S. Jayko, and D.G. Howell, Geology of a subduction complex in the Franciscan assemblage of northern California, *Oceanologic Acta*, 12, 267-272, 1981.

Blake, M.C., Jr., D.G. Howell, and D.L. Jones, Tectonostratigraphic terrane map of California, *U.S. Geological Survey Open-File Report* 82-593, 1:750,000, 1982.

Blake, M.C., Jr., A.S. Jayko, and D.G. Howell, Sedimentation, metamorphism and tectonic accretion of the Franciscan assemblage of northern California, in *Trench-Forearc Geology*, edited by J.K. Leggett, *Geological Society of London Special Publication* 10, 433-438, 1982.

Blake, M.C., Jr., and A.S. Jayko, Geologic map of the Yolla Bolly Middle Eel Wilderness and adjacent roadless areas, *U.S. Geological Survey Miscellaneous Field Studies Map* MF-1595-A, 1:62,500, 1983.

Blake, M.C., Jr., D.G. Howell, and A.S. Jayko, Tectonostratigraphic terranes of the San Francisco Bay region, in *Franciscan Geology of Northern California*, edited by M.C. Blake, Jr., *Pacific Section, Society of Economic Paleontologists and Mineralogists Book* 43, 5-22, 1984.

Blake, M.C., Jr., D.C. Engebretson, A.S. Jayko, and D.L. Jones, Tectonostratigraphic terranes in southwest Oregon, in *Tectonostratigraphic Terranes of the Circum-Pacific Region*, edited by D.G. Howell, *Circum-Pacific Council for Energy and Mineral Resources, Earth Science Series*, 1, 147-157, 1985a.

Blake, M.C., Jr., A.S. Jayko, and R.J. McLaughlin, The Central Belt Franciscan: an oblique transform melange?, *Geological Society of America Abstracts with Programs*, 17, 342, 1985b.

Bosworth, W., The relative roles of boudinage and "structural slicing" in the disruption of layered rock sequences, *Journal of Geology*, 93, 447-456, 1984.

Bourgeois, J., and R.H. Dott, Jr., Stratigraphy and sedimentology of Upper Cretaceous rocks in coastal southwest Oregon: evidence for wrench-fault tectonics in a postulated accretionary terrane, *Geological Society of America Bulletin*, 96, 1007-1019, 1985.

Byrne, T., Early deformation in melange terranes of the Ghost Rocks Formation, Kodiak Islands, Alaska, in *Melanges: their Nature, Origin and Significance*, edited by L.A. Raymond,

Geological Society of America Special Paper 198, 21-52, 1984.

Byrne, T., D. Fisher, and L. DiTullio, Deformation regimes in accretionary prisms: evidence from SW Japan and SW Alaska, *Geological Society of America Abstracts with Programs*, 19, 608, 1987.

Carver, G.A., Late Cenozoic tectonics of the Eel River basin region, coastal northern California, in *Tectonics, Sedimentation and Evolution of the Eel River and Other Coastal Basins of Northern California*, edited by H. Schymiczek and R. Suchsland, *Pacific Section, American Association of Petroleum Geologists Miscellaneous Publication 37*, 61-72, 1987.

Cashman, S.M., P.H. Cashman, and J.D. Longshore, Deformational history and regional tectonic significance of the Redwood Creek schist, northwestern California, *Geological Society of America Bulletin*, 97, 35-47, 1986.

Cloos, M., Flow melanges: numerical modeling and geologic constraints on their origin in the Franciscan subduction complex, *Geological Society of America Bulletin*, 93, 330-345, 1982.

Cowan, D.S., Deformation and metamorphism of the Franciscan subduction zone complex northwest of Pacheco Pass, California, *Geological Society of America Bulletin*, 85, 1623-1634, 1974.

Cowan, D.S., Origin of blueschist-bearing chaotic rocks in the Franciscan Complex, San Simeon, California, *Geological Society of America Bulletin*, 89, 1415-1423, 1978.

Cowan, D.S., Deformation of partly dewatered sediments near Piedras Blancas Point, California, in *Trench-Forearc Geology*, edited by J.K. Leggett, *Geological Society of London Special Publication* 10, 439-457, 1982.

Cowan, D.S., Structural styles in Mesozoic and Cenozoic melanges in the western Cordillera of North America, *Geological Society of America Bulletin*, 96, 451-462, 1985.

Dickinson, W.R., Interpreting detrital modes of graywacke and arkose, *Journal of Sedimentary Petrology*, 40, 695-707, 1970a.

Dickinson, W.R., Relations of andesite, granites, and derivative sandstones to arc-trench tectonics, *Review of Geophysics and Space Physics*, 8, 813-860, 1970b.

Dickinson, W.R., Interpreting provenance relations from detrital modes of sandstones, in *Provenance of Arenites*, edited by G.G. Zuffa, *NATO Advanced Science Institutes Series C*, 148, 333-362, 1984.

Dickinson, W.R., and E.I. Rich, Petrologic intervals and petrofacies in the Great Valley Sequence, Sacramento Valley, California, *Geological Society of America Bulletin*, 83, 3007-3024, 1972.

Dickinson, W.R., R.V. Ingersoll, and S.A. Graham, Paleogene sediment dispersal and paleotectonics in northern California, *Geological Society of America Bulletin*, 90, pt. I, 897-898, pt. II, 1458-1528, 1979.

Dickinson, W.R., R.V. Ingersoll, D.S. Cowan, K.P. Helmold, and C.A. Suczek, Provenance of Franciscan graywackes in coastal California, *Geological Society of America Bulletin*, 93, 95-107, 1982.

Dott, R.H., Jr., Geology of the southwestern Oregon coast west of the 124th meridian, *Oregon Department of Geology and Mineral Industries Bulletin* 69, 1971.

Ernst, W.G., Tectonic contact between the Franciscan and Great Valley sequence: crustal expression of a late Mesozoic Benioff zone, *Journal of Geophysical Research*, 75, 886-901, 1970.

Fisher, D., and T. Byrne, Structural evolution of underthrusted sediments, Kodiak Islands, Alaska, *Tectonics*, 6, 775-793, 1987.

Folk, R.L., *Petrology of Sedimentary Rocks*, Hemphill Publishing Company, Austin, Texas, 1974.

Hamilton, W., Mesozoic California and the underflow of the Pacific mantle, *Geological Society of America Bulletin*, 80, 2409-2430, 1969.

Hansen, E. *Strain Facies*, Springer-Verlag, New York, 1970.

Hossack, J.R., Pebble deformation and thrusting in the Bygdin area (southern Norway), *Tectonophysics*, 5, 315-339, 1968.

Hsü, K.J., Principles of melanges and their bearing on the Franciscan-Knoxville paradox, *Geological Society of America Bulletin*, 79, 1063-1074, 1968.

Hsü, K.J., Melanges and their distinction from olistostromes, in *Modern and Ancient Geosynclinal Sedimentation*, edited by R.H. Dott, Jr., and R.H. Shaver, *Society of Economic Paleontologists and Mineralogists Special Publication* 19, 321-333, 1974.

Hsu, T.C., The characteristics of coaxial and noncoaxial strain paths, *Journal of Strain Analysis*, 1, 216-222, 1966.

Ingersoll, R.V., Petrofacies and petrologic evolution of the Late Cretaceous fore-arc basin, northern and central California, *Journal of Geology*, 86, 335-353, 1978.

Ingersoll, R.V., Evolution of the Late Cretaceous fore-arc basin, northern and central California, *Geological Society of America Bulletin*, 90, 813-826, 1979.

Ingersoll, R.V., Evolution of the Great Valley forearc basin of California, in *Trench-Forearc Geology*, edited by J.K. Leggett, *Geological Society of London Special Publication* 10, 459-467, 1982.

Ingersoll, R.V., T.F. Bullard, R.L. Ford, J.P. Grimm, J.D. Pickle, and S.W. Sares, The effect of grain size on detrital modes: a test of the Gazzi-Dickinson point-counting method, *Journal of Sedimentary Petrology*, 54, 103-116, 1984.

Irwin, W.P., Geologic reconnaissance of the northern Coast Ranges and Klamath Mountains, California, *California Division of Mines Bulletin* 179, 1960.

Irwin, W.P., E.W. Wolfe, M.C. Blake, Jr., and C.G. Cunningham, Jr., Geologic map of the Pickett Peak quadrangle, Trinity County, California, *U.S. Geological Survey Quadrangle Map* GQ-1111, 1:62,500, 1974.

Jayko, A.S., and M.C. Blake, Jr., Geologic terranes of coastal northern California and southern Oregon, in *Tectonics, Sedimentation and Evolution of the Eel River and Other Coastal Basins of Northern California*, edited by H. Schymiczek and R. Suchsland, *Pacific*

Section, *American Association of Petroleum Geologists Miscellaneous Publication* 37, 1-12, 1987.

Kelsey, H.M., and G.A. Carver, Late Neogene and Quaternary tectonics associated with the northward growth of the San Andreas transform, northern California *Journal of Geophysical Research*, 93, in press, 1988.

Kelsey, H.M., and D.K. Hagans, Major right-lateral faulting in the Franciscan assemblage of northern California in Late Tertiary time, *Geology*, 10, 387-391, 1982.

Koch, J.G., Late Mesozoic stratigraphy and tectonic history, Port Orford-Gold Beach area, southwestern Oregon coast, *American Association of Petroleum Geologists Bulletin*, 50, 25-71, 1966.

Kubler, B., Crystallinity of illite. Detection of metamorphism in some frontal parts of the Alps, *Fortschrift der Mineralogie*, 47, 39-40, 1969.

Lowe, D.R., Sediment gravity flows II: depositional models with special reference to deposits of high-density turbidity currents, *Journal of Sedimentary Petrology*, 52, 279-298,1982.

Lucas, S.E., and J.C. Moore, Cataclastic deformation in accretionary wedges: Deep Sea Drilling Project Leg 66, southern Mexico, and on-land examples from Barbados and Kodiak Islands, in *Structural Fabric in Deep Sea Drilling Project Cores from Forearcs*, edited by J.C. Moore, *Geological Society of America Memior* 166, 89-104, 1986.

Massari, F., Resedimented conglomerates of a Miocene fan-delta complex, Southern Alps, Italy, in *Sedimentology of Gravels and Conglomerates*, edited by E.H. Koster and R.J. Steel, *Canadian Society of Petroleum Geologists Memoir* 10, 259-278, 1984.

McCarthy, J., and D.W. Scholl, Mechanisms of subduction accretion along the central Aleutian Trench, *Geological Society of America Bulletin*, 96, 691-701, 1985.

McLaughlin, R.J., and H.N. Ohlin, Tectono-stratigraphic terranes of the Geysers-Clear Lake region, California, in *Franciscan Geology of Northern California*, edited by M.C. Blake, Jr., *Pacific Section, Society of Economic Paleontologists and Mineralogists Book* 43, 221-254, 1984.

Maxwell, J.C., Anatomy of an orogen, *Geological Society of America Bulletin*, 85, 1195-1204, 1974.McLaughlin, R.J., and H.N. Ohlin, Tectonostratigraphic terranes of the Geysers-Clear Lake region, California, in *Franciscan Geology of Northern California*, edited by M.C. Blake, Jr., *Pacific Section, Society of Economic Paleontologists and Mineralogists Book* 43, 221-254, 1984.

Miller, W., III, Discovery of trace fossils in Franciscan turbidites, *Geology*, 14, 343-345, 1986a.

Miller, W., III, New species of *Bathysiphon*, (Foraminiferida: Textulariina) from Franciscan flysch deposits, northernmost California, *Tulane Studies in Geology and Paleontology*, 19, 91-94, 1986b.

Monsen, S.A., and K.R. Aalto, Petrology, structure and regional tectonics: South Fork Mountain Schist of Pine Ridge Summit, northern California, *Geological Society of America Bulletin*, 91, 369-373, 1980.

Moore, J.C., Accretionary prisms: symbiosis of structural and fluid evolution, *Geological Society of America Abstracts with Programs*, 19, 777, 1987.

Moore, J.C., D.S. Cowan, and D.E. Karig, Penrose Conference report: structural styles and deformation fabrics of accretionary complexes, *Geology*, 13, 77-78, 1985.

Murphy, J.M., Franciscan Complex geology at Point St. George, California, B.Sc. thesis, Humboldt State University, Arcata, California, 1982.

Murphy, J.M., and K.R. Aalto, Flow melange: numerical modeling and geologic constraints on their origin in the Franciscan subduction complex, California: discussion, *Geological Society of America Bulletin*, 94, 1241-1243, 1983.

Mutti, E., and F. Ricci Lucchi, Turbidites of the northern Appennines: introduction to facies analysis, *International Geology Review*, 20, 125-166, 1978.

Page, B.M., Franciscan melanges compared with olistostromes of Taiwan and Italy, *Tectonophysics*, 47, 223-246, 1978.

Platt, J.P., Dynamics of orogenic wedges and uplift of high-pressure metamorphic rocks, *Geological Society of America Bulletin*, 97, 1037-1053, 1986.

Ramsay, J.G., *Folding and Fracturing of Rocks*, McGraw Hill, New York, 1967.

Ramsay, J.G., Shear zone geometry: a review, *Journal of Structural Geology*, 2, 83-99, 1980.

Ramsay, J.G., and M.I. Huber, *The Techniques of Modern Structural Geology: Strain Analysis*, Academic Press, New York, 1983.

Raymond, L.A., Classification of melanges, in *Melanges: their Nature, Origin and Significance*, edited by L.A. Raymond, *Geological Society of America Special Paper*, 198, 7-20, 1984.

Sassi, F.P., and A. Scolari, The b_o value of the potassic white micas as a barometric indicator in low-grade metamorphism of pelitic schists, *Contributions to Mineralogy and Petrology*, 45, 143-152, 1974.

Silver, E.A., and E.C. Beutner, Penrose Conference report on melanges, *Geology*, 8, 32-34, 1980.

Silver, E.A., M.J. Ellis, N.A. Breen, and T.H. Shipley, Comments on the growth of accretionary wedges, *Geology*, 13, 6-9, 1985.

Suppe, J., Geology of the Leech Lake Mountain-Ball Mountain region, *University of California Publications in the Geological Sciences*, 107, 1-81, 1973.

Widmier, J.M., Mesozoic stratigraphy of the west-central Klamath Province: a study of eugeosynclinal sedimentation, unpublished PhD dissertation, University of Wisconsin-Madison, 1962.

Young, J.C., Geology of the Willow Creek quadrangle, Humboldt and Trinity Counties, California, *California Division of Mines and Geology Map Sheet* 31, 1:62,500, 1978.

Young, S.W., Petrographic textures of detrital polycrystalline quartz as an aid to interpreting crystalline source rocks, *Journal of Sedimentary Petrology*, 46, 595-603, 1976.

PALEONTOLOGY OF FRANCISCAN FLYSCH AT
POINT SAINT GEORGE, NORTHERN CALIFORNIA

William Miller, III,
Department of Geology and Marine Laboratory,
Humboldt State University, Arcata, California

INTRODUCTION

Other than a few occurrences of ammonoids and bivalve mollusks (see Bailey *et al.*, 1964; Blake and Jones, 1974), flysch sequences within the Late Mesozoic Franciscan Complex are usually considered to be unfossiliferous. A recent exploration of coastal outcrops in northern California, however, has yielded a surprisingly varied and locally abundant assemblage of trace fossils along with peculiar tubular body fossils. These fossils represent a previously undescribed indigenous benthic fauna of Franciscan submarine fan and fan-margin environments. The biostratigraphically important bivalves and ammonoids probably were derived from up-slope settings and from upper (not necessarily local) levels of the oceanic water column, respectively. The autochthonous (in-place to slightly disturbed) benthic fossils have potential in the reconstruction of original depositional environments and in the evaluation of paleobathymetric schemes based on sedimentologic features and morphotectonic scenarios. For the paleontologist the newly discovered fossils provide an important glimpse of life at the seafloor within a Cretaceous submarine fan-trench system.

The purpose of this section is to provide a description of Franciscan megafossils and a brief evaluation of their possible significance, focusing on the locality that has yielded the greatest variety of ichnofossils and the largest number of tube fossils. The locality is a sequence of fine-grained, thin-bedded turbidites (Facies D; terminology of Mutti and Ricci Lucchi, 1978) within the Yolla Bolly terrane at Point Saint George (Figure 1). Sedimentology and structural development of Franciscan rocks in this area, including the Facies D sequence, are described in detail by Aalto and Murphy (1984) and Aalto (this volume). My contribution is an annotated inventory of fossils intended to supplement the geologic account of the Point Saint George exposures.

FRANCISCAN FOSSIL RECORD: AN OVERVIEW

Franciscan fossils receiving the most recent attention are planktonic radiolarians and foraminiferids. These microfossil groups have been crucial in establishing the ages and paleogeographic

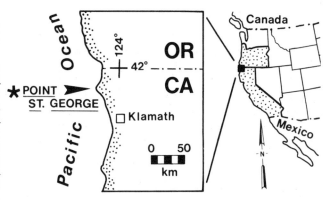

FIGURE 1. Location of Point Saint George in northern California (see Aalto, this volume, for geologic map).

relations of Franciscan terranes (e.g., Murchey and Jones, 1984; Tarduno *et al.*, 1986). They occur in tectonically incorporated blocks of chert and limestone that originated as pelagic oozes in low-latitude areas of high biotic productivity or carbonate deposition, and were then rafted to the western edge of North America on oceanic plate. A few vertebrate remains (pelagic deadfalls?) also have been reported from these open-ocean sediments (e.g., Camp, 1942).

In contrast, the flysch sediments and fossils accumulated in near-continent submarine fan and trench basin settings. Megafossils in the turbidites can be divided into four categories based on their origin and mode of preservation: 1) ichnofossils, mostly the burrows of deposit-feeding infauna and the regularly patterned biogenic structures of uncertain function known as "graphoglyptids"; 2) siliceous, matrix-filled tubes resembling collapsed drinking straws, which are the recrystallized tests of giant foraminiferids; 3) carbonized plant detritus, with concentrations of large fragments apparently restricted to Facies B deposits; and 4) the presumably transported <u>Inoceramus</u> and <u>Buchia</u> bivalves and ammonoids. The trace fossils and giant foraminiferids are the only unequivocally autochthonous remains of benthic communities that inhabited fan lobes and adjacent basin floors.

Burrows are widespread and locally abundant in both coarse-grained ("proximal") and fine-grained ("distal") Franciscan turbidites in northern

TABLE 1. Trace fossils from Facies D sequence at Point Saint George.

ICHNOTAXON	MORPHOLOGY	SUBSTRATE NICHE[1]	ABUNDANCE[2]
Graphoglyptids (specialized search patterns, traps, and microbial "farms"):			
Megagrapton aequale Seilacher	irregular nets	shallow predepositional, horizontal	common
M. irregulare Ksiazkiewicz	irreguler nets	shallow predepositional, horizontal	common
Lorenzinia apenninica DeGabelli	radial tubes	shallow predepositional, tangential	common
?Glockeria ichnsp.	radial tubes	shallow predepositional, tangential	rare
Desmograpton cf. geometricum Seilacher	biramous meanders	shallow predepositional; horizontal	rare?
?Belorhaphe ichnsp.	uniramous meanders	shallow predepositional; horizontal	rare
Helminthorhaphe crassa (Schafhäutl)	continuous meanders	shallow predepositional; horizontal	rare
Paleodictyon (Squamodictyon) petaloideum Seilacher	regular nets	shallow predepositional; horizontal	rare
Pascichnia (infaunal grazing systems):			
Planolites ichnsp.	slightly curved, unbranched tubes	moderate to deep postdepositional; oblique to parallel	abundant
Helminthopsis ichnsp. A (large clay-filled form)	sinuous tubes	shallow (?) postdepositional; oblique	common
Helminthopsis ichnsp. B (smaller sand-casted form)	sinuous tubes	shallow predepositional; parallel	rare
Neonereites ichnsp.	sinuous rows of irregular beads	shallow (?) postdepositional; parallel	common
Clay-filled burrows (large form)	wide vertical tubes	moderately deep postdepositional; vertical to oblique	common
Clay-filled burrows (small form)	narrow vertical tubes	moderately deep postdepositional; vertical to oblique	rare?
Fodinichnia (semi-permanent deposit-feeding structures):			
Phycosiphon ichnsp.	branching U-shaped spreiten	shallow to moderately deep postdepositional; oblique	abundant
Chondrites ichnsp.	branching root-like tubes	moderately deep pre- and postdepositional; vertical to tangential	common
Micatuba ichnsp.	vertical shaft with radial tubes	shallow to moderately deep predepositional; vertical to horizontal	rare
Taenidium serpentinum Heer	solitary meniscate tubes	moderately deep postdepositional; horizontal	rare?
Domichnia (dwelling structures):			
Conichus-like pits containing Bathysiphon tests	small cones	shallow predepositional; vertical	abundant
?Conichus ichnsp.	large cones	shallow predepositional; vertical	rare
Fugichnion (escape structure):			
Disturbed stratification with a central structureless zone	cones surrounded by downturned laminae	moderately deep postdepositional; vertical	rare?

1 - Orientation term is relative to sedimentary bedding; predepositional = pre-turbidite structure

2 - After ~ 40 hr of intensive collecting at the site: abundant, > 10 specimens; common, 3 - 9 specimens; rare, 1 or 2 specimens at most

California. The greatest diversity of biogenic structures has been found in the latter type of desposit (Miller, 1986a). The assemblage of deposit-feeder burrows and graphoglyptids is characteristic of the Nereites ichnofacies, a distinctive grouping of trace fossils found in Phanerozoic flysch and pelagic sediments deposited at bathyal to abyssal depths (Seilacher, 1967; Ekdale et al., 1984; Frey and Pemberton, 1985; cf. Byers, 1982). Ichnofossil diversity trends across submarine fan environments are related to variations in trophic resources, O_2 tension in bottom water and sediments, recruitment dynamics of benthos, and especially to the nature, frequency and intensity of natural disturbances. The pattern is clearly shown in the Franciscan by the zonation of both the entire trace fossil assemblage as well as the diversity changes within ethologic groups. For example, only one type of graphoglyptid, Megagrapton, has been found in Facies B within the Yolla Bolly and Central terranes, but at least eight different graphoglyptids occur in the Facies D sequence at Point Saint George (Table 1). Such interstratal zonation of biogenic structures deserves more study as it is almost certainly a reflection of diversity variation along deep-sea environmental gradients.

Occurring together with the ichnofossils in Facies D beds are straight to slightly curved, unbranched, siliceous tubes. These have been called "Terebellina" by some workers, usually considered to be either annelid body fossils (Howe, 1962) or grain-lined burrows (Chamberlain, 1978). Based on wall microstructure and a comparison with modern abyssobenthic organisms, these tubes are now interpreted as recrystallized tests of the giant foraminiferid, Bathysiphon (Miller, 1986b; Miller, 1988; Miller, in press). At least two species occur in the Yolla Bolly terrane: a large, straight-sided form 100 mm long (B. aaltoi) and a smaller curved, tapering form up to 45 mm long (B. cashmanae). Similar "worm tubes" occur in Mesozoic to Cenozoic basinal deposits throughout the Pacific borderlands and many of these also could prove to be Bathysiphon tests (see Moore, 1988; Miller, in press).

POINT SAINT GEORGE LOCALITY

The Facies D sequence at Point Saint George is perhaps the most important Franciscan fossil locality in northern California. This is because of the variety of well-preserved trace fossils and abundance of Bathysiphon tests found here -- making it one of the only "windows" on a Cretaceous deep-sea ecosystem in western North America.

Biogenic structures are preserved as casts on the soles of turbidite beds (hyporelief), as well as within sandstone and mudstone beds (full relief). The most abundant structures are the deposit-feeder spreiten systems known as Phycosiphon (Figure 2J) and infaunal grazing/locomotion burrows assignable to the ubiquitous ichnogenus Planolites (Figure 2L). Other common structures produced by deposit feeders are the root-like systems of Chondrites (Figure 2G); two forms of the large, sinuous burrow Helminthopsis (Figure 2C, H); and rows of irregular, bead-like bodies known as Neonereites (Figure 2F). Rare burrows include Taenidium (Figure 2B) and Micatuba (Figure 2K). Dwelling structures are large, sand-casted depressions tentatively identified as ?Conichus (Figure 2I) and smaller "wallow pits" containing in situ segments of Bathysiphon (Figure 3B). (The latter is the only example of an ichnofossil-body fossil association known from Franciscan rocks.) Rare escape structures (Figure 2A) were made when organisms in the predepositional community attempted to move through a blanket of newly deposited turbidite sand.

One intriguing aspect of the trace fossil assemblage is the richness of graphoglyptids (Table 1). These structures consist of predepositional burrow systems organized in nets, uniramous and biramous meanders, continuous meanders, and radial tubes. Graphoglyptids have long puzzled ichnologists because they do not appear to have been constructed by deposit-feeders or used merely as domiciles. Instead, they may be the specialized search traces and subsurface traps constructed by small, motile predators, or a kind of microbial "farm" in which the inhabitant "cultivated" microorganisms for food (see

FIGURE 2 (Figure on next page). Trace fossils from Facies D sequence (ruler in A, C, E, F, G, J, M and O graduated in mm's; bar scale in B, D, H, K and N=5 mm, in L=10 mm, and in I=20 mm). A, escape structure in cross-stratified sandstone; B, Taenidium serpentinum (hyporelief); C, Helminthopsis ichnsp. B (hyporelief); D, Desmograpton cf. geometricum (hyporelief); E, Helminthorhaphe crassa (hyporelief); F, Neonereites ichnsp. (full relief); G, Chondrites ichnsp. (full relief); H, Helminthopsis ichnsp. A (full relief); I, ?Conichus ichnsp. (hyporelief); J, Phycosiphon ichnsp. (full relief); K, Micatuba ichnsp. (S = central shaft, B = one of several radiating burrows; hyporelief); L = Planolites ichnsp. (hyporelief?); M, Paleodictyon (Squamodictyon) petaloideum (hyporelief); N, cluster of several Lorenzinia apenninica (hyporelief); O, large specimen of Megagrapton aequale (hyporelief).

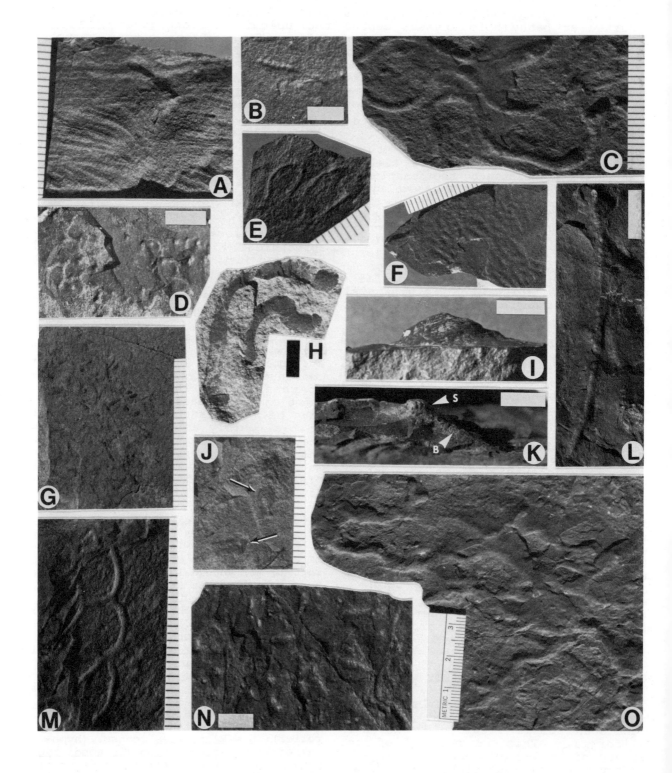

FIGURE 2. Caption on previous page

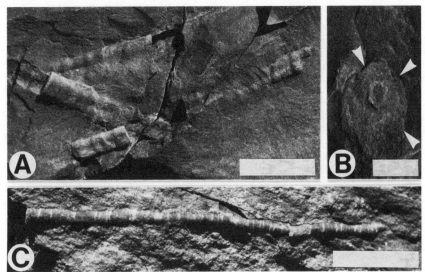

FIGURE 3. <u>Bathysiphon aaltoi</u> (bar scale in A and C = 10 mm, in B = 5 mm). <u>A</u>, cluster of long segments in mudstone (Bouma T_e division); <u>B</u>, test fragment in life orientation within sand-casted pit, exposed on sole of turbidite bed; <u>C</u>, a nearly complete test in silty sandstone turbidite.

Seilacher, 1977). Graphoglyptids have been observed in modern pelagic ooze, but the organisms responsible for the traces were not found in the structures (Ekdale, 1980). Swinbanks (1982) recently pointed out the similarities between graphoglyptids, notably <u>Paleodictyon</u>, and the tests of modern, deep-sea xenophyophore protozoans. Although their function and biologic affinities remain obscure, they are well known in flysch deposits and generally are taken to indicate food poor, bathyal or deeper marine environments. The most common graphoglyptid at Point Saint George is <u>Megagrapton</u> (Figure 2O). Other forms, representing the varied morphologies exhibited by the group, are <u>Lorenzinia</u> (Figure 2N), <u>Helminthorhaphe</u> (Figure 2E), <u>Desmograpton</u> (Figure 2D), and, the most distinctive of all the graphoglyptids, <u>Paleodictyon</u> (Figure 2M).

Perhaps the most unusual fossil found at the site is <u>Bathysiphon aaltoi</u>, named in honor of our field trip leader. <u>Bathysiphon</u> tests occur in varied states of preservation: small, barrel-shaped segments in the turbidite sandstone beds; segments 20-30 mm long in the mudstone beds (Figure 3A); immature ends of tests still in life orientation in small, sand-casted pits on the soles of some of the sandstone beds (Figure 3B); and nearly complete individuals, > 50 mm long, oriented oblique or parallel to bedding in parts of the sequence consisting of thick mudstone beds with thin, discontinuous, "fading-ripple" siltstone laminae (Figure 3C). Paleoecology of Franciscan <u>Bathysiphon</u> is the subject of another paper (Miller, in press).

SIGNIFICANCE

(1) Recognition of the <u>Nereites</u> ichnofacies in Franciscan flysch would appear to confirm the bathyal to abyssal origin of these deposits, and may point to lower trench slope or even trench floor areas as the most likely depositional environments.

(2) Presence of large numbers of postdepositional <u>Phycosiphon</u> and <u>Chondrites</u> may suggest episodic deposition of organic-rich sediments in these settings (see Vossler and Pemberton, 1988). Graphoglyptids probably represent the long background periods of relative stability when resident organisms had to eke out a living from the normally meager trophic resources.

(3) Large tests of <u>Bathysiphon</u> (largest specimen was a <u>broken</u> tube 98 mm long) probably are the result of long intervals characterized by slow metabolism and uninterrupted growth ($10 - 10^2$ yr ?) that were punctuated by turbidity currents. Could these organisms be used to estimate the frequency of such deep-sea disturbances?

(4) Because <u>Bathysiphon</u> probably occurs throughout the Pacific Rim in post-Paleozoic basinal deposits there is the potential for basing time-correlations on the species belonging to this genus. However, the variable preservation and anatomical simplicity of tubes could prevent definitive comparisons.

(5) Trace fossils and foraminiferids taken together represent a complex and diverse benthic community that was subjected to regular, small-scale biogenic and physical disturbances, as well as to infrequent, catastrophic disturbances associated with

turbidity currents. The first type (or scale) of disturbance actually may have promoted diversity (e.g., Grassle and Morse-Porteous, 1987), while the second type probably eliminated the predepositional community over large areas of fan surface or basin floor. Changes in disturbance regime, in space and time, could explain the zonation of trace fossils commonly observed in flysch sequences.

Acknowledgments

I wish to thank Dr. K. R. Aalto for showing me the Point Saint George locality and for editorial directions. Ms. Camellia Armstrong typed the manuscript. Acknowledgment is made to the Petroleum Research Fund, of the American Chemical Society, for support of this research.

REFERENCES

Aalto, K. R., and J. M. Murphy, Franciscan Complex geology of the Crescent City area, northern California, in Franciscan Geology of Northern California, edited by M. C. Blake, Jr., Pacific Section Society of Economic Paleontologists and Mineralogists, 43, 185-201, 1984.

Bailey, E. H., W. P. Irwin, and D. L. Jones, Franciscan and related rocks, and their significance in the geology of western California, California Division of Mines and Geology, Bull. 193, 1964.

Blake, M. C., Jr., and D. L. Jones, Origin of Franciscan melanges in northern California, in Modern and Ancient Geosynclinal Sedimentation, edited by R. H. Dott, Jr. and R. H. Shaver, Society of Economic Paleontologists and Mineralogists, Special Publication 19, 345-357, 1974.

Byers, C. W., Geological significance of marine biogenic sedimentary structures, in Animal-Sediment Relations, edited by P. L. McCall and M. J. S. Tevesz, p. 221-256, Plenum, New York, 1982.

Camp, C. L., Ichthyosaur rostra from central California, Journal of Paleontology, 16, 362-371, 1942.

Chamberlain, C. K., Recognition of trace fossils in cores, in Trace Fossil Concepts, edited by P. B. Basan, Society of Economic Paleontologists and Mineralogists, Short Course 5, 119-166, 1978.

Ekdale, A. A., Graphoglyptid burrows in modern deep-sea sediment, Science, 207, 304-306, 1980.

Ekdale, A. A., R. G. Bromley, and S. G. Pemberton, Ichnology, Trace Fossils in Sedimentology and Stratigraphy, Society of Economic Paleontologists and Mineralogists, Short Course 15, 1984.

Frey, R. W., and S. G. Pemberton, Biogenic structures in outcrops and cores, Bulletin of Canadian Petroleum Geology, 33, 72-115, 1985.

Grassle, J. F., and L. S. Morse-Porteous, Macrofaunal colonization of disturbed deep-sea environments and the structure of deep-sea benthic communities, Deep-Sea Research, 34, 1911-1950, 1987.

Howe, B. F., Worms, in Treatise on Invertebrate Paleontology, Part W, edited by R. C. Moore, p. W144-W177, Geological Society of America and University of Kansas Press, 1962.

Miller, W., III., Discovery of trace fossils in Franciscan turbidites, Geology, 14, 343-345, 1986a.

Miller, W., III., New species of Bathysiphon (Foraminiferida: Textulariina) from Franciscan flysch deposits, northernmost California, Tulane Studies in Geology and Paleontology, 19, 91-94, 1986b.

Miller, W. III., Giant agglutinated foraminiferids from Franciscan turbidites at Redwood Creek, northwestern California, with the description of a new species of Bathysiphon, Tulane Studies in Geology and Paleontology, 21, p. 81-84, 1988.

Miller, W. III., Giant Bathysiphon (Foraminiferida) from Cretaceous turbidites, northern California, Lethaia, in press.

Moore, P. R., Terebellina - sponge or foraminiferid? A comparison with Makiyama and Bathysiphon, New Zealand Geological Survey Record, 20, 43-50, 1988.

Murchey, B. L., and D. L. Jones, Age and significance of chert in the Franciscan Complex in the San Francisco Bay region, in Franciscan Geology of Northern California, edited by M. C. Blake, Jr., Pacific Section Society of Economic Paleontologists and Mineralogists, 43, 23-30, 1984.

Mutti, E. and F. Ricci Lucchi, Turbidites of the northern Apennines: introduction to facies analysis, International Geology Review, 20, 125-166, 1978.

Seilacher, A., Bathymetry of trace fossils, Marine Geology, 5, 413-428, 1967.

Seilacher, A., Pattern analysis of Paleodictyon and related trace fossils, in Trace Fossils 2, edited by T. P. Crimes and J. C. Harper, p. 289-334, Seel House Press, Liverpool, 1977.

Swinbanks, D. D., Paleodictyon: the traces of infaunal xenophyophores?, Science, 218, 47-49, 1982.

Tarduno, J. A., M. McWilliams, W. V. Sliter, H. E. Cook, M. C. Blake, Jr., and I. Premoli-Silva, Southern hemisphere origin of the Cretaceous Laytonville Limestone of California, Science, 231, 1425-1428, 1986.

Vossler, S. M., and S. G. Pemberton, Superabundant Chondrites: a response to storm buried organic material? Lethaia, 21, 94, 1988.

THE REDWOOD CREEK SCHIST - A KEY TO THE DEFORMATIONAL HISTORY OF THE NORTHERN CALIFORNIA COAST RANGES

S. M. Cashman
Department of Geology, Humboldt State University,Arcata, California

P. H. Cashman
Department of Geological Sciences, University of Nevada - Mackay
School of Mines, Reno, NV, 89557

INTRODUCTION

The Redwood Creek schist, previously referred to as the "Redwood Mountain Outlier of the South Fork Mountain Schist" (Irwin, 1960), is a fault-bounded body of metasedimentary and metavolcanic rock that occurs within the Franciscan Complex of the northern California Coast Ranges. The Redwood Creek schist crops out in an elongate belt, 70 km long and 10 km wide, parallel to the north-northwest regional structural trend. The Redwood Creek schist closely resembles the South Fork Mountain schist, which crops out to the east and southeast along the eastern edge of the Franciscan Complex, and the Colebrooke schist, which crops out to the north in the Coast Range province of southwestern Oregon(Fig. 1). These three metamorphic units have been correlated by numerous workers (for example, Irwin, 1960; Blake et al., 1967; Coleman, 1972; Cashman et al.,1986); they constitute the Pickett Peak terrane of Blake et al. (1982). The metamorphic age of the units is thought to be Late Cretaceous (115-120 m.y.) based on K-Ar and Ar39-Ar40 ages for the South Fork Mountain Schist (Lanphere et al.,1978).

The Redwood Creek schist appears to be a klippe which has been modified on both the east and west by right-lateral strike-slip faults (Cashman et al., 1986). On the west, the Bald Mountain fault separates the schist from blueschist bearing melange of the Central Franciscan belt. On the east, the Grogan fault separates the schist from sedimentary rocks of the Franciscan Yolla Bolly belt; Kelsey and Hagans (1982) have documented 80 km of right lateral offset on this fault.

REGIONAL TECTONIC SIGNIFICANCE

Two fundamentally different tectonic styles - convergence and strike-slip faulting - have been documented along the western boundary of North America during late Mesozoic time. The Franciscan Complex has been interpreted as an accretionary prism formed along an east-dipping subduction zone (Hamilton, 1969; Ernst, 1970; Maxwell, 1974; and many others) between Late Jurassic and Miocene time. Recent paleomagnetic studies have

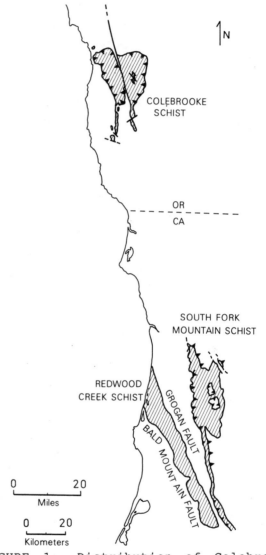

FIGURE 1 Distribution of Colebrooke Schist, South Fork Mountain Schist, and Redwood Creek schist, northern California and southern Oregon.

documented large-scale translation and/or rotation during the same time period in the Franciscan Complex (e.g. Alvarez et al., 1980,) and elsewhere in the western Cordillera. (e.g. Wells and Heller, 1988). In this regional context, the Redwood Creek schist becomes a key to the deformational history of the Franciscan Complex by providing a way to evaluate the relative displacement and timing of thrust

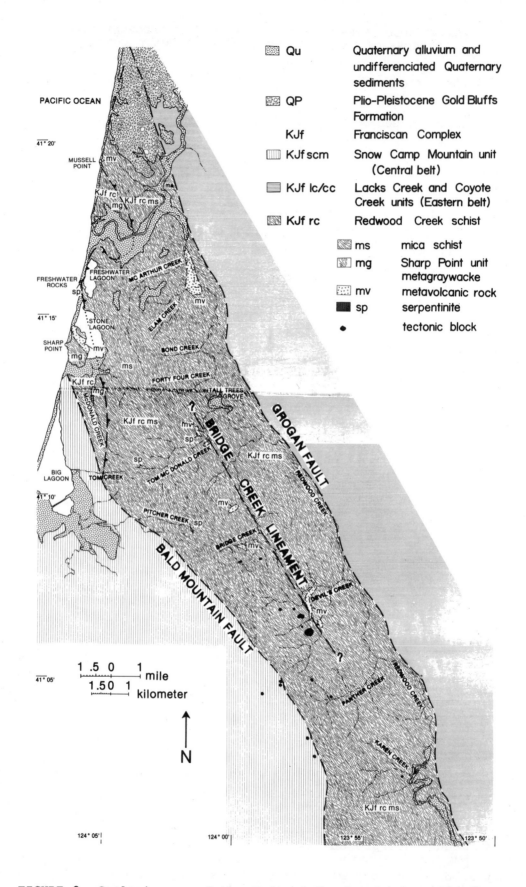

FIGURE 2   Geologic map of the Redwood Creek schist north of latitude 41° 00'.

versus strike-slip faulting. The three correlative schist units - Redwood Creek schist, South Fork Mountain Schist, and Colebrooke Schist - not only record synmetamorphic deformation but also provide a distinctive "marker unit" that can be used to trace postmetamorphic faulting.

LITHOLOGY

The two major rock types in the Redwood Creek schist are clastic metasedimentary rocks and basaltic metavolcanic rocks. Metasedimentary rocks make up the bulk of the Redwood Creek schist and are characteristically light-colored, fissile, fine-grained schist and phyllite. The mineral assemblage of the mica schist is quartz + chlorite + white mica + albite $\pm$ graphitic material $\pm$ lawsonite $\pm$ sphene $\pm$ calcite or aragonite $\pm$ clinozoisite. Other metasedimentary rocks recognized within the schist unit include semischistose metasandstone and foliated, micaceous quartzite. Interbedded with these metasedimentary rocks, there are minor amounts of metavolcanic rock, ranging from massive greenstone to fine-grained foliated and laminated metavolcanic rocks. The mineral assemblage for the several types of metavolcanic rocks is: chlorite + albite + quartz $\pm$ actinolite $\pm$ epidote or lawsonite $\pm$ pumpellyite $\pm$ white mica $\pm$ calcite or aragonite.

A distinctive semischistose metasandstone crops out along the northwestern edge of the Redwood Creek schist. It is informally termed the "Sharp Point unit" of the Redwood Creek schist (Cashman et al., 1986). The mineral assemblage of Sharp Point metasedimentary rocks is the same as that of the mica schist. The mineral assemblage of minor interbedded foliated metavolcanic rocks is quartz + chlorite + albite + epidote or lawsonite $\pm$ pumpellyite $\pm$ actinolite.

Mesoscopic-scale descriptions of Franciscan Complex rocks commonly include textural grade, or zone, to describe the extent of recrystallization or development of metamorphic texture. For the Redwood Creek schist, we follow the usage of Suppe (1973) and Worrall(1981). Sharp Point unit metasandstones contain relict clasts and are therefore textural zone 2, whereas most metasandstones in the rest of the Redwood Creek schist are totally recrystallized with no relict clastic grains and are classified as textural zone 3.

Serpentinite and tectonic blocks occur in clusters within the Redwood Creek schist and along its borders. These tectonic blocks define a broad north-northwest-trending linear zone within the schist belt or occur in proximity to the boundary faults (Fig. 2).

METAMORPHISM

Mineral assemblages in the Redwood Creek schist are indicative of the pumpellyite-actinolite facies, or, more commonly, the lawsonite-albite-chlorite facies (as defined by Turner, 1981). Both facies appear in the blueschist series of regional metamorphism; however, as no blue amphibole is present in the Redwood Creek schist (except in tectonic blocks), this unit does not record blueschist facies metamorphism.

STRUCTURE

The Redwood Creek schist has undergone three periods of penetrative deformation. These deformational events are recorded by foliations and by linear fabric elements in all of the major rock types. All penetrative fabrics predate the last movement on the faults that bound the schist belt. In addition, numerous faults are found within the schist belt; several styles and ages of brittle deformation can be distinguished. Structural analysis of fabric elements indicates that the Redwood Creek schist is divided into domains that have remained rigid internally but have rotated relative to one another. This rotation took place late in the structural history of the schist.

Penetrative Structures

Bedding, S0, has been identified only in the semischistose metagraywacke of the Sharp Point unit, where graded bedding is preserved. Here, the primary foliation, S1, is parallel to bedding except at the hinges of F1 isoclinal folds, where S1 is an axial-planar cleavage.

The dominant foliation, S1, is a schistosity defined by segregation and preferred orientation of metamorphic minerals. At the microscopic scale, S1 is characterized by compositional layering, by preferred orientation of quartz grain boundaries, and by a strong preferred orientation of lawsonite (010) and muscovite and chlorite (001) faces subparallel to S1.

In the Sharp Point unit, isoclinal F1 folds are well exposed in coastal outcrops of interbedded metagraywacke and argillite. The folds are most conspicuous in metasandstone layers and have amplitudes of several meters. Axial-planar cleavage, S1, is well developed in argillite layers but only weakly developed in metasandstone beds. In the mica schist unit, microscopic F1 folds were observed in thin sections, but

F₂ FOLD AXES

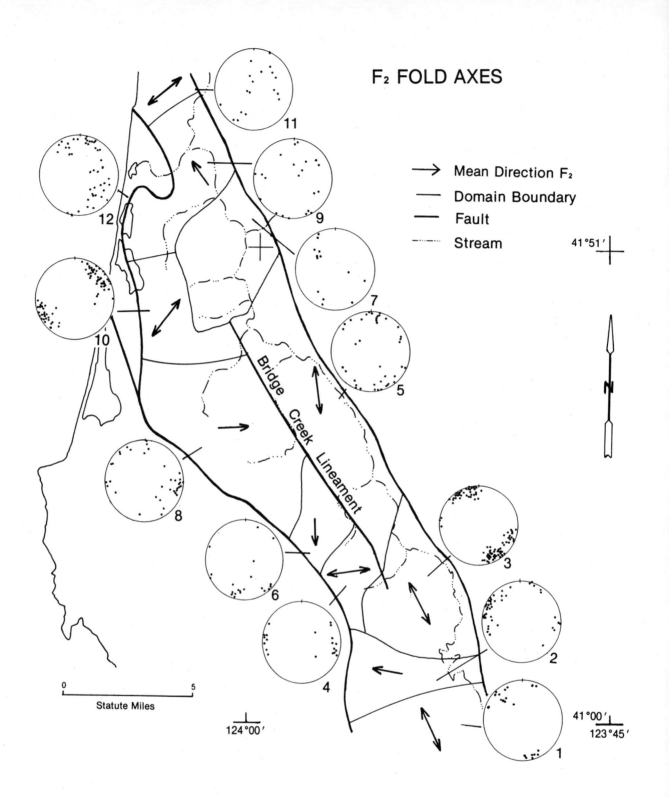

→   Mean Direction F₂

—   Domain Boundary

━   Fault

‑·‑·‑   Stream

0   5
Statute Miles

FIGURE 3   Orientations of F_2 fold axes in the Redwood Creek schist. Number of data points in each domain: 1,17; 2,41; 3,90; 4,20; 5,42; 6,21; 7,24; 8,35; 9,14; 10,77; 11,18; 12,45.

no larger F1 folds were recognized.

Superimposed on S1 is a well-developed crenulation cleavage S2. S2 is defined by alternation of quartz-rich microlithons, in which S1 is preserved, with mica-rich cleavage domains, in which S1 has been deformed and rotated by asymmetric microfolds. S2 is observed in most outcrops of the Redwood Creek schist; locally it is the dominant foliation .

F2 folds, for which S2 is axial planar cleavage, are widely developed in the Redwood Creek schist. F2 folds vary considerably in style and range in scale from microscopic to mesoscopic. In thin section, lawsonite grains (formed synchronously with S1) are seen to be rotated around F2 folds, indicating that crystallization of this important indicator mineral ceased prior to formation of F2.

Mesoscopic folds related to a third deformational event are developed locally. These F3 folds are commonly open chevron, box, or kink folds. Where axial-planar cleavage (S3) accompanies these folds, it tends to be a fracture cleavage.

Faults

Several types and ages of faults occur within the Redwood Creek schist; poor exposure prevents unequivocal interpretation of all of them. Thrust faults exposed in several localities parallel S2 foliation, have the same vergence direction as F2 folds, and appear to have developed concurrently with S2 and F2. Other thrust faults clearly cut, and therefore postdate, F3 folds. The most recent strike-slip movement on the faults that bound the schist belt postdates all of the structures described above.

Structural Analysis

Domains within the Redwood Creek schist belt were defined using orientations of F2 folds. F2 fold hinge orientations show strong local clustering (Fig. 3) but display a wide range of trends throughout the Redwood Creek schist belt. In spite of the wide range of F2 fold trends, fold plunges are consistently shallow. This unusual distribution of fold orientations was interpreted by Cashman et al. (1986) as suggesting rotation of relatively rigid domains about a vertical axis. The rotation of domains within the Redwood Creek schist most probably accompanied dextral strike-slip on the Grogan and Bald Mountain faults as the schist was emplaced in its present position within the Franciscan Complex.

DISCUSSION

Correlation of the Redwood Creek schist

and South Fork Mountain Schist of northern California and Colebrooke Schist of southern Oregon is supported by their similarities in protolith, metamorphic grade, major element geochemistry, deformational history and structural position (Cashman et al., 1986). The shared deformational history of the units includes F1/S1 folding and cleavage formation accompanying peak metamorphism, F2/S2 folding and associated thrusting , local F3 folding, and finally brittle deformation along north-northwest trending strike-slip faults. F2/S2 deformation appears to have accompanied thrusting of the schist units over adjacent Franciscan or equivalent rocks (Coleman, 1972; Jayko, 1984).

Metamorphism and accompanying folding and thrusting of these units is Late Cretaceous in age; the breakup and dispersal of the units therefore occurred after this time. Clasts of Colebrooke Schist in the Eocene Umpqua Formation establish that the Colebrooke Schist was near its present position by Eocene time. Northward strike-slip faulting of the Redwood Creek schist (probably accompanied by rotation of relatively rigid domains within the schist belt) and the Colebrooke Schist therefore occurred between Late Cretaceous and Eocene time. Continuing strike-slip faulting north of the Colebrooke Schist is suggested by Walker et al., 1987, who correlate the Colebrooke Schist with the Shuksan Metamorphic Suite in the Washington Cascade Range and infer an additional 600 km of northward movement for Shuksan rocks.

The sequence of deformational events for the Redwood Creek schist constrains interpretations of Coast Range tectonics by placing both thrust and strike-slip faulting events in a limited time framework. Thrusting (mainly Late Cretaceous) was followed by large-scale right-lateral strike-slip faulting between Late Cretaceous and Eocene time. These events predate Middle Eocene accretion and subsequent right-lateral shear-related rotation of rocks in the Oregon and Washington Coast Ranges north of the study area (Wells and Heller, 1988). They also predate probable right-lateral strike-slip faulting in Coastal belt Franciscan rocks south of the Redwood Creek schist (McLaughlin et al., 1982). Repeated episodes of strike-slip faulting therefore played a part in the tectonic evolution of the western margin of North America in Late Mesozoic and Cenozoic time.

REFERENCES

Alvarez, W., Kent, D. V., Silva, I. P., Schweickert, R. A., and Larson, R. A.,

1980, Franciscan Complex limestone deposited at 17° south paleolatitude: Geological Society of America Bulletin, v. 91, p. 476-484.

Blake, M. C., Jr., and others, 1967, Upside-down metamorphic zonation, blueschist facies, along a regional thrust in California and Oregon: U.S. Geological Survey Professional Paper 575C, p. 1-9.

Blake, M. C., Jr., Howell, D. G., and Jones, D. L., 1982, Preliminary tectonostratigraphic terranes map of California: U. S. Geological Survey Open-File Report 82-593, scale 1:750,000.

Cashman, S. M., Cashman, P. H., and Longshore, J. D., 1986, Deformational history and regional tectonic significance of the Redwood Creek schist, northwestern California: Geological Society of America Bulletin, v. 97, p. 35-47.

Coleman, R. G., 1972, The Colebrooke Schist of southwestern Oregon and its relation to the tectonic evolution of the region: U.S. Geological Survey Bulletin 1339, 61 p.

Ernst, W. G., 1970, Tectonic contact between the Franciscan melange and the Great Valley Sequence, crustal expression of a late Mesozoic Benioff zone: Journal of Geophysical Research, v. 75, p. 886-902.

Hamilton, W., 1969, Mesozoic California and the underflow of the Pacific mantle: Geological Society of America Bulletin, v. 80, p. 2409-2430.

Irwin, W. P., 1960, Geologic reconnaissance of the northern Coast Ranges and Klamath Mountains, California: California Division of Mines and Geology Bulletin 179, 80 p.

Jayko, A. S., 1984, Deformation and metamorphism of the eastern Franciscan belt, northern California [Ph.D. thesis]: Santa Cruz, California, University of California, 217 p.

Kelsey, H. M. and Hagans, D. K., 1982, Major right-lateral faulting in the Franciscan Assemblage of northern California in late Tertiary time: Geology, v. 10, no. 7, p. 387-391.

Lanphere, M. A., Blake, M. C., Jr., and Irwin, W. P., 1978, Early Cretaceous metamorphic age of the South Fork Mountain Schist in the northern Coast Ranges of California: American Journal of Science, v. 278, p. 798-815.

Maxwell, J. C., 1974, Anatomy of an orogen: Geological Society of America Bulletin, v. 85, p. 1195-1204.

McLaughlin, R. J., Kling, S. A., Poore, R. Z., McDougall, K., and Beutner, E. C., 1982, Post-middle Miocene accretion of Franciscan rocks, northwestern Caliofrnia: Geological Society of America Bulletin, v. 93, p. 595-605.

Suppe, J., 1973, Geology of the Leach Lake-Ball Mountain region, California: University of California Publications in Geological Sciences, v. 107, 82 p.

Turner, F. J., 1981, Metamorphic petrology - Mineralogical, field and tectonic aspects (2nd edition): New York, McGraw-Hill, 524p.

Walker, N. W., Plake, T. D., and Brown, E. H., 1987, Tectonic significance of Jurassic protolith ages of meta-plutonic rocks of the Shuksan Metamorphic Suite, Washington, and the Colebrooke Schist, Oregon: Geological Society of America Abstracts with Programs, v. 19, p. 879.

Wells, R. E., and Heller, P. L., 1988, The relative contribution of accretion, shear, and extension to Cenozoic tectonic rotation in the Pacific Northwest: Geological Society of America Bulletin, v. 109, p. 424-435.

Worrall, D. M., 1978, Imbricate low angle faulting in uppermost Franciscan rocks, South Yolla Bolly area, northern California: Geological Society of America Bulletin, jPart I, v. 92, p. 703-729.

NEOTECTONIC DEFORMATION IN THE SOUTHERNMOST CASCADIA FOREARC DUE TO NORTHWARD GROWTH OF THE SAN ANDREAS TRANSFORM FAULT, NORTHERN CALIFORNIA

H. M. Kelsey

Department of Geology, Western Washington University, Bellingham

INTRODUCTION

Strain patterns within the forearc at a convergent margin adjacent to a passing fault-fault-trench triple junction show a systematic evolution. I describe some aspects of deformation within the southernmost part of the Cascadia forearc associated with northward migration of the Mendocino triple junction and the San Andreas transform fault. The Mendocino triple junction was the product of collision of the Pacific-Farallon ridge with the North American Plate about 28 Ma (Atwater, 1970). Late Neogene and Quaternary deformation in northern coastal California is in part the result of geometric instability of the Mendocino triple junction (McKenzie and Morgan, 1969) as it migrates northward.

Deformation in the Cascadia forearc of North America near Humboldt Bay consists of a 30 km wide zone of on-land contraction with a narrower zone of translation to the east (Kelsey and Carver, 1988). The zone of subaerial contraction near Humboldt Bay is part of the relatively wide, southernmost extent of forearc contraction within the Cascadia subduction zone. Net Quaternary northeast-southwest contraction across forearc thrust faults (Fig. 1) is at least 7.9 km, and minimum fault slip rates on the six major thrust faults is 0.8-2.3 mm/a (Kelsey and Carver, 1988). Net right slip within the zone of translation is a minimum of 3 km. South of the forearc, translation is predominant in the North American plate and is accommodated on land along two major right-lateral fault zones of the San Andreas transform boundary, the Lake Mountain and Garberville fault zones (Fig. 1).

Present day deformation near Humboldt Bay thus reflects contraction associated with subduction of the Gorda plate and translation further east associated with oblique convergence at the plate boundary. As the triple junction migrates, the strike-slip faults of the San Andreas transform boundary will extend to the north-northwest into the present forearc, and the forearc strike-slip faults will become faults of the San Andreas system. The forearc strike-slip faults are the main focus of this paper. These faults, collectively referred to as the Eaton Roughs-Grogan-Lost Man Fault Zone (Fig. 1) are the northernmost faults that demonstrate significant dextral shear

associated with the San Andreas transform zone. I will first demonstrate the method used to determine that these faults are translational faults, next discuss the fault system and finally discuss the regional significance of this fault system. I will concentrate on the Grogan and Lost Man faults.

USE OF MESOSCALE STRUCTURES TO DEDUCE REGIONAL STRAIN

Associated with the major faults that cut Neogene and younger sediments in northern California is a broad (400-600 m wide) zone of fractures. These fractures can be divided into three classes: those with evidence of shear origin, called mesoscale (outcrop-scale) faults; those with evidence of extension origin, called extension fractures; and those of indeterminate origin, which I call fractures because they cannot be further classified.

Analysis of mesoscale faults has been used as a field technique to identify megascale strike-slip faults (Kelsey and Cashman, 1983) and to differentiate megascale strike-slip faults from megascale reverse faults (Kelsey and Carver, 1988). The type of strain recorded by mesoscale faults is apparent in outcrop by offset relations and slickenside striae or grooves on fault surfaces. Where fractures and faults occur together in groups of similar orientation (clusters), we assume the same origin for the fractures and the faults.

Mesoscale fractures and faults are ubiquitous in the late Neogene sediments in northern coastal California. Generally, these structures increase in density within 200-300 m either side of megascale faults, with greatest density near the fault trace. We therefore consider mesoscale faults as genetically related to the megascale fault.

Pairs of fault sets are common at many measuring sites. The sets are separated by an acute angle. In cases where the two fault sets formed synchronously (each offsets the other), have opposite senses of displacement, and show displacement perpendicular to the line of intersection of the fault sets, the two fault sets are a conjugate fault pair (Hobbs et al., 1976).

I have found two types of conjugate faults in Neogene sediments near megascale faults. In type one, exemplified by the Eaton Roughs, Grogan and Lost Man faults, all mesoscale faults were vertical or subvertical and all indicators of fault slip were horizontal or sub-horizontal. In type two, all faults dipped less than 60°, all indicators of fault slip showed a down-dip sense, and offsets showed a reverse movement sense. The second case is exemplified by the contractional thrust and reverse faults near Humboldt Bay (Kelsey and Carver, 1988).

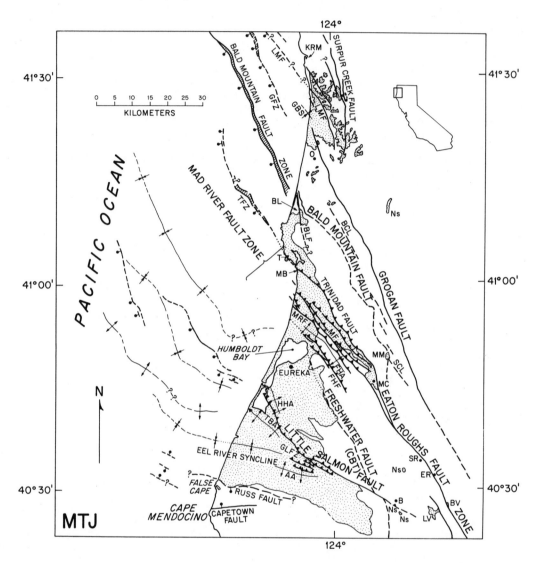

FIGURE 1 Tectonic map of north coastal California north of Cape Mendocino. Offshore data from Clarke, 1987. Faults with solid lines are probably active (for criteria, see Kelsey and Carver, 1988); faults with dashed lines do not meet criteria of probably active faults. Faults--LMF, Lost Man fault; GFZ, Grogan fault zone (offshore only); BCL, Bridge Creek lineament; BLF, Big Lagoon fault; Trinidad fault zone (offshore only); BF, Blue Lake fault; MF, McKinleyville fault; MRF, Mad River fault; FHF, Fickle Hill fault; SCL Snow camp lineament; CBT, coastal belt thrust. Folds--Gold Bluffs syncline; FHA, Fickle Hill anticline; HHA, Humboldt Hill anticline; TBA, Table Bluff anticline; AA, Alton anticline; RS, Redway syncline. Intermontane valleys--MM, Murphy's Meadows; LV, Larrabee Valley; BV, Burr Valley. Localities--O, Orick; T, Trinidad; MB, Moonstone Beach; A, Arcata; E, Eureka; B, Bridgeville; D, Dinsmore. Geologic units--dot pattern: all later Neogene and Quaternary sediments (mid-Miocene to present; excludes Miocene rocks of Kings Range terrane of McLaughlin et al., 1982); no pattern: early Cenozoic and pre-Cenozoic rocks of Franciscan assemblage (includes Kings Range terrane) and Klamath Mountain province; NS: isolated patches of late Neogene sediments, age designation uncertain; MTJ--approximate location of Mendocino triple junction.

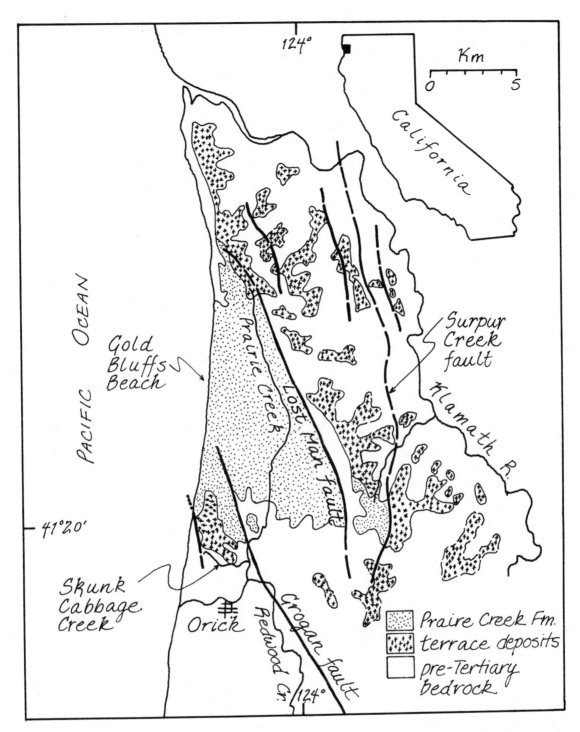

FIGURE 2   Map showing Plio-Pleistocene Prairie Creek Formation and faults cutting this unit.

GROGAN-LOST MAN FAULT ZONE

North of Big Lagoon (BL, Figure 1), the Pliocene and Pleistocene (predominantly Pleistocene) Prairie Creek Formation (Kelsey and Cashman, 1983; Cashman et al., in press) (top of figure 1 and Figure 2) crops out in a 125 km^2 coastal area. The sediments are the onshore part of an extensive sequence of sediments deposited on the continental shelf at the mouth of the Klamath River. A prominent structure in the unit is the northwest-trending, slightly northwest plunging Gold Bluffs syncline (GBS, Figure 2). The northeast limb of the syncline is steeper and cut off by faulting.

Three major megascale faults, as well as a number of minor megascale faults, offset the Prairie Creek Formation (Figure 2). Two of these faults, the Grogan fault and the Surpur Creek fault, are each exposed in a single exposure, and the Lost Man fault is not exposed at all. However, abundant mesoscale faults and fractures occur in Prairie Creek sediment in proximity to the Lost Man fault and to a lesser extent near the Grogan fault. These features are predominantly high angle and form conjugate sets (Kelsey and Cashman, 1983).

High angle mesocale faults with horizontal striae are formed by translational strain. Evidence for this contention is that the suite of mesoscale folds and faults associated with the Grogan-Lost Man fault zone is identical to folds and faults created by deformation of clay cake models by strike-slip faulting (Wilcox et al., 1973). The two dominant fracture directions formed by experimental shearing of the clay, the synthetic and antithetic shear, are equivalent to the conjugate fractures observed in the field in the Prairie Creek Formation sediments. The direction of maximum contraction bisects the acute angle between these shears. Numerous field investigations (e.g. Sylvester and Smith, 1976; Keller et al., 1982) describe a similar pattern of structural deformation for sediments overlying basement that has experienced strike-slip fault displacement. In the early phase of strike-slip fault zone formation, concurrent folding, conjugate fracturing, extensional faulting, and thrust faulting occur in the younger cover sediments at predictable angles relative to the horizontally directed regional stresses. Eventually the master strike-slip fault breaks to the surface and further rock deformation is confined to a relatively narrow zone.

Superimposing the structures produced from the sheared clay cake model (Wilcox et al., 1973) on the observed regional structures in Prairie Creek sediments near the Grogan-Lost Man faults shows a close similarity in relative orientation of all structures (Figure 3). Kelsey and Cashman (1983) therefore concluded that the Prairie Creek Formation sediments were deformed in a right shear couple in

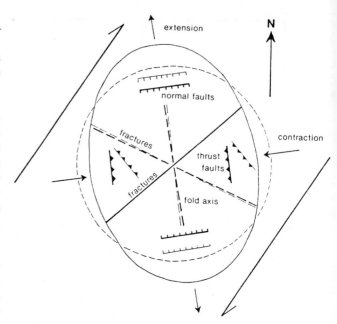

FIGURE 3 Structures produced in clay cake models of strike-slip fault-associated deformation (heavy lines) [after Wilcox et al., 1973] superimposed on the observed structures in the Prairie Creek Formation sediments (lighter lines) demonstrate close similarity in orientation of modeled and observed structures. The northeast trending fracture set of both data sets is arbitrarily superimposed. Orientations of structures are shown in context of the deduced strain ellipse.

Pleistocene time prior to surface rupture of the cover sediments along the major faults. After surface rupture, right slip deformation has been localized along the Grogan and Lost Man faults zones. Analysis by Kelsey and Cashman (1983) further showed that the high angle, dextral Lost Man and Grogan fault have an appreciable, but less significant, component of reverse dip-slip offset as well.

Because of extensive erosion, sense of translational fault offset cannot be inferred from outcrop distribution alone. However, the outcrop distribution of Prairie Creek sediments in relation to the two faults constrains right lateral offset to less than 10 km. Translational fault displacement across this zone is therefore relatively small, reflecting relatively minor translation in the forearc related to oblique subduction of the Gorda Plate.

SIGNIFICANCE OF FOREARC STRIKE-SLIP FAULTS (GROGAN AND LOST MAN FAULTS) TO GROWTH OF THE SAN ANDREAS TRANSFORM BOUNDARY

Present day deformation in the vicinity of Humboldt Bay consists of contractional deformation associated with subduction of the Gorda plate, and translational faulting further to the east associated with oblique convergence at the Gorda-North American plate boundary. This translational faulting is mainly accommodated by the Grogan-Lost Man faults, and by the Eaton Roughs fault zone (Figure 1). As the Mendocino triple junction migrates further northward and the San Andreas transform boundary extends in length, the strike-slip faults of the San Andreas will extend to the north-northwest into the present forearc, and the above forearc strike-slip fault zones will become faults of the San Andreas fault system.

I suggest therefore that there will be an eventual shift of major transform activity to this northernmost and easternmost right-slip fault system and that this shift will coincide with the migration of the triple junction to the approximate latitude where these faults intersect the continental margin. A shift of motion eastward as well as northward is likely because the present plate boundary formed by the San Andreas fault and the Cascadia trench is not colinear on a small circle. A stable triple junction would migrate on a small circle route (McKenzie and Morgan, 1969) and the Grogan-Lost Man fault zone is rupturing along such a trend.

A precedent for my suggested shift of major San Andreas fault to the younger Grogan-Lost Man fault zone is the proposed history of the San Gregorio-Hosgri fault, which lies west of the San Andreas fault in central coastal California. Graham and Dickinson (1978) proposed, based on offset geologic units, that prior to 10 Ma ago, a significant component of right-lateral transform displacement was taken up by the San Gregorio-Hosgri fault. My geologic evidence leads me to suggest that, in similar fashion, another eastward shift to the Lost Man-Grogan fault zone is now in progress as the Mendocino triple junction migrates northward.

REFERENCES

Atwater, T., Implication of plate tectonics for the Cenozoic tectonic evolution of western North America, Geological Society of America Bulletin, 81, 3513-3536, 1970.

Cashman, S. M., Kelsey, H. M., and Harden, D. R., Geology and descriptive geomorphology of the Redwood Creek basin, Humboldt County, California, in Geomorphic Processes and Aquatic Habitat in the Redwood Creek Basin, edited by K. M. Nolan, H. M. Kelsey, and D. C. Marrou, Northwestern California, U. S. Geological Survey Professional Paper 1454, in press.

Clarke, S. H., Geology of the California continental margin north of Cape Mendocino, Geology and resource potential of the continental margin of western North America and adjacent ocean basins-Beaufort Sea to Baja California, edited by D. W. School, D. W., Grantz and J. G. Vedder, Circum-Pacific Council for Energy and Mineral Resources, Earth Science Series, 6, 15A1-15A8, 1987.

Graham, S. A., and Dickinson, W. R., Evidence for 115 kilometers of right slip on the San Gregorio-Hosgri fault trend, Science, 199, 179-181, 1978.

Hobbs, B. E., Means, W. D., and Williams, P. F., An outline of structural geology, 571 pp., John Wiley, New York, 1976.

Kelsey, H. M., and Cashman, S. M., Wrench faulting in northern California and its tectonic implications, Tectonics, 2, 565-576, 1983.

Kelsey, H. M. and Carver, G. A., Late Neogene and Quaternary tectonics associated with the northward growth of the San Andreas transform fault, northern California, Journal of Geophysical Research, 93, 4797-4819, 1988.

Keller, E. A., R. J. Bonkowski, R. J. Korsch, and R. J. Shlemon, Tectonic geomorphology of the San Andreas fault zone in the southern Indio Hills, Coachella Valley, California, Geological Society of America Bulletin, 93, 46-56, 1982.

McLaughlin, R. J., Kling, S. A., Poore, R. Z. McDougall, K. L., and Beutner, E. C., Post-middle Miocene accretion of Franciscan rocks, northwestern, California, Geological Society of America Bulletin, 93, 595-605, 1982.

McKenzie, D. P., and morgan, W. J., Evolution of triple junctions, Nature, U224, 125-133, 1969.

Sylvester, A. G. and R. R. Smith, Tectonic transpression and basement-controlled deformation in the San Andreas Fault Zone, Salton Trough, California, American Association of Petroleum Geology Bulletin, 60, 2081-2102, 1976.

Wilcox, R. E., T. P. Harding, and D. R. Seeley, Basic wrench tectonics, American Association of Petroleum Geology Bulletin, 57, 74-96, 1973.

ACTIVE CONVERGENT TECTONICS IN NORTHWESTERN CALIFORNIA

Gary Carver and Raymond Bud Burke

Department of Geology, Humboldt State University, Arcata, CA

95521

INTRODUCTION

The Pacific northwest coast of the United States is situated on an accreated continental margin developed during repeated episodes of subduction. In northwest California, younger to the west belts of tectonically emplaced oceanic rocks (Aalto and Murphy, 1984; Jayko and Blake, 1987) record a succession of Mesozoic and early Tertiary subduction episodes. These accreted terranes make up the upper crust in northwest California and the forearc basement upon which the modern accretionary margin has developed.

Northward passage of the Farallon-Kula plate boundary in the early Tertiary (Engebretson and others, 1985) initiated the present episode of subduction. During the late Cenozoic, dextral oblique convergence of the Farallon plate with North America has dominated the tectonics and sedimentation in the Pacific Northwest coast region. Tectonic processes and rates have varied as the Farallon plate has been consumed. The rate of convergence has decreased from about 50 km/my in the early Miocene, to the present rate of less than 20 km/my (Engebretson and others, 1985). Concurrently, the age of the subducting plate has decreased. Modern Cascadia subduction in the northern California region involves oceanic crust as young as 6 m.y. (Engebretson, 1987).

In the modern subduction zone, the Gorda and Juan de Fuca plates, remnants of the Farallon plate, descend beneath the accretionary margin of North America at a shallow angle of about 12-15 degrees (Walter, 1986; Taber and Smith, 1985). Offshore the subduction zone is represented as a shallow sediment filled trench. The shallow trench, low subduction angle, moderate to slow convergence rate and young descending oceanic plate are characteristic of strongly coupled subduction zones with actively contracting accretionary prisms and high seismic potentials (Heaton and Kanamori, 1984).

FOREARC BASIN-SEDIMENTARY TECTONICS

Since initiation of Cascadia subduction, oblique convergence has resulted in a northward migration of subducting oceanic plate segments and boundaries along the northwest California coast. Early phases of Cascadia subduction resulted in regional forearc subsidence and accumulation of thick clastic sedimentary basin fills. Sediments of the Wildcat Group (Figure 1; Ogle, 1953; Ingle, 1976, 1987) and Falor Formation (Manning and Ogle, 1950; Carver, 1987; Nilsen and Clarke, 1987) were deposited in a sequence of forearc basins developed across the western portion of the pre-Neogene accretionary complex.

The late Pliocene-early Pleistocene passage of the Blanco fracture zone probably resulted in the present episode of folding and thrusting in northwestern California. Migration of this fracture past the northwest California region resulted in the abrupt decrease in age of the subducting crust. Additionally late Neogene changes in Pacific plate motion motion (Engebretson and others, 1985), and internal deformation of the southern Gorda plate (Riddenhough, 1984) represent major events in the late Cenozoic tectonic evolution of the northern California region. The final stage of convergent margin evolution, the replacement of the contractile tectonic environment resulting from plate

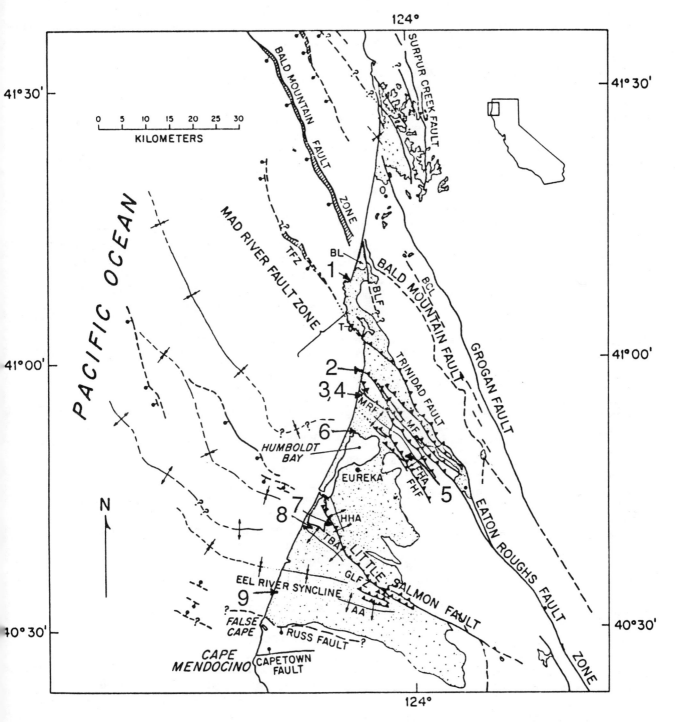

FIGURE 1 General regional map of coastal northern California showing the principal late Quaternary tectonic structures and distribution of Eel River basin sediments (stippled). Planned stops described in the guidebook are numbered 1-9.

convergence by transform tectonics at the propagating north tip of the San Andreas transform system, is progressing in the Mendocino triple junction region (Kelsey and Carver, 1988).

The early history of Farallon plate subduction in the northern California region includes periods of basin formation and intervals of relatively passive tectonism. More than 4 km of late

Cenozoic sediments accumulated in these basins. Several periods of basin development are represented. Bathyal to abyssal mudstones of the Bear River beds near Cape Mendocino are interpreted as trench slope basin sediments of lower to middle Miocene (Ingle, 1976; Nilsen and Clarke, 1987). These sediments are preserved as fault bounded slices within a broad belt of melange in the False Cape and Mattole shear zones south of Cape Mendocino (Ogle, 1950). Paleobathymetric interpretation of these sediments suggests the early Cascadia trench was considerably deeper than the modern trench.

Sediments of equivalent age on land are sparse. However, remnants of a regionally extensive low relief landscape capped by a thick saprolitic weathering profile are well preserved as accordant summit surfaces throughout the Coast Ranges of northern California and southern Oregon (Carver, 1987). The deeper trench, low coastal relief, and prolonged landscape stability indicated by the saprolite suggest early Cascadia subduction was characterized by steeper descent of older denser oceanic crust and weak coupling across the plate boundary.

Pliocene and early Pleistocene sediments, including the Wildcat Group and Falor Formation were deposited in a regionally extensive forearc basin (Nielson and Clarke,1987), named the *Eel River basin* by Ogle (1950) and the *Humboldt basin* by Ingle (1976). The basin sediments show initial deep water deposition followed by gradual basin filling and shoaling during the late stages of formation (McCorey, 1985; Ingle, 1976; Carver, 1987).

Tectonic disruption of the forearc began in the mid-Pleistocene and resulted in the development of a broad accretionary margin fold and thrust belt (Figure 1; Carver, 1987). This contraction ended Eel River basin sedimentation and deformed the basin sediments. In numerous depocenters, localized in the cores of growing synclines and on the footwalls of thrusts, late Quaternary sediments have accumulated on the Neogene basin deposits. Pre-Cenozoic basement rocks have been thrust over the late Cenozoic basin deposits. Glacio-eustatic high stand marine terraces, cut into uplifted portions of the coastline, record the late Pleistocene growth of anticlinal ridges and elevation of the hanging walls of thrusts.

DEFORMATION RATES

Estimates of the deformation rate represented by the faulted and folded terraces suggest much of the plate convergence has been accommodated by contraction of the North American plate margin, indicating strong coupling between the subducting Gorda plate and the overriding accretionary prism. The vigorous thrusting and folding initiated in the mid-Quaternary has remained active to the present. Holocene sediments are locally folded and faulted by major thrusts and raised late Holocene marine terraces are present along portions of the northern California coast. Holocene growth of the *Freshwater syncline*, one of the larger folds in the modern accretionary fold and thrust belt, has been documented at the north end of Humboldt bay (Figure 1).

Field relations and ages of deformed sediments show the growth of folds, uplift of the coast, and displacement on the thrusts has occurred as sudden events occurring at intervals of several hundred years. These deformation events are interpreted to be coseismic and indicate periodic large earthquakes are generated by the Cascadia subduction zone. The accrued late Quaternary deformation has shortened the modern accretionary margin by at least 20 percent.

FIELD TRIP STOPS

Introduction

This field trip starts at Patricks Point State park, about 60 km east of the deformation front of the active accretionary prism. Falor Formation sediments, equivalent in age to the middle member of the Rio Dell Formation of the Wildcat Group at Centerville Beach (Sarna and others, 1987), and late Pleistocene marine terraces are exposed in the modern sea cliff at Agate beach. These nearshore marine and terrestrial sediments fix the location of the eastern margin and paleo-shoreline of the late Pliocene-early Pleistocene Eel River basin, and record the initiation and

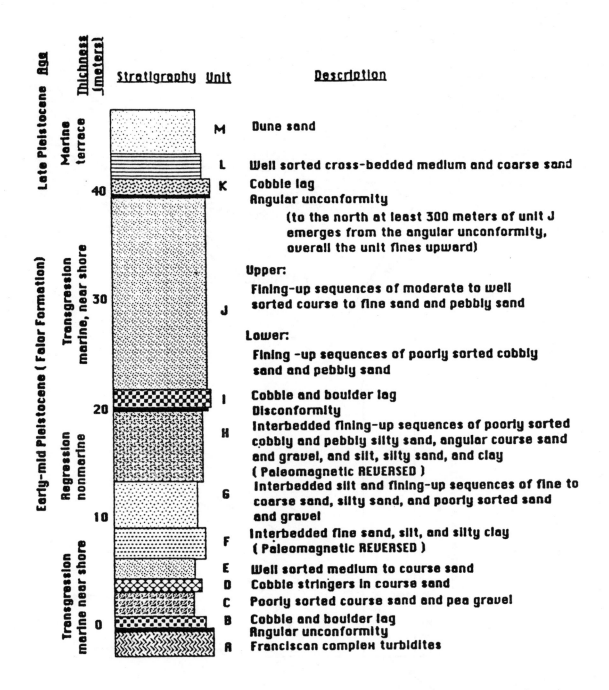

FIGURE 2 Stratigraphy and paleoenvironmental aspects of sediments exposed in the sea cliffs at Agate Beach. The section is described in the area of the Patricks Point State Park trail and stairway to the beach.

structural style of the late Quaternary contractual tectonics.

During the field trip subsequent stops are planned where generally time-equivalent sediments are exposed at sites progressively closer to the active subduction front and the paleo-axis of the Eel River basin. The focus of the trip is to examine structures, stratigraphy and geomorphology related to the development of the Pliocene and early Pleistocene Eel River basin and the mid and late Quaternary thrust and fold belt. The trip ends at Centerville beach, near the axis of the Eel River basin and about 30 km east of the active deformation front of the Cascadia subduction zone. Figure 1 is a generalized regional map showing the location of Eel River basin sediments, principal fold and thrust belt

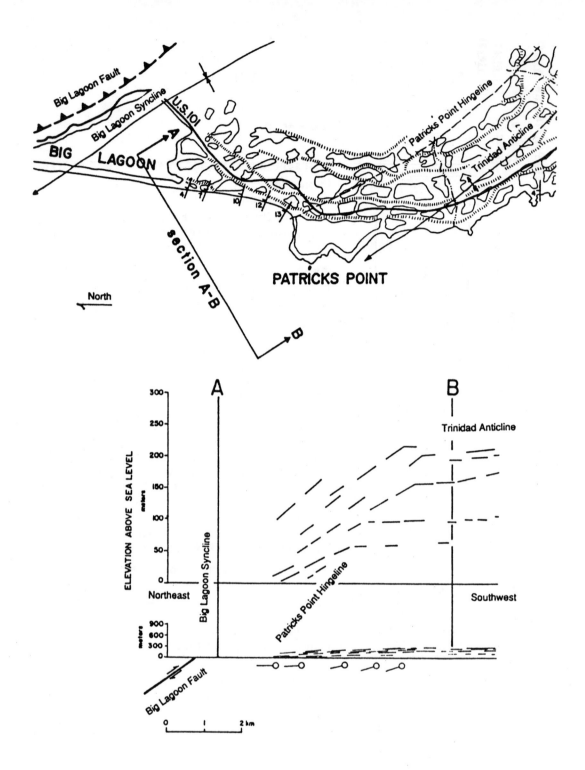

FIGURE 3 Map and cross section of raised and folded marine terraces at Patricks Point. Note that the older terraces have been elevated more on the crest of the Trinidad anticline, but the dip of the terraces on the north limb of the fold is nearly constant regardless of the terrace age. This type of folding is best explained as the result of slip of the hanging wall of a nearly flat thrust across a ramp of bend in the thrust.

structures, and stops planned for this trip.

Stop 1-Agate Beach, Patricks Point State Park

The seacliffs at Agate Beach provide an unusually large continuous exposure of Eel River basin sediments near the basin's eastern margin. Contained in the section is a north-tilted sequence of upper shoreface marine and non-marine sediments resting unconformably on Central belt Franciscan rocks (Figure 2). The sediments include more steeply dipping Falor Formation strata overlain by less steeply dipping late Pleistocene marine terrace sediments. The terrace abrasion surface constitutes a prominent angular unconformity truncating the Falor strata.

Along the length of Agate Beach, more than 330 m of Falor sediments are exposed. At the State Park stairway to the beach, silty and clayey layers near the base of the Falor sequence are paleomagnetically reversed suggesting an age in the 2.3 to 0.7 million year range. The sequence dips about 14-17 degrees to the north-northwest at the south end of the beach, and less, about 4-7 degrees, to the north at the end of the seacliff exposure near Big Lagoon. The Falor strata are conformable and include transgressive and regressive cycles and associated marine and non-marine sequences interpreted as being glacio-eustatic.

A prominent raised marine terrace is cut into the Falor and Franciscan rocks at Patricks Point. The Patricks Point terrace has been assigned to the stage 5A (83 ka) glacio-eustatic high stand (Woodward- Clyde, 1980; Carver and others, 1986; Carver and Burke, 1987). Higher and older raised marine terraces spanning the last several hundred thousand years are present inland (Figure 3). North of Patricks Point, the terraces are inclined to the northeast about 4-6 degrees. At Patricks Point, and inland along a prominent hingeline, the terraces bend sharply and are nearly horizontal to the south.

The terraces are interpreted to record fault bend folding of the hanging wall of a deeply buried, northeast-dipping thrust above a ramp on the fault (Figure 4). The

length of the inclined segment of the Patricks point terrace is about 2 km. The fault bend fold model predicts this length should correspond with the total accrued slip on the buried fault, i.e. about 2 km in 83 ky or about 2.4 cm/yr. To the south, a system of at least 8 southwest verging imbricate thrusts cut the surface. Late Pleistocene slip rates for each of these faults range between about 1 mm/yr and 10 mm/yr, and total across the contractile zone about 2 cm/yr, a rate similar to estimates of the rate of plate convergence.

Stop 2-McKinleyville Airport area

In the vicinity of the Humboldt County airport, the *McKinleyville fault* deforms a flight of late Pleistocene marine terraces. At the airport the lowest terrace, correlated to stage 5A (83ka; Burke and others, 1986), is displaced down to the south about 33 m. The McKinleyville fault is one of at least six principal thrusts in the *Mad River fault zone*. The fault has been mapped inland for about 35 km. The surface expression of the fault varies greatly along strike. At the airport, the fault includes two main traces defined geomorphically as broad scarps and warps of the terrace.

A relatively undeformed segment of the terrace is present between the two prominent traces near the freeway, where displacement is distributed across a width of about 500 m. To the east , near the entrance to the airport, the two traces converge into a narrower and steeper scarp. Similar, but larger, displacement relationships on higher and older terraces (Figure 5) indicate the McKinleyville fault has a late Pleistocene slip rate of about .9 mm/yr (Carver and Burke, 1987).

Stop 3-Clam beach, the mouth of Mad River

We will walk (climb) down to the beach from the vista overlook on US 101. This stop allows examination of recently eroded exposures resulting from the northward migration of the mouth of the Mad River. The erosion is ongoing and consequentially the exposure is constantly changing. This brief summary

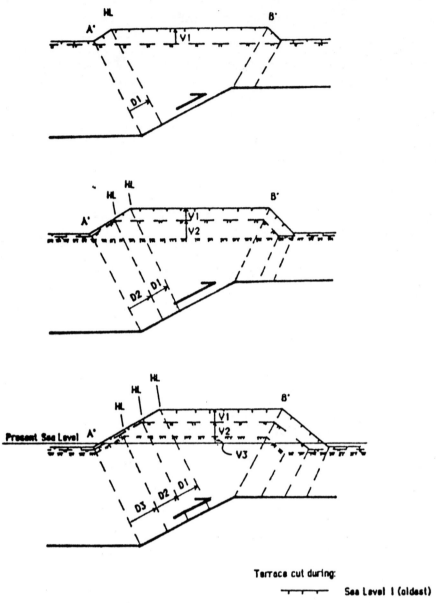

Terrace cut during:

┬─┬─┬─ Sea Level 1 (oldest)

┬▿┬▿┬ Sea Level 2

▿▿▿▿ Sea Level 3

FIGURE 4   This diagram illustrates the geometry and kinematic development of a growth fold of the fault-bend type, as expressed in a sequence of glacio-eustatic marine terraces. The axial surfaces A' and B' are fixed with respect to the position of a ramp on a deeply buried thrust fault. A hingeline forms at the axial surface A', after a terrace cutting event migrates away form the axis as slip occurs on the fault. These hingelines are preserved and carried with the overthrust block as displacement accrues in the interval between high sea level stands (D1, D2, D3). The dip of the limb remains constant and the length of the limb increases at the slip rate on the fault. The terraces are progressively elevated at the fold crest (V1, V2, V3). Because successive glacio-eustatic high sea levels rose to different elevations relative to modern sea level, and the time interval between high stands varied, the altitudinal spacing of the resulting terraces is not uniform.

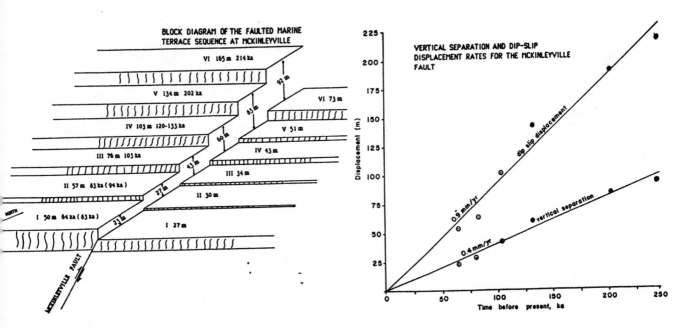

FIGURE 5 Block diagram of the faulted marine terraces and slip rate for the McKinleyville fault.

of the exposure was written in early November, 1988, and discusses the features exposed at that time. Two main stories are interpreted from the seacliff geology at the mouth of the Mad River: 1) cross-section view of a portion of the McKinleyville fault, and 2) raised late Holocene marine terrace and terrace stratigraphy interpreted to reflect recent paleoseismicity of the Cascadia subduction fold and thrust belt.

Story 1: The McKinleyville fault. The recent cliff erosion has exposed in cross-section a portion of the McKinleyville fault, seen at the last stop as a broad warp and vertical offset of the terrace surface. Only the southern part of the fault is exposed as of this writing. Fault displacement has occurred as small amounts of slip (<1 m) on many northeast dipping thrust faults and fractures across a zone at least 100 m wide. Additionally, much of the vertical offset has been accomplished by folding. The dip of late Pleistocene strata in the fault zone changes from nearly horizontal south of the fault, to 40-50 degrees south within the zone of faulting. This style of surface deformation is common for thrusts in the Mad River fault zone.

Detailed observation of individual thrust surfaces shows complex offsets of early-formed faults by later fault surfaces, indicating multiple episodes of slip involving many of the meso-scale faults. Also apparent in the cliff exposure are many southwest-dipping antithetic faults or back-thrusts. The synthetic and antithetic faults are strongly oriented and geometrically fit a strain ellipsoid with a horizontal principal compression axis oriented northeast-southwest and a nearly vertical minimum compression axis.

At this locality the faults are not rotated with the beds in the fault-line fold. Similar cross-cutting relations between thrusts and inclined bedding in fault-line folds are common throughout the Mad River fault zone. Probably the earliest deformation occurred on a blind thrust at depth and resulted in the growth of a fault-propagation fold near the surface. With subsequent displacement, the fault tip propagation through the fold ceased and the detached fold stopped growing. The most recent displacement has accrued as slip on the many distributed fault surfaces.

Story 2. Clam beach, a raised late Holocene marine terrace. Clam beach occupies the seaward edge of a broad terrace elevated a few meters above sea level. The terrace terminates at the base of a geomorphically youthful sea cliff about 60 m high. The terrace is overlain by stabilized and vegetated sand dunes, and the abandoned sea cliff is

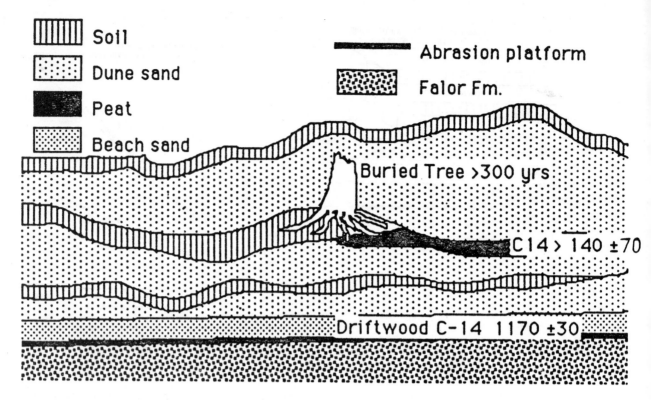

Soil

Dune sand

Peat

Beach sand

Abrasion platform

Falor Fm.

Buried Tree >300 yrs

C14 > 140 ±70

Driftwood C-14 1170 ±30

FIGURE 6 Schematic cross-section of the sea cliff exposure at the mouth of the Mad River in the raised Holocene marine terrace at Clam Beach. The abrasion platform is cut into faulted and folded late Pleistocene sediments and is about 2 m above seas level.

covered with dense vegetation including many very large old growth spruce trees. Erosion at the mouth of the Mad River during the past several years has exposed a cross-section through the clam beach terrace (Figure 6).

The section shows that the terrace abrasion platform and shoreline angle have been raised several meters above present sea level. The terrace cover sediments include a thin (1-2 m) layer of beach sand, similar to that on the modern beach, that is buried under three sequences of dune sand and associated soils. The beach sand contains abundant driftwood including logs and smaller woody debris. A small piece of driftwood tree limb collected from the buried beach sands yielded a C^{14} date of 1170 ±30 yrs .

Above the buried beach are three sequences of cross-bedded dune sand. Weak, but distinct, buried soils cap the lower two buried dune sequences, and a weak modern soil has formed on the stabilized and vegetated surface of the upper set of dunes. Near the abandon sea cliff, the upper buried soil grades laterally into a peat layer containing abundant leaf fossils, spruce cones, and

plant remains. Stumps of large spruce and fir trees are rooted in the upper soil and associated peat. Carbon^{14} dates for the peat and one of the buried stumps and suggest an age of about 300 years.

The preliminary interpretation of the late Holocene section at Clam beach is that the marine terrace was raised during three late Holocene episodes of sudden uplift. Each uplift event resulted in several meters of emergence of the coastline, enough to expose a broad strip of sandy shoreline to wind erosion and generate a landward sequence of coastal dunes. Restabilization of the shoreline and vegetation of the newly formed emergent platform shut off the sand source shortly after each uplift event. During the interval between events, soils formed on the dunes.

The first of the uplift events was about 1200 years ago, and the most recent about 300 years ago. The age of the intermediate event is not known. Each uplift episode is interpreted to be the result of sudden slip on a deeply buried fault. Similarities in the ages of uplift events at Clam Beach and ages of slip events on the Little Salmon fault suggest

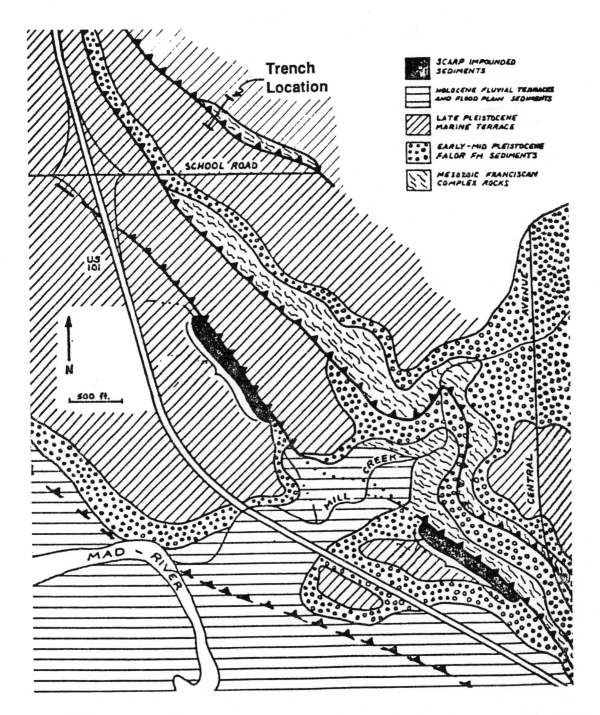

The legend in the figure reads:

- SCARP IMPOUNDED SEDIMENTS
- HOLOCENE FLUVIAL TERRACES AND FLOOD PLAIN SEDIMENTS
- LATE PLEISTOCENE MARINE TERRACE
- EARLY-MID PLEISTOCENE FALOR FM SEDIMENTS
- MESOZOIC FRANCISCAN COMPLEX ROCKS

Trench Location

SCHOOL ROAD

US 101

N

500 ft.

MAD - RIVER

MILL CREEK

CENTRAL

AVENUE

FIGURE 7 General geologic map and trench location on the Mad River fault at School Road near McKinleyville.

that fault may be the causative structure. The uplift at Clam beach was probably coseismic and accompanied by very large earthquakes.

Stop 4-Mad River Fault

The purpose of this brief stop is to look at the *Mad River fault* where it displaces the stage 5A (83 ka) marine terrace. As is characteristic of thrusts in northwestern California, the Mad River fault includes several imbricate traces distributed across a zone as much as a kilometer wide (Figure 7). At this site, four principal thrusts displace the marine terrace down to the southwest a total of more than 40 m. The thrusts are separated by several hundred-meter-wide segments of gently folded and warped terrace. The terrace here is correlated with the Patricks Point terrace at

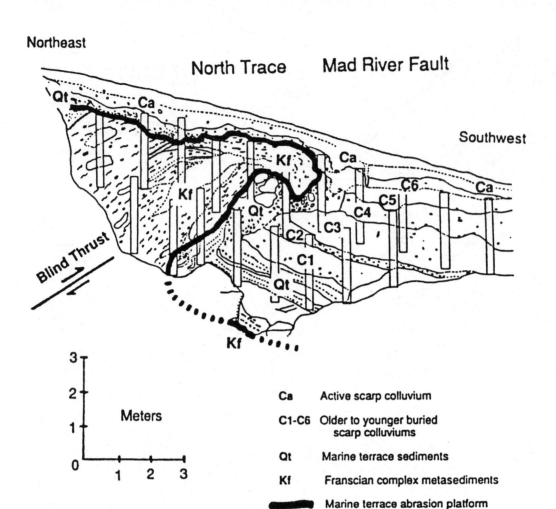

Northeast

North Trace   Mad River Fault

Southwest

Qt
Ca
Kf
Kf
Qt
Ca
C6
Ca
C5
C4
C3
C2
C1
Qt
Kf
Blind Thrust

3
2
1
0
Meters
1   2   3

Ca   Active scarp colluvium

C1-C6   Older to younger buried
             scarp colluviums

Qt   Marine terrace sediments

Kf   Franscian complex metasediments

▬▬▬   Marine terrace abrasion platform

FIGURE 8 A portion of a trench log across the northern trace of the Mad River fault at School Road near McKinleyville. At this location the surface expression of the fault is a well defined scarp and vertical separation of the stage 5A (83 ka) terrace. The trenching showed the scarp is composed of a tight overturned anticline with a sequence of 6 colluvial sheets containing buried soils under the overturned south limb. The anticline is interpreted as a fault propagation fold and each colluvial sheet marks a coseismic slip event. Trenches 300 m to the west and east of this site revealed low angle thrusts extending to the surface.

Trinidad based on similarities in soil development, mineral etching, and map continuity (Carver and others, 1987). The terrace offset indicates the Mad River fault has a late Pleistocene slip rate of about 1.2 mm/yr.

Geologic investigation trenches have been excavated across the upper (northeast) imbricate trace of the Mad River fault at several locations. Near the freeway, a few hundred meters northwest of this site, geotechnical investigation trenches exposed several closely spaced well defined thrusts offsetting gently folded terrace sands.

The faults extend into the surface colluvium along the lower portion of the prominent scarp. Trenches across the same scarp at this location revealed that a tight overturned asymmetrical fold at the surface has accommodated the fault displacement (Figure 8). The fold exhibits characteristics of episodic growth resulting from periods of sudden slip on a shallow buried thrust. Along strike, a few hundred meters to the southeast, the scarp diminishes in height and disappears.

The remarkable variation in structural

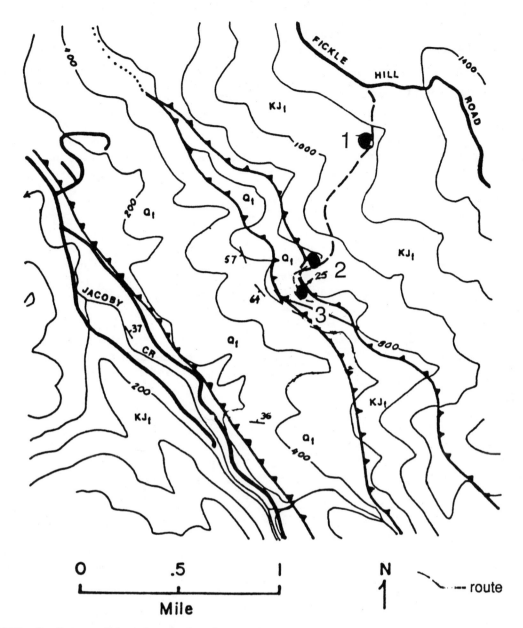

FIGURE 9 Generalized geologic map of the south side of Fickle Hill along the field trip route. The dashed line shows the approximate location of the abandoned logging road followed on the walk. Q_f-Falor formation sediments. KJ_f-Central belt Franciscan complex rocks. The three thrust faults represent the principal traces of the Fickle Hill fault. The large dots are: 1) Central belt Franciscan complex melange in poor exposures, 2) one of the three main traces of the Fickle Hill fault where KJ_f is thrust over Q_f, 3) Coulomb faulting in steeply south dipping Q_f sediments in the footwall of the northern trace of the fault.

style and surface expression exhibited by this fault is typical for thrusts along the Humboldt County coast. Based on surface expression, the principal faults in the fold and thrust belt appear to be composed of imbricate slip surfaces which laterally are segmented. The segments overlap along strike and are bounded by intervening zones where slip occurs as ductile deformation or distributed microfracture shear. Displacement is transferred from one imbricate segment to another, or changed from fault slip (brittle) to folding (ductile) or distributed shear from place to place along the fault.

Stop 5-Fickle Hill anticline and the Fickle Hill fault

Fickle Hill is the prominent northwest-trending ridge which separates the Mad River and Jacoby creek watersheds (Figure 1). About 25 kms long, it rises from sea level at Arcata to 730 m near its eastern end. The ridge, a late Quaternary anticline, is capped and flanked by Falor Formation marine sediments. On the ridge crest, scattered patches of basal Falor sediments are nearly horizontal. Along the broad northeast flank of Fickle Hill Falor strata dip 20-30 degrees to the northeast, more or less parallel to the landscape. The upper half of the steeper south-facing slope is composed of Franciscan rock, predominantly melange. Vertical to steep southwest-dipping Falor sediments make up most of the lower south slope of Fickle Hill (Figure 9). These Falor sediments are in thrust contact with the overlying Franciscan rocks.

The thrust, the *Fickle Hill fault*, also displaces marine terraces in Arcata, similar to the faulted terraces across the Mad River and McKinleyville faults near McKinleyville. Although exposure is limited, outcrop mapping on the south slope of Fickle Hill shows at least 350 m of Falor section have been overthrust by Franciscan melange on three main traces of the Fickle Hill fault. Additionally, the steeply dipping Falor beds on the south slope define the attenuated limb of a fault propagation fold with at least 300 m of vertical relief, and the entire faulted anticline has more than 1000 m of structural relief.

Stop 5 involves walking from the crest of Fickle Hill to the Fickle Hill fault on old logging roads. The walk is about 1.5 km and descends about 200 m. Along the way are poor but representative outcrops of Franciscan melange (Figure 9, location 1).

At location 2 (Figure 9), the Fickle Hill fault is exposed in a small roadcut. Sheared Franciscan melange is in sharp thrust contact with sandy Falor sediments. The fault surface consists of about 25 cm of gray-brown gouge composed of highly polished and oriented lens-shaped clasts derived from both Falor and Franciscan rocks in a strongly foliated and slickensided clayey matrix. Striations and grooves on the fault surface and shear fabric in the gouge indicate nearly pure dip-slip movement has occurred on the fault. The fault strikes about N55W and dips to the northeast at about 25-35 degrees. Within a few meters of the fault the Falor beds are pervasively sheared and appear to be overturned. Numerous shears, fractures, and small faults generally parallel to the main fault also are present in the overriding Franciscan rocks.

At location 3 (Figure 9), distributed imbricate thrusts and conjugate fractures and faults have deformed steeply southeast-dipping Falor sediments in the footwall of the upper trace of the Fickle Hill fault. The geometry and orientation of the meso-scale faults and fractures is the same as those seen in the seacliff exposure of the McKinleyville fault (stop 3). This pattern of distributed conjugate thrust faults and fractures is nearly always present across a wide zone where major thrusts cut late Cenozoic sandy sediments. It is a useful outcrop-scale field indicator of the presence of significant faulting in coastal Humboldt county (Kelsey and Carver, 1988). In general, the near surface thrust deformation in coastal Humboldt County is expresses as a combination of faulting and folding (Figure 10).

Although no thick continuous section of the Falor sediments is exposed on the south side of Fickle Hill, numerous outcrops throughout the Jacoby Creek Valley allow a general characterization of the stratigraphy. The Falor sediments on the south flank of Fickle Hill are predominantly marine, with considerably fewer and thinner fluvial sequences in the section than in the Mad River area to the northeast. The section is marked by frequent vertical changes and probably reflects numerous local transgressive-regressive cycles, a few of which resulted in short non-marine intervals of deposition.

The marine sediments include thick, upper shoreface sequences, predominantly medium to course sand, pebbly sand and locally fossiliferous bay muds, silts, and sands. Also present are well sorted massive fine to medium sand and silty sand which may represent mid-shelf

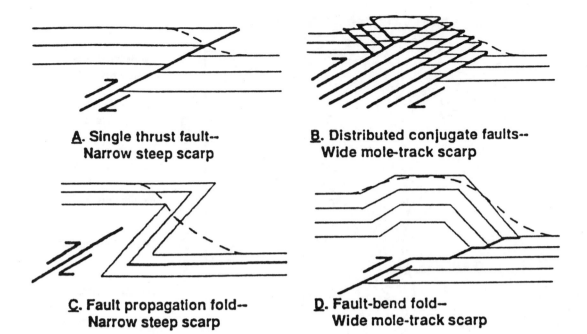

A. Single thrust fault--
Narrow steep scarp

B. Distributed conjugate faults--
Wide mole-track scarp

C. Fault propagation fold--
Narrow steep scarp

D. Fault-bend fold--
Wide mole-track scarp

FIGURE 10 Schematic diagram of the near-surface structural styles for thrust faults in coastal northwestern California and generalized characteristics of the resulting scarps. Heavy lines indicate faults with arrows showing sense of slip; fine lines indicate bedding.

deposition below wave base. The sands and gravels seen at location 2 and 3 are typical of open marine, high energy littoral deposition and represent a common lithofacies of the Falor in the Jacoby creek and Freshwater area. In general, the nearshore and upper shelf sediments in the Jacoby Creek-Freshwater area grade to the north into the fluvial and marginal marine dominated deposits in the Patricks Point-Mad River area (near the basin margin), and into the coeval shelf-slope facies Rio Dell sediments to the south in the Eel River area (near the basin axis).

Stop 6-Mad River Slough, north Humboldt bay

Late Holocene growth of folds in the Cascadia fold and thrust belt has resulted in rapid episodic sedimentation in the axis of the Freshwater syncline, where the fold intersects the northern end of the Freshwater syncline (Vick and Carver, 1988). The signature of sudden subsidence of the fold axis is preserved in the microstratigraphy of the late Holocene sedimentation of intertidal and salt marsh sediments along the bay margin. Described initially by Atwater

(1987), the sudden subsidence marsh stratigraphy consists of high marsh surface peats abruptly overlain by intertidal mud which grade into overlying high marsh peats. At least three buried high marsh peats are present at the north end of Humboldt bay. Carbon 14 ages for the buried marsh peats suggest the upper buried sequences may be correlative with the raised beach and dune deposits at Clam beach (stop 3), and formed during the same coseismic events.

Stop 7, Little Salmon fault, Little Salmon Creek trench site

During the past several years, trenching studies have shown some of the major thrusts in the Eel River-Mad River region have been active during the late Holocene. The *Little Salmon fault* exhibits abundant evidence of Holocene activity. Figure 11 shows the surface geology in the vicinity of the trenches placed across the west trace of the fault. Trenches also show east trace of the fault juxtaposes faulted and over-turned Hookton (late Pleistocene) sediments beneath lower Rio Dell siltstones and generates scarps in late Holocene terraces along Salmon Creek.

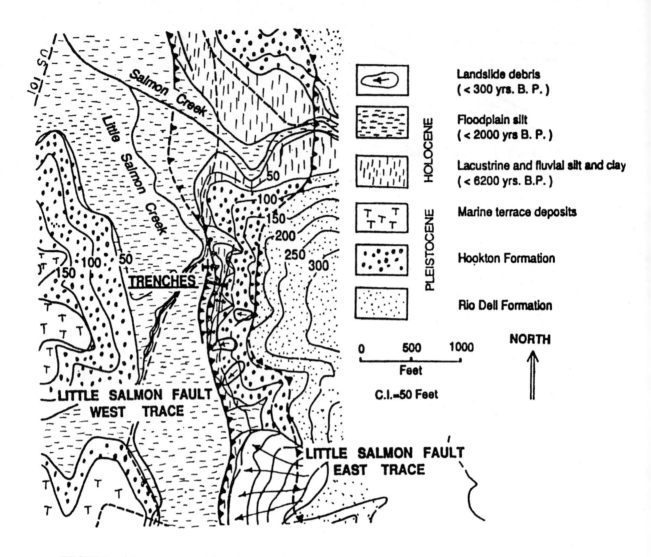

FIGURE 11 Generalized geologic map of the Little Salmon fault trench site and adjacent area.

Figure 12 is a composite interpretative section across the west trace of the fault, based on a series of exploration trenches and shallow bore holes. The trenching shows that the mid-Holocene (about 6.2 ka) valley floor sediments have been offset vertically about 15 m (Carver and Burke, 1987a,1987b). The offset has generated a sharp fault-bend fold (Figure 11-D), as well as a thrust which propagates to within 20 cm of the ground surface. Three latest Holocene overbank floodplain silt sequences and capping soils onlap the flank of the fold and are progressively more deformed, probably reflecting the last three slip events on the fault. Wood and charcoal recovered from a number of horizons in the trenches provide the control on the general ages shown in figure 12. The last three events appear to have occurred

about 300, 1000, and 1800 years ago (Figure 13).
in the following respects:
a) slip rate: at least 6.5mm/yr and possibly more than 10 mm/yr,
b) apparent amount of slip during each event: at least 4 m, and possibly as much as 7 m,
c) recurrence interval: 500 to 700 years.
These vital statistics for the Little Salmon fault indicate it is very active and probably is capable of generating very large earthquakes.

Stop 8-Table Bluff, Southport Landing quarry

The purpose of stop 8 is to look at the Table Bluff anticline. Late Pleistocene growth of the anticline is indicated by

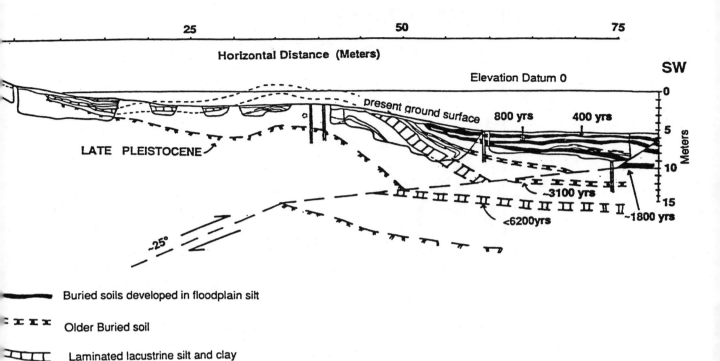

Little Salmon Fault   West Trace   Little Salmon Creek Site

Buried soils developed in floodplain silt

Older Buried soil

Laminated lacustrine silt and clay

FIGURE 12 Interpretive cross section across the west trace of the Little Salmon fault based of exploration trenches and shallow borings at the Little Salmon trench site. General ages for stratigraphic markers is based on C^{14} dates for many horizons exposed in the trenches. The fault shows growth during overbank sedimentation on the floor of Little Salmon Creek valley.

the deformation of a flight of late Pleistocene marine terraces draped across the axis of the fold (Figure 14). Several different ages of terraces can be distinguished on the basis of soils and geomorphology. The western most terrace (large dot pattern, figure 14a) has been correlated to the Patricks Point terrace (stage 5a, 83 ka) at Trinidad (Carver and others, 1986). Progressively older surfaces are present inland. A volcanic ash is exposed in road cuts on the east end of Table Bluff (Tompkins Hill ash, Woodward-Clyde Consultants, 1980) and on the south side of the anticline near Loleta, also near the terrace surface and probably within the terrace cover sediments. The ash has provided an age of about 350 ka (Woodward-Clyde, 1980).

The terraces at Table bluff have been folded into a broad open anticline, with planer limb segments and multiple axial surfaces or hingelines. Two of these hingelines are easily seen in the topography, and are shown on the map in figure 14a. The geometry of the fold indicates the structural style is that of a fault bend fold. Thus the Table Bluff anticline has formed in response to slip on a buried thrust. Fault generated folds such as Table Bluff and Fickle Hill represent the dominant fold type in the Eel River basin region. Stratigraphic and geomorphic evidence suggests all of these structures are youthful and actively forming to the present.

The Southport Landing quarry has been excavated into the axis of one of several small folds developed on the north limb of the Table Bluff anticline. Surface mapping and terrace soil studies indicate that the terrace at this location formed during the stage 5a highstand (83ka). In the quarry exposure, the 5a terrace deposits are gently folded and faulted. They overlay with angular discordance older terrace sediments which are more tightly folded and severely faulted.

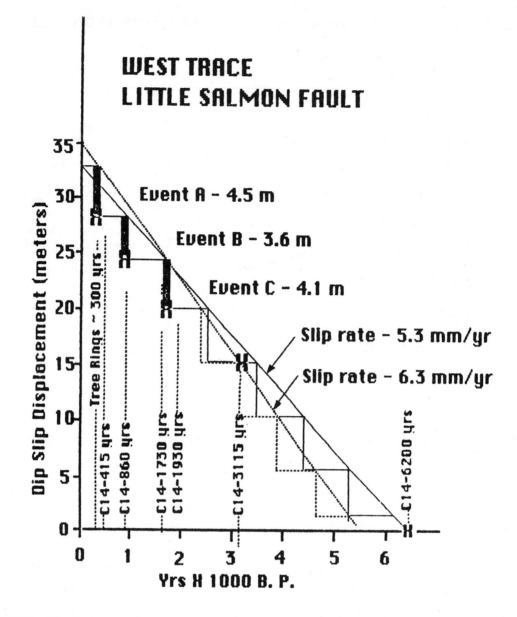

FIGURE 13 Late Holocene paleoseismic history and slip rate for the west trace of the Little Salmon fault based on displacement measurments and C^{14} ages obtained from the trenches at the Little Salmon Creek site. Three slip events of about 4 to 5 m each have occurred in the past 2 ky. The data indicates this trace of the fault has a late Holocene slip rate of about 6 mm/yr. Trenches across the east trace of the fault also show late Holocene displacement, but insufficient age control precludes event correlation.

Thermolumnesence dating of the older terrace sediments has yielded an age of 105 ka (Berger and others, 1988), indicating continued late Pleistocene growth of the fold. Both normal and reverse faults cut the fold. The normal faults parallel the axis of the fold and are interpreted to be rootless bending-moment faults resulting from flexure of the growing anticline.

Stop 9-Centerville beach

The sea cliff exposures along the coast in the Centerville beach area constitute the reference section for the Eel River basin stratigraphy and have been described by many researchers (Ogle,

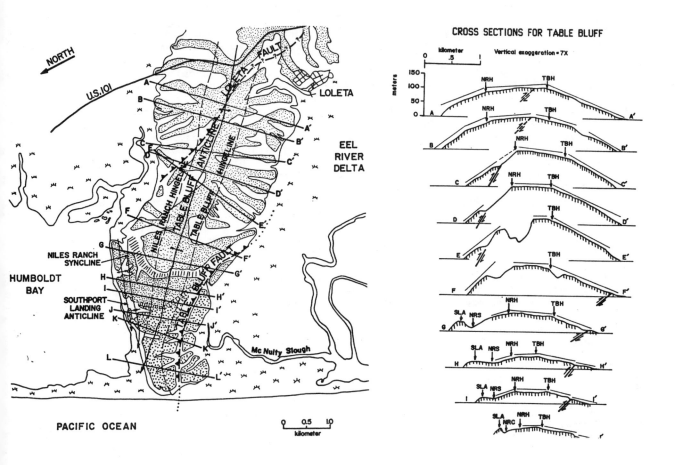

FIGURE 14 Map and topgraphic cross sections of folded late Pleistocene marine terraces at Table Bluff. The Table Bluff anticline is interpreted to be a large fault-bend fold above the Table Bluff fault. The fold geometry consists of planar fold limbs separated by several sharp hingelines. Several smaller folds are present on the north limb of the large anticline.

1950; Ingle, 1976). A thick sequence of marine sediments is exposed, ranging from upper Miocene to late Pleistocene. The sediments are interpreted to reflect filling of the forearc basin and shallowing of the depositional environment from initial bathyal depths to above wave base in the early Pleistocene (Ingle, 1976; McCorey, 1987). Late Pleistocene marine terraces are unconformably cut across the earlier basin sediments. The terraces are tilted 4 to 6 degrees to the north and rise from sea level at Centerville to more than 400 meters near Cape Mendocino.

The basin sediments are in fault contact with melange of the *False Cape shear zone* at the south end of the sea cliff exposure. This melange may represent the accretionary complex of the Cascadia subduction zone. The melange contains tectonically infaulted blocks of Wildcat sediments, and reportedly includes sheared shaley matrix containing foraminifera of Pliocene age (Ogle, 1950). The melange and uplifted late Cenozoic sediments and terraces may reflect rapid uplift of the southern end of the accretionary margin above the tectonically thickened Gorda plate, and thus represent the tectonic bow wave of the northward migrating Pacific plate.

REFERENCES CITED

Aalto, K. R., and Murphy, J. M., 1984, Franciscan Complex geology of the Crescent City area, northern California: in Blake, M. C., Jr., ed., *Franciscan geology of northern California*: Pacific Section, Society of Economic Palentologists and Mineralogists Book 43, p.185-201.

Atwater, B. F., 1987, Evidence for great Holocene earthquakes along the outer coast of

Washington State: Science, v. 236, no. 4804, p. 162-168.

Berger, G., Burke, R. M., Carver, G. A., and Easterbrook, D. J., 1988, Thermolumnescence analysis and paleomagnetic signature-techniques for establishing numerical age estimates on tectonically deformed marine terraces, northern California (abs): Geological Society of America, Abstracts with Programs, v. 20, no. 7, p. A 53.

Burke, R. M., Carver, G. A., and Lundstrom, S. C., 1986, Soil development as a relative dating and correlation tool on marine terraces of northern California: Geological Society of America Abstracts with Programs, v. 18, p. 91.

Carver, G. A., 1987, Late Cenozoic tectonics of the Eel River basin region, coastal northern California: *in* Schymiczek, H. and Sushsland, R., eds., *Tectonics, Sedimentation and Evolution of the Eel River and associated coastal basins of northern California*: San Joaquin Geological Society Miscellaneous Paper no. 37, p. 61-72.

Carver, G. A., 1988, Seismic potential of the Gorda segment of the Cascadia subduction zone (abs): *in Holocene Subduction in the Pacific Northwest*: Abstracts with Programs, Quaternary Research Center symposium, Seattle, p.6.

Carver, G. A., and Burke, R. M., 1987a, Late Holocene paleoseismicity of the southern end of the Cascadia Subduction zone (abs): EOS, v. 68, n. 44, p.1240.

Carver, G. A. and Burke, R. M. 1987b, Late Pleistocene and Holocene paleoseismicity of the Little Salmon and Mad River thrust systems, N. W. California; Implications to the seismic potential of the Cascadia subduction zone (abs): Geological Society of America Abstracts with Programs, v. 19, n. 7, p. 614.

Carver, G. A., and Burke, R. M., 1987c, Investigations of late Pleistocene and Holocene thrust faulting in coastal California north of Cape Mendocino: *in* Jacobsen, M. L., and Rodrigues, T. R., eds., *Summaries of Technical Reports*, v. XXIII, National Earthquake Hazards Reduction Program, p. 165-168.

Carver, G. A., Stephens, T. A., and Young, J. C., Quaternary Geology of the Mad River Fault Zone, Arcata North, Arcata South, Korbel, and Blue Lake 7.5' Quadrangle: U.S. Geological Survey Open File Report (in press).

Heaton, T. H., and Kanamori, H. 1984, Seismic potential associated with subduction in the northwest United States: Bulletin of the Seismological Society of America, v. 74, p. 933-941.

Ingle, J. C., Jr., 1976, Late Neogene paleo-bathymetry and paleoenvironments of the Humboldt basin, northern California: *in* Fritche, A. E., TerBest, H., and Wornardt, W., eds., *The Neogene symposium*, Pacific Section, Society of Economic Palentologists and Mineralogists, p. 53-61

Jayko, A. S., and Blake, M. C., 1987, Geologic terranes of coastal northern California and southern Oregon: *in* Schymiczek, H. and Sushsland, R., eds., *Tectonics, Sedimentation and Evolution of the Eel River and associated coastal basins of northern California*: San Joaquin Geological Society Miscellaneous Paper no. 37, p. 1-13.

Kelsey, H. M., and Carver, G. A., 1988, Late Neogene and Quaternary tectonics associated with the northward growth of the San Andreas Transform Fault, northern California: Journal of Geophysical Research, v.93, p. 4797-4819.

McCory, P.A., 1987, Late Cenozoic history of the Humboldt basin, Cape Mendocino area, California: Unpublished Ph.D. dissertation, Stanford University, 229 p.

Nilsen, T. H., and Clarke, S. H., Jr., 1987, Geologic evolution of late Cenozoic basins of northern California: *in* Schymiczek, H. and Sushsland, R., eds., *Tectonics, Sedimentation and Evolution of the Eel River and associated coastal basins of northern California*: San Joaquin Geological Society Miscellaneous Paper no. 37, p. 15-31.

Ogle, B. A., 1953, Geology of the Eel River valley area, Humboldt County, California: California Department of Natural Resources, Division of Mines, Bulletin 164.

Riddihough, R. R., 1984, Recent movements of the Juan de Fuca plate system: Journal of Geophysical Research, v.93, p. 6980-6994.

Sarna-Wojcicki, A. M., S. D. Morrison, C. E. Meyer, and J. W. Hillhouse, 1987, Correlation of upper Cenozoic tephra layers between sediments of the western United States and eastern Pacific Ocean and comparison with biostratigraphic and magnetostratigraphic age data: Geological Society of America Bulletin, v. 98, p. 207-223.

Taber, J. J., and Smith, S. W., 1985, Seismicity and focal mechanisms associated with the subduction of the Juan de Fuca plate beneath the Olympic Peninsula, Washington: Bulletin of the Seismological Society of America, v. 75, p. 237-249.

Vick, G. and Carver, G. A., 1988, Late Holocene paleoseismicity, northern Humboldt Bay, California (abs): Geological Society of America Abstracts with Programs, v. 20, p. A232.

Woodward-Clyde Consultants, 1980, *Evaluation of the potential for resolving the geologic and eismic issues at the Humboldt Bay Power Plant*: Unit No. 3, Appendices, Woodward-Clyde Consultants, Walnut Creek, California.

The San Andreas Transform Belt

Long Beach to San Francisco, California
July 20–29, 1989

Field Trip Guidebook T309

Leaders:
Arthur G. Sylvester   *John C. Crowell*

with contributions by:
Jon S. Galehouse   *E. A. Hay*   *N. T. Hall*
W. R. Cotton   *Carol S. Prentice*   *John D. Sims*

American Geophysical Union, Washington, D.C.

COVER   Oblique aerial photograph looking north-northwest at the San
Andreas Fault in the Carrizo Plain, Central California.   Photograph
by Robert E. Wallace, May 25, 1965.

GUIDE TO THE SAN ANDREAS TRANSFORM BELT

GUIDE TO THE SAN ANDREAS TRANSFORM BELT

PREFACE

Arthur G. Sylvester
Department of Geological Sciences, University of California, Santa Barbara

John C. Crowell
Institute of Crustal Studies, University of California, Santa Barbara

It is always a moving experience for us here in California to stand on the San Andreas fault - to realize that the ground beneath our feet has shifted 330 km dextrally over the last 24 Ma - to realize that even as we stand there, it may shift laterally as much as 6 m - and to know that in some places it is presently creeping as rapidly as 35 mm/yr.

This tectonic activity mars the utopian image generally evoked by the word "California," but it is just that tectonic activity and associated earthquakes which are primarily responsible for the attractive geographic variety throughout the "Golden State" and which are the subjects of this field trip. The word "California" itself is mythological in origin; it is believed that California was named for Calafia, a legendary Amazon queen renowned for her power and beauty. She reigned over a utopian paradise replete with gold, pearls, sunshine, and fruit. Today, California still boasts the gold, sunshine, and fruit, lacking only pearls and Queen Calafia herself!

California is a natural laboratory to study mechanics of strike-slip faults, and the San Andreas fault is the most thoroughly studied strike-slip fault in the world because of the earthquake hazard it presents to the 25 million inhabitants of the state; because of the intensive geological and geophysical studies by the petroleum industry for oil and gas; because of its proximity to several major university, government, and petroleum industry research laboratories; and because it is well exposed and readily accessible. But its history is only the latest of a complex sequence of superposed tectonic events to mold the present geologic structure of California.

The first step to unravel these events is palinspastic restoration of the San Andreas fault. This task has gained considerable momentum in the last 25 years, and began before the unifying theory of plate tectonics. In addition, the discovery of the frequency of great earthquakes on the San Andreas and related faults has spawned, in turn, the realization of the enormous risk that California faces from future earthquakes as its population and development increase. Considerable time, effort, and dollars have been expended by hundreds of scientists to determine the past history and present activity on the San Andreas fault system in order to predict its future activity.

Highlights of the lessons learned from the San Andreas fault include the great amount of lateral offset that has occurred in Late Cenozoic time which supported the gradual realization nearly 40 years ago that great strike-slip predominates the movement on other major faults worldwide; that strike-slip faults are shallow structures and may penetrate downward only to the seismogenic zone; that the fault may be subdivided into characteristic segments which reflect their long and short-term seismic activity; that parts of the fault continually slip by fault creep; that the return frequency of its major earthquakes (>M 7) is measured in terms of only a few hundred years compared to the thousands to hundreds of thousands of years typical of other kinds of faults; and that parts of the crust within the San Andreas fault system have rotated as much as 90° about a vertical axis. Seismological and geodetic studies have characterized the depth, breadth, and locus of shear strain accumulation and release to a degree approached for few other faults. Seismic reflection and refraction, heat flow, and stress measurements coupled with analyses of P-wave delays from earthquakes and quarry blasts indicate that the San Andreas fault and much of the terrane within the great San Andreas transform belt in southern California is detached on horizontal discontinuities at depth. Much of southern California is therefore allochthonous. A remarkably repetitive and identical set of earthquakes between 1857 and 1966 on the San Andreas fault at Parkfield in central California gives a basis for an unique earthquake prediction experiment which, if successful, will yield a wealth of data about the constitutive properties of earthquake sources as well as the physical changes in the earth that precede and accompany earthquakes.

These fundamental and exciting revelations about one of the master geologic structures of the earth have raised a host of ideas and questions to apply to, and to advance understanding of, less well-studied faults.

The purpose of this field trip of international scholars is to share our accumulated knowledge and ideas about the San Andreas fault system and to pose questions for which we and our California colleagues presently seek answers. Our objective is to foster a broader understanding of the tectonics and mechanics of strike-slip faults which will help our guests learn more from their respective studies elsewhere and which may lead back to answers to our problems.

The literature pertaining to the San Andreas fault is

vast, because it is such an intensively studied fault, and to give proper reference citations for every fact, interpretation, and idea would fill an additional guidebook and unnecessarily clutter this one. This is a guidebook and is therefore not a "stand-alone" scientific document and, accordingly, is not fully referenced. We regret that its purpose and length requirements limited us from acknowledging the significant contributions of many people. We have cited key references, however, and refer readers to the Rubey Volumes (Ernst, 1981; Ingersoll and Ernst, 1987) for more detailed information about the tectonics of California, and to SEPM Special Publication 37 (Biddle and Christie-Blick, 1985) and Sylvester (1988) for discussion of strike-slip fault mechanics and a more comprehensive bibliography.

We might entitle this field trip "The Geology of Western California", because the San Andreas is a profound structure of regional scale, stretching through western California for a distance of about 1100 km, and because the San Andreas fault cuts rocks of all ages, from Precambrian to Recent, and of all types, from high-grade metamorphic rocks to the most recent alluvium. We shall drive that distance and more in nine days. In that time and distance we shall study and discuss the origin and evolution of the great Mesozoic batholiths; of Mesozoic ophiolitic rocks; of Mesozoic and Cenozoic turbidites; of Cenozoic diatomaceous and fluvial strata; of Precambrian anorthosite and related rocks; of Cenozoic and Recent basalt and rhyolite; of Mesozoic thrusts and Cenozoic detachments; of pull-apart basins and uplifts; of oil and gas fields; and of a variety of landforms.

The field trip focuses on the structural styles of faults, folds, basins, uplifts, and associated landforms of the San Andreas fault which are abundant and well-displayed along nearly the entire length of the fault system. Some concepts may be addressed at one location, whereas corollaries to those concepts may be illustrated better a few hundreds of kilometers away and a couple of days later. Field discussions, highlighted by the evidence displayed by the rocks and structures, will address the inception and evolution of the San Andreas fault system, as well as the history of the development of ideas and concepts about it. We shall discuss (and may experience!) its earthquake activity - both historic and paleoseismic - and we shall call frequent attention to the ever-present hazards and risks of future activity all along the fault.

The field trip begins at the edge of the urban sprawl of the Los Angeles metropolitan area and ranges back and forth across the plate boundary from near the International Boundary with Mexico to Pt. Reyes north of San Francisco. We start from the crest of Signal Hill, an enormously productive oil field on a squeezed up structural block in the Newport-Inglewood fault zone. We'll drive along the Pacific Ocean coastline, over the Santa Ana Mountains to the Salton Trough below sea level, where scars of several recent earthquakes along the San Jacinto, Imperial, and related faults are well-preserved in the desert. We'll cross the Transverse Ranges and climb up into the pines in the San Gabriel Mountains where the San Andreas fault reaches its highest elevation - nearly 2500 m above sea level - and where it is eroded to its deepest structural level, then continue along the locus of the great 1857 earthquake at the edge of the vast Mojave Desert to Pallett Creek to study evidence for nine similar earthquakes in the last 1600 years.

At Palmdale, we'll cut back into the Transverse Ranges to study sedimentation related to strike-slip faulting in Ridge basin. We'll continue northward into the Great Valley to consider the prolific oil fields in a remarkable series of en echelon folds adjacent to the San Andreas fault. The route will proceed across the Temblor Range, and we'll drive along the San Andreas fault to Carrizo Plain to view Wallace Creek which might be regarded as the prototype for a stream course displaced by strike-slip. From there, we'll drive along the San Andreas fault to Parkfield to consider the evidence for an on-going prediction experiment. We'll inspect the Franciscan Formation on Table Mountain before crossing the Diablo Range. The following day takes us along the actively creeping segments of the San Andreas fault along State Highway 25 and the Calaveras fault in the town of Hollister. We'll examine the Hayward and San Andreas faults in the San Francisco Bay area on the next day. The final day of the trip takes us across the Golden Gate Bridge and north of San Francisco along the locus of the great 1906 earthquake.

We look forward with pleasure to guiding participants of the 28th International Geological Congress through this geology and structure, as well as through some of California's cultural and social wonders. We hope it will be an inspiring experience for the participants, and perhaps even a "moving" one should we experience an earthquake that adds to the 330 km of displacement already accumulated on the San Andreas fault!

Subsequent users of this guidebook will need the 1:750,000 scale geologic map of California (Jennings, 1977) as well as a standard road map.

Penny Hanshaw, Tom Blenkinsop, Craig Nicholson, and Alan Hull read and edited parts or all of the manuscript, and Karin Sylvester retyped parts of it. Dave Crouch prepared a number of the figures. We are grateful for their help.

INTRODUCTION TO THE SAN ANDREAS TRANSFORM BELT

Arthur G. Sylvester
Department of Geological Sciences, University of California, Santa Barbara

John C. Crowell
Institute of Crustal Studies, University of California, Santa Barbara

REGIONAL TECTONIC SETTING

The San Andreas transform belt is the surface expression of one of the world's best known plate boundaries and, along with the divergent boundary of the Gulf of California and the Salton Trough, is the only exposed, active plate boundary within the continental United States. The transform belt strikes from N35°W to N45°W across western California, separating the relatively northwestward drifting Pacific plate from the North American plate (FIG. 1). This relative motion imparts an overall dextral pattern of motion between the two plates at a rate of 55 mm/yr, based on interpretations of seafloor magnetic anomalies. Movement between the two plates is concentrated across a zone from 80 km to 100 km wide in the central part of California, but it may be from 500 km to 1000 km wide when viewed regionally. Within the narrower zone are a number of right-slip faults which, collectively, constitute the San Andreas fault system. Some of the faults, including the Hayward, Calaveras, San Jacinto, and Elsinore faults, as well as the San Andreas fault itself, are active, having produced significant earthquakes in historic time. Paleoseismic evidence shows that others, including the Hosgri, Rinconada, and San Gabriel faults, have been active in Quaternary time, so they are potentially active.

The northwest structural grain of western California, which is followed by the San Andreas fault, is interrupted in southern California by the Transverse Ranges, a physiographic province of east-west mountains, valleys, faults and folds. The strike of the San Andreas fault changes from N35°W north of the Transverse Ranges to N45°W on the south side. The Transverse Ranges form a relatively youthful physiographic feature, having been uplifted in Pleistocene and Recent time in response to concentrated crustal shortening where an irregular edge of the North America plate impinges against the Pacific plate.

The San Andreas transform came into existence about 29 Ma ago as a result of the rearrangement of the interactions between the Pacific and North American plates when the East Pacific Rise came into contact with the subduction zone that had characterized the edge of the North American plate through Mesozoic time (FIG. 2). Since mid-Miocene time, 330 km of right slip has occurred along the fault, based primarily on correlations of several offset rock units which range greatly in composition, origin, and age. That diastrophism included the opening of several, deep, irregular-shaped basins which accumulated as much as 10,000 m of mainly clastic marine sedimentary strata.

THE MODERN SAN ANDREAS FAULT

The great extent of the San Andreas fault was discovered in 1906 when it abruptly ruptured the surface in the severe San Francisco earthquake. Geologists had mapped short stretches of the fault in various places previously, and although Fairbanks had followed its physiographic expression from San Francisco to San Bernardino, no one realized it was a single, continuous structure until the earthquake.

The historic slip rate of the San Andreas fault, judging from fault creep and repeated geodetic surveys across the central part of its active trace, is 35 mm/yr, about 20 mm/yr less than the geologic rate of slip of 55 mm/yr between the Pacific and North American plates. This discrepancy has been explained by apportioning the 20 mm/yr residual to other active faults of the San Andreas fault system, especially those offshore, and perhaps on faults as far east as those in the Basin and Range Province. Displacements on the north- and northwest-striking faults of the Basin and Range Province are largely normal, but a component of right slip has occurred along several of these faults during earthquakes, and offset Pleistocene alluvial fans and glacial moraines along some of the major faults have a component of right slip that is as great as the component of normal slip.

Segmentation

The San Andreas fault system has been subdivided into five major segments (FIG. 3) based on the size and type of historic earthquakes generated in those segments (Allen, 1968). The central segment is straight, narrow, and parallel to the plate motion vector. The active strand of the San Andreas fault is contained almost entirely within the Mesozoic ophiolitic rocks of the Franciscan Complex in this segment. Aseismic fault creep, having a maximum rate of 35 mm/yr over at least the last 25 years, is characteristic of this segment. Creep diminishes to zero at each end of the creeping segment where it joins the adjacent bent, locked segments. Earthquakes having magnitudes less than 6 are common in the creeping segment, and some

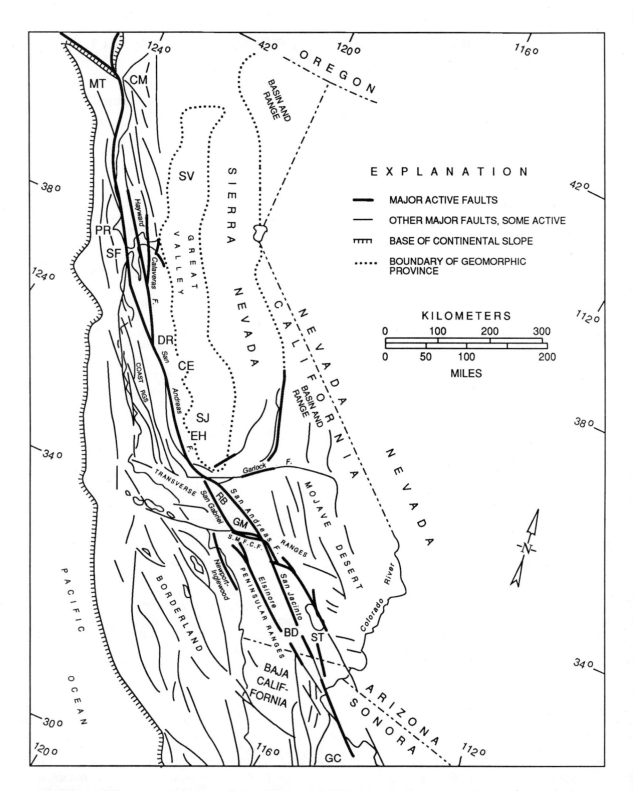

FIGURE 1 The San Andreas Fault System and Physiographic Provinces of California. BD - Borrego Desert; CE - Coalinga; CM - Cape Mendocino; DR - Diablo Range; EH - Elk Hills; GC - Gulf of California; GM - San Gabriel Mountains; MT - Mendocino Triple Junction; PR, Point Reyes; RB - Ridge basin; SC - Santa Cruz Mountains; SF - San Francisco; SJ - San Joaquin Valley ; SM - Santa Monica Mountains; SMFCF - Santa Monica fault-Cucamonga fault trend, including the Raymond and Sierra Madre faults; ST - Salton Trough; SV - Sacramento Valley. From Crowell (1985) with modifications.

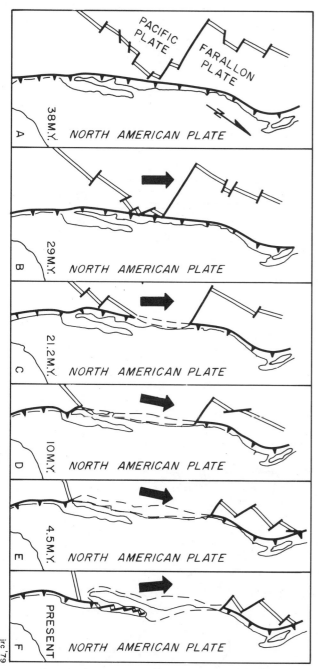

FIGURE 2 Interpretative Diagram Showing Interaction Among the Farallon, Pacific, and North American Lithospheric Plates for Six Intervals of Tertiary Time. The North American plate is considered fixed. The dashed line in diagrams C, D, E, and F outlines the transform region of basins interspersed with high-standing terrace. The present shoreline is shown on each map for reference, although it did not have this configuration at the times of the diagrams. Transform plate boundaries are shown as single lines, divergent as double lines, and convergent as barbed. After Blake et al. (1978).

seismologists believe that the creep prevents sufficient elastic strain energy from being stored here, thus precluding earthquakes greater than M 6.5.

Northwest and southeast of the creeping segment, the fault juxtaposes Mesozoic granitic basement and the oceanic Franciscan Complex, and the fault zone is curved, correspondingly wider, and structurally more complex (FIG. 1). The largest historic earthquakes in the San Andreas fault system, those of 1857 and 1906, occurred on these segments. The return frequency of paleoseismic earthquakes in these segments is about every 150-300 yrs, with a range of about 70-300 yrs. Thus, these segments are thought to "lock" and build up sufficient strain to generate infrequent, major earthquakes.

Northwest and southeast of the locked segments, several major faults splay from the San Andreas fault (FIG. 3). Historic earthquakes are more frequent and smaller in each of these splayed domains relative to the locked segments of the fault system, but they are larger and less frequent than those in the creeping segment. Aseismic horizontal creep has also been documented on several of these faults, but at rates that are from a tenth to a third of the maximum rate measured in the central segment.

The Active Fault Trace

The presently active trace of the San Andreas fault lies in a physiographically distinct zone of crushed crystalline basement rocks and locally highly contorted strata that ranges from a few tens of meters to a several hundreds of meters wide. The fault stretches about 1100 km northwestward from the southern Salton Sea as an almost continuous, but sinuous fault to Cape Mendocino in northern California. Several changes of strike interrupt the relatively straight trace - the San Bernardino bend in the San Bernardino Mountains where the fault's path is sufficiently tortuous that the fault has several, braided strands; the "big bend" in the Transverse Ranges where a 10 km-long stretch of the fault strikes east-west; northwest of the juncture with the Hayward and Calaveras faults; and the Mendocino bend north of San Francisco, most of which is offshore.

Major Splay Faults

Southern Faults. The main fault strands at the south end of the San Andreas fault system are the San Jacinto and Elsinore faults, and the Newport-Inglewood fault zone. All are right slip faults. The San Jacinto fault is the most seismically active fault in California in historic time, having generated six M 6+ earthquakes since 1899, and its slip rate is about 8-12 mm/yr in late Quaternary time. Total right slip since Pliocene time is 40 km.

The Elsinore fault is 215 km long and has been credited with only one M≥6 earthquake in historic time, but stretches of the fault near the International Boundary show geomorphic evidence of very recent activity. The sense of displacement ranges considerably on the Elsinore fault, including normal and reverse in combination with right slip. Total right slip may be as much as 40 km; geomorphic evidence all along the

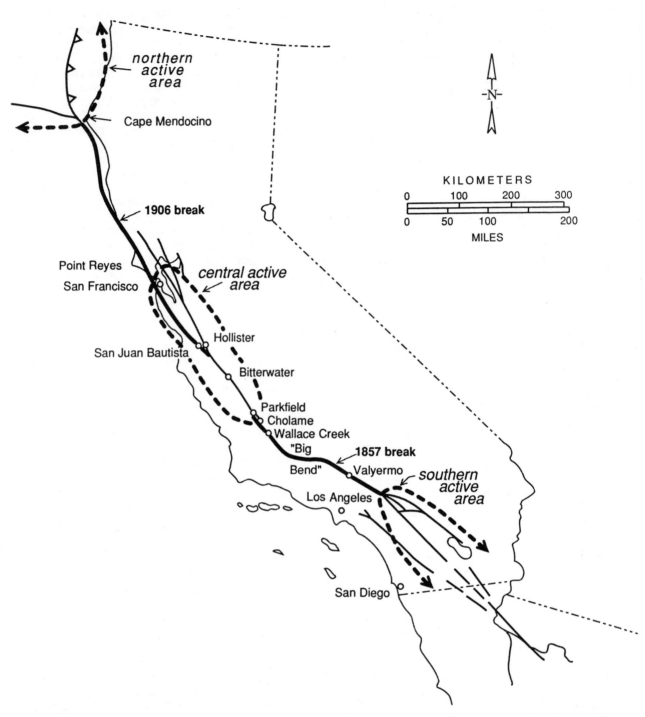

FIGURE 3 Segmentation of San Andreas Fault System According to Contrasting Seismic Behavior (after Allen, 1968). The 1857 and 1906 segments produce infrequent, outstanding earthquakes; the segment of active fault creep is in central California between the ends of the 1857 and 1906 segments.

Elsinore fault indicates oblique-slip movement during late Quaternary time. The Quaternary slip rate is estimated to be 5 mm/yr, but trenching data are conflicting.

The Newport-Inglewood fault zone along the southwest margin of the Los Angeles Basin is known primarily for its associated en echelon folds which have trapped great quantities of petroleum. Although horizontal displacement of Pliocene markers does not

exceed 800 m, the destructive Long Beach earthquake (M 6.3) of 1933 was associated with the Newport-Inglewood fault zone, which juxtaposes oceanic rocks against continental crystalline rocks in the deep subsurface .

Northern faults. The principal splay faults in northern California include the Hayward and Calaveras faults, the Green Valley fault zone, and their discontinuous and incompletely studied northwestern extensions: the Rodgers Creek, Healdsburg, and Maacama fault zones northwest of the Hayward fault, and the Concord and Green Valley faults northwest of the Calaveras fault (FIGS. 1, 21). All are right slip faults with components of vertical displacement. The amount and timing of slip is not clearly established for the Hayward fault, whereas right slip is about 20 km along the Calaveras fault since Pliocene time. Both the Hayward and Calaveras faults creep aseismically at rates of 4-5 mm/yr and 12 mm/yr, respectively (see Galehouse, this volume). The Hayward fault generated major earthquakes (M≥6) with associated surface ruptures in 1836 and 1868, and moderate earthquakes occurred in 1979, 1984, and 1988 on the southern part of the Calaveras fault. Microearthquakes show that lesser faults participate in the dextral motion among the major faults, including parts of the Pilarcitos, Bear Valley, Zayante, Butano, Sargent, and Berrocal faults, but parts of these faults also have reverse or thrust components of displacement.

Other faults. Other major faults comprising the San Andreas fault system are offshore and less well known. They include the Seal Cove-San Gregorio-Hosgri fault zone which projects southward from the San Francisco region to the west end of the Transverse Ranges, and the Rose Canyon fault which some geologists regard as the offshore extension of the Newport-Inglewood fault zone, and which comes onshore north of San Diego (FIG. 1). Assignment of historic earthquakes to these faults is problematic, and recent geologic slip rates are debatable and inconclusive. Still farther offshore are several northwest-striking faults in the Continental Borderland of southern California, including the San Clemente fault, which are also right-slip faults, but for which the amount of offset, recent slip rate, and earthquake history and potential are largely unknown.

GEOMORPHOLOGY AND STRUCTURE

Geologic and geomorphic evidence of the fault, its major splay faults, and individual traces is abundant. The most distinctive aspect of the San Andreas fault is the extreme linearity of the "rift" topography over much of its length: "*It has a curiously direct course across mountains and plains with little regard for gross physiographic features, yet it influences profoundly the local topographic and geologic features within it*" (Noble, 1927, p. 37). The "rift" may be up to 10 km wide with a variety of fault-formed features, including pressure ridges, fault scarps of various heights, closed depressions called sag ponds if filled with water, shutterridges, lines of springs, and deflected streams and canyons. These physiographic features are especially well-developed and preserved along the segments of the fault which ruptured in the 1857 and 1906 earthquakes. Fault gouge is a barrier to the movement of groundwater, so that the subsurface water level is higher on one side of the fault than the other,

thus promoting more abundant growth of vegetation on the high side. This is seen especially well in the Salton Trough.

Elongate structural and topographic depressions called pull-apart basins are developed locally between en echelon or parallel, sidestepping fault strands. One of the most symmetrical of these is the Lake Elsinore pull-apart on the Elsinore fault. The San Jacinto pull-apart on the San Jacinto fault is an active, rapidly-subsiding structure, as are several pull-aparts in the deep subsurface of the Salton Trough which have been documented by seismic refraction studies. San Francisco Bay occupies a large, structurally complex "sag" between the San Andreas and Hayward faults.

Rocks ranging greatly in age and lithology from high-grade Proterozoic metamorphic rocks to poorly-lithified Recent alluvium are found along the San Andreas fault. In many places rocks of very different age and structure are juxtaposed across the fault. In some places, volcanoes have erupted lava flows and breccias; elsewhere, basins have accumulated sediments containing distinctive clasts from unique source areas. Both the volcanoes and the source areas are now separated by strike-slip from their flows and sedimentary detritus, respectively. Correlations of these distinctive rock units as well as geologic lines on either side of the faults have provided the evidence for the sense, timing, and amount of their geologic offsets.

EN ECHELON FOLDS

En echelon folds, which are one of the characteristic elements of the strike-slip structural style, are commonly developed in sedimentary strata along the San Andreas fault and related strike-slip faults. The large en echelon folds along the southwest edge of the San Joaquin Valley and the smaller folds along the Newport-Inglewood fault zone in the Los Angeles area are particularly noteworthy, because they have yielded more than half of California's oil and gas production for nearly 70 years. Other notable areas of en echelon folds are in the Berkeley Hills between the Hayward and Calaveras faults, and along the southern San Andreas and San Jacinto faults in the Salton Trough.

Recent studies have called into question the genesis of the en echelon folds in the San Joaquin Valley, finding that the axis of normal stress, derived from analyses of well-breakouts, is presently oriented perpendicular to the fault strike. This observation is contrary to what is predicted by the Riedel model of simple shear, in which the normal stress is oriented at an angle of 45° to the fault strike (FIG. 4). The paradox may be reconciled by considering that the transform fault displacement of 35 mm/yr is decoupled from the regional shortening deformation of 5 mm/yr, and that they operate simultaneously and largely independently. But that simple conclusion has revealed a host of problems and questions related to strike-slip faults and to mechanical layering and decoupling within the upper crust.

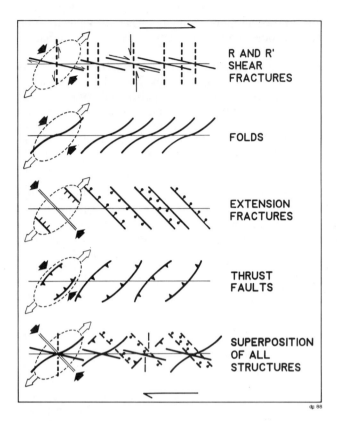

FIGURE 4 Geometry of Fractures, Folds, and Faults that Form in Right Simple Shear. Solid arrows indicate direction of shortening, open arrows indicate direction of extension. From Sylvester (1988) with modifications.

TECTONIC ROTATION AND CRUSTAL DECOUPLING

Interpretations of paleomagnetic declination studies show that large and small crustal blocks in southern California have rotated clockwise and counterclockwise about vertical axes since middle Miocene time in response to right simple shear and oblique extension along the San Andreas fault system (FIG. 5). Mechanically, the phenomenon requires thin blocks, relative to their areal dimensions, that they are relatively rigid, and that they are somehow decoupled from more ductile crust at depth.

Seismic reflection studies in southern California and across the San Andreas fault, analyses of P-wave delays from earthquakes and quarry blasts, and analyses of earthquake focal mechanisms suggest that the upper crust is layered, and perhaps decoupled and allochthonous. The data are best beneath the western Mojave Desert and central Transverse Ranges. Moreover, the recent Coalinga (1983) and Whittier Narrows (1987) earthquakes occurred on blind thrust faults which project under the San Andreas fault and Transverse Ranges, respectively, recalling the concept of "flake tectonics". These recent revelations have led to hypotheses backed up by permissive seismic and geologic evidence that several of the major strike-slip faults do not extend below the seismogenic zone, and

that even the San Andreas fault is locally offset at the base of the seismogenic zone from its extrapolated position in the lower crust and lithosphere.

The concept of tectonic "flakes," which may move abruptly over a broad area of the seismogenic zone and generate earthquakes on blind thrusts, has generated considerable concern about related earthquake hazards. Previously, many seismologists believed that all of the active seismogenic structures in southern California, mostly strike-slip faults, reached the surface and that the hazards could be determined by mapping the faults and by studying their paleoseismic histories in near surface exposures. Now seismologists speculate several large, early historic earthquakes in and around the Transverse Ranges, such as the southern California earthquakes of 1812, may have occurred at the base of one of these flakes rather than on some, as yet undetermined, high angle fault.

As of this writing, geologists and geophysicists are seeking answers to questions regarding the areal extent of the crustal decoupling postulated for parts of southern California along the San Andreas belt. Does it extend farther north into central and northern California? How does the decoupling relate to crustal accretion and denudation processes in western North America. What have been the effects of plate convergence in Mesozoic time such as underplating and consequent crustal thickening, terrane migration, docking, and accretion, followed by Tertiary thermal bulging, crustal thinning and upper crustal extension?

TERRANES AND OROGEN-PARALLEL STRIKE-SLIP

An increasingly popular concept is that much of the western North American plate in California, Oregon, Washington, British Columbia, and Alaska is an amalgamation of 10 to 50 small lithospheric plates called terranes. They are separated from one another by great faults and are characterized by regional domains of rocks and structures whose age and deformation history are greatly dissimilar to adjacent terranes. According to some geologists, these terranes traveled piggyback on an oceanic plate and lodged against the North American continental plate when the oceanic plate subducted. Where subduction was oblique, the terranes slid horizontally along the edge of the continental plate. Other writers postulate that at least some of the terranes were sliced off Mexico or Central America and slid northwestward along an ancient transform fault system, older than the San Andreas system, and formed as the result of a different lithospheric plate arrangement. The juxtaposition of these tectonostratigraphic terranes occurred in Mesozoic and early Paleogene time, long before the inception of the San Andreas fault system, defined as that transform belt originating in post-Oligocene time.

Vigorous controversy concerns the timing of emplacement of the tectonic blocks in western California. In particular, the Salinian block - consisting of granitic rocks and overlying strata, now situated west

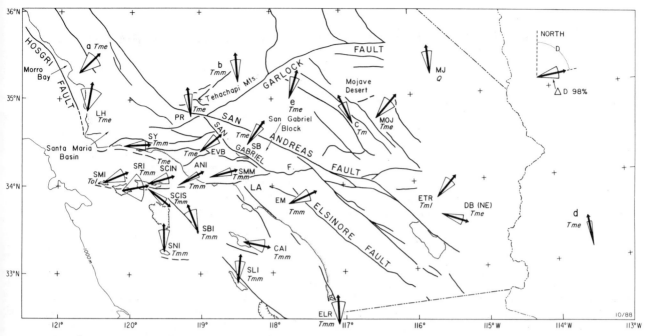

FIGURE 5 Paleomagnetic Declinations Measured in Neogene age Rocks, Southern California (after Luyendyk et al., 1985, with additions from Luyendyk, pers. commun., 1988). The mean declination at each site is shown with the 98% confidence limit on the mean   Tol, Oligocene rocks, Tme, early Miocene rocks; Tmm, middle Miocene rocks; Tml, late Miocene rocks; Tm, Miocene rocks. Site keys are MJ, Quaternary Mojave; MOJ, Mojave; ETR, eastern Transverse Ranges; DB (NE), northeast Diligencia basin; SB, Soledad basin; PR, Plush Ranch Formation; CM, Chico Martinez; ELR, El Rosario; EM, El Modeno; SLI, San Clemente Island; CAI, Catalina Island; SNI, San Nicolas Island; SBI, Santa Barbara Island; SMI, San Miguel Island; SRI, Santa Rosa Island; SCIN, north Santa Cruz Island; SCIS, south Santa Cruz Island; ANI, Anacapa Island; SMM, EVB, east Ventura basin; Santa Monica Mountains; SY, Santa Ynez Range; LH, Lions Head. Site a is from Greenhaus and Cox (1979), b is the Miocene result from Kanter and McWilliams (1982), c is from J. Morton and J. Hillhouse (unpub. ms), d is from Calderone and Butler (1984), and e is from Golombek and Brown (1988).

of the San Andreas fault and underlying much of the Coast Ranges - is interpreted by some geologists as having been emplaced primarily by major strike-slip displacement in pre-San Andreas time, and by others as a result of movement along the San Andreas fault. The far-travelled and ancient interpretation is supported by some paleomagnetic data consisting of low inclinations. The younger and lesser emplacement, viewed as taking place during San Andreas movement, is supported by interpretations that match basement rocks of the Salinian block with similar rocks in the desert of southeastern California. This controversy regarding the history and place of origin of the Salinian block is discussed in another section of this guidebook.

Controversy concerning the pre-San Andreas

emplacement of other tectonic terranes is less vigorous, and indeed some blocks are considered to be far-travelled, based both on the interpretation of paleomagnetic data, and on the mismatch of the rocks across boundary contacts that may prove to be tectonic sutures. The test of this concept, however, involves finding terranes from which the California blocks have been derived. This has not yet been done and involves research that will be underway for many years. Whereas both "schools" advocate displacement during San Andreas times of several hundreds of kilometers on specific major faults, the crux of the problem is to find supporting evidence for displacements older and greater than these amounts.

THE TECTONIC HISTORY OF THE SAN ANDREAS TRANSFORM BELT

John C. Crowell

Institute of Crustal Studies, University of California, Santa Barbara

INTRODUCTION

California has been at or near lithospheric plate boundaries during much of its decipherable geologic history, and at present is undergoing deformation along the broad belt of interaction between the Pacific and North American plates (FIG. 1). This belt of interaction constitutes the San Andreas transform belt which consists of many subparallel faults and folds, and the San Andreas fault itself is only one tectonic element within it. Rocks now exposed within the belt and along the course of the field excursion are made up of many types and ages. These include rocks as old as Proterozoic, and deformation accompanied by sedimentation, volcanism, and plutonism has taken place during this long interval since then. The San Andreas transform belt, however, originated only about 30 Ma, and many of the rocks and structures are older than this. It is essential to sort out and separate those features resulting from the evolution of the San Andreas system from those that are older.

The geology of California is therefore complex because of the progressive superposition of profound tectonic events upon one another from late Proterozoic time to the present (Dickinson, 1981; Ernst, 1981). Eastern California and western Nevada were the site of a passive continental margin from late Proterozoic to middle Triassic time when more than 15,000 m of miogeoclinal strata accumulated on the edge of a craton consisting of older basement rocks. This great prism of Cordilleran rocks was intermittently deformed on the west by the collision of plates coming in against the North American continent from Pacific Ocean regions during Phanerozoic time. The Antler Orogeny took place in Devono-Carboniferous time when a huge and complex mass impinged against the Cordilleran prism along the Roberts Mountain thrust system. Later, in Permo-Triassic time, an island-arc terrane was accreted to the continent during the Sonoma Orogeny along the Golconda thrust. The record of both the Antler and Sonoma orogenies, however, is preserved primarily in Nevada, and only fragments of this record reach as far southwest as the San Andreas fault system. Moreover, only locally along the San Andreas belt are Proterozoic, Paleozoic, and rocks identified that are older than Late Jurassic.

The structural framework of western California is dominated by the remnants of plate convergence and its products between late Jurassic and mid-Tertiary time. An evolving and intermittently changing trench-arc system prevailed during this long interval of about 140 Ma. Principal products of this long and complex series of deformations are the great batholiths of the Sierra Nevada and Peninsular Ranges, very thick sediment accumulations lying upon ophiolitic floors within forearc basins (Great Valley Sequence), melanges and disrupted parts of trenches, trench slope, outer-arc ridges, and accretionary prisms (Franciscan Complex). Blocks within this vast subduction complex include blueschists which record high pressure metamorphism deep within a subduction zone where temperatures were far below normal for those depths. These blocks and tectonic units have since reached the surface tectonically. Other blocks may have been scraped off of subducting oceanic plates far from their resting place today. Some are viewed as coming from seamounts and oceanic plateaus within ancient Pacific oceans. Others may have been cut from the margin of Central America and then carried northwestward by ancient transform displacements (Crowell, 1985).

In fact, much of the margin of western North America consists of tectonostratigraphic terranes that have docked against the continent and constitute regions with distinctive rocks and inferred geologic histories. At places the sutures where such exotic blocks have met the continent are identified but many of the blocks are only "suspect", and their joining sutures are not recognized. The designation of distinctive terranes, characterized by their own geologic history, and the search for sutures bordering them and their regions of origin, constitute challenges intriguing many geologists.

The San Andreas transform belt, consisting of strike-slip faults and associated folds and other structures, originated in mid-Oligocene time, following a period of tectonic adjustment to the waning and ending of subduction. This belt formed along the boundary between lithospheric plates coming in from Pacific regions and the North American plate. At the end of the Miocene epoch, the Peninsula of Baja California was sliced from western Mexico, and the Gulf of California was born. The Baja Peninsula and western California are now moving relatively northwestward along faults of the San Andreas system as the Gulf of California widens. The San Andreas transform belt proper is now about 1200 km long and 500 km wide. At places, transpression prevails within the belt so that crustal blocks are contracted; at others, transtension has resulted in deep and irregular pull-apart basins. Blocks within the splintered system have been rotated (Luyendyk et al, 1985).

Because these tectonic processes along this continental transform belt are at work today, in reviewing briefly the tectonic history of western California, it is logical to work from the present backwards in time.

THE TECTONIC SETTING AT PRESENT

Today the San Andreas transform system extends from the Salton Trough, at the head of the Gulf of California, to the Mendocino Triple Junction in northern California. Well to the southeast, off the coast of southern Mexico, the East Pacific Rise comes into the continent from the floor of the deep Pacific Ocean, and enters the Gulf of California. The gulf constitutes a divergent plate boundary. Associated convergent plate boundaries lie northwest of the Mendocino triple junction and far to the southeast off the coast of central Mexico, where oceanic crust moves beneath the North American plate. The divergent plate boundary within the Gulf of California and the San Andreas transform system provide the tie between these two convergent realms.

Upon the floor of the Gulf of California active spreading centers, here viewed as sidestepped parts of the East Pacific Rise, lie between northwest-trending transform faults. Most of these faults do not extend into continental rocks on either side of the Gulf, that is, into either mainland Mexico or Baja California. On the northwest, the divergent plate boundary narrows and forms the Salton Trough. Here the San Andreas transform system, consisting of several subparallel strike-slip faults, takes over plate movements on to the northwest (FIG. 1).

Along the San Andreas transform belt, coastal California and the adjoining Baja Peninsula move relatively toward the northwest as part of the Pacific plate. Although the San Andreas fault proper is the main transform fault in this displacement, several others (San Jacinto, Elsinore, Hayward, etc.) play an active role. Long slices between these faults - because the faults are not exactly parallel to lithospheric plate movements - are either squeezed or stretched as plate movements continue. Where squeezed, the blocks are raised to make uplands or mountains, and where stretched, they underlie lowlands or basins. The arching upward and bowing downward of these long slices, all caught in a regime of active lateral displacement, is referred to as "porpoise structure" (Crowell and Sylvester, 1979). The transform belt along the field excursion route is therefore broad and complex, and the geology reflects not only active deformation going on today, but also tectonic movements and geological events of the past.

THE GEOLOGIC HISTORY OF WESTERN CALIFORNIA

The San Andreas transform system, and the principal fault itself, originated when the Pacific Plate first nudged against the North American plate, in mid-Oligocene time. Before that impingement, plate convergence existed along the margin of the North American continent. The transform system has lengthened and broadened since mid-Oligocene time, but the regions of convergence remain today off of central Mexico on the southeast and off of northern California, Oregon, and Washington on the northwest. The history of this lengthening and broadening of the transform system through time is illustrated by means of a time (vertical coordinate) and geographic (horizontal coordinate) plot of plate tectonic events in late Cenozoic time (FIG. 6). The diagram is based on fitting together interpretations of magnetic anomalies from the sea floor with those from the geologic record on land. Because of uncertainties in the interpretation of the magnetic patterns, two plots (A and B) are shown. The Pacific plate first reached North America in the general vicinity of the Mexico-United States border about 29 Ma, long before the Gulf of California had opened. The plot shows that convergent boundaries to the northwest and southwest grew apart as the strike-slip regime lengthened, and as the Mendocino and Rivera triple junctions moved apart. The Mendocino triple junction marched steadily northwestward parallel to the coast with only slight changes in its speed. The Rivera triple junction, on the other hand, moved back and forth off the coast, and then very quickly sped far to the southeast about 13 Ma. About 12 Ma it changed directions and moved to the region off the mouth of the future Gulf of California. The rifting of continental rocks that resulted in the gulf began about 5 Ma.

Figure 6 also shows the time and place of origin of several of the Cenozoic basins that evolved within the broad transform belt. In addition, fields of time and space on the diagram depict when and where, with respect to the restored coastline, block rotations, strike-slip and rifting, and other styles of tectonics took place. With the opening of the Gulf of California beginning about 5 Ma, oblique rifting occurred within it. As the Baja Peninsula migrated northwestward, the region of the Transverse Ranges was shortened and the mountains grew. Although strike-slip dominates as the principal style of interplate displacement north of the Transverse Ranges as far as the Mendocino triple junction, upper crustal rocks are shortened in directions nearly normal to the surface trace of the present plate boundary: the San Andreas fault. The Coast Ranges are therefore characterized by folds and thrust faults nearly parallel to the transform plate boundary. These structures are now active, and the style of deformation is also shown by earthquake first motions and geodetic measurements.

The style and timing of tectonics described briefly above with reference to Figure 6 emphasizes those events younger than about 30 Ma. Older tectonic features have been overprinted by these events, but are nonetheless conspicuous in parts of the region traversed by the field excursion. Prior to the Miocene epoch, when convergent tectonics prevailed along the California margin, the Pacific oceanic floor gave way eastward to a subduction zone inclined beneath the continent. At the beginning of the Tertiary period, a diagrammatic tectonic cross-section through central California shows, from west to east, a gently flexed

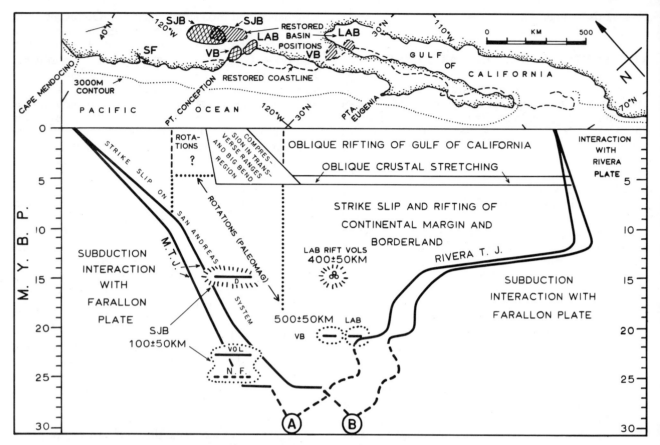

FIGURE 6 Diagram Showing Plate-boundary Regimes and Deformation in Late Cenozoic Time with respect to the California and Mexican coastlines, restored to their approximate positions about 10 Ma. The restoration is based on seafloor magnetic anomalies and derived plate-circuit reconstructions (Atwater, 1970; Atwater and Molnar, 1973; and Atwater, in Crowell, 1987a). The present shoreline is shown for approximate geologic reference. The coordinates for the diagram are geography (horizontally) and time (vertically). Two favored time-position curves of the Mendocino and Rivera Triple Junctions show the timing and possible locations for the onset of interaction between the Pacific and North American plates. The 29 Ma date of this event and the locations of the present-day triple junctions are reasonably well established. Path A suggests that this breakup occurred off what is now southern California. This initial interaction between the Pacific and North American plates was followed by tectonic stretching and subsidence, including basin formation. Only after several millions of years did the San Andreas transform fault begin to lengthen through the activity of the northward-migrating Mendocino and southward-migrating Rivera Triple Junctions. The geologic time and geographic locations lying within these paths of plate interaction imply positions within the evolving strike-slip regime at a given time. In the upper diagram, the Ventura, Los Angeles, and San Joaquin basins are shown as dotted ellipses based on data described by Crowell (1987a) and data presented by Goodman et al (in press). The origin of the San Joaquin basin may have been outside this boundary or, alternatively, the timing of this event may locate the triple junction and thus favor Path A. The approximate time of deepening for the Tejon embayment is denoted by 'D' within the dashed area and lies well within the strike-slip regime. Other symbols: SF - San Francisco; LAB - Los Angeles basin; VB - Ventura basin; SJB - San Joaquin basin; M.T.J. - Mendocino Triple Junction; N.F. - onset of normal faulting at Tejon embayment (estimated), VOL - extrusion of Tunis volcanic rocks at the Tejon embayment. From Goodman et al (in press)

oceanic crust, a trench, an inner trench slope (with some trench-slope basins), an accretionary prism (with a steep slope shedding olistostromes on the inshore or continental side), a forearc basin (at times narrow, at times broad), a magmatic arc, and far to the east, a backarc basin (FIG. 7). In general such a cross section prevailed as far back as Late Jurassic time.

Remnants of these pre-San Andreas tectonic domains are clearly discernible today, even though this tectonic arrangement has been fragmented and blocks displaced for many tens or hundreds of kilometers by superposed strike-slip of the San Andreas system. Blocks between major faults have been rotated. Others

have been stretched to form the floors of deep basins, and still others squeezed and uplifted and as a result deeply eroded. The San Andreas fault at the northwestern end of the Salton Trough, for example, cuts through the deeply eroded foundation of the Cretaceous magmatic arc, so that granitic rocks now predominate. Parts of the old forearc basin remain in the topography today as the broad Great Valley of California, and forearc strata of Paleogene and upper Mesozoic age are preserved beneath it and around its margins. The Franciscan terranes in the Coast Ranges, consisting in part of melanges with associated rocks, are now uplifted and deformed again, perhaps several

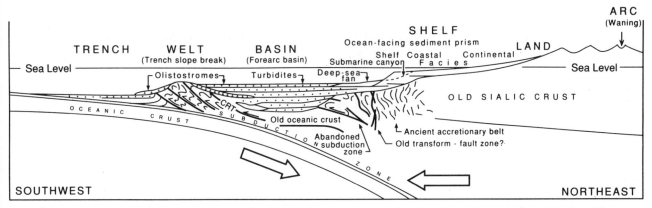

FIGURE 7 Diagrammatic Cross Section across Coastal Central and Southern California at the Beginning of Tertiary Time. CRT - Coast Range thrust; CS - Catalina Schist. From Crowell (1976, Fig. 6).

times. The terranes were originally formed within a succession of accretionary prisms and are now uplifted and exposed.

Much of the excursion route passes through country formed originally in this way. At many places within these regions, however, Neogene beds lie unconformably upon trench-arc units, and these younger beds in turn dip steeply and have been markedly folded and faulted. Clearly the older beds below have been folded and faulted also, so that some of their complexity is the consequence of these later and oft-repeated deformations. The crustal rocks of California have been subjected to many deformations through time.

The tectonic mobility is shown by interpretations of paleomagnetic and paleontological data, and by displacements on major faults such as the main strands of the San Andreas system. For example, within Mesozoic strata along the coast in central California, the remnant paleomagnetic vectors record a very low angle of inclination. This suggests to several workers that slices of terrane have been displaced several thousands of kilometers from low latitude and equatorial sites. The slices are viewed as displaced from their sites of origin and then accreted to the continent along sutures far to the north. Sea-floor spreading mechanisms associated with great transform displacements are visualized as responsible. Other geologists, however, are skeptical of movements of this magnitude because they consider several of the supposed far-travelled blocks as matching satisfactorily continental terranes that are only a few hundred rather than a few thousand

kilometers away. Studies concerning the magnitude of these displacements are now underway by California geologists.

Faunal provinces mapped within Tertiary strata show significant horizontal offsets along the San Andreas fault, but these separations are not larger than those shown by the matching of offset rock units. For some Mesozoic faunas, however, the displacements are interpreted by paleontologists to be several times larger, but the displaced faunas are recovered largely from blocks isolated within melanges or from small fault slices. The rocks do indeed show evidence, at places supported by paleomagnetic data, that they have travelled far across the Pacific Ocean from seamounts or other sources before accretion to the continent.

Rocks exposed at places in western California are older than Upper Jurassic. Tracts of gneiss, schist, granulite, and other metamorphic rocks are known to include Proterozoic, Paleozoic, and pre-Upper Jurassic rocks. Some of these are clearly related to the North American craton, and have been torn from it and displaced within the San Andreas system. Others have been broken from the continent and occur as pendants and septa within the Mesozoic arc rocks. Masses of marble, quartzite, and hornfels, for example, come from miogeoclinal prisms still preserved in eastern California and adjoining states. Major structures, such as the Vincent thrust system are preserved within the crystalline basement and within Mesozoic and Tertiary strata. These document major tectonic events previous to the onset of San Andreas displacements.

TECTONIC PROVINCES, TERRANES, AND REGIONAL STRUCTURES OF WESTERN CALIFORNIA

John C. Crowell
Institute of Crustal Studies, University of California, Santa Barbara

Arthur G. Sylvester
Department of Geological Sciences, University of California, Santa Barbara

California consists of several geomorphic and tectonic provinces (FIG. 1), which are intimately involved in the history of the San Andreas fault system. Some of them will be traversed on the excursion, whereas others will be observed distantly.

THE PENINSULAR RANGES

In Late Jurassic and Early Cretaceous time, an extensive volcanic-plutonic arc spatially associated with ophiolitic rocks similar, and perhaps equivalent to, the Franciscan Complex, formed in the region that is now the Continental Borderland. During Cretaceous time the magmatic axis moved eastward and formed the great Southern California batholith which underlies most of the Peninsular Ranges, one of the largest geologic units in western North America. It extends 1450 km from the Transverse Ranges southward to the tip of Baja California. The province varies in width from 50 to 150 km between the Gulf of California and the Pacific Oceans in Baja California, and between the Salton Trough and Continental Borderland in southern California.

Extensive exposures of plutonic rocks with mineral ages ranging between 120 and 90 Ma contain pendants and septa of metamorphosed miogeoclinal sedimentary rocks. Locally, high temperature contact metamorphism of magnesian limestone produced a remarkable series of 138 minerals which are displayed in museums around the world. Late Jurassic volcanic rocks and Cretaceous marine deposits are well-represented in the northernmost ranges and part of the coastal strip of Baja California. Paleogene volcanic, marine, and nonmarine strata form a restricted shelf sequence on the coastal part of the Peninsular Ranges between Los Angeles and San Diego.

The batholithic rocks range in composition from gabbro to true granite, however, the average composition is tonalitic in contrast to the calc-alkaline composition of the Sierra Nevada batholith. In addition, the Southern California batholith is associated with volcanic rocks which show that eruptions accompanied pluton emplacement. Concurrent and later uplift and erosion yielded abundant clastic volcanic and plutonic detritus that was deposited in thick accumulations along the continental margin.

Tourmaline-bearing pegmatites are prevalent in the Peninsular Ranges between the Transverse Ranges and the International Boundary. They are typically 1-3 m thick and as much as 2 km long. Most of the tourmaline crystals are small and black, but at Pala - half way between San Diego and the central Salton Trough - large, gem-quality crystals of the pink, green, and yellow varieties were mined for many years for the Dowager Empress of China.

THE COLORADO DESERT

East of the divergent lithospheric plate boundary (the Gulf of California and the Salton Trough), lies a broad region extending into Arizona and Sonora, Mexico. It is a part of the Basin and Range Province of North America and topographically is characterized by many mountain ranges trending mainly northwest and separated by broad, alluviated valleys. Rocks of this region include Precambrian gneisses of several ages, Paleozoic and Mesozoic metamorphosed strata similar in facies to those of the Grand Canyon, broad tracts of Jurassic and Cretaceous granitic rocks, and overlying Eocene, Oligocene, and Neogene volcanic and sedimentary strata. The region is complex tectonically, and includes Jurassic and Cretaceous thrust sheets, and Miocene low-angle normal faults (detachment faults). Post-mid-Miocene faults associated with both the San Andreas system and opening of the Gulf of California are present locally and are recognized as far east as the Colorado River region. Minor faults, gentle folding, and tilting of Pliocene beds in easternmost California and western Arizona attest to gentle tectonism. Much of this broad desert region is inaccessible and is still not mapped geologically except at the reconnaissance level.

THE SALTON TROUGH

The divergent lithospheric plate boundary associated with the northward continuation of the East Pacific Rise terminates within southern California as the Salton Trough. The region is part of the Colorado Desert physiographic province and consists of a broad alluviated plain lying between high mountain ranges; those near its northwestern end exceed elevations of 3000 m.

The Salton Sea, with a surface elevation of about 70

m below sea level lies in its midst and above a spreading center as shown by repeated earthquakes, geodetic distortions, recent volcanism, gravity measurements and interpretations, and very high heat flow. The rugged margins of the trough are due to faults of several types. These include major transform faults related to the opening of the Gulf of California, such as the Elsinore, San Jacinto, and San Andreas proper. In addition, normal faults, both low and high angle, are also related to the gulf opening. The bare and rugged mountains also afford excellent exposures of older faults unrelated to the opening, such as mid-Miocene detachment faults, early Tertiary faults of several types, and of Mesozoic regional thrusts of at least two distinctly different ages.

Strata within the Salton Trough consist of two groups: Paleogene marine and nonmarine strata deposited in regional basins which are now displaced by the San Andreas fault from their counterparts in the Transverse Ranges; and Late Cenozoic fluvial and lacustrine strata which represent fillings of the present and proto-basin. Sediments within the trough include detritus eroded from the Colorado Plateau and transported to the Salton Trough by the Colorado River and its ancestors.

The basement rocks on either side of Salton Trough are quite different from one another. Those on the northeast side - that is, on the North American Plate - are Proterozoic gneisses intruded by anorthosite and related plutonic rocks of Precambrian age, intruded by granodiorite of Permian age, intruded by calc-alkaline granitoids of Mesozoic age and Sierra Nevada batholith affinity, all of which are thrust upon Orocopia Schist by the Orocopia and Chocolate Mountains thrust system. The entire lithotectonic complex is intruded by felsite dikes related to the great Neogene volcanic centers in southeastern California and western Arizona. On the southwest side - the Pacific Plate - are the plutonic rocks of the great Southern California batholith, locally mylonitized, intruded into a late Paleozoic/early Mesozoic sequence of schist, marble, and quartzite whose protoliths were miogeoclinal rocks.

THE TRANSVERSE RANGES

The Transverse Ranges province is an exceedingly complex, east-trending region of mountain ranges, valleys, reverse faults, and folds that interrupts the prevailing northwest structural and physiographic trends of the Coast Ranges to the north and the Peninsular Ranges to the south. The province, which is about 500 km long and averages about 50 km in width, has been uplifted to its present elevation during the last million years, but it has had a long, complicated history of formation and deformation. The southern margin is bounded by a poorly-understood, complex array of north-dipping thrust faults, including the Malibu, Santa Monica, Raymond, San Fernando, Sierra Madre, Cucumonga, and Elysian Park faults which carry the Transverse Ranges over the northwest-trending Peninsular Ranges. This array of faults is probably a reactivated ancestral transform that separated segments of the oceanic plate that collided with the North American continental margin in Mesozoic time. The northern margin of the western half of the Transverse Ranges is demarcated mainly by late Cenozoic strike-slip faults: From west to east, the Santa Ynez, Big Pine, San Andreas faults. The north margin east of the San Andreas fault slopes upward from the Mojave Desert and is cut locally by small thrust faults.

The province may be subdivided into three physiographic and tectonic segments which are separated by the San Gabriel and San Andreas faults. West of the San Gabriel fault, the Transverse Ranges consists of a thick and almost continuously deposited marine and nonmarine succession of Cretaceous and Cenozoic sedimentary rocks intruded locally by Miocene volcanic rocks. Part of the western segment includes the long, narrow Santa Barbara Channel and Ventura basin, a major structural and depositional trough which contains more than 8000 m of late Cenozoic deep-water marine strata. Herein is found the largest oil field in California outside of the Los Angeles and San Joaquin basins.

The central segment is an extremely rugged, mountainous region composed of the same unusual suite of crystalline rocks as described above in the mountains on the northeast side of the Salton Trough: Precambrian gneiss, Precambrian anorthosite and related rocks, Permian granodiorite, and Mesozoic granitic rocks all thrust upon the Pelona Schist on the Vincent thrust. In fact, it was correlation of this complex of metamorphic and plutonic rocks between the Transverse Ranges and the Salton Trough across the San Andreas fault that aided in recognition and definition of the 330 km of displacement (Crowell and Walker, 1962). The Soledad and Ridge basins developed in late Cenozoic time in sags along major faults within this segment. The central segment rose more than 2 m in 1971 as a result of the San Fernando earthquake on the San Fernando fault zone along part of its southern margin.

The eastern segment has basement rocks like those of the Colorado Desert and of the northwest side of the Salton Trough, described above: Mesozoic calc-alkaline granitoids intruded Precambrian gneiss, but also late Paleozoic pre-batholithic rocks. The uplifted ranges, including the high San Bernardino Mountains, have subdued, plateau-like summits, in contrast to the peaked summits in the San Gabriel Mountains in the central segment. The uplands are erosion surfaces that slope generally eastward into the Mojave Desert. Their steep, southern flanks are bounded by active faults of the San Andreas system. The range culminates in Mt. San Gorgonio (Old Grayback), the highest peak in southern California (El. 3539 m).

LOS ANGELES-SAN BERNARDINO LOWLAND

Between the Transverse Ranges and the Peninsular Ranges lies a lowland region consisting of the coast Los

Angeles Plain on the west and the Pomona-San Bernardino Plain on the east. This region, also with an east-west trend, is the living place for most of the large population of southern California. Geologically, the region consists of the filled Los Angeles basin, one of the irregularly-shaped pull-apart basins associated with the San Andreas transform belt, and is similar in tectonic origin to the unfilled basins in the California Borderland. Its character is discussed in more detail below. The San Bernardino region, on the other hand, is the depressed and north-tilted part of the Peninsular Ranges. The sedimentary section in this area is relatively thin, and hills of Peninsular Ranges basement rocks locally protrude above broad alluvial fans than extend southward from the rising San Gabriel and San Bernardino Mountains. Hills surrounding the lowland, especially to the west, are underlain by upper Cenozoic sedimentary and volcanic rocks that are locally severely folded and faulted.

THE MOJAVE DESERT

The Mojave Desert is closely connected with the Colorado Desert from a geological point of view, but physiographically, they are regarded as the "high desert" and "low desert", respectively. The Mojave Desert is a great tract of broad, alluviated valleys, playas and low, isolated mountains located structurally between the Garlock and San Andreas faults. The Mojave region was originally part of the Basin and Range province, but since development of the Garlock fault about 10 Ma, the Mojave has had its own tectonic history.

Proterozoic rocks are widespread in the Mojave Desert. They are divided into high grade, gneiss and schist with granitoid intrusions, and a thick succession of Late Precambrian sedimentary strata akin to the Grand Canyon series in northern Arizona. Paleozoic miogeoclinal strata ranging from Cambrian to Permian in age, also correlatable with the succession in the Grand Canyon, crop out extensively in the eastern Mojave Desert, but only in isolated, erosional patches in the western part. The pre-Mesozoic rocks were widely intruded by Mesozoic plutons and dikes swarms related to the Sierra Nevada batholith. The extensive Independence dike swarm cuts northwestward across the central Mojave Desert, is sinistrally offset about 60 km by the Garlock fault, and continues across the southwest corner of the Basin and Range province into the Sierra Nevada.

Much of the pre-Cretaceous sequence of rocks, representing the western edge of the North American plate, has been thrust upon greenschist of the Pelona-Orocopia-Rand association which is interpreted to be metamorphosed Pacific ocean floor that was subducted and metamorphosed beneath the North American crust. The schist is now locally exposed in tectonic windows in the Mojave Desert. Deep seismic reflection profiling by COCORP and CALCRUST reveals several low-angle reflectors in the upper crust of the Mojave Desert. These reflectors are interpreted both as Miocene detachment faults and as parts of the dismembered Mesozoic thrust system. Some of them have been traced to the surface where they coincide with the thrust contact between the greenschist and overlying continental crustal rocks: the Vincent-Chocolate Mountains thrust system.

The central Mojave Desert endured extensive erosion in Cenozoic time, and especially in Paleogene time. Much of the detritus has accumulated in basins that were to the west and west of the San Andreas transform belt. Distinctive clasts of durable Mojave Desert rocks, especially Cambrian quartzite, Jurassic metavolcanic rocks, and some granitic rocks comprise much of the upper Cretaceous through early Miocene detritus in basins of southern and central coastal California.

Great piles of intermediate lava flows and pyroclastic rocks were erupted throughout the central Mojave Desert in Miocene time at the time the East Pacific Rise impinged against, and disrupted the west edge of North America, and when the San Andreas transform system originated. Simultaneously, local intermontane basins accumulated lacustrine, fluvial, and alluvial deposits, some of which are rich in borate salts derived from hot springs.

The western Mojave Desert is cut by a set of active, northwest-striking transcurrent faults having dextral offsets of a few tens of kilometers. Recent basaltic volcanoes dot the region, culminating in a large field of more than 25 cones and attendant lava flows in the eastern Mojave. Some of the lavas contain a rich assemblage of ultramafic xenoliths.

THE GREAT VALLEY

The Great Valley is an asymmetric synclinal trough, 640 km-long and about 80 km-wide, that covers the suture zone of oblique convergence between contemporaneously-formed rocks of the magmatic arc of the Sierra Nevada batholith on the east and the ophiolitic rocks of the Franciscan Complex on the west. It is near sea level in its central part, and about 120 m above sea level at both its northern and southern ends.

Since late Jurassic time, the Great Valley has been a synformal receptacle for marine sediments and volcaniclastic detritus, derived mostly from the east. Great thicknesses of sandstone, shale, and conglomerate collectively known as the Great Valley Sequence were deposited as turbidity flows, locally in abyssal depths, and are now turned up and exposed along the west edge of the valley in the eastern Coast Ranges. The basin was deformed at different times and different places in the Cenozoic era, producing elongate uplifts and depressions in the basin floor that guided and received clastic sediments. By Pliocene time, the basin was filled to sea level. Brackish and freshwater lakes, fed by streams from the west slope of the rising Sierra Nevada, replaced marine waters, and the Central Valley assumed its present form. On the southwest margin of San Joaquin Valley, however, intense shear deformation along the San Andreas transform belt produced a series of remarkable folds that trapped about 30 percent of California's eventual petroleum reserves. Much of California's gas is produced from stratigraphic traps in the Sacramento Valley. At the south, where the Great Valley meets the Transverse Ranges, Pleistocene and active tectonism is especially noteworthy.

THE COAST RANGES

The Coast Ranges stretch northwest nearly 1000 km from the Transverse Ranges to the northern border of California. They comprise several elongate mountain ranges and large structural valleys containing many strands of the San Andreas fault system, and Neogene and late Cenozoic tectonics have shaped the recent structural fabric of the province. Thus, a blanket of Late Cretaceous and Cenozoic sedimentary strata as young as Pliocene and Pleistocene are cut by many strike-slip faults which bound several en echelon basins, and are accompanied by en echelon folds and some east-west thrust faults. Many of the mountain ranges are complex antiforms with mobile and diapiric cores of Franciscan rocks, but some are bounded on at least one side by strike-slip faults having a component of vertical movement.

The southern Coast Ranges consist of three, strike-slip fault-bounded, lithotectonic domains: A subduction zone complex - the Franciscan Complex; forearc basin sedimentary rocks - the Great Valley Sequence; and a magmatic arc - plutonic and metamorphic rocks of the Salinian block. Each is composed of roughly contemporaneous Late Mesozoic rocks. The Salinian block may have been displaced about 330 km from its original position on the southeast and is now separated from the Franciscan Complex on both sides: By the San Andreas fault on the northeast and by the Nacimiento fault on the southwest.

Much of the Franciscan and Salinian basement rock in the Coast Ranges is blanketed by a gently folded supracrustal succession of Paleogene marine and nonmarine, clastic sedimentary rocks, by fine-grained diatomaceous strata of the Miocene Monterey Formation, and by marine and nonmarine sandstone and siltstone of Pliocene age. A prominent and important field of rhyolitic volcanic rocks was extruded in early Miocene time in the Pinnacles area, and parts of this volcanic field were subsequently slivered and strung out along the San Andreas fault. The main slice constitutes the Neenach field, bordering the Mojave Desert. In addition, stumps of Miocene rhyolitic volcanoes remain in the San Luis Obispo area in the southern Coast Ranges. Both of these regions of volcanism are one of the manifestations of the tectonic reorganization of the Pacific-North American plate boundary when it changed to a transform belt upon impingement of the East Pacific Rise.

THE FRANCISCAN COMPLEX

The Franciscan Complex makes up much of the core of the Coast Ranges. Its components were both assembled and dismembered in Late Jurassic, Cretaceous, and Early Tertiary time by oblique subduction of oceanic plates beneath the western margin of North America.

Melanges and large, coherent rock units of oceanic and terrigenous rocks make up the Franciscan Complex. The coherent units include Cretaceous turbiditic sandstone with mudstone intercalations, chert-graywacke sequences, and upper Jurassic chert-greenstone units. Conglomerate is locally present as lenses and channel fillings. The melanges include similar rocks as well as serpentinite, blueschist, eclogite, amphibolite, and conglomerate in a pervasively sheared, argillaceous matrix. Mineral compositions of the clastic sedimentary rocks indicate their derivation from the magmatic arc to the east. Many of the blocks and coherent units have undergone blueschist-facies metamorphism resulting from high pressure, low temperature conditions during rapid subduction and subsequent buoyant uprise, perhaps aided by oblique strike-slip.

THE SALINIAN BLOCK

Much of California north of the Transverse Ranges and west of the San Andreas fault is underlain by granitic rocks. This large tectonic block, termed the Salinian block, lies between the San Andreas fault and the Nacimiento and Rinconada faults (FIG. 1). The visible basement both to the west and east consists mainly of Franciscan Complex and strata of the Great Valley sequence overlying ancient ophiolitic seafloor (FIG. 7). The granitic block, which is about 750 km long and only about 60 km wide, differs markedly from terranes on either side and has long been recognized as anomalous in the tectonic arrangement of California. In fact, Hill and Dibblee (1953) used its position in advocating several hundreds of kilometers on the San Andreas fault. The boundary of the Salinian block on the east is the San Andreas fault itself. The sharp and clearcut boundary on the west, however, is more complex, and consists of a series of faults of different ages. Along this boundary on the northwest, an older fault system (Nacimiento) at places demarcates the edge of the Salinian block. This fault system is in part a thrust system that was active in early Tertiary time and was later overprinted and reactivated by Neogene and later displacements. Other faults along the trend of the western margin of the Salinian block include the San Gregorio on the north, the Hosgri on trend with it to the south and largely offshore, and the Rinconada still farther to the southeast and inshore. These three faults, and others related to them, have been active in late Cenozoic time but have as yet undeciphered beginnings. The attenuated northwestern end of the Salinian block extends offshore beyond Point Arena, but no submarine geophysical data require its continuation to the Mendocino fracture zone. On our excursion, granitic rocks near the onshore northwestern part of the Salinian block will be observed at Point Reyes. They also crop out at Bodega Head and make up the Farallon Islands offshore from the mouth of San Francisco Bay.

The granitic basement rocks of the Salinian block, including associated gneissic and metasedimentary rocks, show strong affinities to the rocks of both the Sierra Nevada and the Peninsular Ranges. Controversy still prevails concerning the original source of the block, and the manner and timing of its emplacement. With the developing recognition of large lateral displacements on the San Andreas fault, geologists (beginning with Hill and Dibblee in 1953) have suggested that it has been moved northwestward along the fault from a site originally between the Sierra Nevada and the Peninsular Ranges. Silver and Mattinson (1986) identified a set of geological lines with the fault within the Salinian physiographic province and in the Mojave Desert and western Arizona formed by a line of Mesozoic granitic rocks having apparently unique isotopic composition and radiometric ages. That line is dextrally offset about 330 km, the same distance that several other geologic lines and their piercing points have been offset along the San Andreas fault in Cenozoic time. Silver and Mattinson concluded that Salinia slid northwestward along the San Andreas fault from a position originally between the south end of the Sierra Nevada and the north end of the Peninsular Ranges. This agrees with the pre-plate tectonics hypothesis of Hill and Dibblee (1953), but it is contrary to interpretations of paleomagnetic inclination data, which suggest that Salinia was sliced off southern Mexico and has travelled 2500 km since Mesozoic time to its present position.

At present, the San Andreas fault is viewed as having a known right slip of about 330 km, and this displacement has taken place since some time in the Early Miocene, beginning about 24 Ma. The length of the Salinian block is more than twice this amount, however, so two explanations for its length exceeding the preferred offset on the San Andreas fault have been investigated. One explanation is that slices of the block have been strewn out relatively to the northwest during Miocene and younger movements; the second explanation is that a granitic terrane was already in place in western California before the San Andreas fault originated. Studies within the central and northern Coast Ranges now show that during Paleogene time, a series of granitic islands and a granitic peninsula extended northwestward from the vicinity of the present Transverse Ranges, but the tectonic reason for their location is as yet unclear. According to this reconstruction, a shorter Salinian block already existed in late Paleogene time which was then lengthened by San Andreas displacements in Neogene and Quaternary time. To account for the Paleogene Salinian block, several investigators have suggested that it was emplaced as a composite of exotic slices from a source region far to the south or southeast. This explanation has been based on interpretations from paleomagnetic vectors which show very low, or near-equatorial, inclinations. Therefore, the block is viewed as a far-travelled terrane not at all related to the Sierra Nevada or Peninsular Ranges basement, but as displaced from a similar batholithic terrane somewhere along the western margin of the continent bordering the Pacific Ocean. After its voyage to the northwest, it was sutured against California in Mesozoic or early Paleogene time. In addition, along with other slices consisting of largely granitic basement, it has been moved even farther to the northwest by post-early-Miocene displacements when the San Andreas system dominated the region.

Although parts of the Salinian block show very close affinities to basement-rock types east of the San Andreas fault system within the Mojave Desert and, therefore, probably originated from this region lying roughly between the Sierra Nevada and Peninsular Ranges, parts of the Salinian terrane, in contrast, have no such relatives so far recognized within California or nearby regions. Moreover, probable sutures between western parts of the block and eastern slices with affinities to the basement of the Mojave Desert have not been recognized.

In summary, the origin of the Salinian block is still incompletely known. Its history apparently involves at least two very different stages in order to account for its emplacement and position today. First, an obscure history during the Paleogene or even earlier time accounts for blocks of granitic basement west of the dying forearc basin which lay west of the Sierra Nevada and Peninsular Ranges volcanic arc with its underlying batholiths. During this first stage, three hypotheses have been advocated to explain the exotic presence of the granitic terrane, but none has yet been adequately documented: 1) Suturing of far-travelled granitic terrane from sources well out of the region (Champion et al, 1984), 2) Displacement along flat faults (either detachments or thrusts) that have brought granitic rocks relatively westward from granitic arcs (Silver and Mattinson, 1986), or 3) Pre-San Andreas strike-slip faulting (McLaughlin et al, 1988; Avé Lallement and Oldow, 1988). The second, distinct and well-documented stage of emplacement history, involves right slip of about 330 km on the San Andreas system beginning in early Miocene time. The Miocene and younger displacements carried the Mojave Desert rocks northwestward and still continue. When the San Andreas transform system began to dominate the tectonics of western California, some right-slip faults within the Salinian block contributed to the length of the block, but as yet, to an unknown degree.

Additional tectonic models are emerging, based largely on the interpretation of paleomagnetic data and the concept that crustal California is broken up into coherent and semi-rigid units rotating under the influence of simple shear upon detachments within the upper crust (Luyendyk et al, 1985). These models propose that some of the major strike-slip faults of the San Andreas system terminate abruptly against other blocks. For example, paleomagnetic measurements augmented by matching the provenance of some stratigraphic units with their source areas suggest that the Transverse Ranges block has pivoted from a roughly north-south orientation to its present east-west orientation (Link et al, 1984). Under this interpretation, some major strike-slip faults within the Salinian block, such as the Rinconada, end against the swinging semi-rigid beam of the Transverse Ranges block and do not cut into it or through it. When the beam swings, however, slices of the Salinian block are required to move relatively toward the northwest upon strike-slip faults, including the San Andreas.

This model also allows major faults south of the present Transverse Ranges to terminate on their northwestern ends, solving one of the long-standing problems of southern California tectonics: How do major faults such as the Elsinore and Newport-Inglewood terminate at their north ends?

The model also allows for different magnitudes of slip on different parts of the same fault. For example, displacement on the San Andreas fault proper northwest of the Transverse Ranges block could have begun earlier than the part of the fault cutting through the Transverse Ranges, and possess, therefore, more kilometers of right-slip. Splays of the San Andreas fault southeast of the Transverse Ranges would be both younger and have lesser displacements. Vigorous research is now underway to evaluate the significance of these concepts involving the interplay between strike-slip faulting and the rotation of crustal tectonic blocks within central and southern California.

VINCENT THRUST SYSTEM AND ASSOCIATED PELONA/OROCOPIA SCHIST

A thrust system underlies much of the Transverse Ranges and the region along the San Andreas system to the southeast. It emplaces basement gneiss and related rocks, along with their sedimentary and volcanic cover, above greenschist rocks of late Mesozoic age. The thrusting took place, according to regional cross-cutting relations and isotopic dating, in latest Cretaceous or earliest Tertiary time. The greenschist is mapped as Pelona Schist in the San Gabriel Mountains, Orocopia Schist in the Colorado Desert, Rand Schist in the Mojave Desert, and perhaps also Catalina Schist offshore. Below the thrust the schist is severely deformed and locally is metamorphosed to the blueschist and lower amphibolite facies. Rock types include quartzo-feldspathic schist, mica schist, albite amphibolite, quartzite and rare marble.

Hanging wall rocks along the thrust are mylonitized, and at places, the belt of mylonites is over 500 m thick. Some minor structures and kinematic indicators show that the hanging wall was displaced relatively toward the present northeast, but it is controversial whether these indicators reveal a regional movement picture or only the latest movements.

The protolith of the footwall schist contains deformed bodies interpreted as mafic sills or lava flows injected into, or laid down within, a several kilometer-thick section of bedded graywacke and mudstone of continental provenance. One interpretation at present holds that the thick mass of metasedimentary and oceanic-type volcanic rocks was deposited in a rift, similar to the present Gulf of California (Haxel and Dillon, 1978). Another, but less likely interpretation, holds that the schist protolith accumulated elsewhere

and has been accreted to the continent by major lateral movements from regions far to the south or southeast (Vedder et al, 1983)

The Vincent thrust system has been both folded and dismembered by Cenozoic faulting. The footwall schist, assigned to three very similar stratigraphic units (Pelona Schist, Orocopia Schist, and Rand Schist), is exposed at the surface as tectonic windows within faulted and breached antiforms. At places the mylonite zones along the thrust have become zones of later movements, including those associated with low-angle normal faulting or detachment during mid-Tertiary time. The time of erosional breaching of the footwall is indicated by the influx of schist debris into mid-Late Miocene stratigraphic units at several localities, such as the Mill Creek pull-apart basin along the southern front of the San Bernardino Mountains (Sadler and Demirer, 1986).

The regional extent of the Vincent thrust system, which underlies or lies within much of the basement terrane of southern California, is not well known at present. A thrust fault with similar rocks both above and below the mylonitic movement zone is north of the Garlock fault within the Transverse Ranges and is about the same age, although several thrust strands that are not necessarily of the same age have been mapped in this region. Major mylonitic zones of similar character and age, and with similarly directed kinematic indicators, extend through the northern part of the Peninsular Ranges (the Santa Rosa mylonite belt) but do not have greenschist units in their footwalls. Instead, the mylonite zones separate granitic and gneissic terranes of different aspect and history. Far to the northwest, within ranges lying west of Salinas city, is a range consisting of similar greenschist as in footwalls farther south, but the relations of this mass to the others is as yet unknown although it may be part of the same widespread unit before metamorphism and tectonic emplacement.

The Vincent thrust system and related thrust faults are presently interpreted as the result of either convergence when lithospheric plates from Pacific Ocean regions came in against the continent, or as the result of a short-lived rifting phase during this convergence. It is a major tectonic element of western California of latest Cretaceous age, primarily, and must be kept in mind during our excursion.

DETACHMENT FAULTS

Low-angle normal faults are exposed in the desert of southeastern California and elsewhere locally (Davis, et al., 1980; Frost and Martin, 1982). Near the Colorado River, gneissic core complexes, overlain tectonically by younger sequences of sedimentary and volcanic rocks, characterize several mountain ranges. The younger beds commonly dip steeply into a fault contact with underlying gneiss and mylonite. The zone of movement consists of fault breccia, often chloritized, and exhibits mullions and grooves attesting to stretching displacements along a northeast-southwest trend.

Similar detachments are mapped within basement blocks, and deep reflection seismic profiles show that they extend into the middle crust with similar gently undulating shapes (Okaya and Frost, in press). Important gold deposits have accumulated near the detachment faults within severely brecciated rocks near the surface.

The low-angle detachments formed in late Miocene time between about 22 and 14 Ma. During this interval the Basin and Range Province underwent extreme extension, at places regionally exceeding 100 percent. Plate convergence was waning along the continental margin, and the San Andreas transform belt had not yet extended broadly, nor reached far eastward into the continent. The East Pacific Rise had not yet entered the Gulf of California. A plausible plate-tectonic explanation involves differential stretching above a flat subducting plate extending far under western North America and even to the eastern front of the Rocky Mountains.

The mid-late-Miocene extension, characterized by these low-angle detachment faults, is also recognized west of the present San Andreas transform system. The detachments are within the rugged mountains bordering the Salton Trough on the west, and detachments of the same age have long been recognized within the western Transverse Ranges. Within the Santa Monica Mountains northwest of Los Angeles, for example, low angle faulting brought younger units over older with displacements of about 25 km during late Miocene time. Whether these detachments are part of the regional extension of the southeastern deserts, however, is debatable. The detachment faults in the Santa Monica Mountains, although of about the right age, may have resulted from major downslope sliding of upper crustal rocks into an opening, major pull-apart basin.

Young, low-angle normal faults are mapped around the margin of the Salton Trough. At one place, for example, steeply dipping beds not older than 5 Ma meet downward at a detachment with basement. These relations are interpreted as extension concomitant with the opening of the Salton Trough, and not as the result of mid-late-Miocene regional extension. In addition, in the northern Peninsular Ranges several low-angle fault zones bring basement over basement. Kinematic indicators on some of the faults have a normal-slip sense of relative displacement, but it is not always known whether these are regional detachments, down-to-basin normal-slip faults, or tipped thrust faults, inasmuch as the age of faulting in such examples is not known. In addition, within the Transverse Ranges and southernmost part of the Sierra Nevada, recent mapping has disclosed flat faults, but of poorly constrained age. They are younger than faulting of the Vincent thrust system, and older than the regional mid-late-Miocene detachments, and may record either low-angle thrusting or detachment faulting of early Tertiary age.

THE SAN ANDREAS TRANSFORM BELT THROUGH TIME

John C. Crowell

Institute of Crustal Studies, University of California, Santa Barbara

INTRODUCTION

The San Andreas fault system, as a continuous series of main faults and splays, has a right slip of about 330 km that began in early Miocene time about 24 Ma ago (Crowell, 1979, Stanley, 1987). This magnitude of slip and date of beginning of displacement are largely based on determining the age of strata and volcanic rocks in western California that show the maximum right slip and where, as well, all older rocks show the same displacement. The most convincing offset correlations are Miocene shorelines, facies changes, and bathymetric interpretations by Addicott (1968), Miocene volcanic centers by Matthews (1976), facies correlations by Huffman (1972), basement-rock characteristics by Crowell (1960; 1968), and Miocene facies by Stanley (1987).

To determine the maximum displacement, geologic features that correlate must be identified on the two offset walls of the fault. These statements involve concepts and caveats that require discussion. The slip on any fault is defined as the vector magnitude, lying in the fault surface, of two points that were adjacent before the faulting and are now displaced so that each of the two points is identified in the two walls. The only practical *points* are *piercing points* where *geological lines* meet the fault surface, in much the same way as one visualizes an arrow piercing a target. Our task in regional and structural geology is to idealize geological features as lines (either straight or crooked) and to locate where these lines are cut by the fault. In addition, the position and elevation of the piercing points so identified need to be known. Some examples of geological lines within pre-faulting blocks are pinch-out lines, isopachous lines, truncation lines where steeply dipping features meet overlying unconformities, intersecting dikes, isotopic and geochemical isopleths, and many others.

It may be difficult to recognize offset lines and to be sure that the offset counterparts were segments of the same line before displacement. This is particularly difficult where displacements are huge and measurable in several tens or hundreds of kilometers. Under such circumstances, the geologist establishing the correlation needs to have a detailed knowledge of the two distant areas and of the region in between. It is largely for this reason that the discovery of great strike slip on California's faults, and on major strike-slip faults elsewhere over the world, has come so late in the development of geology. The recognition that the San Andreas fault had right slip of about 300 km was not possible until much of western California had been mapped by Dibblee, and then Hill and Dibblee (1953) were able to advance their hypothesis. In southern California, a region astride the San Andreas and San Gabriel faults, about 50 km wide and well over 300 km long had to be reasonably well known, with parts much better known, before similar displacements could be documented (Crowell, 1960, 1962).

In addition, true correlatable lines and their piercing points are not always discoverable. The dispersed geologic record is too piecemeal and fragmented, some parts are concealed by younger deposits, and some of the needed record has been eroded away. Instead, displacement may need to be established by finding domains where the complex of rocks and their recorded history is deemed close to unique. Most of the great displacements on faults of the San Andreas system are advocated on the basis of such an approach. For example, the rocks of three regions in southern California are displaced on the San Andreas - San Gabriel faults: The Tejon block on the northwest is interpreted as displaced from the Soledad block which in turn is displaced from the Orocopia block farther to the southeast (FIG. 8).

Another geometric and geologic difficulty faces geologists working with great strike-slip faults. This involves the concept of *regional trace slip*. Strata and some volcanic units are laid down subhorizontally on a regional scale. Where such flat or gently undulating beds are displaced by strike slip, the slip vector is parallel to the trace of the bedding against the fault surface. The offset counterparts of reference beds across from each other will therefore display no vertical separation or throw. Both conceptually and actually, formations can be displaced for hundreds of kilometers and yet, in crossing a major strike slip fault, the same formation will occur in both walls, and with very little vertical separation. Moreover, with regional trace slip, the vertical separation along the fault may differ in sense from place to place if the formation under consideration is slightly folded. This means that at places the reference bed on one side will first be high, and then, farther along the fault, low. Note in passing, that subsurface reference layers which are contoured on the basis of well data or seismic reflections may not reveal the fault where the vertical separation is less, or of the same magnitude, as the contour interval. Such *scissoring* along the trend of a fault is therefore the result of changes in the sense of vertical separation across a major strike slip displaying regional trace slip, and does *not* necessarily represent a change in the kinematic regime so that first one side is moved upward, and then the other.

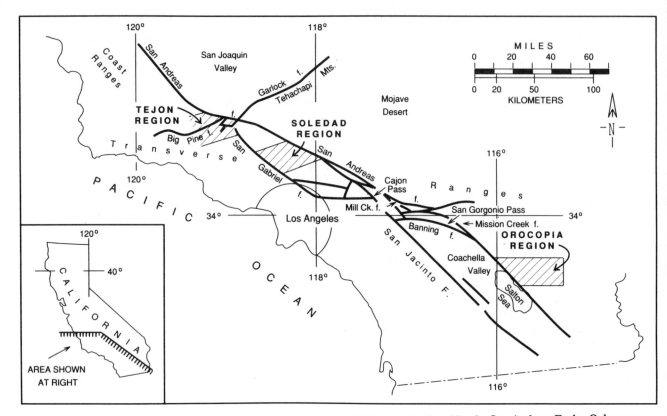

FIGURE 8 Tejon, Soledad, and Orocopia Regions in Southern California, Displaced by the San Andreas Fault. Only some the main faults are shown. From Crowell (1962).

Along the excursion route, these geometric concepts have led to vigorous controversy over the years. The Monterey Formation, of middle and late Miocene age, is widespread in California and mainly consists of siliceous shale. The San Andreas fault along several of its long reaches has juxtaposed units of the formation, so that the fault seems insignificant when crossing it, inasmuch as strata of the same lithology and biostratigraphic content occur in both walls. Many of California's large oil fields are nearby, and the Monterey Formation is an important petroleum unit. It is no wonder, therefore, that geologists, and especially petroleum geologists, were reluctant to accept the notion of major strike-slip.

Documentation of the time of origin of a major strike-slip fault is also difficult and is dependent on access to the right kind of geologic record. Where datable units are replete, as is often the case in western California, it may be possible to date the beginning of faulting rather closely using a time-displacement plot, as mentioned briefly above. Rock units containing geological lines and their piercing points *older* than a determined age will all show the same displacement. Those *younger* than this age will show successively less displacement coming up to the present through geologic time. The usefulness of finding the timing of fault onset, using such an approach, is of course dependent on having dated units that are both slightly older and slightly younger than the time of origin. At some places along the San Andreas system, such circumstances exist. At others, however, the preserved

record may not straddle this critical time of fault origin. On a regional scale, however, the concept is most important. In southeastern California, for example, distinctive Proterozoic units are displaced the same amount laterally as distinctive Miocene units. This information shows only that fault displacement took place sometime during the long interval between the age of the Proterozoic basement and deposition of the Pliocene units lying above it. California geologists have therefore been searching for places where the youngest strata are preserved that show the same offset as the oldest (FIG. 6).

Two other approaches are employed in order to pinpoint the time of origin of major faults of the San Andreas system. First, with faulting at the surface, coarse sedimentary debris may be shed from a rising fault scarp, and be preserved within the datable geologic sequence. The Violin Breccia along the San Gabriel fault is such a deposit which will be examined on the field excursion. In reality, the dating of such a breccia actually dates only the timing when the fault possessed a vertical separation, and trace slip of a flat topographic surface on a pure strike-slip fault would not form a scarp. Fortunately, however, almost all of the major California faults have a slight dip-slip component and developed scarps very early in their histories.

Second, folds may grow along major strike-slip faults, following the simple-shear scheme (FIG. 4). Here the problems are to determine the age of the beginning of the folding, and to relate the folds to the beginning of strike-slip faulting. Along the western margin of the San Joaquin Valley, the interiors of several oil fields display growing folds through time, based on stratal thickness changes and unconformities (bald-headed anticlines) across the folds. The trends of these folds, both at the surface and in the subsurface, show growth through time and an increasing obliquity to the nearby San Andreas, interpreted as the result of development during simple-shear (Harding, 1976). These folds began to grow in the early mid-Miocene, but the record does not allow a more precise dating of their inception. Younger folds, including those growing at present, seem to be responding to contraction nearly at right angles to the San Andreas fault, and do not follow the simple-shear arrangement (Mount and Suppe, 1987).

DISPLACEMENT PARTITIONING

According to earth-girdling plate circuits, undertaken in order to find the total strike slip along the transform boundary between the Pacific and North American lithospheric plates, about 1500 ±120 km of total slip is required (Atwater and Molnar, 1973; Atwater, 1989). This amount of slip, based on interpretations of sea-floor magnetic anomalies, began at about 29 Ma. As described below, not more than 330 km of post-29 Ma slip can be recognized as yet on the San Andreas fault itself. Well over 1000 km is therefore probably partitioned on other faults of the system, including hypothetical major faults such as one speculated as trending along the base of the continental slope. The strike-slip displacement on such a fault is illusive, however, mainly because relatively young sea floor is on the oceanic side, and a borderland underlain by complex geology on the continental side; therefore, there is little likelihood of finding previously existing geological lines to show offset. Attempts have been made, however, to match debris within deep-sea fans lying upon the oceanic plate with identifiable sources coming down offset submarine canyons (Hein, 1973). Although this work suggests progressive displacement through the latest spans of the Cenozoic Era, no information from this technique has divulged information as yet on the total displacement nor on the age of its initiation.

Displacement may have been partitioned on many other faults in the San Andreas transform belt. Several are conspicuous bathymetrically in the borderland, off both central and southern California. Many are prominent onshore, including important splays such as the Hayward and Calaveras faults in northern California, and the San Jacinto and Elsinore in southern California. In addition, components of strike-slip are known on many faults in the desert regions of eastern California and extending eastward into the Basin and Range Province. Some of these faults are active as shown by both geomorphic features along them and geophysical events associated with them. Others are overlapped by upper Cenozoic deposits of different ages and so have not acquired displacement since their deposition. Some of the older ones, but still younger

than mid-Oligocene, have been bent and deformed and abandoned. Many years of work remain to document both the partitioning and timing of displacements on the various faults.

Before the inception of the San Andreas system of strike-slip faults, long faults probably existed within the region, and especially during times of oblique plate convergence (McLaughlin, et al., 1988; Avé Lallement and Oldow, 1988). Basement segments constituting the Salinian block and within the southern California borderland are interpreted as having been assembled by lateral movements during late Mesozoic and early Tertiary time. Some of these faults are sutures where far-travelled blocks docked against coastal blocks. These faults in several instances are suspected to have been reactivated when the San Andreas system prevailed. Thus, some of the long and straight faults of western California may have originated as sutures between far-travelled blocks, others may have come into being during oblique convergence within rocks continentally inboard from the trench, and others may have been born when the Pacific plate met the North American plate. Here we are considering those that acquired displacement as the result of transform displacement between the two plates as the only ones truly belonging to the San Andreas system. These will include those showing post-mid-Oligocene displacements even though the fault originated earlier under an entirely different tectonic regime.

Some of the older strands of the San Andreas system that will be examined or discussed on the excursion include deep displacements on the Newport-Inglewood fault (still active), faults in the Banning Pass region at the northwestern end of the Salton Trough, the San Gabriel fault (it was probably the main strand of the San Andreas system between 12 and 5 Ma and formed the margin of Ridge basin), and the Pilarcitos fault within the San Francisco Peninsula. In addition, many strands of the fault system were recently active but are not now marked by active-fault landforms. Several additional "dead" faults are required by geologic relations but are out of sight beneath alluvium or young deposits that are not ruptured.

DISPLACEMENT ALONG THE SAN ANDREAS FAULT

The recognition of many tens of kilometers of strike slip on the San Andreas fault proper came slowly in California. In 1953 Hill and Dibblee published their landmark paper suggesting that the main strand of the present system (here referred to as the San Andreas fault proper in order to distinguish it from the system of many subparallel faults) had a total displacement of as much as 560 km. Their data indicated that this displacement had taken place since Jurassic time, but since 1953 information on the splays and strands, and on the ages of rock units displaced, now indicates that the difference between 560 km and 330 km, 230 km, includes both older faulting in pre-mid-Oligocene time, and the displacement contribution of splays joining the

San Andreas fault in northern California. The principal splay is the San Gregorio-San Simeon-Hosgri system which is viewed at present as perhaps adding as much as 115 km of right slip (Graham, 1978; Page, 1981). This fault system, as well as others, have played a part in the elongation of the Salinian block.

Prior to the Hill and Dibblee paper of 1953, several geologists had written concerning major strike-slip faulting in California. Noble (1926) was the first to show major strike-slip on the fault, on the reach bordering the Mojave Desert on the south. Here distinctive Miocene conglomerate bodies were offset about 40 km. Wallace (1949) suggested large right-slip, but was unable to find tie points of older rocks to establish total slip, nor its timing. Other papers dealing with right-slip of many kilometers preceded the Hill and Dibblee paper but on faults of the system rather than on the San Andreas proper. These include that by Vickery (1925) on the Calaveras fault and by Crowell (1952) on the San Gabriel fault.

Since the mid-1950s, many studies have added to our knowledge of the displacement history of the San Andreas fault proper. In central and northern California, especially significant contributions have been made by Curtis et al. (1958), Hall (1960), Fletcher (1967), Addicott (1968), Ross (1970), Huffman (1972), Matthews (1976), Nilsen (1984), and Stanley (1987), as well as by many others who have added valuable information and interpretations on local areas. In southern California, contributions have been made on the displacement history of the San Andreas proper by Crowell (1960, 1962), Ehlig et al. (1975), and Powell (1982), although, again, many investigators have contributed vital regional data. Smith (1977) marshalled evidence to propose that a major fault strand was active during Miocene time by connecting faults through the Salinia block (Red Hill, Chimineas, San Juan, and others), to faults within the Transverse Ranges, and thence to the Clemens Well fault in the Orocopia Mountains. Crowell (1982a) has shown that the San Gabriel fault was the main strand of the system

in late Miocene time and was then abandoned about 5 Ma. At that time, corresponding with the opening of the Gulf of California, reinforcement of the "big bend" of the fault, and elevation of the Transverse Ranges, the main San Andreas strand moved to its position bordering the Mojave Desert and connected highly deformed regions between Tejon and Cajon passes. The story is by no means completed, and research on how to link together major faults through time is still underway.

SUMMARY

The displacement histories on the San Andreas fault proper for northern California can be represented by a time-distance diagram such as the one prepared by Stanley (1987). During the decades ahead, such a plot will be improved and extended to the fault system as a whole. As changes in the rate of slip are documented, because the transform belt is complex, we can be sure that different faults of the system will be shown to have been active at different times. We now know that some have changed their rates of displacement through time or have been reactivated following episodes of dormancy. Moreover, recently proposed models, based in part on paleomagnetic data gathered during the past decade by Luyendyk and his colleagues, suggest that different stretches have displaced at different times as the continental margin has been subjected to simple shear (Luyendyk, et al., 1985; Kamerling and Luyendyk, 1985).

The evidence now in hand suggests that the San Andreas fault zone (properly defined, but including a few strands) has a total right slip of about 330 km, and that this slip has accumulated irregularly since the end of the Oligocene epoch. It is also clear that strike-slip faulting of major tectonic significance took place along the margin of the continent at intervals long before that time (Stewart and Crowell, in press).

THE MODERN SAN ANDREAS TRANSFORM BELT

Arthur G. Sylvester

Department of Geological Sciences, University of California, Santa Barbara

INTRODUCTION

Considerable research efforts have focused on the behavior of the San Andreas fault system in the recent past to understand how it may behave in the near future. This intensive research is justified because the San Andreas fault is the longest fault in the state and, therefore, it is capable of generating the largest earthquakes and affecting the greatest number of people and structures. Most of the research has been conducted by scientists of the U.S. Geological Survey and by university and private scientists supported by the National Earthquake Hazards Reduction Program. In just the past decade, dates of major slip events, Holocene slip rates, recurrence intervals, characteristic rupture patterns, and rupture lengths of specific prehistoric earthquakes have been determined for many faults in California, particularly for those of the San Andreas fault system. Indeed, our understanding of its past behavior has reached the point, where experiments have been devised and emplaced to predict future earthquakes along the fault in a few places, the most notable example being that at Parkfield in central California.

That the San Andreas fault system is active is quite clear from the frequency and distribution of earthquakes on it in historic time and from geodetic studies of fault creep and rates of strain accumulation. For many segments of the San Andreas fault system, the average earthquake repeat time is 200 years, but this can be as short as 40 years or as long as 350 years. That it was at least as active during the last millenium as it has been in the last 200 years is becoming more clearly established by painstaking paleoseismic studies of earthquake effects on old trees and in recent sedimentary deposits. That it has been active throughout Quaternary time is shown by the great multitude of fault movement-induced landforms along nearly all of its 1000 km length.

EARTHQUAKES

The first written record of an earthquake in California was provided by the Portolá expedition on July 28, 1769, when both the explorers and the Santa Ana River they camped along in southern California were thrown out of their respective beds by an earthquake. Its exact location is unknown. Seismographic data became available for California as early as 1887, but comprehensive earthquake monitoring commenced only in 1932 for southern

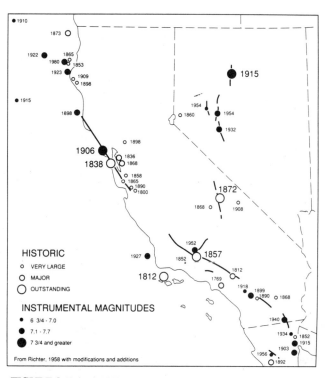

FIGURE 9 Principal Earthquakes and Associated Faulting, California, 1769-1988.

California.

Most of California's destructive earthquakes have not occurred on the San Andreas fault itself (Table 1; FIG. 9), however, two of the three outstanding historic California earthquakes did occur on the San Andreas fault and are the basis for its awesome reputation: The San Francisco earthquake of 1906 in northern California, and the Fort Tejon earthquake of 1857 in southern California. Each earthquake occurred in a "locked" segment of the San Andreas fault, both had surface ruptures measured in hundreds of kilometers, the "magnitude" of each is estimated to have been nearly 8.0, and if either earthquake recurred today, it is estimated there would be upward of 10,000 fatalities, 400,000 serious injuries, and from $30-40 billion damage.

California's earthquake history is quite short (< 200 years), so it is hazardous to generalize about what may be regarded as typical earthquake activity. All major earthquakes, however, have occurred on pre-existing faults, and most earthquakes greater than M 6.5 generate ground ruptures. The modern fault

TABLE 1 Significant California Earthquakes[1]

DATE (Greenwich)	NAME OR REGION	FAULT	MAG	TIME[2]
28 JUL 1769	Santa Ana region	?	?	1300
11 OCT 1800	San Juan Bautista area	San Andreas ?	6+	
22 NOV 1800	San Diego region	?	6.5	1330
8 DEC 1812	Palmdale/Wrightwood area	San Andreas	7.5[3]	0700
21 DEC 1812	Santa Barbara Channel	?	7.1	1100
10 JUN 1836	Hayward	Hayward ?	6.8	0730
JUN 1838	East Bay region	?	7.0	PM
29 NOV 1852	Southern California	?	6+	
9 JAN 1857	Ft. Tejon	San Andreas ?	8+	0800
8 OCT 1865	San Jose	San Andreas ?	6.3	1246
21 OCT 1868	Hayward	Hayward	6.8	0753
26 MAR 1872	Owens Valley	Owens Valley	8+	0230
23 NOV 1873	Oregon border region	offshore ?	6.7	2100
2 FEB 1881	Parkfield	San Andreas	5+	
12 APR 1885	Salinas	?	6.2	2005
9 FEB 1890	SE of Anza?	San Jacinto fault zone ?	6.3	
24 FEB 1892	San Diego/Baja California	Laguna Salada	8+?	2320
19 APR 1892	Winters	?	6.4	0250
21 APR 1892	Winters	?	6.2	0743
28 MAY 1892	SE of Anza?	San Jacinto fault zone ?	6.3	
20 JUN 1897	Gilroy area	San Andreas ?	6.2	1214
31 MAR 1898	Rodgers Creek	Rodgers Creek ?	7	2343
15 APR 1898	Point Arena	offshore ?	6.4	2307
22 JUL 1899	Cajon Pass region	?	6.5	1232
25 DEC 1899	San Jacinto-Hemet	San Jacinto	6.6	0425
3 MAR 1901	Parkfield region	San Andreas	6+	2345
18 APR 1906	San Francisco	San Andreas	8+	0512
19 APR 1906	Imperial region	Imperial ?	6+	1630
4 NOV 1908	Inyo region	?	6.5?	0037
15 MAY 1910	Elsinore region	Elsinore ?	6	0747
1 JUL 1911	Santa Clara	Hayward	6.6	1400
23 JUN 1915	Imperial region	Imperial ?	6.3	1959
22 OCT 1916	Tejon Pass region	San Andreas ?	6	1844
21 APR 1918	San Jacinto-Hemet region	San Jacinto	6.8	1432
10 MAR 1922	Cholame Valley	San Andreas	6.5	0321
22 JAN 1923	Cape Mendocino region	offshore ?	7.2	0104
22 JUL 1923	San Bernardino Valley	San Jacinto fault zone	6.3	2330
29 JUN 1925	Santa Barbara	? offshore	6.3	0642
4 NOV 1927	Pt. Arguello	? offshore	7.3	0551
6 JUN 1932	Humboldt County	? offshore	6.4	0044
11 MAR 1933	Long Beach	Newport-Inglewood	6.3	1754
8 JUN 1934	Parkfield	San Andreas	6.0	2030
25 MAR 1937	Terwilliger Valley	San Jacinto fault zone	6.0	0849
19 MAY 1940	El Centro	Imperial	6.7	2037
3 OCT 1941	Cape Mendocino region	offshore ?	6.4	0813
21 OCT 1942	Borrego Valley region	San Jacinto ?	6.5	0822
15 MAR 1946	Walker Pass	?	6.3	0550
10 APR 1947	Manix	Manix	6.2	0758
4 DEC 1948	Desert Hot Springs	Banning (San Andreas)	6.5	1543
21 JUL 1952	Kern County	White Wolf	7.5	0352
19 MAR 1954	Santa Rosa Mountain	San Jacinto	6.2	0154
25 NOV 1954	Cape Mendocino region	offshore ?	6.25	0317
21 DEC 1954	Eureka area	offshore ?	6.5	1156
22 MAR 1957	Daly City area	San Andreas	5.3	1144
27 JUN 1966	Parkfield-Cholame	San Andreas	5.3	2026
12 SEP 1966	Truckee region	?	6.3	0841
9 APR 1968	Borrego Mountain	San Jacinto	6.4	1829
9 FEB 1971	San Fernando	San Fernando	6.4	0601
6 AUG 1979	Coyote Lake	Calaveras	5.9	0905
15 OCT 1979	Imperial Valley	Imperial	6.4	1417
27 JAN 1980	Livermore	Greenville	5.8	1834
25 MAY 1980	Mammoth Lakes	Long Valley caldera	6.4	0849
25 MAY 1980	Mammoth Lakes	Long Valley caldera	6.4	1145
27 MAY 1980	Mammoth Lakes	Long Valley caldera	6.3	0651
8 NOV 1980	Cape Mendocino region	offshore ?	6.9	0228
2 MAY 1983	Coalinga	?	6.4	1642
24 APR 1984	Morgan Hill	Calaveras	6.2	1315
8 JUL 1986	North Palm Springs	Banning	5.9	0120
13 JUL 1986	Offshore Oceanside	offshore ?	5.3	0547
21 JUL 1986	Chalfant	White Mountain	6.2	0642
1 OCT 1987	Whittier Narrows	Elysian Park ?	6.1	0642
23 NOV 1987	Westmorland	Elmore Ranch	6.2	1754
24 NOV 1987	Superstition Hills	Superstition Hills	6.6	0515

MAG = typically local Richter magnitude or surface wave magnitude when available

[1] after Wesnousky (1986), Coffman et al. (1982), Toppozada et al. (1986) with modifications and additions

[2] Pacific Standard Time

[3] after Jacoby et al. (1988)

morphology is largely the result of slip in large earthquakes. Focal depths of large mainshocks are typically between 10 km and 15 km. Major earthquake sequences tend to cluster in time and space. Nearly 60 percent of the major damaging earthquakes since 1900 occurred within two hours of sunrise or sunset (Table 1) when the rate of change of the solar tide is the greatest. Periods of high earthquake activity are sometimes widely scattered over short periods of time, as, for example, in 1986 when major earthquakes happened within one month at such widely separated places as North Palm Springs, Oceanside, and Chalfant Valley. Some earthquakes occur on "blind" faults, or those at depth that do not reach the surface, such as the 1983 Coalinga earthquake and the Whittier Narrows earthquake of 1987.

San Francisco Bay Region

Earthquakes were common in the San Francisco Bay region before 1906. Several investigators have suggested that this high level of seismicity presaged the 1906 earthquake. The catalogs describe the effects of many earthquakes which would correspond to earthquakes in the local magnitude range from 3 to 6. Three of the largest were in June 1836, June 1838, and October 1868. Details pertaining to these earthquakes are few, but the 1836 and 1868 earthquakes are considered to have occurred on the Hayward fault in the East Bay, whereas the 1838 earthquake generated cracks for many kilometers along a segment of the San Andreas fault that was subsequently the locus of the 1906 earthquake. The 1868 Hayward earthquake was sufficiently impressive and damaging that it was regarded as "the great earthquake" until 1906. After 1906, the numbers of strong earthquakes in onshore northern California dropped nearly to zero for almost 50 years (Table 1).

The Daly City earthquake of 22 March 1957 is significant, because it was the first M≥5 earthquake to occur on the segment of the San Andreas fault that ruptured in 1906. No ground rupture was reported, and damage and injuries were low. The next earthquake in that segment happened near Los Gatos on 27 June 1988 with a provisional magnitude of 5.4. It is the latest, as of this writing, of a series of damaging earthquakes that have struck the south end of the San Francisco Bay south of San Jose. Several earthquakes with magnitudes between 4 and 6 have occurred between 1984 and 1988 at the south end of the Calaveras fault between San Jose and Hollister (Table 1), the most notable of which was the M 6.2 Morgan Hill earthquake of 24 April 1984 which caused one of the strongest horizontal earthquake accelerations ever measured at 1.3 g (Bakun et al., 1984).

San Andreas Fault

1812 Palmdale/Wrightwood. Dendro-chronologic evidence was presented by Jacoby et al. (1988) for the occurrence of a major earthquake on the Mojave segment of the San Andreas fault on 6 December 1812, an earthquake whose epicenter was previously assigned to a coastal location in the vicinity of San Juan Capistrano where one of the Spanish missions was destroyed. They estimated that the rupture segment was at least 12 km-long and perhaps hundreds of kilometers long. Fault dormancy before the earthquake was about 330 years, and the earthquake occurred only 44 years before the greater 1857 Ft. Tejon earthquake which shared much of the same rupture zone.

1857 Fort Tejon. This earthquake was the largest known to have occurred on the San Andreas fault in historic time. It was described as a violent shock that was felt over an area in southern California equivalent to the 1906 felt area. Few details are available for such a great earthquake, because the population was small and scattered in 1857. The earthquake may have been preceded by a few hours by a foreshock (M 6) in the vicinity of Parkfield. The surface rupture extended 400 km from Parkfield southeastward to Cajon Pass near San Bernardino. One anecdote tells of a circular corral, 10 m in diameter, that was offset into an S-shape as a result of slip during the earthquake. Subsequent studies have found many stream and rivelet courses also displaced right laterally from 2 to 5 m.

1906 San Francisco. The 1906 earthquake occurred at 0512 on the morning of 18 April. The surface rupture extended 430 km from Cape Mendocino to San Juan Bautista, connecting several previously mapped fault segments and thus revealing the continuity and great length of the fault. The earthquake and the surface rupture also clearly demonstrated the relation between earthquakes and faults. Surficial displacement was dextral, maximum offset was 5 m, and vertical movement was generally less than 0.5 m. Insurance adjustors estimate that 20 percent of the destruction in San Francisco was due to shaking, and 80 percent to the ensuing fire, but these figures are debatable. The official number of fatalities was 700, however, recent historic research provides evidence to believe that this number may be to low by at least a factor of two.

Among the significant observations on this earthquake were the triangulation surveys in northern California which provided the data for Reid's formulation of the elastic rebound theory. The earthquake also established the fact that strike-slip is the principal slip mechanism along the San Andreas fault, and it caused scientists to realize that strike-slip may be important along other great faults.

The Parkfield Earthquakes. The Parkfield-Cholame area in central California has experienced a series of similar earthquakes since 1857 that is remarkable for its apparent periodicity (FIG. 10). Earthquakes in 1857, 1888, 1901, 1922, 1934, and 1966 apparently comprise the series characterized by having a similar source and source dimension, magnitude, apparent rupture length, and surface offset (Bakun and McEvilly, 1984). They have a return frequency of 21.9 ± 3.1 years which is upset only by the earthquake of 1934. That earthquake occurred "early", but the interval between its bracketing earthquakes in 1922 and 1966 is 44 years.

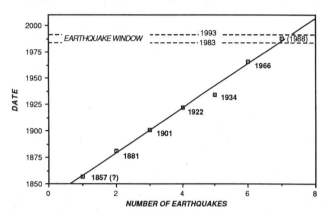

FIGURE 10 Repetition of Characteristic Parkfield Earthquakes, 1857 to 1966. Recurrence interval is 21.9 ± 3.1 years, extrapolated to 1988. Redrawn after Bakun and Lindh (1985).

The remarkable periodicity and similarity in character of the earthquakes in this series has attracted much research attention since 1983 on the Parkfield segment of the San Andreas fault, where the next earthquake in the series was predicted for January 1988 ± 5 years (Bakun and Lindh, 1985; Sims, this volume). The anticipated earthquake has not occurred as of this writing in December, 1988, but the area around Parkfield remains the one of the most heavily instrumented earthquake prediction sites in the world.

The 1966 earthquake is noteworthy for the discovery and documentation of afterslip that followed it. Since then, afterslip has been observed and studied in several strike-slip earthquakes. The earthquake also produced the strongest free field ground acceleration (0.5 g) that had been recorded up to that time.

San Jacinto Fault Zone

The San Jacinto fault is the most active fault in California, having generated at least six and perhaps 10 magnitude 6 earthquakes since 1890 (Sanders and Kanamori, 1984; Table 1). Little is known about the 1899 and 1918 earthquakes, except that their felt areas were comparable, and much damage occurred in the San Jacinto-Hemet area. The 1923 earthquake caused relatively minor damage in San Bernardino area; it is assigned to the San Jacinto fault northwest of the 1899-1918 rupture zone owing to the lack of surface evidence for fault slip on the adjacent segment of the San Andreas fault (Sanders and Kanamori, 1984). The 1937, 1954, and 1968 earthquakes occurred on various branches of the fault zone (Table 1; Clark fault, Coyote Creek fault). A 30 km-long stretch along the central part of the fault zone is called the "Anza seismic gap", because no known recent earthquake has ruptured through it, and its recent background seismicity has been much lower than that on adjacent fault segments.

The 1968 Borrego Mountain earthquake is noteworthy, not only because it had a significant, but complicated history of co-seismic and post-seismic fault slip, but because it also triggered minor fault slip on the Superstition Hills, Imperial, and San Andreas faults as far as 70 km from the epicenter. Major earthquakes have subsequently occurred on the Imperial and Superstition Hills faults in 1979 and 1987, respectively, and in view of the long period of time that has elapsed since the last major earthquake on the southern end of the San Andreas fault, some geologists and geophysicists postulate that the triggered slip there is an indication that the San Andreas fault is primed and ready to generate the next large earthquake in the Salton Trough.

Focal depths of San Jacinto fault earthquakes are clustered between 15 and 20 km between San Bernardino and the southeast end of the Anza gap where the crust is about 25 km thick. Then they decrease to depths between 3 and 10 km in the Salton Trough where the crustal thickness is about 10 km.

The Superstition Hills earthquake of 24 November 1987 (M 6.6) occurred at the southeasternmost end of the mapped San Jacinto fault zone. It was significant because it was preceded 12 hours earlier (and triggered?) by a M 6.2 earthquake on the Elmore Ranch fault perpendicular to it. The surface ruptures for both faults were remarkably well-exposed on the desert floor. The coseismic surface slip on the Elmore Ranch fault was 15 cm left lateral; the coseismic slip on the Superstition Mountains fault was 35 cm right lateral, followed by an addition 45 cm afterslip. Subsequent resurveys of geodetic markers indicated about 0.4 m of left-slip at depth on the Elmore Ranch fault, and about 1.2 m of right-slip on the Superstition Hills fault. A major earthquake was expected on the Superstition Hills fault because of its recent history of triggered sympathetic slip (Sieh, 1982), and because of the high rate of adjacent activity on the Brawley seismic zone in 1981, and the Imperial fault in 1979 (Wesson and Nicholson, 1988).

The structural and tectonic relations of the San Jacinto, Superstition Mountain, and Superstition Hills faults to one another is not well-known, because their surficial connection, if any, is obscured by a blanket of recent lake deposits, and few earthquakes have occurred to fill the information gap.

Imperial Fault

The earthquake history in Imperial Valley is fragmentary prior to 1940. A large earthquake occurred in 1903 and another severe earthquake occurred on April 18, 1906, the same day as the great San Francisco earthquake. Damage occurred in towns that would be similarly affected by later earthquakes on the Imperial fault, but the fault itself was not "discovered" until the 1940 earthquake. The magnitude of the 1906 earthquake is estimated to be greater than 6.

Imperial Valley was struck by major earthquakes on the Imperial fault in 1940 and 1979. The locus of the surface rupture of the 1979 earthquake occurred exactly, or within a few centimeters, of that of 1940, supporting the observation that, 90 percent of the time, surface ruptures of California earthquakes tend to occur almost exactly along the locus of the most recently active geomorphic trace.

There were actually two earthquakes on the Imperial fault in 1940: one having a magnitude of 6.9, followed over an hour later by one almost as large. The surface rupture was 60 km-long - half of it in California, half in Mexico. Maximum offset of 6 m occurred near the middle of the surface break and is well-known because of the photograph of offset trees that was published in several popular magazines of the time. Vertical displacement of up to one meter occurred at the north end of the surface rupture where the fault curves northward from its north-northwest strike.

The epicenter for the 1979 Imperial earthquake was 5 km farther north than 1940, the magnitude was less (M 6.4), and the surface rupture was only on the U.S. side of the International Boundary. It yielded the most comprehensive series of strong motion records ever obtained for a strike-slip earthquake.

Elsinore Fault

The Elsinore fault (FIG. 1) is a major fault, 250 km-long, which lacks much geomorphic evidence for recent movement in the U.S., but it has a large, well-preserved scarp in Mexico along the Laguna Salada strand of the fault that continues at least 60 km southward from the International Boundary. This scarp may have been formed in the earthquake of 1892. Sense and amount of geologic movement are debatable, although the consensus is that its Late Cenozoic slip has been dextral, perhaps as much as 40 km (Lamar and Rockwell, 1986). Only one earthquake greater than M 5 on 15 May 1910 has been recorded on the fault, except on its extension in Mexico.

Paleoseismic studies of the Glen Ivy strand, between Corona and Elsinore, give evidence for several events in the last 1000 years, with a recurrence interval of 250 years. The slip rate is estimated to be from 3 to 7 mm/yr, and typical earthquake magnitudes are in the range of 6 to 7 (Rockwell, et al., 1986).

Newport-Inglewood Fault Zone

The principal earthquake on this fault zone (FIG. 1) occurred on 10 March 1933, and although it was not especially large (M 6.3), damage and loss of life (115) were excessive because it occurred in the thickly populated Long Beach and Compton areas of the greater Los Angeles region. Many poorly constructed buildings combined with the water-saturated alluvium and other unfavorable geological conditions increased the damage. Many kinds of structures, particularly public school buildings, were hard hit. Fortunately, the earthquake occurred near sunset after most of the schools were empty. The main lesson learned from this earthquake was that building construction practices and codes needed to be improved. That realization led to enactment of the Field Act, one of several state laws that

resulted in stricter building codes and earthquake zoning.

No surface displacement was found along the fault zone for this earthquake, nor is there evidence for more than about a maximum of 800 m total right-separation across the fault zone since Late Pliocene time. The fault zone continues southeastward and offshore from Huntington Beach to San Diego. Minor earthquake activity has persisted along the northern half of the zone, leading to concern about the capability of the zone and its extensions farther south to produce damaging earthquakes in the San Diego region where a major damaging earthquake has not occurred in historic time.

Other Important California Earthquakes

The earthquake of 21 December 1812, which rocked all of southern California, was particularly severe in Santa Barbara. Missions in the western Transverse Ranges were strongly damaged or destroyed. Because of the many anecdotes of associated sea waves, seismologists generally assign the earthquake to the Santa Barbara Channel and thus estimate a local magnitude for it of from 7.5 to 8. The occurrence of a tsunami, however, could have been triggered by a submarine slide, and so is not definitive evidence of the earthquake's location. Toppazada el al. (1986) rated its magnitude as 7.1, judging from intensity effects.

The Owens Valley earthquake of 26 March 1872 is considered to be one of the ten strongest earthquakes to strike the North American continent in historic time and is the third of California's three great earthquakes mentioned above. It was felt over most of California, and probably most of Nevada and Arizona. It was very destructive in Owens Valley east of the Sierra Nevada. Well-defined scarps up to 7 m high were formed along the fault accompanied by right-slip of as much as 6 m. The surface rupture was at least 160 km long. Thousands of aftershocks, some severe, occurred over the next few years. The region has been seismically quiet since 1872 except for occasional M 6 earthquakes north of the north end of the 1872 surface break. Notable earthquakes include four M 6 shocks in three days in the Mammoth Lakes area in May 1980, and the M 6.2 earthquake of 21 July 1986 in Chalfant Valley.

Santa Barbara was struck by another devastating earthquake on 29 June 1925 which caused more damage than that in 1812. The local magnitude was only 6.3, with the epicenter either onshore or somewhere in the Santa Barbara Channel. The main reason for the high levels of damage in 1925 was that more buildings, and more poorly-constructed buildings were extant than in 1812. This earthquake was the first to do significant damage to a California city since the 1906 San Francisco earthquake, and it did much to kindle the awareness and smite the conscience of concerned structural engineers, urban planners, and earth scientists about the hazards and risks of earthquakes in a rapidly growing and urbanizing state. Regrettably, it required a still more destructive earthquake, the 1933 Long Beach earthquake, before

Californians began to plan and design systematically for earthquakes and earthquake hazards.

The 1927 Pt. Arguello earthquake (M 7.3) caused damage in the San Luis Obispo region. Its offshore location is controversial, but it may have been on the Hosgri fault and is critical, therefore, in appraising the hazard for the Diablo Canyon nuclear power station.

The largest earthquake to strike California since 1906 was the M 7.5 Kern County earthquake of 21 July 1952. It was felt over all of southern California, and damage was extensive in Bakersfield at the south end of the San Joaquin Valley. Surface ruptures along the White Wolf fault showed a left-oblique reverse-slip mechanism. The aftershock sequence for this event lasted several years, whereas the general level of seismicity in other parts of southern California remained low between 1952 and 1971 relative to the average seismic flux of the previous 30 years.

That seismic quiescence in the Los Angeles region was rudely interrupted at 6 AM on 9 February 1971 by a M 6.4 earthquake in San Fernando. Property damage exceeded $500 million because the earthquake occurred in a heavily populated area, and because strong ground motion ranged from 0.6 g to 0.8 g over a large area. The earthquake also generated the largest accelerations ever recorded for an earthquake to that time (1.25 g). The earthquake was felt by at least 10 million people, 1 million people lived in the damaged area, yet only 64 fatalities were recorded, and 40 of those were in one building - an old hospital that was built in 1918. In spite of the great damage, it could have been much worse had not Californians applied some of the knowledge gained from previous destructive earthquakes, and if the earthquake had occurred later in the day.

The Coalinga earthquake of 2 May 1983 (M 6.4) is significant, because it occurred in part of California thought to have a relatively low earthquake hazard - the west side of the San Joaquin Valley just east of the Parkfield region of the San Andreas fault. The earthquake is generally believed to have occurred on a blind thrust which dips southwestward toward the San Andreas fault. The only surface ruptures occurred with some of the large aftershocks more than one month after the main earthquake. Damage was also severe to unreinforced brick and masonry buildings in the older part of the city.

The most recent destructive earthquake in California was that of 1 October 1987 (M 6.1) near the city of Whittier, a major suburb of Los Angeles. There was no surface rupture, evidently because the earthquake occurred on a blind thrust that dips gently northward beneath the Transverse Ranges, but one-half meter of uplift was detected by precise releveling. Although this thrust was suspected from other evidence, the earthquake gave strong support to the presence of a fault system which may underlie the northern part of the Los Angeles metropolitan area, and may pose a greater hazard to the region than the San Andreas fault or any of its strike-slip relatives.

The damage and destruction caused by earthquakes in urban areas is capricious and wanton, but some

events have actually been beneficial. The charm of Santa Barbara and the safety of many high occupancy buildings throughout the state are direct consequences of earthquake experiences.

Several major faults in California exhibit abundant physiographic evidence for recent geologic activity, however, only sparse seismic activity has occurred along them in historic time. These faults include the 265 km-long, left-slip Garlock fault in the Mojave Desert, which is the second longest fault in California ; the 135 km-long, left-slip (?) Santa Ynez fault in the western Transverse Ranges ; the 360 km-long (right oblique-slip?) San Gregorio-Hosgri fault system which is mostly offshore of central California; the 180 km-long Nacimiento fault zone; and the 200 km-long Rinconada fault, both in central California between the coast and the San Andreas fault. The sense and amount of slip are still largely unknown for most of these faults.

Pre-Historic Earthquakes

The seminal study of pre-historic San Andreas earthquakes is Sieh's investigation of the sag pond deposits at Pallett Creek on the Mojave Desert strand of the San Andreas fault (Sieh, 1978a; 1984). By careful studies of stratigraphic relations, fault offsets, and radiocarbon aging of the peats exposed in deep trenches, Sieh found that nine large earthquakes have occurred at the site since A.D. 500, yielding an average recurrence interval of 140 ± 40 years, however, the dates of individual slip events ranging from 40 to 350 years demonstrates considerable irregularity of recurrence. Nevertheless, the average recurrence interval is at least ten times more frequent than recurrence intervals determined for other kinds of faults in the western United States. For example, normal faults in the Basin and Range province are characterized by return frequencies of from thousands to hundreds of thousands of years.

In order to determine the length of the paleoseismic fault ruptures, Sieh, his colleagues, and other investigators have done similar studies elsewhere on the San Andreas fault, principally at Indio in the Salton Trough; at Lost Lake in Cajon Pass (Weldon and Sieh, 1985); at Three Points between Pallett Creek and the Big Bend (Rust, 1983); at Mill Potrero in the Big Bend (Davis, 1983); at Wallace Creek in the Carrizo Plain (Sieh and Jahns, 1984); at Crystal Springs Reservoir on the San Francisco Peninsula (Hall, 1984); and at Dogtown near Olema in northern California (Cotton et al., 1982).

Sieh has provisionally correlated some of the events among the sites south of the central creeping segment as possibly representing rupture in a single earthquake. Events in 1480 and 1100 appear to be the largest, greatly exceeding even the 1857 earthquake, with ruptures apparently extending 500 km from Cholame southeast to the Salton Sea.

Paleoseismic studies have identified a few individual events on the San Andreas fault in northern California and on other faults of the San Andreas fault system, and in some cases permit estimation of an average slip rate, but sufficient numbers of events have not been identified on any single fault to compile a meaningful recurrence interval or minimum rupture length. Part of the problem is the paucity of suitable study sites with well-preserved, detailed stratigraphy and datable material. Another problem in correlation is the imprecision of radiocarbon dating relative to the apparent frequency of San Andreas fault earthquakes.

QUATERNARY ACTIVITY

The San Andreas fault exhibits the full range of characteristic landforms that originate on, or near, various strike-slip faults (FIG. 11). Indeed, it was the characterization of these landforms on the San Andreas fault after the 1906 earthquake that permitted recognition of other strike-slip faults, especially in tropical areas, which had not experienced a major, ground-rupturing earthquake in historic time. These tell-tale landforms range greatly in scale from small, simple sag ponds to great, compound depressions like the Salton Trough; from elongate ridges in the fault zone to entire, strike-slip fault-bounded mountain ranges within the fault system; from the stubs of ridges offset a few hundred meters as shutterridges to entire mountains ranges that have been transected and dextrally offset hundreds of kilometers from one another.

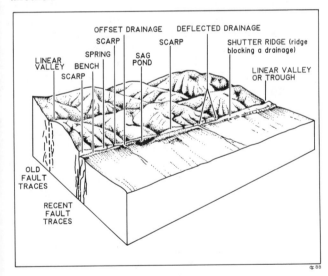

FIGURE 11 Assemblage of Landforms Associated with Strike-slip Faults. Redrawn after Wesson et al. (1975).

Many of the San Andreas landforms have been accentuated by historic earthquakes, especially in 1906 and 1857, leading to the conclusion that previous earthquakes contributed substantially to their formation and development and current morphology. Little quantitative work has been done, however, to measure the rate of growth and development of strike-slip landforms. Notable exceptions to this generalization are the detailed studies of offset stream courses at Wallace

Creek in the Carrizo Plain (Sieh and Jahns, 1984), and on a displaced alluvial fan in the Indio Hills (Keller, et al., 1982). Quantification of various contributing factors almost defies a general solution, however, given variables such as climate, rock type, slip rate, direction and fault mechanism. For example, it is nearly impossible to detect a difference between landforms formed in a fault zone characterized by episodic slip and one characterized by rapid creep (Hay et al., this volume). Moreover, fault blocks can rise and subside as they are transported along the strike in a major strike-slip fault zone over time, and scarps bounding those blocks may not exhibit the rather regular slope-degradation processes and rates characteristic of normal faults. The most reliable proof of age of a given landform, and thus of the timing of various Quaternary events, comes from systematic radiocarbon dating and from soil chronosequences. Without such data, it is hazardous to say *a priori* that a given scarp or landform in a strike-slip fault zone is necessarily young or old, judging from its relative degree of development.

REGIONAL AND NEARFIELD STRAIN

Regional and local strain across faults of the San Andreas fault system is measured by a variety of techniques and instruments, including laser-ranging trilateration, alignment arrays, precise geodetic leveling, creepmeters, borehole strainmeters, Very Long Baseline Interferometry (VLBI), and Global Position Satellite (GPS) surveys (Thatcher, 1986; Sylvester, 1986).

Regional Horizontal Strain

Annual laser-ranging surveys since 1970 show that interseismic straining is spread over a wide zone across the southern San Andreas fault system (Thatcher, 1979). Significant deformation extends out to distances as far as 100 km from the San Andreas fault, but the lack of any strain discontinuity at the fault trace indicates that the San Andreas fault is locked at the surface (Thatcher, 1986). Increasing movement rates away from the fault suggested to Thatcher (1986) that the fault slips freely below some locking depth. The regional interseismic strain build-up does not match the coseismic strain release, suggesting that a component of permanent deformation occurs which can be estimated geodetically.

In general, the geodetic and paleoseismic measurements show an average shear strain rate of about 35 mm/yr along the San Andreas fault. In central California nearfield geodetic measurements show that the shear is concentrated on the active, creeping trace of the San Andreas fault itself, where it is narrow and straight (FIG. 3). In northern California, however, where the fault zone is composed of several main strands, the shear strain is partitioned on some or all of the strands, with maximum slip rates as follows (Hall, 1984; Galehouse, this volume): San Andreas fault - 12 mm/yr; Hayward fault - 4 mm/yr; Calaveras - 12 mm/yr; Concord/Green Valley fault- 4 mm/yr.

Similarly, this rate of strain accumulation is partitioned across the major fault strand in southern California - with 20-25 mm/yr on the southern San Andreas fault, 10-15 mm/yr on the San Jacinto fault, and at least 5 mm/yr on the Elsinore fault. The Imperial fault currently accumulates horizontal shear strain at an average rate of 35 mm/yr.

Geodetically-determined surficial slip rates across the entire San Andreas fault system agree well with Holocene slip rates of about 35 mm/yr, but they do not match the plate tectonic rate of 55 mm/yr determined from analyses of ocean floor magnetic anomalies and space geodesy (Minster and Jordan, 1987). The conventional explanation is that the residual slip of about 20 mm/yr is taken up on other faults in a broad zone of shear that may be as wide as 500 to 1000 km across the Pacific-North American plate boundary. It is also recognized that surficial slip is temporally episodic and heterogeneously distributed in space. Thus, the 6 m of right slip on the Owens Valley fault east of the Sierra Nevada in 1872 would be cited as evidence that right shear is partitioned on faults other than the San Andreas fault across much of California and perhaps into Nevada in the Basin and Range Province where significant components of right-slip commonly accompany dip-slip earthquakes.

Regional Vertical Strain

Repeated precise leveling at ten year intervals in much of California shows, in general, interseismic and coseismic rise of mountains and subsidence of basins relative to sea level, consistent with the youthful topography. But the measured rates are slow, the number of surveys is small, and the time span of the surveys is short, so meaningful trends and rates of elevation changes have not been established. Coseismic vertical movements, however, have been dramatic: The elevation of the central San Gabriel Mountains increased 2 m as a result of uplift related to the 1971 San Fernando earthquake. The mountains on the hanging wall of the White Wolf fault rose almost one meter in the 1952 earthquake.

Much of the basin subsidence in the last 50 years is nontectonic, however, being caused instead by withdrawal of groundwater for irrigation or of oil and gas. Parts of the San Joaquin Valley have subsided as much as 8 m due to withdrawal of groundwater, and parts of the cities of Long Beach and Wilmington have also subsided about 8 m due to withdrawal of oil from the great Wilmington oil field. Subsidence was arrested in both places by cessation of pumping or reinjection of fluids. Locally the subsidence is concentrated along old faults in some basins, giving the false impression that displacement along the faults is tectonic (e.g., Pampeyan et al., 1988).

Creep

The San Andreas fault is unique among faults because it exhibits surficial, aseismic creep, which is a surficial phenomenon limited to the upper 100 m or so of the fault. Only the North Anatolian fault in Turkey has also shown evidence of continuous creep (Aytun, 1980).

Horizontal creep was first recognized in 1956 on the San Andreas fault at the Cienega Almaden winery, 12 km south of Hollister, by the progressive shearing damage to a large winery building built in 1948 upon the active trace of the fault (Steinbrugge and Zacher, 1960). Since then, creep has been documented on several faults of the San Andreas system in the San Francisco Bay area (see Galehouse, this volume) and more recently on the southern end of the San Andreas fault along the northeast side of the Salton Sea. Creep rates vary from 3-4 mm/yr near the Salton Sea (Louie et al., 1985) to maximum rate of 35 mm/yr in central California where the fault is narrow and straight (Burford and Harsh, 1980). The location and rate of creep is measured intermittently by precise surveying (Sylvester, 1986; Galehouse, this volume) and continuously by creepmeters (Schulz et al., 1982).

Creep is believed to be driven by elastic loading of the crust at seismogenic depths. Its occurrence at the surface represents either deep aseismic motion which is expressed as seismicity in one area and aseismic slip in another, or the accumulation of seismic dislocations away from slipping faults (Louie et al., 1985). Some writers believe that creep represents steady state slip which relieves buildup of stress on strike-slip faults, thereby decreasing the maximum size of an earthquake that may occur in that segment (Brown and Wallace, 1968; Prescott and Lisowski, 1983). Considerable horizontal strain data support the hypothesis (Langbein, 1981). Other writers postulate that creep is the first step in progressive failure leading to a large earthquake (e.g., Nason, 1973; 1977), however, creepmeter data show that several moderate earthquakes in southern California were not preceded by surficial, pre-seismic creep (Cohn et al., 1982), although some earthquakes initiated at the ends of creeping segments.

Afterslip, which is fault slip that occurs along a fault in the days, weeks, or years following the main shock, is another phenomenon characteristic of strike-slip faults that was first recognized on the San Andreas fault following the 1966 Parkfield-Cholame earthquake (Wallace and Roth, 1967). There it was made evident by the gradual increase of displacement of a paint-stripe on a highway. The coseismic displacement was only a few centimeters, the afterslip decreased logarithmically following the earthquake, and the surficial displacement totaled about 10 cm a year after the earthquake. Afterslip has been subsequently observed along the surface trace of many earthquakes on faults of the San Andreas fault system.

Triggered slip is another unique phenomenon of strike-slip faults that was discovered on the San Andreas fault (Allen et al., 1972). Triggered slip is coseismic slip or post-seismic delayed slip on a fault outside of the epicentral area of the main shock and at a distance from the causal fault. It has occurred repeatedly in the Salton Trough where up to 30 mm of slip has occurred on the San Andreas, Superstition Hills, and Imperial faults as far as 70 km from the causal fault and its epicenter following earthquakes in 1968, 1979, 1981, 1986, and 1987.

The mechanism of triggered slip is debatable. Some writers think it is induced by dynamic shaking by the main earthquake; others think the regional static strain field is perturbed so that the stress is concentrated on other faults which then yield by creep. Recently seismologists postulated that the occurrence of triggered slip on a fault is an indication that it is close to failure. It is noteworthy that moderate earthquakes (M>6.5) have occurred on the Superstition Hills and Imperial faults after they exhibited triggered slip following the 1968 Borrego Mountain and 1979 Imperial earthquakes. If the hypothesis is correct, then the southern San Andreas fault is primed to generate an earthquake in the Salton Sea region.

MODERN PROBLEMS

Each earthquake in California teaches new lessons. The 1971 San Fernando, 1983 Coalinga, and 1987 Whittier Narrows earthquakes caused seismologists to realize that all of California's earthquakes are not limited to nearly vertical strike-slip faults of the San Andreas system. Those earthquakes and subsequent geological and geophysical studies showed that several if not many active and potentially active faults, capable of generating damaging earthquakes, are gently-dipping thrust faults, some of which do not intersect the surface as discrete, mappable faults. These earthquakes and some lesser ones, together with seismic reflection and refraction studies, heat flow values and analyses of P wave delays, point to the existence of a mechanically layered upper crust, one which is detached, perhaps, on several layers at various depths to the base of the seismogenic zone. These detachments are required to explain not only thin-skinned rotation of crustal blocks, flakes, and slabs about vertical or nearly vertical axes, but also recent hypotheses that the San Andreas fault itself is offset at depth across nearly horizontal detachments or faults. The lack of a peaked heat flow anomaly along the San Andreas fault is also better explained if frictional shear heating is dispersed by the presence of shallow, low-angle thrusts or detachments.

Deep studies of the San Andreas fault have been emphasized in recent years. In 1986 and 1987 a deep hole (3.4 km) was drilled near the San Andreas fault where it transects the Transverse Ranges in Cajon Pass to obtain information about heat flow and stress state related to the fault. Near-surface stress measurements (0-1 km) show an increase of stress with depth, consistent with estimates based on laboratory frictional experiments (about 90 bars/km). However, extensive conductive heat flow measurements made near the fault show no discernible anomalies relative to regional values, suggesting that there is little frictional heating on the fault, and that the upper limit for the average shear stress on the fault is less than 200 bars. Therefore, near surface stress measurements cannot be extrapolated to depths greater than 3 km without violating existing heat flow constraints.

Stress measurements in the Cajon Pass drillhole imply that the present maximum compression axis is nearly perpendicular to the San Andreas fault, contrary to the 45° orientation that would be predicted from model studies and simple shear theory. Well-breakouts, well-bore elongation, and orientations of fold axes also show that the compression axis is nearly normal to the San Andreas fault in western San Joaquin Valley and along the Calaveras fault. Such stresses have to be maintained by the plate motion, otherwise they would be relaxed quickly by earthquakes. Mount and Suppe (1987) proposed that transcurrent and contractile components of deformation are decoupled and operate both simultaneously and largely independently of each other. This hypothesis also implies the presence of a mechanically decoupled upper crust and is significantly different from hypotheses originally proposed to explain such features as the en echelon anticlines in the San Joaquin Valley which have typically been cited as holotypes for fold formation in simple shear. Detailed stratigraphic studies show that the early history of the folds did, indeed, conform to the dictates of the simple shear theory, but an inferred change in the regional stress orientation from 2 to 5 Ma caused a departure from pure strike-slip on the San Andreas fault system to oblique-convergent strike-slip across a broad zone up to 700 km wide. That change is inferred to have coincided with the main episode of crustal shortening and folding which created the California Coast Ranges. Thus, the young thrust faults in the cores of the en echelon anticlines (e.g., Anticline Ridge near Coalinga) accommodate recent crustal shortening of about 5 mm/yr across the California Coast Ranges, whereas the San Andreas fault takes up the much greater transform displacement. It is possible, however, that the fractures observed in well-breakouts are *not* extension fractures, as is commonly assumed, but are shear fractures (Byerlee and Lockner, 1988). If that hypothesis is correct, then the principal stress axes, so derived, are consistent with a Riedel model of simple shear, where the greatest stress would be oriented 45° to the principal displacement zone, normal to the fold axes, and the paradox is explained.

These concepts are recent and profound revelations which have not been completely factored into the tectonic evolution of California, but their implications and significance for a variety of complex problems are being attacked with vigor by many California investigators.

SEDIMENTATION AND TECTONICS
ALONG THE SAN ANDREAS TRANSFORM BELT

John C. Crowell
Institute of Crustal Studies, University of California, Santa Barbara

INTRODUCTION

The anastomosing belt of faults along the San Andreas system has cut the continental crust into long slices and wedges. As movement progresses through time, some of these blocks are shortened in transpression and others are lengthened. Shortened blocks rise as they are squeezed to make uplands and, therefore, constitute source areas that shed debris into nearby depressions. These depressions are stretched blocks which sag to form basins. The rising and falling of the blocks in the laterally moving transform system results in what is termed colloquially in California as *porpoise structure* (Crowell and Sylvester, 1979). In accordance with this movement scheme, some blocks stand high for a time and serve as sediment source areas and then are carried laterally into a transtensional regime where they sag to make a basin floor.

Basins formed in this way are all along the San Andreas belt, ranging in size from small sag ponds conspicuous within the "rift zone" of active displacement to huge pull-apart basins (Crowell, 1974 a; b; 1987; Ingersoll, 1988). Some of these basins originated early in the evolution of the transform system and then have grown and have been structurally overprinted and modified through time. Some, such as the Los Angeles basin, have undergone at least two stages of deepening (Crowell, 1987), and deformation along their margins has been affected by marked rotations of the blocks forming their walls (Luyendyk et al., 1985). As the locus of lateral displacement has moved inland during the evolution of the transform system, coastal or western basins formed by tectonic stretching have then undergone isostatic adjustment. They have largely been abandoned as actively forming basins and new basins originated along new strands of stretching. For example, basins are actively forming and growing within the divergent boundary at the Salton Trough.

SALTON TROUGH

The Salton Trough at the head of the enlarging Gulf of California can perhaps serve, therefore, as one model of the interplay of processes underway at present, and of the processes that formed basins in the past. The tectonic processes that manufacture the basin are accompanied by uplift of mountainous margins. These high-standing regions provide sediment for the growing depression. The Salton Trough (FIG. 12) contains

about 10 km of sediments and originated as a deep, divergent gash in the continental crustal about 5 Ma. Before then, sedimentary and sparse volcanic rocks of Miocene age occupied the region, and in late Miocene time a shallow sea invaded the region (Metzger, 1968; Winker and Kidwell, 1986; Buising, 1988). Strata contained within the trough consist mainly of alluvial debris washed down from rugged bordering uplands, bajada and lacustrine beds in central areas, and rare thin beds deposited during intermittent marine incursions. Ancestral Colorado rivers brought in debris rich in carbonate detritus eroded from thick Paleozoic sections of the Grand Canyon region.

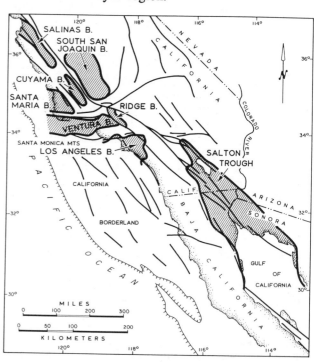

FIGURE 12 Late Cenozoic Basins in Southern California. Hachured line offshore is the 1000 m contour. From Crowell (1987a).

The stages in the development of this type of continental rift, such as the Salton Trough and some other basins in southern California, are illustrated in Figure 13. First, processes at depth succeed in stretching the crust so that the surface subsides. In stage 2 the region is arched above the developing basin, presumably because of crustal expansion resulting from heat from below and isostatic adjustment. As the stretching and heating continue, volcanic rocks may enter the basin at depth. In the Salton Trough, for example, volcanic rocks of at least two suites are present: 1) oceanic basalts at depth that show up as inclusions in younger intrusions and account for high values of gravity along the axis of the trough, and 2) rhyolitic domes which reach the surface. Stretching has gone on long enough in this region, shielded from sedimentation, so that the present surface lies 70 m below sea level. The elevation of the sedimentation surface has varied through time and is the result of the balance between sedimentation rates and tectonic subsidence. The very high heat flow, which drives geothermal power plants in the region, and active seismicity show that the rift is still enlarging. It fits into the scheme of Figure 13 at the beginning of stage 3 - the mature rift stage. Other basins in southern California, such as some offshore in the Continental Borderland and the onshore Los Angeles and San Joaquin basins, are inferred not to have an active heat source at depth any longer: They have entered the post-rift and subsidence state - Stage 5. As thermal and isostatic equilibrium are slowly restored, the surface continues to subside and the basins overfill. The marginal faults and folds, as well as volcanic rocks at depth, sink down so that they lie at great depths. Basins of this type contain a thick section of sediments laid down during the first subsidence and stretching stage, followed by a thick sequence laid down during the second stage after the active stretching of the basin has ceased.

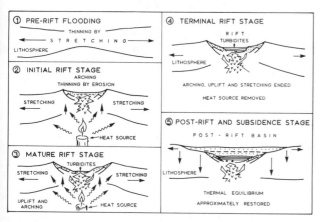

FIGURE 13 Stages of Rift Basin Development. From Crowell (1987a)

LOS ANGELES AND SAN JOAQUIN BASINS

The combination of these processes in the Los Angeles basin has lowered the top of Miocene beds to a depth of about 7 km at its center, but with a huge and unknown thickness beneath. During the total history of the basin, however, deformation has gone on nearly continuously, but with different orientations and styles from time to time. Deep-water turbidites were deposited during the mature stages, but folds and faults were disrupting the basin floor, so that the bottom-

seeking flows of sediment were guided by the tectonically controlled bathymetry. Many of these channel turbidites found their way between anticlinal hills on the basin floor. As the hills grew, some sand bodies were thereby arched across the anticlinal noses, and thus became prolific hydrocarbon traps in time. A similar style is manifest in the San Joaquin Valley where the careful subsurface study of turbidite channels has led to the discovery of several oil fields.

Deformation in both the Los Angeles and San Joaquin basins continued through the post-rift stages, but at different places and at different times, with different paces and styles. In the Los Angeles basin, deformation continues vigorously today in the broad transitional region where the northwest migrating Baja California block impinges against the Transverse Ranges. The Newport-Inglewood fault zone is being reactivated, and a chain of anticlines and associated structures nicely fits a simple-shear scheme where a buried fundamental fault lies beneath a thick cover of strata. Along the Santa Monica fault trend at the northern margin of the Los Angeles basin, another belt of complex folds and faults is also actively deforming, and so are the beds at depth. From geophysical evidence gained from the 1987 Whittier Narrows earthquake, these beds are now inferred to be shortening along flat and blind faults, sliding nearly parallel to bedding, and deep enough to be within the seismogenic zone.

PULL-APART BASINS

The major basins within the San Andreas transform belt are complex pull-apart basins. Large, truly rhombic-shaped basins, are rare, but some rhombic basins can be discerned in the bathymetry of the offshore basins in the Southern California Borderland, and also in growing basins in the Gulf of California. An idealized rhombic basin (FIG. 14) has subparallel margins: Those parallel to the transform direction are straight and consist of braided strike-slip faults, and those across the ends of the basin are oblique to the transform displacements - the pull-apart margins. The margins are oversteepened structurally and fail where high-standing continental rocks are adjacent to low-standing floors. Listric normal faults, slide blocks, detachments, and other complex structures lie along these margins. From the steep, strike-slip margins, coarse sedimentary facies give way to fine-grained sediments over short distances toward the axis of the basin. Features pictured diagrammatically in Figure 14 can be discerned for many of the California basins, but most have a more complicated arrangement, stacking sequence, and history.

In view of the mobility of crustal blocks in the San Andreas transform belt, most of the pull-apart basins do not display a rhombic shape. Some are roughly triangular in outline and may have formed as marginal blocks rotated, approximately following the scheme proposed by Luyendyk et al., 1985). The Los Angeles basin along its northern margin, and the San Joaquin

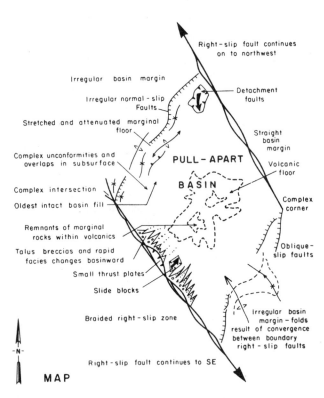

Right-slip fault continues
on to northwest

Irregular basin margin

Irregular normal-slip
Faults

Detachment
faults

Stretched and attenuated marginal
floor

Straight
basin
margin

Complex unconformities and
overlaps in subsurface

PULL-APART

Volcanic
floor

BASIN

Complex intersection

Oldest intact basin fill

Complex
corner

Remnants of marginal
rocks within volcanics

Oblique-
slip faults

Talus breccias and rapid
facies changes basinward

Small thrust plates

Slide blocks

Braided right-slip zone

Irregular basin
margin-folds
result of convergence
between boundary
right-slip faults

-N-

Right-slip fault continues to SE

MAP

FIGURE 14 Sketch Map of Idealized Pull-apart Basin. From Crowell (1974b)

along its southern, have been subject to severe deformation long after their origin. These margins now do not reflect their original shape or orientation. Both basins appear to have initiated about 24 Ma, but uplift of the Transverse Ranges block and complex folding and faulting along the marginal belts have very much modified their original shape. As the tectonics of the basins and their marginal blocks receive more study in the future, it is doubtful that a simple kinematic history of block movement and basin evolution will emerge.

Pull-apart basins seem to lie between two end members in regard to the nature of their floors. At one end are true rifts that extend at depth into hot rocks of the upper mantle, such as those expected above an oceanic spreading ridge. Older rocks, largely continental, are ripped asunder, first by attenuation of the upper crust, and then by actually breaking apart as the mantle material wells up into the widening gash. Under these circumstances, older parts of the basin floor on which sediments are laid down may be missing completely. These basins lack a true basement, and a well drilled to depth would go through sediments into a sill and dike complex of volcanic rocks. Because they intrude the oldest sediments in the basin, the volcanic rocks are younger than sediments at the base of this type of pull-apart. If the well were drilled vastly deeper, it would presumably reach hot and even molten rocks of the lower crust and upper mantle. Salton basin is an example of this kind of basin.

At the other end of the spectrum of pull-apart basin types are those that bottom-out on a detachment or decollement surface within the upper crust, and they may be grouped into two subtypes: 1) those that bottom against flat tectonic surfaces, or the detachments or flat faults themselves, and 2) those that bottom unconformably against older basement where the detachment is deeper still. Seismic profiling by COCORP and CALCRUST and other geophysical studies, along with down-dip extrapolation of surface observations, show that much of southern California is underlain by subhorizontal tectonic surfaces (Cheadle et al., 1986; Frost and Okaya, unpub. ms.). It is quite likely that some of the through-going reflections are from structural discontinuities on which crustal blocks pull apart and rotate. The profiles disclose several suspect detachments, however, so it is not yet clear on which, if any, the rotations and pull-aparts occur. Perhaps the reflections are stacked decollements, and intermediate blocks between them rotate differently from their underlying and overlying neighbors.

SEA LEVEL AND EUSTASY

In addition to the tectonic controls on sedimentation described briefly above, eustatic controls of changing sea levels are also important. For example, most of the major basins in coastal California originated and were flooded at about 22 or 24 Ma (Crowell, 1987). This was a time when relative sea level rose rapidly (FIG. 15). In general, sorting out the influence of sea-level changes along the San Andreas transform belt has barely begun, although sequence boundaries corresponding with world-wide eustatic levels are locally recognized. In the main, tectonic influences of sedimentation have overridden and obscured those due to eustacy.

SUMMARY

Tectonics and sedimentation have preceded hand in hand along the San Andreas transform belt, and the interaction between the two is clearly discernible today. Facies changes downflow from rising source areas are abrupt and range from terrestrial to deep-water marine. These facies changes are guided by the continuing deformation with only a minor influence from eustatic sea-level changes. Tectonic displacements determine the topography or bathymetry, and these, in turn, control the spread and types of sedimentary facies. Source areas range widely in elevation and rock type which, in turn, influence sediment character. Some source areas consist of rocks older than the onset of the transform belt, such as arc-granitic rocks of Mesozoic age. Other source areas consist of rocks originally laid down within a basin formed within the transform belt and later uplifted.

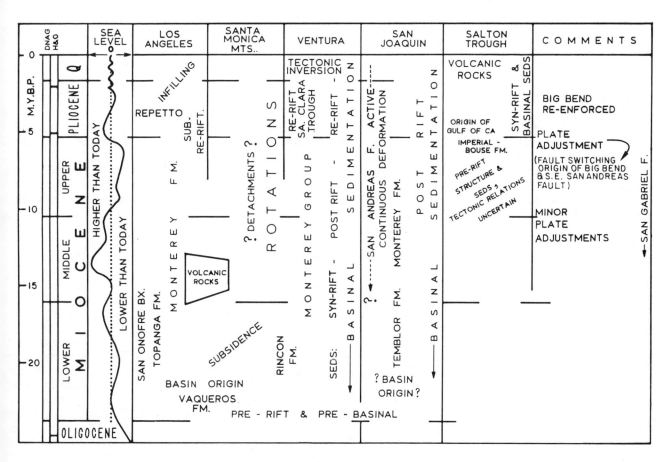

FIGURE 15 Relations between Sea Level Curve and Tectonism in Late Cenozoic Basins of Southern California. From Crowell (1987a).

RAPID CREEP ON THE SAN ANDREAS FAULT AT BITTERWATER VALLEY

Edward A. Hay[1], N. Timothy Hall[2], and William R. Cotton[3]

INTRODUCTION

We have attempted to determine slip rates and also to recognize geomorphic and subsurface criteria that may, in general, serve to distinguish rapid-rupture segments of a fault from those characterized by aseismic creep.

An offset road and the barbed-wire fences on either side of it, at the Flook Ranch site in Bitterwater Valley (FIG. 16) near King City, California, indicate that rapid creep occurred there during the last 75 or 80 years (first reported by Brown and Wallace, 1968). The zone of deformation is only a few meters wide and is topographically expressed by a shallow depression. This site is ideally situated for the accumulation of relatively young, laterally continuous sedimentary units, because it is located at the mouth of a canyon. The Holocene alluvial deposits here are incised by a modern channel (believed to be approximately 100 years old) that is offset right-laterally.

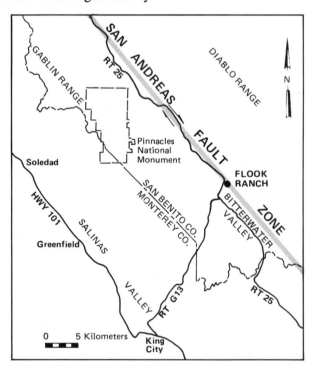

FIGURE 16 Location of Flook Ranch, Bitterwater Valley.

[1]De Anza College, Cupertino, California
[2] Earth Science Associates, Palo Alto, California
[3] William Cotton and Associates, Los Gatos, California

PALEOCHANNELS AND SLIP RATES

Five potential strain markers cross the San Andreas fault at the Flook Ranch site: a fence, the banks of a modern incised stream channel, and three different, buried stream channels (paleostream channels). The fence was constructed around 1908 (Brown and Wallace, 1968), and its offset yields an average creep rate of 35 mm/yr. The stream banks of the modern incised channel are offset a minimum of 3.3 m. Leopold and Miller (1956) stated that a period of widespread gullying began in the western United States in the year 1885. Assuming that the incised channel began to form in that year, then the average creep rate indicated by its offset banks is 34 mm/yr, which is consistent with the rate determined from the offset fence.

Three paleostream channels were located and defined by extensive trenching. The longest one, at a depth between 1.5 and 3.0 m and buried beneath the present alluvial fan surface, was traced laterally more than 60 m (Channel A-A' in FIG. 17). Within a zone approximately 10 m wide, two right bends account for most of the 7 ± 1 m of the channel offset. Rupture surfaces associated with the southwestern offset, which coincides with the locus of modern creep, were exposed in two trenches. Fault attitudes were N30°W, dip 90° and N45°, dip 80°SW.

Downstream from Trench 3, the form of the paleochannel changes abruptly over a short distance (FIG. 17); it is about 1 meter shallower and lacks the lower fine-gravel and pebbly-gravel deposits that characterize it upstream. The contrast of the channel appearance in Trench 3 with that in Trench 17 is dramatic. Trench 15C, in which the lower deposits are exposed, is located 21 m northwest of the deposits that were lacking along A-A'. We believe that this is an offset of tectonic origin which antedated 7 ± 1 m of slip observed along Channel A-A' between Trench 11 and Trench 3. If this reconstruction is correct, then the zone of active faulting stepped eastward following the 21 m of slip indicated along Deep Channel A-A", and the total width of the zone of deformation within the last 500 to 1000 years is 18 m.

The average paleo-slip rates can be calculated by dividing their tectonic offsets by their age. Radiocarbon ages from channel deposits, as well as from associated fan deposits, provide a basis for estimating the age of the paleochannel. However, an important characteristic of radiocarbon dating must be taken into consideration. The age of the sedimentary deposit is younger than the ^{14}C age of its charcoal, assuming that the charcoal is uncontaminated, because of three factors: 1) the

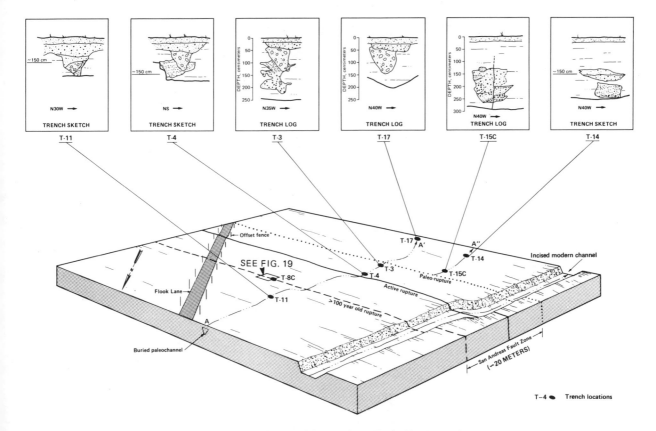

FIGURE 17 Trench Locations and Logs of Creeping Strand of San Andreas Fault, Flook Ranch.

charcoal may have been derived from very old wood (e.g., heartwood from redwoods, etc.), 2) there may have been significant time between burning and sedimentation, and 3) earlier deposits may have been reworked. Even if the radiocarbon age for a deposit is a reasonably accurate assessment of a deposit's age, the channel must be even older than the oldest deposit found in it, but it is younger than the fan deposits it cuts. Thus, the best estimate of a channel's age is bracketed between the youngest ^{14}C age of the oldest deposit in the channel and the youngest ^{14}C age from a fan deposit cut by the channel. Assuming that the charcoal used for the ^{14}C ages is uncontaminated, the channel age thus determined will always be a maximum. These relationships are well demonstrated by consideration of the Deep Channel A-A" at Bitterwater (FIG. 18).

The age of 1080 ± 10 yr B.P. is a maximum channel age because both ^{14}C ages upon which this value is based are maxima.

Dendrochronologic correction (Stuiver, 1978) of the 1080 ± 10 yr B.P. [14]C age indicates that the deep channel was 998 calendar years old in 1983 when we did our study. Because a slip rate of 34 mm/yr since 1885 was determined from offset of the modern stream channel, the average rate of movement for the prior 900 years was at least 27 mm/yr. Either the average slip rate here was slower before the turn of the century than since, or the radiocarbon age of 1080 ± 10 years B.P. is too old.

GEOMORPHIC CHARACTERISTICS AND STYLES OF RUPTURES

The active creep zone is not topographically expressed, even though it is in the middle of a topographic depression which trends parallel to the fault strike. This depression is 22 m wide in the region of our trenches, and it has a maximum relief of about 80 cm. Breaks in slope that define the margins of the depression are outside, but within a few meters of, the subsurface evidence for paleoruptures. In addition, recognizable expression of the relative vertical displacement that was once active for a significant period of time, is documented by structural relief shown on stratigraphic markers seen in Trench 8C (FIG. 19). Although the vertical offset of the surface is only 15 cm, an alluvial sand stratum is offset 55 cm, and an older silty clay is offset 80 cm. This northeastern zone of faulting has been inactive, however, for at least 100 years, because it does not disturb the fence or the modern incised stream. It is approximately 2 m wide and is characterized <u>not</u> by the northwest-striking ruptures seen in the active creep zone, but instead by many northnortheast-striking extension fractures whose average strike is N10°±5°E with near vertical dips. They accommodate both horizontal and vertical offsets in the stratified sequence and in the topography (Trench Log 8C, FIG. 19).

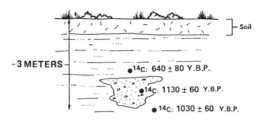

A. DIAGRAMMATIC SKETCH OF DEEP CHANNEL (A-A')
AT BITTERWATER

B. ANALYSIS OF ¹⁴C DATA USED TO DETERMINE AGE
OF DEEP CHANNEL

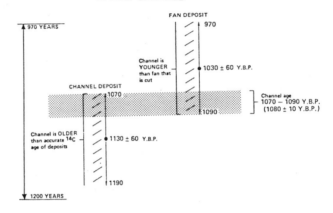

FIGURE 18 Diagrammatic Sketch of Deep Channel (A-A"), Flook
Ranch.

The southwest margin of the trough approximately coincides with a buried rupture zone, which parallels the active creep zone about 8 m to the southwest. It is the oldest expression of faulting at the site, offsetting a channel that was approximately 1000 years old in 1983, based on radiocarbon data. The rupture sharply truncated Deep Channel A-A" and appears to have been offset by northwest-trending shearing similar to current movement in the active zone, rather than by northnortheast-striking extension fractures, such as those along the northeast rupture zone, which is older than 100 years. We cannot judge whether the movement was by aseismic creep or by distinct rapid-rupturing, because both styles of slip seem to be characterized by northwest-trending shears.

We postulate that the great differences among rupture styles in the active creep zone are reflections of significantly different rates of slip. The radiocarbon age for the channel offset by the northeastern zone of faulting is not well constrained. It ranges so greatly that the slip rate varies between 17 mm/yr and 43 mm/yr. We favor the slower of these slip rates.

If the northeastern fault zone developed its north- to northeast-striking extensional fractures while the slip rate was in the 18-20 mm/yr range, during the time interval of perhaps 300 to 100 years ago, then the earlier 21 m of dislocation indicated by the offset of

Deep Channel A-A" would have accumulated at an average slip rate in excess of 30 mm/yr. Because this rate is comparable to the current slip rate, and because the style of rupturing in the offset of Deep Channel A-A" is also similar to the style in the one of active movement, we believe that the slip rate at Bitterwater slowed for about 200 years to 50-60 percent of the characteristic rate for the last 900-1000 years.

CONCLUSIONS

1) Can the style of rupturing in the subsurface of a creeping fault segment be distinguished from that produced by rapid rupture?

Our preliminary conclusion is that a zone of aseismic creep is characterized by fractures oriented about 45° clockwise from the strike of the fault-trace orientation if the creep rate is relatively slow. Just what rate constitutes "relatively slow" is not well established, and a quantitative estimate awaits further ¹⁴C age determinations. In the zone of creep that experienced an average rate of 34 mm/yr for the last 100 years, however, extension fractures oriented about 45° clockwise from the fault strike were not observed.

Distinction between rapidly creeping segments and rapid rupturing faults cannot be made confidently with

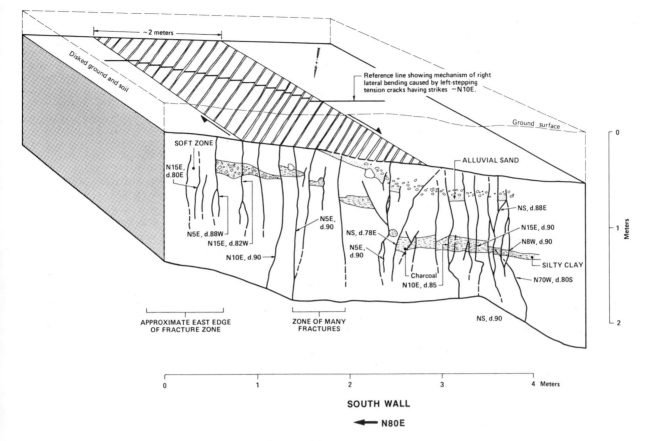

FIGURE 19 Diagram of Fault Ruptures and Log of Trench 8C: abrupt rupture segment of San Andreas fault, Flook Ranch.

our current information. Comparisons of the fractures seen at Dogtown (Cotton, et al., 1982), where rapid rupturing is known to have occurred, and the creeping segment at Bitterwater, however, indicate that creeping segments may be characterized by a zone of diffuse ruptures many meters wide that often strike several tens of degrees clockwise from the strike of the fault zone itself. This style of subsurface fractures was also documented along the creeping segment of the Hayward fault near the Fremont City Hall (Cotton et al., 1986). When enough strain accumulates in deep bedrock for it to rupture rapidly, the break through overlying near-surface materials will be sharply defined and confined to a narrow zone. In contrast, if strain is relieved in very small increments of movement, then a broad and diffuse zone of many fractures emerges, almost like plastic deformation.

2) Can long-term slip rates be determined along creeping segments of a fault?

Slip rates are determined just as easily for creeping segments as for those experiencing rapid rupturing. Paleochannels in Bitterwater Valley are superb markers for measuring up to 28 m of right slip. Charcoal samples collected from offset paleochannels, together with several samples from various stratigraphic markers

in the alluvial fan that antedate the paleochannel, yielded an average slip rate for the last 100 years of at least 28 mm/yr. We believe that the slip rate was significantly slower than the average rate during an undefined period prior to 100 years ago.

3) Are fault-related geomorphic features along creeping fault segments recognizably different from those of rapid rupture segments?

The rate of fault slip and variations in the strike-direction of ruptures are more important factors for geomorphic expression of a fault zone than whether or not the fault is creeping. In this region we have not recognized any geomorphic criteria to distinguish between segments of the fault zone experiencing different styles of slip; however, if reaches of a fault undergo relatively slow creep for long durations, then they develop subsidence depressions parallel to the fault. Thus, in a slowly creeping zone, extension fractures may develop in order to accommodate strain, causing small vertical movements to accompany the dominantly horizontal creep.

Research supported by U.S. Geological Survey Earthquake Hazards Reduction Program contract #14-08-0001-21335.

INVESTIGATIONS OF THE SAN ANDREAS FAULT
AND THE 1906 EARTHQUAKE, MARIN COUNTY, CALIFORNIA

N. Timothy Hall[1], Edward A. Hay[2], William R. Cotton[3]

INTRODUCTION

Following the great earthquake of 1906, G.K. Gilbert spent considerable time in 1906 and 1907 investigating fault features along the San Andreas fault southeast of Tomales Bay. We have reoccupied many sites that he photographed and described in his field notes. We have mapped the location of surface ruptures described by Gilbert and located aerial photo lineations and ground observed geomorphic features that are indicative of fault origin. The picture that emerges is one of a varied and complex fault zone.

LOCATION

Gilbert described surface ruptures at the Beisler and Bondietti Ranches (ranch names in 1906), which are approximately 8 km northwest of the town of Bolinas, and 10 km southeast of Olema (FIG. 20). The Bondietti house and barn are located on the southwest side of Highway 1. Walk southwest a few hundred meters through this ranch on a dirt road to reach the rift zone.

Gilbert also visited the Boucher Ranch approximately 2 km southeast of Olema. Entrance can be obtained by permission from the Vedanta Society, reached by turning southwest from Highway 1 at 300 m southeast of the intersection of Highway 1 with Sir Francis Drake Highway in Olema. From the Vedanta buildings, the Boucher site is approximately 2 km southeast on a dirt road.

BEISLER AND BONDIETTI RANCHES

The Beisler and Bondietti ranch sites lie in the San Andreas rift zone approximately 5 miles northwest of the town of Bolinas within the boundary of Point Reyes National Seashore. The Bondietti ranch buildings still exist, whereas all signs of the Beisler ranch, with the exception of a row of Monterey cypress tress, have been removed. The Beisler buildings are shown, however, on U.S. Geologist Survey W.R.D. photos taken 6/24/66 and their approximate positions are plotted on this map.

In his study of the 1906 earthquake in Marin County, G.K. Gilbert referenced several of his field

[1] Earth Sciences Associates, Palo Alto, California
[2] De Anza College, Cupertino, California
[3] William Cotton and Associates, Los Gatos, California

observations and photographs to these two ranches:

Bondietti Ranch--5/11/06 (G.K.G. Note Book No. 108, p. 31-32). "Bondietti's is east of main crack 200 rods. House shifted 3' toward the fault. Barn tilts. Men milking the cows thrown from the fault. Secondary cracks run through ponds. Bondietti infers that there were earthquakes before. This theory I verify by visiting a pond which is margined by the main fault. The up throw of earlier faults -- like this one has interrupted a drainage hollow and dammed the water. The catchment is small and a pond is the result. So this faulting is on an old line, recently used. I see 3 other lakes on same line and am told of 4 more. They are features of a faulted valley on the slope of a hill (G.K. Gilbert's section A, see GKG photo 2850):

West   East

"Displacement here works against erosion and the topography reflects both. Erosion has enough strength to determine the general slope, but displacement retains its ridge where the erosional force is at least-in drainage lines. In these slopes I see no soil differences of the type of the new rupture so cannot well estimate the trend lines of the last break."

Beisler Ranch--5/11/06 (G.K.G. Note Book No. 108, p. 30-31). "Beisler's is midway between Olema and Bolinas. The fault is there divided. The west part had a throw of 6 feet, the east 2 feet. The west passed under the barn, the east between house and barn. Mr. Beisler was milking within 6 feet of main fault, on the west side, and was thrown to west--A tank near the fault shifted slightly. Main part of barn between 2 faults shifted 2 feet north parallel to crack. Other buildings also shifted."

Bondietti and Beisler Ranches--3/30/07 (G.K.G. Note Book No. 110, p. 35-37). "Lunch at Bondietti's. He has lived here 17 years and has felt several noteworthy quakes. On one occasion a store was so shaken the loss from breakage was considerable--about 7 years ago. Several times has milk in pans had been spilt.

"The rift profile a little N. of Beisler's is (G.K.

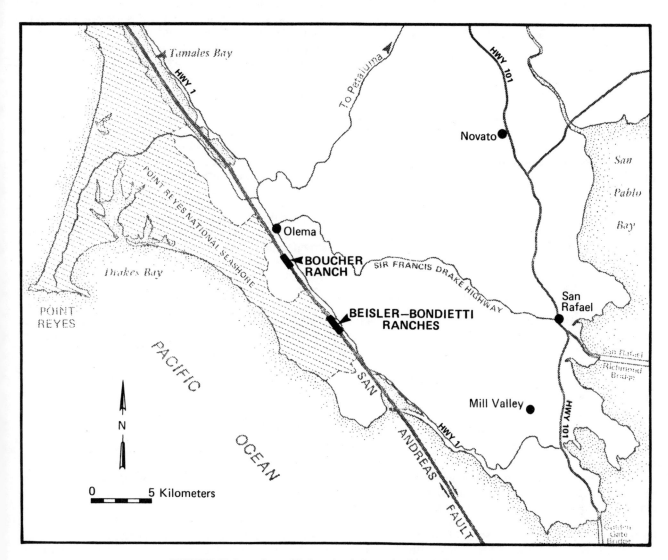

FIGURE 20 Locations of Paleoseismic Investigations in Olema Valley.

Gilbert's section B, see GKG photo 3047):

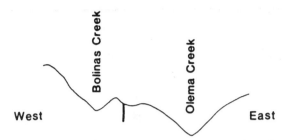

"A half mile south of Bondietti's it is (G.K. Gilbert's section C):

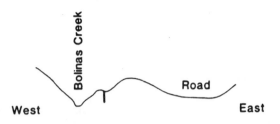

"Between Bondietti's and Beisler's (G.K. Gilbert's section D):

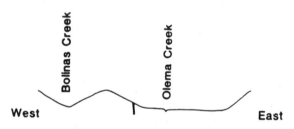

"Bondietti now understands the ponds, for the crack sends a branch to each. Approaching Beisler's the fault crack (fault crack) swerves a little to east and a parallel starts west of center course. The two come together. Each starts slides on opposite sides of a sag.

"From Beisler's the fault crack climbs a hill following a groove in the end of it. The hill is the end of a ridge. On the east side, is a side hill sag and this the fault crack finds (sketch B above) and follows for a 1/4 mile or more. There I leave it for the night. . ."

The Bondietti and Beisler ranch sites lie on a structurally complex, dissected medial ridge within the San Andreas rift valley. This ridge is flanked on the southwest by Pine Gulch Creek, which drains southeastward into Bolinas Lagoon, and on the northeast by Olema Creek, which flows to the northwest into Tomales Bay. The topographic expression of the active fault traces is quite clear along this reach of the San Andreas and the photo coverage of the 1906 activity by G.K. Gilbert is extensive and excellent. With the exception of a substantial increase in vegetation, which made relocation of Gilbert's photo stations difficult, this area has remained essentially unchanged since 1906.

Southeast of the Bondietti Ranch

Here a single 1906 trace ties along the southwestern side of the medial ridge and is marked by a series of elongate sag ponds impounded by segments of a dissected sidehill ridge. The ponds are bounded on the southwest by northeast-facing scarps. Gilbert observed that the 1906 faulting elevated the sidehill ridge with respect to the sags and thus increased their closure. It was also evident to Gilbert that the sidehill ridge and the ponds trapped behind it had formed by pre-1906 slip on the San Andreas fault.

Between 100 and 200 feet southwest of the 1906 trace, two knobs and a series of northwest-southeast-trending subsequent gullies suggest the existence of a subsidiary fault trace. Since Gilbert did not mention this possible trace, or photograph it, it probably was not active in 1906. The channel of Pine Gulch Creek lies between 500 and 600 feet southwest of the active trace. Because this channel is so linear and parallel to the 1906 trace, Pine Gulch Creek probably follows another strand within the San Andreas fault zone.

The crest of the medial ridge lies 500 to 600 feet northeast of the 1906 trace. The ridge is cut by several north-northwest to north-trending topographic lineaments suggestive of faulting. Two are well defined branch fractures, the western one of which is marked by a narrow linear valley and a sag pond. Gilbert photographed this branch fracture (photo 3037) but did not indicate it was active in 1906. The trend of these features is more northerly than the trend of the 1906 rupture, thus they probably experienced an extensile component of strain in addition to right shear and are, therefore, right-normal slip faults.

Between Bondietti and Beisler Ranches

Here the topographic expression of the rift valley and the 1906 faulting changes dramatically. The 1906 fault splays into two or three branches which lie along the northeast side, or base of the medial ridge which is confined to the western side of the fault. Two large sag ponds lie between the eastern side of the medial ridge and Olema Creek. In contrast to the linear path of Pine Gulch Creek, Olema Creek follows a sinuous path between the Bondietti and Beisler ranches across a rather broad alluvial valley floor. Here, east of the 1906 faulting, the medial ridge is interrupted, possibly by lateral erosion of Olema Creek, but more probably by downwarping along the axis of an inferred syncline. The topographic low caused by this apparent downwarping has created the closed depressions now occupied by the ponds. These ponds are presently being filled on their northern and eastern margins with alluvium from Olema Creek.

The two traces which Gilbert reported active at the Beisler Ranch are difficult to locate with much confidence, but we show three possibilities. The eastern break which passed between the house and the barn and which showed two feet of right-slip probably traverses the low marshy ground east of the medial ridge. Just as today, it was apparently not very photogenic during Gilbert's visits, for he left no record of its precise location. Two possibilities are shown for the western and major strand which displayed six feet of right slip in 1906. One follows the obvious topographic break along the northeastern margin of th central segment of the medial ridge, a path which passes beneath the site of the Beisler barn. The southeastern end of the most westerly strand was definitely active in 1906 as shown in Gilbert's photograph 3031, but its northwestern extension toward the Beisler barn is not well defined topographically. In between these two strands, we recognized a third strand by anomalous changes in slope.

Northwest of the Beisler Ranch

Here the medial ridge broadens and is once again a prominent feature on the northeastern side of the San Andreas fault. Two parallel traces ascend this ridge just southwest of its crest and form a prominent groove in it. Although the eastern trace is marked by an abrupt linear break in slope and a small sag, Gilbert did not indicate it was active in 1906 (see his sketch B above). The western trace was active, however, and is clearly shown in photo 3947. Apparently ground cracks from the 1906 faulting promoted a slope failure during the winter of 1907.

BOUCHER RANCH

The Boucher Ranch study area lies between 2 and 3.2 km southeast of Olema and is on land belonging to the Vedanta Society. We gratefully acknowledge Mr. Clair Scott of the Vedanta Society whose interest and assistance made our field investigation on this site possible.

After first visiting the Skinner Ranch (now the headquarters of Point Reyes National Seashore) and the Payne Shafter Ranch (now the Vedanta Retreat), G.K. Gilbert followed the San Andreas fault southeastward toward the Boucher Ranch.

3/29/07 (G.K.G. Note Book No. 110, p. 25-26). "The fault crack breaks up for a space and disappears, showing again in a pasture as a group of gentle furrows. It touches a summit pool in a minute sag near a hill top. The profile is (see GKG photo 3089):

"A little farther (see GKG photo 3043):

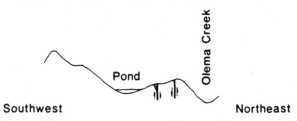

"Here for the first time I see features (within the fault crack belt) which might have been made by a similar fault 10 or 20 years ago. They are inflections of the surface quite similar to the ones recently made.

"The numbering of ponds includes only the larger, 3 and 4 are close to the Boucher Ranch (unoccupied). 5 is 1/4 mile south, but there are four others nearby--At the Boucher ranch and beyond the lake country is a plateau or terrace somewhat lumpy and only obscurely ridged:

"The fault crack is here divided and not rectilinear. There is an obscure echelon on a large scale:

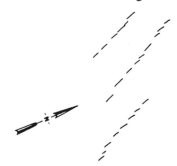

"This division of the fault crack continues for a mile south of the Boucher ranch and the fault line then enters forest, where I do not attempt to follow."

At the Boucher Ranch study area, the San Andreas fault has formed a broad sidehill bench or terrace between Olema Creek on the east and the northeast-facing slope of Inverness Ridge. A higher bench west of the Boucher ranch site probably also has a tectonic origin. With the exception of the removal of all traces of the Boucher buildings and a significant growth of vegetation, this area has remained virtually unchanged since 1906.

The San Andreas fault displays two distinctly different patterns of traces within this study area. Northwest of the site of the Boucher ranch buildings, the fault is characterized by multiple traces all of which are more or less parallel to the regional trend of the fault zone. A well-defined tectonic ridge separates these fault traces from Olema Creek. On the hillside north of the northernmost elongated sag pond, near tree "X",

Gilbert sketched as many as five closely spaced fault breaks (see profile A above). Near tree "Y" Gilbert sketched three 1906 breaks (see profile B above). It was on this hillside, which is underlain by Franciscan rocks, that Gilbert saw evidence for fault traces similar to the 1906 breaks that he thought could have been made 10 to 20 years prior to 1906.

Southeast of the Boucher Ranch site, the San Andreas fault crosses a broad "lumpy" terrace which lacks a well-defined bounding ridge on its northeastern side. There are no through-going strands parallel with the general trend of the San Andreas fault zone. Instead, the fault is characterized by multiple, N10-20°W trending, left-stepping en echelon breaks that were active in 1906. Gilbert photographed one of those breaks (photo 3045). Additionally, several vegetation lineations, sag pond margins and other topographic inflections have a more westerly direction than the regional fault trend. These two sets of lineaments divide the terrace into a cross-hatched pattern of subtle mounds, alluviated flats, knobs, saddles and sags.

Research supported by U.S. Geological Survey Earthquake Hazards Reduction Program, contract #14-08-0001-21242.

SAN FRANCISCO BAY REGION FAULT CREEP RATES MEASURED BY THEODOLITE

Jon S. Galehouse, F. Brett Baker, Oliver Graves, and Theresa Hoyt
Department of Geological Sciences, San Francisco State University, California

INTRODUCTION

We began to measure creep rates on San Francisco Bay region faults in September 1979. Amount of slip is determined by noting changes in angles between sets of measurements taken across a fault at different times. This triangulation method uses a theodolite to measure the angle formed by three fixed points to the nearest tenth of a second of arc. Each day that a measurement set is done, the angle is measured 12 times and the average determined. The amount of slip between measurements can be calculated trigonometrically using the change in average angle.

We presently have theodolite measurement sites at 20 localities on 10 active faults in the San Francisco Bay region (FIG. 21). Most of the distances between our fixed points on opposite sides of the various faults range from 50-275 m. The precision of our measurement method is such that we can detect with confidence any movement more than a millimeter or two between successive measurement days. We remeasure most of our sites about once every two to three months. The following is a brief summary of our results thus far.

SAN ANDREAS FAULT

Since March 1980 when we began our measurements across the San Andreas fault in South San Francisco (Site 10), no net slip has occurred. Our Site 14 at the Point Reyes National Seashore Headquarters has also shown virtually no net slip since we began measurements in February 1985. Our Site 18 (not shown on FIG. 21) in the Point Arena area has averaged about one millimeter per year of right-slip since January 1981. These results indicate that the northern segment of the San Andreas fault is virtually locked, with very little, if any, creep occurring.

HAYWARD FAULT

The average rate of right-slip on the Hayward fault is 5.4 mm per year in Fremont and 4.3 mm per year in Union City (FIG. 22) since we began our measurements in September 1979 in Fremont (Site 1) and Union City (Site 2). Since we began measuring two sites within the City of Hayward in June 1980, the average annual rate of right-lateral movement is 5.0 mm at D Street (Site 12) and 4.6 mm at Rose Street (Site 13). Since we began measurements in San Pablo (Site 17) near the northwestern end of the Hayward fault in August 1980, the average rate of movement has been about 4.2 mm per year in a right-lateral sense. However, superposed on the overall slip rate are changes between some measurement days of up to nearly a centimeter in either a right-lateral or a left-lateral sense. Paradoxically, right-slip tends to be measured during the first half of a calendar year and left-slip during the second half.

In summary, our data show that since 1979 the average rate of right-lateral creep on the Hayward fault is about 4 to 5 mm per year.

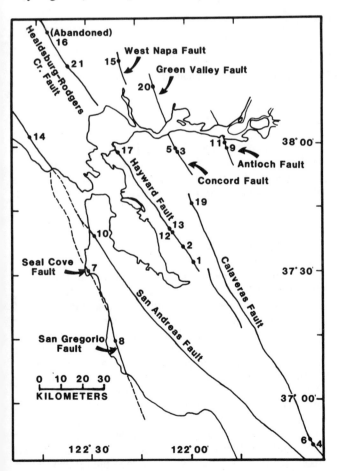

FIGURE 21 San Francisco State University Theodolite Measurement Sites.

HAYWARD FAULT

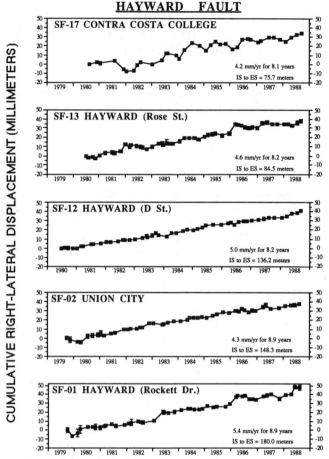

CUMULATIVE RIGHT-LATERAL DISPLACEMENT (MILLIMETERS)

SF-17 CONTRA COSTA COLLEGE
4.2 mm/yr for 8.1 years
IS to ES = 75.7 meters

SF-13 HAYWARD (Rose St.)
4.6 mm/yr for 8.2 years
IS to ES = 84.5 meters

SF-12 HAYWARD (D St.)
5.0 mm/yr for 8.2 years
IS to ES = 136.2 meters

SF-02 UNION CITY
4.3 mm/yr for 8.9 years
IS to ES = 148.3 meters

SF-01 HAYWARD (Rockett Dr.)
5.4 mm/yr for 8.9 years
IS to ES = 180.0 meters

FIGURE 22 Graphs of Cumulative Right-lateral Displacement of Hayward Fault, 1979 to 1988.

CALAVERAS FAULT

We have three measurement sites across the Calaveras fault, and the rates of movement are different at all three (FIG. 23). Since we began monitoring our Site 4 within the City of Hollister in September 1979, the average rate of movement has been about 7.0 mm per year right-laterally. Slip at this site is episodic, with most times of relatively rapid slip occurring early in a calendar year and little net movement occurring during the rest of the year.

At our Site 6 just 2.3 km northwest of Site 4, the slip rate was rather constant from late 1979 to June 1984. Since then, however, fault displacement has been episodic. Virtually no net movement occurred for more than a year between late 1985 and late 1986 (FIG. 23). Overall, the average rate of right-slip since October 1979 is 12.5 mm per year. This rate is the fastest of any of our sites in the San Francisco Bay region and is nearly twice as fast as that at our nearby Site 4 on the Calaveras fault. Inasmuch as the two sites are so close and the rates are so different, undetected movement may be occurring somewhere outside of our 89.7 m-long survey line at Site 4 within the City of Hollister.

In contrast to the relatively high creep rates in the Hollister area, our Site 19 in San Ramon near the northwesterly terminus of the Calaveras fault has shown virtually no net movement since we began measuring it in November 1980.

CONCORD FAULT

We began our measurements at Site 3 and Site 5 on the Concord fault in the City of Concord in September 1979 (FIG. 24). We measured about one centimeter of right-slip at both sites during October and November 1979. Following this, both sites showed relatively slow slip for the next four and one-half years at a rate of about one millimeter per year. In late spring-early summer 1984, both sites again moved relatively rapidly, slipping about 7 mm in a right-lateral sense in a few months. The rate again slowed to about a millimeter per year for about the next three years, beginning in late August 1984. Between late November 1987 and late February 1988, the Concord fault moved about 8 mm right-laterally. Since then the rate has again slowed down. Therefore, it appears that characteristic movement on the Concord fault since at least 1979 is

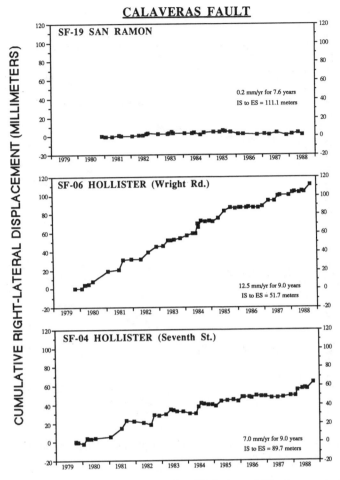

FIGURE 23 Graphs of Cumulative Right-lateral Displacement of Calaveras Fault, 1979 to 1988.

relatively rapid displacement over a period of a few months alternating with relatively slow displacement over a few years. Overall, the average rate of movement since late 1979 is 4.1 mm per year of right-slip at Site 3 and 3.3 mm per year at Site 5.

OTHER FAULTS

The Seal Cove fault (Site 7) and the San Gregorio fault (Site 8) have shown very little net slip since November 1979 and May 1982, respectively. However, both sites often show large variations in the amounts and directions of movement from one measurement day to another.

Much subsidence and mass movement creep appear to be occurring both inside and outside the Antioch fault zone, and it is probable that these nontectonic movements obscure any tectonic slip that may be occurring. Our Site 9 has shown about 1.7 mm per year of right-slip since November 1982, and Site 11 has

shown about 1.0 mm per year of left-slip since May 1980.

Seasonal and/or gravity-controlled mass movement effects are also present at our Sites 16 and 21 on the Rodgers Creek fault. Both sites show large variations from one measurement day to another. Site 16 had to be abandoned in early 1986 because our line of sight became obscured. Net movement had been virtually nil for the previous 5.4 years. Our replacement site on the Rodgers Creek fault (Site 21) has also shown no net slip since measurements began in September 1986.

Our Site 15 on the West Napa fault also shows large variations, but virtually no net movement has occurred since we began measurements in July 1980.

Since we established Site 20 on the Green Valley fault in June 1984, measurements show right-slip at an average rate of about 5.5 mm per year. Large variations also tend to occur here between measurement days. Preliminary results suggest that the Green Valley fault may behave similarly to the Concord fault to the southeast, that is, relatively rapid movement in a short

CONCORD FAULT

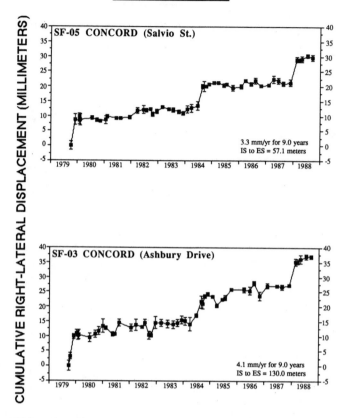

FIGURE 24 Graphs of Cumulative Right-lateral Displacement of Concord Fault, 1979 to 1988.

period of time (months) alternates with relatively slow movement for a longer period of time (years). Continued monitoring of the Green Valley fault will help to confirm its movement characteristics and rate and determine if the Green Valley fault is actually the northwestward continuation of the Concord fault.

This research is supported by the U.S. Geological Survey, Department of the Interior, under award number 14-08-0001-G1186. However, the contents of this report do not necessarily represent the policy of the agency, and the reader should not assume endorsement by the Federal Government.

THE NORTHERN SAN ANDREAS FAULT:
RUSSIAN RIVER TO POINT ARENA

Carol S. Prentice

Division of Geological Sciences, California Institute of Technology, 170-25
Pasadena, California

INTRODUCTION

The segment of the San Andreas fault, between the mouth of the Russian River and Point Arena, last broke at the time of the great San Francisco earthquake in 1906. It was part of a much longer rupture segment from near San Juan Bautista, south of San Francisco, to the vicinity of Cape Mendocino (about 430 km). Since then, this segment of the fault has been locked and essentially aseismic (Bolt and Miller, 1975; Prescott et al, 1981). Although this segment has not been studied as extensively as segments farther south, two references are invaluable to those wishing to become familiar with the fault zone in this area: Lawson (1908), and Brown and Wolfe (1972). U.S.G.S. 7.5' quadrangle maps covering this area are also listed at the end of the paper. Another very useful map is the Automobile Club Association of America's North Bay Counties map.

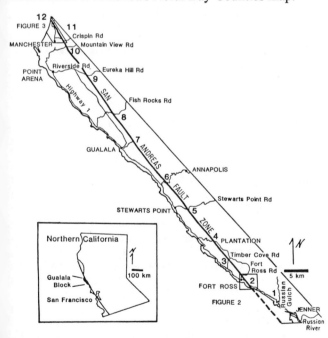

FIGURE 25 Map of the Gualala Block with Roads that Access the San Andreas Fault. Numbers refer to areas mentioned in text. Locations of figures 26 and 27 are indicated.

The region west of the San Andreas fault between Fort Ross and Point Arena is known as the Gualala block (FIG. 25). The San Andreas fault juxtaposes rocks of very different origin along most of its length,

and in the area of the Gualala block, rocks of the Franciscan Complex on the east are faulted against a >10,000 m thick sequence of Upper Cretaceous through Tertiary sedimentary and minor volcanic rocks that contain no Franciscan-derived clasts. These rocks were studied by Wentworth (1967). Many interesting geologic problems are associated with the bedrock of the Gualala block, including paleomagnetic interpretations of several thousands of kilometers of displacement since Cretaceous time (Kanter and Debiche, 1985), the origin of the granitic clasts in the Gualala Formation (Wentworth, 1967; James et al, 1986), and the nature of the Gualala basement. This paper only summarizes the Quaternary history and geomorphology associated with the San Andreas fault in this region.

PALEOSEISMIC AND SLIP-RATE RESEARCH

Excavations in a small, late Holocene alluvial fan near Point Arena exposed evidence of five or six seismic events. Because the stratigraphic sequence consists primarily of very coarse fluvial gravel, and deposition of younger deposits clearly involved erosion of underlying units, a minimum number of earthquakes is recorded at this site. Radiocarbon dating of charcoal collected from a layer that has been faulted by all of the events indicates that the layer is less than about 2000 yr old, indicating a <u>maximum</u> average recurrence interval of about 400 yr. The most recent earthquake at this site prior to 1906 occurred after AD 1524 and probably after AD 1635. A buried channel was located on both sides of the fault; a small branch in the gravel that fills the channel yielded a radiocarbon age of 2350-2710 yr.b.p. The maximum offset of this channel is 64 ± 2 m, giving a maximum slip rate of 25.5 ± 2.5 mm/yr. At this rate, the fault would take a minimum of about 200 ± 20 yr to accumulate the five meters of slip that occurred here in 1906. If this segment of the fault is characterized by five-meter slip events, then a great earthquake may not occur here for at least another one hundred years. These data suggest that a long average recurrence interval (from about 200 to 400 yr) characterizes this segment of the fault, leading to the tentative conclusion that a repeat of the 1906 earthquake is not likely to occur within the next century. Results of further radiocarbon analyses will better determine the dates of the individual paleoearthquakes.

Several Pleistocene marine terrace risers have been offset across the San Andreas fault near Point Arena. Preliminary correlation and age estimates of these terraces suggest an average slip rate since the late Pleistocene of about 18-20 mm/yr.

The Pliocene Ohlson Ranch Formation is a shallow marine formation that caps flat, to gently rolling surfaces on ridge tops east of the San Andreas fault near and south of Annapolis (Higgins, 1960). As pointed out by Higgins, no Pliocene marine strata are found on the surfaces preserved on the coastal ridge of the Gualala block west of the San Andreas fault. The Pliocene sea must have had an inlet; deposits near Point Arena that carry mollusks tentatively identified as Pliocene (Peck, in Boyle, 1967) may represent the offset inlet of the Ohlson Ranch Formation sea. Foraminifera collected from strata in both areas allow, but do not prove, this correlation. Zircons collected from an ash in the Ohlson Ranch Formation have a fission track age of 3.3 ±.8 Ma (Naeser, personal communication, 1988). If the correlation and proposed offset (50 km) are correct, then the average slip rate across the San Andreas fault since the retreat of the Pliocene sea is at least 12-20 mm/yr.

These tentative slip rates, if correct, imply three important conclusions about this segment of the San Andreas fault: 1) the slip rate has remained fairly constant over the past several million years; 2) a significant part of the Pacific-North American plate motion must be accommodated on other structures in this region; and, 3) a repeat of the 1906 earthquake is unlikely during the next 100 years.

GEOMORPHOLOGY AND CULTURAL FEATURES

Geomorphic and cultural features associated with the San Andreas fault lie along Highway 1 between Jenner and Point Arena. Some of the features are accessible only through private or state lands, and so permission from the owners is required for entry. These features, from south to north, are keyed to the maps of figures 26 and 27 by numbers and letters.

1. Uplifted Pleistocene marine terraces are found along most of the California coastline, and the highway north of Jenner crosses such a terrace. However, just north of Russian Gulch, the highway climbs through a set of switchbacks well above the terrace surface. The next several miles of highway, known to local residents as the Jenner Grade, is spectacular because the cliffs drop precipitously, with no intervening marine terraces, into the ocean from a height of about 200 m. Part of the reason for this conspicuous lack of terraces is that right-lateral displacement along the San Andreas fault has offset the terraces and has juxtaposed an area of land that was inland at the time of terrace formation next to the modern shore. The fault lies under the ocean less than 2.5 km west of the highway here; northwestward up the coast, marine terraces lie west of the San Andreas fault where it intersects the coastline south of

Fort Ross. Note the deformation of the terraces near the fault zone in this region.

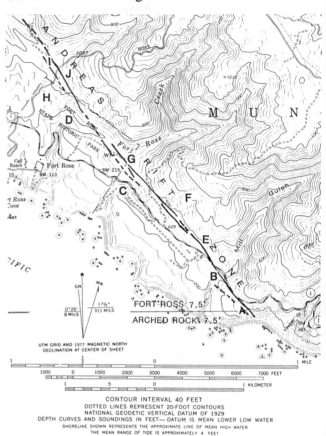

FIGURE 26 Topographic Map of the Fort Ross Area. Letters refer to features discussed in text.

2. The San Andreas fault intersects the coastline about 2.5 km southeast of Fort Ross (FIG. 26). Several small shutterridges (A) can be seen within a few tens of meters west of the highway as it descends from the Jenner Grade onto the terrace. Between here and the ranch houses north of Mill Gulch, the highway is immediately east of and adjacent to the fault zone; it crosses the fault within about 100 m of the ranch buildings and remains on the west side of the fault zone to a point about 10 km north of the town of Point Arena. Several drainages are beheaded or offset within a few tens of meters of the highway in this area (B, C, D). A small stream that was offset and ponded after the 1906 earthquake is a few meters east of the highway (E) (see Lawson, 1908, plate 37, p. 65, and map 3; this is the old Doda ranch referred to in his text). The offset here in 1906 was reported to be about 3.7 m; the stream is actually offset considerably more than this, so the 1906 offset was only the most recent movement to affect this channel. Within 200 m east of the highway in this area the fault traverses a large landslide (F); part of the toe of this slide has been offset nearly 2 km (H, FIG. 26).

A fence that was offset in 1906 and carefully surveyed (see p. 64 of the Lawson report) can still be seen (FIG. 26, G). Although many of the fenceposts have fallen, the original offset is still very clear. A resurvey of this fence in 1987 suggests that less than 0.5 m of afterslip has occurred since the original survey in 1906. Approximately 25 m south of the fence, the old Russian road that led to Fort Ross (in the 1800s) was also offset in 1906, although because the road crosses the fault at such a low angle, the offset is more obscure. The low, fresh, northeast-facing scarp in this area is probably the result of movement in 1906. The offset fence and road are on state land behind a locked gate; contact the ranger at Fort Ross for access.

Turning east on Fort Ross Road will lead to the fault again in about 0.6 to 0.8 km (FIG. 26). The road traverses a large landslide deposit 0 to 600 m southwest of the fault (H); the source of this deposit is probably the landslide headscarp east of the fault mentioned above. A charcoal sample collected from the base of this deposit was older than the range of radiocarbon dating (i.e. older than 43,700 yr), indicating that the average slip rate of the fault over this time period is less than 39 mm/year. The fault is expressed especially well geomorphically to the northwest (I) and southeast (J) of the point where it crosses Fort Ross Road: there are several sag ponds, linear troughs and ridges, and small scarps. This is also state land, so contact the Fort Ross rangers for access.

3-10. Between Fort Ross and Point Arena the fault is up to several kilometers inland from the Coast Highway, and access is limited to the roads indicated on figure 25. Heavy vegetation along much of this section conceals fault features. Of the available fault crossings, the best expressions of the fault are where the roads to Plantation, Stewart's Point and Annapolis cross the fault zone (4,5,6).

11. About one-half km north of Manchester, turn east on Crispin Road (FIG. 27). A linear trough is evident where the fault crosses this road, and a modified sag pond lies a few hundred meters southeast of the road (K).

Well-developed fault geomorphology (sag ponds, scarps, and shutterridges) exists on the Skaramella ranch (FIG. 27) (this is private land, permission must be secured before visiting the area). In addition, a row of cypress trees that was offset in 1906 is still present less than 100 m northwest of the ranch house (L). An excavation site, where data documenting prehistoric earthquakes and the Holocene slip rate were collected, is about 500 m northwest of the ranch house (M).

12. The fault crosses Highway 1 again under the bridge across Alder Creek (FIG. 27). Within a few hundred meters southwest of Alder Creek, about 200 m northeast of the highway, is a series of sag ponds and scarps (N). The unnamed, paved side road leading northwest from the highway, about one-half km south of Alder Creek, (at mile marker 22.48), leads to the mouth of the creek. This is the site of the old bridge that was destroyed by fault rupture in 1906 (O) (see Lawson, p. 59 and plate 32). The bridge was later rebuilt and the cement abutments of this later structure still stand. In the stream bed within a few meters of the

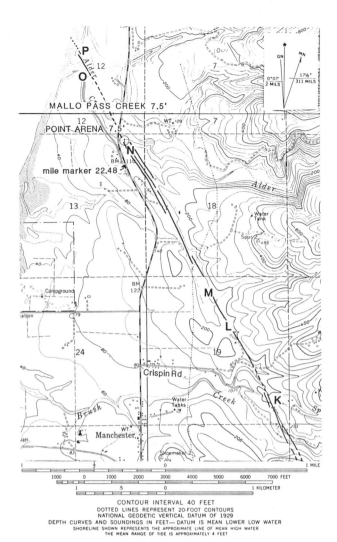

FIGURE 27 Topographic Map of the Point Area Area. Letters refer to features discussed in text.

western bank, (a few tens of meters northwest of the bridge site), an exposure of fault gouge can be seen when the tide and stream flow are low. A sag pond is present between the stream valley and the modern sea cliff above the northern bank of the stream (P). The cliffs along the beach for several hundred meters north of the stream mouth give an idea of how wide and highly sheared the bedrock fault zone is, but the 1906 trace is obscured by slumping and other erosional processes. Franciscan bedrock is exposed *west* of the 1906 fault trace here, but whether this represents basement west of the fault or merely material caught up in the fault zone is not clear.

U.S.G.S. 7.5' QUADRANGLE MAPS

Arched Rock	Annapolis	Saunders Reef
Fort Ross	Stewart's Point	Point Arena
Plantation	Gualala	Mallo Pass Creek

Acknowledgments

This work was funded by U.S. Geological Survey contracts 14-08-0001-61370, 14-08-0001-61098, and 14-08-0001-22011, a Grant-in-Aid of Research from Sigma Xi, The Scientific Research Society, and the Geological Society of America. Field accommodations were generously provided by Mendocino College at its Point Arena Marine Institute.

ROADLOG

LONG BEACH TO BORREGO SPRINGS

SIGNAL HILL-DANA POINT-LAKE ELSINORE-SALTON TROUGH

The route proceeds along the southwest edge of the Los Angeles basin, parallel to the Newport-Inglewood trend. The first stop is on Signal Hill in the Long Beach oil field, which is a squeezed-up block between two faults in the Newport-Inglewood fault zone. Then the route progresses south along the Pacific coast to San Clemente, thence east up the gentle flank of the Santa Ana Mountains to Lake Elsinore and the Elsinore pull-apart basin. It winds then through exposures of batholithic rocks in the Peninsular Ranges to the head of the Salton Trough at Borrego Springs.

INTRODUCTION TO THE LOS ANGELES BASIN

The Los Angeles basin (FIG. 28), in the geologic context, is one of several Tertiary depositional basins in California, each of which received very thick deposits of clastic marine and non-marine sediments in Miocene and Pliocene time. These basins were deformed in late Pliocene and Pleistocene time. The deformation is so recent that it is vividly expressed in the geomorphology.

Many of the hills are structural highs. The elevation of coastal areas is shown by numerous wave-cut marine terraces. The highest terrace yet recognized is in the Palos Verdes Hills, south of Los Angeles International Airport, at an elevation of 400 m above sea level. Remnants of fluvial terraces are found along all the major streams around the basin. The streams have produced surfaces that are partly erosional and partly depositional. The most notable depositional features are the coalesced alluvial fans along the flanks of the

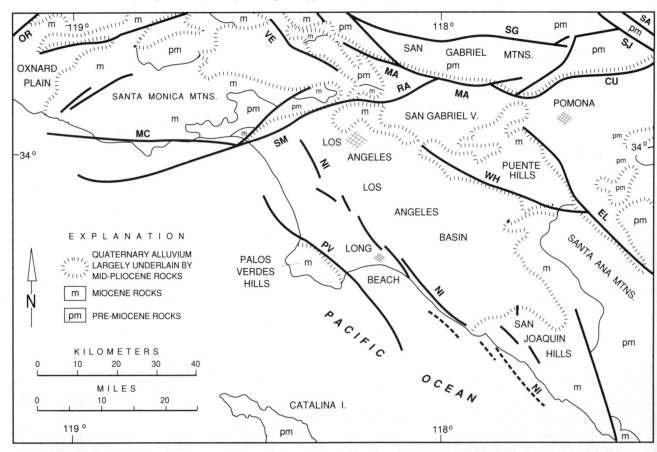

FIGURE 28 Generalized Geologic and Geographic Map of the Los Angeles Basin. OR, Oak Ridge fault; SI, Simi fault; MC, Malibu Coast fault; SM-RA, Santa Monica-Raymond fault; VE, Verdugo fault; MA, Sierra Madre fault; SG, San Gabriel fault; SA, San Andreas fault; SJ, San Jacinto fault; CU, Cucumonga fault; CH, Chino fault; WH, Whittier fault, EL, Elsinore fault; NI, Newport-Inglewood fault zone; PV, Palos Verdes fault. From Crowell (1987a).

young, uplifted mountains.

Although the late Tertiary Los Angeles basin probably took form in middle Miocene time, parts of the area it occupies received sediments during early Tertiary time and also during Late Cretaceous time as shown by strata of Late Cretaceous and Paleogene age in mountains surrounding the basin. During Pliocene and Pleistocene time the basin became progressively shallower and smaller. Foraminiferal studies indicate that the depth of the sea gradually decreased from more than 1300 m to about 250 m in the interval from early to late Pliocene time.

Marine Pliocene and Quaternary sedimentary rocks have a known thickness of at least 10,000 m in the central part of the Los Angeles basin, judging from seismic reflection studies. Basement has never been penetrated in the central part of the basin, but the deepest well, southwest of downtown Los Angeles, reached a total depth of 6528 m into upper Miocene strata. Within oil field drainage areas, the Los Angeles basin contains 6550 cubic kilometers of sedimentary strata.

The basin has 62 oil fields from which over 7.7 billion barrels of oil and 7.1 MMCF of gas have been produced. Twelve of the fields have produced between 100 million and one billion barrels of oil. Huntington Beach field and the giant Wilmington field have produced about 1 billion and 2 billion barrels, respectively, since 1932 and are still major producers. A total of 25,811 wells had been drilled in the basin as of January 1986.

The majority of the oil fields produce from multiply stacked Miocene and Pliocene turbidite sandstone reservoirs. Oil generation occurs below 2500 to 3000 m subsea at temperatures exceeding 164° C from upper Miocene shale beds of the Puente and Monterey Formations, and where present, of the Topanga Formation. Vertical migration occurs largely along faults and fractures, although many of the thin, turbidite sand beds are virtually self-sourced by intercalated shale beds. Oil production columns are locally as thick as 2000 m.

Los Angeles basin is bounded by several strike-slip fault zones which also control the main hydrocarbon-bearing structural traps in the basin. The northwest-trending structural trends - the Palos Verdes, the Newport-Inglewood, Whittier-Elsinore - are truncated by faults along the southern edge of the Transverse Ranges - the Malibu Coast-Santa Monica-Raymond-Hill and Sierra Madre fault zones. Complex subsurface extensions of the northwest-trending fault zones, such as the northwest end of the Newport-Inglewood fault zone, are now deeply buried by thick alluvial fan deposits which were shed southward from the Santa Monica and San Gabriel Mountains.

STOP: LONG BEACH OIL FIELD (also called Signal Hill field)

Signal Hill and the surrounding area looked much different when oil exploration was booming here in the 1920s. Today Long Beach, the fifth largest city in California and the 35th largest city in the United States, is one of southern California's major industrial and oil producing centers. Before the discovery of oil, the Long Beach field consisted of several large, undeveloped farms and small, undeveloped subdivisions. These lands were converted to the exclusive use of the oil operations and industry when more and more oil was found and produced. Today, residential construction is gradually replacing abandoned wells and oil facilities.

The discovery well, Alamitos #1, was completed by Shell Oil Company in 1921 to a depth of 958 m. It came in flowing 600 barrels of 22° API gravity oil per day. More than 1240 oil wells have been drilled in the Long Beach oil field since 1921, and 520 wells still produce a total of nearly 10,000 barrels per day. Estimated reserves in 1981 were 41.1 million barrels with the ultimate recovery expected to be 930 million barrels. As of 1989, the total production exceeded 900 million barrels.

Structure

The oil field occupies the crest of an elongate, northwest-trending, uplifted and arched block between the Cherry Hill and Northeast Flank faults, which are parts of the Newport-Inglewood fault zone (Yeats, 1973; Harding, 1973). (FIG. 29). The fault zone has been active since Miocene time, but Signal Hill itself was uplifted in Pleistocene time (Barrows, 1973). It is estimated that 800 m of right separation and 300 m of vertical separation have occurred along the fault zone at this location. Numerous cross faults are production barriers and limit production in some cases.

At basement level, the Newport-Inglewood fault zone separates "continental" igneous and metamorphic rocks on the northeast from "oceanic" Franciscan rocks on the southwest. Some writers maintain that the fault zone at basement level is a vestige of the Mesozoic subduction zone that has been reactivated as a strike-slip fault in Tertiary time (e.g., Hill, 1971). Others view it as a buried, pre-mid-Miocene strike-slip fault. The destructive Long Beach earthquake (M 6.3) occurred on the Newport-Inglewood fault zone in 1933.

Stratigraphy

More than 4000 m of deep marine sediments of Miocene and Pliocene age were deposited here in a rapidly subsiding trough. The oldest rocks penetrated in this field are mid-Miocene (?) black, resistant shales, probably equivalent to the Monterey Formation. Productive zones many tens of meters thick are composed of individual sand members from 0.5 to 3 m thick, separated by shale breaks which act as seals. Each of the sand members is probably composed of one or more graded beds, and each sand member is virtually self-sourced. Overall, the productive zones at Long Beach are 50-60% net sand. Sand porosities range from 25 to 30% with permabilities up to 850 md.

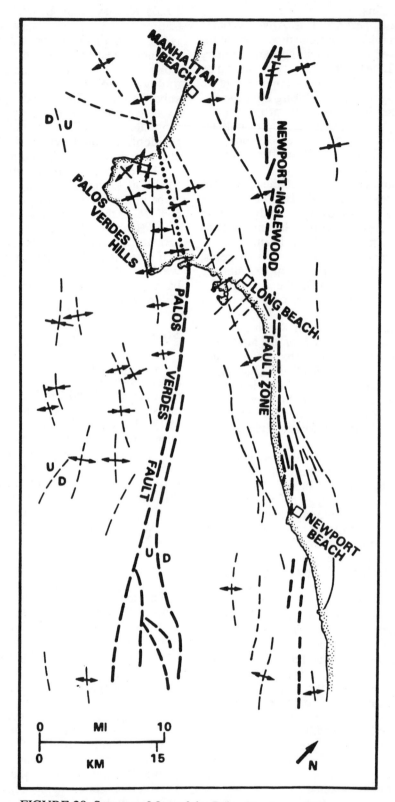

FIGURE 29 Structure Map of the Palos Verdes and Newport-Inglewood Fault Zones (courtesy of T.P. Harding from Jennings, 1977).

Proceed from Long Beach southward along Interstate Highways 405 and 5 to San Clemente and to the park at Dana Point Harbor.

STOP: SUBMARINE CHANNELS AT DANA POINT

Strata exposed in the cliff along the north side of Dana Point Harbor include San Onofre Breccia on the west, faulted against sedimentary rocks of the upper Miocene and Pliocene Capistrano Formation on the east. Details of three submarine channels filled with conglomerate and sandstone are especially well exposed.

The marine, mid-Miocene San Onofre Breccia accumulated at the base of steep escarpments interpreted as fault scarps. Debris flows and alluvial fans extended eastward from this scarp, consisting mainly of Catalina Schist, the basement terrane underlying part of the region now under the Pacific Ocean on the west. Rocks of this complex are now exposed on Catalina Island, in the Palos Verdes Hills, and are recovered from a number of wells and dredgings. Blueschist, greenschist, amphibolite, and gabbro are among the rock types represented as clasts. The San Onofre Breccia documents an episode of mid-Miocene deformation that apparently included faulting with marked vertical separations, but all are within the San Andreas transform block. Aspects of these tectonic and sedimentation events may be comparable to those that occurred onshore in southern California, especially in the Salton Trough.

The submarine channels, exposed in the seacliff and now filled with sandstone and conglomerate, flowed southeastward and are interpreted as scoured within the near-contemporaneous Capistrano Formation by bottom currents leading from the shoreline of the Los Angeles basin into deep water. Coarse rip-up breccias, where clasts of Miocene shale are embedded in coarse sandstone, and overhanging walls of channels are exposed in the cliff face. Nested submarine fan-channels are also present about 10 km to the southeast.

The route proceeds from San Clemente across the Santa Ana Mountains via State Highway 78 and about 2 km beyond the crest of the range to a vista point into the Elsinore basin.

SANTA ANA MOUNTAINS

The Santa Ana Mountains lie at the northwestern end of the Peninsular Ranges province, but their geologic history is tied more closely to that of the Los Angeles basin which developed during Miocene time. A homoclinal sequence, largely marine strata about 1500 m (4500 ft) thick and ranging in age from Upper Cretaceous to early Pliocene, unconformably overlies Cretaceous granitic rocks and Jurassic volcanic rocks (Santiago Peak Volcanics) and a sequence of incipiently metamorphosed, thin-bedded argillite, graywacke, and quartzite, with local pods of marble (Bedford Canyon Formation). The metasedimentary rocks are regarded as early and middle Jurassic in age. Farther east and southeast, the Peninsular Ranges are composed chiefly of Cretaceous granitic and metamorphosed pre-batholithic rocks that comprise the great Southern California batholith. The northeast side of the Santa Ana Mountains is bounded by the Elsinore fault.

Silverado Formation

Of particular interest in the Santa Ana Mountains is the Silverado Formation of Paleocene age, because it provides a basis for estimating horizontal separation for the Elsinore fault (FIG. 30). The formation ranges in thickness from 225 to 500 m and is characterized by distinctive strata near its base, including claystone, lignite, and sub-bituminous coal beds. These nonmarine units grade upward into marine sandstone and siltstone, which were deposited in a coastal belt at the margin of a deeply eroded terrane that had a humid climate. Inasmuch as this shoreline belt of distinct facies is repeated in the Elsinore trough, right separation of about 40 km is indicated, but discrete geological lines and their piercing points with the Elsinore fault have not yet been identified.

Elsinore Fault

The Elsinore fault zone has a known length of 215 km, extending from the Whittier Narrows area, east of Los Angeles, southeastward through the Peninsular Ranges to at least within a few kilometers of the International Boundary (FIG. 1). Across the boundary its name changes to the Laguna Salada fault, and it continues many more kilometers into Mexico. For most of its length, it is parallel to the San Jacinto fault to the northeast and to the Newport-Inglewood fault to the southwest, and thus it is regarded as part of the San Andreas transform belt. At its northwest end in the Santa Ana Mountains, the Elsinore fault splays into the Whittier fault (west) and the Chino fault (east). Locally, structural data apparently indicate right-lateral reverse displacement of about 5 km on the Whittier fault since middle Miocene time.

The sense of vertical separation varies considerably on the Elsinore fault, including normal and reverse in combination with right separation. The magnitude of right separation on the northern segment of the Elsinore fault is perhaps as much as 40 km since Paleocene time, judging from offset facies changes and pinch-out lines in the lower Silverado Formation, but that figure has not been confirmed by similar offsets in widely exposed basement rocks. Geomorphic evidence all along the Elsinore fault zone indicates oblique-slip movement during late Quaternary time. Weber (1977) found from 9 to 11 km of right separation along the fault based on offset of thin bodies of gabbro, aberrant foliation in the Bedford Canyon Formation, and the contact between the Santiago Peak Formation and the Bedford Canyon Formation. Vertical separations are variable and are at least 3 km at Lake Elsinore.

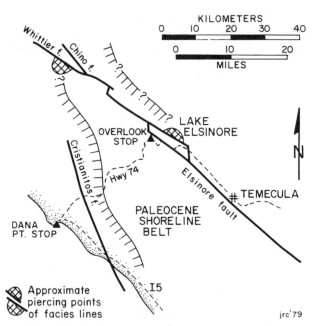

FIGURE 30 Sketch Map of Geometry of Probable Right slip on Elsinore Fault Zone. Offset geological lines in Paleocene beds include facies-change lines, pinch-out line, and isopachous lines. Refer to text for explanation. From Crowell and Sylvester (1979).

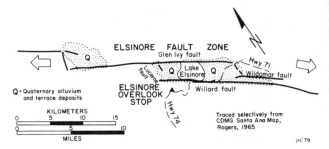

FIGURE 31 Sketch Map of Lake Elsinore Region with Right-stepping of Elsinore Fault Strands. Depressions are inferred at pull-apart junctions. From Crowell and Sylvester (1979).

STOP: LAKE ELSINORE OVERLOOK

This stop affords an opportunity to view the geometry of a relatively small pull-apart basin, to appreciate the relations between separation and slip in a strike-slip zone, and to note the relations between tectonics and sedimentation. In a flexible crust, such as that envisioned for this region, domains of the crust are arched up at places and stretched and depressed at others within a strike-slip fault regime.

From the walled parking area and overlook (FIG. 31), one views Lake Elsinore, the rectangular alluviated area around it, and the Perris Plain beyond it to the northeast. The lake is the main tourist attraction in the area, but in the early days Elsinore, named in 1884 for the Danish castle made famous by Shakespeare's *Hamlet*, was known for its mineralized hot springs and resorts. Many hot springs issued along the northeast side of the lake prior to 1890 when an irrigation canal disrupted the water table. Now hot water is obtained only from wells. The City of Elsinore utilized water from thermal wells until a decade or so ago when 5 ppm fluoride was found in the water, an amount equal to more than five times the recommended limit. The lake is presently filled with water imported by aqueduct from the Colorado River and with runoff from the San Jacinto River.

The lake basin is bounded by a rectangular arrangement of active faults which are easily distinguished from the overlook by their prominent scarps and which, in combination, constitute a pull-apart depression under right slip between two main, right-stepping, parallel strands of the Elsinore fault system: the Glen Ivy and Willard faults. In this

arrangement, the Lucerne fault at the northwest end of the lake, and the north-striking, unnamed fault at the southeast end of the lake are dip-slip faults. About 6 km of horizontal separation is indicated by the size and shape of the alluviated depression, if the pull-apart is closed up and allowance is made for some crustal stretching and sagging. Although vertical separations here are great and probably amount to several kilometers, the right-slip component amounts to much more. If the total right-slip on the fault zone is 40 km as suggested by the offset Paleocene shoreline facies, then the slip required for the Elsinore pull-apart must be a partial slip only, and therefore, considerably younger.

The route continues from Elsinore to Interstate 15, proceeds southward to Temecula, thence eastward on State Highway 79 across granitic rocks of the Southern California batholith in the Perris Plain and Peninsular Ranges and through Warner Hot Springs.

EN ROUTE: SOUTHERN CALIFORNIA BATHOLITH (after Woyski and Howard, 1987)

The Southern California batholith extends from the Transverse Ranges through southern California and the length of Baja California in a SSE-trending belt 1450 km long and spreads 110 km wide from the foothills along the Pacific coast to the Salton Trough along the San Andreas fault (FIG. 32). The batholith consists of numerous igneous plutons that were intruded side by side, leaving a few screens of metamorphosed country rocks between them.

The country rocks on the west consist of low-grade metamorphic Triassic-Jurassic flysch-type strata (dark gray argillite or slate interbedded with fine arkosic or lithic quartzite) of the Bedford Canyon Formation overlain unconformably by andesitic volcanic rocks (interlayered with quartz latite and minor rhyolite) of the Santiago Peak Formation, whereas to the east they are medium-grade, metamorphosed Mesozoic and Paleozoic clastic sedimentary rocks with minor limestone which form large masses, screens between plutons, roof pendants, and abundant inclusions in plutons.

The Southern California batholith is interpreted as the root system of a Late Jurassic- Early Cretaceous volcanic arc off the southwest edge of the North

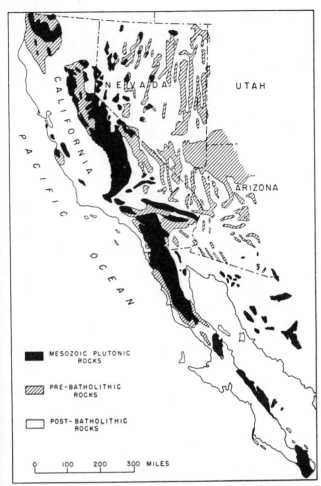

FIGURE 32 Generalized Distribution of Mesozoic Plutonic Rocks in Southern California and Baja California. After Larsen (1954).

American craton. The extrusive rocks of this arc (Santiago Peak Volcanics) were erupted upon the earlier Mesozoic flysch deposits of the Bedford Canyon Formation. In Late Jurassic time, a plate west of the North American plate (possibly the Kula or Farallon plate) began to subduct beneath the North American plate creating a volcanic arc that formed southwest of the Precambrian craton. Rising magmas intruded and erupted onto the Triassic-Jurassic flysch. From 130 to 105 Ma the island arc system remained static while plutons intruded across the 50 km width of the arc. From 105 to 80 Ma the locus of plutonism changed and migrated steadily eastward to the craton, possibly due to an increase in the rate of plate convergence.

By Late Cretaceous time, isostatic uplift and erosion exposed the root system of the arc. The Southern California batholith was separated from its initial position contiguous to mainland Mexico when the Gulf of California developed as an extension of the East Pacific Rise, beginning in late Miocene time.

A few kilometers beyond Warner Hot Springs, the route turns east on highway S2 and then bears left on highway S22. Park in a vista point in the winding downgrade that leads into Borrego Valley.

STOP: SALTON TROUGH OVERLOOK

View eastward into the Salton Trough with Borrego Valley in the foreground. Note the fault-controlled topography, the granitic and gneissic rocks of the Peninsular Ranges basement in the vicinity, and the sedimentation pattern now prevailing in the trough below. Stream courses into the Borrego Valley are much shorter and steeper than those flowing westward into the Pacific Ocean. The large alluvial fans grade distally into playa deposits. With good visibility the Salton Sea and the crest of the Chocolate Mountains may be seen about 100 km due east.

The Salton Trough or graben is structurally complex (FIG. 33), and its width between the marginal rims at this latitude is about 100 km. Faults with vertical separations, including those with normal slip, are partly responsible for the difference in topographic relief in the vicinity of the overlook. For example, aligned notches in the basement rocks below the overlook to the southeast are along a fault apparently within homogeneous granite. Some of these faults are interpreted as related to collapse under gravity as the graben widened, but others may be related to the formation of the proto-Gulf of California or even older basin-range faulting.

Displacements along braided strands of the San Jacinto fault zone are largely responsible for the topography within Borrego Valley and at its northwest

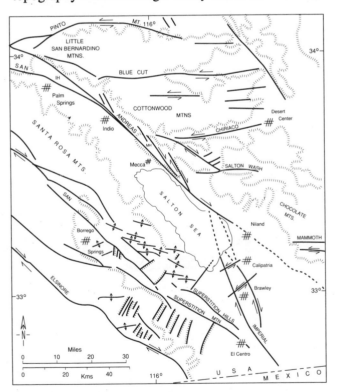

FIGURE 33 Sketch Map of Salton Trough Region with the Orientation of Faults and Folds and the Inferred Shape of the Pull-apart Basins near Brawley. MH - Mecca Hills; SHF - Superstition Hills fault; SMG - Split Mountain Gorge; IH - Indio Hills; 7D - Seven Levels Detachment fault locality. After Crowell (1985).

and southeast margins. Note the faceted spurs along the Coyote Creek fault and beyond Clark Valley along the southwest face of the Santa Rosa Mountains. Alluvium, folded into a complex broad arch, constitutes the Borrego Badlands lying southeast of the plunging end of Coyote Mountain. Borrego Mountain lies along trend farther southeast. Structural details in this region were mapped and studied intensively following the Borrego Mountain earthquake of April 9, 1968 (M 6.8).

Beyond the overlook are the community of Borrego Springs, the southern end of the Santa Rosa Mountains, and Coyote Mountain. Note the depression occupied by Clark Lake and the gentle swell of the Borrego Badlands consisting of Upper Cenozoic sedimentary strata deformed and antiformally arched along the trend of the San Jacinto fault zone. Clark Valley lies between the two major strands of this system, as shown topographically by the faceted spurs and linear trend of the ranges. Huge blocks of basement rocks, of probable landslide origin, lie along the base of the Santa Rosa Mountains.

Rocks in Coyote Mountain include the Santa Rosa mylonite, a tectonic-movement zone of Late Cretaceous age that has been displaced dextrally by Late Cenozoic faults. The Santa Rosa mylonite belt extends from Palm Springs southeastward through the Santa Rosa Mountains to their southeastern tip where it is displaced dextrally to Coyote Mountain; it is traced farther southward into the eastern Peninsular Ranges. The zone consists of pervasively sheared mylonitic rocks of mid-Cretaceous and older basement types, deformed under ductile conditions at depths estimated between 11-23 km and temperatures approaching the minimum melting temperature of granite (650° - 700° C). The thickness of the mylonitic zone as now known reaches 8 km south of Indio. It is associated at many places with gently cross-cutting thrust faults with northeast dips, inferred to be only slightly younger than the mylonitization. At places these thrust faults have been reactivated with normal-slip where they have provided sites for faulting in connection with the formation of the Salton Trough. The movement zone, now deeply eroded, is probably a manifestation of events when a convergent plate boundary existed in western North America.

BORREGO SPRINGS TO INDIO

SPLIT MOUNTAIN GORGE - IMPERIAL FAULT - MECCA HILLS

This part of the trip focuses on stratigraphy and structure of strata deposited in the Salton Trough during Late Cenozoic time together with deformation of those strata by strike-slip faulting. The surficial geology related to active tectonics at the northeast tip of the East Pacific Rise where it cracks continentward between the Pacific and North American Plates is displayed at the south end of the Salton Sea.

Drive southeast from Borrego to Anza-Borrego State Park via State Highway 78 to Ocotillo Wells, thence south via Split Mountain Road to the mouth of Split Mountain Gorge in Anza-Borrego State Park.

Note the pattern of the fan sedimentation while driving across Borrego Valley. Older fan surfaces were arched along the general trend of the San Jacinto fault zone to block drainage and so form a depression now occupied by the Borrego Sink. Fanglomerate overlies unconformably tilted, similar strata. Elongate hills, such as Borrego Mountain and those southeast of Ocotillo Wells, are interpreted as anticlines, up-tilted blocks, and squeeze-ups with the San Jacinto fault zone. The linear margins of the Ocotillo Badlands and Half Hill are fault-controlled, and the depression occupied by Halfhill Dry Lake probably owes its origin to a combination of tectonic blocking of drainage by these hills and to the building of the broad Fish Creek fan.out toward the northeast.

EN ROUTE: THE OCOTILLO BADLANDS

The strata of the Ocotillo Badlands consist of the Pleistocene Ocotillo Conglomerate that unconformably overlies folded and faulted sandstone and siltstone layers of the nonmarine Plio-Pleistocene Palm Spring and Borrego Formations. The deformation is the result of shortening in an uplifted pull-apart in the San Jacinto fault zone.

STOP: SPLIT MOUNTAIN GORGE

A spectacular sequence of Late Cenozoic marine turbidites, nonmarine fanglomerate, and landslide deposits is exposed in the near vertical canyon walls of Split Mountain Gorge near Borrego Springs. These deposits represent the lower part of the Salton Trough fill and record the initial tectonic subsidence of the trough before and during the origin of the proto-Gulf of California. The turbidite deposits represent the principal incursion of marine water into the Salton Trough.

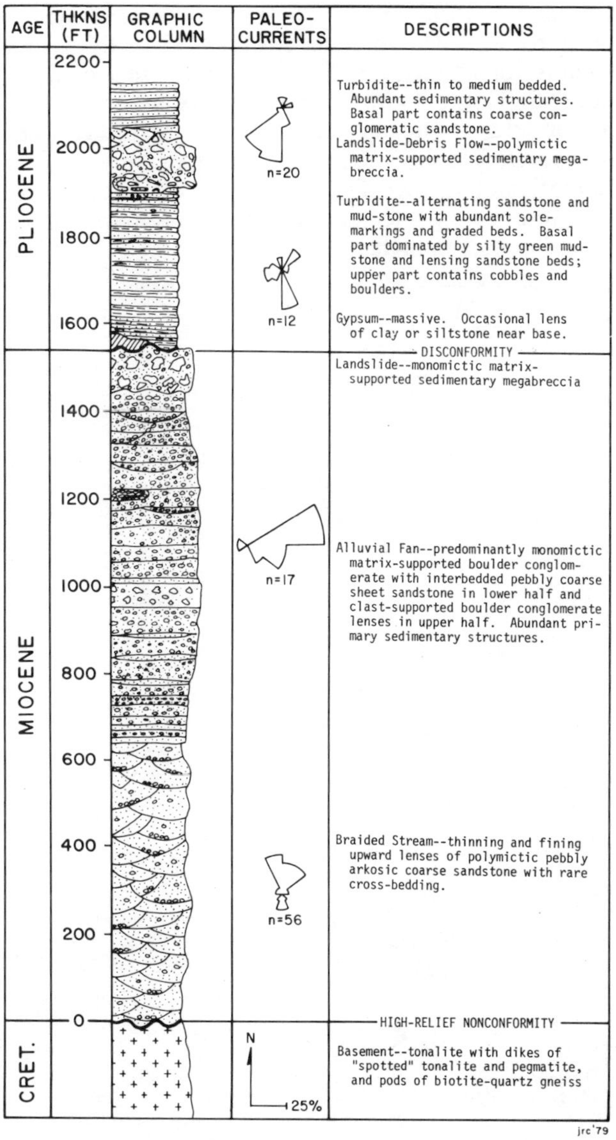

Stratigraphy

From oldest to youngest, Miocene braided stream deposits of polymictic pebbly arkosic sandstone overlie Cretaceous granite (FIG. 34). Nearly unidirectional paleocurrent indicators point toward a source about 40 km to the south, but the region has probably been

AGE	THKNS (FT)	GRAPHIC COLUMN	PALEO-CURRENTS	DESCRIPTIONS
PLIOCENE	2200–1600		n=20 / n=12	Turbidite--thin to medium bedded. Abundant sedimentary structures. Basal part contains coarse conglomeratic sandstone. Landslide-Debris Flow--polymictic matrix-supported sedimentary megabreccia. Turbidite--alternating sandstone and mud-stone with abundant sole-markings and graded beds. Basal part dominated by silty green mudstone and lensing sandstone beds; upper part contains cobbles and boulders. Gypsum--massive. Occasional lens of clay or siltstone near base. — DISCONFORMITY —
MIOCENE	1400–0		n=17 / n=56	Landslide--monomictic matrix-supported sedimentary megabreccia. Alluvial Fan--predominantly monomictic matrix-supported boulder conglomerate with interbedded pebbly coarse sheet sandstone in lower half and clast-supported boulder conglomerate lenses in upper half. Abundant primary sedimentary structures. Braided Stream--thinning and fining upward lenses of polymictic pebbly arkosic coarse sandstone with rare cross-bedding. — HIGH-RELIEF NONCONFORMITY —
CRET.			N / 25%	Basement--tonalite with dikes of "spotted" tonalite and pegmatite, and pods of biotite-quartz gneiss.

jrc'79

FIGURE 34 Columnar Section for Split Mountain Gorge. From Kerr et al. (1979).

rotated clockwise about 45° as shown by paleomagnetic studies (Johnson, et al., 1983). A thick alluvial fan sequence, also of Miocene age, overlies the braided stream sequence. Debris-flow, sheetflood, and fluvial-channel fill lithofacies may be distinguished. Sediment

T309: 59

size analyses and rare paleocurrent indicators point to the paleo-fanhead having been west of the present canyon. These strata are overlain by a monomictic sedimentary megabreccia (landslide) which probably required an air cushion to permit a long distance of flow, but its source is not known.

A disconformity separates the top of the megabreccia from massive gypsum of Pliocene age, now being mined by the U.S. Gypsum Company. Maximum thickness of the gypsum is about 35 m in the synclinal valley northeast of Split Mountain Gorge. It interfingers with alluvial fan and marine deposits at the extremities of its outcrop area. The gypsum is overlain by a succession of sandstone-mudstone turbidites into which slid a remarkable landslide-debris flow of megabreccia. This breccia is about 35 m thick and extends for about 3 km both east and west of Split Mountain Gorge. Some of the boulders are up to 10 m in diameter. Rip-up clasts of the turbidites are incorporated in the basal part of the megabreccia. Its source is not known. The megabreccia is overlain by more turbidites which locally contain enigmatic "lonestones" up to 1 m in diameter.

Structure

The sedimentary strata are broadly folded into a WNW-trending anticline, but within the canyon, note that the geology cannot be extrapolated from one canyon wall to the other. This is because most of the canyon has been carved along a steep fault which cuts the anticline and dips steeply eastward (FIG. 35). Examine the tightly folded turbidites where they are in contact with the polymictic debris flow, and decide if the fold formed before, during, or after the debris flow!

Return to Highway 78, proceed east to Highway 86, thence southeast to Westmorland.

From near Split Mountain Gorge to Westmorland the route proceeds along the floor of ancient Lake Cahuilla, a lake that occupied the Salton Trough from time to time between 1600 and 300 years ago. Most of Imperial Valley is covered by a thick blanket of Recent lacustrine and fluvial sediment. Shorelines are conspicuous as well as bottom features formed in shallow water, such as spits, bars, and mantling of

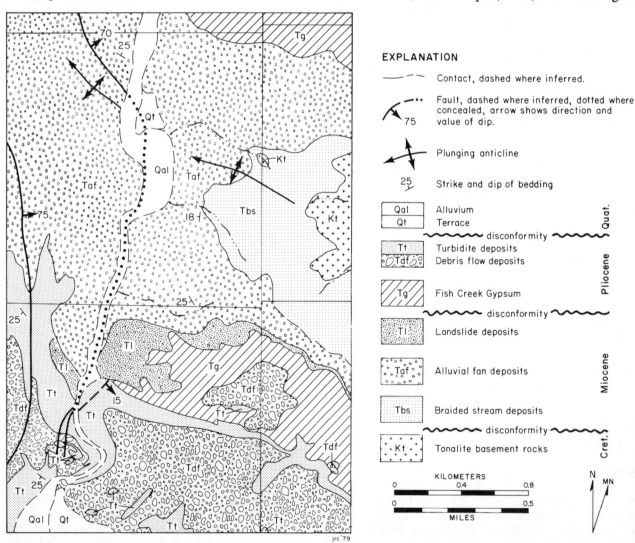

FIGURE 35 Geologic Map of Split Mountain Gorge Area, Eastern San Diego County. From Kerr et al. (1979).

sediment containing abundant small fossils. The shoreline is near the present sea level and looks like a bathtub ring where the lake surface was against steep outcrops of bedrock.

The present Salton Sea was formed during the interval between 1905 and 1907 when levees along the Colorado River broke during floods, and water entered the deepest part of the Salton Trough along New River, inundating the railway and farmland. The field trip route crosses New River north of Westmorland. Today the sea surface is slowly rising and its salinity increases due to the influx of irrigation water.

INTRODUCTION TO THE IMPERIAL VALLEY

Imperial Valley is the widest and deepest part of the Salton Trough, and most of it is below sea level as is clearly shown by the "sea level" stripes 25 m high on silos between Brawley and El Centro. Gravity and seismic refraction surveys show that the deepest part of Imperial Valley is 8 km, but surprisingly it is characterized by a gravity high (-20 mgal) centered beneath the Obsidian Buttes volcanic field.

Structurally the Gulf of California is generally regarded as a modern oceanic divergent plate boundary, the opening of which is propagating northwestward toward and into the Salton Trough. At its southern end the Gulf is floored by normal oceanic lithosphere; there the Mohorovicic discontinuity is interpreted from seismic data to be about 12 km below sea level. The discontinuity is deeper beneath the Salton Trough, but its exact depth is not known. From gravity data, the discontinuity is inferred to dip northwestward from 14 km at the International Boundary to 20 km beneath Coachella Valley. Thus the peculiar nature of a deep sedimentary basin with a gravity high may be viewed as a rift where low-density sediments cover young, high-density rocks, such as basalt, that have welled up at depth along the axis of the rift.

Seismic refraction studies show that the southern half of the trough is a westward tilted half-graben (FIG. 36) with considerable structural relief beneath the cover of Recent sediments which were deposited largely by the Colorado River when it occasionally drained into the "Salton Sink". On the west and deepest side of the trough are small basement blocks downfaulted from the Peninsular Range. A sequence of sediments up to 6 km thick is present locally in the center of the trough in the vicinity of Mesquite Lake. The topographic depression there and the large delay times can be interpreted as showing a deep rift in the crust or very low velocity sediment fill, both of which might be expected over a pull-apart basin bounded by the Brawley and Imperial faults. The subsurface relief on the east side of the trough is smoother and shallower.

Active faults bound some of the sub-basins of Salton Trough. Occasionally they slip and generate earthquakes which rupture the ground surface, giving valuable clues to the on-going tectonics of the basin. Important strands of the Elsinore, San Jacinto, San Andreas, and Imperial faults extend into Imperial Valley and have been active at least since 1940, as follows (see also Table 1):

1940 El Centro (M 6.9); on the Imperial fault with as much as 6 m right-slip
1968 Borrego Mountain (M 6.5) on the Coyote Creek strand of the San Jacinto fault; 1 m right-slip
1979 Imperial (M 6.4) on the Imperial fault with 1 m right-slip
1987 Westmorland and Superstition Hills (M 6.2; 6.6) on the Elmore Ranch and Superstition Hills faults with 15 cm left-slip and 80 cm right-slip, respectively

Several historic earthquakes have also occurred at the head of the trough in Coachella Valley:
1948 Desert Hot Springs (M 6.5) on Mission Creek strand (?) of San Andreas fault
1986 North Palm Springs (M 5.9) on Banning strand of San Andreas fault with 10 cm right-slip

Significantly, no earthquakes have been documented since AD1680 by historic or paleoseismic means on the San Andreas fault between the south and north ends of the Salton Trough, although the geomorphology clearly shows that the fault has been active in Holocene time. With so much historic activity at each end of this great fault, as well as along much of the remainder of its length in California, seismologists are understandably concerned that the probability of a major earthquake on the fault in Salton Trough is quite high, perhaps as high as 60 percent in the next 30 years.

Five rhyolite domes at the south end of the Salton Sea, collectively called the Salton or Obsidian Buttes, attest to the active rift tectonics occurring here. The lavas contain inclusions of basalt, and lesser quantities of sedimentary rocks and granite which suggest the nature of the basement. Their eruptive ages range from 55,000 to 16,000 years BP. Steam still discharges from cracks in three of the domes, and subsurface geothermal energy is actively exploited by several operators today.

From Westmorland, the route proceeds to one of the areas of recent surface rupture. The specific area will depend on recency of faulting, access, preservation of fault features, and time.

ACTIVE FAULTING

Two major earthquakes occurred in Imperial Valley in November 1987. The first (M 6.2) was on a fault which strikes northeast across the south end of the Salton Sea. It was followed about 6 hours later by a larger earthquake (M 6.6) on the northwest-striking Superstition Hills fault. Subsequent field studies revealed 15 cm left slip on a previously mapped but unnamed fault - now called the Elmore Ranch fault - and 25 cm right slip on the Superstition Hills fault. Over the next two months, the right-slip increased to a maximum of 80 cm by afterslip. The earthquakes were significant because they marked the first time in California that seismologists found that one earthquake

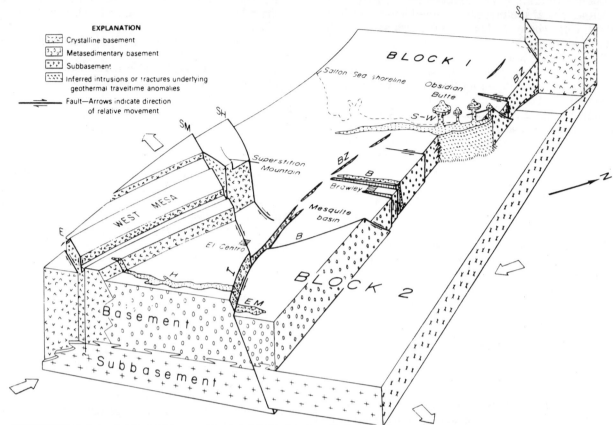

FIGURE 36 Schematic Block Diagram of Imperial Valley Region. Sedimentary rocks are removed and basement is cut away along a surface nearly parallel to the Brawley seismic zone. Geographic names are projected downward on to basement for reference. B - Brawley fault zone; BZ - Brawley seismic zone; E - Elsinore fault; I - Imperial fault; SA - San Andreas fault; SH - Superstition Hills fault; SM - Superstition Mountain fault. Geothermal areas: B - Brawley; EM - East Mesa; H - Heber; S - Salton Sea; W - Westmorland. Blocks 1 and 2 move away from Brawley seismic zone and inferred spreading center in direction parallel to southern section of Imperial fault. Rhombochasms or pull-apart basins form at Mesquite basin and Salton sink which is presently occupied by the Salton Sea. In addition, ductile thinning of both blocks may be occurring as they move away from the Brawley seismic zone. From Fuis et al. (1982).

triggered another. Now they are concerned that another earthquake on any one of a myriad of northeast-striking faults in Imperial Valley might similarly trigger the San Andreas fault along the opposite margin of the Salton Trough.

A M 6.4 earthquake occurred on the Imperial fault in October 1979, centered a few kilometers east of the town of Imperial. In many aspects it mimicked a M 6.9 earthquake on the same fault in 1940. The latter earthquake offset the International Boundary up to 6.3 m right laterally, and the break extended about 15 km on both sides of the boundary. In 1979, the break was limited to the U.S. side, and the maximum horizontal displacement was about 1 m, also right-laterally. Later in 1980, another M 6 earthquake occurred on the Mexican segment of the Imperial fault without noteworthy surface rupture. Later in 1980, a M 5.6 earthquake and related swarm occurred on the Brawley fault between the Imperial and San Andreas faults. Left-lateral displacement has been noted recently on northeast-striking faults and has been interpreted as seismic support for interpretations of paleomagnetic data for clockwise rotation of blocks in a zone of right

simple shear (Nicholson et al., 1986).

From Westmorland, proceed north on route S30 past the geothermal operations to Red Hill Marina at the south end of the Salton Sea. Park in the saddle between the two domal hills of the "island"; walk to the top of the southern hill to view the geothermal area and the Salton Sea, to collect black obsidian, and to straddle the East Pacific Rise.

STOP: SALTON BUTTES VOLCANOES AT RED HILL

Red Hill consists of two of the five rhyolite domes that intrude Quaternary sediments near the southeast shoreline of the Salton Sea (FIG. 37). The domes are composed of low-calcium alkali rhyolite, gray pumice, and black obsidian with from 1 to 2 percent crystals. Inclusions of low-potassium tholeite basalt and partly melted granite, many showing strong hydrothermal alteration, are present. The bimodal basalt-rhyolite assemblage formed by partial fusion of mantle peridotite in two stages, yielding successive rhyolitic and basaltic melts. Compositions and textures suggest that the granite inclusions are fragments of basement rocks rather than the crystallized equivalents of the rhyolitic magma.

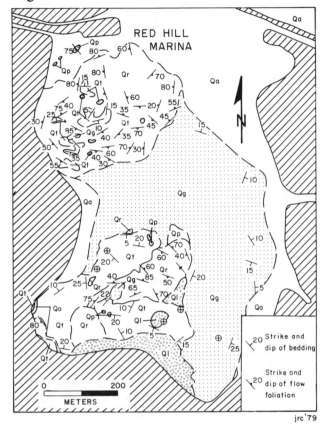

FIGURE 37 Geologic Map of Red Hill. Qr - rhyolite (dominantly lithoidal) that forms the north and south protrusions; Qo - obsidian and flow breccia; Qt - subaqueous pyroclastic material; Qg - gravel and sand; Ql - lag gravel; Qa - alluvium. Diagonal lines indicate area covered by Salton Sea in October 1964. Redrawn from Robinson et al. (1976).

The domes are interpreted to lie above a hot spreading center underlain at great depth primarily by a mantle diapir that has broken and fragmented the attenuated basement of older continental rocks when the Pacific and North American plates diverged.

The route continues via farm roads from Red Hill Marina past carbon dioxide wells to Niland.

OPTIONAL STOP: CO$_2$ GAS WELLS AND MUD VOLCANOES

Northeast of Red Hill lies an abandoned CO$_2$ gas field that produced from 1934 to 1954 from 54 wells with depths ranging from 150 m to 250 m at pressures exceeding hydrostatic. The gas was used mainly for manufacturing "dry ice". Abundant CO$_2$ is released by decarbonation accompanying low grade metamorphism of the young, carbonate-rich sediments; then the CO$_2$ migrates upward into the gas field. About 2 km south of the CO$_2$ gas field is field of interesting mud

volcanoes, also driven by CO$_2$, which erupt warm water and mud.

The initial source of the carbonate is probably detrital carbonate grains eroded from older sedimentary formations of the Colorado Plateau region and brought to the Salton Trough by ancestral Colorado Rivers. Battered Cretaceous forams are found in Pliocene and Pleistocene sedimentary strata in Imperial Valley, having been derived originally from the Mancos Shale in Utah and Colorado and transported to Salton Trough by the Colorado River.

The route continues northwest from Niland to Mecca via Highway 111.

EN ROUTE: BOMBAY BEACH

The southeastern end of the San Andreas fault terminates at Bombay Beach at the southeast corner of the Salton Sea. There is speculation and a little geophysical evidence that a fault or "structural discontinuity" continues southeastward approximately beneath the trend of the Algodones sand dunes which stretch across the International Boundary into Mexico. The fault is continuous for 1100 km down the length of California from the point where it comes onshore north of San Francisco.

OPTIONAL STOP: EN ECHELON FOLDS AT DURMID

A remarkable sequence of en echelon folds is developed in Quaternary lacustrine strata near Durmid and Frink (Babcock, 1974) (FIG. 38). The folds are best seen from the air, because they have been beveled by wave erosion of ancient Lake Cahuilla, and topographic relief is very low. Many of the outcrops are covered with a thin patina of calcareous tufa, and fossil shells from the latest of these lakes are ubiquitous on the ground surface. The folds trend generally east, oblique to the northwest-trending San Andreas fault, and they verge south away from the fault toward the Salton Sea. Interbeds of volcanic ash and pumice were erupted from the Salton Buttes (55,000 to 16,000 yrs BP), and from the Long Valley caldera (710,000 yrs BP) in eastern California.

OPTIONAL STOP: SAN ANDREAS FAULT AT SALT CREEK

Non-marine beds of Plio-Pleistocene silt, sand, and clay are tightly folded and faulted adjacent to the San Andreas fault zone beneath and upstream from the railroad trestle across Salt Creek. Microstratigraphic studies show that this part of the San Andreas fault has not ruptured in the last 350 years.

A few kilometers up Salt Creek are several distinctive volcanic plugs of Oligocene age that fed lava flows which crop out today near Lake Palmdale on the

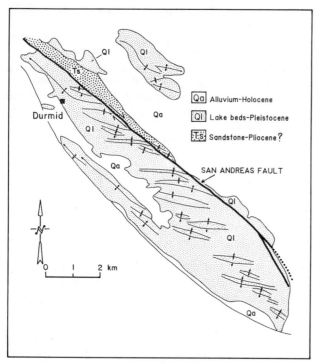

FIGURE 38 En Echelon Folds at South End of San Andreas Fault near Durmid. Redrawn from Dibblee (1977).

southwest side of the San Andreas fault, about 330 km northwest of Salt Creek. This is one of the excellent cross-fault correlations which demonstrate great strike-slip along the San Andreas fault.

At Mecca, proceed east on highway 195 across the Coachella Valley Canal to a gravel road which proceeds northwest 5 km along a powerline to the mouth of Painted Canyon.

MECCA HILLS

The Mecca Hills, along with the Indio and Durmid Hills, are one of three tectonic culminations along the San Andreas fault in the Salton Trough (FIG. 39). Each culmination is adjacent to a segment of the San Andreas fault which strikes only a few degrees west of the inferred direction of transform plate movement in this region and, thus, is a site of convergent right-slip and resultant uplift at a gentle restraining bend (Bilham and Williams, 1985). The areas of low topographic and structural relief between the culminations are where San Andreas fault segments are parallel to the the relative plate motions. Structurally, the Mecca Hills represent a "palm-tree structure" (Sylvester and Smith, 1976), almost all of which is well-exposed in the walls of Painted Canyon.

Lithology

Dark brown, gravel-strewn slopes and hills along the southwest edge of the Mecca Hills are underlain by Pleistocene Ocotillo Conglomerate which contains schist detritus derived from the Orocopia Mountains to

the east, and which was subsequently offset 20 km northwestward along the San Andreas fault. The higher, ruggedly sculptured part of the Mecca Hills is underlain by Plio-Pleistocene sandstone and conglomerate with interbeds of greenish-gray siltstone. In the core of the Mecca Hills, the lower part of the Palm Spring Formation is interbedded with and underlain by the late Miocene and Pliocene Mecca Formation, a coarse, dark reddish-brown, thickly-stratified unit of torrentially deposited breccia and conglomerate. The Mecca Formation, in turn, lies nonconformably on a heterogeneous basement of Precambrian gneiss, Mesozoic granitic rocks, and Tertiary felsic dikes which, in the subsurface, is thrust upon Orocopia Schist. The entire sedimentary section is replete with local unconformities and abrupt facies changes, which are evidence of sedimentation during tectonism.

Structure

The Mecca Hills are warped and uplifted between the San Andreas fault on the southwest and the Painted Canyon and Hidden Spring faults on the northeast (FIG. 40). Precambrian and Mesozoic basement rocks are exposed in the core of the Mecca Hills, and the 30° NW plunge of the structures offers a deep structural profile. The San Andreas fault itself strikes northwest parallel to the gravel road leading to Painted Canyon from Hwy 195. Northeast of the road the surface trace of the fault is marked in the low hills by streaks of red-brown clay gouge in talus-strewn slopes and hillocks. The bounding faults are steep at depth but flatten upward like the fronds of a palm tree, and carry rocks of the central block outward short distances upon the adjacent blocks.

Painted Canyon

The general structure traversed by the road up Painted Canyon is that of an asymmetric antiform with crushed basement and a complex zone of faulting in its core (FIG. 41). However, because of strike-slip on the core faults, the structure is much more complex. From the San Andreas fault zone at the mouth of Painted Canyon, the road up Painted Canyon crosses the Skeleton Canyon syncline which, because of a local structural depression, is not well expressed in the canyon walls except for a few gentle reversals of dip. About 300 m up the canyon, however, the dip of the beds is consistently down-canyon (SW) on the southwest flank of the main antiform. The generally thin-bedded, gently-dipping, tawny-colored sandy and conglomeratic beds constitute the upper part of the Palm Spring Formation. About 800 m up the canyon, the sandstone and conglomeratic strata have interbeds of greenish-gray siltstone. These strata, which are overturned locally, constitute the lower member of the Palm Spring Formation. All of the Palm Spring Formation in the Mecca Hills represents coalescing alluvial fan deposits, derived from mountains northeast of the Mecca Hills. On the southwest limb of the

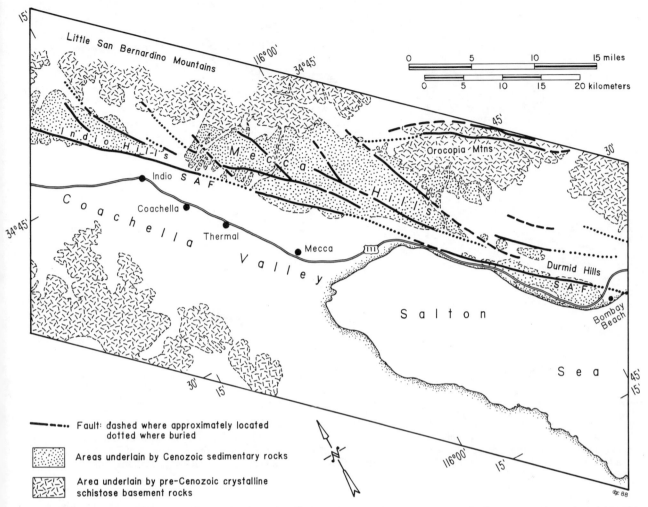

FIGURE 39 Generalized Geologic Map of San Andreas Fault Zone in Coachella Valley and Salton Sea Region, with Locations of Indio, Mecca, and Durmid Hills. From Sylvester (1988).

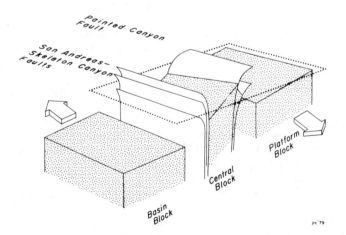

FIGURE 40 Idealized Block Diagram of Basement and Principal Faults in the Central Mecca Hills. Dotted parallelogram represents the ground surface. Non-parallel arrows indicate relative convergent, right oblique-slip of basin and platform blocks resulting in crustal shortening manifested by folding of sedimentary rocks and palm-tree geometry of the faults. Modified from Sylvester and Smith (1976).

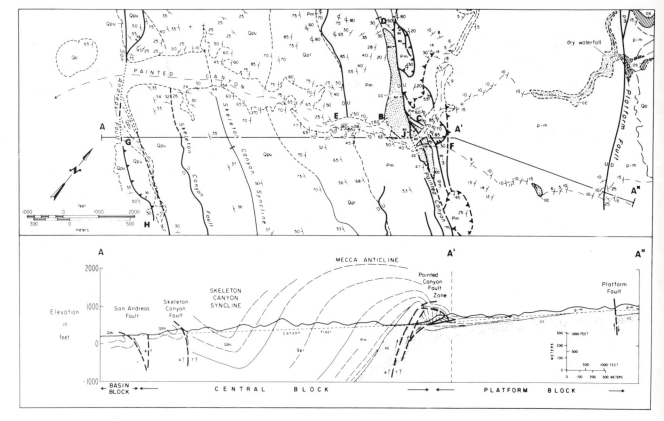

FIGURE 41 Geologic Map and Structural Profile of Painted Canyon, Mecca Hills. cc - crystalline basement rocks of Chuckawalla Complex; os - Orocopia Schist; Pm - Mecca Formation; Qpl, Qpu - Palm Spring Formation, lower and upper members; p-m - Palm Spring and Mecca Formations, undifferentiated; Qo - Canebrake and Ocotillo Conglomerate, undifferentiated. Base map from U.S. Geological Survey 1:24,000 Mecca (1955), Thermal Canyon (1956), Mortmar (1958). From Sylvester and Smith (1975).

antiform, the Palm Spring Formation is more than 1300 m thick, whereas, it is less than 100 m thick northeast of the Painted Canyon fault.

Up-canyon, the Palm Spring Formation is underlain by the dark red-brown Mecca Formation, exposing about 100 m of coarse breccia and conglomerate of locally derived gneiss, schist, and granitic rocks. The contact between the Mecca and Palm Spring Formations on the northwest canyon wall is a minor fault, but not on the southeast wall. About 300 m farther up-canyon from the Mecca-Palm Spring contact are vari-colored outcrops of highly fractured basement, including black gneiss, white granitic rocks, and orange felsic dikes. The contact between the Mecca Formation and the basement is a buttress unconformity which has been folded and faulted. The Painted Canyon fault cuts through the central part of the basement outcrop on the northwest side of the canyon, but on the southeast side, one of the main strands of the fault forms the contact between the basement and the Mecca Formation.

Northeast of the basement outcrops, the nearly vertical and overturned vermillion and tawny-colored strata belong to the Mecca and Palm Spring Formations, respectively, and are on the northeast limb of the antiform. High on the northwest canyon wall, flat-lying beds, which are structurally and stratigraphically continuous with beds farther up-canyon, tectonically overlie the nearly vertically dipping beds. The contact, which is somewhat obscured by talus, is one of a family of low-angle faults associated with the Painted Canyon fault.

Return to Hwy 195 in Box Canyon - proceed northeast on the paved road up Box Canyon.

EN ROUTE: BOX CANYON WASH

The San Andreas fault is crossed near the mouth of Box Canyon Wash. The fault is exposed on the northwest side of the highway where low outcrops of weakly resistant, brown sandstone and siltstone are in a poorly exposed contact with tan fanglomerate which dips gently down-canyon.

At the 30 mph right curve in the road and on the northwest side of the highway, a gentle angular unconformity is clearly exposed between the lower and upper members of the Palm Spring Formation. The unconformity is widespread within that formation throughout the Mecca Hills and is evidence of basin margin deformation and erosion during deposition.

The last of the highly-folded strata of the Palm Spring Formation and the beginning of the gently-dipping beds of fanglomerate marks the position of the

Painted Canyon fault in Box Canyon. The fanglomerate is deposited on the platform, or northeastern block (FIG. 41).

STOP: OROCOPIA SCHIST AT SHAVER'S WELL

The Orocopia Schist is one of the principal basement rocks on the northeast side of the Salton Trough. Here the schist is overlain nonconformably, with a small fault complication, by Pleistocene fanglomerate which dips gently downcanyon and was derived from the mountains to the north.

At Shaver's Well the Orocopia Schist is a greenschist facies albite-chlorite-mica schist with an abundance of minor folds. This rock unit, probably ensimatic in origin, is confined to the footwall of the great Vincent-Orocopia-Chocolate Mountains thrust system of probable latest Mesozoic age, and underlies much of southern California. The thrust sheet was folded in mid-Tertiary time and later disrupted so that folds were displaced from the Transverse Ranges region to this area by Late Cenozoic right-slip.

EN ROUTE: MARINE EOCENE ROCKS IN THE OROCOPIA MOUNTAINS

About 1700 m of Early and Middle Eocene marine strata nonconformably overlie granitic basement in the northeastern Orocopia Mountains. Called the Maniobra Formation, these strata comprise interbedded fossiliferous siltstone, sandstone, some sandy limestone, breccia, and massive conglomerate with large, polished boulders up to 10 m in diameter. They are interpreted as having been deposited at the east edge of a forearc basin, later displaced by San Andreas movements.

The Eocene strata are overlain unconformably by the Diligencia Formation which comprises about 1700 m of volcanic and nonmarine clastic rocks of early Miocene and perhaps Oligocene age, based on a single vertebrate fossil find, and K-Ar isotopic ages on interlayered volcanic rocks from 18.6 ± 1.9 my to 22.4 ± 2.9 Ma. Neither the Maniobra nor the Diligencia Formations, nor rocks equivalent in age and lithology, have been found elsewhere in the Salton Trough, but have probable correlatives in the central Transverse Ranges from 210 to 330 km to the northwest across the San Andreas fault.

Maniobra is the Spanish word for manoever and for a wash in the Orocopia Mountains near where General George C. Patton practiced desert tank warfare tactics before fighting in North Africa in World War II. Tracks from those tanks can still be seen in parts of the desert today, and a museum in Patton's honor opened in 1988 at Chiriaco Summit. Diligencia is the Spanish word for stagecoach. The route of the Butterfield Stage passed along the present route of Interstate Highway 10 north of the Orocopia Mountains.

Proceed northeast along State Highway 195 to its intersection with Interstate Highway 10, turn left and proceed west about 40 km to Indio.

EN ROUTE: BOX CANYON TO INDIO

The south front of the Little San Bernardino Mountains on the north side of the highway is controlled by the east-striking Chiriaco fault, which is largely buried along its entire length. The Chiriaco fault is one of several east-striking, left-slip faults northeast of the San Andreas fault. Occasional tailings at the mountain front are from the underground Coachella Valley aqueduct which carries Colorado River water to the valley to support its agricultural industry.

Over the next 30 km the highway traverses dissected alluvial fans shed from the mountains on the north to an irregular pediment surface carved on granitic basement rocks which locally project to the surface from beneath the fans.

As the route descends the grade, good views may be had of the narrow northwest end of the Salton Trough and the rugged Santa Rosa Mountains on the far side of the trough to the southwest.

At the base of the grade and of the mountain slope, the route crosses the Coachella Valley aqueduct and the San Andreas fault and proceeds onto the Pacific lithospheric plate. The fault trace is indistinct here from ground level, but is marked by a relatively lush growth of vegetation resulting from groundwater impounded upstream of the fault, and by a change in the topographic style from the dissected hills on the drier, northeast side of the fault from the flat valley floor.

INDIO TO SAN BERNARDINO

COACHELLA VALLEY - SEVEN LEVELS - SAN GORGONIO PASS - SAN JACINTO FAULT - MILL CREEK

The route proceeds west from Indio on Hwy 111 to State Highway 74, the Palms-to-Pines Highway, to a chlorite-mylonite zone of rocks representing a detachment fault in granitic basement. The route crosses Coachella Valley to the San Andreas faultt and to a deflected stream course in Pushawalla Canyon. The route continues west on I-10 to a side trip up Whitewater Canyon, thence through San Gorgonio Pass to Beaumont where the route turns south for a sidetrip across the San Timoteo badlands to an actively-subsiding pull-apart on the San Jacinto fault. The route returns to I-10 and continues to San Bernardino with an optional sidetrip to the Mill Creek pull-apart basin.

Follow Highway 111 westward from Indio to Palm Desert and Highway 74, proceed south about 7 km to a conspicuous parking area on the right (west side of road, square, paved). Park and carefully cross the road to examine the fault on the hillside about 50 m above the highway.

STOP: SEVEN LEVELS FAULT - CLOSE-UP VIEW

A low-dipping fault surface, dipping toward the Coachella Valley on the north, is well-exposed across several small spurs and in gullies. This, the Asbestos Mountain fault, emplaces upper Cretaceous granodiorite above metasedimentary tectonites and gneisses along a discrete, meter-thick zone of movement. Rocks above and below the low-angle fault are crushed and deformed. Kinematic indicators along the zone, also locally characterized by a chlorite-breccia zone, show that the hanging wall has moved relatively toward the northeast. Controversy concerns the timing of the faulting and whether it has been reactivated. It may be (1) part of the Santa Rosa mylonite zone of late Cretaceous age, or (2) an element of the widespread early Miocene detachment fault system of the desert regions northeast of the Salton Trough, or (3) a late Cenozoic listric surface within basement rocks where the upper plate slides into the opening Gulf of California, and therefore, nor older than 5 Ma. Fault grooves, polished surfaces, and slickensides are oriented with northeast trends. Paleomagnetic work has not been done on rocks in this region.

Return to vehicle and proceed up switchbacks about another 5 km to large viewpoint on left (north side of the road, paved, prepare to turn around; watch traffic carefully upon swinging into parking area). Gather on rock point at edge of parking area, look north across the Coachella Valley approximately down the dip of the fault.

STOP: SEVEN LEVELS VISTA

From the viewpoint and large parking lot at the summit of the Seven Levels Grade, fine views are afforded of the northwest end of the Coachella Valley, disrupted and displaced landforms along the San Andreas fault zone (about 21 km to the north), the Salton Sea to the southeast, and the high mountains (San Gorgonio (3503 m) and San Jacinto (3293 m)). In this very active region, the landforms closely reflect tectonic movements because the Gulf of California opened only 5 Ma, and this, the northwest part, is presumably much younger. Transpression across the valley is now responsible for elevating the higher mountain flanks.

The Seven Levels fault is well exposed in the hills and canyons in the foreground and is easily traced in the desert region by the color changes in the weathered rocks both above and below the rust-colored fault zone.

Return to Highway 111, proceed north on Bob Hope Drive to Interstate Highway 10, thence east on Thousand Palms Road.

INDIO HILLS

The Indio Hills are another tectonic culmination, or "porpoise", like the Mecca and Durmid Hills in the San Andreas fault zone (FIG. 39). The hills are composed of folded and faulted strata belonging to the Palm Spring and Imperial formations; a few isolated outcrops of granitic basement rocks are exposed at the southeast end of the hills, but their relation to the overlying sedimentary rocks is not clear. Two faults bound the uplifted part of the Indio Hills: the Banning fault on the southwest side and the Mission Creek fault on the northeast side. The two faults merge at the southeast end of the Indio Hills where they continue southeastward as the San Andreas fault. The San Andreas fault crosses Dillon Road 1 km north of I-10 and is marked by a prominent growth of vegetation due to impounding of southward-flowing groundwater. Within the tectonic depression between the Indio and Mecca Hills, the San Andreas fault is marked by south-facing scarps in the alluvium and vegetation contrasts reflecting other groundwater barriers beneath the surface.

Groundwater barriers are particularly prominent in the Indio Hills; they are marked by lush growth of vegetation, especially palm trees: The moisture-loving native California fan palm Washingtonia filifera is a "trademark" of these oases. Thousand Palms, about 2 km south of the intersection of Thousand Palms Canyon and Dillon Road, is one such oasis along the northeast-facing scarp of the Mission Creek fault. The water table slopes south from the Little San Bernardino Mountains, and pulverized rock along the fault trace dams subsurface water to the extent that the ground along the northeast side of the fault is moistened; active surface springs are present locally.

Thousand Palms Canyon Road crosses the trace of the Banning fault at Willis Palms, about 2 km down-canyon from the Mission Creek fault. The fault is marked by a straight and prominent southwest-facing scarp with abundant palm trees at its base. The structure of the Indio Hills along Thousand Palms Canyon Road between the two faults is a gentle syncline in alluvial fan deposits of the Late Pleistocene Ocotillo Formation. Light-colored outcrops west of Willis Palms are nearly vertical strata of the Imperial Formation. Exposures of the Imperial Formation are also present in the northwest part of the Indio Hills, but it is not known to crop out northeast of the Mission Creek fault.

STOP: DEFLECTED STREAM IN PUSHAWALLA CANYON (after Meyer, 1979; Keller et al., 1982)

An excellent view of an offset and abandoned stream course can be observed in Pushawalla Canyon in the southeast Indio Hills by walking up the gentle floor of a dry wash to a linear oasis (FIG. 42). The oasis ends abruptly near the edge of a major canyon which is 25 m below the floor of the canyon.

The thin layer of alluvium in the dry wash overlies steeply-dipping layers of Late Pliocene and Pleistocene fanglomerate and fluvial sandstone of the lower Ocotillo Formation. These layers are also exposed in the opposite wall of Pushawalla Canyon to the southeast. The layers exposed in Pushawalla Canyon are bounded by two left-stepping segments of the Mission Creek strand of the San Andreas fault. The southern segment is a wide gouge zone with discrete shears. Locally, sets of minor normal faults having vertical separations of about 50 m are between two major shears. The northern segment is also a wide gouge zone, and within the zone, beds are steepened to nearly vertical attitudes. On the southwest side of Pushawalla Canyon, nearly horizontal layers of upper Ocotillo Formation overlie lower Ocotillo Formation in angular unconformity.

The older part of Pushawalla Canyon parallel to Mission Creek fault was a progressively lengthened dog-leg offset as the right-lateral movement progressed on the Mission Creek fault. However, at some time in the past few thousand years, its course was captured by another canyon moving northwest with the southwestern block. With capture, the offset part of Pushawalla Canyon became abandoned, and that part upstream became rejuvenated, because the new route presented a more direct, steeper path to the desert floor. The evidence of the rejuvenation is the sets of abandoned stream terraces parallel to the presently

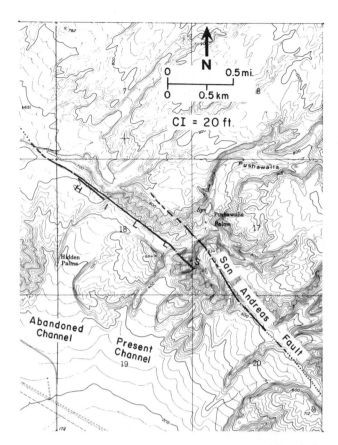

FIGURE 42 Topographic Map Showing the Left Step of the San Andreas Fault and Abandoned Channel of Pushawalla Canyon, Myoma 7.5-min Quadrangle, Indio Hills, California. From Keller et al. (1982).

active creek.

"A definitive slip rate for the San Andreas fault in the Indio Hills is not available. An estimated range for the slip rate of 10 to 35 mm/yr is based on a 0.7 km cumulative offset of an alluvial fan whose age is estimated by soil profile development to as much as 70,000 yr, but which most likely is about 20,000 to 30,000 yr. The latter age provides an estimated slip rate of 23 to 35 mm/yr" (Keller et al, 1982, p. 56).

OPTIONAL STOP: BANNING FAULT GEOMORPHOLOGY, BISKRA PALMS

Geomorphic features related to, and characteristic of, strike-slip faulting are well-exposed along the southeast part of the Banning fault in the Indio Hills. Offset and beheaded alluvial fans and stream courses, pressure ridges, sags, and fault scarps have been been mapped and studied. Of particular significance at this stop is a beheaded alluvial fan that was offset northwestward several hundred meters right-laterally from the canyon feeding the present fan. Individual stream courses on the surface of the fan are also beheaded, and the spacing between the drainages led to the inference that offsets along the fault take place episodically, from 4 m to 7 m at a time.

Return to I-10 and proceed west up the Coachella Valley.

EN ROUTE: COACHELLA VALLEY

The route gently ascends the northwest end of Coachella Valley via I-10 across the desert floor which is covered by wind-blown sand, and alluvial sand and gravel derived largely from the Whitewater River. Heavy stands of tamarisk trees line the railroad to prevent sand from burying the tracks. The northwest-trending mountains on the left, the San Jacinto Mountains, are in the Peninsular Ranges province and are composed of high-grade, locally mylonitic metamorphic rocks intruded by Cretaceous calc-alkaline granitic rocks. The mountain ranges on the north - the Little San Bernardino and Cottonwood Mountains - are in the E-W trending Transverse Ranges province and are composed of Precambrian gneiss and schist intruded by Mesozoic granitic rocks unlike those in the Peninsular Ranges.

Here in the middle of Coachella Valley, the upper end of Salton Trough, the sedimentary fill is more than 3200 m thick, judging from detailed gravity interpretations and from the Texas Company's "Edom" Stone 1 well (T.D. 2321 m) in one of the big, whale-backed anticlinal hills on the north side of the highway. The well bottomed in red-brown conglomeratic sandstone, 634 m beneath the base of the marine Imperial Formation of late Miocene or early Pliocene age. The well penetrated 511 m of the Imperial Formation, the thickest section of that formation (if it is not repeated by faulting) found anywhere in Salton Trough. Seven other oil wildcats have been drilled in Salton Trough: six of those were before 1927. All were dry holes.

The thickness of the sedimentary fill increases southeastward along the axis of the trough to 4200 m at Mecca and to a maximum of nearly 6450 m near the International Boundary.

EN ROUTE: WHITEWATER RIVER FAN

The southward-flowing Whitewater River debouches into the east end of Banning Pass, forming a constricted alluvial fan that has built out southeastward into Coachella Valley. Huge boulders and coarse detritus attest to the flooding and erosive power of this river which drains the highest parts of the San Bernardino Mountains. Powerful winds blow frequently through the pass and down into the valley. As they blow over the Whitewater River fan, they pick up fine sand and dust and carry them southeast, principally along I-10, causing intense sand and dust storms that have sandblasted the paint on countless cars and frosted innumerable windshields. An extensive field of low sand dunes covers the area between I-10 and Highway 111 almost all of the way from Thousand Palms to Whitewater. A plethora of wind machines take advantage of the abundant wind energy.

The optional sidetrip proceeds north 8 km up Whitewater Canyon to the end of the road.

OPTIONAL SIDE TRIP AND STOP: WHITEWATER CANYON AND COACHELLA FANGLOMERATE

In Whitewater Canyon the Banning fault strikes N85°W across the canyon, placing Cabazon Fanglomerate (Pleistocene) on the south against Precambrian gneiss and granite on the north (FIG. 43). A prominent vegetation line in the canyon bottom marks the trace of the fault. On the east side of the canyon the Whitewater fault (striking N30°W, almost parallel to the canyon) juxtaposes old alluvium against late Miocene Coachella Fanglomerate. A M 5.9 earthquake occurred on the Banning fault in 1986. The thrust mechanism of the earthquake conflicted with the minor right-slip ground rupturing.

At the Rainbow Rancho Trout Farm (FIG. 43) the Upper Miocene Coachella Fanglomerate forms the abrupt east wall of Whitewater Canyon. The Mission Creek strand of the San Andreas fault crosses the canyon about 5 km farther upstream. The Coachella Fanglomerate lies unconformably on Precambrian gneiss between the Banning and Mission Creek faults. It consists of up to 1600 m of coarse conglomerate and breccia with interbedded minor sandstone lenses. Interbedded andesite flows near the base of the formation have a K-Ar age of 10 Ma. The formation is subdivided into two stratigraphic units: a light gray upper unit which is predominantly fluvial, and a lower, dark-colored unit which is mainly composed of debris-flow deposits. Metamorphic, granitic and volcanic rocks are in the fanglomerate. Paleocurrent indicators, thickness changes and downflow diminution of stone size show that transport was largely from northeast to southwest from a source across the Mission Creek fault. Particularly distinctive clasts include a slightly metamorphosed potassium-feldspar porphyritic quartz monzonite with large K-feldspar megacrysts, the so-called "Peterson porphyry", and magnetite. A possible source area with rocks matching the distinctive quartz monzonite and magnetite clasts has been recognized near the Cargo Muchaco Mountains near the International Boundary, a suggested correlation which requires about 200 km of right slip since deposition of the fanglomerate.

Continue northwest on Interstate 10 toward Banning and Beaumont.

EN ROUTE: SAN GORGONIO PASS AND THE BANNING FAULT

The towns of Beaumont and Banning are in San Gorgonio Pass which lies between two of the loftiest peaks in southern California, and which is the locus of the contact between the Transverse Ranges and the northwest end of the Peninsular Ranges. Because the pass is fairly wide and straight, and because the San Jacinto Mountains rise abruptly from the floor of the pass as if fault-controlled, some geologists have postulated that San Gorgonio Pass is a graben bounded

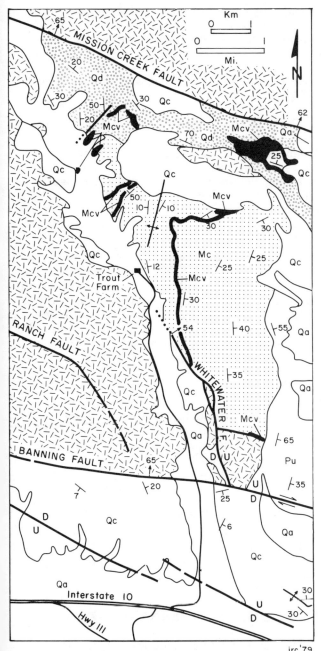

FIGURE 43 Geologic map of Whitewater Canyon. Hackled pattern - gneiss; Mc - Coachella Fanglomerate; Mcv - volcanic rocks intercalated with Coachella Fanglomerate; Pu - Imperial Formation; Qd - gravels of Whitewater River; Qc - terrace; Qa - alluvium. From Crowell and Sylvester (1979).

by reverse faults; however, the so-called "South Pass" fault has no geophysical or geological expression, and if present, it is buried deep below the Whitewater alluvial fan.

The north side of the pass is bounded structurally by the Banning fault which dips steeply north and is considered to be the extension of the Sierra Madre fault zone of the south front of the San Gabriel Mountains, now offset right laterally about 20 km. by the San Jacinto fault.

"The Banning fault zone had an initial period of 55
km of left-lateral displacement prior to late Miocene time, followed by right lateral displacement during latest Miocene and early Pliocene time. The left lateral displacement juxtaposed Transverse Ranges basement against Peninsular Ranges basement. Clasts within Tertiary conglomerates in the San Gorgonio Pass area have been right-laterally displaced from 10 to 24 km from their source areas. Stratigraphic successions in the area suggest right-lateral displacement occurred between 4 and 7.5 Ma when the Banning fault was the active strand of the San Andreas transform system.

"During Quaternary time, the eastern two-thirds of the Banning fault was reactivated to form once again an active strand of the San Andreas fault system. The western third of the fault appears to have been dormant since Pliocene time. The renewed displacement on the eastern third of the Banning fault appears to be mostly right lateral, with a total of from 1.6 to 3.2 km of displacement. The central part of the Banning fault in the San Gorgonio Pass area underwent reverse displacement during Quaternary time, producing a complex of low-angle thrust faults. The lateral component of this renewed displacement is not known."

(italicized parts with modification from Morton et al. (1987).

At Beaumont the route proceeds 12 km south on highway 79 to Gilman Spring Road, thence southeast along the base of the steep scarp of the San Jacinto fault to Soboba Hot Springs.

EN ROUTE: SAN TIMOTEO BADLANDS

The route from Beaumont to Gilman Hot Springs through Lamb Canyon crosses a badlands area of Pliocene nonmarine rocks comprising the San Timoteo badlands northeast of the San Jacinto fault. The basal strata are generally red beds of coarse, arkosic sandstone and conglomerate. These are overlain by gray and greenish-gray sandstone, conglomerate beds and lenses, and siltstone with interbeds of white tuff. Some conglomerate lenses are monolithologic, composed of quartz diorite clasts up to 6 m in diameter. The lower part of the sequence (Mt. Eden Formation) is mainly middle Pliocene in age, judging from a Hemphillian fauna and flora. The upper strata (San Timoteo Formation) are upper Pliocene and early Pleistocene in age.

Locally the contact with the granitic basement is exposed. Gently inclined buttress unconformities show that the strata filled an area of granite hills.

STOP: SAN JACINTO FAULT AND PULL-APART BASIN

The San Jacinto fault is the only fault of the San Andreas system, including the San Andreas fault itself, to cross the Transverse Ranges without deviating from its general northwest strike. At its northwest end, the San Jacinto fault splays into a number of faults in the core of the San Gabriel Mountains. Along its length southeast of San Bernardino, it comprises several en echelon strands including, from northwest to southeast, the Casa Loma, Claremont, Coyote Creek, Buck Ridge, Clark Valley, San Felipe Hills, Superstition Hills, and Superstition Mountain faults.

Total right separation on the San Jacinto fault is about 40 km and accumulated since mid-Pliocene time; vertical separation exceeds 3 km. Where the fault zone splays upon entering the Borrego Desert, high-standing blocks within the zone are flanked by deep, alluvium-filled depressions as where Coyote Mountain stands high between Borrego and Clark Valleys. This pattern suggests that within a strike-slip fault zone, some blocks are squeezed up and others are depressed in accordance with the principles of fault convergence and divergence along a braided strike-slip fault. Several deep basins along the fault, such as the San Jacinto Valley between the Casa Loma fault and the main San Jacinto fault, are pull-apart basins (FIG. 44).

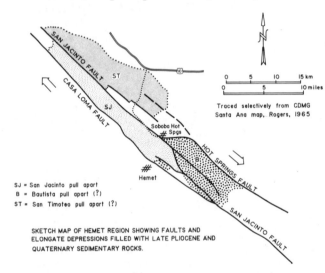

FIGURE 44 Sketch Map of Hemet Region Showing Faults and Elongate Depressions Filled with Late Pliocene and Quaternary Sedimentary Rocks.

In San Jacinto Valley, the depth to basement in this alluvium-filled pull-apart is about 2500 m below the valley floor. The basin is about 20 km long and 3 km wide. The pull-apart probably developed and filled since early Pleistocene time and is still undergoing rapid subsidence.

The San Jacinto fault is the most active fault in California in historic time, having generated at least six earthquakes greater than magnitude 6 since 1899, including that of 1918 (M 6.5) along the segment between Gilman and Soboba Hot Springs. Sense and amount of slip in those earthquakes were not determined but are presumed to have been right oblique slip, in accord with the geologic displacements. At a slip rate of 10 mm/year, determined from paleoseismic studies, sufficient strain should have accumulated by now for a repeat of the 1918 earthquake (M 6.8).

Return to Interstate 10, continue west and northwest toward

Redlands and San Bernardino. Proceed via surface streets in Yucaipa to the mouth of Mill Creek for an optional stop. Otherwise continue northwest on Interstate 10 to San Bernardino.

OPTIONAL STOP: MILL CREEK PULL-APART BASIN

Right-slip on the San Andreas fault has caused the rocks of the modern San Gabriel and San Bernardino Mountains to pass alongside one another en route to their present positions (FIG. 45). Clast suites, including Pelona Schist, record that passage by the infilling of Mill Creek basin (FIG. 46), a small pull-apart basin, now present in the San Andreas fault zone in the San Bernardino Mountains (Sadler and Demirer, 1986).

The south flank of the basin exposes a breccia facies dominated by Pelona Schist clasts probably derived from the Mount Baldy area of the San Gabriel Mountains, approximately 50 km to the northwest. Sediments entering the basin from the north and east <u>lack</u> clasts characteristic of the San Bernardino Mountains and indicate that the basin developed farther southeast (Sadler and Demirer, 1986).

Stratigraphy

Fault-line breccia, mostly Pelona Schist, intertongues with fluvial and lacustrine sandstone, siltstone of Miocene (?) age. The stratigraphic thickness exceeds 2000 m.

Structure

Pull-apart basin (10 km long, 5 km wide) bounded by the main active strand of the San Andreas fault (southwest); the Mill Creek strand of the San Andreas fault (northeast); Wilson Creek fault (east); unnamed fault (west).

EN ROUTE: SAN BERNARDINO VALLEY

The city of San Bernardino, named for the 15th century Franciscan preacher Saint Bernardino of Siena, Italy, is situated on a plain of deep Quaternary alluvium shed southward from the eastern San Gabriel Mountains and western San Bernardino Mountains These rise abruptly to the northwest and northeast, respectively, to elevations locally greater than 3000 m. Close to the mountains, the alluvium is spread in the form of large, coalescing alluvial fans on which are constructed entire cities, such as Pasadena, 40 km west of San Bernardino, and lesser communities of Monrovia, Arcadia, Duarte, Azusa, Glendora, Pomona, Claremont, Upland, Etiwanda, Cucumonga, Rialto, Colton, as well as San Bernardino itself.

The alluvial fans effectively obscure the nature of the abrupt northwest termination of the Peninsular Ranges against the southern edge of the Transverse Ranges. Virtually nothing is known of this contact at

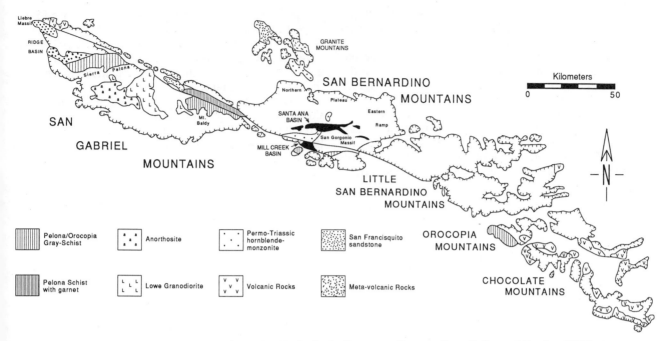

Legend for Figure 45:

Pelona/Orocopia Gray-Schist

Pelona Schist with garnet

Anorthosite

Lowe Granodiorite

Permo-Triassic hornblende-monzonite

Volcanic Rocks

San Francisquito sandstone

Meta-volcanic Rocks

FIGURE 45 Location of the Mill Creek and Santa Ana Basins in the Transverse Ranges. From Sadler and Demirer (1986).

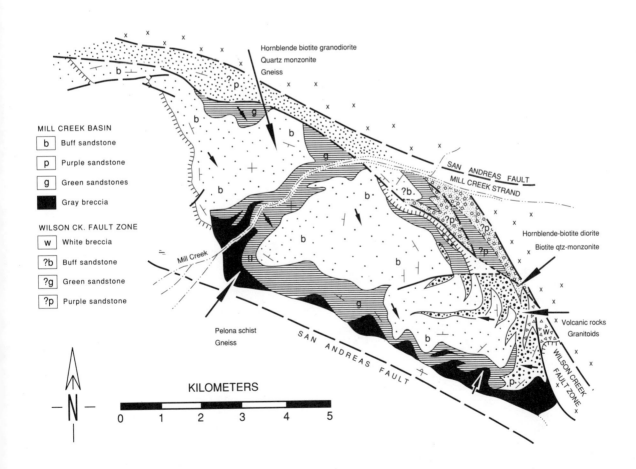

MILL CREEK BASIN

b   Buff sandstone

p   Purple sandstone

g   Green sandstones

■   Gray breccia

WILSON CK. FAULT ZONE

w   White breccia

?b   Buff sandstone

?g   Green sandstone

?p   Purple sandstone

FIGURE 46 Facies and Provenance of the Mill Creek Basin. Arrows show representative paleocurrent directions. Larger arrows have list of key clast types. From Sadler and Demirer (1986).

depth, except locally where active reverse, left-oblique slip faults are along the south front of the San Gabriel Mountains. East of Cajon Pass, this boundary is displaced 20 km right-laterally along the San Jacinto fault to near the town of Redlands. From this area the boundary continues eastward into the Salton Trough as the active Banning fault.

The Whittier Narrows earthquake (M 6.5) of October 1987 apparently occurred on a blind thrust fault that dips northward beneath the Sierra Madre fault and the San Gabriel and Santa Monica Mountains. The earthquake and others similar to it along the south front of the Transverse Ranges provide some evidence for the hypothesis that the central Transverse Ranges province is allochthonous. This system of blind thrusts would reach the surface from 5 to 10 km south of the Sierra Madre fault zone, but their tips are evidently buried by the great alluvial fans.

In early days, oranges and grapes were grown on the fans, and gravel and sand were quarried. Urbanization has almost completed obliterated much of the citrus and grape industry, but may of the quarries are still in operation   Some quarries are refuse dumps for that urbanization.

SAN BERNARDINO TO VALENCIA

SAN GABRIEL MOUNTAINS - SAN ANDREAS FAULT - MOJAVE DESERT

The route follows the San Andreas fault northwestward along the trace of the 1857 surface rupture from Cajon Pass to Palmdale. Then it turns westward down the axis of the Soledad basin to Valencia.

EN ROUTE: SAN JACINTO FAULT OVERCROSSING

The San Jacinto fault is encountered at the interchange of I-15 and I-10. The fault comes out of the San Timoteo Hills on the southeast, heads northwest through the grain elevator and the graceful interchange, and bisects a former motel on the west. This major right-slip fault has been very active in the 20th century. The segment through San Bernardino is particularly dangerous, because it transects a major urban area, and because it has not moved in historic time, earthquake activity would seem to be overdue. Southeast of San Bernardino the fault has generated at least six M 6 earthquakes in historic time.

The San Jacinto fault contributed substantially to the founding of the city of San Bernardino. A subsurface gouge zone formed an aquaclude (locally known as the Bunker Hill Dike) that traps abundant groundwater northeast of the fault and beneath the city.

EN ROUTE: FAULT AND DISSECTED ALLUVIAL FAN MORPHOLOGY IN CAJON WASH

At the north edge of the city are low hills that are the remains of the Muscoy Terrace, a thick, deeply dissected alluvial fan truncated at its lower end by Cajon Creek and along the mountain front by a graben in the San Andreas fault zone. The hills behind the golf course were laid bare by a disastrous brush fire, which burned into the northern limits of San Bernardino in the fall, 1980, consuming more than 300 homes.

Cable Canyon is on the northeast side of the highway, a few more kilometers up Cajon Pass, where the highway meets the base of Cajon Creek alluvial fan. The head of Cable Canyon is sharply displaced a few hundred meters right-laterally by the San Andreas fault.

Here in the Cajon Pass gorge, Interstate Highway 15 cuts a wide swath across the San Andreas fault zone. Crushed and distorted rocks, which typify the fault zone, are evident in the road cuts. Precambrian crystalline rocks form the basement along the northeast side of the fault; Pelona Schist is on the southwest side.

Exit Interstate 15 at Kenwood Road; proceed west beneath I-15 to the old highway and follow it northeast, upcanyon, tabout 3 km to a point where the highway bends east to the Blue Cut.

OPTIONAL STOP: PELONA SCHIST IN THE BLUE CUT (Old Highway)

Blue Cut takes name from the bluish hue of the Pelona Schist at this locality. The San Andreas fault crosses Cajon Creek at the northeastern edge of the gray-green outcrops of Pelona Schist in the canyon bottom. Cajon Creek is offset about 2 km right laterally.

The heavy fence was constructed along part of Blue Cut to prevent fractured quartz diorite from falling on the road. The quartz diorite is in a fault slice with Pelona Schist on each side.

Note the unfaulted, nearly horizontal nonconformity at the base of the pink alluvial terrace above the railroad tracks on the north side of the highway. Pelona Schists is the gray rock at railroad grade level.

The character of the Pelona Schist may be examined more readily in the roadsides on the south side of the highway and in fresher exposures behind the cottonwood trees several hundred feet back down the highway. Small folds and horizontal slickensides are present in the latter exposure.

After leaving Blue Cut and the San Andreas fault, the route crosses the edge of the North American lithospheric plate. Upstream and north on the old highway, the route traverses stratigraphically upsection from Precambrian gneissic rocks, through Eocene and Paleocene marine sandstone and shale, to pinkish and grayish nonmarine sandstone of Miocene age. Precambrian crystalline rocks of the San Bernardino Mountains complex form the slopes on the east side of the highway.

Proceed northeast up Cajon Wash past the intersection of I-15 with State Hwy 138.

EN ROUTE: STREAM PIRACY, DOSECC DRILL SITE, MOJAVE DESERT

The DOSECC deep drilling site was located on the east side of I-15 about 1 km north of the intersection with Highway 138. Continue another 6 km up the canyon to the Mojave Desert at Cajon Summit. Note the major streams which drain internally into the Mojave Desert and whose drainage heads have been captured by Cajon Creek which flows eventually into the Pacific Ocean.

The rim of bluffs along the skyline to the northeast marks the southwest edge of the Mojave Desert. The south face of the so-called "inface bluff" is formed in loose, gravelly beds of Pleistocene age which dip gently northward beneath the desert floor. The gravels are on the northeast side of the San Andreas fault and were derived from the San Gabriel Mountains southwest of the fault. The upper gravels consist almost entirely of Pelona Schist clasts; the lower part of the section contains clasts of rocks that crop out only in the central and western San Gabriel Mountains. Rare clasts of "polka-dot granite", in part of second cycle derivation from the Devil's Punchbowl area, show that from 30 to 40 km of right-slip has occurred on this strand of the San Andreas fault since deposition of the lower beds. These gravels provide the first indication of uplift of the San Gabriel Mountains during Pleistocene time.

Return down the canyon along I-15 to State Hwy 138 and turn west. Drive about 1 km to a turnout and dirt road on the northeast. Park and walk across the wash to inspect the bold sandstone outcrops.

STOP: MIOCENE SANDSTONE AT MORMON ROCKS

Lithology. Conglomeratic arkose and sandstone; nonmarine, fluvial; display crossbedding as well as cut-and-fill structures; 3300 m thick.

Structure. North-dipping homocline. Most of the deformation probably occurred in Late Miocene time and thus may not be associated with the San Andreas fault.

Age. Middle to Late- Miocene.

The rocks here are remarkably similar to strata at Devil's Punchbowl County Park, about 40 km to the northwest. The two localities are on opposite sides of the San Andreas fault, and their separation has been cited as evidence for large-scale right-slip on the fault since Miocene time. This interpretation poses difficulties, however, because considerable evidence from other rock suites shows that this part of the San Andreas fault should have about 250 km of post-Miocene right-slip, not just 40 km. In fact, the correlation of the two localities is more apparent than real, underscoring the necessity of careful and thorough geologic work. First, collections of fossil mammals from the two localities show that the strata at Devil's Punchbowl are from Early to Middle Pliocene in age, whereas the strata here at Mormon Rocks are Middle to Late Miocene in age. The strata at Devil's Punchbowl were deposited by streams flowing from the east and northeast and contain granite and volcanic clasts and distinctive clasts of "polka-dot granite". At Mormon Rocks, the "polka-dot granite" is lacking, however, and in its place is phyllite which is not present at the Devil's Punchbowl. Rock units in the Cajon Pass area are slices between several major faults, including the San Andreas proper and the San Jacinto. Deformation is complex and has occurred during several episodes since Early Miocene time.

The route from Mormon Rocks up Lone Pine Canyon to Wrightwood is a deceptively steep, straight ascent of 700 m over 10 km in the San Andreas "rift" zone. A vista at the head of the canyon provides an opportunity to look back along the route.

PHOTO STOP: SAN ANDREAS FAULT IN LONE PINE CANYON

On smog-free days a magnificent view is gained along the San Andreas fault in Lone Pine Canyon southeast of the hamlet of Wrightwood. This "rift valley," a deeply eroded, 300 m wide zone of intensely crushed and broken rock, lies between two prominent ridges. Blue Ridge to the south (right) is composed of typically foliated rocks of the Pelona Schist, and on the north is a narrow, uplifted wedge of granitic rock lying between the San Andreas and Cajon Valley faults. Large pods of white, coarsely crystalline marble (late Paleozoic) are included in the granite and are quarried locally for cement.

Note how the straight lines of the canyon are quite unlike the meandering nature of most creeks. This linearity is caused by erosion along the crush zone of the fault. The few ranches in Lone Pine Canyon depend on water from springs that are controlled by faulting. The Clyde Ranch, a few kilometers down the canyon, is located on a spring resulting from the damming of underground drainage from the canyon to the southwest. Three kilometers farther down-canyon, recent scarps are on the south side of the canyon behind the Sharpless Ranch. In the far distance, on opposite sides of a line of projection of the rift zone, two prominent peaks may be observed, smog or fog permitting. On the left is Mt. San Gorgonio (El. 3539 m; Old Grayback), the highest peak in southern California; the peak on the right is Mt. San Jacinto (El. 3333 m). Between them at the head of the Salton Trough is San Gorgonio Pass.

EN ROUTE: WRIGHTWOOD MUD FLOWS AND SAN ANDREAS FAULT

The vacation hamlet of Wrightwood (El. 2000 m) is spread across Swarthout Valley, which is eroded along the San Andreas fault zone. The 1857 earthquake surface rupture and scarp are upslope of the south side of town. Alluvium covers most of the valley floor, so that little is known about the condition of the underlying bedrock. About 50 houses in Wrightwood are situated on the active trace of the fault and will probably be strongly damaged during the next major earthquake here.

On the south side of the highway, high up on the slope of Blue Ridge, is a large landslide scar. It is the periodic source of great mudflows, such as that of 1941, which caused considerable damage in Wrightwood. As we proceed through town, notice the broad, deep channels and levees constructed to contain future mudflows. The gray, monolithologic Pelona Schist detritus is carried northward along Sheep Creek to the floor of the Mojave Desert where it is deposited on a large, gray alluvial fan which stands out clearly on Landsat imagery of this region.

On the western outskirts of Wrightwood, evidence of faulting can be seen south of the highway where a line of trees and swampy ground make a large area of impounded groundwater on the south side of the fault. Several trenches have been dug across the fault zone here in the past year

Continue NW along the paved road from Wrightwood, past the Holiday Hill Ski Area to Big Pines junction.

BIG PINES JUNCTION: HIGH POINT OF THE SAN ANDREAS FAULT

The San Andreas fault reaches its highest elevation (2111 m) at Big Pines Summit, 6 km west of Wrightwood. The saddle at Big Pines is fault-controlled; the 1857 scarp is on the north side of the intersection and extends behind the ranger station to the east. The fire station west of the junction is on saturated ground within the fault zone, south of the actual 1857 break.

Time and road conditions permitting, turn southeast at Big Pines Junction, drive past deep road cuts in Pelona Schist and note the layering in the schist, and proceed 3 km to the crest of Blue Ridge and the viewpoint into the core of the San Gabriel Mountains, the headwaters of the San Gabriel River, and the high peaks of the mountain range.

OPTIONAL STOP: BLUE RIDGE VISTA POINT

The view is south down the East Fork of the San Gabriel River into the core of the San Gabriel Mountains. On a clear day the east San Gabriel Valley and points beyond are visible. The San Gabriel River comes north from the valley, upstream toward the viewpoint and splits into the Prairie and Iron Forks whose courses are controlled by the Punchbowl fault. Pelona Schist is exposed on both sides of the Punchbowl fault.

The Punchbowl fault is an old, inactive strand of the San Andreas fault with about 40 km of right slip. It is not a continuation of the young, highly active San Jacinto fault as was frequently mentioned in the older literature.

The Vincent thrust, exposed in the middle distance in the San Gabriel Canyon beyond the confluence of Prairie and Iron Forks, is a major thrust fault in southern California, part of the Vincent-Orocopia-Chocolate Mountains thrust system, which is now segmented and right-laterally offset by the younger San Andreas fault system. Here the Vincent thrust places Precambrian gneiss and Mesozoic plutonic rocks of continental character upon Pelona Schist with a zone of mylonitic and retrograded rocks at the base of the upper plate (Vincent thrust zone). The thrust is exposed on the northeast flank of Mt. Baden-Powell (El. 2892 m), named for the founder of the Boy Scouts of America, to the southwest at about eye level, about 150 m above a light colored swarm of Miocene dacite and quartz latite sills. Thrusting occurred synchronously with prograde metamorphism of Pelona Schist during Late Cretaceous or early Cenozoic time. The thrust "vees" down San

Gabriel Canyon and crosses the north flanks of Iron Mountain (El. 2464 m) to the south and Mt. San Antonio (Old Baldy; 3097 m) to the southeast.

If road conditions permit, proceed west 5 km from Big Pines summit to Vincent Gap to exposures of the Punchbowl fault.

OPTIONAL STOP: FAULT SLICE IN VINCENT GAP

The Punchbowl fault strikes N65°W through Vincent Gap. Red sandstone and conglomerate of the Punchbowl Formation (?) exposed south of the parking lot are in fault contact with basement rocks upslope. A fault slice of granitic rocks intervenes between sandstone and Pelona Schist on the north side of the gap. A fault slice of sedimentary breccia, exposed on the ridge spur between Vincent Gulch and Prairie Fork to the southeast contains blocks of syenite probably derived from outcrops west of Soledad Pass (50 km northwest near Palmdale) and now offset by right-slip from their source. These relations show how faulted slivers of exotic rocks are distributed along the San Andreas fault.

Return to Big Pines Junction and proceed northwestward toward Palmdale. Between Big Pines and Pallett Creek, drive northwest along the San Andreas fault, descending from an elevation of about 2300 m to 1000 m, and from the conifer forest to the "high desert".

STOP: SQUEEZED UP FAULT WEDGES AT APPLETREE CAMPGROUND

Structure - San Andreas fault zone; pressure ridges (squeeze-ups)

Lithology - fault contact between Pelona schist and Precambrian (?) gneiss and granite

One of the best bedrock exposures of the San Andreas fault anywhere along its length is near Appletree Campground. A deep canyon is eroded along the fault which juxtaposes Precambrian (?) gneiss and granite on the northeast side (North American plate) against Mesozoic Pelona Schist on the southwest side (Pacific plate). Both basement rocks are extremely fractured and crushed. Long, isolated, northwest-trending ridges in the fault zone are pressure ridges of crushed basement, squeezed up like watermelon seeds by contractile strike-slip in the fault zone. Colluvium has partly filled the gap between the ridges and canyon walls, making pseudo-"sidehill graben" (FIG. 47). The significance of these exposures is that the crushed basement is probably typical of the nature of the fault to depths of up to 4 km in this area.

Continue northwest along roadway past Mile High, Jackson and Caldwell Lakes to Big Rock Creek.

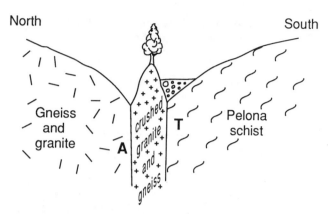

North   South

Gneiss and granite   *crushed granite and gneiss*   A   T   Pelona schist

FIGURE 47. Diagrammatic Cross Section across San Andreas Fault Zone at Appletree Campground.

EN ROUTE: SAN ANDREAS FAULT ZONE TOPOGRAPHY

The topography along this segment of the fault is characteristic of strike-slip faults in general (Schubert, 1982). The many elongate ridges and swales and flats are the result of thousands and thousands of years of mainly horizontal movement along the fault. Neither Caldwell Lake nor Jackson Lake are true sag ponds, that is, ponds formed as pull-aparts. Both are formed in part by the shutterridge of Shoemaker gravel on the north, which blocks drainage from canyons to the south. Jackson Lake has been dammed and modified substantially for recreational use, and Caldwell Lake to increase its water capacity. In the northern roadcut just opposite the lake, crushed basement rock is exposed, and a zone of gouge separates the crushed basement from Shoemaker gravel.

EN ROUTE: DISPLACED GRAVEL IN SHOEMAKER CANYON

Southeast of the 1857 scarp is Shoemaker Canyon, a straight valley along the San Andreas fault. From Caldwell Lake to Big Rock Creek (8 km), Pleistocene gravels are present along and north of the fault zone. These are well exposed in switchbacks which descend the ridge to the fault zone and to Big Rock Creek. Distinctive clasts in these gravels were derived from the Pallett Creek drainage, 12 km northwest and on the opposite side of the San Andreas fault, showing right-lateral displacement of these deposits through time relative to their source.

PAUSE: FAULT SCARP AT BIG ROCK CREEK

The scarp north of the road was not here in 1852 when Dutton and his fellow railroad route surveyors camped here. It was uplifted in 1857 during the great earthquake, and it diverted the creek from a course southeast of the scarp to its present position northwest of the scarp.

Proceed about 3 km northwest along the San Andreas fault to Pallett Creek trench site.

STOP: PALLETT CREEK TRENCH SITE

The Pallett Creek record of paleoseismicity spans at least 1600 years. It provides a history of surface-rupturing earthquakes, including at least six earthquakes since the 9th century similar to the 1857 event.

Structure - Uplifted and drained sag pond in San Andreas fault zone

Stratigraphy.
Gravel, sand, siltstone, and peat (Holocene marsh deposits

angular unconformity
Sandstone, siltstone (Anaverde Formation - Plio-Pleistocene)

Two kilometers west of St. Andrew's Priory, west of Valyermo, where the drainage of Pallett Creek is deflected eastward by the San Andreas fault, peat-bearing sag pond deposits were cut by surface ruptures associated with the great earthquake of 1857 on the San Andreas fault. By careful studies of stratigraphic relations, fault offsets, and radiocarbon dating of the peats exposed in deep trenches, Sieh (1978a; 1984) determined that approximately nine equivalent fault movements occurred at the site since 500 AD, yielding an average earthquake recurrence interval of 140 years ± 40 years. However, the range in the intervals between large events, from 40 to 350 years, casts considerable doubt on the usefulness of the concept of "recurrence intervals" for the purposes of earthquake risk analysis or earthquake prediction. Interpretation of the record here and elsewhere suggests that major earthquakes may occur in temporal clusters. The probability that this segment will slip again by the year 2020, however, is estimated to be 60 percent.

Continue northwest along road from Pallett Creek to Longview Road, turn south and follow signs to Devil's Punchbowl Park.

STOP: DEVIL'S PUNCHBOWL SYNCLINE

The Devil's Punchbowl is a picturesque, bowl-shaped area among smooth, massive exposures of colorful Pliocene sandstone and crushed fault rocks (Noble, 1954).

Structure. West-plunging asymmetric syncline, bounded by Punchbowl fault and Precambrian metamorphic rocks and Mesozoic plutonic rocks of the San Gabriel Mountains to the southwest, and by the San Andreas fault and the Mojave Desert to the northeast.

Stratigraphy.

Quaternary fluvial terrace
angular unconformity
Pliocene fluvial fanglomerate, sandstone
(Punchbowl Formation)
angular unconformity
Paleocene marine shale and sandstone
(San Francisquito Formation),
nonconformity
Mesozoic granite

Devil's Punchbowl exposes bold outcrops of rounded, buff arkosic sandstone and conglomerate similar to those at Mormon Rocks in Cajon Pass. Although superficially identical, significant lithologic and age differences exist between outcrops here in Devil's Punchbowl and those in Cajon Pass. Here the deposits are 1230 m thick, half as thick as those in Cajon Pass, and here the strata are uniformly highly indurated, whereas induration is much more selective at Cajon Pass.

The Punchbowl fault is an ancient, inactive (?) strand of the San Andreas fault zone that separates the high, mountainous terrain of Precambrian basement to the south from the more gentle topography underlain largely by Mesozoic granitic rocks to the north. Physiographically the Punchbowl fault is the boundary between the San Gabriel Mountains of the Transverse Ranges and the Mojave Desert. In surface exposures the fault is a reverse fault dipping 70° south. According to recent geologic studies by the California Division of Mines and Geology, the fault originated after deposition of the Punchbowl Formation, moved as recently as 5 my ago, and shows field evidence for about 44 km of right slip.

Retrace route from Devil's Punchbowl Park to Mt. Emma Road. Proceed northwest across Littlerock Creek to Cheseboro Road, turn left on to Barrel Springs Road. Drive in San Andreas fault zone to 47th Street East, turn north across the Littlerock and Cemetery faults to State Hwy 138. Turn west and proceed to Highway 14, thence north to a broad turnout at Lamont-Odet Vista Point to view the Palmdale Reservoir. Proceed next northward to the famous Palmdale roadcut.

STOP: LAMONT-ODET VISTA POINT

From this vantage point one looks over the California Aqueduct, the Palmdale reservoir, the San Andreas fault, Antelope Valley, and the western Mojave Desert. With exceptional visibility, the southern end of the Sierra Nevada may be seen to the north, and Telescope Peak at the crest of the Panamint Range, west of Death Valley, is about 150 km to the northeast.

The reservoir was originally a large sag pond which was dammed in the 1890s to increase its capacity for water storage and recreation. The San Andreas fault zone here is about 2 km wide, extending from the vista point to a point beyond the low ridge north of the reservoir. The most recently active trace of the San Andreas fault is along the north shore of the reservoir. The last major surface displacement here was at the time of the great Fort Tejon earthquake of 1857 (M 8), amounting to as much as 6 m of right slip.

The aqueduct in the foreground is the California Aqueduct which brings water from the Feather River in northern California to several thirsty districts in southern California.

The ridge extending northwesterly from the low hills on the opposite side of the reservoir is Portal Ridge, underlain chiefly by Pelona Schist of Mesozoic age. The dark ridge to the west is the Sierra Pelona, also underlain by Pelona Schist. The Pelona Schist supports little vegetation, and its name in Spanish means "bald".

The western Mojave Desert is a wedge-shaped structural block bounded on the northwest by the Garlock fault along the base of the Tehachapi and El Paso Mountains, and on the southwest by the San Andreas fault along the northeast edge of the Transverse Ranges. Except for local areas bordering these faults, the block was eroded to a low-relief surface in late Cenozoic time. Low mountains and hills poke up through a thick alluvial blanket derived from erosion of the same mountains and hills. Most of the rocks are Mesozoic granite and Tertiary volcanic rocks that have intruded and extruded upon late Paleozoic and Proterozoic metamorphic rocks. Locally nonmarine Tertiary basins have accumulated as much as a few hundred meters of clastic and evaporitic sedimentary rocks.

The Mojave Desert is cut by active, northwest-striking right-slip faults. Quaternary volcanism has occurred only in the eastern Mojave Desert. Windows of Pelona and Rand Schists are present in the western Mojave Desert, leading to the inference that the entire Mojave Desert is allochthonous and underlain by schist.

Continue north 2 km on Highway 14. Exit at Avenue S, turn west and park on the roadside 100 m west of the highway. Walk north 300 m to the crest of the hill and an overlook into the freeway roadcut.

STOP: HIGHWAY 14 ROAD CUT

The large roadcut exposes highly folded and faulted Pliocene nonmarine evaporitic and sandy strata of the Anaverde Formation. Their deformation is a consequence of contraction between the San Andreas fault zone, which lies immediately to the south, and the Littlerock fault on the north end of the roadcut. Careful mapping and matching sections on each side of the highway shows that the axes of the folds trend nearly east-west, oblique to the NW-SE strike of the San Andreas fault. When fresh, the structures in the roadcut were quite impressive and detailed, but in 20 years, the soft rocks have weathered and eroded deeply.

PALMDALE BULGE

The Palmdale bulge, the Palmdale uplift, the southern California uplift - no matter by what name it is known - has stimulated great debate in the scientific literature with the ultimate result that the sources of errors in geodetic leveling are better understood. Thus, tectonicists know how to be more careful and critical in their search for historical crustal movement.

"*Castle et al. (1976) identified the region of southern California surrounding Palmdale as uplifted when they analyzed repeated leveling surveys performed between 1955 and 1975. Most of the uplift reportedly occurred between early 1961 and 1965. A map was produced depicting the pattern, extent and magnitude of uplift, showing 25 cm of uplift centered at Palmdale. Further analysis by Castle (1978) raised the estimated uplift at Palmdale to 35 cm. Castle's original map of uplift was constructed by studying profiles of relative vertical crustal movement traversing a region bounded by Bakersfield and Barstow to the north; and Ventura, San Pedro, and Riverside to the south*" (Holdahl, 1983).

Following the revelations of Jackson et al (1980) that much of the uplift may be erroneous and attributable to inadequate calibration of leveling rods, Strange (1981) presented the idea that the appearance of an uplift was created by the different refraction error accumulations in successive surveys. He said that the primary cause of this difference was the reduction of standard sight lengths starting in 1964 after the uplift was thought to have occurred.

"*The leveling data used by Castle et al (1976) did not have the benefit of refraction corrections . . . . . Of primary importance to Castle's derivation of uplift at Palmdale is the analysis of direct levelings between Saugus and Palmdale. The Saugus to Palmdale route is special in several respects. It has a gentle average gradient of 0.012, allowing average sight lengths as long as 83 m. Secondly, this segment of leveling is along the Southern Pacific Railroad. The ballast of railroads, and the tracks, cause large vertical temperature gradients to result. Early leveling surveys between Saugus and Palmdale tended to have longer sight lengths than later surveys. Also, the older levelings were usually carried out on the track (i.e., on the ballast), while later levelings were made to the side of the tracks. All of the levelings between Saugus and Palmdale suffer from refraction error, but the older ones have significantly more error. It is the difference in these accumulations of refraction error which give appearance of uplift at Palmdale*" (Holdahl, 1983).

Thus, "*the uplift at Palmdale, relative to San Pedro, is not large enough to warrant a confident declaration that the motion is real. On the other hand, the motion parameters . . . . . are not all close enough to zero to justify a statement that no significant local motions occurred* " at Palmdale. In fact, "*in view of what is now know about leveling procedure changes after 1961, and what constitutes a physical setting which is favorable for large accumulations of refraction error, there is some justification to suspect that*" what little remains (7.5 cm vs Castle's 25-35 cm) "*of the apparent uplift at Palmdale may be caused by remaining systematic error* " (Holdahl, 1983).

Castle (1987) challenged these analyses: "*Examination of thousands of kilometers of geodetic leveling developed over nearly a century, coupled with reanalysis of the two cited refraction experiments, indicate that routinely constrained and corrected geodetic levelings have remained remarkably free of systematic error - rarely above 2×10^{-5}. The selectivity and special pleading to which various latter day critics of geodetic leveling have resorted challenge the judgment if not the credibility of these critics rather than the existence of the uplift.*"

Proceed southwestward on Highway 14 from Palmdale through Soledad basin to Valenica

EN ROUTE: SOLEDAD BASIN

The Soledad basin is a small, northeast-trending basin that began to open in Oligocene time and accumulated a west-dipping sequence about 8000 m thick of proximal, nonmarine detritus, most of which was shed from the ancestral San Gabriel Mountains to the south. The basement rocks of the basin and the adjacent San Gabriel Mountains consist of gneiss, anorthosite, quartz diorite, and quartz-bearing syenite of Precambrian age which have been intruded locally by granodiorite of late Permian age, Cretaceous granitic rocks, as well as andesite and basalt of Oligocene age. These rocks form the gentle hills north and east of Red Rover Mine Road as well as the north flanks of the San Gabriel Mountains. More importantly, this suite of distinctive rocks is also present in the Orocopia Mountains across the San Andreas fault, about 240 km southeast of Soledad basin. This correlation is one of the main lines of evidence for great strike-slip on the San Andreas fault.

Colorful Oligocene and early Miocene redbeds, sandstone, breccia, and conglomerate with intercalated volcanic flows are exposed in Vasquez County Park. Some of the boulders in the breccia and conglomerate of this, the Vasquez Formation, are as large as 2 m in diameter and were deposited by debris flows. The offset equivalent of the Vasquez Formation is also exposed in the Orocopia Mountains.

Near the sprawling urbanization of Canyon Country, roadcuts on both sides of the highway expose an unconformity between west-dipping, late Miocene terrestrial Mint Canyon Formation and horizontal Pleistocene fluvial gravels.

VALENCIA TO MARICOPA

RIDGE BASIN - PULL-APART SEDIMENTATION - BIG BEND

The route begins at the southeast end of Ridge basin at the intersection of Interstate Highway 5 and State Highway 126. It proceeds northwestward along Interstate 5 with several stops to the hamlet of Gorman which is located astride the San Andreas fault. There the route turns left onto Frazier Mountain Road and follows the San Andreas fault through its "big bend" to Lake of the Woods, thence it bears right onto Cuddy Valley Road to the Mt. Piños Road, thence it bears right onto the Cerro Noroeste Road through Mill Potrero, to Reyes Station at the intersection of State Highway 33. The distance along the San Andreas fault from Gorman to Reyes Station is about 65 km. Then the route leaves the San Andreas fault and turns right onto Highway 33 and proceeds 22 km northeastward to Maricopa.

INTRODUCTION TO RIDGE BASIN

Ridge basin (FIG. 48), about 40 km long and 10 km wide, was filled with about 13,500 m of clastic sediments, both marine and nonmarine, during Late Miocene and Early Pliocene time. The basin formed in the San Andreas transform belt when the San Gabriel fault, bordering it on the southwest, was the principal active strand of the belt. In Pliocene and Quaternary time, the region was deformed, uplifted, and deeply eroded so that rapid facies changes and many types of sedimentary structures and features are now well-displayed. The central part of the basin (FIG. 48) contains the best exposures and affords an opportunity for a visitor to appreciate sedimentation within a strike-slip regime where depocenters migrate laterally with respect to source areas. The sedimentary facies and tectonic relations are described in Crowell and Link (1982) and in maps and cross sections in Crowell et al. (1982a).

The southwest edge of the basin is bordered by the San Gabriel fault, which was continually active from about 11 to 4 Ma, as shown by the belt of coarse talus and debris-flow fanglomerate constituting the Violin Breccia (Crowell, 1982b) (FIG. 49). This formation has a total aggregate thickness of more than 11,000 m, but extends laterally into the basin no more than 1500 m where it interfingers with sandstone and shale along the basin trough. Near the fault, boulders and blocks up to 2 m in diameter are embedded within an earthy matrix. The clasts consist of gneissic and granitic rocks derived from basement outcrops across the fault on the southwest, but the source area is now displaced relatively northwestward. This source area moved laterally alongside the basin during Late Miocene time, dumping debris across the fault scarp into it. The result is a shingling or overlapping of younger units toward the northwest (FIG. 50), and we shall traverse the thick stratal section as we drive along Interstate 5. We shall examine the facies changes of the coarse Violin Breccia - to debris-flow deposits and fanglomerate and into sandstone and shale near the basin axis - at Stop PG. Dating of units comes from their stratigraphic position, from vertebrate and invertebrate fossils, from foraminifers in the marine section, and from magneto-stratigraphic studies.

Most of the sediment in Ridge basin was derived from sources to the north and northeast, as shown by paleocurrent indicators and facies and thickness changes; only about 15 percent was derived from the southwest or Violin Breccia margin. These basinal facies are well-displayed at stops TOR and CC (FIG. 48), where they consist of lacustrine and fluvial beds. At Stop TOR the transition upward from marine beds, including shallow-water turbidites, into nonmarine fluvial-deltaic deposits is exposed. Slump structures, breccia beds, and intestiniform layers here are interpreted as caused by earthquake shaking during deposition. The unconformity at the base of the Ridge Basin Group is exposed at Stop CC, where a coarse conglomerate lies on sandstone, conglomerate, and shale of the Cretaceous and Paleocene San Francisquito Formation.

From Valencia, drive north on Interstate 5 about 25 km to Templin Highway turnoff from freeway . Turn right (east) and drive 2 km to intersection with Old Ridge Route. The first stop has a short walkabout near this road intersection.

STOP: TEMPLIN HIGHWAY AT OLD RIDGE ROUTE

Overview of Ridge basin. Note structure and overlap relations to the northeast near the bottom of Castaic Canyon. Roadcuts display good directional current structures and minor faults and folds. Of particular interest are unusual deformed, intestiniform layers, interpreted as "seismites" related to sliding on subaqueous slopes during either tectonic oversteepening of the depositional slope or during earthquakes. The facies at this locality are shown in Figure 51. The carbonate marker bed passes under the highway crossing (FIG. 52).

Stratigraphically we stand in the upper part of the Marple Canyon Sandstone Member, the lowermost unit of the Ridge Basin Group. We are near the transition between marine beds stratigraphically below us and nonmarine beds above us, exposed in the russet-colored outcrops along the Old Ridge Route north of Templin Highway (FIG. 51). Deformed beds near this transition are conspicuous. Although most of the Ridge

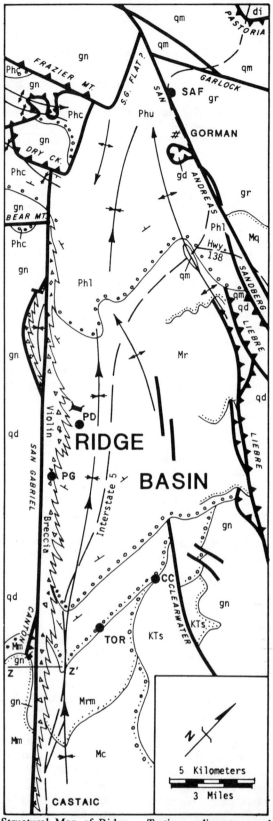

FIGURE 48. Simplified Structural Map of Ridge Basin. Stops: SAF, San Andreas fault; PD Pyramid Dam; PG, Piru Gorge; CC, Castaic Creek; TOR, Templin Highway-Old Ridge Route. Lithologic and formation symbols: gn, Proterozoic gneiss; gd, Mesozoic quartz diorite; di, diorite; qm, quartz monzonite; gr, granite; KTs, Cretaceous and early Tertiary sedimentary rocks; Miocene Mq, Quail Lake Formation; Mc, Castaic Formation; Mr, Ridge Basin Group; Mrm; Marple Canyon Member; Mm, Mint Canyon Formation; Phl, Hungry Valley Formation, lacustrine member; Phc, Hungry Valley Formation, nonmarine member. From Crowell (1982a).

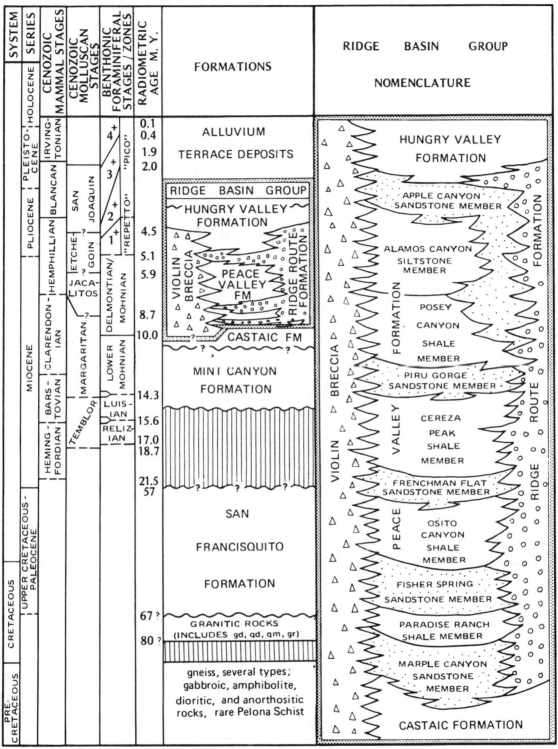

+ 1. REPETTIAN,   2. VENTURIAN,   3. WHEELERIAN,   4. HALLIAN.

FIGURE 49. Stratigraphic Column, Ridge Basin. From Link (1982).

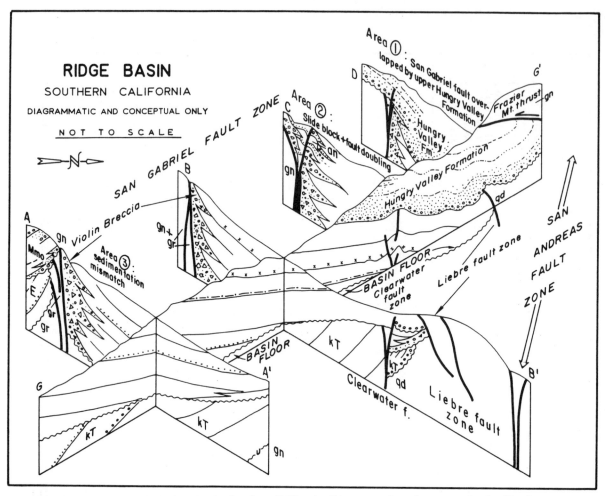

FIGURE 50. Isometric Sketch of Ridge Basin, Southern California. Diagrammatic and conceptual only. Not drawn to scale. Mmo - Modelo Formation. From Crowell (1986).

basin section is out of sight to the northwest, marginal relations are in view northward in the middle distance to the left (west) of the green surge tank. The notch in the ridge near where power lines cross is the unconformable contact between the San Francisquito Formation (mainly of Paleocene age) on the east and the rugged ledges of conglomerate and sandstone of the Castaic Formation (Miocene) on the west.

Stratigraphy and Depositional Environments

The Marple Canyon Sandstone Member of the Ridge Route Formation (FIG. 49) is well-exposed at the Templin Highway/Old Ridge Route intersection. The nonmarine-marine transition in Ridge basin is in this unit at approximately the intersection of the two roads. Marine rocks crop out along the Old Ridge Route south of Templin Highway and contain marine mollusks and foraminifers indicating moderate water depths. Slope facies and channel and interchannel turbidites are exposed here. The slope facies consists of large-scale sequences of slump-folded strata, slide blocks, and growth faults cut and filled by channels and bounded by laterally continuous strata (FIG. 51). Composite channel deposits consist mainly of sandstone beds

which form laterally discontinuous thinning- and fining-upward sequences. Graded beds, displaced mollusks, dish structures, sole marks, and Bouma intervals Ta and Tab are common. Associated with the channel sequence are interchannel deposits of mudstone and thin-bedded sandstone. These deposits wedge-out laterally and are highly slump-folded and locally brecciated. Many of the sandstone interbeds are graded and contain rip-up clasts, dish structures, Bouma intervals Tabcde, Tabe, Tbcde, and Tbde, and are interpreted as turbidites forming from overbanking processes from major distributary channels. They are deposited in interchannel areas adjacent to the channels.

The upper part of the marine section contains a carbonate marker horizon, which can be traced across Ridge basin and consists of interbedded micrite and dolomite with gypsum crystals. Many of the beds in this horizon are brecciated and contain abundant peloids. Mudcracks have not been observed. This horizon is interpreted as a sabka in a paralic setting, or as an evaporitic phase in a restricted marine embayment similar to, but on a much smaller scale to that of the Messinian event in the Mediterranean area.

Overlying the marine rocks is the nonmarine fluvial-deltaic sequence of the upper part of the Marple Canyon

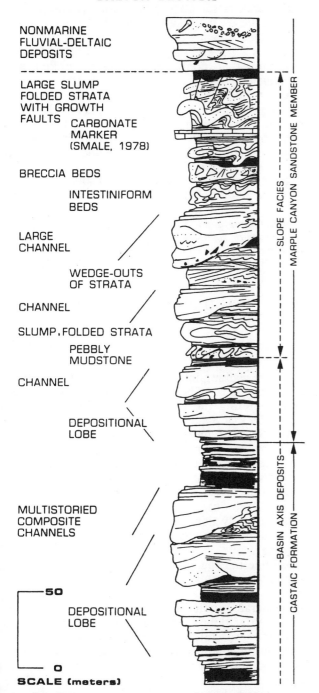

TEMPLIN HIGHWAY/OLD RIDGE ROUTE SKETCH SECTION

NONMARINE FLUVIAL-DELTAIC DEPOSITS

LARGE SLUMP FOLDED STRATA WITH GROWTH FAULTS
CARBONATE MARKER (SMALE, 1978)

BRECCIA BEDS

INTESTINIFORM BEDS

LARGE CHANNEL

WEDGE-OUTS OF STRATA

CHANNEL

SLUMP, FOLDED STRATA
PEBBLY MUDSTONE

CHANNEL

DEPOSITIONAL LOBE

MULTISTORIED COMPOSITE CHANNELS

50

DEPOSITIONAL LOBE

SCALE (meters)

0

SLOPE FACIES — MARPLE CANYON SANDSTONE MEMBER

BASIN AXIS DEPOSITS — CASTAIC FORMATION

FIGURE 51. Columnar Section along Old Ridge Route near Templin Highway. Diagram shows the transition from marine to nonmarine beds and the associated depositional features. From Advocate et al. (1982).

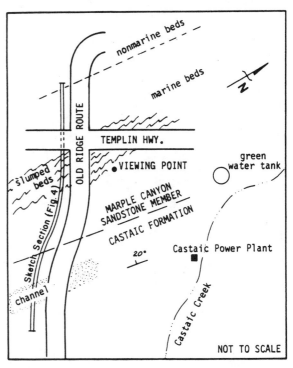

FIGURE 52. Sketch Map of Templin Highway and Old Ridge Route Intersection. Map shows the vantage point, location of columnar section, and geologic features. From Advocate et al. (1982).

laminae, and scour channels are common. These fluvial-deltaic or fan delta deposits mark the start of nonmarine sedimentation in Ridge basin and apparently prograded over deeper-marine deposits as the basin subsided.

Proceed down grade ahead to north about 3 km to concrete bridge, cross it and park. Walk up fireroad about 300 m along east side of Castaic to viewpoint amidst pine trees.

OPTIONAL STOP: CASTAIC CREEK

The irregular floor of Ridge basin, where the Castaic Formation lies unconformably on steeply dipping beds of the San Francisquito Formation of uppermost Cretaceous and Paleocene age, is exposed along Castaic Creek (FIG. 53). Two kilometers north of this point, the Castaic Formation interfingers with the Ridge Route Formation and lies unconformably on the San Francisquito Formation (FIG. 54).

From the concrete bridge crossing the creek, walk about 200 m north to the alluvial flat. From this vantage point examine the cliff on the west side of the canyon and trace the unconformity across the creek and up the eastern side, noting some small faults. Beds above the unconformity include a coarse basal conglomerate laid down on an irregular surface and crowded with large boulders and cobbles of the underlying San Francisquito Formation. In this area, these lower beds consist primarily of shale and siltstone but with interbedded resistant layers of sandstone and

Sandstone Member (FIGS. 49, 51). It is well-exposed along the Old Ridge Route north of Templin Highway. It consists of interbedded sandstone, conglomerate, and mudstone which form laterally continuous beds. Vertebrate remains, nonmarine mollusks and ostracods, and charophytes are in this upper section. Trough- and planar-cross bedding, climbing ripples, parallel

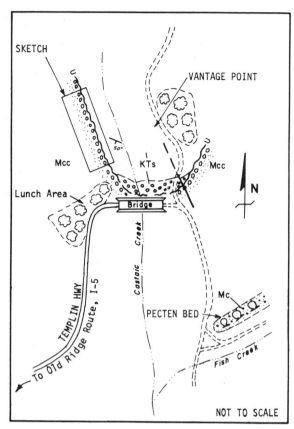

FIGURE 53. Sketch Map of Castaic Creek Field Trip Stop Locality. Map shows location of the unconformity, vantage point, and fossil bed. From Advocate et al. (1982)

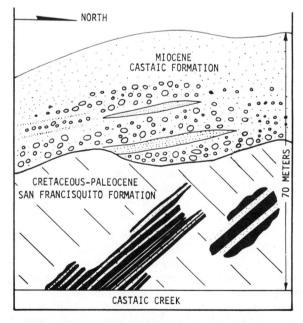

FIGURE 54. Sketch Map of the Angular Unconformity between Castaic Formation (Mc) and the San Francisquito Formation (KTs) along Castaic Creek. From Advocate et al. (1982).

conglomerate. In the cliff these resistant layers mark ridges on the unconformity, so that the basal conglomerate of the Castaic Formation is thinner over them and thicker in the swales carved by erosion during the Miocene epoch. South of the bridge, near Fish Creek, lies a gravelly conglomerate bed of the Castaic Formation that contains abundant, broken and disarticulated Pectens.

Good exposures of the San Francisquito Formation are along the stream course, including a conglomerate bed near where the basal unconformity of the Castaic Formation crosses the creek. The San Francisquito Formation here is mainly flysch-like and includes thin turbidites. Rounded stones in the conglomerate probably came from regions now far displaced laterally on the San Andreas fault.

Retrace route along Templin Highway to I-5, and follow frontage road on southwest side of freeway for about 10 km to Frenchman Flat stop. Park in campground on west side of highway. Walk down gorge along the left bank of Piru Creek.

STOP: PIRU GORGE AND VIOLIN BRECCIA

The Violin Breccia and its transition into the Osito Shale Member of the Peace Valley Formation may be studied by walking into Piru Gorge downstream from the parking area at Frenchman Flat (FIGS. 55, 56). Before entering the gorge, however, note that the Frenchman Flat Sandstone Member overlies the Osito Canyon Shale Member and extends to the ridge on the northwest held up by resistant Violin Breccia. The sandstone intertongues with breccia high on the ridge to the northwest. Stratigraphically below this marker bed, Violin Breccia is exposed along the western ridge, whereas Osito Shale is exposed along the highway to the east, directly along strike. Several steeply plunging but gentle folds are present in shale in the approach to the gorge.

Examine the change in bedding characteristics and sedimentary structures in crossing the transition from shale to breccia. Tongues of sandstone and conglomerate, including sedimentary breccias are interpreted as debris flow deposits, and they extend northeastward into the shale. Oncolites and stromatolitic coatings on some boulders and cobbles show that these flows extended into Ridge Basin Lake. Other fossils preserved in the Violin Breccia include rare vertebrate bones, animal tracks, and burrows. As the San Gabriel fault is approached, breccia clasts become much larger and the bedding less distinct. Some of the boulders are about 2 m in diameter, and most are composed of varieties of gneiss and granitic rocks.

The San Gabriel fault itself is exposed in the wall of the gorge where about 6 cm of consolidated gouge and cataclasite separate severely broken gneiss from sheared Violin Breccia. The fault zone in the gneiss terrane is wide and consists of several strands.

Proceed northward along old paved road about 2 km to turnaround at the Pyramid Dam Stop.

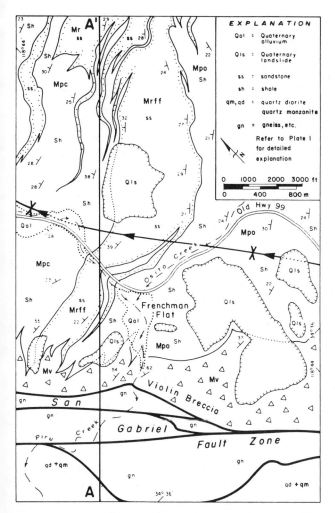

FIGURE 55. Geologic Map of Frenchman Flat area at Piru Gorge Stop. Map symbols: Mv, Violin Breccia; Mpo, Mrff, Mpc, Mr, are members of the Peace Valley Formation (see Fig. 49). From Advocate et al. (1982).

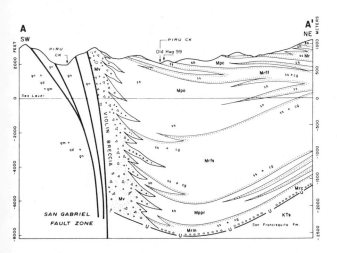

FIGURE 56. NE-SW Cross Section in the Frenchman Flat Area at Piru Gorge Stop. Symbols as in FIG. 55. From Advocate et al. (1982).

STOP: STRATIFIED SEQUENCE AT PYRAMID DAM

Outcrops of the Piru Gorge Sandstone Member of the Ridge Route Formation and the Posey Canyon Shale Member of the Peace Valley Formation are at the base of Pyramid Dam (FIG. 48). The Piru Gorge Sandstone Member is one of the best exposed of the fluvial-deltaic sequences characterize the axial part of Ridge basin. The Piru Gorge Sandstone Member is 185 m thick here and consists of two sandstone bodies separated by an interval of shale (FIG. 57). Each of the sandstone sequences is up to 60 m thick and thickens and coarsens upward. These sandstone packages consist of thin-bedded, graded sandstone at the base, overlain by thicker-bedded, cross-bedded sandstone. The cross-bedded sandstone is channeled into the underlying strata and shows lateral accretionary bedding. These thick, cross-bedded units are locally slump-folded and transitional laterally and vertically into finer-grained strata. These beds consist of interbedded blue-gray mudstone and sandstone and are inclined away from the channels and wedge-out laterally. They contain abundant organic material, mudcracks, rootlets, burrows, animal tracks, and are locally slump-folded. These sandstone bodies are interpreted as major, meandering, fluvial-deltaic lobes which prograded into a shallow lake. Thin-bedded turbidites or pro-deltaic deposits were generated off the front and flanks of the prograding fluvial system.

These deposits are overlain in turn by fluvial channel and interchannel deposits. The fluvial channel deposits are conspicuously cross-bedded and form meandering, laterally accretionary sequences. The interchannel deposits consist of well-developed levees and interchannel areas (ponds, marshes, floodplains). The levees are inclined away from channels and wedge-out laterally and consist of alternating sandstone and mudstone beds which are locally slump-folded. Interchannel areas consist of blue-gray mudstone and sandstone rich in organic material which were subject to desiccation and burrowing and reworking by organisms.

The shale interval separating the two sandstone bodies consists of organic-rich, laminated, black shale, limestone, chert, and thin-bedded sandstone. It is 30 m thick and contains locally abundant ostracods and plant fossils. Rare animal tracks are in the shale, and mudcracks and ripple marks are common sedimentary structures. This shale is interpreted as a shallow lake deposit rich in organic material. It was locally desiccated with anoxic bottom conditions as suggested by the mudcracks, chemical deposits, preservation of the organic material, and the laminated nature of the deposits.

The Posey Canyon Shale Member of the Peace Valley Formation (FIG. 49) conformably overlies the Piru Gorge Sandstone Member and consists of indurated mudstone and shale, soft mudstone, pyritic shale and siltstone, and gypsiferous siltstone. These rocks contain from 30 to 60 percent analcime, 20 to 50 percent ferroan dolomite, pyrite, micrite, and lesser

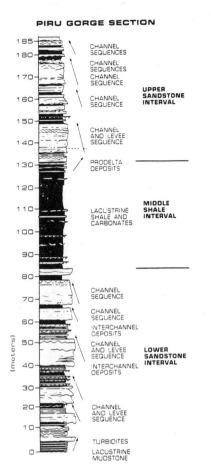

PIRU GORGE SECTION

- CHANNEL SEQUENCES
- CHANNEL SEQUENCES
- CHANNEL SEQUENCE
- CHANNEL SEQUENCE — UPPER SANDSTONE INTERVAL
- CHANNEL AND LEVEE SEQUENCE
- PRODELTA DEPOSITS
- LACUSTRINE SHALE AND CARBONATES — MIDDLE SHALE INTERVAL
- CHANNEL SEQUENCE
- CHANNEL SEQUENCE
- INTERCHANNEL DEPOSITS
- CHANNEL AND LEVEE SEQUENCE — LOWER SANDSTONE INTERVAL
- INTERCHANNEL DEPOSITS
- CHANNEL AND LEVEE SEQUENCE
- TURBIDITES
- LACUSTRINE MUDSTONE

(meters)

FIGURE 57. Pyramid Dam Columnar Section in Piru Gorge below the Dam. From Advocate et al. (1982).

amounts of detrital quartz, feldspar, clay, mica, and organic material. Analcime and dolomite are in both peloids and as fine-grained disseminated crystals. Clay includes illite, smectite, and kaolinite. The organic content of shale ranges from 0.44 to 2.18 percent and is considered one of the best potential source rocks for petroleum in Ridge basin. Sedimentary structures include even, varve-like laminae, soft sediment deformation features, syneresis cracks and injection features, small-scale faults, gypsum and pyrite nodules, graded beds, rip-up clasts, displaced fossils, and breccia beds. The Posey Canyon Shale Member has been interpreted as a closed, brackish lake deposit with locally anoxic bottom conditions. The Posey Canyon Member may have been deposited in relatively deep (20-25 m) water in comparison to the other shale members of the Peace Valley Formation, because of the paucity of beetle burrows, distinctive mudcracks, rare wave ripples and bioturbation. Turbidites and debris-flow deposits are locally present in these fine-grained chemical deposits.

Backtrack along road to Templin Highway onramp to I-5 and drive northwest about 30 km along freeway to Gorman exit. Drive up frontage road on southwest side of freeway for about 3 km to the top of Tejon Pass to San Andreas fault stop.

STOP: SAN ANDREAS FAULT AT TEJON PASS

Roadcuts along the western side of the frontage road at the top of Tejon Pass (El. 1263 m) expose the main active strand of the San Andreas fault (FIG. 58). Here over a meter of black gouge, tar-like in consistency, separates sheared and comminuted basement granitic rocks on the southwest from buff Pleistocene gravel on the northeast. The fault, which last broke the surface here in the 1857 Fort Tejon earthquake, can be traced across the freeway toward the southeast and through the notch in the hillside. One of the strands is parallel to this strand southwest of the rounded hill close to the freeway. The hill between the notch and the freeway is a landslide mass that was emplaced before the last movement on the fault. It apparently came down from the high hill northeast of Gorman, a hill consisting of pink granite on top with several long slices of hornfels, gabbro, diorite, marble, schist, and volcanic rocks - and then was displaced right laterally. These slices have been tectonically assembled to form a lens-shaped region in the hillside.

INTRODUCTION TO BIG BEND REGION

The Garlock fault meets the San Andreas fault north of the San Andreas fault stop (FIG. 58). The main Garlock fault zone lies beneath the alluvium near the intersection of I-5 with Frazier Park Road, but a splay with a crushed zone passes in notches in hills north of the road. In this vicinity, displacements along the San Andreas zone have disrupted the topography, but the Garlock fault has not. Only saddles weathered out along the belt of crushed rock mark its course. From the San Andreas fault the Garlock fault can be traced eastward about 260 km to the area southeast of Death Valley, where its structural and tectonic relation to major faults and other structures is still under investigation. The Garlock fault zone forms a "transfer fault" between major blocks in the Mojave Desert which have extended differently across it. Although its western end is apparently inactive, judging from the lack of fault landforms, it is indeed active along its central reaches and on to the east.

Within the Tehachapi Mountains the Garlock fault zone includes several strands, some of which may not be related at all to late Cenozoic tectonic events. For example, the North Garlock fault is probably a strand of the Rand thrust (part of the Vincent thrust system) and is largely of late Mesozoic age. In addition, the Pastoria fault is characterized by thick mylonite, is somewhat younger, but has not been active since early Tertiary time. Much of the basement block north of the intersection between the Garlock and San Andreas faults, and extending for many kilometers both east and west, has not yet been mapped in detail. The region is exceptionally steep, brush covered, and subject to severe landsliding.

The San Andreas fault through the Big Bend region separates very different rocks. North of the fault

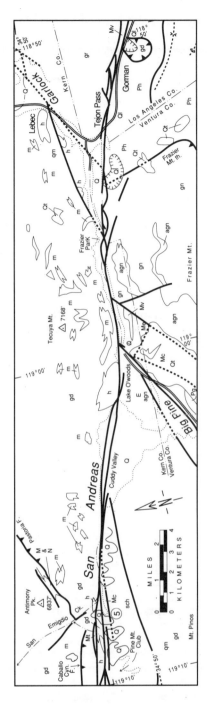

FIGURE 58. Generalized Geologic Map
of Field Trip Route between Gorman and
Pine Mountain.   From Crowell (1975),
with modifications.

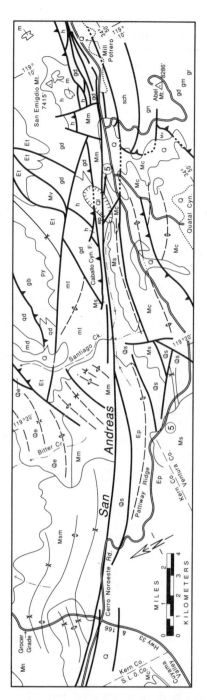

FIGURE 59. Generalized Geologic Map
of Field Trip Route between Mill Potrero
and State Highway 33. From Crowell
(1975), with modifications from Davis
and Duebendorfer (1987).

basement rocks consist of granite, granulite, and
gneiss, mainly with Sierra Nevada affinities and largely
of Mesozoic age. Bodies and pendants of marble and
schist, correlated with Paleozoic metasedimentary rocks
elsewhere, are intruded by granitic rocks. On the west
the basement is composed of mafic metamorphic rocks,
many of which are meta-ophiolitic and are interpreted as
uplifted parts of ancient oceanic crust of Mesozoic age.

The basal strata in the sedimentary section north of
the fault are marine Eocene beds, assigned to the Tejon
Formation, which lies unconformably upon the
basement complex. These beds are succeeded by a
nearly complete sequence, including some intercalated
volcanic rocks, up into the Holocene section. Many of
these northern units change facies from continental beds
on the east into deep water marine facies on the west.

The total stratigraphic thickness along the northwest flank of the San Emigdio Mountains is about 8500 m. About 10 km farther north, beneath the southwestern corner of the San Joaquin Valley, the sedimentary section is very much thicker, and the depth to seismically acoustic basement is over 12 km. The nature of the sedimentary rocks beyond the reach of the drill within this very deep basin is not well known. The basin may have originated in Early Miocene time as a pull-apart basin.

Southwest of the San Andreas fault zone in the Big Bend region, Precambrian gneiss and migmatite, gneiss of unknown age, and Mesozoic granitic rocks of several types and ages are exposed. The oldest sedimentary rocks consist of sandstone of Paleocene and Eocene age near the fault, but within 10 km a prism of Cretaceous strata up to 8 km in thickness lies within the southern Coast Ranges.

Along our route, mid-Tertiary marine and nonmarine units with intercalated volcanic rocks make up a very different sequence from the beds of the same age across the fault to the north. Miocene units in the western part of this region came from sources east of the San Andreas fault, but no possible source area is now present directly across the fault. The source for these strata has now been offset to the southeast by right-slip on the fault zone. In fact, in this region the contrast between facies of the same age requires major right slip and is one of the classic regions for reaching such a conclusion in California. It was first recognized by Hill and Dibblee (1953) but has been studied subsequently in more detail by many geologists. Rocks of the eastern Caliente Range are important, because intercalations between nonmarine and marine Miocene strata, with intercalated volcanic rocks, have provided a sequence for calibrating time scales from foraminifers, marine mollusks, vertebrate remains, and isotopic dating.

The tectonics of the Big Bend region are complex (FIGS. 58, 59). Several ages of faulting have been documented, and the sedimentary sequence is replete with unconformities and facies changes. The Big Pine fault meets the San Andreas fault along this stretch near the hamlet of Frazier Park. The Big Pine fault has at least three distinctly different episodes of movement. In Oligocene time the southern wall stood high so that coarse sedimentary breccias, including boulders up to 2 m in diameter, were shed northward into nonmarine talus cones and steep alluvial fans. The fault was reactivated in post-Late Miocene time with left-slip of about 13 km as shown by the offset of volcanic bodies and major faults. It was then apparently inactive until recently. At present subdued fault scarps are along it, but judging from the disruption of landforms, it is not as vigorously active as the San Andreas fault. The Oligocene Big Pine fault is interpreted as the offset counterpart of faults with similar histories in the Soledad and Orocopia regions far to the southeast.

In driving from Interstate 5 to Maricopa, many of the geologic features en route can be seen from the bus windows, although a few pause stops for photographs may be desirable in addition to the three scheduled stops. The landforms are shown in detail on a large-scale map by Davis and Duebendorfer (1987).

From the intersection of Interstate Highway 5 with Frazier Park Road, drive westward through the hamlets of Frazier Park and Lake of the Woods, and follow the Cuddy Valley Road along the San Andreas fault. Several good pause and photo stops are along the way, but the first major stop is at the Ward Street intersection with Mill Potrero Road. Park at intersection and walk up Ward Street about 100 m to a lookout into San Emigdio Canyon.

STOP: SAN EMIGDIO CANYON OVERLOOK

The San Andreas fault zone follows along the canyon to the west, and long slices of several rock types can be observed on the mountain to the northwest. These slices include dark shale of mid-Miocene age (Temblor Formation) and several varieties of basement rocks. The Caballo Canyon fault, of mid-Tertiary age, cuts several of the canyons and spurs and places shattered basement rocks above these sedimentary slices. Northward down San Emigdio Canyon are views of the San Joaquin Valley in the distance, and of steeply dipping Tertiary units lying unconformably on basement.

Continue westward about 11 km along San Andreas fault zone through Apache Saddle to a stop on right about 200 m west of intersections with unsurfaced roads to Quatal Canyon and Caballo Camp. Walk up the ridge 150 m to view the deep canyon of Santiago Creek cut into the San Andreas fault zone.

STOP: SANTIAGO CANYON OVERLOOK

From the insecure lip of a steep amphitheater on the west edge of Santiago Creek (FIG. 59), views along the San Andreas fault zone to east show several slices of Tertiary strata. The fault trace is also discernible where it crosses a large landslide. Strands of the San Andreas dip steeply in the deep part of the canyon and flair upward and outward, constituting "palm-tree structure". The up-dip lips of some of these fault slices merge into landslides where they curve over and dip outward from the fault zone and downward into the valley to the south. These are combination reverse faults, thrusts, and landslides, and are colloquially termed "slusts" or "thrides".

Drive westward and northwestward about 16 km and park 100 m west of stone monument marking the exit from Los Padres National Forest.

STOP: UNCONFORMITY OVERLYING TWO FORMATIONS

At the east end of a long roadcut, near a stone monument marking the border of the Los Padres National Forest, nonmarine conglomerate beds of the Pliocene Quatal Formation (also termed Paso Robles

Formation), dip about 30° to the west. These beds lie unconformably upon east-dipping, nonmarine sandstone and conglomerate of the Oligocene and lower Miocene Simmler Formation which, in turn, is conformably overlain by marine brown siltstone of the Miocene Soda Lake Formation. The San Andreas fault zone crosses ridges and valleys to the north, and if the weather is clear, the San Joaquin Valley far below may be in view. In the far, far distance may loom the peaks of the Sierra Nevada, about 200 km to the north-northeast.

Continue to drive along the highway, meeting Highways 33 and 166 in about 15 km. Turn east, cross the San Andreas fault at Reyes Station, and drive about 22 km to town of Maricopa.

Good views may be seen of the San Andreas fault and distant vistas of the Cuyama Valley and bare foothills of the southern San Joaquin Valley. These hills are underlain by folded and faulted siliceous shale of the Monterey Formation, locally called the Maricopa Shale. Fractured shale of this type is a prolific oil producer in offshore areas of California in the Santa Barbara and Santa Maria basins.

SAN JOAQUIN VALLEY ANTICLINES - TEMBLOR RANGE - WALLACE CREEK
BITTERWATER VALLEY - CHOLAME - PARKFIELD - TABLE MOUNTAIN

The route proceeds northward from Maricopa to the Buena Vista and Elk Hills to study the great anticlines beneath them. Then it continues northwestward over the Temblor Range to Wallace Creek to see a remarkable set of deflected and offset stream courses. The route continues up the Carrizo Plain through Bitterwater Valley to Cholame Valley where the San Andreas fault makes a major right step. At Parkfield the data for the Parkfield earthquake prediction experiment will be reviewed together with the geologic evidence for 330 km of dextral slip in this region. The route proceeds over Table Mountain to view the Franciscan Complex with its extrusions of serpentinite.

INTRODUCTION TO THE SOUTHERN SAN JOAQUIN VALLEY

The southwest corner of the San Joaquin Valley (FIG. 60) is underlain by a very deep Neogene basin which has been repeatedly infilled and deformed. In Late Pleistocene time, this part of the valley had internal drainage collected in Buena Vista Lake, and rivers from the southern Sierra Nevada and the Tehachapi and San Emigdio Mountains did not flow northward through the valley to the Pacific Ocean.

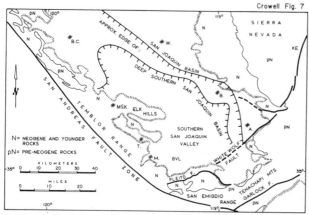

Crowell Fig. 7

FIGURE 60. Generalized Geologic and Geographic Map of Southern San Joaquin Valley. A, Arvin; B, Bakersfield; BC, Blackwells Corner; BVL Buena Vista Lake; KE, Kern Canyon fault; M, Maricopa; McK, McKittrick; T, Taft; W, Wasco. From Crowell (1987a)

Basement rocks are not known in this region, except in the San Emigdio Mountains to the south. The oldest sedimentary strata in the mountains to the northwest are of Upper Cretaceous age and are overlain by a very thick Cenozoic section. Facies changes, unconformities, as well as folds and faults have trapped great quantities of oil and gas in some of California's most prolific oil fields: Midway-Sunset, Buena Vista Hills, Elk Hills, and Kettleman Hills. Production in this region began in the early part of the century, so that many of the oil fields are now being produced by utilizing secondary and tertiary production methods. In 1987 the deepest well ever drilled in California reached a depth of 7445 m in the Elk Hills oil field. Structure at

depth is reported as complex, and beds as old as Eocene or Cretaceous may have been reached.

Cretaceous and Paleogene strata crop out in the Temblor Range to the west and are reached by the drill in some areas peripheral to the deep part of the basin. These beds were deposited in a forearc environment and are overlain by huge thicknesses of Neogene and younger units. A few kilometers east of Taft, the stratigraphic section is at 12 km thick, but the character of deeper rocks is not yet known. The deeper part of the basin, known as the Maricopa sub-basin, may have originated as an irregular pull-apart in early Miocene time (about 22 Ma); since then, however, its shape and character have been modified by continuing deformation. In fact, the latest deepening may be due to the superposition of a foreland basin in front of the rising San Emigdio Mountains, which are being actively thrust northward.

During Miocene time, turbidites spilled into the basin from sources both to the east and southeast. They followed the seafloor, weaving and winding between topographic highs caused by folding of the substrate (FIG. 61). The sub-basin was open to the ocean on the west during much of its history, but beginning in late Miocene time, displacements along the San Andreas fault on the southwest margin closed it off. Today, pre-late Miocene outlets to the sea have been identified and are dextrally offset about 300 km.

The orientation of Miocene structures along a wide belt east of the San Andreas fault appears to follow the simple shear scheme. In Plio-Pleistocene time and continuing to the present, however, folds and faults are oriented in such a way that they indicate that shortening in upper crustal rocks is occurring nearly normal to the San Andreas fault. The 1983 Coalinga earthquake is interpreted as a result of this shortening in which thrust displacements occurred on "blind" faults at depth that are not related directly to strike slip on the San Andreas fault itself, 40 km southwest of the epicenter.

Ranges bordering the valley on the west and south are actively deforming today. Regional tectonic subsidence of the basin probably continues today, but man-induced subsidence has been extensive and overshadows the tectonic component. Pumping of groundwater and petroleum especially since 1940 has caused local subsidence as great as 7 m in some parts of

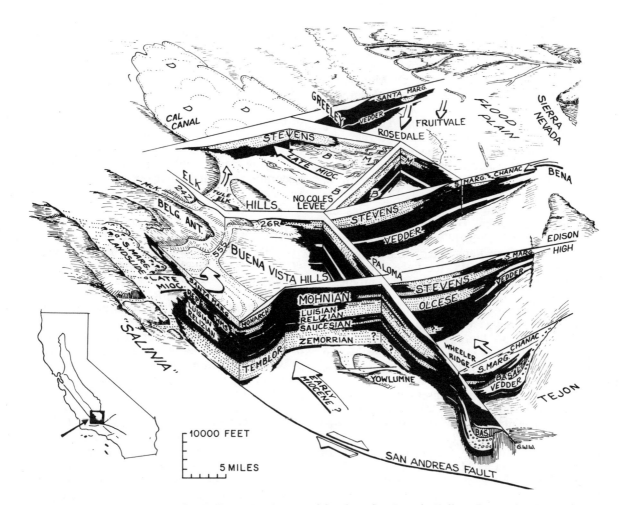

FIGURE 61. Combined Fence and Physiographic Diagram of Southern San Joaquin Valley. Restored to Late Miocene Time. Water is white, marine shale black, turbidite sands are stippled, shallow-marine and nonmarine sands white. Land surfaces and shorelines shown around basin and basement surface locally within it. Turbidity channel directions shown by arrows. Facies: M - meandering channel; B - braided channel (upper suprafan); D - distal (lower suprafan to basin plain). From Webb (1981).

the valley. The field trip route passes over some of the subsidence areas between Maricopa and the Temblor Range, but the effects are subtle and documented only by precise spirit leveling.

Easier to see are uplifted areas, expressed as the great oil fields in en echelon anticlines along the southwest edge of the San Joaquin Valley. The Elk Hills anticline is a prominent physiographic feature and typical of the surface folds in the valley. It is a doubly-plunging anticline about 25 km long and 8 km wide which, from a distance, looks like a giant whale beached on the dry sands of the desert. The field was discovered in 1915 and was designated Naval Petroleum Reserve No. 1. It covers more than 20,000 surface hectares, and subsurface petroleum reserves exceed one billion barrels.

From Maricopa, the optional sidetrip proceeds east on Hwy 166 about 12 km to Green Leaf Road.

OPTIONAL SIDETRIP AND STOP: SAN EMIGDIO MOUNTAINS

The north slopes of the San Emigdio Mountains are clearly in view from State Highway 166 about 12 km east of Maricopa. Note the huge alluvial fans debouching from the mountain front, incised flights of terraces, and numerous landslides. Deformation is occurring along several strands of the Pleito thrust and is progressing outward into the valley with time. The hypocenter of the 1952 Kern County earthquake (M 7.5) was beneath the eastern part of the mountain range, and about 22 km farther to the northeast, the ground was broken along the surface trace of the White Wolf fault.

Return to Maricopa and drive 2 km north on State Highway 33 to its intersection with Petroleum Club Road. Follow it about 1 km to the Lakeview Gusher Historical Monument.

STOP: LAKEVIEW GUSHER

Union Oil Company's. Lakeview No. 1 was spudded on January 1, 1909, and took 14 months to reach a depth of 615 m by cable tools. On March 14th or 15th, the bailer stuck, and the driller yanked at it with his cable. This effort blew the bailer through the crown block, and the well was out of control. The gushing column could be seen for 45 km, and the roar was heard for many kilometers. Sand flowing from the well soon buried the engine house, bunk houses, and coal shack; the derrick was demolished. A "trout stream" of oil flowed down the hill from the well. Four hundred men with mule teams and fresnos built barricades and prepared earthen reservoirs. Three pumps moved some of the oil to a pair of 55,000 bbl tanks. A month after the blowout, it still flowed at a rate of 40,000 bbl/day. Finally, on September 9, 1911, the Lakeview gusher caved in, 544 days after the blowout. It had produced about 9 million bbls of which 4 million were saved. During its flow the price of oil dropped from $1 per bbl to 30 cents per bbl. It was redrilled in 1913 by Union, but it produced at 30 bbl/day and died not long afterwards. In 1930 it was redrilled again, this time by General Petroleum (now Mobil), but no production was realized.

Production in the Midway-Sunset area (discovered in 1894) comes from folded Miocene sandstone bodies, many of which are lenticular, that are closed against unconformable, overlying Pliocene and Pleistocene rocks, and by up-dip pinchout. Many of the upper Miocene "Santa Margarita" sandstone beds are subsea debris flows that moved eastward and downslope from a source terrain west of the San Andreas fault before much of the major strike-slip on that fault system. The source area for the debris probably lies now about 300 km to the northwest in the vicinity of the Pinnacles National Monument, but this correlation is still under study by a number of workers. The geology of the region is so complicated both structurally and stratigraphically, and the subsurface records and logs from the ancient wells are so poor, that exploration for overlooked petroleum accumulations is difficult and challenging.

Continue into Taft city, bear right 3 km on Airport Road to a point about 150 m before its intersection with Honolulu Road and where the road bends from a northeast to a north trend on the south flank of the Buena Vista Hills. The Buena Vista thrust fault strikes nearly east-west across Airport Road and through the bend in the road. Park and walk west along the fault scarp.

STOP: BUENA VISTA OIL FIELD

The Buena Vista Hills are another of the anticlinal uplifts in southwest San Joaquin Valley. It is one of the "giant" oil fields in San Joaquin Valley, having produced 646 million bbls of about 28° API gravity oil to 1986 since its discovery in 1910.

The north and south flanks of the Buena Vista Hills are cut by north-dipping thrust faults (FIG. 62). The southern fault has buckled pipe-lines, cut well-casings at shallow depth, and produced a scarp in the surficial Pliocene and Pleistocene alluvium. The surficial shortening is readily seen along the scarp where slabs of asphalt roads have been thrust upon one another, by tilted utility poles and oil derricks, and by slackened utility wires. The damaged wells show that the fault dips north about 25° and strikes about N75°W. Thrusting has occurred aseismically at a rate of 20 mm/yr at least since 1930. A finite strain analysis of surveying data concluded that the displacement is underthrusting, rather than overthrusting, in response to surficial subsidence due to fluid withdrawal at depth in the oilfield.

In detail, however, the cause of subsidence in oil fields is much more complicated. According to Strehle (1987): *"With a reduction in reservoir fluid pressure, grain to grain loading of the sand increases and compaction begins through 1) repacking and rearrangement of grains, 2) plastic flow of softer intergranular material, 3) crushing of sand grains, and 4) elastic deformation of grains. The above processes take place where neither the formation nor the overburden have any self-supporting structure. Initial pressure reduction, and therefore compaction, takes place in the more porous and permeable sands. As pressures are depleted, a pressure differential develops across the sand/shale interface, and the shale (or clay or siltstone) begins to dewater to the sand. The amount of time needed for the sand to reach equilibrium with the increased overburden pressure is relatively short. The shales and siltstones, however, dewater very slowly due to their very low permeability. Depending on thicknesses, shales may take many years to reach pressure equilibrium".*

Continue on Airport Road, turn left (west) on Honolulu, cross Highway 119 on to Midway Road about 5 km to Fellows and on north on Mocal Road. Drive through the oil field at the end of this road, making way north and east to Highway 33 near the hamlet of Derby and thence north to McKittrick and Highway 58. Good views of tar sands and oil seeps may be had by driving on some side roads in McKittrick.

EN ROUTE: ELK HILLS OIL FIELD

North of the Buena Vista Hills are the Elk Hills with the great Elk Hills Naval Petroleum Reserve. About 35 wells had been drilled into the structure before oil was produced commercially in 1919. By the end of 1986, 811 million bbls of about 34° API gravity oil had been produced. It ultimate recovery is estimated to be about 1500 million bbls.

Surface outcrops, like those in the Buena Vista Hills, consist of Pleistocene nonmarine sand, silt, and conglomerate of the Tulare Formation. Locally the Tulare Formation is considerably deformed, and along the north flank of the San Emigdio Mountains, it is overturned. Both the Elk Hills and Buena Vista Hills have a right-stepping, en echelon orientation relative to

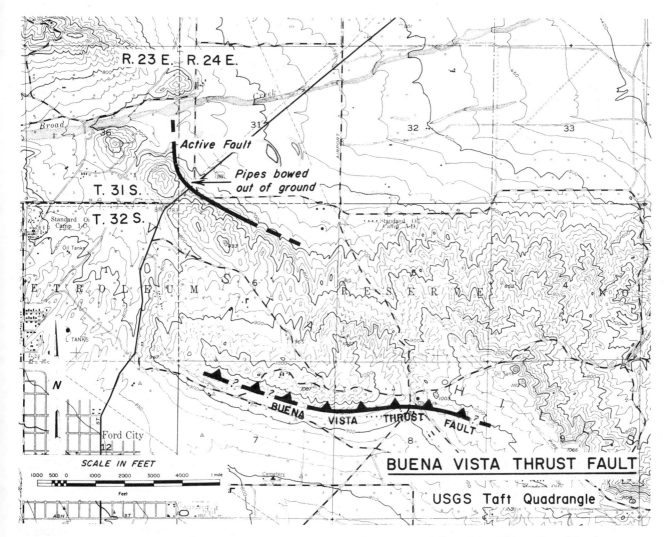

FIGURE 62. Topographic Map of Part of the Buena Vista Hills and Surface Trace of Buena Vista Thrust. From Manning (1973).

the San Andreas fault.

From McKittrick, proceed WNW on State Highway 58 across the Temblor Range to the Carrizo Plains.

TEMBLOR AND CALIENTE RANGES

The Temblor and Caliente Ranges bound the northeast and southwest sides, respectively, of the Carrizo Plain, a broad, undrained depression in the central Coast Ranges (Dibblee, 1973). The San Andreas fault marks the structural and physiographic boundary between the Temblor Range and Carrizo Plain, and it juxtaposes two different basement terranes. Basement beneath the Temblor Range is Franciscan Complex, whereas Mesozoic granitic rocks of Salinia form the basement beneath the Carrizo Plain and Caliente Range. Miocene marine strata overlie marine Eocene and lower Cretaceous strata in the Temblor Range; similar, but non-correlative Miocene marine strata overlie Oligocene nonmarine strata beneath the

Carrizo Plain and Caliente Range. Detailed biostratigraphic and paleogeographic studies of the Miocene sequences east and west of the San Andreas fault show that they were deposited in different basins, perhaps hundreds of kilometers apart, which subsided unevenly adjacent to rising highlands that shed different types of detritus into each basin.

Post-Miocene thrust faults beneath the Temblor Range and the Carrizo Plain dip toward the the San Andreas fault, posing an interesting structural question of their relation to the San Andreas fault. Are they cut by the fault and thus have offset counterparts in the subsurface tens or hundreds of kilometers along the fault, or do they cut the fault at depth?

The two mountain ranges take their names from the Spanish words for earthquake (temblor) and for hot (caliente). It is evident in July that both ranges are aptly named!

Temblor Range

Up to 5000 m of Miocene marine strata lie

unconformably on as much as 800 m of marine Eocene strata which, in turn, lie unconformably on more than 2000 m of lower Cretaceous marine strata in the Temblor Range. This sequence lies on Franciscan Complex in a northeast-dipping homocline.

In the central and southern Temblor Range, the sedimentary strata are deformed into numerous, northwest-trending, southeast-plunging folds which are increasingly tightened near the San Andreas fault. The folds are cut by a series of southwest-dipping thrust faults which, if projected to the San Andreas fault, would intersect the fault at a depth of about 700 m.

An interesting geological problem exists in the Recruit Pass area of the Temblor Range. Locally, thin, flat masses of pre-Cretaceous mica schist, coarse marble, quartzite, and coarse-grained granodiorite - underlain by gouge and breccia - overlie lower Miocene marine strata and are overlain, in turn, by upper Miocene marine strata. These crystalline rocks are unlike any nearby crystalline rocks on this side of the San Andreas fault where the prevailing basement is Franciscan Complex. One interpretation holds that they are the vestiges of an almost completely eroded thrust sheet. Another interpretation says that they are erosional remnants of a massive landslide. No matter which interpretation is correct, the lithology requires that they be derived from across the San Andreas fault, but their source has not yet been confidently identified.

Carrizo Plain

In contrast to the Cretaceous, Eocene, and Miocene marine strata of the Temblor Range, the Carrizo Plain is underlain by a 4000 m-thick sequence of Oligocene and Miocene strata lying on gneissic and granitic basement of the Salinian block. The sedimentary section is broadly folded into a great, asymmetric anticline, which extends into the southern Caliente Range, and which is cut by a major northeast-dipping thrust. If projected to the San Andreas fault, the thrust would intersect it at a depth of about 12,000 m, although it probably bottoms into a "flat" or decollement.

Caliente Range

Directly west of the San Andreas fault, the Miocene marine/Oligocene nonmarine succession lies directly on granitic gneiss and Mesozoic granitic basement, or on Cretaceous strata in the La Panza area of the Caliente Range, and on Paleocene marine strata in the southeast Caliente Range. The stratified succession thickens from 1200 m in the La Panza area to 4000 m in the Caliente Range and Carrizo Plain. Miocene arkosic strata must have been derived from a granitic source east of the San Andreas fault, because they are coarsest near the fault. The nearest such source without a cover of Miocene rocks, is the San Emigdio Mountains, 30 km to the southeast.

EN ROUTE: PHYSIOGRAPHY OF THE CARRIZO PLAIN

The San Andreas fault passes along the northeast edge of Carrizo Plain, a nearly level, undrained valley - the only one in the central Coast Ranges - lying between the Temblor and Caliente Ranges. Carrizo Plain is about 80 km long, 10 km wide, at an altitude of about 600 m. The climate is arid, therefore, streams are intermittent and flow only in springtime. Most of the water flows into a desert-type of playa, Soda Lake, which is encrusted with sodium carbonates and sulfates when dry.

Carrizo means "reed grass, bunch grass, and cane" which were important commodities for the California Indians who made a sweetening substance from it called panoche (Gudde, 1969). During the hot summer, about all that may be seen growing on the Carrizo Plain, aside from the parched weeds, are solar panels which generate electric power by a photo-voltaic process.

At the summit of Highway 58 across the Temblor Range, the highway bends sharply southwest down San Diego Creek. At the mouth of the creek, the road branches into three forks. Follow the left fork, Elkhorn Road, southeast 7 km along the southwest front of the Temblor Range to the second cattleguard and fence. Walk the fenceline uphill to the San Andreas fault and the Wallace Creek fault crossing.

STOP: DISPLACED DRAINAGE AT WALLACE CREEK

Wallace Creek is an ephemeral stream dextrally offset by the San Andreas fault in Pleistocene and Recent time. It was named in honor of U.S. Geological Survey geologist Robert E. Wallace who has been one of the principal students of the San Andreas fault for nearly 50 years. Sieh and Jahns (1984) worked out the evolution of the stream channel and its associated deposits to learn about the Holocene slip history of the fault (FIG. 63). They found that the average rate of slip has been 33.9 ± 2.9 mm/yr for the past 3700 years and $35.8 \pm 5.4/-4.1$ mm/yr for the past 13,250 years. The slip during the past three great earthquakes ranged from about 9.5 to 12.3 m. They used these values to determine that the dormancy period between major earthquakes is from 240 to 350 years which is from 100 to 200 years more than what Sieh determined at Pallett Creek. They argued, however, that the 90 km-long segment of the San Andreas fault northwest of Wallace Creek, which experienced up to 3.5 m of slip in 1857, may generate a major earthquake before the end of the century.

Sieh and Johns noted also that the long term slip rates are the same as those determined from short term creep and geodetic studies along the creeping segment farther north, and from that they concluded that elastic strain is not accumulating in the creeping segment. They also noted that the Wallace Creek slip rate is much less than the 55 m/yr rate determined from plate motions in the last 3 Ma, so they concluded that the 20 mm/yr residual must be taken over by other faults of the San Andreas fault system.

The geologic history of the channel involves several episodes of entrenchment, lengthening along the fault

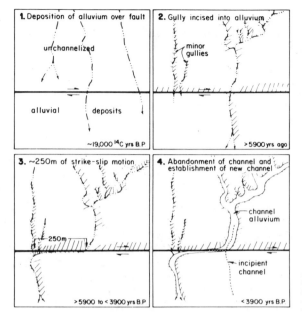

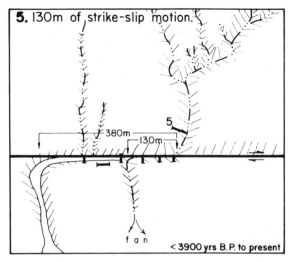

FIGURE 63. Holocene-Late Pleistocene Evolution of Wallace Creek. Heavy bars in map 5 are locations of trench excavations. From Sieh (1981).

trace as the fault slips episodically, abandonment, and establishment of a new channel by entrenchment. The history is rather easy to decipher at Wallace Creek, because this history is rather simple, it evolved quickly, and the exposures are excellent.

Wander along the fault trace, note the narrow zone of faulting, measure the channel offsets, both large and small, and decide which way and how much are the offsets of small gulches.

Return to Highway 58 and proceed 25 km westward across the Carrizo Plain to a narrow paved road that bears right and north to Bitterwater, Choice Valley, and Palo Prieto Pass.

EN ROUTE: CHOICE VALLEY AND PALO PRIETO PASS

Bitterwater is a common name in California and the arid parts of the southwestern United States where the potability and taste of water was of great importance to herdsmen, prospectors, and surveyors . Here it is applied to a creek in Choice Valley between the virtual wasteland of Carrizo Plain and the oak and grassland of Cholame Valley. The San Andreas fault follows Choice Valley for most of its length. Scarps, sidehill graben, offset stream courses, springs, and landslides are abundant and frequently visible on the east side of the valley.

Palo Prieto, meaning black tree, may derive from the fine stand of California live oak trees at the head of the valley.

Proceed to intersection with Hwy 46, turn east and drive about 2 km to Cholame.

PAUSE STOP: HAMLET OF CHOLAME

Cholame's principal claim to fame is its memorial to actor James Dean who died tragically in an automobile crash in 1955 a few kilometers down the highway. It also marks the northwest end of the locked segment of the San Andreas fault, the southeast end of the actively creeping segment, and the southeast end of the Parkfield-Cholame segment which is typified by a periodic repetition of characteristic earthquakes.

Proceed about 1 km east from Cholame, turn left (northwest) up a narrow, paved road and continue toward Parkfield. The following section of the roadlog describes the geology of the Parkfield area together with six excursion stops.

FIELD GUIDE TO THE PARKFIELD-CHOLAME SEGMENT
OF THE SAN ANDREAS FAULT, CENTRAL CALIFORNIA

John D. Sims
U.S. Geological Survey, 345 Middlefield Road, Menlo Park, California

SUMMARY OF GEOLOGY

The Parkfield-Cholame segment of the San Andreas fault is west of the southern Diablo Range and northern Temblor Range in central California. Displacement on the San Andreas fault here juxtaposed dissimilar tectonic terranes separated by a belt of melange (FIG. 64). The active main trace of the San Andreas fault in the Parkfield-Cholame area is characterized by a 1 km right stepover in Cholame Valley and a 5° left bend in the fault trace on Middle Mountain in the northern part of the area. This 40-km-long segment of the San Andreas is situated between the creeping segment to the northwest and the locked segment to the southeast (FIG. 3).

The area northeast of the San Andreas fault is characterized by complexly folded and faulted rocks of the Franciscan Complex, Coast Range ophiolite, sedimentary rocks of the Great Valley sequence, and upper Cenozoic marine and nonmarine sedimentary rocks (Dickinson, 1966a). Deformation of Upper Cretaceous and Cenozoic rocks northeast of the San Andreas fault zone varies in general with their age and proximity to the fault. The Upper Cretaceous rocks are strongly deformed and overlain by upper Cenozoic rocks with angular unconformity. Lower(?) and middle Miocene strata are less deformed than upper Miocene to lower Pliocene rocks. Upper Pliocene and Pleistocene rocks generally are the least deformed.

The area southwest of the San Andreas fault consists of granitic basement rocks of the Salinian block (Ross, 1984), which are overlain by Miocene and Pliocene marine sedimentary rocks which, in turn, are capped by Pliocene and Pleistocene nonmarine gravel and sand. In general, rocks west of the San Andreas fault are less deformed than similar-aged rocks east of the fault.

The active trace of the San Andreas fault lies along the southwest side of a belt of highly deformed rocks separating the Salinian block from the region to the northeast (FIG. 64). The deformed belt consists of sheared and deformed rocks of the Franciscan Complex and of upper Cenozoic units and exotic blocks of granite, gabbro, and marble. These exotic blocks are along a 25 to 30 km stretch of the Parkfield segment of the San Andreas fault zone (Sims, 1986) and were derived from Jurassic and Cretaceous crystalline basement rocks and upper Miocene volcanic units exposed to the northwest and southeast adjacent to the San Andreas fault (FIG. 65).

The Parkfield-Cholame segment also contains exotic blocks of crystalline basement and associated Tertiary volcanic and sedimentary rocks. North of the 1-km right stepover in Cholame Valley are two fault-bounded blocks of particular interest that have been displaced over 150 km from their counterparts. One block includes the hornblende quartz gabbro of Gold Hill, which correlates with a petrographically similar body of the same age near Eagle Rest Peak, about 145 km to the southeast, and with rocks near Logan about 165 km to the northwest (Ross, 1970). The second exotic block lies southwest of the fault (FIG. 66) and consists of a northwest trending belt of the Miocene volcanic rocks of Lang Canyon (Sims, 1986). These volcanic rocks correlate with compositionally similar rocks of the Neenach Volcanics about 150 km to the southeast and with the Pinnacles Volcanics about 160 km to the northwest. Both exotic blocks have Tertiary sedimentary rocks associated with them that are correlated with similar strata associated with the parent bodies on the east side of the San Andreas in the Big Bend area. These correlations of sedimentary units support the correlation of the gabbro and Miocene volcanic rocks (FIG. 65).

The Gold Hill block of gabbro is an elongate tabular body about 2 km wide, 8 km long, and less than 1 km thick. It is bounded by near-vertical faults on the northeast and south (R. C. Jachens, pers. commun., 1986) and lies northeast of the main trace of the San Andreas fault (FIG. 66). In depositional contact with the gabbro are sedimentary rocks of the Eocene Tejon Formation that consist of sandstone and conglomerate that bears pebbles, cobbles, and boulders of the gabbro. These rocks are derived from the Tejon Formation in the San Emigdio Mountains and equivalent rocks of the lower San Juan Bautista Formation of Kerr and Schenk (1925) near San Juan Bautista.

The volcanic rocks of Lang Canyon are part of an elongate, northeast-dipping block 16 km northwest of Gold Hill (FIG. 64). The volcanic rocks are overlapped on the southwest side by upper Miocene sandstone of the Santa Margarita Formation (FIG. 67).

The relative positions of the hornblende quartz gabbro of Gold Hill and the volcanic rocks of Lang Canyon are reversed with respect to the positions of their counterparts northwest and southeast of the Parkfield-Cholame area. The reversed relative positions reveal some details of the history of movement on the San Andreas fault. The restored Eagle Rest Peak-Gold Hill-Logan body, in its initial, prefaulting position at the margin of the Mojave Desert, was about 55 km northwest of the reconstructed Neenach-Lang Canyon-

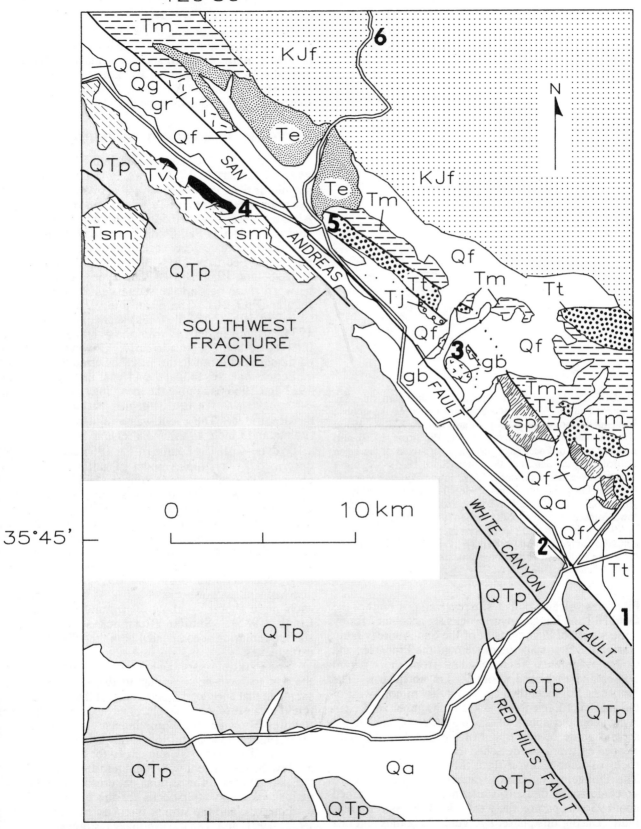

FIGURE 64. Geologic Map of the Southern Diablo Range and Northern Temblor Range. The Parkfield-Cholame area is indicated and the field trip stops are shown by bold numbers. Map units are: Qa- alluvial deposits (Quaternary); Qf - alluvial fan deposits (Quaternary); Qg - nonmarine gravel and sand deposits (Quaternary); QTp - Paso Robles Formation (Pleistocene? and Pliocene); Tsm - Santa Margarita Formation (upper Miocene); Te - Etchegoin Formation (upper Miocene); Tj - Tejon Formation (Eocene); Tm - Monterey Formation (middle Miocene); Tv - volcanic rocks of Lang Canyon (lower Miocene); Tt - Temblor Formation (lower Miocene); KJf - Franciscan Complex and serpentinite; sp - serpentinite; g - hornblende quartz gabbro of Gold Hill (Ross, 1970); gr - granitic rocks. See figures 66 and 67 for detailed maps of the Gold Hill and Parkfield areas.

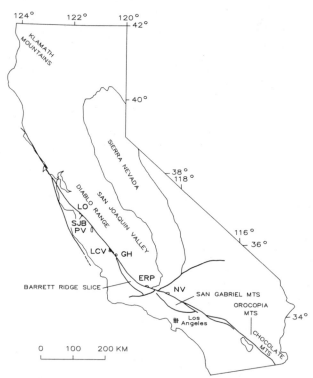

FIGURE 65. Map of Exotic Blocks Correlated across the San Andreas Fault in Central and Southern California. Symbols: ERP, quartz hornblende gabbro of Eagle Rest Peak (Ross, 1970); GH - hornblende quartz gabbro of Gold Hill (Ross, 1970), also includes rocks of Eocene age correlated with rocks of the Tejon Formation in the San Emigdio Range (Sims, 1986); NV, lower Miocene Neenach Volcanics; LCV, lower Miocene volcanic rocks of Lang Canyon (Sims, 1986); LO, quartz hornblende gabbro of Logan; PV, early Miocene Pinnacles Volcanics; SJB, lower Miocene volcanic rocks near San Juan Bautista and Eocene to Miocene stratigraphic section correlated with rocks in the San Emigdio Range (Nilsen, 1984); TF, lower Miocene volcanic rocks of the Tecuya Formation and nearby rocks of the Eocene Tejon Formation.

Pinnacles volcanic body. The reversed positions of the Gold Hill and Lang Canyon blocks are accounted for by three stages of development of the San Andreas fault. During the first stage of movement the Pinnacles and Logan fragments were detached from their parent bodies and displaced about 95 km northwest. The Pinnacles fragment then lay about 40 km northwest of the gabbro of Eagle Rest Peak, and the gabbro of Logan lay at about the latitude of Chico Martinez Creek. In the second phase of movement of the San Andreas fault stepped eastward and detached the sliver of the volcanic rocks of Lang Canyon from the Neenach Volcanics. The Pinnacles Volcanics, detached earlier, and the volcanic rocks of Lang Canyon thereafter remained about 95 km apart on the west side of the San Andreas fault. During this phase, two volcanic bodies and one gabbro body were moved away from their parent bodies. Movement during this phase lasted until the gabbro of Logan was at the latitude of Gold Hill. When it reached this latitude, the Gold Hill block and its associated Eocene sedimentary rocks were sliced off and remained on the east side of the San Andreas fault.

Following the positioning of the Gold Hill block, all the detached fragments maintained their relative positions, and the gabbro of Logan and the Pinnacles Volcanics were displaced an additional 160 km northwest to their present positions. A total of about 310 km of movement is recorded by these displacements.

HISTORIC PARKFIELD SEISMICITY

The first recorded earthquakes in the Parkfield area were on January 9, 1857. Two earthquakes of about magnitude 6 are thought to have occurred near Parkfield just prior to the great Fort Tejon earthquake of that date. All three epicenters are thought to have been on the San Andreas fault near Parkfield. Since 1857, earthquakes of about magnitude 6 have occurred on the San Andreas fault on 2 February 1881, 3 March 1901, 10 March 1922, 8 June 1934, and 28 June 1966. The intervals between these earthquake sequences are remarkably regular (FIG. 10). The mean interval is 21.9 ± 3.1 (standard deviation of the mean) years. The time of the 1934 earthquake departs from the average recurrence interval by being a decade early. However the 1966 earthquake conforms to the sequence established by the earlier earthquakes in that the 44-year interval between 1922 and 1966 is double the mean interval. The 1966 earthquake was the last damaging earthquake in the Parkfield area. The event was assigned a $M_L = 5.6$ (Bakun and Lindh, 1985) and a seismic moment of 1.4 x 10^5 dyne-cm. The source of the 1966 earthquake is described by a simple model of unilateral rupture propagation to the southeast along the fault.

All the M~6 earthquakes along the Parkfield-Cholame segment in this century are thought to have had similar characteristics, but only the 1934 and 1966 events were instrumentally recorded. The 1934 and 1966 earthquake sequences are remarkably similar. Both had main shocks with similar epicenters, magnitudes and fault plane solutions. Both sequences also had similar foreshocks of $M_L \sim 5.1$ which preceded each main shock by about 17 minutes (Bakun and Lindh, 1985). Similar aftershock sequences and surface rupturing accompanied both the 1934 and 1966 earthquakes.

The Parkfield earthquakes of 1922 and 1901 are thought to have been similar to the 1934 and 1966 events in that anecdotal reports suggest that cracks were found in some of the same places as the later earthquake ruptures (Brown et al., 1966). Intensity patterns for the 1901, 1922, 1934, and 1966 earthquakes are similar (Sieh, 1978a). Reports on intensity for the 1881 event are few but consistent with the intensities reported for the later events. The similar patterns of foreshocks, mainshocks, and aftershocks for the 1934 and 1966 earthquakes, and the similar intensity patterns for the 1922, 1901 and 1881 earthquakes suggest that the Parkfield segment of the San Andreas fault behaves in a characteristic fashion. The U.S. Geological Survey has therefore designed an earthquake predication experiment around the recurrence of the characteristic Parkfield earthquake (Bakun and Lindh, 1985: Bakun et al, 1987).

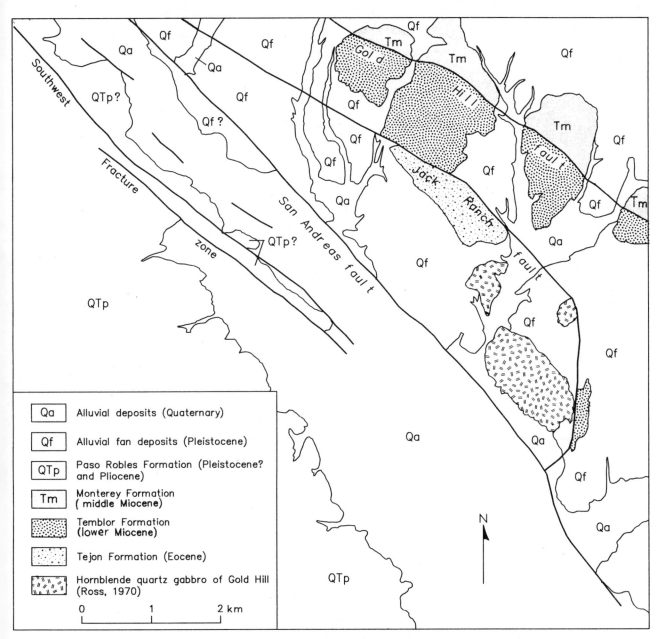

FIGURE 66. Geologic Map of the Gold Hill Block and Adjacent Area (after Sims, 1988).

LOG OF FIELD TRIP STOPS

STOP 1: SITE OF 1966 FAULT RUPTURE

Fault scarps and offset streams are visible at this site (Fig. 68). One stream is reported to be offset 3.2 ± 0.2 m right laterally (Sieh, 1978b), and a shallow trench contains the San Andreas fault. The broad alluvial plain of Cholame Valley stretches northwestward with Gold Hill in the distance (see STOP 3).

The offset stream at this location (FIG. 69) is typical of offset streams in general. Here we see a modification of the topography by man's activities. The amount of offset is ambiguous, and considerable disagreement exists as to the true offset. Sieh (1978b) considered the stream to have been offset in the 1857

Fort Tejon earthquake. However, later study of this and other offset streams in this area by Lienkaemper (1987) suggests that it may have been blocked by fault movement which ponded water and sediment behind a small ridge. The ridge perhaps was later overtopped and the new stream course was established. Subsequent to the establishment of the new stream course, some of the ponded sediment may have slumped to deflect the stream to its present position.

STOP 2: SCARPS AND OFFSET ALLUVIAL FAN

A fence, built in 1908, has been offset 65 cm at this site by the Parkfield earthquakes of 1922, 1934, 1966.

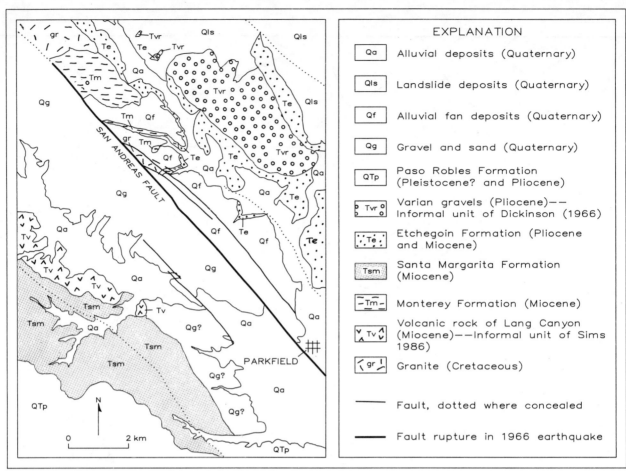

FIGURE 67. Geologic Map of the Volcanic Rocks of Lang Canyon and Adjacent Area (after Sims, in press).

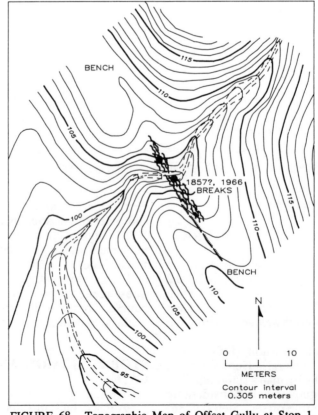

FIGURE 68. Topographic Map of Offset Gully at Stop 1. After Sieh (1978b).

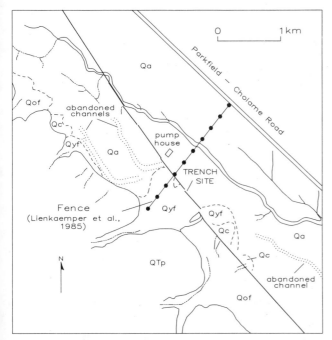

FIGURE 69. Watertank Site with Offset Fence, Scarps in Alluvial Fans and Trench Site of Sims (1987). Lithologic units: Qc, colluvium (Holcene); Qyf, younger alluvial fan deposits (Holocene); Qof, older alluvial fan deposits (Holocene); Qa, alluvium (Holocene); QTp, Paso Robles Formation (Pleistocene? and Pliocene).

Creep is currently measured at 4 ± 1 mm/yr. Nearby scarps in alluvial-fan deposits are cut by the San Andreas fault. This is also the site where a trenching study of the San Andreas was carried out by Sims (1987). The small alluvial fan here was offset 47.4 ± 2.7 m over the last $1,773 \pm 143$ cal yr B.P. This yields an average slip rate of 26.7 ± 2.6 mm/yr (Sims, 1987).

STOP 3: GOLD HILL EXOTIC BLOCK

This exotic block was separated from its parent bodies, the gabbro of Logan and at Eagle Rest Peak (FIG. 64).

STOP 4: EXOTIC BLOCK OF VOLCANIC ROCKS

The exotic block of volcanic rocks of Lang Canyon (FIG. 65) is composed of flow banded rhyolite, rhyolitic agglomerate, obsidian and perlitic obsidian. These rocks are identical in age, composition, and general stratigraphy to the Pinnacles Volcanics 165 km to the northwest and to the Neenach Volcanics 140 km to the southeast. The unit is fault bounded and unconformably overlain by friable quartz sandstone of the upper Miocene Santa Margarita Formation. On the

west side of the synclinal axis, the Santa Margarita Formation is composed of coarse conglomerate containing clasts of flow-banded rhyolite, purple amygdaloidal andesite, and granite.

STOP 5: PARKFIELD BRIDGE

This bridge crosses Little Cholame Creek about one-half kilometer south of the hamlet of Parkfield. The bridge, constructed in April, 1932, of a steel framework of I-beam truses that rest on seven pairs of vertical I-beams or bents with a wooden deck, was replaced by the present concrete deck in 1960. The alinement of the bridge beams is offset by displacement on the main fault trace. According to a survey in 1985 the beams are offset 76 cm which represent 118 ± 2 cm of slip along the strike of the fault since its construction in 1932 (J. J. Lienkaemper, written commun., 1988).

Proceed through Parkfield up Little Cholame Creek between Middle Mountain and Table Mountain. As the road winds up the hill, notice mounds and masses on the hillsides of isolated, exotic blocks of red-brown chert, metabasalt, and blueschist in greywacke matrix of the Franciscan Complex.

OPTIONAL STOP 6: TABLE MOUNTAIN SERPENTINITE EXTRUSION

Table Mountain is a remarkably smooth and narrow ridge capped by a subhorizontal sheet of serpentinite breccia of the Franciscan Complex. The breccia is weakly foliated and composed of massive serpentinized peridotite in a matrix of crushed serpentinite. The breccia extruded by plastic flow from steeply dipping feeders that tap its parent body at depth in a complex anticline that makes up the core of this part of the southern Diablo Range (Dickinson, 1966b). These feeders probably developed in the compressive stress regime that has dominated the tectonics of this area since Miocene time. In particular the increase in strain during the latest phase of movement on the San Andreas fault in Pliocene and Pleistocene time either initiated or reactivated the feeders. The breccia moves downslope from the crest of Table Mountain in large, lobate, landslide-like masses and are mapped on the lower slopes as landslides. Pleistocene fan deposits in the Cholame Valley and Parkfield areas postdate the serpentinite extrusion, because they contain abundant serpentinite debris, and in some outcrops, felted masses of asbestos minerals are found as single laminae.

Continue on the gravel road over the summit of the central Diablo Range to Highway 198, and then continue east to Coalinga.

COALINGA TO HOLLISTER

ANTICLINE RIDGE - PINNACLES NATIONAL MONUMENT
FAULT CREEP - HOLLISTER

The route proceeds through Coalinga, site of a major earthquake in 1983, through Priest Valley, and over the Diablo Range to rejoin the San Andreas fault. State Highway 25 follows the creeping segment of the fault and crosses it several times from the foot of Mustang Grade to Hollister. On a brief sidetrip to the Pinnacles National Monument, a distinctive suite of Miocene volcanic rocks is seen which correlate with those at Lang Canyon near Parkfield, and with the Neenach Volcanics farther southeast near Palmdale.

THE 1983 COALINGA EARTHQUAKE

Prior to 1983, seismologists gave little thought to the possibility of major earthquakes in the San Joaquin Valley, especially on faults that do not reach the surface. The prevailing opinion seemed to be that active faults would also have been sufficiently active in recent geologic time that they would be expressed physiographically by readily evident scarps and landforms that characterize active faults elsewhere in California. Geologists were certainly aware that the folds along the west edge of the San Joaquin Valley are quite young and may be growing today. Petroleum industry geologists were also aware of the presence of blind, fold-propagation faults in the cores of some of the folds. But the connection between active folding and activity on the faults in the cores of those folds escaped those scientists entrusted with defining earthquake hazards in California. Thus, the earthquake of May 2, 1983 (M 6.4), which was centered 12 km beneath Anticline Ridge, 35 km northwest of the San Andreas fault, came as a "shock" to more than just the good people of Coalinga.

Coalinga was known as Coaling Station A on the branch line that the Southern Pacific Railroad built into the district in 1888 after lignite deposits were widely promoted as great coal seams. The "Coalinga coal boom" petered out like all other California coal booms and was replaced in later years by a more substantial petroleum boom, and the name of the town was changed to the more euphonious Coalinga (Gudde, 1969). Coalinga is located in the Pleasant Valley syncline at the southern end of the Diablo Range. This northwest-trending range of mountains is sharply bounded on the northeast by the San Joaquin Valley and on the southwest by the San Andreas and Calaveras faults. Anticline Ridge is northeast of the syncline and is a major structural trap for petroleum.

Seismic reflection and refraction data suggest that northeast-dipping reverse faults and a southwest-dipping thrust fault lie beneath the anticline. Most of the focal-mechanism studies prefer movement on one of the steep reverse faults, because one of the nodal planes closely fits their orientations, however, geodetic data are compatible with both the reverse faults and the southwest-dipping thrust fault. The lack of a surface rupture with the main shock is explained by surficial folding which accompanied the subsurface faulting. Minor surficial faulting occured with a M 5.2 aftershock on a normal fault five weeks after the mainshock.

Drive southwest then northwest from Coalinga up Waltham Canyon on State Highway 198 past Curry Mountain to Priest Valley, thence westward across the Diablo Range via Mustang Ridge to State Highway 25.

EN ROUTE: PRIEST VALLEY AND MUSTANG RIDGE

Waltham Canyon cuts an unusually straight course parallel to the San Andreas fault through Pliocene marine strata in Waltham Canyon, but no fault has been mapped.

Mustang Ridge is underlain by ophiolite of the Franciscan Complex. Exposures of graywacke, red chert, and serpentine are prevalent in roadcuts, and hummocky topography typical of landslides is ubiquitous. Exposures and landslides are best observed on the downgrade into Bitterwater Valley.

At the base of Mustang Grade, turn right on State Highway 25. The Bitterwater oil field is about 5 km beyond the intersection of Highway 25 and roadway G13 which goes to King City.

EN ROUTE: BITTERWATER OIL FIELD

The Bitterwater oil field may be the smallest oil field along the San Andreas fault. Only two small pumpers are visible from the highway. The discovery well was completed in 1952 from a total depth of 600 m for 52 bbls/day of 26° gravity oil. The deepest well was drilled to a total depth of 2000 m and bottomed in granite, however, Franciscan rocks are exposed in the San Andreas fault zone only 400 m away. Production is from upper Miocene arkose in a faulted anticlinal trap. Since its discovery to 1986 it has produced about 330,000 bbls.

Turn southeast about 500 m east of Bitterwater oil field, drive parallel to the San Andreas fault about 1 km to a point where the road bends eastward into a canyon. The entrance into Flook Ranch is on the west side of the road at the bend in the road.

STOP: FAULT CREEP IN BITTERWATER VALLEY

This stop at the Flook Ranch offers the opportunity to photograph an offset road and an 80 year-old fence line on each side of the road, which demonstrate that aseismic creep here has averaged 35 mm/yr (Hay, et al., this volume), congruent with geodetically determined rates in this creeping segment of the San Andreas fault during the last 20 years.

The site is an ideal place to determine fault slip rates, because the deformation zone is only a few meters wide, and it cuts an active alluvial fan whose Holocene alluvial deposits are incised by a modern channel believed to be about 100 years old. Its banks are dextrally offset a minimum of 3.3 m, yielding a slip rate of 34 mm/yr. Offset paleo-channels exposed in trenches excavated across the fault yielded a slip rate of 28 mm/yr for the last 1000 years

The deformation zone near the fence is expressed as a narrow, shallow depression which, about 100 m to the northwest, transforms into a north-facing scarp about 5 m high. Although the tectonic landforms are smoothed by plowing, they are still readily recognizable.

EN ROUTE: THE SAN ANDREAS FAULT ALONG STATE HIGHWAY 25

The route follows several northwest-trending valleys which parallel the San Andreas fault and other faults of the San Andreas fault system. Stops of opportunity will be made to study en echelon fractures in asphalt pavement which are manifestations of the active fault creep that characterizes the segment of the San Andreas fault between Cholame and San Juan Bautista.

En Route: Little Rabbit Valley

Hills west of the highway consist of upper Miocene arkose that probably correlates with the producing formation in the Bitterwater oil field. Hills to the east consist of undifferentiated Pliocene sandstones.

En Route: Bear Valley

The Bear Valley and San Andreas faults strike parallel to Highway 25, separating a sliver of Plio-Pleistocene nonmarine rocks from white outcrops of upper Miocene arkose on the west, equivalent to the producing sandstone in the Bitterwater oil field, from middle and lower Pliocene marine strata on the east. Both of the faults are active as shown by frequent, small earthquakes in this area.

OPTIONAL SIDETRIP: PINNACLES NATIONAL MONUMENT

The Pinnacles area is a picturesque region of jagged, cavernous, volcanic peaks composed of Late Miocene rhyolitic breccia, agglomerate, flow-banded lava, tuff, and obsidian intrusions and flows, with minor andesite and basalt flows. These rocks were intruded in and extruded above a wide range of granitic rocks containing roof pendants of biotite schists and marble of the Salinian block. The suite of volcanic rocks in the Pinnacles on the southwest side of the San Andreas fault have been correlated with the lithologically and temporally identical volcanic rocks of Lang Canyon near Parkfield (Sims, this volume) and the Neenach Volcanics (Matthews, 1976).

OPTIONAL STOP: FAULT CREEP AT THE CIENEGA WINERY

The route proceeds up a valley which, before being drained, was the "cienega" or swamp for which the land grant "Cienega del Gabilan" (Swamp of the Hawk) was derived. A remnant of the original swamp is present north of the winery. The swamp and others like it in Cienega Valley are sag ponds in the fault zone.

The Cienega-Almaden winery is the site of a plaque proclaiming the San Andreas fault as Registered National Landmark. Look for it on an exterior brick wall of the tasting room. The main building of the winery, which houses the fermenting tanks, is located directly on an active trace of the San Andreas fault (FIG. 70) It was constructed in 1948 and has been offset progressively at a rate of about 12 mm/yr since 1958. The fault damage was first noticed in 1956 when concrete floor slabs and walls had been dextrally offset nearly 100 mm. Creepmeters show that most of the displacement occurs as "creep events" lasting from several days to a week.

Two earthquakes, M 5.6 and 5.5 occurred in Cienega Valley on 8 April 1961, causing major damage to the winery buildings. A 15 m-long fissure was observed about 5 km from the winery.

Several springs exist on the active fault trace north and south of the winery building, and a series of planks covering one of them has been rotated about 15° counterclockwise, analogous to the way Luyendyk et al. (1985) visualized that the Transverse Ranges rotated in simple shear in southern California. Drainage ditches and vineyard rows are also dextrally offset along the fault trace.

Proceed to the town of Hollister. Park on Eighth Street and search streets between Eighth and Fourth Streets and between Powell and West for curb offsets and sidewalk bends.

STOP: FAULT CREEP IN HOLLISTER

The City of Hollister is situated on the south end of the Calaveras fault near its junction with the San Andreas fault (FIG. 71). The surface trace of the Calaveras fault can be readily discerned by mapping dextral bends and offsets in concrete curbs, sidewalks, street slabs, walls, and building foundations. Look for

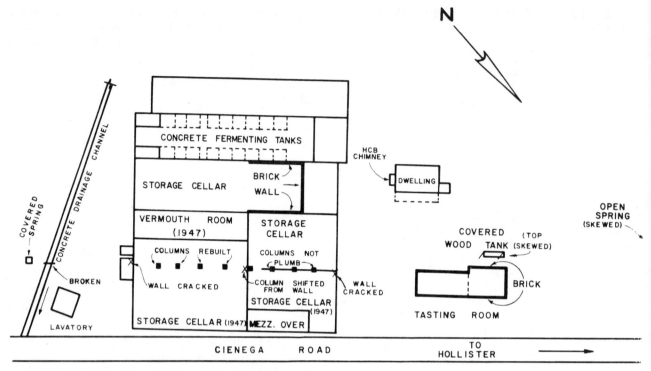

FIGURE 70. Site Map of Cienega-Almaden Winery Storage Cellar and Associated Features on Active Trace of the San Andreas Fault. From Tocher and Nason (1967).

narrow zones of en echelon fractures which are prevalent in asphalt street surfaces over the fault trace. Notice how many sidewalks are buckled, and several curbs are rotated or pushed out into the street at corners and at house walkways where sidewalks are oriented parallel to the shortening direction.

The slip rates are variable in time and space (Rogers and Nason, 1971). Movement did not occur between 1910 and 1929, judging from the amount of offset in two sidewalks that were laid in 1910 and 1929, and in a pipeline laid in 1929. Creep commenced sometime after 1929 and averaged 8 mm/yr. Between 1961 and 1967, the slip rate was about 15 mm/yr. Since 1979 two sites have been monitored in Hollister, one showing 6.6 mm/yr and the other, only 2.3 km to the northwest, creeps 12 mm/yr - the fastest rate of movement measured across any fault in the San Francisco Bay region (Galehouse, this volume).

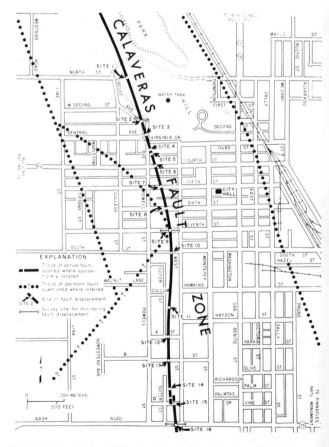

FIGURE 71. Map of Calaveras Fault Zone in City of Hollister. From Rogers and Nason (1971).

HOLLISTER TO SAN FRANCISCO

SAN JUAN BAUTISTA FAULT SCARP - CALAVERAS FAULT - HAYWARD FAULT SAN ANDREAS LAKE - DALY CITY - SAN FRANCISCO

The route proceeds from San Juan Bautista across "Silicon Valley" to examine the Calaveras and Hayward faults on the east side of San Francisco Bay. Then it crosses the Bay to the peninsula to view the long narrow lakes along the San Andreas fault. The route continues to Daly City where the fault goes out to sea. The day concludes with a spectacular view of San Francisco and its geological and environmental setting from Telegraph Hill.

STOP: FAULT SCARP AT SAN JUAN BAUTISTA

The pleasant little town of San Juan Bautista sits on a low hill overlooking San Juan Valley (FIG. 72). The northeast side of the hill is a 15-m high fault scarp along the San Andreas fault; the rodeo grandstand east of the mission is built on the face of the scarp. The first church was established on the scarp by the Spanish in 1797, but it was extensively damaged by earthquakes. The present structure was completed in 1812. Parts of the outer walls collapsed in 1906, and although the remaining structure was braced with concrete buttresses and other supports, the broken walls were not restored and can still be seen.

San Juan Bautista lies at the southeastern terminus of the surface rupture of the 1906 earthquake and near the northwest end of the creeping segment (FIG. 3). Some seismologists argue that San Juan Bautista was the site of the initial rupture of the 1906 earthquake and that it propagated northwestward. Others argue equally strongly that the earthquake initiated in the center of the ruptured segment, and still others say that it propagated southeastward from an initiation point at the northwest terminus of the surface rupture.

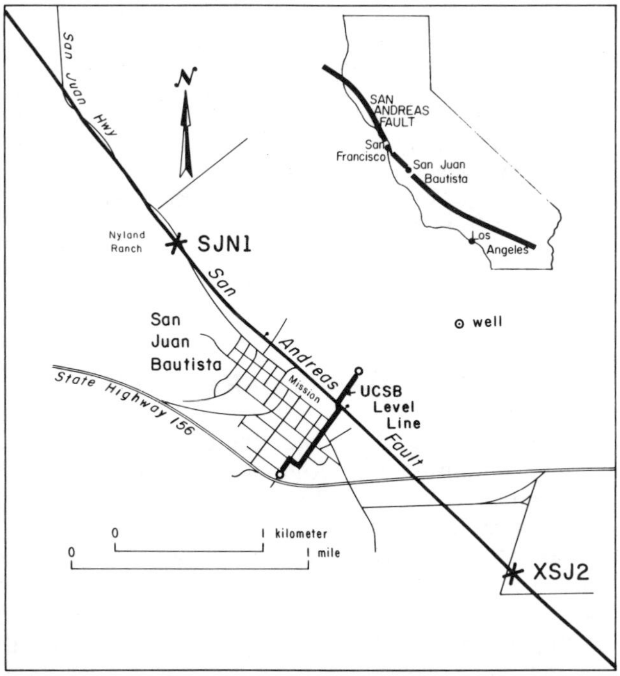

The low hill and the town are underlain by granite of the Salinian block that has been squeezed up alongside of the San Andreas fault in a segment where its creep rate abruptly diminishes from 12 mm/yr southeast of the town to 7 mm/yr two kilometers to the northwest at Nyland Ranch. These rates are determined by U.S. Geological Survey creepmeters and by periodic resurveys of aligned arrays of nails that were emplaced in 1967. The height fault scarp relative to San Juan Valley is apparently increasing at a rate of 10 mm/yr; however, it is more likely that the height change is due to nontectonic subsidence of the valley rather than to vertical tectonic fault creep, because the water table has been dropping substantially in response to withdrawal of groundwater for irrigation.

Between San Juan Bautista and the south end of the San Francisco Peninsula, the San Andreas fault traverses a chain of low, but rugged and heavily vegetated mountains. That stretch of the fault was inactive since 1906 until June 1988 when a M 5.5 earthquake occurred at the intersection of the San Andreas and Sargent faults.

Drive northwest 2 km from the center of San Juan Bautista to the Nyland Ranch which is marked by an array of white fences. Park at the driveway entrance to the ranch.

STOP: ACTIVE FAULT CREEP AT NYLAND RANCH

A line, nearly 100 m-long, of 20 nails was placed along the driveway across the 1906 trace of the San Andreas fault in 1967. Since then, it and the fences on each side of the drive way have been offset about 20 cm right laterally. The zone of slip is only about 1 m-wide.

The route proceeds from San Juan Bautista northward via U.S. Highway 101 to Morgan Hill.

FIGURE 72. Index Map of UCSB Level Line, USGS Creepmeters, San Andreas Fault, and the Town of San Juan Bautista. From Sylvester et al. (1980).

OPTIONAL STOP: SILVER CREEK FAULT AT ANDERSON RESERVOIR

Highly sheared rocks of the Franciscan Complex are well-exposed in the Silver Creek fault zone near the spillway of the Anderson Reservoir. The fault strikes nearly east-west and meets the main trace of the northwest-striking Calaveras fault north of the town of Morgan Hill. The spillway location offers excellent exposures of a major fault in the Franciscan Complex.

Continue northward on Highway 101 to San Jose, parallel to the Calaveras fault, and thence by Highway 580 to the Hayward fault.

EN ROUTE: HAYWARD FAULT

The steep front of the East Bay Hills on the east side of San Francisco Bay is caused by uplift along the Hayward and Mission faults. The Hayward fault is a major structural boundary between the Franciscan Complex on the west and two rock sequences on the east: The Great Valley Sequence of Jurassic and Cretaceous age and Miocene/Pliocene sedimentary and volcanic rocks. The Great Valley Sequence is a turbidite sequence consisting of conglomerate, sandstone, and shale deposited in a deep ocean trench, contemporaneously with but separately from the Franciscan Complex. The Miocene/Pliocene rock sequence has marine and non-marine sedimentary rocks intercalated with basaltic lava flows and rhyolitic tuff. Both rock sequences are intensely folded about northwest-trending axes (Aydin and Page, 1984).

Proceed to the center of the city of Hayward to A, B, C, and D streets.

STOP: ACTIVE FAULT CREEP IN HAYWARD

Surface movement occurred along the Hayward fault in 1836 and 1868, and slow creep takes place today (Galehouse, this volume). En echelon cracks are particularly well-developed in pavement where active strands of the fault extend through the city of Hayward. Curbs and building walls are offset from 5 to 10 cm across each of two active strands (FIG. 73).

Cross San Francisco Bay via high 92 to the San Francisco Peninsula, continue southeast about 15 km on Highway 101 to University Ave., turn southeast and proceed about 10 km to the community of Portola Valley.

OPTIONAL STOP: THE SAN ANDREAS FAULT IN PORTOLA VALLEY

The community of Portola Valley, named after the Spanish explorer who established the first Spanish settlement in what is now California, is unique because it has been planned and has grown around the San Andreas fault. The Alquist-Priolo Act, a statute of California state law, dictates the set-back distances for construction near known, active faults. These statutes are the basis for the plan of Portola Valley, and state building codes guide the type of construction.

Where the San Andreas fault is mapped in careful detail, it is not a single continuous trace but consists of overstepping, parallel or en echelon segments having characteristic lengths of from 10 to 15 km, a length that is probably related to the depth to the seismogenic zone. In Portola Valley, the San Andreas fault makes a left step from the Trancos strand to the Woodside strand (FIG. 74). The 1906 surface rupture coincided with segments of fault traces having the strongest geomorphic and topographic characteristics.

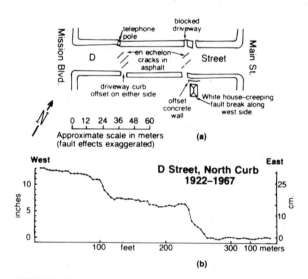

FIGURE 73. Active Creep in Hayward. a) Sketch map, made November 28, 1976, of D Street between Main Street and Madison Boulevard, showing location of offset curbs and buildings of Hayward fault zone in City of Hayward. b) Record of precisely surveyed creep displacement. The line with dots represents the position of the curb in 1967, which was placed perfectly straight along D Street in 1922. From Nason (1971) in Wahrhaftig (1984).

Proceed to Hwy 280, turn northwest and drive about 10 km to the first of three vista stops.

STOP: SAN ANDREAS LAKE OVERLOOK

The presence of a major fault through the "rift" valley containing San Andreas and Crystal Spring reservoirs was known to local geologists at the turn of the 20th century, but that it was an active fault which transects the length of California was widely recognized only after the 1906 surface rupture connected the various segments of the great fault. The segment through the reservoirs was called the San Andreas fault, and that name was applied to the entire fault after 1906.

Palou named the valley Cañada de San Andres on November 30, 1774, the feast day of Saint Andrew, and the reservoir was called San Andreas Reservoir when it was created in 1875 (Gudde, 1969).

Crystal Springs Dam was one of the largest concrete dams in the world when it was constructed in 1888. The 1906 faulting sheared the east abutment of the dam from 2 to 3 meters, according to historic records and large-scale mapping, across a 30 m-wide zone. Paleoseismic studies (Hall, 1984) showed that the 1906 fault traces have been active for much of Holocene time, with five 1906-type events in the last 1130 ± 160 years, yielding a minimum slip rate of 12 mm/yr and an average return frequency of 224 ± 25 years. The zone of active faulting is characterized by slow upwelling of highly sheared rock and clay gouge which, together with differential right shear and stream erosion, has produced most of the distinctive landforms in the rift valley, including closed depressions, elongate ridges, shutterridges, deflected stream courses, and ponded alluvium (Hall, 1984).

Continue northwest on highway 280 to Highway 1, turn left, drive to Skyline Blvd., turn northwest toward the refuse station. Turn east into the residential area and proceed uphill to a landslide vista.

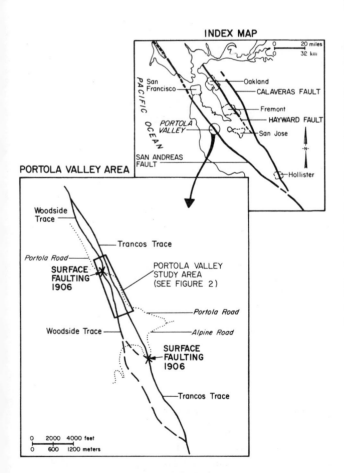

FIGURE 74. Index Map and Mapped Fault Traces in the Portola Valley Area with Reported Locations of Surface Faulting in 1906. From Taylor et al. (1980).

STOP: LANDSLIDE AND SAN ANDREAS FAULT AT DALY CITY

The San Andreas fault goes out to sea at Mussel Rock at Daly City. In contrast to Portola Valley, Daly City was developed with little if any regard for the presence of the San Andreas fault, in spite of the fact that the presence and location of the fault was clearly established by 1906 surface rupture. Ironically, Daly City was settled on John Daly's dairy pasturelands by refugees from the 1906 San Francisco earthquake and fire (Gudde, 1969).

Perhaps even more of a continual hazard to this seacliff community than the San Andreas fault are the landslides which have nipped into the residential areas and destroyed several houses in recent years. One of the largest landslides is in the San Andreas fault zone where marine Pleistocene sedimentary rocks of the Merced Formation are sheared and water-saturated.

Return to Highway 280, proceed from Daly City to San Francisco and to Telegraph Hill for an overview of The City.

STOP: TELEGRAPH HILL VISTA POINT

Coit Tower, at the top of Telegraph Hill, is a gift to the city in 1933 by Lillie Hitchcock Coit as a memorial to volunteer fire fighters whom she admired all her life. The tower is supposed to have the form of a fire hose nozzle. The view is spectacular from the top of the 70-m high tower. Murals on the ground floor were painted in the late 1930's by artists of the WPA Arts Project and show economic and social life during the Great Depression.

The present land surface beneath San Francisco is an irregular surface cut across the Franciscan Complex and veneered by Quaternary dune sand, and lagoonal, and alluvial deposits. Within the city, the Franciscan Complex may be divided into five northwest-trending, fault-bounded litho-tectonic units (FIG. 75). Most of the hills which give the city its distinctive character are exotic blocks of Franciscan sandstone, gabbro, basaltic greenstone, chert, serpentinite, and schist in a highly-sheared, argillaceous matrix. Telegraph Hill is situated on an enormous sandstone block, which was quarried extensively for fill along harbor area.

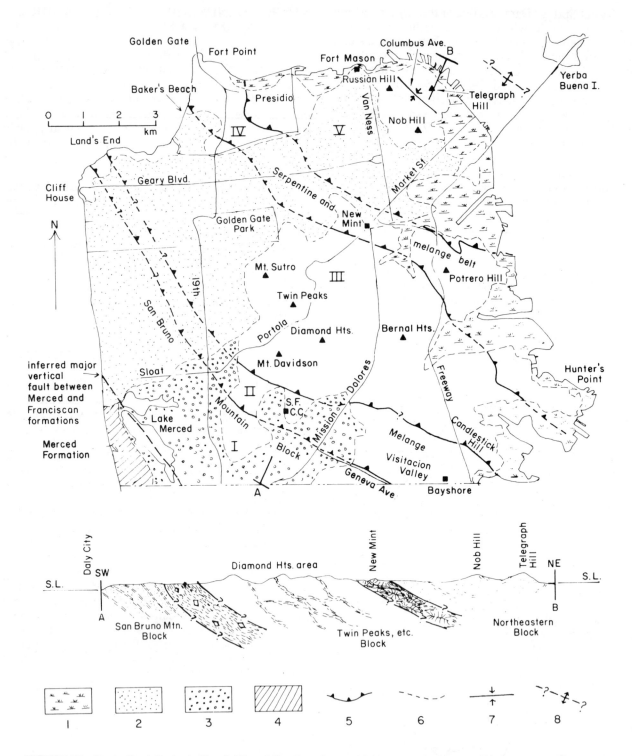

FIGURE 75. Generalized Geologic Sketch Map of San Francisco and Diagrammatic Geologic Section. 1 - filled land; 2 - area cover mostly by Holocene dune sand; 3 - area covered most by Pleistocene Colma Formation (dune sand and lagoonal and alluvial deposits); 4 - area underlain by the Plio-Pleistocene Merced Formation (blank areas have Franciscan rocks at the surface); 5 - boundary between major blocks in the Franciscan Complex (dashed where covered by surficial deposits; blocks are indicated by Roman numerals; boundaries are probably thrust faults; black triangles on down-dip side of faults); 6 - contacts of surficial (Quaternary) deposits; 7 - synclinal hinge line; 8 - inferred anticlinal hinge line. From Wahrhaftig (1984).

SAN FRANCISCO TO POINT REYES

GOLDEN GATE AND MARIN HEADLANDS - BOLINAS LAGOON
POINT REYES - TOMALES BAY

Cross the Golden Gate Bridge to the Marin Headlands to view the structural and tectonic setting of the San Francisco area. Continue via State Highway 1 to Bolinas Lagoon where the San Andreas fault comes back on shore on its route from Daly City. Examine the fault's landforms and recent earthquake history between Dogtown and Olema. Walk the Earthquake Trail at Point Reyes National Seashore Visitors' Center. Continue northwest along the shore of Tomales Bay through Inverness for a walk along Kehoe Beach to see spectacularly faulted granitic basement of the Salinian block. Return to San Francisco via a remarkable sequence of pillow basalts in the Franciscan Complex.

INTRODUCTION

North of San Francisco the San Andreas fault separates the Franciscan Complex on the North American lithospheric plate from granitic rocks of the Salinian block and its cover of upper Cenozoic sedimentary rocks in the Pacific plate. Most of the field trip route is on the Franciscan Complex. The granitic rocks and their abundant roof pendants of quartzite, schist, and marble, are exposed on Point Reyes Peninsula and on the Farallon Islands about 30 km offshore.

Point Reyes Peninsula is a regional of low, rolling grassland underlain by upper Cenozoic marine strata between two ridges of granite: the 15 km long promontory of Point Reyes itself, and Inverness Ridge.

The upper Cenozoic rocks on Point Reyes are marine Miocene and Pliocene sandstone and siltstone that have been correlated with several formations exposed at least 150 km to the southeast across the Seal Cove-San Gregorio fault. Quaternary alluvium and aeolian deposits cover both the sedimentary and granitic rocks over much of the Point Reyes Peninsula. The peninsula was broadly warped in Late Pleistocene time, causing the south end at Bolinas to be uplifted as much as 70 m and causing valleys at the north end to be drowned. The southern, coastal slope of Inverness Ridge is a large area of landslides with a full assemblage of landslide landforms.

It was this segment of the San Andreas fault between Bolinas Lagoon and Tomales Bay on which the greatest surficial displacements were recorded in 1906, averaging about 4 m, and locally exceeding 5 m. Vertical displacement was negligible. The great American geologist, Grove Karl Gilbert, spent several days here taking photographs and notes and making topographic profiles of the rupture zone. His notes were so clear that, in spite of the growth of vegetation and culture, it is easy to retrace his footsteps more than 80 years later. The sites of some of his photographs and descriptions were revisited and reevaluated in the 1980's, and trenches have been excavated across some of the fault strands at Dogtown to determine times and sizes of previous earthquakes, and to compare that record with those determined at Wallace Creek and Pallett Creek. The results of some of these studies are presented in this volume by Hall et al.

From San Francisco, drive northward on Highway 101 across the Golden Gate Bridge, take turnoff to the Golden Gate National Recreation Area. Proceed to the top of Hawk Hill for an unparalled vista of the Golden Gate.

STOP: GOLDEN GATE VISTA

If the fabled fog permits, this stop affords an incredible view of the Golden Gate, the famed bridge across it, the City of San Francisco, Alcatraz Island, and San Francisco Bay, as well as an opportunity to study the intense, chevron-folds in chert outcrops of the Franciscan Complex. Because of its strategic importance, a number of forts and fortifications were constructed around the Golden Gate more than 50 years ago, but shots were never fired from them.

The Golden Gate is a drowned river valley eroded by the ancestral Sacramento-San Joaquin River. The present channel beneath the bridge is more than 100 m deep and is famed for its strong currents.

Lower Jurassic oceanic crust is exposed here in the rugged mountains and cliffs of the Marin Headlands north of the Golden Gate. It is one of the least disrupted, most structurally coherent blocks of the Franciscan Complex anywhere in California (FIG. 76). Within the block, geologists have mapped a sequence of eight or ten imbricated, northwest-striking, southwest-dipping, thrust-bounded slabs of oceanic crust, now folded into a broad, southwest-plunging syncline. The slabs are an ophiolitic assemblage composed typically of lenticular to tabular bodies of pillow basalt, ribbon chert, sandstone, and shale. Each of the slabs is from a few to a few hundreds of meters thick and from a few hundred meters to 3 km or more long. The slabs are stratigraphically upright, and biostratigraphic studies of radiolarians in the chert show tectonic repetition of the slabs.

Paleomagnetic analyses of chert and clastic rocks in the Marin Headlands block indicate that it has rotated clockwise nearly 130°, similar to other Franciscan-aged terranes along western North America. Unlike other west coast terranes, however, it lacks a significant component of poleward displacement.

The nature of the contact between the southern

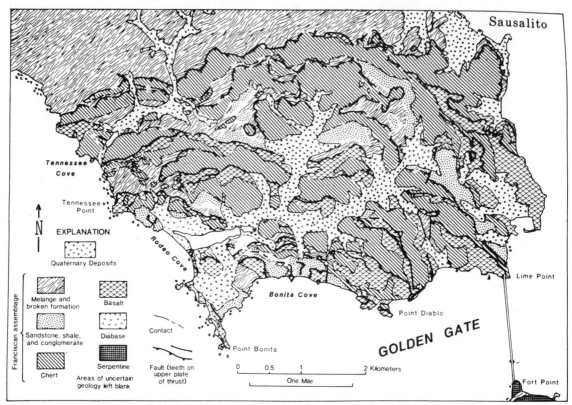

FIGURE 76. Geologic Map of Marin Headlands. From Wahrhaftig and Murchey (1987).

Marin Headlands and the San Francisco Peninsula beneath the Golden Gate has been a topic of considerable debate since 1917 when two of California's leading geologists, J. C. Branner of Stanford University and A. C. Lawson of the University of California, took opposite sides in a heated debate about the feasibility and advisability of constructing the Golden Gate Bridge. The contact must be a profound tectonic dislocation, because a sequence of east-dipping serpentine, melange, and sandstone in San Francisco does not match the west-dipping sequence in the Marin Headlands that lacks serpentine but is dominated by chert and basalt. The south pier of the bridge is on relatively low strength serpentine, whereas the north pier is on stronger basalt, chert, and sandstone. The orientation, sense and age of movement of the dislocation is unknown.

Because Pleistocene marine deposits or surfaces are not present above subaerial Sangamonian (latest interglacial) deposits that crop out along the shore south of the Golden Gate, sea level has not been higher in Quaternary time than it is now. In fact, Wahrhaftig (1984) thinks that the Marin Headlands may be tectonically subsiding, unlike much of the California coast.

The route winds through the Marin Headlands back to Highway 101, thence onto State Highway 1 which wends its way through the Franciscan Complex back to the rugged coastline at Stinson Beach and to a highway turnout that affords a fine view of Bolinas Lagoon and its bay-mouth bar.

PHOTO STOP: BOLINAS LAGOON

The San Andreas fault is about 500 m offshore from the view stop and comes onshore along the southwest edge of Bolinas Lagoon. Northwest of Bolinas, the San Andreas fault and its 1906 surface rupture follow the long, straight Olema Valley for 21 km, then along the narrow Tomales Bay which is 25 km long, before reentering the sea. Recall that the 1906 surface rupture extended 450 km from Shelter Cove 30 km northwest of here, to San Juan Bautista, 150 km southeast of here. Therefore, much of the rupture was subsea. The dextral offset at Bolinas in 1906 was about 4 m.

From San Francisco to this stop the route has been across Jurassic and Cretaceous rocks of the Franciscan Complex. But on the far side of Bolinas Lagoon, juxtaposed against the Franciscan Complex by the San Andreas fault, are yellow cliffs of soft, marine siltstone and sandstone of Pliocene age.

Oil-filled fractures and asphaltic clastic joints, typically of the Monterey Formation, stirred up enough enthusiasm in Bolinas in 1865 to build the Petroleum Hotel. Massive tar and oil sands of middle Miocene age are also present farther up the coast, but commercial oil or gas has not been found in the region.

Continue northwest on Highway 1.

EN ROUTE: OLEMA VALLEY AND THE 1906 RUPTURE ZONE

The 21 km-long stretch of Olema Valley separates the Point Reyes Peninsula from "mainland" California, and is the locus of the central part of the subaerial rupture of the 1906 earthquake (Hall et al., this volume). The valley is a "rift valley" eroded along the San Andreas fault zone between the resistant granitic rocks of Inverness Ridge to the southwest and the Franciscan Complex beneath Bolinas Ridge northeast of the fault. The ground rupture passed through several ranches, offsetting lines of trees, fences, and roads as much as 4.3 m. One of the ranchers was milking a cow at the time of the earthquake, and a crack developed 2 m from him. In the milking yard of the Skinner Ranch, G. K. Gilbert noted that "men and cows were thrown downhill in a heap".

The fault trace is well-marked in Olema Valley by scarps, closed depressions, sag ponds, benches, elongate ridges, linear valleys, and offset stream courses (Hall, et al., this volume). About half way between Bolinas and Olema village, the topography is so affected by the San Andreas fault that two parallel streams, 300 m apart, actively erode along strands of the San Andreas fault zone, but they flow in opposite directions. The eastern one flows northwest and empties into Tomales Bay; the western one flows southeast and empties into Bolinas Lagoon.

Proceed to Olema and the visitor's center for Point Reyes National Seashore.

STOP: EARTHQUAKE TRAIL

The Point Reyes National Seashore is a 26,000 hectare wilderness preserve, covering most of the Point Reyes Peninsula, created by an act of Congress in 1962. It is a magnificent national treasure because of its spectacular scenery, natural history, and unique geology. It is primarily a "trail park" with more than 250 km of hiking trails through woods, across rolling grasslands, and along sandy beaches.

The self-guiding, 1 km-long Earthquake Trail is the only outdoor exhibit in California that focuses on the San Andreas fault (FIG. 77). It was created by Professor Tim Hall and his students in 1972 and preserves the location and some of the dextral offsets related to the 1906 earthquake where those offsets reached a maximum of 5 m.

Proceed northwest from the Point Reyes visitor's center to the hamlet of Inverness on the edge of Tomales Bay. Then drive west across Inverness Ridge and follow the road signs that lead to McClure's Beach. Park at the trailhead to Kehoe Beach and walk to and north along the shore about 600 m to the Kehoe fault.

STOP: KEHOE BEACH

A magnificent outcrop of a forced, or drape fold is exposed in the 100 m-high seacliff on Kehoe Beach. The basement rocks are composed of highly-fractured, gray granodiorite and granite complexly intruded by leuco-adamellite and pegmatite. Miocene sandstone and siltstone nonconformably overlie the granitic rocks.

The stratified rocks are folded and downthrown to the south over the faulted edge of the basement in the style of a drape fold, except that close inspection of the nearly vertical fault surface reveals the presence of horizontal slickensides. The fault strikes east-west at a high angle to the San Andreas fault, parallel to the steep, south face of the Point Reyes promontory, which may be a fault-line scarp. Both are oriented properly to be an R' shear in a right-simple shear scheme, and the slickensides along the Kehoe fault indicate that the most recent slip is sinistral, which is the appropriate sense of slip for an R' fault in right simple shear.

Measured rates of cliff retreat on the Point Reyes Peninsula range from 0.3 to 0.8 m/yr and vary according to the type of rocks being attacked by powerful storm waves.

Return to San Francisco via Lucas Valley and San Rafael.

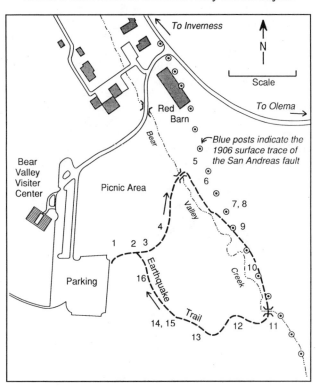

FIGURE 77. Earthquake Trail at Point Reyes National Seashore. Key to trail site numbers:
1) Start of trail;
2) Introduction panel;
3) Directional sign;
4) 1906 earthquake photos;
5) San Andreas fault panorama;
6) Can the San Andreas fault swallow cities?
7) Fence offset in 1906;
8) Fractured barn;
9) Rock exhibit;
10) Continental drift and moving plates;
11) Continental drift timetable;
12) Plate boundary;
13) Small quakes won't stop big ones
14) How the geology performed during 1906
15) Active faults in the San Francisco Bay area
16) Earthquake preparedness

REFERENCES CITED

Addicott, W. O., Mid-Tertiary zoogeographic and paleogeographic discontinuities across the San Andreas fault, California, pp. 144-165, *in* Dickinson, W. R., and A. Grantz, *eds.*, Proceedings of Conference on Geologic Conference of the San Andreas Fault System, Stanford Univ. Pubs. Geol. Sci. 11, 374 p., 1968.

Advocate, D. M., J. C. Crowell, and M. N. Link, Road log and field trip guide for Ridge basin, southern California, pp. 277-291, *in* Crowell, J. C., and M. N. Link, *eds.*, Geologic History of Ridge Basin, Southern California, Pacific Section, Soc. Econ. Paleontologists and Mineralogists Guidebook, Bakersfield, 304 p., 1982.

Allen, C. R., The tectonic environments of seismically active and inactive areas along the San Andreas fault system, pp. 70-82, *in* Dickinson, W. R., and A. Grantz, *eds.*, Proceedings of Conference on Geologic Conference of the San Andreas Fault System, Stanford Univ. Pubs. Geol. Sci. 11, 374 p., 1968.

Allen, C. R., The modern San Andreas fault, pp. 511-534, *in* Ernst, W. G., *ed.*, The Geotectonic Development of California, Prentice-Hall Inc., 706 pp., 1981.

Allen, C. R., M. Wyss, J. N. Brune, A. Grantz, and R. E. Wallace, Displacements on the Imperial, Superstition Hills and the San Andreas faults triggered by the Borrego Mountain earthquake, U.S. Geol. Surv. Prof. Paper 787, 87-104, 1972.

Atwater, T., Implications of plate tectonics for the Cenozoic tectonics of western North America, Geol. Soc. Am. Bulletin 81, 3513-3536, 1970.

Atwater, T., Plate tectonic history of the northeast Pacific and western North America, *in* Winterer, E. L., D. M. Hussong, and R. W. Decker, *eds.*, DNAG Vol. N: The Eastern Pacific Ocean and Hawaii, Geol. Soc. Am., 1989.

Atwater, T., and P. Molnar, Relative motion of the Pacific and North American plates deduced from sea-floor spreading in the Atlantic, Indian, and south Pacific oceans, pp. 136-148, *in* Kovach, R. L. and A. Nur, *eds.*, Proceedings of Conference on Tectonic Problems of the San Andreas Fault System, Stanford Univ. Pubs. Geol. Sci. 13, 494 p., 1973.

Avé Lallemant, H. G., and J. S. Oldow, Early Mesozoic southward migration of Cordilleran transpressional terranes, Tectonics 7, 1057-1075, 1988.

Aydin, A., and B. M. Page, Diverse Pliocene-Quaternary tectonics in a transform environment, San Francisco Bay region, California, Geol. Soc. Am. Bulletin 95, 1303-1317, 1984.

Aytun, A., Creep measurements in the Ismetpasa region of the North Anatolian fault zone, *in* Isakara, A. M., and A. Vogel, *eds.*, Multidisciplinary Approach to Earthquake Prediction 2, Friedrich Vieweg & Sohn, Braunschweig/Weisbaden, 279-292, 1982.

Babcock, E. A., Geology of the northeast margin of the Salton Trough, Salton Sea, California, Geol. Soc. Am. Bull. 85, 321-332, 1974.

Bakun, W. H., M. M. Clark, R. S. Cockerham, W. L. Ellsworth, A. G. Lindh, W. H. Prescott, A. F. Shakal, and P. Spudich, The 1984 Morgan Hill, California, earthquake, Science 225, 288-291, 1984.

Bakun, W. H., and A. G. Lindh, The Parkfield, California, earthquake prediction experiment, Science 229, 619-624, 1985.

Bakun, W. H., and T. V. McEvily, Recurrence models and Parkfield, California, earthquake, J. Geophys. Res. 89, 3051-3058, 1984.

Bakun, W. H., and many others, Parkfield, California, Earthquake predictions scenarios and response plans, U.S. Geol. Surv. Open-File Rept., 87-192, 59 p., 1987.

Barrows, A. A., A review of the geology and earthquake history of the Newport-Inglewood structural zone, southern California, Calif. Div. Mines & Geol. Spec. Rept. 114, 115 p., 1974.

Bennett, J. H., and R. W. Sherburne, *eds.*, The 1983 Coalinga, California, earthquakes, Calif. Div. Mines & Geol. Spec. Pub. 66, 335 p., 1983.

Biddle, K. T., and N. Christie-Blick, *eds.*, Strike-slip Deformation, Basin Formation, and Sedimentation, Soc. Econ. Paleontologists and Mineralogists Spec. Pub. 37, 375-385, 1985.

Bilham, R., and P. Williams, Sawtooth segmentation and deformation processes on the southern San Andreas fault, California, Geophys. Res. Lett. 12 (9), 557-560, 1985.

Blake, M. C., Jr., R. H. Campbell, T. W., Dibblee, Jr., D. G. Howell, T. H. Nilsen, W. R. Normark, J. G. Vedder, and E. A. Silver, Neogene basin formation in relation to plate-tectonic evolution of San Andreas fault system, California, Amer. Assoc. Petrol. Geol. Bull. 62, 344-372, 1978.

Bolt, B. A., and R. D. Miller, Catalog of earthquakes in northern California and adjoining areas: 1910-1972, Spec. Pub. Seismographic Stations, Univ. Calif., Berkeley, 1975.

Boyle, M. W., Stratigraphy, sedimentation, and structure of an area near Point Arena, California, unpublished M. S. thesis, Univ. Calif., Berkeley, 1967.

Brown, R. D., Jr., and E. W., Wolfe, Map showing recently active breaks along the San Andreas fault between Point Delgada and Bolinas Bay, California, scale 1:24,000, Geol. Invest. Map I-692, U.S. Geol. Surv., Washington, D.C.,1972.

Brown, R. D., and R. E. Wallace, Current and historic fault movement along the San Andreas fault between Paicines and Camp Dix, California, *in* Dickinson, W. R., and A. Grantz, *eds.*, Proceedings of Conference on Geologic Conference of the San Andreas Fault System, Stanford Univ. Pubs. Geol. Sci. 11, 22-41, 1968.

Brown, R. D., Jr., J. G. Vedder, R. E. Wallace, E. F. Roth, R. F. Yerkes, R. O. Castle, A. O. Waananen, R. W. Page, and J. P. Eaton, The Parkfield-Cholame, California, earthquakes of June-August 1966: Surface geologic effects, water-resources aspects, and preliminary seismic data, U.S. Geol. Surv. Prof. Paper 579, 66 p., 1967.

Buising, A. V., Contrasting subsidence histories, northern and southern proto-Gulf of California: Implications for proto-Gulf tectonic models, Cordilleran Sec. Geol. Soc. Am. Abstr. with Progs. 20, 146-147, 1988.

Burford, R. O., and P. W. Harsh, Slip on the San Andreas fault in central California from alinement array surveys, Seism. Soc. Am. Bull. 70, 1233-1262, 1980.

Byerlee, J. D., and D. A. Lockner, An alternative explanation for borehole breakouts in a strike-slip regime, Eos Trans. AGU 69, p. 1455, 1988.

Calderone, G., and R. F. Butler, Paleomagnetism of Miocene volcanic rocks from southwestern Arizona: Tectonic implications, Geology 12, 627-630, 1984.

Castle, R. O., Leveling surveys and the Southern California uplift, Earthquake Inform. Bull. 10, 88-92, 1978.

Castle, R. O., Bulge bashing: Modern sophistry in action, Eos Trans. AGU 68, 1506, 1987.

Castle, R. O., J. P. Church, and M. R. Elliott, Aseismic uplift in southern California, Science 192, 251-253, 1976.

Champion, D. E., D. G. Howell, and C. S. Grommé, Paleomagnetic and geologic data indicating 2,500 km of northward displacement for the Salinian and related terranes, J. Geophys. Res. 89, 7736-7752, 1984.

Cheadle, M. J., B. L. Czuchra, T. Byrne, C. J. Ando, J. E. Oliver, L. D. Brown, S. Kaufman, P. E. Malin, and R. A. Phinney, The deep crustal structure of the Mojave Desert, California, from COCORP seismic reflection data, Tectonics 5, 293-320, 1986.

Coffman, J. L., C. A. von Hake, and C. W. Stover, Earthquake History of the United States, U.S. Dept. Comm. Pub. 41-1, Boulder, Colorado, 1982.

Cohn, S. N., C. R. Allen, R. Gilman, and N. R. Goulty, Pre-earthquake and post-earthquake creep on the Imperial fault and the Brawley fault zone, in The Imperial Valley, California, Earthquake of October 15, 1979, U.S. Geol. Surv. Prof. Paper 1254, 161-168, 1982.

Cotton, W. R., N. T. Hall, and E. A. Hay, Holocene behavior of the San Andreas fault at Dogtown, Point Reyes National Seashore, California, Final Tech. Rept. for U.S. Geol. Surv. Contract No. 14-08-0001-19841, 1982.

Cotton, W. R., N. T. Hall, and E. A. Hay, Holocene behavior of the Hayward-Calaveras fault system, San Francisco Bay area, California, Final Tech. Report for U.S. Geol. Surv. Contract No. 14-08-0001-20555, 1986.

Crowell, J. C., Probable large lateral displacement on the San Gabriel fault, southern California, Am. Assoc. Petrol. Geol. Bull. 36, 2026-2035, 1952.

Crowell, J. C., The San Andreas fault in southern California, Int. Geol. Congress, XXI Session, Norden, Part XVIII, 45-52, 1960.

Crowell, J. C., Displacement along the San Andreas fault, California, Geol. Soc. Am. Spec. Paper 71, 61 p, 1962.

Crowell, J. C., Movement histories of faults in the Transverse Ranges and speculations on the tectonic history of California, in Dickinson, W. R., and A. Grantz , eds., Proceedings of Conference on Geologic Conference of the San Andreas Fault System, Stanford Univ. Pubs. Geol. Sci. 11, 374 p., 1968.

Crowell, J. C., Sedimentation along the San Andreas fault, California, in Dott, R. H., Jr., and R. H. Shaver, eds., Modern and Ancient Geosynclinal Sedimentation, Soc. Econ. Paleontologists and Mineralogists Spec. Pub. No. 19, 292-303, 1974a.

Crowell, J. C., Origin of Late Cenozoic basins in southern California, in Dickinson, W. R., ed., Tectonics and Sedimentation, Soc. Econ. Paleontologists and Mineralogists Spec. Pub. No. 22, 190-204, 1974b.

Crowell, J. C., The San Andreas fault between Carrizo Plains and Tejon Pass, southern California, pp. 223-233, in Crowell, J. C., ed., San Andreas Fault in Southern California, A Guide to San Andreas Fault from Mexico to Carrizo Plain, Calif. Div. Mines & Geol. Spec. Rept. 118, 272 p., 1975.

Crowell, J. C., Implications of crustal stretching and shortening of coastal Ventura basin, California, pp. 365-382, in Howell, D. G., ed., Aspects of the Geologic History of the California Continental Borderland, Pacific Sec., Am. Assoc. Petrol. Geol. Misc. Pub. 24, 561 p., 1976.

Crowell, J. C., The San Andreas fault system through time, Quart. Jour. Geol. Soc. London 136, 293-302, 1979.

Crowell, J. C., The tectonics of Ridge basin, southern California, pp. 25-41, in Crowell, J. C., and M. H. Link, eds., Geologic History of Ridge Basin, Southern California, Pac. Section, Soc. Econ. Paleontologists and Mineralogists Guidebook, Bakersfield, 304 p., 1982a.

Crowell, J. C., The Violin Breccia, Ridge basin, southern California, pp. 89-97, in Crowell, J. C., and M. H. Link, eds., Geologic History of Ridge Basin, Southern California, Pac. Section, Soc. Econ. Paleontologists and Mineralogists Guidebook, Bakersfield, 304 p., 1982b.

Crowell, J. C., The recognition of transform terrane dispersion within mobile belts, pp. 51-61, in Howell, D. G., ed., Tectonostratigraphic Terranes of the Circumpacific Region, Circum-Pacific Council for Energy and Mineral Resources, Houston, Texas, 585 p., 1985.

Crowell, J. C., Geologic history of the San Gabriel fault, central Transverse Ranges, Kern, Los Angeles, and Ventura Counties, California, Calif. Geol., 276-281, 1986.

Crowell, J. C., Late Cenozoic basins of onshore southern California: Complexity is their hallmark of tectonic history, Chapter 9, pp. 207-241, in Ingersoll, R. V., and W. G. Ernst, eds., Cenozoic Development of Coast California, (Rubey Volume VI), Prentice-Hall, Inc., Englewood Cliffs, New Jersey, 496 p., 1987.

Crowell, J. C., and J. W. R. Walker, Anorthosite and related rocks along the San Andreas fault, southern California, Univ. Calif. Pubs. Geol. Sci. 40, 219-288, 1962.

Crowell, J. C., and M. H. Link, eds., Geologic History of Ridge Basin, Southern California, Pac. Section, Soc. Econ. Paleontologists and Mineralogists Guidebook, Bakersfield, 304 p., 1982.

Crowell, J. C., and A. G. Sylvester, Introduction to the San Andreas-Salton Trough juncture, pp. 1-14, in Crowell, J. C., and A. G. Sylvester, eds., Tectonics of the Juncture between the San Andreas Fault System and the Salton Trough, Southeastern California - A Guidebook, Dept. Geol. Sci., Univ. California, 193 p., 1979.

Curtis, G., J. Evernden, J. Lipson, Age determination of some granitic rocks in California by the potassium-argon method, Calif. Div. Mines & Geol. Spec. Rept. 54, 16 pp., 1958.

Davis, G. A., J. L. Anderson, E. G. Frost, and T. J. Shackelford, Mylonitization and detachment faulting in the Whipple-Buckskin-Rawhide Mountains terrane, southeastern California and western Arizona, pp. 79-129, in Cordilleran Metamorphic Core Complexes, Geol. Soc. Am. Mem 153, 1980.

Davis, T. L., Late Cenozoic structure and tectonic history of the western "Big. Bend" of the San Andreas fault and adjacent San Emigdio Mountains, Ph.D disser., 580 pp., Univ. Calif., Santa Barbara, July 1983.

Davis, T. L., and E. M. Duebendorfer, Strip map of the western big bend segment of the San Andreas fault, Map 60, Geol. Soc. Am. Map and Chart Ser., Scale 1:31,682, 1987.

Dibblee, T. W., Jr., Regional geologic map of San Andreas and related faults in Carrizo Plain, Temblor, Caliente, and La Panza ranges and vicinity, California, scale 1:125,000, Geol. Invest. Map I-757, U.S. Geol. Surv., Washington, D.C.,1973.

Dibblee, T. W., Jr., Strike-slip tectonics of the San Andreas fault and its role in Cenozoic basin evolvement, pp. 26-38 in Nilsen, T. H., ed., Late Mesozoic and Cenozic Sedimentation and Tectonics in California, San Joaquin Geol. Soc., Bakersfield, 145 p., 1977.

Dickinson, W. R., Structural relationship of San Andreas fault system, Cholame Valley and Castle Mountain Range, California, Geol. Soc. Am. Bull. 77, 707-736, 1966a.

Dickinson, W. R., Table Mountain serpentinite extrusion in California Coast Ranges, Geol. Soc. Am. Bull. 77, 51-978, 1966b.

Dickinson, W. R., Plate tectonics and the continental margin of California, pp. 1-28, in Ernst, W. G., ed., The Geotectonic Development of California, Prentice-Hall Inc., 706 pp., 1981.

Ehlig, P. L., K. W. Ehlert, and B. M. Crowe, 1975, Offset of the Upper Miocene Caliente and Mint Canyon Formations along the San Gabriel and San Andreas faults, in Crowell, J. C., ed., San Andreas Fault in Southern California, Calif. Div. Mines & Geol. Spec. Rept. 118, 83-92.

Ernst, W. G., ed., The Geotectonic Development of California, (Rubey Volume I), Prentice-Hall Inc., 706 pp., 1981.

Ernst, W. G., Summary of the geotectonic development of California, pp. 601-613, in Ernst, W. G., ed., The Geotectonic Development of California, Prentice-Hall Inc., 706 pp., 1981.

Fletcher, G. L., Post late Miocene displacement along the San Andreas fault zone, central California, pp. 74-80, in Marks, J. G., chairman, Gabilan Range and Adjacent San Andreas Fault, Guidebook, Joint Pacific Section Am. Assoc. Petrol. Geol. and Pacific Section Soc. Econ. Paleontologists and Mineralogists, 110 p., 1967.

Frost, E. G., and D. L. Martin, eds., Mesozoic-Cenozoic Tectonic Evolution of the Colorado River Region, California, Arizona, and Nevada, Cordilleran Publishers, San Diego, 608 pp., 1982.

Frost, E. G., and D. A. Okaya, Continuity of exposed mylonitic rocks to middle- and lower-crustal depths within the Whipple Mountains detachment terrane, southeastern California, unpub. ms.

Fuis, G. S., W. D. Mooney, J. H. Healey, G. A. McMechan, and W. J. Lutter, Crustal structure of the Imperial Valley region, pp. 25-50, in The Imperial Valley, California, Earthquake of October 15, 1979, U.S. Geol. Surv. Prof. Paper 1254, 451 p., 1982.

Golombek, M. P., and L. L. Brown, Clockwise rotation of the Mojave Desert, Geology 16, 126-130, 1988.

Goodman, E. D., P. E. Malin, E. L. Ambos, and J. C. Crowell, The southern San Joaquin Valley as an example of Cenozoic basin evolution in California, in Price, R. A., ed., The Origin and Evolution of Sedimentary Basins and Their Energy and Mineral Resources, Am. Geophys. Un. Monograph Series, in press.

Graham, S. A., Role of the Salinian Block in the evolution of the San Andreas fault system, Am. Assoc. Petrol. Geol. Bull. 62, 2214-2231, 1978.

Greenhaus, M. R., and A. Cox, Paleomagnetism of the Morro Rock-Islay Hill complex as evidence for crustal block rotations in central coastal California, J. Geophys. Res., 84, 2393-2400, 1979.

Gudde, E. R., California Place Names, Third Edition, Univ. Calif. Press, Berkeley and Los Angeles, California 416, 1969..

Hall, C. A., Jr., Displaced Miocene molluscan provinces along the San Andreas fault, California, Univ. Calif. Pubs. Geol. Sci. 34, 281-308, 1960.

Hall, N. T., Holocene history of the San Andreas fault between Crystal Springs Reservoir and San Andreas Dam, San Mateo County, California, Seism. Soc. Am. Bull. 74, 281-300, 1984.

Harding, T. P., Newport-Inglewood trend, California - An example of wrenching style of deformation, Am. Assoc. Petrol. Geol. Bull. 57 (1), 97-116, 1973.

Harding, T. P., Tectonic significance and hydrocarbon trapping consequence of sequential folding synchronous with San Andreas faulting, San Joaquin Valley, California, Am. Assoc. Petrol. Geol. Bull. 58 (7), 356-378, 1976.

Haxel, G., and J. Dillon, The Pelona-Orocopia Schist and Vincent-Chocolate Mountain thrust system, southern California, pp. 453-470, in Howell, D. G., and K. A. McDougal, eds., Mesozoic paleogeorahy of the western United States, Soc. Econ. Paleontologists and Mineralogists, Pac. Sec., Los Angeles, Ca, 573 p., 1978.

Haukssen, E., and others, The 1987 Whittier Narrows earthquake in the Los Angeles metropolitan area, California, Science 239, 1409-1412, 1988.

Hein, J. R., Deep-sea sediment source areas: Implications of variable rates of movement between California and the Pacific plate, Nature 241, 40-41, 1973.

Higgins, C. G., Ohlson Ranch Formation, Pliocene, northwestern Sonoma County, California, Univ. Calif. Pub. in Geol. Sci., v. 36, n. 3, pp. 199-232, 1960.

Hill, M. L., Newport-Inglewood zone and Mesozoic subduction, California, Geol. Soc. Am. Bull. 82, 2957-2962, 1971.

Hill, M. L., and T. W. Dibblee, Jr., San Andreas, Garlock, and Big Pine faults, California - a study of the character, history, and tectonic significance of their displacements: Geol. Soc. Am. Bull. 64, 443-458, 1953.

Holdahl, S. R., Recomputation of vertical crustal motions near Palmdale, California, 1959-1975, Tectonophysics 97, 21-38, 1983.

Huffman, O. F., Lateral displacement of Upper Miocene rocks and the Neogene history of offset along the San Andreas fault in central California, Geol. Soc. Am. Bull. 83, 2913-2946, 1972.

Ingersoll, R. V., Tectonics of sedimentary basins, Geol. Soc. Am. Bull. 100, 1704-1719, 1988.

Ingersoll, R. V., and W. G. Ernst., eds., Cenozoic Basin Development of Coastal California, (Rubey Volume VI), Prentice-Hall, Englewood Cliffs, New Jersey, 1987.

Jackson, D. D., W. B. Lee, and C. C. Liu, Aseismic uplift in southern California: An alternate interpretation, Science 210, 534-536, 1980.

Jacoby, G. C., Jr., P. R. Sheppard, and K. E. Sieh, Irregular recurrence of large earthquakes along the San Andreas fault: Evidence from trees, Science 241, 196-199, 1988.

James, E. W., D. L. Kimbrough, and J. M. Mattinson, Evaluation of pre-Tertiary displacements on the northern San Andreas fault using U-Pb zircon dating, initial Sr and common Pb isotopic ratios, Geol. Soc. Am. Abstr. with Prog., v. 18, n. 2, p. 121, 1986.

Jennings, C. W., Geologic Map of California, scale 1:750,000, California Division of Mines and Geology, Sacramento, Ca.,1977.

Johnson, N. M., C. B. Officer, N. D. Opdyke, G. D. Woodward, P. K. Zeitler, and E. H. Lindsay, Rates of late Cenozoic tectonism in the Vallecito-Fish Creek basin, western Imperial Valley, California, Geology 11, 664-667, 1983.

Kamerling, M. J., and B. P. Luyendyk, Paleomagnetism and Neogene tectonics of the northern Channel Islands, California, J. Geophys. Res. 90, B14, 12,485-12502, 1985.

Kanter, L., and M. O. McWilliams, Rotation of the southernmost Sierra Nevada, J. Geophys. Res., 87, 3819-3830, 1982.

Kanter, L. R., and M. Debiche, Modeling the motion histories of the Point Arena and Central Salinia terranes, in Howell, D. G., ed, Tectonostratigraphic Terranes of the Circum-Pacific Region, pp. 227-238, 1985.

Keller, E. A., M. S. Bonkowski, R. J. Korsch, and R. J. Shlemon, Tectonic geomorphology of the San Andreas fault zone in the southern Indio Hills, Coachella Valley, California, Geol. Soc. Am. Bull. 93, 46-56, 1982.

Kerr, D. R., S. Pappajohn, and G. L. Peterson, Neogene stratigraphic section at Split Mountain, eastern San Diego County, California, pp. 111-123, in Crowell, J. C., and A. G. Sylvester, eds., Tectonics of the Juncture between the San Andreas Fault System and the Salton Trough, Southeastern California, A Guidebook, Univ. Calif. Santa Barbara Dept. Geol. Sci. Pub., 193 p., 1979.

Kerr, P. F., and H. G. Schenck, Active thrust faults in San Benito County, California, Geol. Soc. Am. Bull. 36, 465-494, 1925.

Lamar, D. L., and T. K. Rockwell, An overview of the tectonics of the Elsinore fault zone, pp. 149-158, in Ehlig, P. L., compiler, Neotectonics and Faulting in Southern California, Guidebook and Volume, 82nd Annual Meeting, Cordilleran Section, Geol. Soc. Am., Los Angeles, Calif., March 25-28, 1986.

Langbein, J. O., An interpretation of episodic slip on the Calaveras fault near Hollister, California, J. Geophys. Res. 86, 4941-4948, 1981.

Larsen, E. S., Jr., The batholith of southern California, pp. 25-30, Chap VII, Calif. Div. Mines Bull. 170, 1954.

Lawson, A. C., The California Earthquake of April 18, 1906, Report of the State Earthquake Investigation Commission, Carnegie Inst. Washington, Pub. 87, vol. 1 & 2, and Atlas of Maps and Seismograms, 1908.

Leopold, L. B., and J. P. Miller, Ephemeral streams - hydraulic factors and their relation to the drainage net, U.S. Geol. Surv. Prof. Paper, 282-A, 36 p., 1956.

Lienkaemper, J. J., 1857 slip on the San Andreas fault southeast of Cholame, California, Eos Trans. AGU 68, 1345, 1987.

Link, M. H., R. L. Squires, and I. P. Colburn, Slope and deep-sea fan facies and paleogeography of Upper Cretaceous Chatsworth Formation, Simi Hills, California, Am. Assoc. Petrol. Geologists Bull. 68, 850-873, 1984.

Louie, J. N., C. R. Allen, D. C. Johnson, P. C. Haase, and S. N. Cohn, Fault slip in southern California, Seism. Soc. Am. Bull. 75, 811-834, 1985.

Luyendyk, B. P., M. J. Kamerling, R. R. Terres, and J. S. Hornafius, Simple shear of southern California during Neogene time suggested by paleomagnetic declinations, J. Geophys. Res. 90 (B14), 12, 454-12, 466, 1985.

Manning, J. C., Field trip to areas of active faulting and shallow subsidence in the southern San Joaquin Valley, Natl. Assoc. Geol. Teachers, Far West Sec. Guidebook, 22 p., 1973.

Matthews, V., Correlation of Pinnacles and Neenach volcanic formations and their bearing on San Andreas fault problem, Am. Assoc. Petrol. Geol. Bull. 60, 2128-2141, 1976.

Mayer, L., Subsidence analysis of the Los Angeles basin, pp. 299-320, in Ingersoll, R. V., and W. G. Ernst, eds., Cenozoic Basin Development of Coast California (Rubey Vol. VI), Prentice-Hall, Englewood Cliffs, 1987.

McLaughlin, R. J., M. C. Blake, Jr., A. Griscom, C. D. Blome, and B. Murchey, Tectonics of formation, translation, and dispersal of the Coast Range ophiolite of California, Tectonics 7, 1033-1056, 1988.

Metzger, D. G., The Bouse Formation (Pliocene) of the Parker-Blythe-Cibola area, Arizona and California, U.S. Geol. Surv. Prof. Paper 600-D, 126-136, 1968.

Meyer, G. L., Aeolian features of northern Coachella Valley and landforms and tectonic features of the San Andreas fault zone in the Indio Hills, pp. 1-16, in Field Guide, Far Western Section, Nat. Assoc., Geol. Teachers, 25th Ann. Meet., Palm Desert, 1979.

Minster, J. B., and T. H. Jordan, Vector constraints on western U.S. deformation from space geodesy, neotectonics, and plate motion, J. Geophys. Res. 92 (B6), 4798-4804, 1987.

Morton, D. M., J. C. Matti, and J. C. Tinsley, Banning fault, Cottonwood Canyon, San Gorgonio Pass, southern California, pp. 191-192, in Hill, M. L., ed., Centennial Field Guide Volume 1, Cordilleran Sec. Geol. Soc. Am., 490 p., 1987.

Mount, V. S., and J. Suppe, State of stress near the San Andreas fault: Implications for wrench tectonics, Geology 15, 1143-1146, 1987.

Namson, J. S., and T. L. Davis, Seismically active fold and thrust belt in the San Joaquin Valley, central California, Geol. Soc. Am. Bull. 100, 257-273, 1988.

Nason, R. D., Investigation of fault creep slippage in northern and central California, Ph.D. disser., Univ. Calif., San Diego (University Microfilms, no. 12,785, Ann Arbor, Mich.), 1971.

Nason, R. D., Fault creep and earthquakes on the San Andreas fault, in Kovach, R. L., and A. Nur, eds., Conference on Tectonic Problems of the San Andreas Fault System, Proceedings, Stanford Univ. Pubs. in Geol. Sci. 13, 275-285, 1973.

Nason, R. D., Observations of premonitory creep before earthquakes on the San Andreas fault, pp. 543-550, in Evernden, J. F., ed., Proceedings of Conference II - Experimental Studies of Rock Friction with Application to Earthquake Prediction, U.S. Geol. Surv., Menlo Park, 701 p., 1977.

Nicholson, C. L. Seeber, P. Williams, and L. R. Sykes, Seismic evidence for conjugate slip and block rotation with the San Andreas fault system, southern California, Tectonics 5, 629-648, 1986.

Nilsen, T. H., Offset along the San Andreas fault of Eocene strata from the San Juan Bautista area and western San Emigdio Mountains, California, Geol. Soc. Am. Bull. 95, 599-609, 1984.

Noble, L. F., The San Andreas rift and some other active faults in the Desert Region of southeastern California, Carnegie Inst. Washington Year Book 25, 415-428, 1926.

Noble, L. F., Geology of the Valyermo Quadrangle and Vicinity, California, scale 1:24,000, Geol. Invest. Map GQ 50, U.S. Geol. Surv., Washington, D.C., 1954.

Okaya, D. A., and E. G. Frost, Seismic profiling in the Mojave-Sonoran extensional terrane - CALCRUST reprocessing and interpretation of industry seismic data, Tectonics, in press.

Page, B. M., The southern Coast Ranges, pp. 329-417, in Ernst, W. G., ed., The Geotectonic Development of California, (Rubey Volume I), Prentice-Hall Inc., Englewood Cliffs, New Jersey, 706 pp., 1981.

Pampeyan, E. H., T. L. Holzer, and M. M. Clark, Modern ground failure in the Garlock fault zone, Fremont Valley, California, Geol. Soc. Am. Bull. 100, 677-691, 1988.

Powell, R. E., Crystalline basement terranes in the southern eastern Transverse Ranges, California, in Cooper, J. D., compiler., Geologic Excursions in the Transverse Ranges, Cordilleran Section, Geological Society of America Guidebook, 109-151, 1982.

Prentice, C. S., and K. E. Sieh, Prehistoric seismic events on the northern San Andreas fault near Point Arena, California, Eos Trans. AGU 69, n. 16, p. 492, 1988.

Prescott, W. H., and M. Lisowski, Strain accumulation along the San Andreas fault system east of San Francisco Bay, California, Tectonophysics 97, 41-56, 1983.

Prescott, W. H., M. Lisowski, and J. C. Savage, Geodetic measurement of crustal deformation on the San Andreas, Hayward, and Calaveras faults near San Francisco, California, Jour. Geophys. Res. 86, pp. 10,853-10,869, 1981.

Robinson, P. T., W. A. Elders, and L. P. J. Muffler, Quaternary volcanism in the Salton Sea geothermal field, Imperial Valley, California, Geol. Soc. Am. Bull. 87, 347-360, 1976.

Rockwell, T. K., R. S. McElwain, D. E. Millman, and D. L. Lamar, Recurrent Late Holocene faulting on the Glen Ivy north strand of the Elsinore fault at Glen Ivy Marsh, pp. 167-175, *in* Ehlig, P. L., *compiler*, Neotectonics and Faulting in Southern California, Guidebook and Volume, 82nd Annual Meeting, Cordilleran Sec., Geol. Soc. Am., Los Angeles, Calif., March 25-28, 1986.

Rogers, T. H., and R. D. Nason, Active fault displacement on the Calaveras fault zone at Hollister, California, Seism. Soc. Am. Bull. 61, 399-416, 1971.

Ross, D. C., Quartz gabbro and anorthositic gabbro: Markers of offset along the San Andreas fault in the California Coast Ranges, Geol. Soc. Am. Bull. 81, 3647-3662, 1970.

Ross, D. C., Possible correlations of basement rocks across the San Andreas fault, San Gregorio-Hosgri and Rinconada-Reliz-King City faults, California, U.S. Geol. Surv. Prof. Paper 1317, p. 37, 1984.

Rust, D. J., Evidence for uniformity of large earthquakes in the "big bend" of the San Andreas fault, Eos Trans. AGU 63, 1030, 1982.

Sadler, P. M., and A. Demirer, Pelona Schist clasts in the Cenozoic of the San Bernardino Mountains, southern California, pp. 129-146, *in* Ehlig, P. L., *compiler*, Neotectonics and Faulting in Southern California, Guidebook and Volume, 82nd Annual Meeting, Cordilleran Section, Geol. Soc. of Am., Los Angeles, Calif., March 25-28, 1986.

Sanders, C. O., and H. Kanamori, A seismotectonic analysis of the Anza seismic gap, San Jacinto fault zone, southern California, J. Geophys. Res. 89, 5873-5890, 1984.

Schubert, C., Neotectonics of a segment of the San Andreas fault, southern California (USA), Eiszeitalter und Gegenwart 32, 13-22, 1982.

Sieh, K. E., Prehistoric large earthquakes produced by slip on the San Andreas fault at Pallett Creek, California, J. Geophys. Res. 83 (B8), 3907-3939, 1978.

Sieh, K. E., Slip along the San Andreas fault associated with the great 1857 earthquake, Seism. Soc. Am. Bull. 68, 1421-1448, 1978.

Sieh, K. E., A review of geological evidence for recurrence times of large earthquakes, Earthquake Prediction - An International Review, Maurice Ewing Series 4, Am. Geophys. Un., 181-207, 1981.

Sieh, K. E., Lateral offsets and revised dates of large earthquakes at Pallett Creek, California, J. Geophys. Res. 89, 7641-7670, 1984.

Sieh, K. E., and R. H. Jahns, Holocene activity of the San Andreas fault at Wallace Creek, California, Geol. Soc. Am. Bulletin 95, 883-896, 1984.

Silver, L. T., and J. M. Mattinson, "Orphan Salinia" has a home (abst): Eos Trans. AGU 67, 1215, 1986.

Sims, J. D., The Parkfield shuffle: Displaced rock bodies as a clue to the post-Eocene history of movement on the San Andreas fault in central California, Geol. Soc. Am. Abstr. with Prog. 18, 185, 1986.

Sims, J. D., Late Holocene slip along the San Andreas fault near Cholame, California, Geol. Soc. Am. Abstr. with Prog. 19, 415, 1987.

Sims., J. D., Geologic map of the San Andreas fault in the Cholame Valley and Cholame Hills Quadrangles, San Luis Obispo and Monterey Counties, California, scale 1:24,000, Misc. Field Studies Map MF-1955, U.S. Geol. Surv., Menlo Park, Ca.,1988.

Sims, J. D., Geologic map of the San Andreas fault in the Parkfield Quadrangle, Monterey County, California, scale 1:24,000, Misc. Field Studies Map, U.S. Geol. Surv., in press.

Smith, D. P., San Juan-St. Francis fault - hypothesized major middle Tertiary right-lateral fault in central and southern California, Calif. Div. Mines & Geol. Spec. Rept 129, 41-50, 1977.

Stanley, R. G., Implications of the northwestwardly younger age of the volcanic rocks of west-central California: Alternative interpretation, Geol. Soc. Am. Bull. 98, 612-614, 1987.

Steinbrugge, K. V., E. G. Zacher, D. Tocher, C. A. Whitten, and C. N. Clair, Creep on the San Andreas fault: Fault creep and property damage, Seism. Soc. Am. Bull. 50, 389-404, 1960.

Stewart, J. H., and J. C. Crowell, Strike-slip tectonics in the Cordilleran region, western United States, *in* Burchfiel, B. C., P. W. Lipman, and M. L. Zoback, *eds.*, The Cordilleran Orogen: Conterminous United States, Boulder Colo., Geol. Soc. Am., The Geology of the United States, vol. G3, in press.

Strange, W. E., The impact of refraction corrections on leveling interpretations in southern California, J. Geophys. Res. 86, 2809-2894, 1981.

Strehle, R., Subsidence in Long Beach, pp. 69-80, *in* Clarke, and C. P. Henderson, *eds.*, Geologic Field Guide to the Long Beach Area, Pacific Section, Am. Assoc. Petrol. Geol., Los Angeles, 159 p., 1987.

Stuiver, M., Radiocarbon timescale tested against magnetic and other dating methods, Nature 273, p. 271 ff, 1978.

Sylvester, A. G., Near-field tectonic geodesy, pp. 164-180, *in* Wallace, R. E., *ed.*, Active Tectonics, National Acad. Press, Washington D. C., 266 p., 1986.

Sylvester, A. G., Strike-slip faults, Geol. Soc. Am. Bull. 100, 1666-1703, 1988.

Sylvester, A. G., and R. R. Smith, Structure section across the San Andreas fault zone, Mecca Hills, pp. 111-118, *in* Crowell, J. C., *ed.*, San Andreas Fault in Southern California, A Guide to San Andreas Fault from Mexico to Carrizo Plain, Calif. Div. Mines & Geol. Spec. Rept. 118, 272 p., 1975.

Sylvester, A. G., and R. R. Smith, Tectonic transpression and basement-controlled deformation in the San Andreas fault zone, Salton Trough, California, Am. Assoc. Petrol. Geol. Bull. 60 (12), 2081-2102, 1976.

Sylvester, A. G., A. H. Brown, and N. R. Riggs, Vertical movements and aseismic horizontal creep on the San Andreas fault at San Juan Bautista, California, pp. 91-109, *in* Streitz, R., and R. Sherburne, *eds.*, Studies of the San Andreas Fault Zone in Northern California, Calif. Div. Mines & Geol. Spec. Rept. 140, 187 p., 1980.

Taylor, C. L., J. C. Cummings, and A. P. Ridley, Discontinuous en echelon faulting and ground warping, Portola Valley, California, pp. 59-70, *in* Streitz, R., and R. Sherburne, *eds.*, Studies of the San Andreas Fault Zone in Northern California, Calif. Div. Mines & Geol. Spec. Rept. 140, 187 p., 1980.

Thatcher, W., Systematic inversion of geodetic data in central California, J. Geophys. Res. 84, 2283-2295, 1979.

Thatcher, W., Geodetic measurement of active-tectonic processes, pp. 155-163, *in* Wallace, R. E., *ed.*, Active Tectonics, National Acad. Press, Washington D. C., 266 p., 1986.

Tocher, D., and R. D. Nason, Fault creep at the Almaden-Cienega winery, San Benito County, p. 99, *in* Guidebook to the Gabilan Range and Adjacent San Andreas Fault, Pacific Section, Am. Assoc. Petrol. Geol., 1967.

Toppozada, T. R., Earthquake history of Parkfield and surroundings, Eos Trans. AGU 68, 1345, 1987.

Toppazada, T. R., C. R. Real, and D. L. Parke, Earthquake history of California, Calif. Geol. 39, 27-33, 1986.

Vedder, J. G., D. G. Howell, and H. McLean, Stratigraphy, sedimentation, and tectonic accretion of exotic terranes, southern Coast Ranges, California, Am. Assoc. Petrol. Geologists Memoir 34, 471-496, 1983.

Vickery, F. P., Structural dynamics of the Livermore region, Jour. Geol. 33, 608-628, 1925.

Wahrhaftig, C., A Streetcar to Subduction and Other Plate Tectonic Trips by Public Transport in San Francisco, Am. Geophys. Un., Washington, D. C., 76 p., 1984.

Wahrhaftig, C., and B. Murchey, Marin Headlands, California: 100-million-year record of sea floor transport and accretion, Centennial Field Guide Volume 1, Cordilleran Sec., Geol. Soc. Am., Boulder, 490 p., 1987.

Wallace, R. E., Structure of a portion of the San Andreas fault in southern California, Geol. Soc. Am. Bull. 60, 781-806, 1949.

Wallace R. E., and E. F. Roth, Rates and patterns of progressive deformation, pp. 23-40, in Brown, R., Jr., ed., Parkfield-Cholame, California, earthquakes of June-August 1966 - Surface geologic effects, water resources aspects, and preliminary seismic data, U.S. Geol. Surv. Prof. Paper 579, 66 p, 1967.

Webb, G. W., Stevens and earlier Miocene turbidite sandstones, southern San Joaquin Valley, California, Am. Assoc. Petrol. Geol. Bull. 65, 438-465, 1981.

Weber, F. H., Jr., Seismic hazards related to geologic factors, Elsinore and Chino fault zones, northwestern Riverside County, California, Calif. Div. Mines & Geol. Open File Rpt 77-4, 96 p., 1977.

Weldon, R. J., and K. E. Sieh, Holocene rate of slip and tentative recurrence interval for large earthquakes on the San Andreas fault, Cajon Pass, southern California, Geol. Soc. Am. Bull. 96, 793-812, 1985.

Wentworth, C. M., The Upper Cretaceous and Lower Tertiary rocks of the Gualala block, northern Coast ranges, California, Ph. D. dissertation, Stanford Univ., 1967.

Wesnousky, S. G., Earthquakes, Quaternary faults, and seismic hazard in California, J. Geophys. Res. 91, 12,587-12,631, 1986.

Wesson, R. L., E. J. Helley, K. R. Lajoie, and C. M. Wentworth, Faults and future earthquakes, pp. 5-30 in Borcherdt, R. D., ed., Studies for Seismic Zonation of the San Francisco Bay Region, U.S. Geol. Surv. Prof. Paper 941A, 1975.

Wesson, R. L., and C. Nicholson, Intermediate-term, pre-earthquake phenomena in California, 1975-1986, and preliminary forecast of seismicity for the next decade, Pure and Applied Geophys. 126, 407-446, 1988.

Winker, C. D., and S. M. Kidwell, Paleocurrent evidence for lateral displacement of the Pliocene Colorado River delta by the San Andreas fault system, southeastern California, Geology 14, 788-791, 1986.

Woyski, M. S., and A. H. Howard, A section through the Peninsular Ranges batholith, Elsinore Mountains, southern California, pp. 185-190, in Hill, M. L., ed., Centennial Field Guide Volume 1, Cordilleran Sec. Geol. Soc. Am., 490 p., 1987.

Yeats, R. S., Newport-Inglewood fault zone, Los Angeles Basin, California, Am. Assoc. Petrol. Geol. Bull. 57 (1), 117-135, 1973.

Tectonic Evolution of Northern California

Sausalito to Yosemite National Park, California
June 28–July 7, 1989

Field Trip Guidebook T108

Leaders:
M. C. Blake, Jr.   D. S. Harwood

Associate Leaders:
R. J. McLaughlin   A. S. Jayko
W. P. Irwin   F. C. W. Dodge   D. L. Jones
M. M. Miller   T. Bullen

American Geophysical Union, Washington, D.C.

Published 1989 by American Geophysical Union

2000 Florida Ave., N.W., Washington, D.C. 20009

ISBN: 0-87590-614-1

Printed in the United States of America

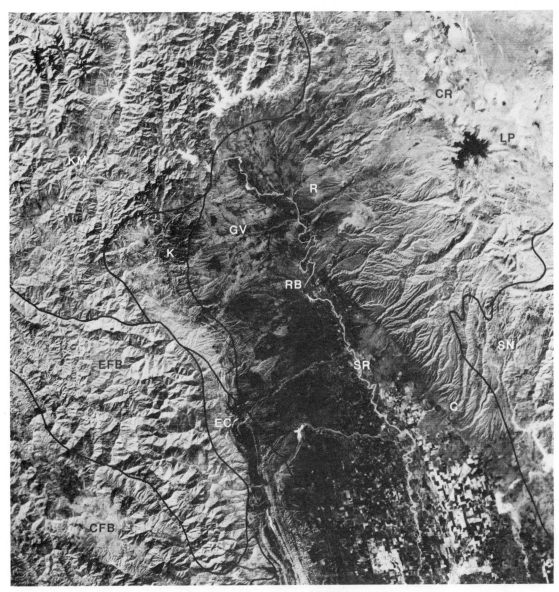

COVER: LANDSAT image centered on northern Sacramento Valley, California. CFB= Central Franciscan belt, EFB= Eastern Franciscan belt, KM= Klamath Mountains, GV= Great Valley, K= Cretaceous overlap sequence, EC= Elder Creek terrane, SN= Sierra Nevada, CR= Cascade Range, C= Chico, R= Redding, RB= Red Bluff, SR= Sacramento River, LP=Lassen Peak.

Leaders:

M. C. Blake, Jr. and D. S. Harwood
U.S. Geological Survey
345 Middlefield Road
Menlo Park, CA 94025

Associate Leaders:

R. J. McLaughlin, A. S. Jayko,
W. P. Irwin, and F. C. W. Dodge
U.S. Geological Survey
345 Middlefield Road
Menlo Park, CA 94025

D. L. Jones
Department of Geology and Geophysics
University of California
Berkeley, CA 94720

M. M. Miller
Division of Geological and Planetary Sciences
California Institute of Technology
Pasadena, CA 91125

T. Bullen
Department of Geology
California State University
Hayward, CA 94542

IGC FIELD TRIP T108:
TECTONIC EVOLUTION OF NORTHERN CALIFORNIA:
OVERVIEW AND INTRODUCTION

M. C. Blake, Jr. and D. S. Harwood
U. S. Geological Survey, Menlo Park, California

The present tectonic regime of northern California includes transform faulting south of the Mendocino triple junction, eastward-directed subduction, north of the junction and extensional rifting and recent volcanism in the Cascade Range to the east. The complex structures that formed during the last few million years and the variety of igneous rocks that were generated in these three different plate tectonic settings provide us with testable models to compare with the more ancient history.

During the first two and one-half days, we will be looking at the rocks and structures of the northern Coast Ranges (Figure 1), including the well-known Franciscan Complex. Evidence will be presented for a complicated history of subduction punctuated by periods of transform and extensional faulting.

After this, we will cross the Western Paleozoic and Triassic belt of the Klamath Mountains (Figure 1), an older accretionary complex containing a greater variety of

rock types and much more complicated and controversial history than the Coast Ranges. The following day will be a more detailed look at the Eastern Klamath terrane with the emphasis on the arc-related volcanic rocks.

The remainder of the trip will be spent in the Sierra Nevada province (Figure 1), beginning in the presently active Cascade arc at Lassen Peak and progressing southward to Yosemite. The trip through Lassen Volcanic National Park on day 5 provides an overview of Quaternary volcanic rocks erupted at the southern end of the Cascade Range (Figure 1) and focuses on products of the 1915 eruption of Lassen Peak and the Lassen geothermal system. From this setting of active continental-arc volcanism, we step back in time nearly 400 m.y. and view Paleozoic and lower Mesozoic island-arc volcanic rocks of the Northern Sierra terrane. The rocks in the Northern Sierra terrane seen on day 6 and 7 invite comparison with the volcanic assemblage seen on day 4 in the Eastern Klamath terrane. Did these

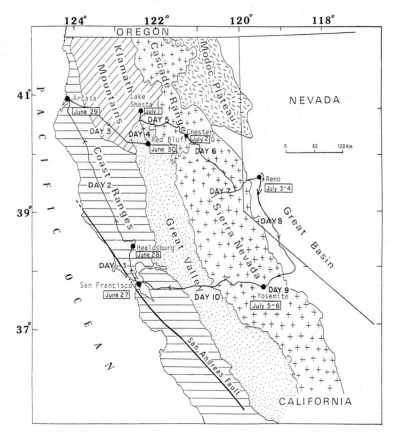

FIGURE 1 Map of northern California showing physiographic provinces and route of field trip.

accreted arc terranes evolve independently at some remote paleolatitude or did they form parts of a long-lived arc complex with North American affinity?

From the northern Sierra Nevada, the trip progresses southeastward along the steep, faulted boundary between the Sierra Nevada and the Great Basin (Figure 1). On days 8 and 9 the trip focuses on the composition, texture and intrusive evolution of granotoid plutons of the Sierra Nevada batholith in Yosemite National Park. Part of day 10 will be spent looking at metamorphic rocks in the western foothills of the Sierra Nevada which host significant gold deposits in the famed Mother Lode.

FIGURE 2 View east toward Mt. Shasta (elev. 4316 m); a Quaternary stratovolcanoe in the Cascade Range.

TERRANES OF THE NORTHERN COAST RANGES

M.C. Blake, Jr. and R.J. McLaughlin
U.S. Geological Survey, Menlo Park, California

D.L. Jones
University of California, Berkeley, California

INTRODUCTION

The Franciscan Complex, the Coast Range ophiolite and Great Valley sequence, and the Salinian block are the three major basement complexes of the northern California Coast Ranges (Figure 1). They overlap in age but differ greatly in lithology, structural state, and metamorphic grade. According to early plate tectonic models, the Great Valley sequence was deposited throughout the late Mesozoic and early Cenozoic in a forearc basin adjacent to the Sierran-Klamath magmatic arc. At the same time, the Franciscan Complex was being accreted by subduction in a trench to the west (Hamilton, 1969; Bailey and Blake, 1969). About ten million years ago the northward migrating Mendocino triple junction passed the latitude of San Francisco (Atwater, 1970; Engebretson et al., 1985), progressively replacing the subduction zone with a right-lateral transform along which the continental Salinian block moved northward. Numerous recent studies (Blake and Jones, 1981; Jayko and Blake, 1987; McLaughlin et al., 1988) make this simple two-stage model obsolete, as alternating periods of oblique subduction, terrane accretion, and right-lateral faulting, and an important event involving crustal extension (Jayko et al., 1987) can now be demonstrated to have occurred in the Cretaceous and early Tertiary.

The purpose of the first two and one-half days of this field trip is to look at a number of far-traveled accreted terranes that have been mapped within the Franciscan Complex and the Coast Range ophiolite in order to contrast the geologic history recorded in these rocks with that implied by the early plate tectonic models.

FRANCISCAN COMPLEX

The Franciscan Complex (Figures 1A to C) has been divided into three broad tectonic belts, the Eastern, Central, and Coastal (Irwin, 1960). More recently, each of these belts has been subdivided into a number of fault-bounded tectonostratigraphic terranes (Blake et al., 1982), each having a stratigraphy that is different from its neighbor. Figure 1C shows the relationship of the tectonostratigraphic terranes discussed in this field trip guide to the belts of the Franciscan Complex. South of about Covelo (Figure 1), the earlier late Mesozoic and early Cenozoic history of the terranes is strongly overprinted by active and recently active structures related to the San Andreas transform system. To the north, however, the older history is better preserved and it is here that most of the terranes and their structural and metamorphic histories have been described (Blake et al., 1984a; 1985a; 1988; Jayko et al., 1986).

Eastern Belt

Pickett Peak terrane. The easternmost and structurally highest terrane is the Pickett Peak terrane, which consists of two units, the South Fork Mountain Schist and the Valentine Spring Formation (Worrall, 1981). These two units are separated by an east-dipping thrust fault.

The South Fork Mountain Schist consists largely of quartz-veined mica schist (metamudstone) with minor metavolcanic rock (including the Chinquapin Metabasalt Member) and scarce metachert. All of the metavolcanic rock is of basaltic composition and includes both tholeiitic and alkalic varieties (Jayko et al., 1986; Cashman et al., 1986). Some of the thinner metabasalt layers appear to have formed from tuffs interbedded with the metasedimentary rocks; however, the larger bodies include relict pillows and pillow breccias that are overlain depositionally by thin (1-5 m) beds of iron- and manganese-rich metachert. At several localities, metachert is depositionally overlain by quartz-mica schist and, less commonly, metagraywacke. Worrall (1981) concluded that the Chinquapin Metabasalt Member was the basement to the South Fork Mountain Schist. This inference combined with the tholeiitic to alkalic geochemistry of the basalts suggests that the basement was probably a seamount or an oceanic island rather than oceanic crust.

The structurally lower Valentine Spring Formation consists largely of schistose to gneissic metagraywacke with scarce metavolcanic rocks and rare metachert. As in the South Fork Schist, the metavolcanic rocks are of basaltic composition and include both tholeiitic and alkalic varieties (Brothers et al., in preparation). Massive metabasalt, up to 100 m thick, locally overlain by thin (1-2 m) metachert lenses, could represent oceanic crust upon which the Valentine Spring metaclastic sediments were deposited.

Fossils have not been found in either the South Fork Mountain Schist or Valentine Spring Formation. Whole-rock K-Ar and Rb-Sr radiometric ages measured on both units range from 110 to 143 Ma, but cluster around 125 Ma (Lanphere et al., 1978; Suppe, 1973; McDowell et al., 1984) suggesting an Early Cretaceous age of metamorphism.

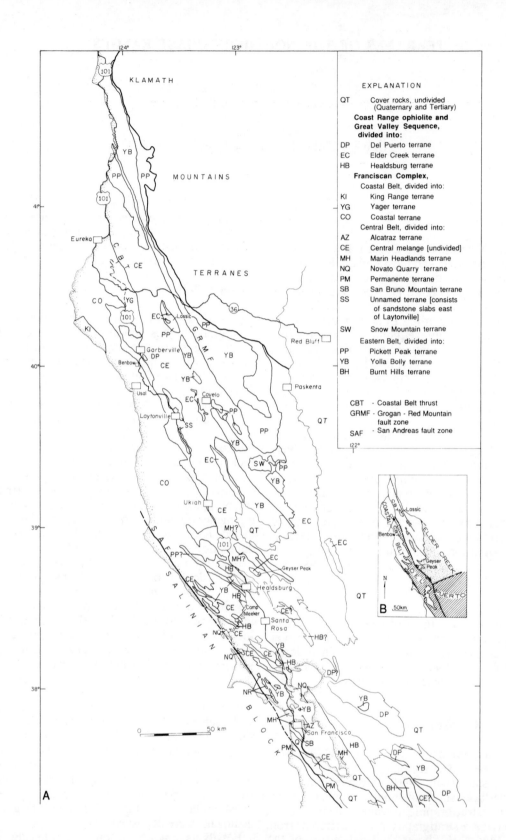

FIGURE 1 A.) Map showing distribution of belts and terranes of the California Coast Ranges and Klamath Mountains (modified from Silberling et al., 1987) B.) Map showing minimum displacement of the Del Puerto and (or) Healdsburg terrane outliers from the west side of the Great Valley. Displaced outliers are incorporated into melange terrane of the Central belt (CE, Figure 1A) C.) Diagram showing correlation of terranes with belts of the Franciscan Complex, and the Coast Range ophiolite and lower part of the Great Valley sequence.

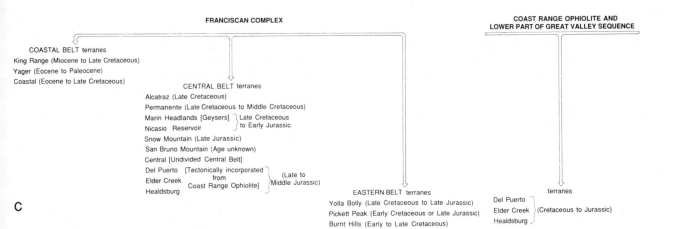

C

All of the rock units assigned to the Pickett Peak terrane contain evidence for three periods of penetrative deformation. The first period is characterized by segregation layering, and the second and third phases produced crenulation cleavages. Blueschist-facies metamorphic conditions persisted during the first two deformational events.

Yolla Bolly terrane. The Yolla Bolly terrane lies west of, and structurally below, the Pickett Peak terrane (Figure 1). It consists of several thrust-bounded units, each characterized by various proportions of quartzofeldspathic metagraywacke, argillite, radiolarian chert, and minor greenstone (altered volcanic rock). These thrust sheets (nappes), appear to represent a dismembered oceanic crustal sequence of rarely-preserved basalt overlain by radiolarian chert that was covered by graywacke and argillite derived from a continental margin. The rocks are locally intruded by Ti-rich alkalic gabbro sills and dikes. Scarce blocks of amphibolite and blueschist, together with serpentinite are also present, particularly along the faulted upper contact with the Pickett Peak terrane.

Radiolarian microfossils and scarce megafossils indicate a Late Jurassic and Early Cretaceous protolith age for most of the Yolla Bolly terrane. However, at one locality southeast of Covelo, mid-Cretaceous (Cenomanian) pelecypods (*Inoceramus*) occur in Yolla Bolly shale and metagraywacke that contains lawsonite and aragonite (Blake and Jones, 1974). K-Ar and Rb-Sr whole-rock metamorphic apparent ages on metagraywacke from the Yolla Bolly terrane range from about 90 Ma to 115 Ma (Suppe, 1973; Lanphere et al., 1978). However, the presence of remnant detrital white mica suggests an inherited age. Therefore, a U/Pb sphene age of 92 Ma from the Yolla Bolly terrane in the Diablo Range (Mattinson and Echeverria, 1980; Mattinson, 1986) may reflect more accurately the timing of metamorphism. Thus, it appears that sedimentation was terminated in the Cenomanian by a tectonic event that caused rapid high-pressure metamorphism. We infer that the Yolla Bolly terrane represents a collapsed continental-margin basin that formed in a transtensional or highly oblique convergent setting, and was subsequently subducted.

The Yolla Bolly terrane retains evidence for two phases of penetrative deformation that were coaxial with the second and third phases of deformation in the Pickett Peak terrane. The first phase of deformation (parallel to the second phase in the Pickett Peak terrane) was accompanied by blueschist facies metamorphism. In the second phase of deformation the metamorphosed basalt-chert-graywacke sequence was complexly folded and imbricated.

Terranes of the Central belt

The Central belt (Figure 1) is a tectonic melange containing numerous blocks and slabs of resistant graywacke, greenstone, chert, limestone, ultramafic rock, and high-grade metamorphic rocks in a sheared matrix of argillite, lithic graywacke, and green cherty tuff. Some of the blocks and slabs, particularly in the San Francisco Bay region, are big enough to be considered as separate terranes and are described later.

Some of the unnamed terranes of resistant graywacke east of Laytonville (as on Figure 1B, labelled "SS") have yielded fossils of mid-Cretaceous (Albian) age, whereas most fossils from the enclosing sheared argillite matrix are Late Jurassic (Tithonian) to Early Cretaceous (Valanginian) (Blake and Jones, 1974). At another locality, southeast of Covelo, Berry (1982) described Cenomanian foraminifers that occur in limestone concretions. Blocks of red pelagic limestone in the Central belt, in particular those near Laytonville, have been the object of much recent study. Both paleontologic and paleomagnetic data suggest this limestone was originally deposited at equatorial latitudes in the southern hemisphere between mid-Cretaceous (Albian) and Late Cretaceous (Coniacian) time (Alvarez et al., 1980; Sliter, 1984; Tarduno et al., 1986), and transported on an oceanic plate to the continental margin where it was tectonically incorporated into the Central belt.

Snow Mountain, located southeast of Covelo, is one of many large slabs of volcanic rocks in the Central belt, but it is distinctive enough to be shown as a separate terrane (Figure 1). These rocks consist of a thick pile of alkalic volcanic rocks with scarce Upper Jurassic chert. They were previously thought to be part of the Coast

Range ophiolite (Bailey and Blake, 1974) but a more recent study (Macpherson, 1986), indicates that the volcanic rocks contain high P/T minerals such as lawsonite and jadeite and they are now interpreted to represent a seamount that was subducted with the Franciscan Complex.

Although both tectonic and sedimentary processes have been invoked to explain how the block-in-matrix fabric that characterizes the Central belt melange formed (Cowan, 1985; Aalto and Murphy, 1984; Cloos, 1985; Aalto, 1986), most workers agree that the melange formed within an accretionary wedge at the interface between an oceanic plate and the North American margin.

At least three sets of faults have contributed to, or have overprinted, the early-phase melange fabric of the Central belt. In the northern part of the transect area, a set of high-angle faults that trend N20°W truncates the Yolla Bolly terrane, and has transposed slabs of it, as well as fragments of the Pickett Peak terrane, Coast Range ophiolite, and the Great Valley sequence as far north as southwest Oregon (Blake et al., 1985b). The easternmost fault of this system is shown as the Grogan-Red Mountain fault on Figure 1. Detailed mapping in the Pickett Peak quadrangle (Irwin et al., 1974) suggests that these faults dip about 60° to the east, implying oblique strike-slip faulting. Timing of this system of faulting is broadly constrained to postdate the age of the youngest fossils and blueschist-facies metamorphism in the Yolla Bolly terrane (Cenomanian, 90 Ma) and to predate early Eocene (50 Ma) strata that overlap the displaced terrane fragments in southwest Oregon, although significant Neogene reactivation has undoubtedly occurred locally (Blake et al., 1985a; Cashman et al. 1986).

A second set of high- to low-angle faults, with reverse and right-slip components, trends about N30°50°W in the vicinity of Covelo (Figure 1). This set of faults appears to truncate or disrupt the earlier N20°W trend. Similar-trending faults with prominent vertical slip components, bound Miocene and Pliocene sedimentary rocks near Garberville (Figure 1).

A third set of faults which trend N20°-40°W includes the active Maacama, and Green Valley-Bartlett Springs-Lake Mountain fault zones (DePolo and Ohlin, 1984; McLaughlin et al., 1985a), and the Middle Mountain-Konocti Bay fault zone (Hearn et al., 1988).

Small terranes in the Central belt of the San Francisco Bay region. Within the San Francisco Bay region, a number of small terranes occur as lenticular slabs dispersed within the Central belt (Blake et al., 1984). The dispersion was previously thought to have taken place during propagation of the presently-active San Andreas transform system. However, most of these terranes lie to the east of the San Andreas fault and several of them as shown by sedimentological, paleontological, and paleomagnetic data, have been displaced greater distances than the 305 km of right slip usually attributed to the San Andreas (Sliter, 1984; Tarduno, et al., 1985).

Of the seven Franciscan terranes shown on Figure 1,

and not discussed earlier, three: Marin Headlands, Nicasio Reservoir, and Permanente are fragments of Early Jurassic to Early Cretaceous oceanic crust or oceanic islands. They consist of pillow basalt overlain by either radiolarian chert or pelagic limestone which is covered by Cretaceous graywacke derived from a continental margin. An additional three, Novato Quarry, Alcatraz, and San Bruno Mountain, are made up almost entirely of graywacke, argillite and conglomerate. San Bruno Mountain is locally intruded by basalt and some chert is also present. These latter three terranes do not show the degree of stratal disruption seen in the surrounding Central belt melange and might in part represent slope deposits formed on previously accreted terranes and subsequently deformed during later transform faulting. The Novato Quarry and Alcatraz terranes contain rare fossils of Late Cretaceous (Campanian) and Early Cretaceous (Valanginian) ages, respectively. The age of the San Bruno Mountain terrane is unknown. Another terrane, found in the Diablo Range, is the Burnt Hills terrane. It is made up of a basal basalt-radiolarian chert sequence of Albian age overlain by arkosic graywacke of Cretaceous (in part Coniacian) age (Blake et al., 1984b; Murchey and Jones, 1984). All of the rocks in the Burnt Hills terrane contain blueschist-facies minerals and, thus, the terrane is probably part of a subduction complex that is somewhat younger than the structurally-overlying Yolla Bolly terrane subduction complex.

Coastal Belt

Coastal terrane. The Coastal terrane is characterized by sandstone, mudstone, and conglomerate that range in age from Paleocene (~60-55 Ma) to late Eocene (~40 Ma). The Coastal terrane is highly disrupted by brittle fracturing, shearing, and folding, with the fabric varying from partially disrupted sandstone-rich rocks in the eastern part to melange in the western exposures. Sandstone and argillite, which form the largest component of the Coastal terrane, commonly contain sheared, argillite-rich zones with carbonate concretions containing dinoflagellate and (or) foraminifer assemblages of Paleocene through late Eocene age (Sliter, et al., 1986). The more chaotic parts of the Coastal terrane contain rare far-travelled blocks of pillow basalt and pelagic foraminiferal limestone, which probably are accreted scraps of Late Cretaceous (Campanian to Maastrichtian) mid-ocean seamounts or plateaus (Sliter, 1984; Sliter et al., 1986; Harbert et al., 1984). Rare blocks of medium-grade blueschist (Type III, Coleman and Lee, 1963) also occur locally (McLaughlin et al., 1982). The foraminiferal limestone in the Coastal terrane in most places overlies or is intercalated with the basalt. In one critical area near Usal, manganiferous basalt and limestone are locally interbedded with arkosic sandstone and tachylitic tuffaceous argillite, suggesting: 1) that the interbedded sandstones were tectonically transported an unknown distance with the basalt and pelagic limestone; and 2) that the active volcanic edifice from which the basalt and limestone were derived, was near enough to the North American Plate margin to receive terrigenous deposits, prior to tectonic interaction with that margin (McLaughlin et al. 1986). Foraminifers from the pelagic limestones at Usal also are indicative of a Campanian to Maastrichtian low-latitude depositional setting.

Yager terrane. The Yager Formation (Ogle, 1953) is the sole component of the Yager terrane (McLaughlin, 1983; McLaughlin et al., 1983b). Rocks of the Yager terrane consist largely of well-bedded, little sheared but locally highly folded mudstone and sandstone, with thick lenses of polymict conglomerate. Sandstone in the Yager and Coastal terranes characteristically is uniformly arkosic to feldspathic, with abundant detrital K-feldspar; compared to the much more variable arkosic to volcaniclastic sandstone compositions in the Coastal terrane. Volcanic rock fragments in sandstones from both terranes are dominantly porphyritic to aphanitic and intermediate to silicic in composition, and also include abundant plutonic detritus, suggesting a magmatic arc source (Underwood and Bachman, 1986). A continental source is also suggested by the presence of detrital muscovite, garnet, epidote, and biotite, which are locally prominent sandstone constituents in both terranes, and by minor fragments of white-mica schist and quartzite and the consistent presence of trace amounts of detrital tourmaline (shorlite).

The Yager terrane is lithologically, structurally and chronologically more homogeneous than the Coastal terrane. Shearing is prominent only locally, particularly near the western and eastern contacts with the Coastal terrane and the Central belt; however, complex folding is widespread (Underwood, 1985). Dinoflagellates and foraminifers indicate that the Yager is no older than Paleocene and as young as late Eocene. Thus, the age of the Yager terrane overlaps that of the younger part of the Coastal terrane. Concretions from a Yager conglomerate along the Eel River, north of Garberville (Figure 1), are richly fossiliferous and include abundant well-preserved dinoflagellates of Paleocene age identical to those described from the upper part of the Moreno Formation (Great Valley sequence) in central California (W. R. Evitt, written commun., 1985). The strata are considered to represent an uplifted and deformed slope or slope-basin deposit that accumulated above partly older, but largely coeval accreted slope and trench-slope deposits of the Coastal terrane (Bachman, 1978; Underwood, 1983, 1985).

Most deformation of the Coastal and Yager terranes pre-dated deposition of the late Miocene "Bear River beds" or Bear River Formation (Gester, 1951) and the upper Miocene and lower Pleistocene Wildcat Group of Ogle (1953). The Central belt and the Yager and Coastal terranes were juxtaposed along the Coastal belt thrust (Figure 1) prior to deposition of the Bear River beds and the Wildcat Group (15 Ma) and after the deposition of the Yager terrane (~38 Ma). Suturing was probably coincident with an Oligocene hiatus in the depositional record. North of the Mattole River mouth rocks of the Wildcat Group and Bear River beds are locally sheared into the Coastal terrane as the result of on-going NE-

SW-oriented compression between the Pacific and North American plates and NW-SE-oriented compression between the Gorda and North American plates (Ogle, 1953; McLaughlin et al., 1983b).

King Range terrane. The youngest Franciscan terrane is the King Range terrane (McLaughlin et al., 1982). The King Range terrane is a composite of two units (the Point Delgada and King Peak subterranes) that are stitched together across steep faults by cross-cutting adularia-bearing veins dated at 13.8 Ma (McLaughlin et al., 1985b). The Point Delgada subterrane consists of basaltic pillow flows, pillow breccias, and tuffs, intruded by diabase sills. This igneous section is overlain by folded and sheared, pumpellyite-bearing, quartz-rich to arkosic sandstone and argillite devoid of detrital K-feldspar. Melange zones containing rare blueschist blocks cut the igneous and overlying sedimentary rocks. Radiolarians from red argillite interbedded with pillow breccias are Late Cretaceous (Campanian or Coniacian).

The King Peak subterrane is characterized by complexly folded and broken calcareous argillite and interbedded quartzofeldspathic to volcanoclastic sandstone, probably deposited in a lower-slope or inner-trench setting. These strata are associated locally with melange containing minor pods and lenses of pillow basalt and pelagic limestone, or basalt plus chert containing radiolaria and diatoms (McLaughlin et al., 1982; McLaughlin, 1983). Both the well-bedded pelagic and hemipelagic rocks and the melange blocks of pillow basalt-chert contain radiolarians and (or) foraminifers of middle Miocene or younger age. A few poorly preserved foraminifers identified as Paleocene or Eocene in age are most likely reworked, although it is possible that part of the King Peak subterrane could be Paleogene in age (McLaughlin et al., 1982).

Aeromagnetic data (Griscom, 1980) indicate that the King Range terrane dips southwestward and overlies the Coastal terrane. Thus, the King Range terrane was overthrust or obducted northeastward onto the North American margin in post-middle Miocene time. This accretion may have resulted from NE-SW-oriented compression and uplift associated with late Neogene convergence between the Pacific and North American plates at the Mendocino triple junction (McLaughlin et al., 1982, 1983a). Vitrinite reflectance data (Blake et al., 1988) show the entire King Range terrane underwent a major thermal event which was confined to the King Range and was most intense at Point Delgada, where the hydrothermal system is dated at ~14 Ma. This is interpreted to indicate that accretion postdated the stitching of the King Peak and Point Delgada subterranes at 14 Ma.

COAST RANGE OPHIOLITE-GREAT VALLEY SEQUENCE

Along the eastern margin of the Coast Range and in outliers to the west, is a distinctive sequence of Upper Jurassic to Upper Cretaceous sandstone, shale, and conglomerate that, in general, lacks the stratal disruption seen in the Franciscan Complex, is relatively unmetamorphosed and fossiliferous. The basement of these sedimentary rocks throughout the Coast Ranges is the Coast Range ophiolite, consisting of ultramafic and mafic rocks upon which the clastic sedimentary rocks of the Great Valley sequence were deposited (Bailey et al., 1971; Hopson et al., 1981). The boundary between the Coast Range ophiolite and (or) the Great Valley sequence and the Franciscan Complex is everywhere a fault. Whereas this fault was earlier called the Coast Range thrust (Bailey et al., 1969) and interpreted to be a paleosubduction zone (Ernst, 1970), recent work suggests that any original thrust has been overprinted by low-angle normal faults of Late Cretaceous to earliest Tertiary age, or younger high-angle faults (Platt, 1986; Jayko et al., 1987; Harms et al., 1987; Jones, 1987).

Like the Franciscan Complex, the Coast Range ophiolite - and the lowermost Great Valley sequence has been subdivided into several terranes based on differences in stratigraphy. There are major differences in the nature of the ophiolitic basements of the Elder Creek and Del Puerto terranes, although their clastic sedimentary rocks have a similar lithology and fossils. The Healdsburg terrane is distinguished from the Del Puerto and Elder Creek terranes by the distinct lithologic character of the Lower Cretaceous sedimentary section which overlies ophiolitic rocks of the Healdsburg terrane.

Del Puerto terrane. The Del Puerto terrane (Figure 1) crops out along the east side of the Diablo Range from north of Mount Diablo south to a termination west of Bakersfield (south of Figure 1). The basal part of this terrane consists of a Middle Jurassic (~166 Ma) ophiolite capped by a thick accumulation of Upper Jurassic (~152 to 146 Ma) siliceous volcanic rocks (extrusive and intrusive keratophyre and quartz keratophyre referred to the Lotta Creek Formation). The silicic volcanic rocks probably represent an island arc formed above an oceanic basement (Bailey and Blake, 1974; Evarts, 1977; Blake and Jones, 1981). The Upper Jurassic Lotta Creek Formation (Raymond, 1970) in the upper part of the ophiolite is overlain by mudstone of Late Jurassic (Tithonian) and Early Cretaceous (Albian) age. These rocks are overlain by sandstone, mudstone, and conglomerate which extend eastward onto Sierran basement and range in age from Late Cretaceous to Paleocene.

The western margin of the Del Puerto terrane is the late Cenozoic Tesla-Ortigalita fault (Page, 1981). The eastern boundary with Sierran basement is buried beneath the Upper Cretaceous and lower Cenozoic overlap deposits, but is believed, from geophysical evidence, to be a fault (Wentworth et al., 1984).

Sixty kilometers south of Figure 1, near Llanada (Hopson et al., 1981, 1986), paleontological and paleomagnetic data have been obtained from the ophiolite and Upper Jurassic basal strata overlying the silicic volcanic section of the Del Puerto terrane. These data suggest that the ophiolitic rocks formed near the equator, were covered at low latitudes by siliceous volcanic rock and radiolarian tuff (Lotta Creek Formation), and

subsequently carried northward to about their present latitude. Here, they were depositionally overlain by Upper Jurassic (Tithonian) strata of the Great Valley sequence. Their arrival at middle latitudes occurred by about 146 Ma. The timing of this arrival corresponds in part to metamorphism and other deformation (150 to 157) Ma in rocks of the Sierra Nevada and Klamath Mountains. This deformation has been attributed to the Nevadan orogeny (Wright and Fahan, 1988).

Elder Creek terrane. North of the Diablo Range, along the west side of the Sacramento Valley, the Coast Range ophiolite has a more oceanic character. The siliceous volcanic rocks of the Del Puerto terrane are absent and the uppermost part of the ophiolite instead consists of a breccia that locally contains well-rounded clasts made up entirely of pieces of the underlying ophiolite (Bailey et al., 1971). The overlying Jurassic sedimentary rocks consist largely of tuffaceous mudstone, sandstone, and chert-pebble conglomerate. Sedimentary serpentinite breccias occur locally in the stratigraphically higher Lower Cretaceous strata Carlson, 1981).

Structurally below the ophiolite of the Elder Creek terrane, and above the Franciscan Complex (Pickett Peak and Yolla Bolly terranes), is a serpentinite-matrix melange that contains blocks and slabs of mafic volcanic rocks, Upper Jurassic radiolarian chert, diabase, minor gabbro, and rare amphibolite. Along the western margin of this melange are slabs of Upper Jurassic Galice Formation, a displaced part of the Western Klamath terrane of the Klamath Mountains (Jayko and Blake, 1986). The melange is interpreted to be the result of extensional deformation which occurred during unroofing of the deeply subducted Franciscan Complex. This extensional event and unroofing of the Franciscan Complex is thought to have occurred during Late Cretaceous to earliest Tertiary time and was accompanied by attenuation of the structurally overlying Western Klamath and Elder Creek terranes (Jayko et al., 1987).

A paleomagnetic study of the Coast Range ophiolite and overlying Upper Jurassic and Lower Cretaceous sedimentary strata from the Elder Creek terrane near Paskenta indicates that these rocks have a magnetic overprinting that post-dates both deposition and folding (Frei and Blake 1987).

Healdsburg terrane. Near Healdsburg (Figure 1) and to the south is a dismembered ophiolite structurally overlain by a sequence of turbidites of Late Jurassic and Early Cretaceous age which, in turn, are overlain by a thick pile of Lower Cretaceous conglomerate. Because the Lower Cretaceous section is so different from coeval strata to the east in the Elder Creek and Del Puerto terranes we describe it in more detail.

At the Dry Creek Dam site 15 km north of Healdsburg, sandstone facies C and B (Mutti and Ricci-Lucci, 1972), are interbedded with thin-bedded turbidites (facies E) and concretion-bearing mudstone (facies G). Carbonaceous debris is conspicuous in the thick-bedded graywackes. In a few places bedding thickness patterns suggest upward thickening sequences.

Upsection from these probable submarine fan deposits, the bedding style changes abruptly to lenticular beds of alternating sandstone and mudstone lacking sedimentary structures. Grading is restricted to thin (about 1 cm thick) sandstone beds and carbonaceous debris is ubiquitous in the sandstone. This association is not indicative of a turbiditic submarine fan facies and more likely represents a shallow marine, possibly deltaic, environment of deposition.

Up to 3000 m of conglomerate overlies and locally fills channels in the shallow marine strata. The conglomerate clasts are principally well-rounded pebbles and cobbles of light-colored rhyolite porphyry, rhyolitic welded ash-flow tuff, and minor amounts of chert. Coeval conglomerate in the Great Valley sequence to the east contains mainly chert pebbles. Species of *Buchia* of Early Cretaceous (Valanginian) age are found throughout the turbidites and the conglomerate of the Healdsburg terrane. Biotite from conglomerate clasts give a K-Ar age of about 145 Ma (Berkland, 1969).

Also shown on Figure 1 is a small area of Healdsburg(?) terrane east of San Francisco. These rocks are clearly part of the Great Valley sequence and contain a thick Lower Cretaceous conglomerate (the Oakland conglomerate of Crittenden, 1951), however, unlike the Healdsburg area, these rocks include an overlying section of Upper Cretaceous and Tertiary strata.

Other outliers of the Coast Range ophiolite and Great Valley sequence. Numerous small composite slabs of rocks assignable to the Del Puerto or Elder Creek terranes of the Coast Range ophiolite and the Great Valley sequence overlie or are faulted into the Central belt of the Franciscan Complex in central and northern California. Many of the ophiolite slabs also include rocks of the Pickett Peak and (or) Yolla Bolly terranes, indicating they were juxtaposed with the Central belt sometime following metamorphism and accretion of the Yolla Bolly terrane about 90 to 92 Ma (Mattinson and Echeverria, 1980). Depositional overlap of the Central belt and Coast Range ophiolite by early Eocene marine strata in southwest Oregon and the first appearance of reworked Franciscan and ophiolitic detritus in the depositional record of the Coast Ranges in rocks of this age, suggests that incorporation of the outliers into the Central belt largely pre-dated 52 Ma.

The ophiolite outliers form two subparallel, northwest-trending belts in northern California (Figure 1). The northeastern belt of outliers all include sedimentary ophiolitic breccia and are assigned to the Elder Creek terrane (examples include the Lassic outlier east of Garberville, and Geyser Peak east of Healdsburg). Outliers of the southwestern belt lack ophiolite breccia but contain keratophyric volcanic rocks and Late Jurassic siliceous tuffs (152-146 Ma) in their upper part, indicating affinities to the Del Puerto and Healdsburg terranes (examples include the Camp Meeker outlier near Occidental, the composite Loma Prieta and Mount Umunum outlier near San Jose, and the Benbow outlier

south of Garberville). The northernmost of the arc-related outliers in California, which is at Benbow, was offset a minimum of 260 km from the northern end of the Del Puerto terrane in the Diablo Range mostly along pre-52 Ma right-lateral faults (McLaughlin et al., 1988). Maximum offset of these outliers possibly exceeds 1200 km if the recent correlation of rocks in the San Juan Islands of western Washington with the Coast Range ophiolite and Great Valley sequence of California is correct (Garver, 1986; McLaughlin et al., 1988.

TECTONIC INTERPRETATION

About 150 Ma, at about the beginning of the Nevadan orogeny, the Coast Range ophiolite, Western Klamath terrane, and the older terranes of the Klamath-Sierra Nevada were imbricated by eastward-directed thrusting (Blake et al., 1988; also, see Irwin, this field trip guidebook). This was followed, in the Early Cretaceous (~125 Ma), by eastward subduction of the Pickett Peak terrane and, somewhat later (~92 Ma), by subduction of the Yolla Bolly terrane. During, or at the end of this subduction event, lower-plate Franciscan terranes were involved in west-directed thrusting or else were subjected to east-directed underthrusting from younger Franciscan rocks to the west. At about the same time, the upper-plate Western Klamath and Elder Creek terranes were attenuated by low-angle normal faulting. This detachment faulting appears to have occurred during the latest Cretaceous or earliest Tertiary, based on map relations northeast of Healdsburg (Swe and Dickinson, 1970; McLaughlin and Ohlin, 1984).

During the early Cenozoic, the tectonic regime changed to one of right-lateral transform faulting, which resulted in the tectonic erosion of the western margin of the previously accreted terranes, and the northward transport of blocks and slabs of Coast Range ophiolite-Great Valley sequence, Pickett Peak, and Yolla Bolly terranes.

By about 40 Ma, eastward-directed subduction resumed and resulted in the accretion and imbrication of the Coastal and Yager terranes and perhaps in some of the young thrusting along the eastern margin of the Coast Ranges (Suppe, 1979). Finally, about 10 Ma at the latitude of San Francisco, the Mendocino triple junction migrated northward, terminating subduction at that latitude and initiating right-lateral strike-slip faulting. Beginning about 15 Ma, and as the result of early transtensional deformation, northwardly younging Neogene volcanism occurred and small extensional structural basins formed in conjunction with propagation of the San Andreas transform (Dickinson and Snyder, 1979; Fox et al., 1985; McLaughlin and Nilsen, 1982). This was followed, about 3 Ma, by the present transpressive phase that elevated the Coast Ranges (Sarna-Wojcicki, 1976; Harbert and Cox, 1986).

In summary, the tectonic history of the northern Coast Ranges was not simply a long period of Mesozoic east-directed subduction followed by San Andreas right-lateral faulting in the late Cenozoic. Instead, there were alternating periods of subduction and right-lateral transform faulting, including an important episode of "proto-San Andreas" faulting (80-50 Ma) when oceanic fragments from as far away as the southern hemisphere were carried northward and accreted to California (Nilsen, 1978; McLaughlin et al., 1988.

ROAD LOG-- DAY 1, June 28, 1988-- SAN FRANCISCO TO HEALDSBURG

The road log begins at the Toll Plaza, Golden Gate Bridge (see Figure 2).

Mileage

1.5 STOP 1-1: Golden Gate Bridge, Vista Point parking area. Just north of the parking area is a well-exposed section of west-dipping pillowed greenstone overlain by about 82 meters of radiolarian chert. The basalt contains high titanium and probably represents a type of off-ridge volcanism, such as a seamount. The basal contact is everywhere a low-angle thrust fault (Wahrhaftig, 1984). Samples from the lower and middle portions of the chert give radiolarian ages of Early to Middle Jurassic (Murchey, 1984). Samples from the upper, gray, more-massive portion are Early Cretaceous (Valanginian), while radiolarians of Cenomanian age have been collected from the uppermost beds.

   Thin-bedded red, radiolarian chert conformably overlies the basalt. However, the contact is locally faulted. Note the local concentration of Mn-oxides near the base of the chert and the irregular bedding described in Bailey et al., 1964. Structural study of these cherts, indicates that the small-scale chevron folding, seen in nearly every outcrop, is related to larger folds with an amplitude of several hundred meters. Overlying the chert is about 300 meters of graywacke and mudstone. A piece of graywacke found along the beach south of here yielded an ammonite of Cenomanian age.

10.7 STOP 1-2: Greenbrae Quarry. Structurally below the basalt-chert graywacke sequence of the previous stop is typical melange of the Central belt consisting of numerous resistant blocks, or "knockers," of greenstone, chert, and graywacke in a highly sheared matrix of argillite and graywacke. At this stop, the matrix of the melange is particularly well-exposed, and the presence of angular, unsorted pebbly mudstone suggests an origin by submarine landsliding (olistostrome). Note also the difference in composition between the well-bedded quartzo-feldspathic graywacke and the lithic graywacke of the matrix with abundant chert and mafic volcanic detritus. The former is probably a slab of Campanian graywacke (referred to the Novato Quarry terrane), which forms a nearly continuous unit a few kilometers to the north. Within the melange matrix, numerous megafossils and microfossils (radiolaria, and palynomorphs)

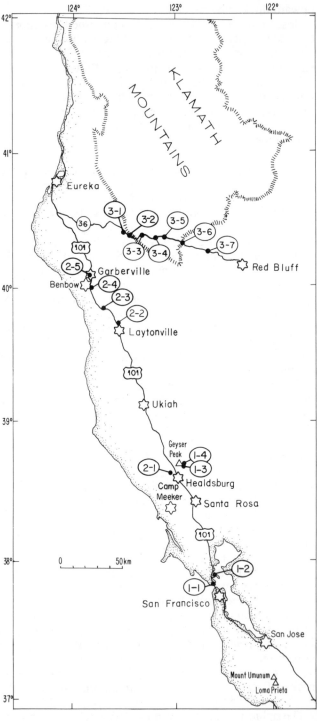

FIGURE 2 Map showing field trip stops for days 1 to 3.

indicate a Tithonian age.

66.9 STOP 1-3: Ophiolite sequence of Geyser Peak and Black Mountain. At this locality, the upper part of a fragmented ophiolite is present (Figures 3 and 4). The ophiolite structurally overlies the Franciscan Complex here, but to the southeast it is incorporated into melange of the Central belt. A remnant of basal strata of the Great Valley sequence is also present in depositional contact with the underlying ophiolite. The basal contact of the ophiolite upon the Franciscan Complex has been interpreted as a deformed thrust fault considered to be an outlier of the upper plate of the Coast Range thrust of Bailey et al., (1971). The Coast Range "thrust" was recently reinterpreted to be a low angle normal (detachment) fault along which section has been removed. Furthermore, the outlier section of the Coast Range ophiolite exposed here on Geyser Peak and Black Mountain was right-laterally translated away from the western edge of the Great Valley province and incorporated into the Central terrane of the Franciscan Complex in earliest Tertiary time. Among the features to observe at this locality are: 1) Basal strata of the Great Valley sequence. These strata vary from steeply tilted to nearly horizontal and are cut by at least two steeply-dipping faults. The basal strata consists of thin-bedded, flysch-like sandstone and mudstone composed of basaltic detritus with sporadic carbonate concretions and thick sedimentary lenses of coarse diabase and basalt-clast breccia in the lower-most part. The age of strata that overlie the diabase breccia at this locality, based on scant fossils, is Early Cretaceous (probably Valanginian). Sandstone, mudstone, and ophiolitic sedimentary breccia stratigraphically below the Lower Cretaceous strata are Late Jurassic (Tithonian and Kimmeridgian). This age determination is based partly on Tithonian to Kimmeridgian radiolaria in a tuffaceous chert lens which unconformably overlies the breccia and basalt on Black Mountain and partly on contact relations near Paskenta (Figure 1), where the same breccia unconformably overlies a gabbroic intrusive radiometrically dated at 155 Ma. 2) Basaltic pillows, pillow breccia, and tuff, exposed along the Geysers road on Black Mountain, appears to be the structurally highest part of the ophiolite succession exposed in this area. A Late Jurassic (Kimmeridgian) or older age for these volcanic rocks is indicated by the radiolaria in the tuffaceous chert which unconformably overlies them on Black Mountain. 3) Diabase and microgabbro compose a sill or concordant dike complex that underlies the pillow flows. Some of this diabase is exposed along the road just southeast of Cold Creek, where it is in fault contact with the basalt flows; here the diabase is overlain depositionally by the mafic ophiolitic breccia at the base of the Great Valley sequence. The diabase and microgabbro sills are thickest beneath Geyser Peak, northwest of the road, where they form a 460-meter-thick southwest-dipping slab. Pyroxene in this diabase has been extensively replaced by brown and green hornblende. Epidote is an abundant replacement mineral in the intrusive and extrusive rocks and presumably formed during low-grade green-schist-facies meta-morphism. Later, regional zeolite-grade (laumontite) metamorphism locally overprinted the greenschist facies assemblage in the volcanic rocks and is prominent in the overlying clastic rocks.

Although cumulate pyroxene gabbro is present in ophiolitic rocks elsewhere in the region, it occurs only locally beneath the basaltic flows and diabase sills in the Geyser Peak-Black Mountain area and is not exposed along the Geysers-Healdsburg Road. Serpentinized and sheared harzburgite peridotite is present as a sheet up to 200 meters thick along the base of the ophiolite complex in this area.

Well-bedded Franciscan graywacke is exposed in Little Sulphur Creek. As you climb out of Little Sulphur Creek, more of this graywacke is exposed in the road cuts, where it is pervasively broken and faulted. Geothermal well logs show that this graywacke slab is the structurally lowest unit exposed in the Geysers region. It is characterized lithologically by abundant chert detritus. The sandstone unit lacks any associated chert and volcanic rock blocks, and it has a relatively unmetamorphosed sedimentary fabric. As you climb higher on the north side of Little Sulphur Creek canyon, note the prominent northwest-trending belt of serpentinite that follows the steep north-dipping Mercuryville fault zone.

73.9 STOP 1-4: Vista of The Geysers steam field and the Geysers chert bed. A zone of intensely acid-leached, hydrothermally altered Franciscan rocks (largely graywacke) can be seen on the north side of Big Sulphur Creek. This area is part of the principal vent zone for The Geysers vapor-dominated hydrothermal system. The Geysers steam field currently produces over 1755 megawatts of electricity, with a projected capacity greater then 2000 megawatts; more than enough to supply the electrical needs of San Francisco. The alteration zone consists of fumaroles and sporadic hot springs that vent along steep-dipping, N30°-40°W-trending, en echelon faults that diagonally transect Big Sulphur Creek. Venting of steam and hot water also occurs to the southeast of this area, beneath a major northeast-dipping sheet of ultramafic rock. Most near-surface hydrothermal alteration in that area appears to result from hydrothermal fluids that have migrated upward along steep-dipping faults and fractures and spread laterally beneath the impermeable lower contact of the north-dipping ultramafic sheet on the northeast side of Big Sulphur Creek.

The Geysers steam field is also the epicentral region for innumerable small earthquakes and microearthquakes. Recent research suggests that much of this seismic activity is induced by steam production and reinjection of geothermal fluids.

In the roadcuts at this locality, the following relationships in the Geysers chert terrane can be observed: (1) Undulatory basaltic flows of the

GEOLOGY OF THE GEYSER PEAK OUTLIER OF THE COAST RANGE
OPHIOLITE, SONOMA COUNTY, CALIFORNIA

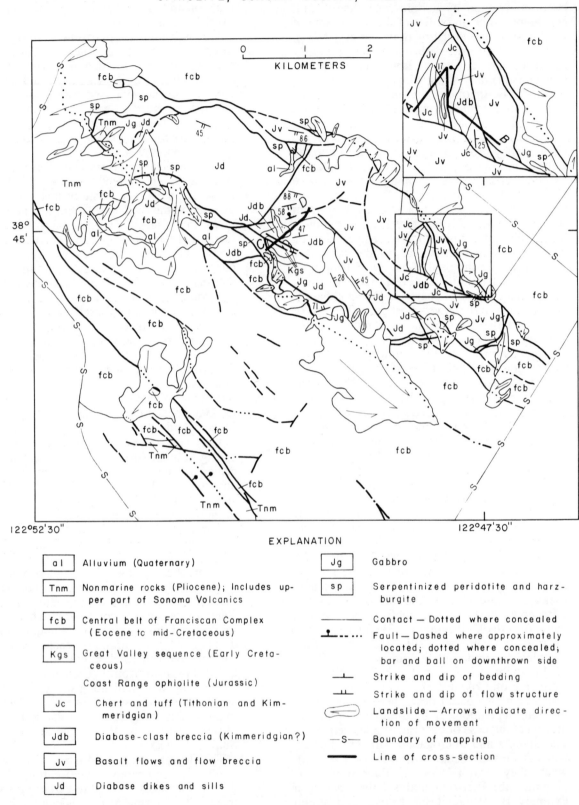

EXPLANATION

al	Alluvium (Quaternary)	Jg	Gabbro
Tnm	Nonmarine rocks (Pliocene); Includes upper part of Sonoma Volcanics	sp	Serpentinized peridotite and harzburgite
fcb	Central belt of Franciscan Complex (Eocene to mid-Cretaceous)		Contact — Dotted where concealed

Great Valley sequence (Early Cretaceous) — Kgs

Coast Range ophiolite (Jurassic)

Jc — Chert and tuff (Tithonian and Kimmeridgian)

Jdb — Diabase-clast breccia (Kimmeridgian?)

Jv — Basalt flows and flow breccia

Jd — Diabase dikes and sills

Fault — Dashed where approximately located; dotted where concealed; bar and ball on downthrown side

Strike and dip of bedding

Strike and dip of flow structure

Landslide — Arrows indicate direction of movement

—S— Boundary of mapping

Line of cross-section

FIGURE 3 Geologic map of the Geyser Peak outlier of the Coast Range ophiolite showing locations of cross sections illustrated in Figure 4 (from McLaughlin et al., 1988).

Franciscan Complex that dip steeply to the east, locally with red siliceous shale between the flows. The high titanium, alkali composition of these basaltic rocks suggests they formed in an oceanic seamount or plateau setting. (2) Overlying the basalt flows is a prominent 67-meter-thick lens of radiolarian chert. The Geysers chert section was the first long-ranging siliceous pelagic section documented in the Franciscan Complex. Radiolaria extracted from this chert were studied in considerable detail by E. A. Pessagno, of the University of Texas at Dallas, and later by B. Murchey (U.S.G.S.), who found that the section, like that at Marin Headlands, ranges in age from Early or Middle Jurassic at the base, to Late Cretaceous (early Cenomanian) at the top. Pelagic sedimentation rates for the chert (disregarding compaction) range from .4 cm/1000 yr. for the Valanginian section, to .1 cm/1000 yr. for the Jurassic and Cretaceous section. Two zones of recrystallized chert separate major radiolarian age assemblages. The lower zone separates Tithonian from overlying Valanginian chert and the upper zone separates Valanginian from overlying Cenomanian chert. Minor manganese-stained radiolaria are reworked into the Cenomanian chert section, suggesting that an unconformity may be associated with the upper zone of recrystallization. Overlying the chert is steep-dipping, medium-grained Franciscan graywacke, locally with prominent carbonaceous debris and detritus derived from the underlying chert.

ROAD LOG-- DAY 2, June 29, 1989-- HEALDSBURG-ARCATA

Mileage

15.8 STOP 2-1: Lower Cretaceous, Healdsburg Conglomerate. Note abundance of siliceous porphyritic volcanic rocks and scarce granitic detritus. Pebble imbrication near here indicates West to East current direction. Buchia-bearing mudstone (Valanginian ?) occurs largely below the conglomerate but Buchias also occur locally within and above.

91.1 Laytonville Quarry. On the right side (east) of 101 is a small quarry in a large composite blueschist block. The block consists of metabasalt, metachert, metaironstone, probably metamorphosed about 160 Ma. Three new blueschist minerals, deerite, howieite, and zussmanite, were discovered here by Stuart Agrell of Cambridge University in 1965.

98.9 STOP 2-2: Laytonville limestone. This block of pink pelagic limestone is one of many found near Laytonville. The foraminifers from these blocks have been extensively studied by Isabella Premoli Silva and William Sliter who have determined an age of Albian to Coniacian. Paleomagnetic studies by Alvarez and others (1980) and

Tarduno and others (1986) suggest that the limestone formed at about 14⁰ south latitude.

124.7 STOP 2-3: Yager terrane, Cummings. At this locality is well-exposed and little-sheared arkosic sandstone, mudstone, and minor conglomerate of the Yager terrane of the Coastal belt. Palynomorphs collected near here are reportedly of Paleogene age (O'Day, 1974). The section exposed along the old highway, consists of thick-bedded, up to 50 m-thick fining-upward sequences of sandstone and minor conglomerate, alternating with lenticular thin-bedded flysch-like sandstone and predominant mudstone. The association and sedimentary structures are consistent with deposition near the channelized portion of a deep-sea fan complex. These rocks are juxtaposed with the Central belt of the Franciscan Complex along a major fault referred to as the Coastal belt thrust.

148.2 STOP 2-4: Lunch stop, Richardson Grove State Park

156.0 STOP 2-5: Yager terrane, Garberville. At this locality we again view rocks assigned to the Yager terrane and some of the complex folding and faulting associated with these rocks. In the vicinity of Garberville City Park along the Eel River, note complexly folded thin-bedded sandstone and argillite exposed beneath terrace deposits. Note, also, that upside-down strata of the Yager terrane are refolded into tight outcrop-scale plunging folds within a few hundred meters of the Coastal belt thrust. Steep-dipping faults associated with the late Cenozoic Garberville fault zone overprint the Coastal belt thrust here and vertically offset near-shore marine siltstone and mudstone of middle Miocene to Pliocene age.

Southwest of Garberville City Park, in a quarry on the opposite (northwest) side of the Eel River, a well-bedded upside-down sequence of sandstone and argillite of the Yager terrane are well-exposed. Abundant sedimentary structures show that these rocks are overturned. The sandstone contains conspicious detrital muscovite, K-feldspar, and silicic volcanic and plutonic detritus, indicative of a continental source. These rocks have been subjected to zeolite facies (laumontite-grade) metamorphism, and vitrinite reflectance data indicate that they have been heated to temperatures mostly in the range 120° to 160° C., but locally to as high as 240° C. adjacent to the Coastal belt thrust. Palynomorphs from carbonate concretions in argillitic sections of the Yager terrane indicate these rocks range in age from late Paleocene to late Eocene.

The Yager terrane was deformed and juxtaposed with the Central terrane in latest Eocene to late early Miocene time, based on the age relations of the Yager terrane and the post-accretion marine rocks which overlap the Coastal belt thrust near Garberville.

A

GEYSER PEAK OUTLIER SECTION A-B, BLACK MOUNTAIN, SONOMA COUNTY, CALIFORNIA

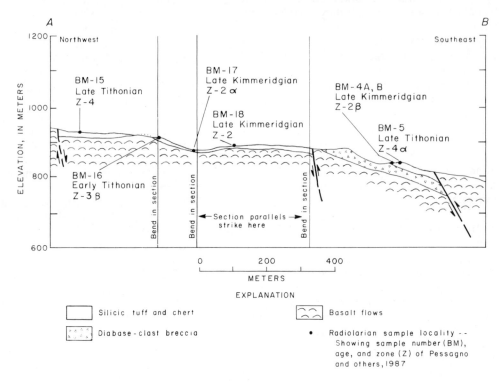

EXPLANATION

	Silicic tuff and chert			Basalt flows
	Diabase-clast breccia		•	Radiolarian sample locality -- Showing sample number (BM), age, and zone (Z) of Pessagno and others, 1987

B

SKETCH OF GEYSER PEAK OUTLIER SECTION C-D ALONG THE GEYSERS-HEALDSBURG ROAD, SONOMA COUNTY, CALIFORNIA
(Viewed toward the southeast)

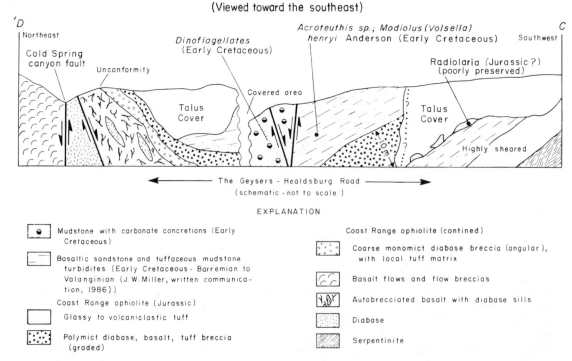

EXPLANATION

Mudstone with carbonate concretions (Early Cretaceous)

Basaltic sandstone and tuffaceous mudstone turbidites (Early Cretaceous- Barremian to Valanginian (J.W. Miller, written communication, 1986))

Coast Range ophiolite (Jurassic)

Glassy to volcaniclastic tuff

Polymict diabase, basalt, tuff breccia (graded)

Coast Range ophiolite (contined)

Coarse monomict diabase breccia (angular), with local tuff matrix

Basalt flows and flow breccias

Autobrecciated basalt with diabase sills

Diabase

Serpentinite

FIGURE 4 Structure sections through the Geyser Peak outlier of the Coast Range ophiolite. A.) Section A-B B.) Section C-D, viewed from the southeast. See Figure 3 for section locations.

REFERENCES

Aalto, K.R., Structural geology of the Franciscan Complex of the Crescent City area, northern California, in *Cretaceous stratigraphy of western North America, Pacific Section, Society of Economic Palentologists and Mineralogists*, v. 46, edited by P. Abbott, pp. 197-209, 1986.

Aalto, K.R. and J.M. Murphy, Franciscan Complex Geology of the Crescent City area, northern California, in *Franciscan geology of northern California, Pacific Section, S.E.P.M.*, v. 43, edited by M.C. Blake, Jr., pp. 185-201, 1984.

Alvarez, W., D.V. Kent, I. Premoli Silva, R.A. Schweickert, and R. Larson, Franciscan Complex limestone deposited at 17° south paleolatitude, *Geol. Soc. Am. Bull.*, v. 96, pp. 476-484, 1980.

Atwater, T., Implications of plate tectonics for the Cenozoic tectonic evolution of western North America, *Geol. Soc. Am. Bull.*, v. 81, pp. 3513-3536, 1970.

Bachman, S.B., A Cretaceous and early Tertiary subduction complex, Mendocino Coast, northern California, in *Mesozoic paleogeography of the western United States, Pacific Section, S.E.P.M., Pacific Coast Paleogeography, Symposium 2,* edited by D.G. Howell, and K.A. McDougall, pp. 419-430, 1978.

Bailey, E.H., and M.C. Blake, Jr., Major chemical characteristics of Mesozoic ophiolite in the California Coast Ranges, *J. Res. U.S. Geological Survey,* v. 2, no. 6, pp. 637-656, 1974.

Bailey, E.H., and M.C. Blake, Jr., Tectonic development of western California during the late Mesozoic, *Geotektonika*, no. 3, pp. 17-30 and no. 4, pp. 24-34, 1969.

Bailey, E.H., M.C. Blake, Jr., and D.L. Jones, On-land Mesozoic ocean crust in California and Coast Ranges, in *U.S. Geol. Surv. Prof. Pap. 700-C*, pp. C-70-C81, 1971.

Bailey, E.H., W.P. Irwin, and D.L. Jones, Franciscan and related rocks and their significance in the geology of western California, *California Div. Mines and Geol. Bull. 183,* 177 p., 1964.

Berkland, J.O., Geology of the Novato quadrangle, Marin County, California, M.S. thesis, San Jose State University, San Jose, California, 1969.

Berry, K.D., New age determinations in Franciscan limestone blocks, northern California (abstract), *American Association of Petroleum Geologists Bulletin*, v. 66, no. 10, p. 1685, 1982.

Blake, M.C., Jr., A.S. Jayko, R.J. McLaughlin, and M.B. Underwood, Metamorphic and tectonic evolution of the Franciscan Complex, northern California, in *Metamorphism and crustal evolution of the western United States, Rubey vol. VII,* edited by W.G. Ernst, pp. 1036-1059, Prentice-Hall, Inc., Englewood Cliffs, N.J., 1988.

Blake, M.C., Jr., Jayko, A.S., and McLaughlin, R.J., Tectonostratigraphic terrane*s of northern California, in Earth Science Series, Circumpacific Council for Energy and Mineral Resources,* edited by D.G. Howell, v. 1, p. 159-171, 1985a.

Blake, M.C., Jr., D.C. Engebretson, A.S. Jayko, and D.L. Jones, Tectonostratigraphic terranes in southwestern Oregon, in *Earth Science Series, Circumpacific council for Energy and Mineral Resources,* edited by D.G. Howell, v. 1, pp. 147-157, 1985b.

Blake, M.C., Jr., D.S. Harwood, E.J. Helley, W.P. Irwin, A.S. Jayko, and D.L. Jones, Geologic map of the Red Bluff 1:100, 000 quadrangle, California, *U.S. Geol. Surv. Open File Rep. 84-105*, 33 p., 1 map, scale 1:100,000, 1984a.

Blake, M.C., Jr., D.G. Howell, and A.S. Jayko, Tectonostratigraphic terranes of the San Francisco Bay region, in *Franciscan Geology of Northern California, Pacific Section, S.E.P.M.*, v. 43, edited by M.C. Blake, Jr., pp. 5-22, Bakersfield, Calif, 1984b.

Blake, M.C., Jr., D.G. Howell, and D.L. Jones, Tectonostratigraphic terrane map of California, *U.S. Geol. Surv. Open File Rep. 82-593* scale 1:750,000, 1982.

Blake, M.C., Jr., and D.L. Jones, The Franciscan assemblage and related rocks in northern California, a reinterpretation, in *The Geotectonic Development of California,* edited by W.G. Ernst, Englewood Cliffs, N.J., Prentice Hall, pp. 306-328, 1981.

-----Origin of Franciscan melanges in northern California, *Society of Economic Paleontologists and Mineralogists Special Paper 19*, pp. 255-263, 1974.

Brothers, R.N., P.M. Black, M.C. Blake, Jr., and A.S. Jayko, Geochemistry of Franciscan metabasites, Pickett Peak terrane, northern California (in preparation).

Carlson, Christine, Sedimentery serpentinite of the Wilbur Springs area--a possible Early Cretaceous structural and stratigraphic link between the Franciscan Complex and the Great Valley sequence: M.S. thesis, Stanford University, 105 p., 1981.

Cashman, S.M., P.H. Cashman, and J.D. Longshore, Deformational history and regional tectonic significance of the Redwood Creek schist, northwestern California, *Geol. Soc. of Am. Bull.*, v. 97, pp. 35-47, 1986.

Cloos, M., Thermal evolution of convergent plate margins: thermal modeling and re-evaluation of isotopic Ar-ages for blueschists in the Franciscan Complex of California, *Tectonics*, v. 4, pp. 421-434, 1985.

Coleman, R.G., and D.E. Lee, Metamorphic aragonite in the glaucophane schists of Cazadero, California, *American Journal of Science*, v. 260, pp. 577-595, 1963.

Cowan, D.S., Structural styles of Mesozoic and Cenozoic melanges in the western cordillera of North America, *Geol. Soc. of Am. Bull.*, v. 96, pp. 451-462, 1985.

Crittenden, M.D., Jr., Geology of the San Jose-Mount Hamilton area, California: *California Division of Mines and Geology Bull.* 157, 74 pp., 1951.

DePolo, C.M., and H.N. Ohlin, The Bartlett Springs fault zone: an eastern member of the California plate boundary system, *Geol. Soc. of Am. Abstr. Programs*, v. 16 no. 6, p. 486, 1984.

Dickinson, W.R., and W.S. Snyder, Geometry of triple

junctions related to San Andreas transform, *J. of Geophys. Res.*, v. 84, pp. 561-572, 1979.

Engebretson, D.C., A. Cox, and R.G. Gordon, Relative motions between oceanic and continental plates in the Pacific basin, *Geol. Soc. of Am. Spec. Pap. 206*, p. 59, 1985.

Ernst, W.G., Tectonic contact between the Franciscan melange and the Great Valley sequence--crustal expression of a Late Mesozoic Benioff zone, *J. of Geophys. Res.*, v. 75, pp. 886-901, 1970.

Evarts, R.C., The geology and petrology of the Del Puerto ophiolite, Diablo Range, central California Coast Ranges in *North American ophiolites, Oregon Dept. Geol. and Mineral Ind. Bull.*, v. 95, pp. 121-140, 1977.

Fox, K.F., Jr., R.J. Fleck, G.H. Curtis, and C.E. Meyer, Implications of northwestwardly younger age of the volcanic rocks of west-central California, *Geol. Soc. of Am. Bull.*, v. 96, no. 5, pp. 647-654, 1985.

Frei, L.S., and M.C. Blake, Jr., Remagnetization of the Coast Range ophiolite and lower part of the Great Valley sequence in northern California and southwest Oregon, *J. Geophy. Res.*, v. 92, B-5, pp. 3487-3499, 1987.

Garver, J.I., Great Valley Group/Coast Range ophiolite fragment in the San Juan Islands of Washington, *Geol. Soc. of Am. Abstr. Programs*, v. 18, no. 2, p. 108, 1986.

Gester, G.C., Northern Coast Ranges; American Association of Petroleum Geologists, v. 35, p. 200-208, 1951.

Griscom, A., Aeromagnetic map and interpretation maps of the King Range and Chemise Mountain instant study areas, northern California, *U.S. Geol. Surv. Misc. Field Stud. Map MF-1196-B*, 1980.

Hamilton, W., Mesozoic California and the underflow of Pacific mantle, *Geol. Soc. of Am. Bull.*, v. 80, pp. 2409-2430, 1969.

Harbert, W., and A. Cox, Late Neogene motion of the Pacific plate, abstract, *EOS*, v. 67, no. 44, p. 1225, 1987.

Harbert, W.P., R.J. McLaughlin, and W.V. Sliter, Paleomagnetic and tectonic interpretation of the Parkhurst Ridge limestone, coastal belt Franciscan, northern California, in *Franciscan geology of northern California, Pacific Section, S.E.P.M.*, v. 43, pp. 175-183, 1984.

Harms, T., A.S. Jayko, and M.C. Blake, Jr., Kinematic evidence for extensional unroofing of the Franciscan Complex along the Coast Range Fault, northern Diablo Range, *California, Geol. Soc. of Am. Abstr. Programs*, v. 18, p. 694, 1987.

Hearn, B.C., Jr., R.J. McLaughlin, and J.M. Donnelly-Nolan, Tectonic framework of the Clear Lake Basin, California, in Late Quaternary climate, tectonism, and sedimentation in Clear Lake, northern California Coast Ranges, *Geol. Soc. of Am. Spec. Pap. 214*, edited by J.D. Sims, pp. 9-20, 1988.

Hopson, C.A., W. Beebe, J.M. Mattinson, E.A. Pessagno, Jr., and C.D. Blome, California Coast Range ophiolite: Jurassic tectonics, *EOS Trans. AGU*, v. 67, no. 44, p. 1232, 1986.

Hopson, C.A., M.J. Mattinson, E.A. Pessagno, Jr., Coast Range ophiolite, western California, in *The geotectonic development of California*, edited by W.G. Ernst, Englewood Cliffs, N.J., Prentice Hall, pp. 418-510, 1981.

Irwin, W.P., Geologic reconnaissance of the northern Coast Ranges and Klamath Mountains, California, with a summary of the mineral resources, *Bull. Calif. Div. Mines Geol. 179*, 80 pp., 1960.

Irwin, W.P., E.W. Wolfe, M.C. Blake, Jr., and C.G. Cunningham, Jr., Geologic map of the Pickett Peak quadrangle, Trinity County, California, *U.S. Geological Survey Geologic Quadrangle Map GQ-1111*, scale 1:62,500, 1974.

Jayko, A.S., M.C. Blake, Jr., and T. Harms, Attenuation of the Coast Range ophiolite by extensional faulting and nature of the Coast Range "thrust," California, *Tectonics*, v. 6, pp. 475-488, 1987.

Jayko, A.S., M.C. Blake, Jr., and R.N. Brothers, Blueschist metamorphism of the Eastern Franciscan belt, northern California, in Blueschists and Eclogites, *Geol. Soc. of Am. Memoir 164*, edited by B.W. Evans, and E.H. Brown, pp. 107-123, 1986.

Jones, D.L., Extensional faults in collisional settings: comparison of California and Japan, *Geol. Soc. Am. Abstr. Programs*, v. 19, no. 7, p. 720, 1987.

Lanphere, M.A., M.C. Blake, Jr., and W.P. Irwin, Early Cretaceous metamorphic age of the South Fork Mountain Schist in the northern Coast Ranges of California, *Am. J. Sci.*, v. 278, pp. 798-815, 1978.

Macpherson, G.J., The nature of blueschist facies metamorphism in the Snow Mountain volcanic complex, northern California Coast Range, *Geol. Soc. Am. Abstr. Programs*, v. 18, p. 128, 1986.

Mattinson, J.M., Geochronology of high pressure-low temperature Franciscan metabasites - a new approach using the U-Pb system, in Blueschists and Eclogites, Geol. Soc. of America Memoir 164, edited by B.W. Evans and E.H. Brown, p. 95-105, 1986.

Mattinson, J.M., and L.M. Echeverria, Ortigalita Peak gabbro, Franciscan complex: U-Pb date of intrusion and high-pressure-low temperature metamorphism, *Geology*, v. 8, pp. 589-593, 1980.

McDowell, F.W., D.H. Lehman, P.R. Gucwa, D. Fritz, and J.C. Maxwell, Glaucophane schists and ophiolites of the northern California Coast Ranges: Isotopic ages and their tectonic implications, *Geol. Soc. of Am. Bull.*, v. 93, pp. 595-605, 1984.

McLaughlin, R.J., M.C. Blake, Jr., A. Griscom, C.D. Blome, and B.L. Murchey, Tectonics of formation, translation, and dispersal of the Coast Range ophiolite of California, *Tectonics*, v. 7, no. 5, pp. 1033-1056, 1988.

McLaughlin, R.J., W.V. Sliter, and N.O. Frederiksen, Plate motions recorded by tectonostratigraphic terranes adjacent to the Mendocino triple junction (abstract): *EOS Trans. AGU*, v. 67, no. 44, p. 1219, 1986.

McLaughlin, R.J., H.N. Ohlin, D.J. Thormahlen, D.L. Jones, J.W. Miller and C.D. Blome, Geologic map and structural sections of the Little Indian Valley-Wilbur Springs geothermal area, northern Coast

Range, California, *U.S. Geol. Surv. Open File Rep. 85-285*, 1985a.

McLaughlin, R.J., D.H. Sorg, J.L. Morton, T.G. Theodore, C.E. Meyer, and M.H. Delevaux, Paragenesis and tectonic significance of base and precious metal occurrences along the San Andreas fault at Pt. Delgada, California, *Econ. Geology*, v. 80, p. 344-359, 1985b.

McLaughlin, R.J., and H.N. Ohlin, Tectonostratigraphic framework of the Geysers-Clear Lake region, California, in *Franciscan Geology of Northern California, Pacific Section, S.E.P.M*, v. 43, edited by M.C. Blake, Jr., pp. 221-254, Bakersfield, Calif., 1984.

McLaughlin, R.J., Post-middle Miocene accretion of Franciscan rocks, northwestern California: Reply to discussion by S.G. Miller and K.R. Aalto, *Geol. Soc. Am. Bull*, v. 94, pp. 1028-1031, 1983.

McLaughlin, R.J., S.D. Ellen, S.G. Miller, K.R. Lajoie, and S.D. Morrison, Terrane boundary relations and tectonostratigraphic framework, south of Eel River Basin, northwestern California: *American Association of Petroleum Geologists, Program and Abstracts, 58th Annual Meeting, Pacific Section, A.A.P.G., S.E.G., S.E.P.M*, Sacramento, pp. 112-113, 1983a.

McLaughlin, R.J., H.N. Ohlin, and C.D. Blome, Tectonostratigraphic framework of the Franciscan assemblage and lower part of the Great Valley sequence in the Geysers-Clear Lake Region, California, *EOS Trans. AGU* v. 64, p. 868, 1983b.

McLaughlin, R.J., and T.H. Nilsen, Neogene nonmarine sedimentation in small pull-apart basins of the San Andreas fault system, Sonoma County, California, *Sedimentology*, v. 29, p. 865-867, 1982.

McLaughlin, R.J., S.A. Kling, R.Z. Poore, K. McDougall, and E.C. Beutner, Post-middle Miocene accretion of Franciscan rocks, northwestern California, *Geol. Soc. Am. Bull.*, v. 93, pp. 595-605, 1982.

Murchey, Benita, Biostratigraphy and lithostratigraphy of chert in the Franciscan Complex, Marin Headlands, California, in *Franciscan Geology of Northern California, Pacific Section S.E.P.M.*, v. 43, edited by M.C. Blake, Jr., pp. 51-70, 1984.

Murchey, B.L., and D.L. Jones, Age and significance of chert in the Franciscan complex in the San Francisco Bay Region, in *Franciscan Geology of Northern California, Pacific Section S.E.P.M.*, v. 43, edited by M.C. Blake, Jr., pp. 23-30, 1984.

Mutti, E., and F. Ricci-Lucchi, Le turbiditi dell' Appennio settentrioriale: introduzione ali' analisi di facies (Turbidites of the northern Appenines: Introduction to facies analysis), Memorie della Societa Geologica Italiana, v. 11, pp. 161-199, translated by T.H. Nilsen, *International Geological Review*, v. 20 no. 2, pp. 125-166, 1972.

Nilsen, T.H., Late Cretaceous geology of California and the problem of the proto-San Andreas fault, in Mesozoic paleogeography of the western United States, Pacific Section, S.E.P.M., Pacific Coast Paleogeography, Symposium 2, edited by D.G. Howell, and K.A. McDougall, pp. 559-573, 1978.

O'Day, M.S., The structure and petrology of the Mesozoic and Cenozoic rocks of the Franciscan Complex, Leggett-Piercy area, northern California: Ph.d. dissertation, *Univ. of California, Davis*, 152 p., 1974.

Ogle, B.A., Geology of the Eel River Valley area, Humboldt County, California, *California Division of Mines and Geology Bulletin 164*, p. 128, 1953.

Page, B.M., The southern Coast Ranges, in *The Geotectonic Development of California*, edited by W.G. Ernst, Englewood Cliffs, New Jersey, Prentice-Hall, Inc., pp. 329-417, 1981.

Platt, J.P., Dynamics of orogenic wedges and the uplift of high-pressure metamorphic rocks, *Geol. Soc. Am. Bull.*, v 97, pp. 1037-1053, 1986.

Raymond, L.A., Cretaceous sedimentation and regional thrusting, northeastern Diablo Range, California: *Bull. Geol. Soc. America*, v. 81, pp. 2132-2128, 1970.

Sarna-Wojcicki, A.M., Correlation of the Late Cenozoic tuffs in the central coast ranges of California by means of trace- and minor-element chemistry, *U.S. Geol. Survey Prof. Paper 972*, p.30, 1976.

Silberling, N.J., Jones, D.L., Blake, M.C., Jr., and Howell, D.G., Lithotectonic terrane map of the western conterminous United States: U.S. Geological Survey Misc. Field Studies Map MF-1874-C, 1987.

Sliter, W.V., Foraminifers from Cretaceous limestone of the Franciscan Complex, northern California, in *Franciscan Geology of Northern California, Pacific Section, S.E.P.M.*, edited by M.C. Blake, Jr., v. 43, pp. 149-162, 1984.

Sliter, W.V., R.J. McLaughlin, G. Keller, W.R. Evitt, Paleogene accretion of Upper Cretaceous oceanic limestone in northern California, *Geology*, v. 14, pp. 350-352, 1986.

Suppe, J., Structural interpretation of the southern part of the northern Coast Ranges and Sacramento Valley, California: Summary, *Bull. Geol. Soc. America*, v. 81, pp. 165-188, 1979.

Suppe, J., Geology of the Leech Lake Mountain-Ball Mountain region, California; A cross section of the northeastern Franciscan belt and its tectonic implications, *University of Calif. Pub. Geol. Sci.*, v. 107, 82 p., 1973.

Swe, W., and W.R. Dickinson, Sedimentation and thrusting of Late Mesozoic rocks in the Coast Ranges near Clear Lake, California, *Geol. Soc. Am. Bull.* v. 81, pp. 165-188, 1970.

Tarduno, J.A., M. McWilliams, W.V. Sliter, H.E. Cook, M.C. Blake, Jr., and I. Premoli-Silva, Southern hemisphere origin of the Cretaceous Laytonville Limestone of California, *Science*, v. 231, pp. 1425-1428, 1986.

Tarduno, J.A., M. McWilliams, M.G. Debiche, W.V. Sliter, and M.C. Blake, Jr., Franciscan Complex Calera Limestones-accreted remnants of Farallon Plate oceanic plateaux, *Nature*, v. 317, pp. 345-347, 1985.

Underwood, M.B., Sedimentology and hydrocarbon potential of the Yager structural complex, possible Paleogene source rocks in the Eel River basin,northern California, *American Association of Petroleum Geologists Bulletin*, v. 69, pp. 1088-1100, 1985.

-----Depositional setting of the Paleogene Yager formation, northern coast ranges of California, in *Cenozoic marine sedimentation, Pacific Margin, U.S.A., Pacific Section, S.E.P.M.*, edited by D. Larue and R. Steel, pp. 81-101, 1983

Underwood, M.B., and S.B. Bachman, Sandstone perrofacies of the Yager Complex and the Franciscan Coastal belt, Paleogene of northern California, *Geol. Soc. Am. Bull.*, v. 97, pp. 809-817, 1986.

Wahrhaftig, Clyde, Structure of the Marin Headlands block, California: A progress report, in *Franciscan Geology of northern California, Pacific Section S.E.P.M.*, edited by M. C. Blake, Jr., v. 43, pp. 31-50, 1984.

Wentworth, C.M., M.C. Blake, Jr., D.L. Jones, A.W. Walter, and M.D. Zoback, Tectonic wedging associated with emplacement of the Franciscan assemblage, California coast ranges, in *Franciscan Geology of northern California, Pacific Section S.E.P.M.*, edited by M.C. Blake, Jr., v. 43, pp. 163-173, 1984.

Worrall, D.M., Imbricate low-angle faulting in uppermost Franciscan rocks, south Yolla Bolly area, northern California, *Geol. Soc. Am. Bull.*, v. 92, pp. 703-729, 1981.

Wright, J.E., and Fahan, M.R., An expanded view of Jurassic orogenesis in the western United States Cordillera: Middle Jurassic (pre-Nevadan) regional metamorphism and thrust faulting within an active arc environment, Klamath Mountain, California, *Geol. Soc. of Am. Bull.*, v. 100, p. 859-876, 1988.

TERRANES OF THE KLAMATH MOUNTAINS,
CALIFORNIA AND OREGON

William P. Irwin
U. S. Geological Survey, Menlo Park, CA 94025

Abstract. The Klamath Mountains province is an accumulation of tectonic fragments (terranes) of oceanic crust, volcanic arcs, and melange that amalgamated during Jurassic time prior to accretion to North America. The nucleus of the province is the Eastern Klamath terrane, which consists of oceanic volcanic and sedimentary rocks ranging from early Paleozoic to Middle Jurassic in age and includes the Trinity ophiolite of Ordovician age. The volcanic arc and melange terranes lying west of the nucleus contain three additional dismembered ophiolites that range in age from late Paleozoic to Late Jurassic and that are sequentially younger oceanward (westward). The volcanic arc terranes also contain preamalgamation plutons that are of Devonian, Permian, and Jurassic age and that are cogenetic with the volcanic rocks they intrude. Following amalgamation, the terranes were stitched together by postamalgamation plutons during Middle Jurassic to earliest Cretaceous time. Accretion of the amalgamated Klamath terranes to North America seems to have occurred during intrusion by plutons of the Shasta Bally belt in earliest Cretaceous time, following large-angle clockwise rotation of the terranes during Triassic and Jurassic time. The accreted terranes are unconformably overlapped along much of the southeastern and eastern perimeter of the province by relatively weakly deformed Cretaceous and Tertiary strata. Along the arcuate western boundary, the rocks of the province are in fault contact with accreted Jurassic and Cretaceous rocks of the Franciscan Complex and related formations of the Coast Ranges province.

INTRODUCTION

The Klamath Mountains province is a small part of the vast collage of suspect terranes that make up the western margin of North America from Mexico to Alaska (Coney et al, 1980). The province is a west-facing arcuate region of approximately 30,645 km sq and is a composite of a number of individual terranes, each of which is an allochthonous tectonic fragment of oceanic crust, volcanic island arc, or melange. None of the terranes appears to be a fragment of a former continent, and there is no evidence that any were ever underlain by continental rocks. An uncommon aspect of the Klamath Mountains is the number of ophiolites of widely differing ages that constitute parts of some of the terranes of the province. The ophiolites, which consist partly of upper mantle and oceanic crust, represent oceanic spreading centers that existed during early Paleozoic, late Paleozoic, Triassic, and Jurassic time, and are now distributed in a sequence that is successively younger from east to west (oceanward). Some of the ultramafic rocks of the ophiolites have been mined for their chromite content during times of national emergency since the 1800's. One locality, where lateritic soil formed on the ultramafic rock, is the site of the only significant production of nickel in the United States.

The volcanic arc terranes are dominantly andesitic in composition, but range from basalt to rhyolite, and are the hosts for several economically important massive-sulfide deposits. Similar to the ophiolites, the various volcanic-arc terranes tend to be successively younger westward from the Eastern Klamath terrane. This common pattern of westward younging suggests that the joining together (amalgamation) of the terranes also was an east to west sequence of events. Important limestone formations are present at several stratigraphic horizons in the long-standing arc represented by the volcanic rocks of the Eastern Klamath terrane, and are valuable, particularly the Permian limestones, for the manufacture of cement. The fossil faunas contained in the limestones greatly facilitate determination of the age and stratigraphic succession of the numerous formations that constitute the Eastern Klamath terrane. However, determination of the precise age of the strata in the volcanic-arc and melange terranes that lie to the west is difficult, as limestone bodies are relatively scarce, small, and discontinuous, and their enclosed fossils commonly are destroyed by metamorphism. The faunas of a few limestone bodies are thought to indicate a site of deposition that was far distant from the present locality. Much of the dating of the western terranes is based on fossil radiolarians in thin-bedded chert that is locally interlayered with the volcanic rocks.

The amount of deformation of the terranes varies widely. In the Eastern Klamath terrane the stratigraphy and structure are fairly coherent, but the western terranes are more tectonically deformed and commonly consist of broken formation and melange. The mapping of many parts of this geologically complex province has been only in reconnaissance, particularly in the region beyond the Eastern Klamath terrane, and this has resulted in many imprecise and uncertain correlations. Much additional study is needed before the geology of the Klamath Mountains can be accurately told.

EASTERN KLAMATH TERRANE

The Eastern Klamath terrane is the oldest terrane of the province and also contains the oldest rocks. It consists of three subdivisions---the Trinity, Yreka, and Redding subterranes (Figure 1). The Trinity subterrane is a large broadly-arched sheet of ultramafic and gabbroic rocks thought to be at least in part a dismembered ophiolite. The ultramafic sheet crops out over a broad area in the central part of the Eastern Klamath terrane and extends northwest under the Yreka subterrane and southeast under the Redding subterrane. Evidence from magnetic, gravity, and seismic refraction studies indicate that the ultramafic sheet extensively underlies the Yreka and Redding subterranes (LaFehr, 1966; Griscom, 1973; Fuis and Zucca, 1984; Blakely et al., 1985). The western edge of the ultramafic sheet is upturned and is exposed for a length of 160 km where it forms the western boundary of the Eastern Klamath terrane. The

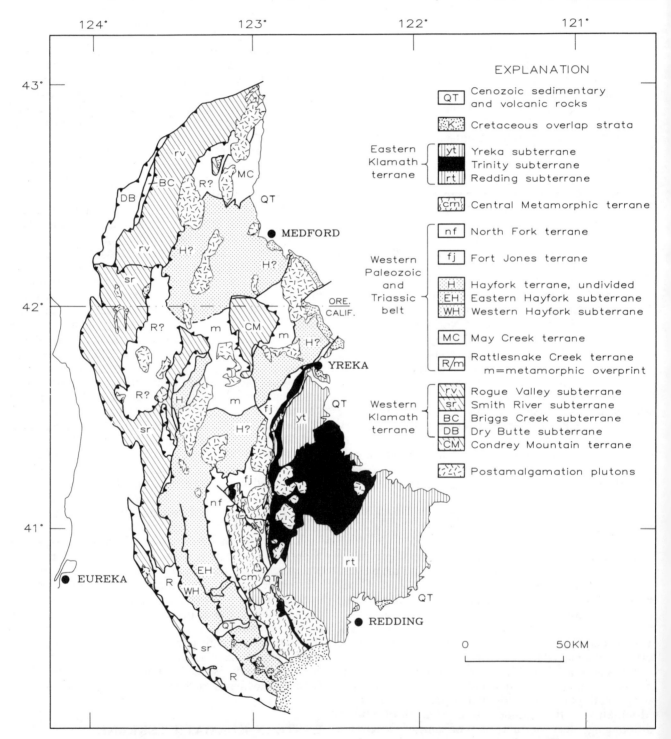

FIGURE 1 Map showing terranes, postamalgamation plutons, and overlap assemblages of the Klamath Mountains. Geology compiled mainly from Smith et al (1982), Wagner and Saucedo (1988), and Fraticelli et al (1987).

ultramafic rocks of the sheet are dominantly serpentinized tectonitic harzburgite, lherzolite, and minor dunite (Lindsley-Griffin, 1982). The gabbroic rocks include cumulate and layered hornblende gabbro grading into hornblende diorite and quartz diorite, and pegmatitic gabbro that may be younger. Locally the gabbro and diorite appear to grade upward into a complex of mafic dikes and sills and finally into mafic volcanic rocks (Lindsley-Griffin, 1977). U-Pb isotopic ages of 455-480 Ma measured on the layered gabbro and diorite, and 430 Ma on the pegmatitic gabbro (Lindsley-Griffin, 1977; Mattinson and Hopson, 1972), indicate an Ordovician age for the Trinity ophiolite. A more recent study reports the presence of tonalitic rocks that crystallized about 565-570 Ma (earliest Cambrian) and suggests that the Trinity ultramafic sheet is polygenetic (Wallin et al., 1988).

The Yreka subterrane is a structurally complex stack of several thrust(?) slices of Ordovician to Devonian strata that rest on the Trinity ultramafic sheet (Potter et al., 1977). The thrust(?) slices variously consist of volcanic and clastic sedimentary strata, locally abundant bedded chert and limestone, quartzite, phyllite, and melange. One melange (schist of Skookum Gulch) contains blueschist blocks, and also contains tectonic(?) blocks of Early Cambrian tonalite similar to that in the Trinity ultramafic sheet (Wallin et al., 1988).

The Redding subterrane consists of sixteen formations of volcanic and sedimentary strata that include representatives of all the Paleozoic and Mesozoic systems from the Devonian to the Middle Jurassic. It also includes preamalgamation plutons of Devonian and late Paleozoic age, which will be discussed on later pages. The volcanic and sedimentary strata are deformed, but they constitute a structurally coherent sequence that generally dips southeastward away from the Trinity ultramafic sheet and has an exposed thickness of more than 10 km (Kinkel et al., 1956; Albers and Robertson, 1961; Miller, 1986; and Sanborn, 1953). The presence of andesitic volcanic rocks and reefal limestones at many places throughout suggests that the sequence represents a long-standing volcanic arc that perhaps was built on Ordovician and older oceanic crust represented by the Trinity ultramafic sheet. However, the present northeast-trending contact between the Redding subterrane and the main area of exposure of the Trinity ultramafic sheet is thought to be a southeast-dipping extensional detachment fault along which some of the stratigraphically lower parts of the Redding subterrane locally have been cut out (Schweickert and Irwin, 1986). Part of the Redding subterrane is described in greater detail by M.M. Miller (this volume).

CENTRAL METAMORPHIC TERRANE

Exposed along the western boundary of the Eastern Klamath terrane, the Central Metamorphic terrane lies generally west of and structurally below the Trinity ultramafic sheet. It consists of mafic volcanic and sedimentary rocks that were metamorphosed during eastward subduction beneath the Trinity ultramafic sheet in Devonian time. The metamorphosed volcanic rocks (Salmon Schist) are the structurally lowest unit of the terrane and consist mostly of fine- to medium-grained hornblende-epidote-albite schist that is locally phyllonitic and retrograde (Davis et al., 1965). The metamorphosed sedimentary rocks (Abrams Schist) overlie the metavolcanic rocks, probably depositionally, and consist of quartz-mica schist, calc schist, and micaceous marble. The widest exposure of the Central Metamorphic terrane is in the southern part of the province. There the terrane forms a nearly detached synformal thrust plate that probably is 3-5 km thick at the axis. The eastern limit of the synform is sharply folded and thinned where it dips eastward beneath the Trinity ultramafic sheet. Klippen of the Eastern Klamath terrane rest on the Central Metamorphic terrane at three widely-spaced localities. The normal outcrop relation between the Central Metamorphic terrane and the Trinity ultramafic sheet is structurally reversed for 35 km along the western boundary of the Yreka terrane, where the Central Metamorphic terrane is represented by a relatively narrow sliver that crops out mostly along the east side rather than west of the northern extension of the Trinity ultramafic sheet. Assignment of a Devonian age of metamorphism is based on K-Ar ages of 390-399 Ma measured on amphibolite correlative with the Salmon Schist in the Yreka area (Hotz, 1977) and on Rb-Sr ages of approximately 380 Ma measured on the Abrams Schist (Lanphere et al., 1968). The Devonian age of metamorphism during subduction beneath the Trinity ultramafic sheet suggests that the Devonian volcanic and plutonic rocks (Copley Greenstone, Balaklala Rhyolite, and Mule Mountain stock) of the Redding subterrane are genetically related to that event.

TERRANES OF THE WESTERN PALEOZOIC AND TRIASSIC BELT

Lying west of the Central Metamorphic terrane is an extensive complex area that was called the Western Paleozoic and Triassic belt during early studies of the region (Irwin, 1960) and which includes the Applegate Group in the Klamath Mountains of Oregon. In the southern part of the province, the belt has been subdivided from east to west into the North Fork, Hayfork, and Rattlesnake Creek terranes (Irwin, 1972), and the Hayfork terrane into the Eastern Hayfork and Western Hayfork subterranes (Wright, 1982). Equivalents of some and perhaps all of these terranes may be present in the central and northern parts of the province (see Wagner and Saucedo, 1987; and Smith et al., 1982), but the correlations are not clearly known and the generalizations shown in figure 1 are speculative. The northern part of the Western Paleozoic and Triassic belt also includes the May Creek Schist which is now considered to be a separate terrane, and, at the California-Oregon border, a window of metamorphic rocks is called the Condrey Mountain terrane.

The North Fork terrane is exposed in the southern part of the province where it occupies a zone 2-10 km wide for a distance of 100 km along the western edge of the Central Metamorphic terrane. The structurally lowest part consists of dismembered ophiolite which is succeeded upward by generally eastward-dipping mafic volcanic rocks that are interlayered with argillite, thin-bedded chert, and discontinuous limestone lenses. Blocks of blueschist occur in a horizon of disrupted argillite (Ando et al., 1983). The age of the ophiolite is late Paleozoic based on the U-Pb isotopic age of a small plagiogranite body (265-310 Ma)(Ando et al., 1983) and on an uncertain relationship with nearby red radiolarian chert of Permian age (Blome and Irwin, 1983). The chert and tuff interlayered with the mafic volcanic rocks contains Permian, Triassic , and Early Jurassic radiolarians. The few limestone bodies that have yielded useful fossils are late Paleozoic in age.

A short distance north of the Salmon River the North fork terrane appears to end, and its position west of the Central Metamorphic terrane is occupied by the Fort Jones terrane of Blake et al. (1982). The rocks of the Fort Jones terrane, called the Stuart Fork Formation by Davis et al. (1965), are similar to those of the North Fork terrane except for a strong metamorphic overprint and the absence of ophiolitic rocks and may well be correlative. The rocks were metamorphosed to blueschist facies (Borns, 1980) and later overprinted to greenschist facies during the emplacement of nearby plutons (Jayko and Blake, 1984) in Jurassic and earliest Cretaceous time. K-Ar isotopic ages measured on blueschist blocks near Fort Jones are ~220 Ma (Triassic)(Hotz et al., 1977), which may indicate that the depositional ages of the volcanic and sedimentary protoliths are not younger than Triassic.

The Hayfork terrane, as exposed in the southern part of the province, consists of the Western and Eastern Hayfork subterranes. The Western Hayfork subterrane is the structurally lower of the two units and is interpreted to represent a Middle Jurassic volcanic arc. It consists mainly of the Hayfork Bally Meta-andesite and the preamalgamation Ironside Mountain batholith and related plutons. The meta-andesite is mainly augite-bearing crystal-lithic tuff and tuff breccia of andesitic to basaltic composition and is interbedded locally with thin-bedded chert and argillite. The meta-andesite has yielded isotopic ages of 156 Ma (M.A. Lanphere, oral commun., 1977) and 168-177 Ma (Fahan, 1982). The Ironside Mountain batholith is largely pyroxene diorite that yields isotopic ages of approximately ~170 Ma (revised constant, Lanphere et al., 1968; Wright, 1982) and is thought to be cogenetic with the meta-andesite.

The Eastern Hayfork subterrane is a melange that is separated from the structurally underlying Western Hayfork subterrane by the Wilson Point thrust fault (Wright, 1982). The melange includes serpentinite, argillite, chert, quartzose sandstone, mafic and locally silicic volcanic rocks, limestone pods, and a few amphibolite blocks. The serpentinite commonly occurs as slivers along the Wilson Point thrust. The chert bodies variously contain Triassic and Jurassic radiolarians. Some of the limestone contains Permian fossils of Tethyan faunal affinity (Irwin and Galanis, 1976; Nestell et al., 1981).

The Rattlesnake Creek terrane is west of the Western Hayfork subterrane, from which it is separated by an east-dipping fault (Salt Creek thrust). It is a melange that consists of dismembered ophiolitic rocks, mafic to silicic volcanic and subvolcanic rocks, plagiogranite, thin-bedded chert, argillite, sandstone and conglomerate, and minor limestone. Bedrock relations are obscured over much of the terrane by widespread landslides and earthflows. The chert contains Triassic and Jurassic radiolarians, and the limestone bodies contain sparsely distributed fossils of Devonian(?), late Paleozoic, and Late Triassic age. U-Pb isotopic ages ranging from 193 to 207 Ma have been measured on the plagiogranitic rocks (Wright, 1982). In the central and northern parts of the province, the Rattlesnake Creek terrane may be represented by rocks of the Preston Peak area (Snoke, 1977), the Takilma melange and Sexton Mountain ophiolite (Smith et al., 1982), and by metamorphosed melange of the Marble Mountain area (Donato et al., 1982), the Condrey Mountain quadrangle (Hotz, 1967), and the Dutchmans Peak area (Smith et al., 1982).

The May Creek terrane consists of the May Creek Schist. The schist is divided into two units (Donato, 1987)---a lower unit of hornblende-plagioclase schist that may represent metamorphosed dikes, sills, and tholeiitic basalt, and an upper unit of metamorphosed quartzose sandstone, tuffaceous sandstone, calcareous metasedimentary rocks, and rare lenses of marble. The age of the schist is unknown. The schist was earlier considered gradational with the volcanic and sedimentary rocks of the Applegate Group (Wells and Peck, 1961), but now is thought to be separated from the Applegate rocks by faults (Smith et al., 1982). The May Creek Schist is herein considered a separate terrane because of its anomalous lithology and amphibolite grade of metamorphism and its fault contacts with the Applegate rocks.

The Condrey Mountain terrane consists of the Condrey Mountain Schist which is mostly metasedimentary rocks, consisting of generally well foliated quartz-albite-muscovite schist, and greenschist-facies metavolcanic rock (Coleman et al., 1983). Several large lenses of flaggy blueschist, consisting of alternating crossite-rich and chlorite-epidote-rich layers, are present in the dominantly metasedimentary unit. Isotopic ages ranging from Middle Jurassic to Early Cretaceous have been measured on various components of the terrane (see Helper, 1986a). The Condrey Mountain Schist is correlated with the Galice Formation of the Western Klamath terrane by some geologists (e.g. Klein, 1977) and with the South Fork Mountain Schist of the Coast Ranges province by others (Helper, 1986b; Brown and Blake, 1987).

WESTERN KLAMATH TERRANE

The Western Klamath terrane (Blake et al., 1985) consists of rocks earlier called the western Jurassic belt (Irwin, 1960). It is mainly a large ultramafic body overlain by volcanic arc and clastic sedimentary rocks, and also includes the Dry Butte and Briggs Creek subterranes. The volcanic arc deposits, assigned to the Rogue Formation (Garcia, 1979), are mostly in the northern third of the terrane, where they interfinger and generally are overlain with weakly slaty shale and sandstone of the Galice Formation. In the central part of the terrane, just south of the Oregon-California border, the ultramafic body forms the base of the Josephine ophiolite which is unusually complete and well exposed along the Smith River (Harper, 1984). There the Galice is deposited on the ophiolite rather than on the volcanic Rogue Formation. In the southern part of the terrane the ultramafic and other ophiolitic rocks are missing, except for a few relatively small fault slices, and volcanic strata associated with the Galice are rare. A continuation of the Josephine ophiolite near the southern end of the terrane is called the Devils Elbow ophiolite remnant by Wyld and Wright (1988) and is overlain by the Galice Formation (Wyld and Wright, 1988).

The age of the Josephine ophiolite is Late Jurassic, based on U-Pb isotopic ages of ~162 Ma of plagiogranite (J.B. Saleeby, oral commun., 1988), with an upper limit of 150-151 Ma based on the ages of dikes that cut the ophiolite and the overlying Galice Formation (Harper, 1984). A section of thinly interbedded argillite, chert, and limestone that directly overlies the Josephine ophiolite along the Smith River contains Late Jurassic radiolarians (Pessagno and Blome, in press). At a few localities elsewhere, clastic strata of the Galice contain sparse pelecypods of Late Jurassic age (mid-middle Oxfordian to Kimmeridgian) (Pessagno and Blome, in press). The Galice was weakly metamorphosed to lower-greenschist facies ~150 Ma (Lanphere et al., 1978), probably while being overridden by rocks of the Western Paleozoic and Triassic belt.

The Galice in the northern part of the terrane, which contains the type locality at Galice Creek and overlies the volcanic rocks of the Rogue Formation, is thought to be in a different thrust plate than to the south where the Galice lies on the Josephine ophiolite (Harper, 1984). The sedimentary rocks of the Galice that overlie the Rogue are volcanogenic in composition whereas those that overlie the Josephine are more quartzofeldspathic Blake et al., 1985). The thrust relationship and the differences in composition of the Galice led Blake et al. 1985) to call the Josephine ophiolite-Galice sequence and the ultramafic rocks-Rogue-Galice sequence separate subterranes---the Smith River and Rogue Valley subterranes, respectively (Figure 1). An anomalous occurrence of Triassic radiolarian chert in the Rogue Valley subterrane (Roure and De Wever, 1983) is difficult to explain. The chert presumably lies below the Rogue and may indicate that the basement beneath the Rogue is older than previously considered (Blake et al., 1985), or that the chert may even be part of an unrecognized tectonic outlier of Rattlesnake Creek terrane.

The Briggs Creek and Dry Butte subterranes lie west of the ultramafic rocks-Rogue-Galice sequence, separated by northeast-trending faults thought to be southeast-dipping thrusts. The Briggs Creek subterrane consists of the Briggs Creek Amphibolite of Garcia 1979), which is mainly amphibolite but includes micaceous quartzite, quartz schist, and recrystallized manganiferous chert, all of unknown age. The original relationship of the Briggs Creek subterrane to the other parts of the Western Klamath terrane is not known. The Dry Butte subterrane is a complex of ultramafic and gabbroic rocks, tonalite and quartz diorite, and rhyodacite flows and tuffs with local andesite and basalt flows (see Smith and others, 1982). The tonalite and quartz diorite have yielded K-Ar isotopic ages of ~153 Ma (Late Jurassic)(Hotz, 1971). The rocks of the Dry Butte subterrane may represent the root zone of the Rogue volcanic arc.

PLUTONIC BELTS

Plutons are present in all the terranes of the Klamath Mountains. They vary widely in composition and range in age from early Paleozoic to Early Cretaceous. Their distribution is not uniform. Few plutons intrude the Rogue and Galice Formations, Yreka subterrane, and Redding subterrane. Many are strikingly elongate parallel to the regional structural trends, and the orientation of some, such as the Ironside Mountain batholith, may represent the orientation of the host volcanic arc. Important constraints on the ages of amalgamation of the terranes are provided by the plutons that are truncated by terrane boundaries and by those that intrude terrane boundaries. The plutons are catagorized as preamalgamation if they intruded before their host terrane joined to an adjacent terrane, and as postamalgamation if the plutons intruded after the joining.

Some preamalgamation plutons are similar in age and composition to the volcanic strata they intrude, forming plutonic-volcanic pairs, and probably represent the roots of volcanic arcs. Examples are the the Mule Mountain stock intruding the Balaklala Rhyolite (Devonian) of the Redding subterrane, plutons of the McCloud belt intruding the Dekkas Andesite (Permian) of the Redding subterrane, and the Ironside Mountain batholith intruding the Hayfork Bally Meta-andesite (Jurassic) of the Western Hayfork subterrane. In contrast, the postamalgamation plutons are significantly younger than their host rocks, seem genetically unrelated, and presumably intruded the host terrane as a result of crustal subduction.

The pattern of distribution of the plutons is important to an understanding of the tectonic history of the province. Isotopic ages have been measured on many of the plutons, and, although some of the isotopic ages measured by different methods are conflicting, the data indicate that plutons of similar age tend to occur in belts that are generally parallel to trends of other regional lithic and structural features (Figure 2). Most of the isotopic dating has been by M.A. Lanphere, J.M. Mattinson, J.B. Saleeby, and J.E. Wright (for listing of age data see Irwin, 1985). Plutonic belts in the northwestern part of the province trend northeast, at large angles to plutonic belts in the southwestern part. The Paleozoic belts are all in the Eastern Klamath terrane and trend northeast-southwest. Jurassic plutons are remarkably rare in the Eastern Klamath terrane. West of the Eastern Klamath terrane, the northeast trending belts are of successivley younger Jurassic and Cretaceous ages except for the

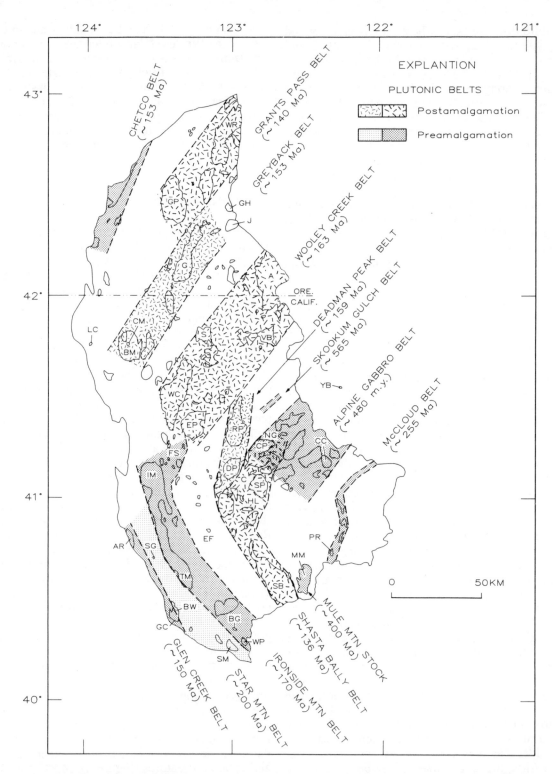

FIGURE 2 Map of Klamath Mountains province showing outlines of major plutons and trends of plutonic belts (modified from Irwin, 1985). Ultramafic rocks are not shown. Letter symbols on map correspond to names of isotopically dated plutons: A, Ashland; AR, Ammon Ridge; BG, Basin Gulch; BM, Bear Mountain; BW, Bear Wallow; C, Caribou Mountain; CC, Castle Crags; CM, Cracker Meadow; CP, Craggy Peak; DP, Deadman Peak; EF, East Fork; EP, English Peak; FS, Forks of Salmon; G, Greyback; GC, Glen Creek; GH, Gold Hill; GP, Grants Pass; HL, Horseshoe Lake; IM, Ironside Mountain; J, Jacksonville; LC, Lower Coon Mountain; MM, Mule Mountain; NG, unnamed gabbro, PR, Pit River; RP, Russian Peak; S, Slinkard; SB, Shasta Bally; SG, Saddle Gulch; SM, Star Mountain; SP, Sugar Pine; VB, Vesa Bluff; WC, Wooley Creek; WP, Walker Point; WR, White Rock; YB, Yellow Butte.

westernmost belt (Chetco belt) which is Late Jurassic. The northwest trending belts also are of Jurassic and Cretaceous age, but the youngest (Shasta Bally belt) is furthest from the ocean and partly overprints an early Paleozoic belt (Alpine gabbro belt).

TECTONIC DOMAINS

The change in orientation of the plutonic belts occurs across a vague northwest-trending zone of discontinuity that divides the province into northeastern and southwestern domains (Irwin, 1985). The change is accompanied by differences in the ages and numbers of plutonic belts and by differences in other regional features. The Paleozoic plutonic belts are all in the northeast domain. Jurassic and Cretaceous belts are in both domains but differ somewhat in age and terrane relations. The number of plutonic belts in the northeast domain is nearly twice that of the southwest domain. In the northeast domain the Western Paleozoic and Triassic belt contains the greatest number of plutons, including nearly all of the postamalgamation Jurassic plutons of the province, but seems to be devoid of preamalgamation plutons (Figure 3). In contrast, most plutons of the Western Paleozoic and Triassic belt in the southwest domain are preamalgamation in age.

Differences in the age of plutonic belts are particularly puzzling where the host terrane seems to be virtually continuous across the zone of discontinuity. The Ironside Mountain and Wooley Creek plutonic belts intrude the Hayfork Bally Meta-andesite on opposite sides of the zone of discontinuity, and virtually coincide at the zone. The Ironside Mountain batholith (~170 Ma) is considered preamalgamation because of its genetic relation to the meta-andesite and its truncation by terrane boundaries. The Wooley Creek belt (~163 Ma) is only a few million years younger than the Ironside Mountain belt, but is postamalgamation because the Wooley Creek, Slinkard, and Vesa Bluffs plutons cross terrane boundaries (Donato et al., 1982; Barnes, 1983; and Mortimer, 1984). Cretaceous plutons of the Grants Pass belt intrude the boundary between the Western Klamath terrane and the rocks of the Western Paleozoic and Triassic belt in the northeastern domain, but comparable plutons and relations are unknown in equivalent terranes in the southwestern domain.

The distribution of the terranes and other features suggest that rocks of the Klamath Mountains are regionally deformed by broad open folding and by doming (Figure 4). The axes of the folds generally are parallel to the overall trends of the terranes and to the trends of the plutonic belts. In the southwest domain, an open synform and a tighter complementary antiform on its eastern limb account for the unusually wide exposure of the Central Metamorphic terrane. The axes of these folds trend northwest to the zone of discontinuity, which at that point is the Browns Meadow fault of Davis et al. 1965). In the northeast domain, the western part of the Western Paleozoic and Triassic belt appears to be a thin synformal thrust plate that rests on the Galice Formation over a broad area. The axis of the synform trends

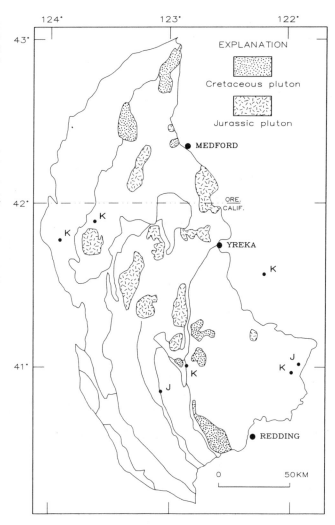

FIGURE 3 Distribution of isotopically-dated postamalgamation plutons. Letter symbols indicate age of small dated plutons: J, Jurassic; K, Cretaceous.

northeast and is parallel a narrow corridor of the Galice formation to the southeast that represents a complementary antiform and which nearly reaches the Condrey Mountain window. The Trinity ultramafic sheet is a broad northeast-trending antiformal upwarp with a complementary northeast-trending synform represented by the Yreka subterrane. The southeast limb of the antiform includes the southeast-dipping stratigraphic section of the Redding subterrane. The Condrey Mountain window represents the core of a structural dome (Coleman and Helper, 1983). The rocks in the core of the dome may well include metamorposed Galice and Rogue Formations. The window is not far from the Galice exposed in the narrow antiformal corridor, and in the usual order of structural stacking the Galice should be overlain by the Rattlesnake Creek terrane. Although the domal structure is complicated by faulting, most of the rocks surrounding the core are indeed metamorphosed melange that is correlative with the Rattlesnake Creek terrane. Generally further from the core of the dome are volcanic rocks that may be correlative with the Hayfork terrane.

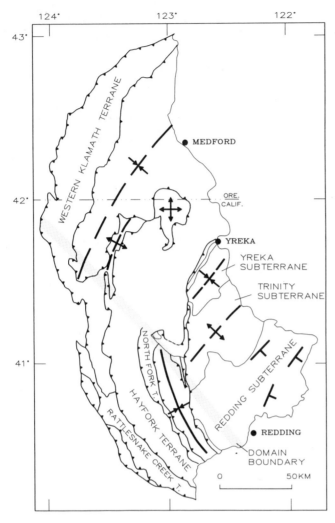

FIGURE 4 Some major structural feature of the Klamath Mountains. Shown are the axes of postulated regional synforms and antiforms, the Condrey Mountain dome, the generally easterly-dipping strata of the Redding subterrane, and the approximate boundary between the southwest and northeast domains.

TERRANE AMALGAMATION

The sequence and timing of terrane amalgamation is established by various kinds of evidence including (1) the age of the youngest rock of the subducted terrane, (2) the age of metamorphism that in some instances occurs in the lower plate adjacent to the terrane boundary during subduction, (3) the age of postamalgamation plutons, and (4) a minimum limit imposed by the age of strata that overlap the terrane boundary. The boundary between the Eastern Klamath and Central Metamorphic terranes is Devonian in age based on the age of the metamorphism that occurred when the protoliths of the Abrams and Salmon Schists were overridden by the Trinity ultramafic sheet (Lanphere et al., 1968). The boundary is intruded by Early Cretaceous plutons of the Shasta Bally belt, and, as with other terranes of the southern Klamath Mountains, it is overlapped by Lower Cretaceous strata of the Great Valley sequence.

The boundary between the Central Metamorphic terrane and the rocks of the North Fork and Fort Jones terranes is intruded by Middle and (or) Late Jurassic plutons (East Fork and Deadman Peak plutons). The North Fork terrane contains radiolarian chert as young as Early Jurassic in age (Pliensbachian)(Blome and Irwin, 1983). Therefore, the age of the terrane boundary must be in the range of late Early to Late Jurassic.

The age of the Eastern Hayfork melange and it boundary with the North Fork terrane is equivocal. The Western Hayfork subterrane includes the cogenetic Middle Jurassic (~170 Ma) Ironside Mountain batholith in the southwest domain. The batholith is cut by the boundary fault between the the Western Hayfork subterrane and Rattlesnake Creek terrane. In the northeast domain, Middle Jurassic (~163 Ma) plutons of the Wooley Creek belt cut the boundary between correlatives of the Western Hayfork subterrane and the structurally underlying Rattlesnake Creek terrane. This suggests that the Western Hayfork subterrane and Rattlesnake Creek terrane were juxtaposed during a interval of ~7 m.y. during Middle Jurassic time.

The youngest part of the Western Klamath terrane the Galice Formation, which contains Late Jurassic (Oxfordian to Kimmeridgian) pelecypods and radiolarians (see Harper, 1984). In the southern part th of the terrane, the Galice was metamorphosed to lowe greenschist facies in Late Jurassic time (~150 Ma presumably during overriding by the Rattlesnake Cree terrane. In the northern part, the boundary between th Galice and rocks of the Western Paleozoic and Triass belt is intruded by Early Cretaceous plutons of th Grants Pass belt, which places an upper limit on the ag of the boundary and supports the Late Jurassic age amalgamation indicated by the metamorphism. Th suturing of the Briggs Creek and Dry Butte subterrane to one another and to the other rocks of the Wester Klamath terrane probably postdates the Late Jurass (~153 Ma) plutons of the Dry Butte subterrane an predates the Early Cretaceous fault boundary with th Coast Ranges to the west.

The foregoing data indicate that the amalgamation the Klamath terranes was a series of Jurassic event except for the Devonian suturing of the Eastern Klama and Central Metamorphic terranes, and appears to ha been sequential westward.

ACCRETION TO NORTH AMERICA

The time of accretion of the Klamath terranes North America is not known with certainty. Nor is clear whether the Eastern Klamath terrane accreted to t continent before being joined by the other terranes, whether all the terranes of the province amalgamated an oceanic setting and then accreted to the continent as composite body. The kinds of evidence used determine the timing of amalgamation of the vario terranes are not available for use in determining the tir of accretion of the Eastern Klamath terrane to t continent, mainly because the terrane boundaries alo the eastern and southeastern border of the province a concealed by a broad cover of Cretaceous and younger sedimentary and volcanic strata. However, indirect evidence for the time of accretion is given by the results of paleomagnetic studies. Most of the studies have been of the middle Paleozoic to Middle Jurassic strata of the Redding subterrane and of some of the plutons. Although the results of some of the studies are conflicting, most indicate that the rocks of the province have experienced large clockwise rotation, but none indicates significant latitudinal displacement.

Large clockwise rotations are reported for the Galice Formation of the Western Klamath terrane (Bogen, 1986), for some plutons of the northeast domain (Schultz, 1983), and for Devonian and Permian strata of the Redding subterrane (Achache et al., 1982; Fagin and Gose, 1983). Paleomagnetic measurements made by Mankinen et al., (1984) in the Redding subterrane were along a northern and a central transect of the Permian and younger strata, and the results of their study indicate clockwise rotations of ~100° for the Permian and Triassic rocks, and clockwise rotation of ~60° for the Lower and Middle Jurassic rocks, relative to stable North America. The data gathered along both transects are consistent with the concept that the Eastern Klamath terrane rotated essentially as a rigid block (Mankinen et al., 1984). However, this is disputed by Renne et al (1988) whose paleomagnetic study of Permian rocks of the central transect indicates no rotation, and who interpret their results as indicating oroclinal bending rather than block rotation.

Paleomagnetic measurements on the postamalgation plutons of the Shasta Bally belt indicate clockwise rotation of ~25° following the intrusion event in Early Cretaceous time (Mankinen et al., 1988). Lower Cretaceous strata of the Great Valley sequence, which overlap the Klamath terranes and Shasta Bally batholith (136 Ma) at the south end of the province, have rotated clockwise ~14° (Mankinen et al., 1988), as have the Tertiary volcanic rocks of the Cascade Range that lie just east of the Klamath Mountains (Beck et al., 1986). This suggests that the Klamath terranes virtually ceased to rotate for a period of ~100 m.y. while the onlapping Cretaceous and Tertiary volcanic strata were deposited. The emplacement of Shasta Bally batholith, which preceeded the deposition of the onlapping Cretaceous (Hauterivian) strata and the long pause in rotation by no more than a few million years, may well mark the time of accretion of the Eastern Klamath terrane.

If the Eastern Klamath terrane accreted to North America in Early Cretaceous time as the preceeding scenario suggests, it would have been as part of a composite body that included the other terranes of the province, because the sutures between the terranes are established as Jurassic and older in age. This concept is also supported by the general similarity of the amount of rotation of the Jurassic postamalgamation plutons northwest of the Eastern Klamath terrane and of the amount of rotation of the Jurassic strata of the Redding subterrane, which suggests that the rocks of both areas probably rotated as a single unit.

ROAD LOG-- DAY 3, June 30, 1989 --- EUREKA TO RED BLUFF

The traverse from Eureka to Red Bluff crosses parts of three major geologic provinces---the northern Coast Ranges, the south end of the Klamath Mountains, and the northern part of the Great Valley. Point-to-point distances listed in the road log are measured in miles to facilitate the use of odometers in American vehicles. The roadlog begins at the junction of State Highway 36 and U.S. Highway 101 at Alton, about 20 mi south of Eureka. From the junction, Highway 36 follows the general course of the Van Duzen River for 47 mi. At mile 55 it crosses the crest of South Fork Mountain, which is the drainage divide between the Coast Ranges and Klamath Mountains. The route continues eastward in the drainage of the Trinity River to the boundary between Trinity and Shasta Counties---about mile 85---where it passes into the Sacramento River drainage of the Great Valley. The traverse ends at old U.S. Highway 99, just north of Red Bluff, for a total length of about 137 mi (220 km).

Mileage
0.0 Junction of State Highway 36 and U.S. Highway 101 at Alton. Travel east on Highway 36 on the floodplain of the Eel and Van Duzen Rivers. The prominent bluff to the northeast exposes Pliocene and Pleistocene nonmarine gravel, sand, and clay, overlain by Pleistocene or Holocene nonmarine terrace gravels.
2.8 Right-angle turn in highway at Hydesville.
4.9 Carlotta. Most of highway for the next 10 mi. is on alluvial terraces of the Van Duzen River.
10.7 Entering redwood grove.
12.7 Van Duzen River bridge.
13.3 Van Duzen River bridge.
15.7 For the next few miles, occassional exposures are seen of well-bedded marine sandstones and shales of the Yager Formation which probably ranges from Late Cretaceous to Eocene in age.
16.2 Enter Grizzley Creek State Park.
22.3 Crossing a northwest trending fault zone, from the Yager Formation into melange of the Franciscan Complex. For the next 30 mi, the Franciscan melange is the predominant unit traversed by Highway 36. The melange terrane is generally characterized by erratic knobby outcrops, surficial creep and landslide, and lumpy grass-covered hillsides with scattered brush, oak trees and patches of timber. Most of the knobby outcrops are blocks of either volcanic rock, graywacke, chert, or rarely blueschist, and generally are in a matrix of sheared sandstone and siltstone. In contrast, the areas underlain by relativley coherent stratigraphic sections of Franciscan sandstone and siltstone are characterized by angular ridges with steep hillsides and heavy timber cover.
23.1 A patch of coherent Franciscan sedimentary strata is exposed along the south bank of the Van Duzen River. The beds are nearly vertical and

strike across the river.

23.8 Bridgeville and the Van Duzen River bridge.

24.6 The many large blocks of rock in the river are residual debris from the hummocky hillside and are a common feature in streams that drain regions of Franciscan melange. Hillside creep and landslide are the main causes of rapid erosion of Franciscan melange and result in extremely high sediment loads in the rivers. For the next 10 mi, excellent views of typical Franciscan melange terrane are seen, with mottled grass-covered hillsides and occasional knobby outcrops of tectonic blocks.

30.5 McClellan Rock--a prominent block of pillow basalt with unusually large pillows exposed near the base.

35.1 Van Duzen River bridge.

41.3 Dinsmore.

46.7 Crest of Mad River Ridge, with view of South Fork Mountain skyline across the Mad River valley.

47.7 Mad River Post Office.

48.2 Mad River bridge.

55.1 STOP 3-1: Summit of South Fork Mountain. Here, Franciscan graywackes show various degrees of penetrative cataclasis caused by regional tectonism and are assigned to Textural Zone 2. Eastward, toward the Klamath Mountains, the degree of cataclasis and recrystallization increases until the graywackes are completely metamorphosed to quartz-albite-white mica-chlorite schist of Textural Zone 3. The schist of Textural Zone 3, called the South Fork Mountian Schist, is considered a regional blueschist owing to the presence of minor lawsonite and aragonite. It locally includes metavolcanic interlayers (Chinquapin Butte Metabasalt) that contain sodic amphiboles. The schist forms a thin structural marker that can be traced for hundreds of miles along the western edge of the Klamath Mountains and Great Valley, and is thought to be a tectonic rind of Franciscan rocks that formed when the Franciscan was overridden by the Klamath terranes and other rocks about 120 Ma (Early Cretaceous). The transition from Textural Zone 2 to Textural Zone 3 is not well exposed along Highway 36, but samples of the gradation can be seen at occasional intervals in low road cuts and as float, beginning in Textural Zone 2 metagraywacke just west of the buried gas line at mileage 55.4 and traveling eastward into Textural Zone 3 metagraywacke (South Fork Mountain Schist) at mileage 55.7 The highway continues downgrade eastward in the schist to the fault boundary of the Western Klamath terrane.

60.4 STOP 3-2: The South Fork fault, which is the boundary between the South Fork Mountain Schist of the Coast Ranges and the Galice Formation of the Western Klamath terrane, is concealed in a gully that is crossed by the highway a few hundred feet east of Clear Creek.

The high roadcut to the east of the gully is in the Galice Formation. The South Fork fault separates the Galice from the South Fork Mountain Schist to the west. A thin slice of serpentinite along the fault is now concealed by debris, and presumably represents a sliver of the Josephine ophiolite. The Galice Formation consists of graywacke, siltstone, and conglomerate of Late Jurassic (middle Oxfordian to Kimmeridgian) age and is the youngest subjacent formation of the Klamath Mountains. Most of the Galice at this latitude belongs to Textural Zone 2. However, the Galice at this stop is unusual because some of it is metamorphosed to Textural Zone 3 and is virtually indistinguishable, except in thin section, from the South Fork Mountain Schist. The South Fork Mountain Schist is seen at this locality only as debris on the west side of the gully.

62.5 Bridge over South Fork of Trinity River at Forest Glen.

62.8 Cross fault into Bear Wallow pluton of the Rattlesnake Creek terrane.

63.4 Turnoff to Hellgate Campground at mouth of Rattlesnake Creek on South Fork of Trinity River.

65.0 Bridge near mouth of North Rattlesnake Creek. A prominent limestone block nearby contains late Paleozoic or early Mesozoic foraminifers. From here to mileage 68.1 are many good exposures of mafic volcanic rocks of the Rattlesnake Creek terrane, as well as red and green chert. Some of the chert contains Triassic or Jurassic radiolarians.

66.7 STOP 3-3: Interlayered mafic volcanic rocks and thin-bedded chert of the Rattlesnake Creek terrane, exposed in high roadcuts.

70.0 Sheared serpentinite and altered volcanic rocks in the Rattlesnake Creek fault zone. Note the extensive landslide topography in the area of the Rattlesnake Creek terrane south of the highway while traveling upgrade for the next mile.

71.4 Crest of grade and junction of Bramlet Road.

72.1 Melange, poor exposures. Mostly sheared mafic volcanic rocks and minor chert and serpentinite to next mileage.

72.8 Junction of Highway 36 and County Route 3 Continue traveling easterly along Highway 36 through metavolcanic and metasedimentary rocks of the Rattlesnake Creek terrane.

78.1 Exposures of pillow lava and volcanic breccia.

78.4 STOP 3-4 (OPTIONAL): Coarse vesicular volcanic breccia with fragments of red radiolarian chert.

78.5 Serpentinized peridotite.

79.2 Crossing into phyllitic volcanic rocks.

79.6 Crossing nose of ridge in metavolcanic rocks and chert. The chert contains Triassic or Jurassic radiolarians.

80.0 Contact between serpentinized peridotite of the Wildwood ultramafic body to east and metavolcanic rocks to the west. The peridotite is exposed along the highway for the next 0.5 mi to a sheared east-dipping contact with metavolcanic rocks. Note occasional small body of light-colored rodingite. To the north, a distant view of the Trinity Alps in the Central Metamorphic terrane.

81.0 Muldoon notch. Approximate location of the suture between the Rattlesnake Creek terrane and the Hayfork Bally Meta-andesite of the Western Hayfork subterrane.

82.4 Hayfork Creek bridge. Highway continues through Hayfork Bally Meta-andesite of Middle Jurassic age.

83.7 Exposures of the Goods Creek pluton, a small Middle(?) Jurassic quartz-diorite body that intrudes the Hayfork Bally Meta-andesite.

85.2 Crest of grade--boundary of Trinity and Shasta Counties. View of Great Valley in distance. Downhill grade in the Hayfork Bally Meta-andesite for the next 2 mi.

87.1 STOP 3-5: Roadside drinking fountain. In the Hayfork Bally Meta-andesite of the Western Hayfork subterrane. Seen in the distance are the peaks of North Yolla Bolly and Black Rock Mountain, which are along the southern extension of the belt of the South Fork Mountain Schist.

87.3 Crossing into moderately-deformed basal conglomerate and sandstone of a depositional outlier of Lower Cretaceous Great Valley sequence. Traverse continues mostly in strata of the Great Valley sequence for several miles to beyond Platina.

89.3 Harrison Gulch Ranger Station.

93.6 Platina.

93.9 Redding road junction, at crest of upgrade through exposures of the outlier of Great Valley sequence. Passing into the Sacramento River drainage basin. For the next two miles, the highway is mostly along rocks of the Eastern Hayfork subterrane.

96.9 STOP 3-6: Junction of road to Beegum. Exposure of Lower Cretaceous (Hauterivian) basal conglomerate of the Great Valley sequence lying unconformably on the Eastern Hayfork terrane.

97.0 Beegum Creek.

99.0 Tedoc Road junction near crest of upgrade. For the next 16 mi the highway passes up-section through generally easterly-dipping beds of the Great Valley sequence that range in age from Early Cretaceous (Hauterivian) to Late Cretaceous (Turonian). Here the stratigraphic thickness of the Great Valley sequence is approximately 4 mi (6.5 km).

107.3 Crossing Dry Creek.

111.3 STOP 3-7: Exposed in the cliff on the north side of Dry Creek is a steep sandstone dike that cuts across gently-dipping beds of the Great Valley sequence. This is one of more than a hundred similar dikes in this part of the Great Valley that

were described before the turn of the century in a classic paper by Diller (1889) and more recently by Peterson (1966). The dikes generally trend northeast and are steeply dipping. They are thought to be tensional features that are post-Turonian and pre-Pliocene in age. Their orientation is similar to that of the Tertiary extensional faults of the southern Klamath Mountains, including the La Grange and related detachment faults.

111.8 Ranch house at sharp curve in road.

112.7 More sandstone dikes cutting the Great Valley sequence are seen in the stream-cut cliff to the north (left) of highway.

119.9 Fire Station at junction of Cottonwood road.

122.1 Crossing South Fork of Cottonwood Creek, then upgrade through exposures of weakly consolidated sand and gravel of the Pliocene Tehama Formation.

122.8 Traveling on a remnant of an old pediment surface covered by gravels of the Pleistocene Red Bluff Formation.

126.0 Roadcuts occasionally expose nearly flat-lying Tehama Formation for the next few miles.

139.0 End of traverse at junction of Highway 36 with old U.S. Highway 99. Turn right for travel to Red Bluff.

REFERENCES CITED

Achache, J., A. Cox, and S. O'Hare, Paleomagnetism of the Devonian Kennett limestone and the rotation of the eastern Klamath Mountains, California, *Earth Planet. Sci. Lett., 61*, p. 365-380, 1982.

Albers, J.P., and J.F. Robertson, Geology and ore deposits of the East Shasta copper-zinc district, Shasta County, California, *U.S. Geol. Surv. Prof. Paper 338*, 107 p., 1961.

Ando, C.J., W.P. Irwin, D.L. Jones, and J.B. Saleeby, The ophiolitic North Fork terrane in the Salmon River region, central Klamath Mountains, California, *Geol. Soc. America Bull., 94*, 236-252, 1983.

Barnes, C.G., Petrology and upward zonation of the Wooley Creek batholith, Klamath Mountains, California, *Jour. Petrology, 24*, 495-537, 1983.

Beck, M.E., Jr., R.F. Burmester, D.E. Craig, C.S. Gromme, and R.E. Wells, Paleomagnetism of middle Tertiary volcanic rocks from the Western Cascade Series, northern California, *Jour. Geophys. Res., 91*, 8219-8230, 1986.

Blake, M.C., Jr., D.G. Howell, and D.L. Jones, D. L., Preliminary tectonostratigraphic terrane map of California, *U.S. Geol. Survey Open-file Rept. 82-593*, 3 sheets, scale 1:750,000, 1982.

Blake, M.C., Jr., D.C. Engebretson, A.S. Jayko, and D.L. Jones, Tectonostratigraphic terranes in southwest Oregon, *in* Tectonostratigraphic terranes of the Circumpacific region, edited by D.G. Howell, *Circumpacific Council for Energy and Mineral Resources, Earth Science series, No. 1*, 147-157, 1985.

Blakely, R.J., R.C. Jachens, R.W. Simpson, and R.W. Couch, Tectonic setting of the southern Cascade Range as interpreted from its magnetic and gravity fields, *Geol. Soc. America Bull., 96,* 43-48, 1985.

Blome, C.D., and W.P. Irwin, Tectonic significance of late Paleozoic to Jurassic radiolarians in the North Fork terrane, Klamath Mountains, California, *in* Pre-Jurassic rocks in western North American suspect terranes, edited by C.H. Stevens, *S.E.P.M., Pacific Section,* 77-89, 1983.

Bogen, N.L., Paleomagnetism of the Upper Jurassic Galice Formation, southwestern Oregon: Evidence for differential rotation of the eastern and western Klamath Mountains, *Geology, 14,* 335-338, 1986.

Borns, D.J., Blueschist metamorphism in the Yreka-Fort Jones area, Klamath Mountains, northern California, Ph.D. thesis, Univ. of Wash., Seattle, 167 p., 1980.

Brown, E.H., and M.C. Blake, Jr., Correlation of Early Cretaceous blueschists in Washington, Oregon and northern California, *Tectonics, 6,* 795-806, 1987.

Coleman, R.G., and M.D. Helper, Significance of the Condrey Mountain dome in the evolution of the Klamath Mountains, California and Oregon, *Geol. Soc. America, Abstracts with Programs, 15,* 294, 1983.

Coleman, R.G., M.D. Helper, and M.M Donato, Geologic map of the Condrey Mountain Roadless Area, Siskiyou County, California, *U.S. Geol. Survey, Misc. Field Studies Map MF-1540-A,* scale 1:50,000, 1983.

Coney, P.J., D.L. Jones, and J.W.H. Monger, Cordilleran suspect terranes, *Nature, 288,* 329-333, 1980.

Davis, G.A., M.J. Holdaway, P.W. Lipman, and W.D. Romey, Structure, metamorphism, and plutonism in the south-central Klamath Mountains, California, *Geol. Soc. America Bull., 76,* 933-966, 1965.

Diller, J.S., Sandstone dikes, *Geol. Soc. America Bull., 1,* 411-442, 1889.

Donato, M.M., The May Creek Schist, southwestern Oregon: Remnant of an incipient back arc basin?, *Geol. Soc. America, Abstracts with Programs, 19,* p. 373, 1987.

Donato, M.M., C.G. Barnes, R.G. Coleman, W.G. Ernst, and M.A. Kays, Geologic map of the Marble Mountain Wilderness, Siskiyou County, California, *U.S. Geol. Survey, Misc. Field Studies Map MF-1452-A,* scale 1:48,000, 1982.

Fagan, S.W., and W.A. Gose, Paleomagnetic data from the Redding section of the eastern Klamath belt, northern California, *Geology, 11,* 505-508, 1983.

Fahan, M.R., Geology and geochronology of a part of the Hayfork terrane, Klamath Mountains, northern California, M.S. thesis, Univ. of California, Berkeley, 127 p., 1982.

Fraticelli, L.A., J.P. Albers, W.P. Irwin, and M.C. Blake, Jr., Geologic map of the Redding 1 X 2 degree quadrangle, Shasta, Tehama, Humboldt, and Trinity Counties, California, *U.S. Geol. Survey Open-file Report 87-257,* scale 1:250,000, 1987.

Fuis, G.S., and J.J. Zucca, A geologic cross section of northeastern California from seismic refraction results, *in* Geology of the Upper Cretaceous Hornbrook Formation, Oregon and California, edited by T.H. Nilsen, *S.E.P.M., Pacific Section, 42,* 203-209, 1984.

Garcia, M.O., Petrology of the Rogue and Galice Formations, Klamath Mountains, Oregon; identification of a Jurassic island arc sequence, *Jour. Geol., 87,* 29-41, 1979.

Griscom, A., Bouguer gravity map of California; Redding Sheet, *Calif. Div. Mines and Geol.,* scale 1:250,000, 1973.

Harper, G.D., The Josephine ophiolite, northwestern California, *Geol. Soc. America Bull., 95,* 1009-1026, 1984.

Helper, M.A., Deformation and high P/T metamorphism in the central part of the Condrey Mountain window, north-central Klamath Mountains, California and Oregon, *Geol. Soc. America Memoir 164,* 125-141, 1986a.

Helper, M.A., Early Cretaceous metamorphic ages for high P/T schists in the Condrey Mountain window, Klamath Mountains, northern California: An inlier of Franciscan?, *Geol. Soc. America, Abstracts with Programs, 18:6,* 634, 1986b.

Hotz, P.E., Geologic map of the Condrey Mountain quadrangle and parts of the Seiad Valley and Hornbrook quadrangles, California, *U.S. Geol. Survey, Geol. Quad. Map GQ-618,* scale 1:62,500, 1967.

Hotz, P.E., Plutonic rocks of the Klamath Mountains, California and Oregon, *U.S. Geological Survey Professional Paper 684-B,* 20 p., 1971.

Hotz, P.E., Geology of the Yreka quadrangle, Siskiyou County, California, *U.S. Geol. Survey Bull. 1436,* 72 p., 1977.

Hotz, P.E., M.A. Lanphere, and D.A. Swanson, Triassic blueschist from northern California and north-central Oregon, *Geology, 5,* 659-663, 1977.

Irwin, W.P., Geological reconnaissance of the northern Coast Ranges and Klamath Mountains, with a summary of the mineral resources, *Calif. Div. Mines Bull. 179,* 80 p., 1960.

Irwin, W.P., Terranes of the western Paleozoic and Triassic belt in the southern Klamath Mountains, California, in Geological Survey research, 1972, *U.S. Geol. Survey Prof. Paper 800-C,* C103-C111, 1972.

Irwin, W.P., Age and tectonics of plutonic belts in accreted terranes of the Klamath Mountains, *in* Tectonostratigraphic terranes of the Circumpacific region, edited by D.G. Howell, *Circumpacific Council for Energy and Mineral Resources, Earth Science Series, No. 1,* 187-199, 1985.

Irwin, W.P., and S.P. Galanis, Jr., Map showing limestone and selected fossil localities in the Klamath Mountains, California and Oregon, *U.S. Geol. Survey, Misc. Field Studies Map MF-749,* scale 1:500,000, 1976.

Irwin, W.P., D.L. Jones, and C.D. Blome, Map showing sampled radiolarian localities in the Western Paleozoic and Triassic belt, Klamath Mountains, California, *U.S. Geol. Survey, Misc. Field Studies*

Map *MF-1399*, scale 1:250,000, 1982.

Jayko, A.S., and M.C. Blake, Jr., Geologic map of part of the Orleans Mountain Roadless Area, Siskiyou and Trinity Counties, California, *U.S. Geol. Survey, Misc. Field Studies Map MF-1600-A*, scale 1:48,000, 1984.

Kinkel, A.R., W.E. Hall, and J.P. Albers, Geology and base-metal deposits of West Shasta copper-zinc district, Shasta County, California, *U.S. Geol. Survey Prof. Paper 285*, 156 p., 1956.

Klein, C.W., Thrust plates of the north-central Klamath Mountains near Happy Camp, California, *California Division of Mines and Geology Special Report 129*, 23-26, 1977.

LaFehr, T.R., Gravity in the eastern Klamath Mountains, California, *Geol. Soc. America Bull., 77*, 1177-1190, 1966.

Lanphere, M.A, W.P. Irwin, and P.E. Hotz, Isotopic age of the Nevadan orogeny and older plutonic and metamorphic event in the Klamath Mountains, California, *Geol. Soc. America Bull., 79*, 1027-1052, 1968.

Lanphere, M.A., M.C. Blake, Jr., and W.P. Irwin, Early Cretaceous metamorphic age of the South Fork Mountain Schist in the northern Coast Ranges of California, *Am. Jour. Sci., 278*, 798-815, 1978.

Lindsley-Griffin, N., The Trinity ophiolite, Klamath Mountains, California, *in* North American Ophiolites, edited by R.G. Coleman and W.P. Irwin, *Oregon Dept. Geol. and Mineral Industries, Bull. 95*, 107-120, 1977.

Lindsley-Griffin, N., Structure, stratigraphy, petrology, and regional relationships of the Trinity ophiolite, Ph.D. thesis, Univ. of California, Davis, 436 p., 1982.

Mankinen, E.A., W.P. Irwin, and C.S. Gromme, Implications of paleomagnetism for the tectonic history of the Eastern Klamath and related terranes in California and Oregon, *in* Geology of the Upper Cretaceous Hornbrook Formation, Oregon and California, edited by T.H. Nilsen, *S.E.P.M., Pacific Section, 42*, 221-229, 1984.

Mankinen, E.A., W.P. Irwin, and C.S. Gromme, Paleomagnetic results from the Shasta Bally plutonic belt in the Klamath Mountains province, northern California, *Geophys. Res. Lett., 15*, 56-59, 1988.

Mattinson, J.M., and C.A. Hopson, Paleozoic ophiolitic complexes in Washington and northern California, *Carnegie Inst., Ann. Rept to Director, Geophys. Lab., 1971-1972*, 578-583, 1972.

Miller, M.M., Tectonic evolution of late Paleozoic island arc sequences in the western U. S. cordillera; with detailed studies from the eastern Klamath Mountains, northern California, Ph.D. thesis, Stanford Univ., 201 p., 1986.

Mortimer, N., Petrology and structure of Permian to Jurassic rocks near Yreka, Klamath Mountains, California, Ph.D. thesis, Stanford Univ., 84 p., 1984.

Nestell, M.K., W.P. Irwin, and J.P. Albers, Late Permian (Early Djulfian) Tethyan Foraminifera from the southern Klamath Mountains, California, *Geol.*

Soc. America, Abstracts with Programs, 12, 519, 1981.

Pessagno, E.A., Jr., and C.D. Blome, Biostratigraphic, chronostratigraphic, and U/Pb geochronometric data from the Rogue and Galice Formations, Western Klamath terrane (Oregon and California): Their bearing on the age of the Oxfordian-Kimmeridgian boundary and the Mirifusus first occurrence event, *in Proceedings of the 2nd International Symposium on Jurassic Stratigraphy*, I.U.G.S., Lisbon, Portugal, in press.

Peterson, G.L., Structural interpretation of sandstone dikes, northwest Sacramanto Valley, California, *Geol. Soc. America Bull., 77*, 833-842, 1966.

Potter, A.W., P.E. Hotz, and D.M. Rohr, Stratigraphy and inferred tectonic framework of lower Paleozoic rocks in the eastern Klamath Mountains, northern California, *in* Paleozoic paleogeography of the western United States, edited by J.H. Stewart and others, *S.E.P.M., Pacific Section, Pacific Coast Paleogeography Symposium 1*, 421-440, 1977.

Renne, P.R., G.R. Scott, and D.R. Bazard, Multicomponent paleomagnetic data from the Nosoni Formation, Eastern Klamath Mountains, California: Cratonic Permian primary directions with Jurassic overprints, *Jour. Geophys. Res., 93*, B4, 3387-3400, 1988.

Roure, F., and P. De Wever, Decouverte de radiolarites du Trias dans l'unite occidental des Klamath, sudouest de l'Oregon, U. S. A.: Consequences sur l'age des peridotites de Josephine, *Acad. Sci Comptes Rendus (Paris), 297*, p. 161-164., 1983.

Sanborn, A.F., Geology and paleontology of the southwest quarter of the Big Bend quadrangle, Shasta County, California, *Calif. Div. Mines Special Rept. 63*, 26 p., 1953.

Schultz, K.L., Paleomagnetism of Jurassic plutons in the central Klamath Mountains, southern Oregon and northern California, M.S. thesis, Oregon State Univ., Corvallis, M.S. thesis, 153 p., 1983.

Schweickert, R.A., and W.P. Irwin, Tertiary detachment faulting in the Klamath Mountains, California: A new Hypothesis, *Geol. Soc. America, Abstracts with Programs, 18*, 742, 1986.

Smith, J.G., N.J. Page, M.G. Johnson, B.C. Moring, and F. Gray, Preliminary geologic map of the Medford 1° X 2° quadrangle, Oregon and California, *U.S. Geol. Survey Open-file Rept. 82-955*, scale 1:250,000, 1982.

Snoke, A.W., A thrust plate of ophiolitic rocks in the Preston Peak area, Klamath Mountains, California, *Geol. Soc. America Bull., 88*, 1641-1659, 1977.

Wagner, D.L., and G.J. Saucedo, Geologic map of the Weed quadrangle, *Calif. Div. Mines and Geol., Regional Geol. Map Series, No. 4A*, scale 1:250,000, 1987.

Wallin, E.T., J.M. Mattinson, and A.W. Potter, Early Paleozoic magmatic events in the eastern Klamath Mountains, northern California, *Geology, 16*, 144-148, 1988.

Wells, F.G., and D.L. Peck, Geologic map of Oregon west of the 121st meridian, *U.S. Geol. Survey, Misc.*

Invest. Map I-325, scale 1:500,000, 1961.

Wright, J.E., Permo-Triassic accretionary subduction complex, southwestern Klamath Mountains, northern California, *Jour. Geophys. Res., 87,* 3805-3818, 1982.

Wyld, S.J., and J.E. Wright, The Devils Elbow ophiolite remnant and overlying Galice Formation: New constraints on the Middle to Late Jurassic evolution of the Klamath Mountains, California, *Geol. Soc. America Bull., 100,* 29-44, 1988.

STRATIGRAPHY AND STRUCTURE OF AN ANCIENT ISLAND ARC: LATE PALEOZOIC AND EARLY MESOZOIC EVOLUTION OF THE EASTERN KLAMATH TERRANE, NEAR MCCLOUD LAKE, NORTHERN CALIFORNIA

M. Meghan Miller
Division of Geological and Planetary Sciences,
California Institute of Technology, Pasdadena

INTRODUCTION

Volcanic arc and related rocks of the eastern Klamath terrane (Redding section) comprise the easternmost and structurally highest fault block in the Klamath Mountains (Figure 1). These rocks form a coherent, moderately to steeply eastward-dipping sequence that extends for 100 km along strike (Figure 2), and records long-lived oceanic-arc magmatism and sedimentation that spanned Early Devonian to Middle Jurassic time (Sanborn, 1960; Albers and Bain, 1985; Miller, 1986, in press; Renne, 1986). The eastern Klamath terrane provides a cross-sectional, along-strike view of the arc-related sequence, recording temporal and spatial variations in deposition, magmatism, and deformation within an ancient island arc setting. This guide reviews the stratigraphic and structural evolution of the eastern Klamath terrane as a whole, and provides a road log for stratigraphy and structure of upper Paleozoic (Miller, 1986, in press) and lower Mesozoic rocks (Sanborn, 1960) in the northern part of the terrane, in the vicinity of McCloud Lake (Figure 2, 3), and eastward above Hawkins Creek.

The volcanic arc represented by rocks of the eastern Klamath terrane plays a critical role in tectonic models for the late Paleozoic and early Mesozoic evolution of the western North American Cordillera. This terrane belongs to a regionally extensive belt of terranes which are thought to have formed a single northeast Pacific fringing arc during the late Paleozoic (Miller, 1987). Ideas concerning its paleogeography and subduction polarity are controversial. If the eastern Klamath arc evolved off the edge of western North America above a

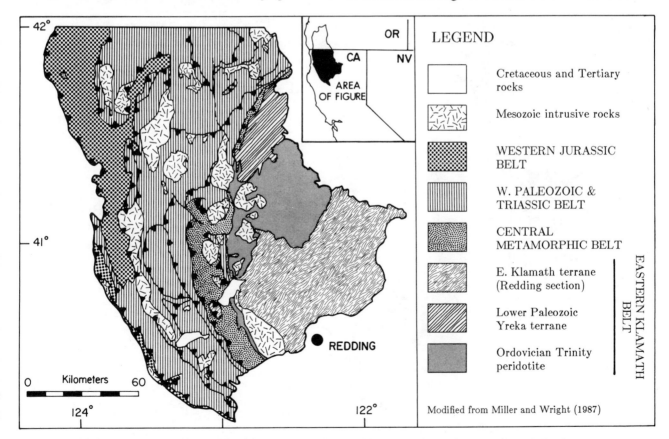

LEGEND

- Cretaceous and Tertiary rocks
- Mesozoic intrusive rocks
- WESTERN JURASSIC BELT
- W. PALEOZOIC & TRIASSIC BELT
- CENTRAL METAMORPHIC BELT
- E. Klamath terrane (Redding section) }
- Lower Paleozoic Yreka terrane   } EASTERN KLAMATH BELT
- Ordovician Trinity peridotite   }

Modified from Miller and Wright (1987)

FIGURE 1 Lithotectonic belts of the Klamath Mountains. The eastern Klamath terrane lies within the easternmost and structurally highest lithotectonic belt of the Klamath Mountains.

dominantly east-dipping subduction zone (Wright, 1982; Miller and Wright, 1987), then rocks to the east, in the Golconda allochthon were deposited in a back-arc setting, and thrust over the continental margin during Permo-Triassic compressional deformation (Silberling, 1973; Churkin, 1974; Burchfiel and Davis, 1981; Miller and others, 1984). Alternatively, if the eastern Klamath terrane was exotic to North America during the late Paleozoic, and formed part of a far-traveled, east-facing arc (Sonomia), then rocks of the Golconda allochthon

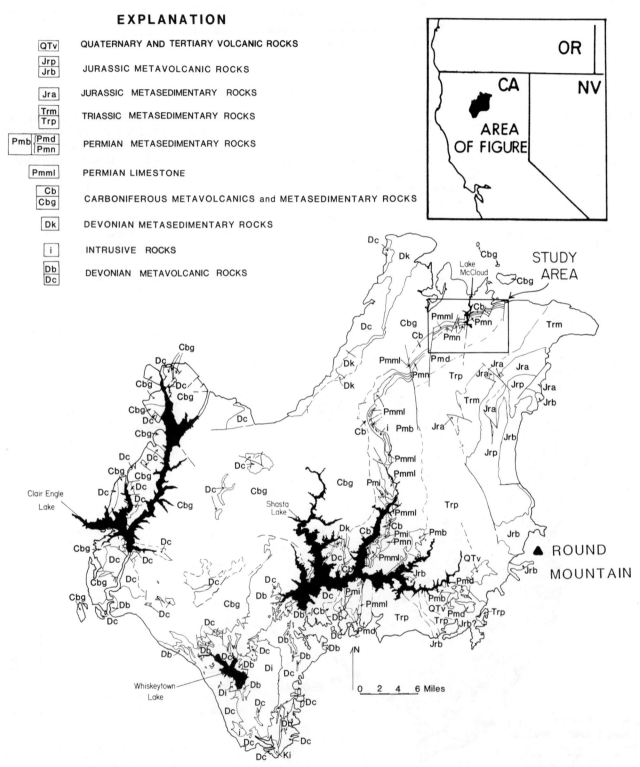

FIGURE 2 Distribution of upper Paleozoic and lower Mesozoic rocks of the eastern Klamath terrane. From compilation in Miller (in press). Location is shown by area of Redding section in Figure 1.

represent a fore-arc accretionary wedge (Speed, 1979; Brueckner and Snyder, 1985). Accordingly, the accretionary wedge was emplaced over the continental margin during a Permo-Triassic collisional orogeny, and the eastern Klamath terrane has no earlier paleotectonic relation to North America.

The paleogeographic relation of the eastern Klamath arc sequence to Paleozoic and Mesozoic rocks further to the west and structurally lower in the Klamath Mountains is also controversial. For example, Wright (1982; see also Miller and Wright, 1987) suggested that the Permian and Triassic arc volcanism in the eastern Klamath terrane was related to eastward subduction recorded by rocks to the west, in the eastern Hayfork subduction complex (western Paleozoic and Triassic belt, Figure 1). In contrast, Mortimer (1984) suggested that Lower Jurassic rocks in the intervening North Fork terrane were not paleogeographically related to the eastern Klamath terrane, and that Klamath terranes were assembled at a subsequent time, during the Middle or Late Jurassic.

Within volcanic arc terranes, direct evidence for paleogeographic ties is sparse (Miller, 1987, in press). Sedimentary and volcanogenic debris generally originates from a relatively high-standing arc edifice and is dispersed to surrounding basinal terranes. However, some clues to the paleogeographic relations between the eastern Klamath and coeval terranes occur within the succession. These include provenance of sparse basinal epiclastic units, evidence for biogeographic provincialism, tectonic and structural history, and stratigraphic relations to coeval terranes.

REDDING SECTION: STRATIGRAPHIC REVIEW

The oldest rocks in the eastern Klamath terrane belong to the Lower and Middle Devonian Copley Greenstone. The Copley has been subdivided into two members; a lower unit of massive lava and pyroclastic debris overlain by an upper unit of pillow lavas, which includes high-Mg andesite (Lapierre and others, 1985a; Figure 4). Sandstone with granitic debris and shaley tuff are locally interbedded with the pillow lavas. The geochemistry of the Copley lavas suggests that they are typical of primitive, oceanic island arc volcanic rocks that formed in an extensional tectonic setting (Lapierre and others, 1985a).

In the vicinity of Lake Shasta, the Copley Greenstone interfingers with and is overlain by tuff and lava of the keratophyric Balaklala 'Rhyolite' (Kinkel and others, 1956; Albers and Robertson, 1961; Lapierre and others, 1985 b; Albers and Bain, 1985). A Middle Devonian fish plate dates submarine tuffs in the Balaklala (Boucot and others, 1974). The trondhjemitic Mule Mountain stock intrudes, and is inferred to be cogenetic with, part of the Balaklala. The Mule Mountain stock yields a concordant U/Pb zircon age of 400 Ma (no error ranges are given) and a K/Ar hornblende age of 392±3 Ma (Albers and others, 1981). Intra-arc extension was recorded by sea-floor graben formation that accompanied silicic magmatism and Kuroko-type massive sulfide mineralization in the Devonian volcano-plutonic association (Lindberg, 1985; Figure 4).

The Middle Devonian (Eifelian) Kennett Formation conformably overlies and interfingers with the Balaklala. Shallow marine volcaniclastic rocks are succeeded by slope turbidites, basinal mudstone, and carbonate debris aprons that were deposited in multiple, small, extension-related basins (Watkins and Flory, 1986). Volcanogenic breccia is also present (Albers and Robertson, 1961). An angular unconformity locally occurs above the Kennett, and clasts of limestone from the Kennett occur in the overlying Bragdon Formation (Albers and Robertson, 1961). The amount of time represented by the hiatus is unknown and, in places, the Bragdon Formation conformably overlies the Kennett Formation, and thus may be as old as Late Devonian (Watkins, 1986).

The Upper Devonian(?) and Mississippian Bragdon Formation contains turbiditic sandstone and argillite, and interstratified channel-filling conglomerate and massive sandstone. Epiclastic and pyroclastic turbidites occur in the same stratigraphic succession (Figure 4). In epiclastic sandstone, constituent grains include sedimentary, metasedimentary (greenschist facies pelitic and quartzose lithic fragments), reworked intermediate volcanic debris, and sparse fragments of felsic tuff and mafic lava (Miller and Cui, in review). This succession records deposition in a 'hybrid', sand-rich submarine fan characterized by apical lobes and discrete, down-cutting channels. Fine-grained, distal facies occur in the base of the section, and are transitional to coarser, channel and interchannel deposits. These are in turn overlain by shallow-water shelf deposits (Watkins, 1986; Miller and Cui, in review).

The Bragdon Formation is gradationally and conformably overlain by volcanogenic debris flows, water-laid tuff, and lava of the Baird Formation (Figure 4). Locally, shallow-water facies are present (Miller, in press). In the central part of the eastern Klamath terrane, the section is dominated by lavas; to the north, coeval strata form volcaniclastic dispersal facies (Watkins, 1985). The facies and lithology of the Baird Formation reflect a transition to shallow-water deposition and an accompanying increase in volcanogenic sediment supply, suggesting localized uplift and variable outbuilding of volcanic debris aprons. Based on brachiopods and corals, the age of the Baird Formation ranges from Late Mississippian to Late Pennsylvanian or Early Permian(?); however, locally within the terrane, the entire Pennsylvanian record may be missing (Watkins, 1973, 1985; see also Yancy, 1986).

The Baird Formation locally interfingers with the conformably overlying Lower Permian McCloud Limestone, although in places the contact is disconformable (Watkins, 1985). Fossiliferous, shallow-water platformal limestone records two stages of carbonate platform deposition (Watkins, 1985). A lower Wolfcampian carbonate platform developed in the southern and central parts of the terrane, and an upper Wolfcampian to lower Leonardian platform overlapped it, extending further northward (Figure 4). In both stages of platform development, shallow-water facies

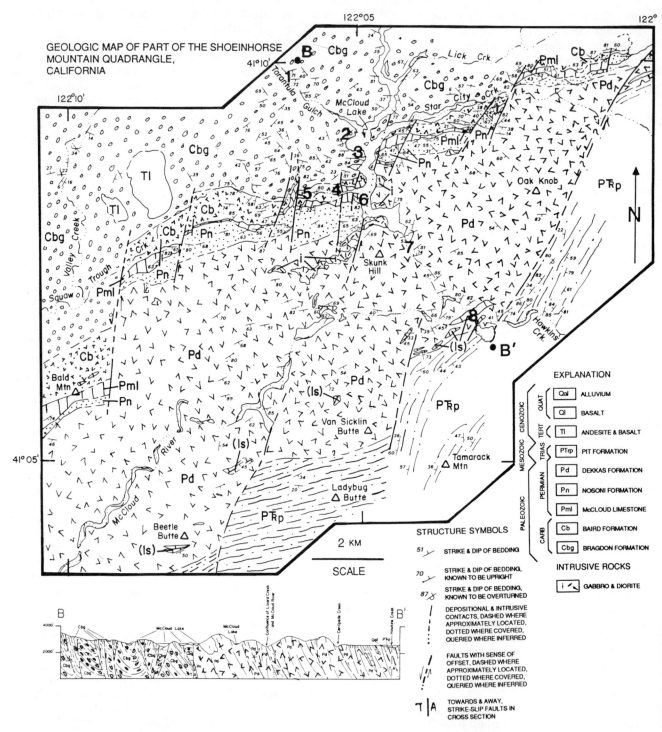

FIGURE 3 Geologic map and cross section of the McCloud Lake area simplified from Miller (1986). Location shown in Figure 2. Large numbers designate field trip stops (1-8).

occur to the south, and more distal, steepened carbonate ramps occur to the north (Watkins, 1985; Miller, 1986, in press). Age constraints and biogeographic indicators come from abundant fusulinids and corals which were comprehensively studied and described by Skinner and Wilde (1965).

In most of the eastern Klamath terrane, the upper Lower Permian rock record is missing across a disconformity that separates the McCloud Limestone from the overlying Upper Permian (Guadalupian) volcanogenic Bollibokka Group (Nosoni and Dekkas Formations, Coogan, 1960). Vent proximal lavas characterize the Nosoni and Dekkas at their southern extent (Albers and Robertson, 1961); more distal facies, such as tuffaceous debris flows, and other volcanogenic sediment gravity flows predominate to the north and were deposited in a more basinal setting. The northerly section may be slightly older, spanning most of Leonardian time (Miller, 1986, in press). Limestone lenses throughout the Bollibokka Group contain Late Permian (Guadalupian) fusulinids (Coogan, 1960) and corals (Stevens and others, 1987). In the south, the Dekkas is overlain by keratophyric tuff and lava of the Bully Hill 'Rhyolite' (Albers and Robertson, 1961; Figure 4). The Pit River stock has yielded a Late Permian U/Pb age of 261 Ma, and is believed to be cogenetic with the Dekkas Formation and Bully Hill 'Rhyolite' (Fraticelli and others, 1985).

The Middle Triassic Pit Formation overlies the Dekkas Formation (Coogan, 1960) and, in the south, is intercalated with the Bully Hill 'Rhyolite' (Albers and Robertson, 1961). The Pit Formation contains argillite, silicic tuffaceous and crystal-rich turbidites, and radiolarian-bearing hemipelagic sedimentary rocks, recording basinal sedimentation (Sanborn, 1960; Curtis, 1983). No Early Triassic fauna have been found in the Redding section. Late Permian radiolaria have been recovered from medium- and thin-bedded white and grey chert beds which overlie the Dekkas Formation in the central part of the terrane. The fossiliferous strata have been assigned to the Pit Formation (Silberling and Jones, 1982), and based on their age, stratigraphic position and lithology, they may be lateral equivalents (silicified tuff)

of the Upper Permian Bully Hill 'Rhyolite'. On Shasta Lake, Middle Triassic fossils occur tens of meters above the base of the Pit Formation (Silberling and Jones, 1982). The Permo-Triassic boundary and the Lower Triassic may thus be represented by unfossiliferous beds in the lower Pit Formation. Alternatively, a hiatus may be present in the lower Pit Formation. If such is the case, the Lower Triassic may be missing from the rock record; as suggested by the lack of Early Triassic fauna, and by structural relations in the northern part of the terrane, discussed below. Furthermore, relatively short intervals of geologic time are represented by thick sequences of water-laid pyroclastics elsewhere within the succession. It seems unlikely that the much of the Upper Permian and entire Lower Triassic are represented by tens of meters of pyroclastic strata at the base of the Pit Formation.

The Pit Formation is conformably overlain by the Upper Triassic (Karnian) Hosselkus Limestone, which contains abundant shallow water fauna (Sanborn, 1960). The Hosselkus is succeeded by black argillite of the Brock Shale and upward-fining, water-laid pyroclastic debris in the Upper Triassic Modin Formation. The Modin Formation has been subdivided into three members, all of which are Upper Triassic (Sanborn, 1960). These include a basal pyroclastic and quartz- and chert-rich conglomeratic section (Hawkins Creek

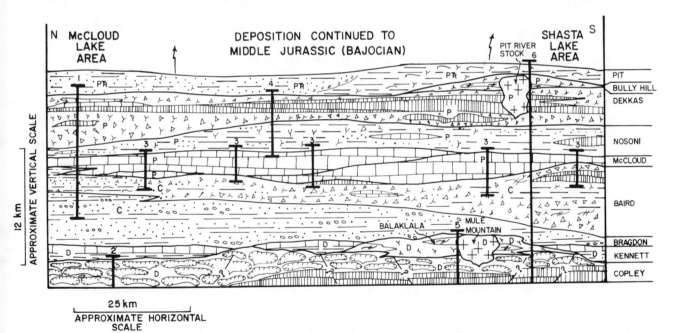

FIGURE 4 Schematic regional stratigraphy of upper Paleozoic rocks within the eastern Klamath terrane. This compilation of stratigraphic thickness and lithologies along 100 km of strike illustrates the strong differences in facies, ages and lithologic units which can be documented within an ancient volcanic-arc stratigraphy. Variations in thickness within formations reflect real variations reported by various workers. Letters designate ages as follows. D: Devonian, C: Carboniferous, P: Permian, PTR: Permian(?) and Triassic. Numbered bars designate the sources of data. 1: Miller, 1986, in press; 2: Lapierre and others, 1985b; 3: Watkins, 1973, 1985, 1986; Skinner and Wilde, 1965; 4: Coogan, 1960; 5: Albers and others, 1981, Lapierre and others, 1985a, c; 6: Albers and Robertson, 1961. Mesozoic history is not depicted, see Sanborn (1960), Albers and Robertson (1961), and Renne (1986).

member), an intermediate tuffaceous limestone and fine-grained calcareous sandstone (Devils Canyon member), and an upper thin-bedded tuffaceous and pyritiferous argillite (Kosk Creek member).

In the northern part of the terrane, the Modin Formation is unconformably overlain by the Lower Jurassic Arvison Formation (Sanborn, 1960); further south the contact is considered conformable (Renne, 1986). Arvison strata record deposition of pyroclastic debris and mafic lavas in a near-shore environment (Sanborn, 1960), marking the reinitiation of voluminous volcanism. The formation is dated by shallow-water marine fauna; it also contains terrigenous plant debris (Sanborn, 1960). The Arvison Formation is overlain by the Potem Formation, which primarily consists of tuffaceous argillite with minor conglomerate and coarse pyroclastic debris (Sanborn, 1960). Fossils from the Potem range from Early to Middle Jurassic (Bajocian) in age. The lower part of the Potem is intercalated with the Bagley Andesite, which consists of intermediate pyroclastic rocks and lavas (Sanborn, 1960). The Bagley was thought to represent a lateral volcanic facies of the Potem Formation, and thus was considered equivalent in age. However, much or all of what has been mapped as Bagley Andesite is now known to be parts of hypabyssal intrusive bodies (Sanborn, 1960; Renne, 1986) which yield younger Jurassic and Cretaceous radiometric ages (Renne, 1986).

In summary, stratigraphic relations in the eastern Klamath Mountains record a long-lived history of island arc development. Extension-related, primitive arc volcanism occurred during the Early and Middle Devonian and was succeeded by carbonate, epiclastic and volcanogenic sedimentation. During the early Carboniferous, an influx of sedimentary and meta-sedimentary detritus was recorded by turbiditic sandstone and related channel deposits, reflecting uplift of an adjacent sedimentary and meta-sedimentary terrane. Mid-Carboniferous volcanism resulted in progradation of volcaniclastic aprons. An Early Permian lull in volcanism led to isolation of volcaniclastic aprons by differential uplift, allowing localized development of carbonate platforms. Leonardian uplift and tilting preceded rapid subsidence and voluminous volcanism during the Late Permian. In the Middle Triassic, episodic volcanism recurred and was followed by Late Triassic shallow water carbonate bank deposition, in turn succeeded by voluminous volcanism, both in the Late Triassic and again in late Early to early Middle Jurassic. These relations record periods of rapid subsidence punctuated by periods of relative tectonic stability and local uplift, suggesting that episodic extensional tectonism controlled deposition within the volcanic arc setting.

STRUCTURAL EVOLUTION

Until recently, rocks of the eastern Klamath terrane have been considered a weakly deformed, homoclinal stratigraphic sequence (e.g., Irwin and Dennis, 1979), consistent with the relatively simple map patterns and the absence of thrust faults or folds which repeat portions of the stratigraphic section on a regional scale. However, in the last decade, the importance of widespread intraformational disruption has been recognized (Albers and Bain, 1985; Miller, 1986, 1988; Renne, 1986). Complex folding, faulting and cleavage formation are variably developed throughout the terrane. This section reviews mesoscopic scale structures developed in the northern part of the terrane (Miller, 1986, 1988) and compares them to structures described elsewhere within the terrane (Albers and Robertson, 1961; Albers and Bain, 1985; Renne, 1986).

McCloud Lake Area

In the vicinity of McCloud Lake, at least three episodes of deformation are recorded by upper Paleozoic rocks (Miller, 1986). Reconnaissance work on Mesozoic rocks suggests a less complex deformational history for the younger rocks (Miller, 1988). The three deformational events are recorded by 1) northwest-vergent, southwest-plunging folds with subvertical to steeply southeastward-dipping axial planes, cleavage and related intraformational thrust faults, 2) subsequent refolding of the early cleavage around moderately west-dipping axial planes, and 3) subvertical left-lateral strike slip faults. The first two fabrics are heterogeneously developed and are strongly controlled by lithology.

D1: Northwest-vergent folding, cleavage formation and faulting. The most pervasive structural grain in the vicinity of McCloud Lake is characterized by folding, cleavage formation and intraformational thrust faulting (Figure 3). Although these features cannot be traced on a map scale due to lack of natural outcrop, road cuts afford excellent exposures of mesoscopic structures formed during D1. Deformation is locally associated with poorly developed low greenschist and sub-greenschist facies metamorphic assemblages.

Mesoscopic folds are most common in thin-bedded sandstone and argillite of the Bragdon Formation. They are also present in thin- to medium-bedded portions of the McCloud Limestone, and are poorly developed in the basal silicified siltstones of the Nosoni Formation. D1 fold axes trend NE-SW and plunge shallowly or moderately to the southwest, or, less commonly, shallowly to the northeast. Folds have NE striking, steeply SE-dipping or subvertical axial planes. They are differentiated from younger structures by their orientation, morphology, apical angles, relation to mesoscopic scale faults and their associated cleavage.

The long limbs of asymmetric folds are predominantly southeast-dipping and upright (Figure 3). Short limbs of asymmetric folds are steeply dipping or overturned, and are commonly truncated by faults. Based on fold asymmetry and fold-fault relations, deformation is northwest-vergent. In bedded sandstone-argillite sequences, folds are associated with steeply-dipping cleavage which is parallel to axial planes. In places, cleavage is coaxially refolded, and the later folds are inferred to be part of a single progressive deformation.

Sparse boudinage of medium- to thick-bedded sandstone occurs in the limbs of folds.

Cleavage associated with D1 generally dips vertically or steeply to the southeast, is axial planar in the cores of mesoscopic folds, and is fanned in the limbs. Refraction occurs within graded beds. Cleavage is ubiquitous in fine-grained rocks throughout the upper Paleozoic section and continuous in hand specimen. On a thin section scale, cleavage is disjunctive and defined by anastomosing seams in sandstone and siliceous argillite or stylolitic surfaces in limestone (e.g., Powell, 1979; Engelder and Marshak, 1985). Concentrations of opaque, insoluble material form cleavage surfaces at sub-millimeter spacing; new mineral growth is generally absent, although exceptions are discussed below. These characteristics suggest a solution origin for cleavage (Engelder and Marshak, 1985); the spacing and geometry of cleavage planes indicate "strong" development, reflecting 25% to 35% shortening (e.g., Alvarez and Engelder, 1982). Bedding-cleavage intersection lineations parallel the SW-NE plunging fold axes, confirming a relation between folding and cleavage formation.

Steeply-dipping, NE-striking faults are spatially associated with D1 folds, and occur throughout the upper Paleozoic section. These faults juxtapose differing facies within individual formations, but rarely cut formation boundaries, implying offsets of tens to hundreds of meters. Several meter wide fault zones contain internally coherent phacoids of differing lithology from within single formations. Massive lithologic units, such as lavas, are complexly and irregularly faulted. Folding and cleavage development are not observed in these units; shortening is accommodated by faulting. Tension gashes are commonly associated with mesoscopic faults and suggest reverse displacement.

Although the disjunctive cleavage associated with D1 is generally of solution origin, new growth of tabular minerals within the cleavage plane has occurred in sparse felsic tuffaceous units. Lepidoblastic minerals include chlorite, sericite and stilpnomelane, and co-exist with albite and quartz, forming a low greenschist assemblage. Sparse biotite within anastomosing cleavage planes occurs only in rocks containing detrital biotite, and is inferred to be a relict phase.

Cleavage and mesoscopic scale folds are absent from rocks of the Middle (and Upper?) Triassic Pit Formation and younger strata in the northeastern part of the terrane, however, thrust faults are present. A regional strain gradient may coincide with this stratigraphic boundary; or alternatively, D1 in the vicinity of McCloud Lake may represent an older, pre-Middle Triassic deformation.

D2: Southwest and northwest plunging folds.
A second folding event is poorly developed jn the vicinity of McCloud Lake. It is characterized by folding of cleavage associated with D1 around shallowly NW-dipping axial planes. D2 is observed on a mesoscopic scale, and has no associated metamorphism.

F2 folds are best developed in fine-grained rocks, and are characterized by open folding about shallowly plunging fold axes. They are most abundant in argillites of the Bragdon Formation and the basal siltstones of the Nosoni Formation. Axial planes dip moderately to the northwest, and fold axes plunge gently to the NNW or SSW. The SSW trending fold axes are distinguished from F1 structures by moderately westward dipping axial planes and greater apical angles. Some F2 structures fold the cleavage formed during D1.

D3: North-northeast striking high-angle faulting.
Left-lateral apparent strike-slip faults form D3 structures, and are best observed on a map scale (Figure 2, 3), as they occur at high angles to stratigraphic trends in the northern part of the terrane. These faults cross cut other deformational fabrics in the study area. Although apparent offset is primarily strike-slip, some component of dip-slip is inferred from juxtaposition of differing depositional facies.

Summary. In summary, three phases of deformation are recorded by rocks in the vicinity of McCloud Lake. Northwest-vergent folding, faulting and cleavage formation affects rocks as young as Upper Permian. Cleavage is subsequently refolded around shallowly westward dipping axial planes. Late steeply dipping strike-slip faults offset older fabrics and regional stratigraphic trends.

WEST SHASTA DISTRICT:

Structures similar to those developed during the oldest deformation in the McCloud Lake area occur in middle to upper Paleozoic rocks in the vicinity west of Shasta Lake (Figure 2). These features were originally described by Kinkel et al. (1956), and have been re-evaluated by Albers and Bain (1985). There, the oldest generation of structures (aside from syndepositional normal faulting) is characterized by northwest-vergent folding, slaty cleavage formation and minor associated faulting. Fold axes plunge shallowly or moderately to the NE, axial planes and coplanar cleavage dip steeply or moderately to the SE. Major fabric elements parallel those of D1 from the McCloud Lake area, described above, except that fold axes are oppositely plunging. A second, nearly coaxial deformation affects rocks in the West Shasta district. It is characterized by crenulation cleavage, and second generation fold axes which parallel F1 axes (Albers and Bain, 1985). These workers note that D2 structures strike and trend in a slightly more easterly direction (ENE) than those attributed to D1 although considerable overlap of the data sets is shown. They speculate that D1 fabrics may be related to soft sediment deformation. This seems unlikely in light of the fact that axial planar slaty cleavage development accompanied deformation. Alternatively, D1 and D2 fabrics may be attributed to a single progressive deformation, where strain accommodated by cleavage development was surpassed, and further deformation occurred by development of crenulation cleavage and F2 folds. The relation of these structures to those in younger rocks to the east is little known. I tentatively correlate both D1 and D2 fabrics in the West Shasta district to D1 structures in the vicinity of McCloud Lake. Structures equivalent to D2 and D3 of McCloud Lake are not known from the West Shasta district.

East Shasta District

Further east, near the central part of Shasta Lake (Figure 2), northeast trending folds affect rocks as young as the Bully Hill 'Rhyolite' (Upper Permian) and are tentatively recognized at one locality in the Pit Formation (Albers and Robertson, 1961). These features parallel the penetrative fabrics developed in the West Shasta district, and probably represent the same period of deformation; hence they may be correlative to D1 in the vicinity of McCloud Lake. In contrast, a younger set of northwestward plunging folds is associated with subvertical to moderately southwestward dipping, axial planar cleavage. These folds are well developed in Mesozoic rocks and post-date the northeastward trending folds (Albers and Robertson, 1961). Based on regional fold asymmetry and cleavage-bedding relations (Albers and Robertson, 1961), these structures may be eastward-vergent. They may correspond to D2 structures in the vicinity of McCloud Lake.

Round Mountain Area

In the east central part of the terrane, the structural development of the Mesozoic section has been described by Renne (1986) and Renne et al. (1986). There, the pervasive and oldest structural fabric is characterized by east-vergent folding around moderately to steeply westward dipping axial planes (Renne, 1986); foliation parallels axial planes. No younger mesoscopic scale structures are recognized. Deformation occurred subsequent to deposition of Middle Jurassic (Bajocian) strata and prior to emplacement of cross-cutting intrusive rocks (166±2 to 168±3 Ma, K-Ar plagioclase, Renne and others, 1986). The age of east-vergent deformation is thus constrained as late Middle Jurassic. This deformation appears to correspond to D2 in the vicinity of McCloud Lake, D2 in the East Shasta district, and is not recognized in the West Shasta district. A second generation of folds in the vicinity of Round Mountain is defined by gentle warping of stratigraphic trends and the older fabric on a several kilometer wavelength scale around subvertical fold axes and E-W to WNW striking axial planes, and may be related to oroclinal bending (Renne, 1986).

Summary of Structural Evolution

In summary, structural relations throughout the terrane collectively suggest that northwest-vergent folding, cleavage formation and faulting in the upper Paleozoic section may have occurred prior to deposition of the Mesozoic section. Although Lower Triassic strata are not recognized in the eastern Klamath terrane, no major angular Permo-Triassic unconformity has been described. Alternatively, D1 structures may abruptly die out along a strain gradient that fortuitously coincides, or nearly coincides, with the Permo-Triassic stratigraphic boundary in both the northernmost and southernmost extent of the terrane. A second, eastward-vergent folding event is variably expressed throughout the terrane, and affects rocks of all ages. It is associated with cleavage development only in the central and southern parts of the terrane. Radiometric age constraints indicate that east-vergent deformation occurred during the late Middle Jurassic (Renne and others, 1986), and was coeval with widespread compressional tectonism elsewhere in the Klamath Mountains. Coeval deformation elsewhere in the Klamath Mountains, however, was west-vergent. Both fold-related fabrics are cross cut by high-angle faults which control the present day exposures of stratigraphic boundaries in the northern part of the terrane.

SUMMARY AND DISCUSSION

Rocks of the eastern Klamath terrane record a protracted history of mid-Paleozoic to mid-Mesozoic intra-arc magmatism, sedimentation and deformation. Deposition was largely controlled by intra-arc extension and related basin subsidence, and punctuated by periods of relative tectonic stability. Pulses in volcanism were not always coincident with rapid subsidence, and may in part reflect migration of the magmatic axis within an evolving oceanic arc. Compressional tectonism rarely affected the terrane, perhaps once during the latest Paleozoic or earliest Mesozoic, as discussed here, and again during the late Middle Jurassic (Renne and others, 1986), when folding, thrust faulting and metamorphism were widespread in the Klamath Mountains province (Wright and Fahan, 1988).

Based on stratigraphic development and the composition of volcanic units, the eastern Klamath terrane volcanic arc was constructed on oceanic crust. Island arcs are generally physically separated from continental or convergent margins by complex basin systems, as in the southwest Pacific today. Such intervening basins obscure original paleogeographic relations. The absolute distance between the Paleozoic eastern Klamath arc and the North American continental margin may never be known; however, relatively large separations are implied by Permian biogeographic relations (Ross and Ross, 1983; Stevens, 1985; Stevens and Rycerski, 1988).

In contrast, several lines of evidence suggest a paleogeographic relation to a continental source or, in cases, specifically to a North American marginal basin system. Second-cycle quartzose debris in sandstones of the Upper Devonian(?) and Mississippian Bragdon Formation was probably derived from uplifted arc basement or an adjacent uplifted basinal terrane. Yet such detritus suggests an ultimate sialic crustal source. Mature, recycled supracrustal sediments are widespread in 'microcontinental' island arcs which fringe continental margins, but are constructed on oceanic crust (Cas and Wright, 1987). Furthermore, the age, depositional setting, and provenance of the Bragdon Formation are similar to those of regionally extensive turbiditic strata of the Earn Sequence (Gordey and others, 1987) in the Canadian Cordillera (discussed in Miller and Cui, in review). These chert-rich strata were deposited in an extensional tectonic setting and have depositional ties to the North American continental margin. The Bragdon Formation may represent the southernmost extent of a marginal graben basin system recorded by the Canadian turbidites (Miller and Cui, in review).

The provenance of Mississippian and Permian clastic units in basinal deposits of the eastward Golconda allochthon suggest a paleogeographic relation to a westward-lying volcanic arc terrane (e.g., Snyder and Brueckner, 1983; Whiteford, 1984; Tomlinson and Wright, 1986). The eastern Klamath arc contains rocks of appropriate age and composition to have supplied such detritus, and may have been the source for these clastic units.

Further evidence suggests a paleotectonic relation between the eastern Klamath terrane and terranes to the west in the western Paleozoic and Triassic belt (Figure). For example, Permian biogeographic ties between ancient seamounts, now preserved in the eastern Hayfork subduction complex, and the eastern Klamath arc (Stevens and others, 1987), imply that the subduction complex and the island arc were both related to the same early Mesozoic (and Paleozoic?) convergent margin (Miller and Wright, 1987).

These relations collectively suggest that the eastern Klamath ensimatic arc evolved primarily above an eastward-dipping subduction zone (present coordinates), flanked by a convergent margin to the west and a marginal basin system of unknown, and likely variable, width to the east. Episodic and short-lived subduction polarity reversals may have occurred, but an ongoing and indirect paleogeographic relation to the continent is implied.

ROAD LOG-- DAY 4, July 1 1989-- McCLOUD LAKE AREA

The field trip begins at exposures of the Upper Devonian(?) and Mississippian Bragdon Formation in the vicinity of McCloud Lake. Take I-5 north from Redding about 60 miles, and exit on route 89 east just before Mount Shasta City. Proceed 10 miles east on route 89 to the town of McCloud and turn right on the road to the reservoir, at the Shell station. The road log begins at this intersection. Stops 1-8 are shown on Figure 3, stops 9-11 are shown on Figure 5.

Mileage

0   Follow the road south from McCloud through the upper Squaw Creek drainage. The valley is filled with recent volcanics from Mount Shasta which is visible to the north. The low hills on the flanks of the valley are underlain by Upper Devonian (?) and Mississippian turbidites and related strata of the Bragdon Formation. As the road rises out of Squaw Valley it cuts through the thick weathering mantle on the north facing slope. Across the pass, in Tarantula Gulch, the exposure improves marginally in the road cuts.

5.9   Junction with Squaw Creek road, keep going straight.

7.4   Top of pass into Tarantula Gulch.

8.7   STOP 4-1: Facies C turbidites of the Bragdon Formation are folded and complexly faulted on a mesoscopic scale. Facing directions, based on graded beds (also indicated by cleavage refraction), vary in this exposure implying isoclinal folding and/or bedding sub-parallel faulting. The turbiditic sandstones contain detrital chert, monocrystalline quartz, feldspar, other sedimentary and low-grade metasedimentary lithic grains. Ongoing isotopic studies on detrital zircon populations indicate a component of early to mid- Proterozoic radiogenic lead (~2.0 or 2.1 Ga; Miller and others, 1988).

In general, bedding faces to the southeast; cleavage dips more steeply than bedding, suggesting northwestward vergence. Cleavage is developed throughout the turbiditic section and is defined by anastamosing septa of opaque insoluble residue, visible in thin section. No new mineral growth is associated with cleavage development. Cleavage is coplanar with axial planes in the cores of folds, and fanned in the limbs. Northwestward vergence is confirmed where mesoscopic folds are asymmetrical, and by fault-fold relations.

9.0   Junction, stay right. Just above the stop sign, cleavage is refolded around moderately westward dipping axial planes (F2).

9.7   STOP 4-2: Channel-fill sequence in the Bragdon Formation. This overturned section consists of interbedded pebbly and boulder-bearing mudstone, thick-bedded massive and amalgamated sandstone, medium-bedded turbidites and hemipelagic rocks. Boulder-sized fragments of thick sandstone beds occur in the pebbly mudstone and are interpreted as slumps from oversteepened channel walls. The thick sandstone and turbiditic units represent channel-fill sequences.

10.3   STOP 4-3 (OPTIONAL): Crystal-rich channel-fill sandstone. This massive sandstone contains primary volcanogenic debris which was re-deposited in the channel of a submarine fan. Framework constituents are primarily volcanic quartz and feldspar crystals set in a fine-grained siliceous matrix. Sedimentary features include argillite rip-up fragments, and a nested channel sequence eroded into the top of the bed.

11.0   STOP 4-4: McCloud Limestone & Baird Formation north of the Battle Creek arm of McCloud Lake. The Baird Formation has two principal contrasting facies types in the vicinity of McCloud Lake: 1) shallow-marine and strand-related facies composed of volcanic debris, and

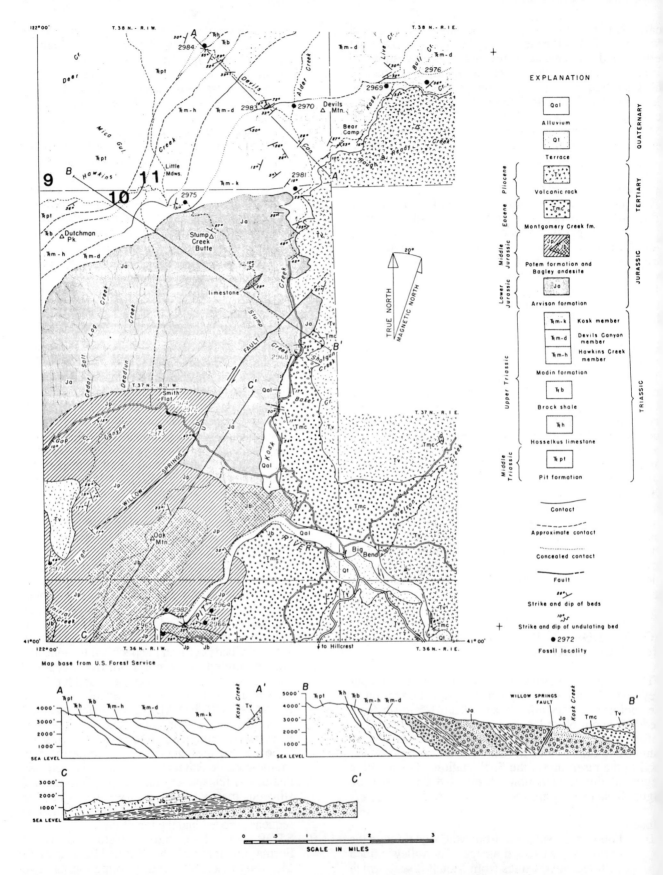

FIGURE 5 Geologic map of part of Big Bend quadrangle (Sanborn, 1960) with field trip stops shown by large numbers (9-11).

2) medium to thick volcanogenic turbidites (exposed here) and tuffaceous argillites which interfinger with the basal McCloud Limestone. The first group is inferred to be stratigraphically lower, although relations in this area are not unequivocal.

The McCloud Limestone in this area is characterized by three differing depositional facies. These include 1) biomicrite and silty micrite, 2) fusulinid and crinoid bearing calcarenite, and 3) carbonate breccia. These lithologies are inferred to represent the distally steepened portions of carbonate platform 'ramp and slope' facies (e.g. Read, 1985). The McCloud Limestone has been dated here by fusulinids and corals by C.H. Stevens. Ages range from zone D to zone H, equivalent to the youngest parts of the McCloud Limestone further to the south (late Wolfcampian and early Leonardian). Just to the east of the Baird exposure, and across a small displacement high-angle fault (D3), beds of the McCloud Limestone are exposed. In places, old cave walls are exposed in the road cut.

11.3   STOP 4-5 (OPTIONAL): McCloud Limestone at Battle Creek. Interbedded limestone and argillite (probable limestone turbidites) at the base of the McCloud Limestone. Fusulinids recovered from this locality are from zone F; the base of the McCloud here is mid-Wolfcampian.

12.3   Back to Junction, go right towards dam and Big Bend.

12.7   STOP 4-6 (OPTIONAL): Basal Nosoni silicified tuffs.

14.0   Cross dam, turn right, following signs to Big Bend.

14.7   STOP 4-7: Well stratified Dekkas.

14.9   Massive amygdaloidal lavas.

15.3   (Ash Camp) red crystal tuff breccia. We will walk through a 0.6 mile road section of the Dekkas Formation beginning with fossiliferous, well stratified polylithologic tuff and tuff-breccia. These strata were deposited as debris flows and turbidites. Upsection lies a thick sequence of mafic amygdaloidal massive lavas. Trace element geochemistry suggests that these are island arc tholeiites. Individual flows can be recognized by concentrations of amygdules along flow tops. Further upsection, towards Ash Camp are more water-laid crystal tuffs and pyroclastic debris flows.

16.0   STOP 4-8: Limestone of upper Dekkas Formation. This is one of several fossiliferous, locally dolomitized limestone lenses that are interbedded with volcaniclastic rocks in the top of the Dekkas Formation, and record shoaling at the end of Dekkas deposition. Guadalupian fusulinids and corals have been recovered from this locality. *Waagenophyllid klamathensis* (Stevens and others, 1987) comes from laterally equivalent carbonates to the southwest.

17.0   Thin- to medium-bedded black siliceous Pit Formation.

18.2   STOP 4-9: Crystal-rich and tuffaceous turbidites of the Pit Formation (Deer Creek). Well-exposed section of siliceous and tuffaceous turbidites and interstratified lithic- and crystal-rich, massive, laminated and amalgamated sandstone and granule-bearing sandstone of the Pit Formation. These strata are more siliceous and finer grained than volcanogenic rocks in the Permian section. Note that mesoscopic folds and penetrative cleavage are not developed in these rocks, however, thrust faults are present.

20.1   Junction.

21.7   Cross Hawkins Creek.

22.7   STOP 4-10 (OPTIONAL): Hosselkus Limestone. The Hosselkus is not shown by Sanborn (1960) in this region, however it appears to be represented here by poorly exposed, silty, fossiliferous limestone beds.

23.1   STOP 4-11: Basal Upper Triassic Modin Formation (Hawkins Creek member of Sanborn, 1960). Well rounded feldspathic and chert-rich sandstone of the basal Modin Formation. Associated conglomeratic facies are not exposed along the road section. Walk to next stop.

23.4   Modin Formation (Devil's Creek member of Sanborn, 1960). Tuffaceous limestone and calcareous, fine-grained sandstone of the middle Modin Formation.

24.0   Arvison: deeply weathered argillite of Arvison

27.5   Pit 5 turnoff. Deeply weathered pyroclastic beds of the Lower Jurassic Arvison Formation exposed in roadcut.

REFERENCES CITED

Albers, J.P., and J.H.C. Bain, Regional setting and new information of some critical geologic features of the West Shasta District, California: *Economic Geology, v. 80*, p. 2072-2091, 1985.

Albers, J.P., R.W. Kistler, and L. Kwak, The Mule Mountain stock, an early Middle Devonian pluton in northern California: *Isochron/West, v. 31*, p. 17, 1981.

Albers, J.P., and J.F. Robertson, Geology and ore deposits of East Shasta copper-zinc district, Shasta County, California: *U.S. Geological Survey Professional Paper 338*, 107 p., 1961.

Alvarez, W. and T. Engelder, Solution cleavage and estimates of shortening, Umbrian Appenines: *in* Atlas of Deformational and Metamorphic Rock Fabrics, edited by G.J. Borradaile, M.B. Bayly, and C. McA. Powell, *Springer-Verlag, Berlin Heidelberg New York*, p. 178, 1982.

Boucot, A.J., D.H. Dunkle, A. Potter, N.M. Savage, and C. Rohr, Middle Devonian orogeny in western North America?: a fish and other fossils: *Journal of Geology, v. 82*, p. 691-708, 1974.

Brueckner, H.K. and W.S. Snyder, Structure of the Havallah sequence, Golconda allochthon, Nevada: Evidence for prolonged evolution in an accretionary prism: *Geological Society of America Bulletin, v. 96*, p. 1113-1130, 1985.

Burchfiel, B.C., and G.A. Davis, Triassic and Jurassic tectonic evolution of the Klamath Mountains-Sierra Nevada geologic terrane: *in* The Geotectonic Development of California, Volume Rubey Volume No. 1, edited by W.G. Ernst, *Prentice-Hall Publishing Company*, p. 50-70, 1981.

Cas, R.A.F. and W.G. Wright, Volcanic successions, modern and ancient: *Allen and Unwin (Publishers) Ltd., London*, 528 pp., 1987.

Churkin, M., Paleozoic marginal ocean basin-volcanic arc systems in the Cordilleran foldbelt: *Society of Economic Paleontologists and Mineralogists Special Publication 19*, p. 174-192, 1974.

Coogan, A.H., Stratigraphy and Paleontology of the Permian Nosoni and Dekkas Formations (Bollibokka Group): *University of California Publications in Geological Sciences, v. 5*, p. 243-316, 1960.

Curtis, D.R., Stratigraphy and origin of the Pit Formation, northern California: *[M.Sci. thesis] University of California, Berkeley, California*, 56 p., 1983.

Engelder, T. and S. Marshak, Disjunctive cleavage formed at shollow depths in sedimentary rocks: *Journal of Structural Geology, v. 7*, p. 327-343, 1985.

Fraticelli, L. A., J.P. Albers, and R.E. Zartman, The Permian Pit River stock of the McCloud plutonic belt, eastern Klamath terrane, northern California: *Isochron/west, n. 44*, p.6-8, 1985.

Gordey, S.P., J.G. Abbott, D.J. Tempelman-Kluit and H. Gabrielse, 'Antler' clastics in the Canadian Cordillera: *Geology, v. 15*, p. 103-107, 1987.

Harland, W.B., A.V. Cox, P.G. Llewellyn, C.A.G. Pickton, A.G. Smith, and R. Walters, A Geologic Time Scale: *Cambridge University Press*, 1982.

Irwin, W.P. and M.D. Dennis, Geologic structure section across the southern Klamath Mountains, Coast Ranges, and seaward of Point Delgado, California: *Geological Society of America, Map and Chart Series MC-28D*, 1979.

Kinkel, A.R., Jr., W.E. Hall, and J.P. Albers, Geology and base-metal deposits of West Shasta Copper-Zinc district, Shasta County, California: *U.S. Geological Survey Professional Paper 285*, 156p., 1956.

Lapierre, H., F. Albarede, J. Albers, C. Coulon, and B. Cabanis, Early Devonian volcanism in the eastern Klamath Mountains, California: evidence for an immature island arc: *Canadian Journal of Earth Sciences, v. 22*, p. 214-227, 1985a.

Lapierre, H., M. Brouxel, F. Albarede, C. Coulon, and B. Cabanis, The Paleozoic and Mesozoic volcanism in the eastern Klamath Mountains (N. California, USA) - Part I: the early Devonian intra-oceanic island-arc series: *Terra Cognita - the Earth, v. 5*, p. 320, 1985b.

Lapierre, H., B. Cabanis, C. Coulon, M. Brouxel, and F. Albarede, Geodynamic setting of Early Devonian Kuroko-type sulfide deposits in the eastern Klamath Mountains (northern California) inferred by the petrological and geochemical characteristics of the associated island-arc volcanic rocks: *Economic Geology, v. 80*, p. 2100-2113, 1985c.

Lindberg, P.A., A volcanogenic interpretation for massive sulfide origin, West Shasta district, California: *Economic Geology, v. 80*, p. 2240-2254, 1985.

Miller, E.L., B.K. Holdsworth, W.B. Whiteford, and D. Rodgers, Stratigraphy and structure of the Schoonover sequence, northeastern Nevada: Implications for Paleozoic plate-margin tectonics: *Geological Society of America Bulletin, v. 95*, p. 1063-1076, 1984.

Miller, M.M., Tectonic evolution of late Paleozoic island arc sequences in the western U.S. Cordillera; with detailed studies from the Klamath Mountains, northern California: *[Ph.D. Thesis], Stanford University, Stanford, California*, 201 p., 1986.

Miller, M.M., Dispersed remnants of a northeast Pacific fringing arc -- Upper Paleozoic island arc terranes of Permian McCloud faunal affinity, western U.S.: *Tectonics, v. 6*, p. 807-830, 1987.

Miller, M.M., Permo-Triassic(?) deformation within the eastern Klamath terrane, northern California: *Geological Society of America Abstracts with Programs, v. 20*, p. 216, 1988.

Miller, M.M., Intra-arc sedimentation and tectonism: late Paleozoic evolution, eastern Klamath terrane, California: *Geological Society of America Bulletin*, in press.

Miller, M.M., and Bingquan Cui, Carboniferous arc-related sedimentation, Klamath Mountains, California: hybrid fan characteristics and dual sediment provenance: *Canadian Journal of Earth Sciences*, in review.

Miller, M. M., J. B. Saleeby, and B. Cui, Provenance of the Carboniferous Bragdon Formation, eastern Klamath terrane, California: Proterozoic detrital zircon and quartzose detritus as continental tracers in oceanic convergent margin systems: *Fourth International Circum-Pacific Terranes Conference, Nanjing, People's Republic of China, Proceedings (in press)*, 1988.

Miller, M.M., and J.E. Wright, Paleogeographic implications of Permian Tethyan corals from the Klamath Mountains, California: *Geology, v. 15* p. 266-269, 1987.

Mortimer, N., Petrology and structure of Permian to Jurassic rocks near near Yreka, Klamath Mountains, California: *[Ph.D. Thesis], Stanford University, Stanford, California*, 83p., 1984.

Powell, C. McA., A morphological classification of rock cleavage: *Tectonophysics, v. 58*, p. 21-34, 1979.

Read, J.F., Carbonate platform facies models: *American Association of Petroleum Geologists Bulletin, v. 69* p. 1-21, 1985.

Renne, P.R., Permian to Jurassic tectonic evolution of the eastern Klamath Mountains, California: *[Ph.D Thesis], University of California, Berkeley*, 127 p., 1986.

Renne, P.R., G. Curtis, and G.R. Scott, Nature and timing of deformation in the Redding section, eastern Klamath Mountains, California: *Geological Society of America Abstracts with Programs, v. 18*, p. 175, 1986.

Ross, C.A., and J.R.P. Ross, Late Paleozoic accreted terranes of western North America: *in* Pre-Jurassic Rocks in Western North American Suspect Terranes, edited by C.H. Stevens, *Pacific Section Society of Economic Paleontologists and Mineralogists, Los Angeles, California*, p. 7-22, 1983.

Sanborn, A.F., Geology and paleontology of the southwest quarter of the Big Bend quadrangle Shasta County, California: *California Division of Mines and Geology Special Report 63*, 26 p., 1960.

Silberling, N.J., Geologic events during Permian-Triassic time along the Pacific margin of the United States: *in* The Permian and Triassic Systems and their Mutual Boundary, Memoir 2, edited by A. Logan, and L.V. Hills, *Alberta Society of Petroleum Geologists*, p. 345-362, 1973.

Silberling, N.J., and D.L. Jones, Tectonic significance of Permian-Triassic strata in northwestern Nevada and northern California: *Geological Society of America Abstracts with Programs, v. 4*, p. 234, 1982.

Skinner, J.W., and G.L. Wilde, Permian biostratigraphy and fusulinid faunas of the Lake Shasta area, northern California: *University of Kansas Paleontological Contributions, v. Protozoa, article 6*, 98 p., 1965.

Speed, R.C., Collided Paleozoic microplate in the western United States: *Journal of Geology, v. 87*, p. 279-292, 1979.

Snyder, W.S., and H.K. Brueckner, Tectonic evolution of the Golconda allochthon, Nevada: problems and perspectives: *in* Pre-Jurassic Rocks in Western North American Suspect Terranes, edited by C.H. Stevens, *Pacific Section Society of Economic Paleontologists and Mineralogists, Los Angeles, California*, p. 103-123, 1983.

Stevens, C.H., Reconstruction of Permian paleogeography based upon distribution of Tethyan faunal elements: *in* Neuvieme Congres international de Stratigraphie et de Geologie du carbonifere: Compte Rendu, Washington and Champaign-Urbana, May 17-26, 1979, 5: Paleontology, paleoecology, paleogeography, editcd by J.T. Dutro and H.W. Pfefferkorn, *Southern Illinois University Press, Carbondale, Illinois*, p. 383-393, 1985.

Stevens, C.H., M.M. Miller, and M. Nestell, A new Permian waagenophyllid coral from the Klamath Mountains, California: *Journal of Paleontology, v. 61, n. 4*, p. 690-699, 1987.

Stevens, C.H., and B. Rycerski, The eastern Klamath Mountains Permian coral province, western North America: *Geological Society of America Abstracts with Programs, v. 20*, p. 235, 1988.

Tomlinson, A.J., and J.E. Wright, The Klamath-Sierran arc: a possible source region for Permian volcaniclastic and conglomeratic units within the Golconda allochthon: *Geological Society of America Abstracts with Programs, v. 18*, p. 193, 1986.

Watkins, R., Carboniferous faunal associations and stratigraphy, Shasta County, northern California: *American Association of Petroleum Geologists Bulletin, v. 57*, p. 1743-1764, 1973.

Watkins, R., Volcaniclastic and carbonate sedimentation in late Paleozoic island-arc deposits, Eastern Klamath Mountains, California: *Geology, v. 13*, p. 709-713, 1985.

Watkins, R., Late Devonian to Early Carboniferous turbidite facies and basinal development of the eastern Klamath Mountains, California: *Sedimentary Geology, v. 49*, p. 51-71, 1986

Watkins, R. and R.A. Flory, Island arc sedimentation in the Middle Devonian Kennett Formation, eastern Klamath Mountains, California: *Journal of Geology, v. 94*, p. 753-761,1986.

Whiteford, W.B., Age relations and provenance of upper Paleozoic sandstones in the Golconda allochthon: the Schoonover sequence, northern Independence Mountains, Elko Co., Nevada: *[M.S. Thesis], Stanford University, Stanford, California*, 126 p., 1984.

Wright, J.E., Permo-Triassic accretionary subduction complex, southwestern Klamath Mountains, northern California: *Journal of Geophysical Research, v. 87*, p. 3805-3818, 1982.

Wright, J.E. and M.R. Fahan, An expanded view of Jurassic orogenesis in the western U.S. Cordillera: Middle Jurassic (pre-Nevadan) regional metamorphism and thrust faulting within an active arc environment, Klamath Mountains, California: *Geological Society of America Bulletin, v. 100*, p. 859-876, 1988.

Yancey, T.E., Volcaniclastic and carbonate sedimentation in late Paleozoic island arc deposits, Eastern Klamath Mountains, California: Comment and Reply: *Geology, v. 14*, p. 538-539, 1986.

GEOLOGY OF LASSEN VOLCANIC NATIONAL PARK

Michael A. Clynne, L. J. Patrick Muffler, and Thomas D. Bullen
U.S. Geological Survey, Menlo Park, California

INTRODUCTION

This section of the field trip provides an overview of Quaternary volcanic rocks in and around Lassen Volcanic National Park, with a stop at each of the major stratigraphic units of the Lassen Volcanic Center. Additional stops focus on the 1915 eruption of Lassen Peak and on the Lassen geothermal system.

On a regional scale, Quaternary volcanism in the southernmost Cascade Range is predominantly basaltic

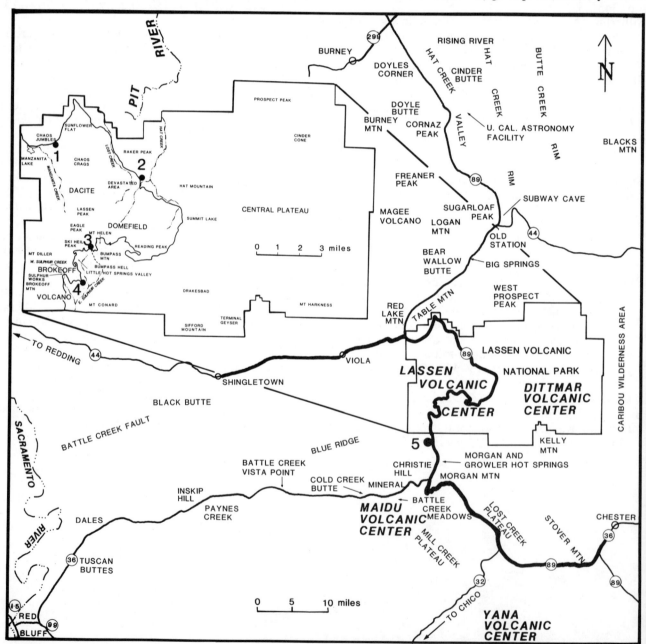

FIGURE 1 Map of the Lassen region showing fieldtrip route.

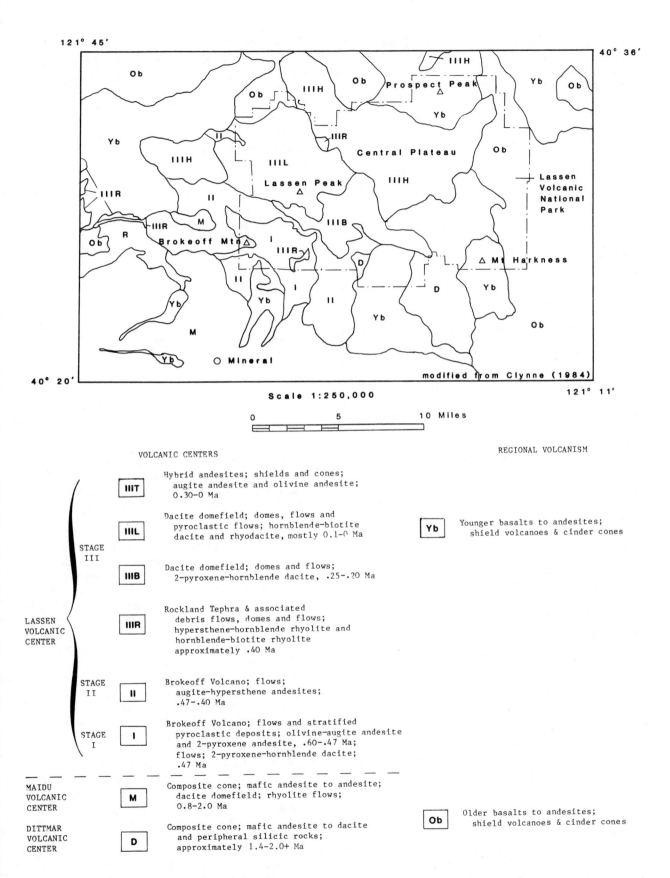

FIGURE 2 Generalized Geologic Map of Lassen Volcanic National Park and vicinity.

to andesitic and consists of hundreds of coalescing monogenetic volcanoes of small volumes (10^{-3} to 10^2 km³) and relatively short lifetimes (10^0 to 10^3 yr). Superposed on this regional mafic volcanism are a few long-lived, much larger volcanic centers that have erupted products ranging from basaltic andesite to rhyolite. Four such centers, each younger than about 3 m.y., have been recognized in the Lassen area (Figure 1; Clynne, 1988).

Each of the larger centers consists of an andesitic composite cone and flanking silicic domes and flows. Each center, of which the Lassen volcanic center (LVC) is typical, was built in three stages (Figure 2):
* **Stage I**, cone-building basaltic andesite to andesite lava flows and pyroclastics;
* **Stage II**, thick cone-building andesite to silicic andesite lava flows;
* **Stage III**, silicic domes and flows flanking the main cone.

The silicic magma chamber of Stage III provides a heat source for a hydrothermal system that develops within the core of the main cone. Alteration of permeable rocks of the cone facilitates glacial and fluvial erosion of the central part of the volcano. The result is selective preservation of a resistant rim of Stage-II lavas and flanking silicic rocks around a central depression. The three older volcanic centers in the Lassen area, the Dittmar, Maidu, and Yana volcanic centers (Figure 1; Clynne, 1988) have reached this stage, and their hydrothermal systems are extinct. LVC, the fourth and youngest center, hosts active Stage III-type volcanism and an active, well-developed hydrothermal system.

Stages I and II of LVC produced the Brokeoff Volcano, an 80 km³ andesitic stratocone. The bulk of Brokeoff Volcano (BV) consists of Stage-I deposits (olivine-augite and hypersthene-augite andesite lava flows and stratified pyroclastic deposits) that erupted from a central vent 0.60-0.47 Ma. Stage I culminated in eruption of a small volume of hornblende-pyroxene dacite lava. During Stage II, which lasted about 0.075 m.y., thick flows of porphyritic augite-hypersthene silicic andesite, generally lacking interbedded pyroclastic material, were erupted. Stage III was initiated by eruption of at least 50 km³ of rhyolitic magma (the Rockland air fall and ash flows) at about 0.40 Ma. This eruption is thought to have produced a caldera that is now filled by a dacite domefield, which consists of three groups of rocks totaling about 30-50 km³:
* **Group 1**: Hornblende-biotite rhyodacite lavas related to the Rockland magma (e.g. Raker Peak);
* **Group 2**: 6 domes and flows of 2-pyroxene-hornblende dacite lavas erupted between 0.25 and .2 Ma (e.g. Bumpass Mt.; Ski Heil Peak);
* **Group 3**: Hornblende-biotite dacite and rhyodacite erupted as domes, lava flows and pyroclastic flows in at least ten episodes, mostly during the past 0.1 m.y. (e.g. Lassen Peak, Chaos Crags).

Throughout Stage III, large flows of hybrid andesite totaling 10 km³ and consisting of thoroughly mixed mafic and silicic magma were erupted peripherally to the dacite domefield, primarily on the Central Plateau.

Porphyritic andesite and dacite with high Al_2O_3, low TiO_2, and medium K_2O contents and FeO/MgO ratios of 1.5-2.0 are the most abundant rock types in LVC. Sparsely porphyritic Rockland rhyolite pumice is the single most voluminous unit. Basaltic andesite, rhyodacite, and hybrid andesite are subordinate in abundance. Rocks of LVC resemble other calc-alkaline volcanic rocks emplaced on continental margins overlying sialic crust. Major-element Harker-variation diagrams of LVC show smooth trends from 55 to 75 % SiO_2 (Clynne, 1984). The general compositional evolution is from mafic to silicic with time, although the evolution is not strictly sequential (Figure 3).

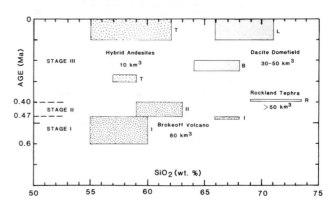

FIGURE 3 Time-composition diagram for rocks of the Lassen Volcanic Center. Some units in Stage III with poorly constrained ages may fall outside the indicated fields. Volume estimates are approximate.

Petrographic characteristics and geochemical systematics indicate a complex origin for mafic and intermediate LVC magmas, involving mixing and crystal fractionation of heterogeneous mantle-derived parental melts. Continued processing of intermediate magmas in a continually evolving magma chamber, and/or partial melting of young mafic crust are likely to be responsible for development of the more silicic magmas. Hybrid lavas and ubiquitous quenched mafic inclusions provide abundant evidence for the interaction of mafic and silicic magma. The long span of volcanic activity at LVC and the presence of a hydrothermal system support the existence of an evolving magma chamber.

LVC differs from most other Cascade volcanoes by having a larger volume of silicic rocks. We envisage the present magma system of the Lassen volcanic center (Figure 4) to be an evolving body of magma 5-8 km in diameter in the middle crust (10-20 km depth) under the northwestern part of Lassen Volcanic National Park. The upper portion of the chamber contains crystal-rich

rhyodacite, which has varied little in composition over at least the last 50,000 years. The system is probably zoned to more mafic compositions at depth. Regional mafic magmas provide heat and material input to maintain the system in its partially molten state.

Despite the volcanologic and petrologic evidence for a complex magmatic plumbing system, teleseismic (Berge and Monfort, 1986) and seismic-refraction (Berge and

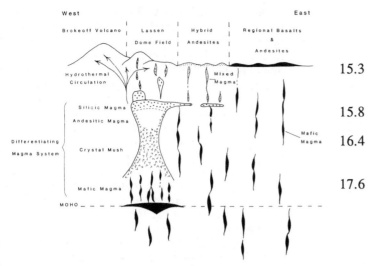

FIGURE 4 Model of Lassen volcanic center.

Stauber, 1987) studies have failed to show the presence of a magma chamber beneath LVC. A 25-km, oval, 50 mGal negative gravity anomaly is centered on the dacite domefield and central plateau (hybrid andesites) of LVC. The gravity low is probably an expression of both near-surface low-density volcanic rocks and Quaternary plutonic rocks beneath the volcanic field. NNW-oriented normal faults in the Lassen region reflect the impingement of Basin and Range tectonics on the Cascade arc (Guffanti and Weaver, 1988).

ROAD LOG-- DAY 5, July 2, 1989-- LASSEN VOLCANIC NATIONAL PARK

Mileage
00.0   Shingletown, at the intersection of California Highway 44 and Wilson Hill Road. Proceed east up the dip slope of the mid-Pleistocene Shingletown basalt, a low-K_2O, high-alumina olivine tholeiite.
3.1   Highway rest area.
5.5   Starlight Pines. Straight ahead to the east one can see dome 1 of the Chaos Crags, a sequence of dacite domes erupted 1,000 years ago.
9.3   Rock Creek Road. Road passes onto glacial gravel related to a major set of moraines to the southeast. Glacial deposits in LVNP and vicinity are thought to represent at least five periods of glaciation which have been tenuously correlated to Sierra Nevadan glacial stratigraphy by Crandell (1972) and Kane (1982).
10.1   Pass out of forest into a meadow with excellent view to southeast (at 12:30) of Lassen Peak, a Stage-III dacite dome emplaced approximately 18,000 BP. To the right are Mt. Diller and Brokeoff Mountain, remnants of Brokeoff Volcano. As the road curves to the left, the 5 major dacite domes of Chaos Crags come into view.
11.5   Road bends to left at Lassen Pines Christian

Center. For the next 3 miles the highway passes through the andesite of Viola, a 0.3 m.y. mixed lava from the Lassen Volcanic Center. Vent for the lava is to the southeast.
15.3   Flat just beyond 5000-foot-elevation sign is underlain by 1000 BP pyroclastic flow of Chaos Crags.
15.8   Roadcuts of a 0.4 m.y. Stage-II Brokeoff andesite.
16.4   Large roadcut just before an area of dense manzanita is basaltic andesite from a vent at Red Lake Mountain (3.5 km. to north).
17.6   Roadcuts for the next 1/2 mile expose a complex volcanic stratigraphy spanning the last 0.4 m.y. No single exposure contains the entire sequence of deposits. The prominent blocky lava flow at the base of the cuts is a 2-pyroxene andesite from Brokeoff Volcano (0.4 m.y.). It is overlain by lithic airfall tephra composed of andesite of Viola (0.3 m.y.), emplaced by formation of Deep Hole, a nearby explosion crater. Then followed a hornblende-biotite rhyolite ash flow from a nearby but unknown source. Crystal-lithic surge deposits comprise the dark stratified deposit near the top of the roadcuts. The last prominent roadcut on the north side of the highway exposes a hot lithic pyroclastic flow initiated by collapse of a dacite dome near Lassen Peak. A 57,000 year old hornblende-biotite ash flow from Eagle Peak preceded emplacement of a pumiceous hornblende-biotite rhyodacite ash flow from the Chaos Crags 1,000 years ago.
18.2   Intersection of Highway 44 with Highway 89. Turn right to Lassen Volcanic National Park. Roadcuts at the junction are in the pyroclastic flow from Eagle Peak.
18.8   Entrance station to Lassen Volcanic National Park. Please note that collecting or disturbing rock, mineral, or other natural specimens in Lassen Volcanic National Park is prohibited except by special permit. Just beyond the entrance station is a view to the south (to the right) across Manzanita Lake to Lassen Peak, Eagle Peak, and Loomis Peak, all Stage-III dacite or rhyodacite domes. The oldest of the three Chaos Jumbles rockfall avalanches (see Stop 5-1) crossed and dammed Manzanita Creek. Radiocarbon ages of wood samples taken from trees submerged in the lake that formed behind the dam confirm the conclusion of Crandall et al. (1974) that all three avalanches occured in quick succession about 300 years ago.
21.2   Pull off on left. STOP 5-1: Chaos Crags, Chaos Jumbles, and quenched mafic inclusions. Detailed geologic mapping (Christiansen and Clynne, 1989) has added significant new detail to the general stratigraphy of the Chaos Crags given by Crandell et al. (1974). The initial event of the Chaos Crags eruptive sequence was the eruption of pumice and lithic debris and the formation of a tuff cone (Table 1). Eruption of two pyroclastic flows, A and B, quickly followed; these were

confined to the valleys of Lost and Manzanita Creeks where they flowed to a distance of 5 km. Dome 0 grew in the vent of the pyroclastic eruptions and then was partly destroyed by a large pyroclastic eruption that emplaced the widespread pyroclastic flow C. At 3 km from the vent, pyroclastic flow C became confined to the drainages of Lost Creek and Manzanita Creek where it flowed to a distance of 18 km and 28 km, respectively. A coignimbrite fall deposited a lobe of pumice that can be traced for 40 km to the northeast. Near the vent, the coignimbrite fall deposit contains up to meter-sized blocks of dome-0 lava. A series of domes were then emplaced in the numerical order shown in Table 1. Dome 3 of Crandell et al. (1974) is now recognized as two domes: 3a and 3b. Domes 2, 3a, and 4 had hot collapse events that produced short pyroclastic flows. Approximately 300 years ago, the Chaos Jumbles were formed when dome 2 partially collapsed in three cold rockfall avalanches.

All the Chaos Crags units are porphyritic hornblende-biotite dacites or rhyodacites containing 66-70% SiO_2. Non-vesicular rocks contain about 40% crystals of plagioclase, biotite, hornblende, quartz, and xenocrystic olivine. The total volume of the Chaos Crags deposits is on the order of 2 km^3.

Quenched inclusions ("blobs") occur in all the rocks, but are especially abundant in the later domes. The quenched inclusions consist of a microvesicular network of predominantly acicular, but sometimes swallowtail or hollow plagioclase, pyroxene, and hornblende microphenocrysts, with abundant interstitial glass. Euhedral grains of an Fe-oxide mineral are a conspicuous accessory. The inclusions have a variety of textures ranging from coarse-grained aphyric to fine-grained porphyritic. Where textural variation occurs in a single inclusion, the fine-grained porphyritic material concentrically surrounds the coarse-grained aphyric material. The phenocrysts in the porphyritic material are resorbed felsic phenocrysts derived from mixing with the silicic host magma at the time of inclusion formation. Some quenched inclusions also contain olivine, calcic plagioclase, and clinopyroxene phenocrysts from their mafic parent. The inclusions display a wide range of chemistry representing the spectrum of calc-alkaline mafic magmas erupted in the Lassen region.

The Chaos Jumbles consists of three separate rockfall avalanches, covering 6.8 km^2, that can be distinguished on the basis of flow margins and grain characteristics. The deposits consist of a monolithologic breccia of Chaos Crags dacite

TABLE 1 Summary of events in the formation of the Chaos Crags. Unpublished data of R.L. Christiansen and M.A. Clynne with ^{14}C and tree-ring correlation ages from the USGS Menlo Park Radiocarbon Laboratory.

--3 cold rockfall avalanches from dome 2 formed Chaos Jumbles, ^{14}C age 275 +
    25 years (weighted average of 3 samples), tree-ring correlation age 1619 +
    63 A.D.

hiatus of approximately 700 years

--hot dome-collapse pyroclastic flow from dome 4
--emplacement of dome 4
--emplacement of dome 3b
--warm dome-collapse avalanche from 3a
--emplacement of dome 3a
--hot dome-collapse pyroclastic flow from dome 2
--emplacement of dome 2

hiatus ??

--emplacement of dome 1
--explosive disruption of dome 0 by eruption of pyroclastic flow C,
    accompanied by a coignimbrite fall deposit and formation of a tuff cone,
    [14]C age 1062 + 14 years (weighted average of 7 samples), tree-ring
    correlation age 984 + 15 A.D.

hiatus of approximately 75 years

--emplacement of dome 0,
--eruption of two column-collapse pyroclastic flows, A and B, ^{14}C age 1124 +
    15 years (weighted average of 7 samples), tree-ring correlation age 911 + 27
    A.D.
--initial vent opening, air-fall pumice and lithic deposit and formation of a
    tuff cone

blocks in a matrix of pulverized dacite. The deposits have steep distal and lateral margins 3 to 5 m in height. The first rockfall avalanche was the largest and traveled 650 m downslope for 4.5 km from the breakaway scar on dome 2 of the Chaos Crags. The other avalanches were successively smaller and shorter, but thicker. The avalanche paths were controlled by existing topography. The intial direction of each avalanche was west-northwest toward Table Mountain. The first avalanche deposit rode up onto Table Mountain, where it is found up to 100 m above the valley floor, before being deflected to the west. Groovelike linear features are interpreted by Eppler et al. (1987) to be strike-slip faults caused by compression during flow as the deposit was deflected by Table Mountain. Regularly spaced surface ridges are oriented perpendicular to the direction of flow. The avalanches were apparantly emplaced as a high yield-strength material capable of deforming and shearing rather than a plug of nondeforming material being carried on a deforming basal layer (Eppler et al. 1987).

21.9 Talus on right is one of the Sunflower Flat dacite domes, a group of seven 35,000 year old hornblende-biotite rhyodacite domes that were preceded by eruption of airfall pumice and a pyroclastic flow. Mature forest obscures the relatively youthful morphology.

22.8 Two-pyroxene andesite of Table Mountain at 7:00 came from a vent to the southwest. Just beyond the sharp bend to the right is a good view to the right of the Sunflower Flat dacite domes.

23.4 Flat is underlain by Sunflower Flat pyroclastic flow.

23.8 On the right is talus from one of the Sunflower flat dacite domes. For the next two miles, the road passes through Tioga till (15,000 to 20,000 years old) that contains boulders of dacite from Lassen Peak.

26.0 Lost Creek. Here, debris flows from the 1915 eruption (see Stop 5-2) overlie pyroclastic flows from the 1000 year-old Chaos Crags eruption; excellent exposures of both deposits can be seen in quarries 1 mile upstream. For the next 1.8 miles, the road crosses 1915 mudflows, climbs onto a Tioga moraine, and then passes over the mudflows again.

27.7 The cliff on the left (northeast) side of the road is the rhyolite of Raker Peak. This rhyolite belongs to Group 1 of the Stage-III lavas and has a composition identical to that of the most silicic Rockland ash sample. The glaciated front face of the dome may have been a caldera-bounding fault related to eruption of Rockland material. Note the internal flow structure of the dome revealed by truncation.

28.4 STOP 5-2, Devasted Area parking lot: New field studies by Christiansen and Clynne (1986; 1989) and a reanalysis of previous work provide a geologic map at 1:24,000 and a revised

scenario of the May 1915 eruptions of Lassen Peak. After a year of intermittent steam blasts and formation of a summit crater 350 m in diameter, in mid-May of 1915 lava welled into the crater to form a small dome of black glassy dacite. The dome grew for 5 or 6 days and on the night of May 19-20 was disrupted by a single explosion. No juvenile material was ejected by this explosion, but fragments of the still-hot dome fell on the snow-covered upper flanks of Lassen Peak, generating an avalanche that flowed 5 km down a 0.8 km-wide path. Fragmentation of the avalanching lava blocks melted enough snow to generate a large debris flow that went 15 km further down Lost Creek. Blocks of the black, glassy dacite carried by the debris flow can be seen in the area just to the northeast of the parking lot.

Later, on the night of May 19-20, dacite lava erupted from the vent opened by the dome-disrupting explosion spilled over the low western and northeastern crater rims, and flowed approximately 300 m down the steep slopes. Then, after two quiet days, an explosive subplinian eruption on the afternoon of May 22 produced a new crater and erupted compositionally layered (banded) dacite-andesite pumice and unlayered dacite to rhyodacite pumice. Fallback onto the still partly snow-covered east slope of Lassen Peak generated a pyroclastic flow that incorported snow. Melting of this snow transformed the pyroclastic flow into a highly fluid debris flow which traveled down Lost Creek. Continued eruption of pumice deposited a fallout lobe to the east and produced several smaller viscous debris flows on the slopes of Lassen Peak.

In the May 1915 eruptions, four volcanic lithologies were produced from Lassen Peak: (1) a black lava dome (64-65% SiO_2), formed between May 14 and 19, and a lava flow (64-65% SiO_2) emplaced on May 19; (2) dark bands in banded pumice (60-61% SiO_2); (3) light bands in banded pumice and unlayered white pumice (64-69% SiO_2) erupted on May 22; and (4) quenched inclusions of andesitic magma (58-60% SiO_2) that occur in the lavas and light pumice.

The black lava dome and flow and the light pumice contain large plagioclase, biotite, hornblende, and quartz phenocrysts in a glassy groundmass that contains plagioclase, clino-pyroxene, hypersthene, and Fe-oxide micro-phenocrysts. Two populations of each pheno-cryst type are present, one that is unresorbed and one that is strongly resorbed. The dark pumice bands have small olivine phenocrysts in a dark glassy groundmass that contains tiny pyroxene and plagioclase microlites. All degrees and scales of mixing between the light and dark bands can be found. Moreover, olivine phenocrysts (Fo_{84}, 2400 ppm Ni) are present in all the lithologies.

Quenched inclusions up to 50 cm in diameter comprise up to 5% of each of the lithologies except the dark pumice bands. The gray to black inclusions have phenocrysts of olivine in a microvesicular network of acicular and occasionally swallowtail or hollow plagioclase, clinopyroxene, and hypersthene microphenocrysts up to 0.5 mm, contained in abundant interstitial light- to dark-brown glass. The dark pumice bands and the quenched inclusions also contain resorbed felsic phenocrysts.

All the 1915 lithologies are mixed. End members are hornblende- biotite rhyodacite (70% SiO_2) and olivine basalt or basaltic andesite (52-54% SiO_2). The quenched andesitic inclusions contain about 35% silicic component. The black lavas were formed by disaggregation of quenched inclusions in rhyodacite and contain about 30-40% mafic component. Dark bands in banded pumice consist of magma similar to quenched inclusions, and light bands consist of magma similar to black lava.

Continue SE on Highway 89.

29.1 Directly ahead is Hat Mountain, a 25,000 year old hybrid andesite. For the next 2.5 miles, the highway passes through till with sporadic outcrops of Stage-III rhyodacite.

31.7 Summit Lake. For the next 3 miles, the highway passes through monolithologic till derived from Reading Peak.

32.5 Lupine picnic area. For the next mile, volcanic vents of the Central Plateau are visible in the near distance to the east (at about 8:00) through the trees. Rocks of the Central Plateau are all hybrid andesites; the youngest is Cinder Cone, which erupted 450 years ago and possibly again in 1851. Continuing clockwise, the skyline is in the Caribou Wilderness Area, a highly glaciated area of many small late Pleistocene basalt and andesite vents. Further clockwise is Mt. Harkness, a late Pleistocene basaltic andesite shield built on the eroded edifice of the 1.5 Ma Dittmar volcanic center. In the distance, beyond Mt. Harkness, is Lake Almanor.

35.2 Kings Creek. To the right is the dacite dome of Reading Peak. To the left is the large dacite flow which vented at Bumpass Mountain.

36.7 Roadcuts in pyroxene-hornblende dacite of Reading Peak, a deeply glaciated, 212,000 year old dome complex.

37.5 View of Lassen Peak to the west (at 12:00), Mt. Helen to the southwest (at 10:30), and Bumpass Dome to the south (at 9:30); all three are Stage-III dacite domes. The shoulder to the left of Bumpass dome is Stage II andesite of Brokeoff Volcano. Further to the left is Mt. Conard, composed of Stage-I andesites.

39.5 Conspicuous switchback to right with roadcuts exposing the Kings Creek flow, a 30,000 year old flow of hornblende-biotite rhyodacite erupted from a vent now covered by Lassen Peak. Much of the pumiceous glassy carapace and structure of the flow is preserved because it has not been heavily glaciated. View to the south and west of Stage-III dacite domes (left to right): Bumpass Mountain (230,000 year old, with its large flow extending SE), Mount Helen, (249,000 year old, with a stubby flow extending SE), and Lassen Peak.

40.1 Roadcut at right exposes thick flow breccia at the base of the the Kings Creek flow. To the left is the Mount Helen dome.

40.4 Pass at 8512 feet.

40.6 The large parking lot on the right marks the start of a good trail that switchbacks for 4 km up the southeast ridge for 600 m to the summit of Lassen Peak, a dacite dome complex formed approximately 18,000 years ago. At the summit one can see the 1915 to 1917 eruption craters and deposits and, on a clear day, can have good views of the northern Sierra Nevada, Sacramento Valley, Coast Ranges, Klamath Mountains, Mount Shasta, and the Medicine Lake Highlands.

Leaving the parking lot, the roadcuts on the left are in the quenched inclusion-rich dacite of Mount Helen (Group 2 of Stage III).

41.3 Sharp bend to right. To the northwest (at 10:30) is Ski Heil Peak, a 244,000 year old pyroxene-hornblende dacite dome. To the north (at 12:00) is Eagle Peak, a 57,000 year old rhyodacite dome that was preceded by a flow (the cliffs ringing the mountain), airfall pumice (that mantles Ski Heil Peak), and a pyroclastic flow. To the right beyond Lake Helen is Lassen Peak.

41.8 STOP 5-3, 4-km round-trip walk to Bumpass Hell: First, walk out to the south end of the parking lot for excellent views of the deposits of all three stages of the Lassen Volcanic Center. The parking lot is built on the contact between Stage-II Brokeoff Volcano flows and Stage-III dacites of Group 2. Note the prominent glacial striations and glacial boulders; the prominent 2-m boulder is hornblende-pyroxene dacite from Bumpass Mountain.

The gentle, level trail to Bumpass Hell winds along this contact, crossing it several times. The massive dacite of Bumpass dome is marked by well-preserved glacial striae. All the dacite domes of the Group 2 have suffered extensive glacial erosion. Although their shape has not been greatly modified, no trace of an outer glassy or pumiceous carapace or talus mantle is preserved. The middle part of the trail has been blasted from the massively jointed devitrified interior of the dome. At the viewpoint where the trail turns east 0.85 km from the parking lot, Stage-II andesite is exposed. The upper flow surface of the andesite was eroded before deposition of the overlying dacite breccia.

The Lassen geothermal system (Figure 5) consists of a central vapor-dominated reservoir at a temperature of 235°C underlain by a reservoir

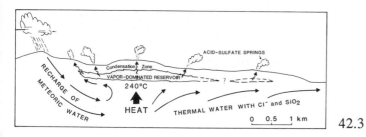

FIGURE 5 Schematic cross-section of the Lassen geothermal system (from Muffler and others, 1982).

of hot water (Muffler et al., 1982; Ingebritsen and Sorey, 1985). The focus and major thermal upflow is at Bumpass Hell, which is located at the vent for Bumpass Mountain, along the contact between BV and the Stage-III dacite domefield. Natural discharge from the deep hot-water part of the Lassen geothermal system occurs only at Morgan Hot Springs and Growler Hot Spring, both located in the canyon of Mill Creek nearly 1000 m below Bumpass Hell. These springs, which discharge near-neutral chloride water and deposit silica, occur at the contact between BV rocks and rocks of the underlying Maidu volcanic center. Part of the deep hot water system also flows laterally to the southeast where it is encountered in well Walker "O" No. 1 at Terminal Geyser.

Bumpass Hell contains numerous superheated fumaroles, one of which had a temperature in 1976 of 159°C. Approximately 75 major fumaroles, acid-sulfate hot springs, and mudpots plus myriad smaller features occur in an area of approximately 0.13 km² (Muffler et al., 1983) that is intensely altered to an aggregate of opal and kaolinite+alunite. Much of the surface of the active part of Bumpass Hell is covered with orange and yellow sulfates. Pyrite is common in many of the hot springs, occurring as linings of the vents and discharge channels, as scum floating on the surface of pools, and as dispersions in gray or black mudpots.

The acid-sulfate water from Bumpass Hell is typical of hot springs related to a vapor-dominated reservoir in having low pH, high sulfate, and no significant Cl. Most other thermal areas in LVNP have thermal features and chemistry similar to Bumpass Hell and are also manifestations of the vapor-dominated reservoir. Some springs near Sulphur Works and in Little Hot Springs Valley and at Drakesbad are relatively rich in HCO_3^- and deposit travertine. These springs are interpreted to be surface discharge from the zone of steam condesate that overlies the vapor-dominated reservoir (Muffler et al., 1983).

42.3   Just beyond the south end of Emerald Lake the road crosses the contact between the 244,000 year old pyroxene-hornblende dacite dome of Ski Heil Peak and the Stage-II rocks of Brokeoff Volcano. The yellowish color of the andesites of Brokeoff Volcano is due to extensive alteration by hydrothermal activity.

43.0   White, kaolinite-rich altered Stage-I Brokeoff andesite at sharp bend in road.

45.5   STOP 5-4, Diamond Peak in the center of Brokeoff Volcano (BV): Virtually all the rocks that can be seen from this viewpoint (Figure 6) are at least incipiently altered, and most areas in the core of the cone are strongly to totally altered. Flows and breccias on Diamond Peak dip steeply eastward, indicating that the Stage-I summit vent of BV was nearby to the west. To the south, in the glacial valley of Mill Creek, the base of BV is exposed where it overlies rocks of the Maidu volcanic center. Mill Creek Canyon and the flanks of Brokeoff Mountain expose nearly the entire BV stratigraphy. Thick flank flows of olivine-augite and hypersthene-augite andesite can be correlated across Mill Creek Canyon with no offset by faulting. The triangle-shaped peak of Brokeoff Mountain to the west is resistant to erosion by virtue of a thick flow of dacite that ended Stage I. To the north, Mt. Diller consists almost entirely of Stage-II flows. To the east, Little Hot Springs Valley displays numerous thermal features. Cliffs on the east side of the valley expose thin lava flows of Stage I intercalated with hydrothermally altered flow breccias and pyroclastic deposits. The cliffs are capped by Stage-II flows. To the southeast, Stage-I flows and breccias form most of Mt. Conard. The cliff that forms the prominent shoulder above Mill Canyon is a single Stage-II lava flow overlain by a thick canyon-filling rhyodacite lava flow of early Stage III. Both flows followed the course of a canyon cut in Stage-I rocks of BV. The rhyodacite flow is

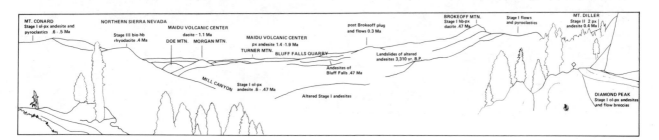

FIGURE 6 Panorama of the Brokeoff Volcano from Diamond Peak.

about 400,000 years old and illustrates the amount of erosion in the core of BV since that time.

Howel Williams, in his classic 1932 paper on Lassen geology, attributed the amphitheater in the central area of BV to a caldera-forming pyroclastic eruption. Detailed mapping, however, has failed to uncover caldera-bounding faults. On the contrary, although the cone is deeply eroded, the stratigraphy is coherent. The Rockland tephra, originally thought to be older than BV, has been shown to be younger than the cone and contains BV rocks as lithic fragments; its source appears to be on the northern flank of the BV. Therefore, the origin of the central depression of BV is probably related to fluvial erosion greatly enhanced by hydrothermal alteration and glaciation. Altered valley walls, oversteepened by glacial erosion, are further destabilized by rapid fluvial downcutting. A large proportion of the terrain below Brokeoff Mountain and Mt. Diller consists of landslide deposits, and the valley of Sulphur Creek below Sulphur Works is filled by reworked landslide material. During interglacial times, therefore, landslides were a significant mass wasting agent.

6.8 Sulphur Works, the most accessible of the thermal areas in LVNP. A short boardwalk guides visitors safely through a portion of the several hundred acres of drowned fumaroles, boiling pools, and steaming ground.

7.7 Lassen Chalet, with a restaurant, rest rooms, and Park information. The buildings and the parking lot are located on a large landslide that moved 7 km down Mill Creek from the cliff at the base of Brokeoff Mountain 3,310 years ago.

7.8 Entrance station for Lassen Volcanic National Park.

49.2 STOP 5-5, Bluff Falls Quarry: Walk along service road 0.2 mile to the quarry, which exposes nearly the entire thickness of a typical silicic andesite lava flow from Stage II of Brokeoff Volcano. The base of the flow (not exposed here) consists of thin reddened scoriaceous flow breccia, which is overlain by a massively jointed, faintly flow-banded, glass-rich zone (Figure 7). This zone grades upward into a thick zone with thin platy jointing. In the flow interior exposed here, the platy jointing is nearly horizontal to slightly wavy. In contrast, at flow edges it is commonly ramped upward or highly contorted. On the glaciated flanks of Brokeoff Volcano, the orientation of the platy jointing can be used to map flow edges and flow direction. The platy interior of the flow grades upward into a zone with columnar joints superposed on the platy jointing. The upper surface of the flow is glassy and vesicular.

This flow is a porphyritic, 60%-SiO_2 rock containing abundant plagioclase, hypersthene, and augite phenocrysts in a cryptocrystalline groundmass. Sparse olivine xenocrysts are derived from the disaggregation of abundant glomeroporphyritic clots that have both gabbroic composition and texture and consist of the same minerals as the phenocrysts plus olivine. Patchy recrystallization of the plagioclase, vague exsolution lamallae in the pyroxenes, and solid-state replacement of olivine by orthopyroxene suggest that the clots represent cumulate material removed from an andesite magma similar to the andesite of Bluff Falls Quarry and subsequently incorporated into the present host. These clots are abundant in all Stage-II lava flows.

51.6 Exposure in roadcut to right of till of middle Tioga-age glaciation (approximately 18,000-

FIGURE 7 Photograph illustrating the features of the silicic andesite lava flow exposed in the Bluff Falls quarry.

25,000 years ago).

51.9 To the south (at 12:00) is Morgan Mtn., one of several dacite domes emplaced at 1.1 m.y. as part of Stage III of the Maidu volcanic center. To the right, across the valley, is Christie Hill, another dacite dome of the Maidu volcanic center. The valley is filled by an olivine- bearing basaltic andesite flow that erupted 300,000 years ago from a vent high on the flank of the Brokeoff Volcano. This flow marks the extinction of Brokeoff volcano and a return to regional mafic volcanism in the Brokeoff volcano area.

53.9 Intersection of Highway 89 with California Highway 36. Turn left. Roadcuts for the next 3.8 miles are in dacite of the Maidu volcanic center.

57.7 Mill Creek. Excellent view to the north (to the left) of the Lassen volcanic center. Brokeoff Mountain is on the left, and Lassen Peak in the center. Three km north of the highway are Morgan and Growler hot springs, near-neutral Cl-bearing waters that deposit siliceous sinter. The springs are the southern manifestation of the Lassen hydrothermal system, centered at Bumpass Hell (the pale yellow, altered area in front of Lassen Peak).The cliffs ahead on the east side of the valley are Stage-II andesites of the Maidu volcanic center.

59.9 Interesection with graveled road to east. For the next 9 miles, the cliffs on the left are formed by a gigantic flow of dacite. Similar huge dacite and rhyolite flows occur to the southwest and west of the Maidu volcanic center. One of these flows, the Blue Ridge rhyolite, has been dated at 1.2 m.y.

63.9 Fire Mountain Lodge. Sporadic outcrops of upper Pleistocene basalt occur in the bottom of the valley, generally obscured by glacial gravels and by extensive colluvium from the dacite flows that form the valley walls.

65.6 Intersection with California Highway 32 from right. Continue to the left on Highways 36 and 89. End of Roadlog.

REFERENCES CITED

Berge, P.A., and M.E. Monfort, Teleseismic residual study of the Lassen Volcanic National Park region in California, *U.S. Geol. Sur. Open-File Rpt. 86-252,* 71 pp, 1986.

Berge, P.A., and D.A. Stauber, Seismic refraction study of upper crustal structure in the Lassen Peak area, northern California, *J. Geophys. Res., 92,* B10, 10,571-10,579, 1987.

Bullen T.D., and M.A. Clynne, [Redbook Conference Paper on isotopic and trace-element geochemistry of Lassen volcanic center), in prep.

Christiansen, R.L., and M.A. Clynne, The climactic eruptions of Lassen Peak, California, in May, 1915 (abs.), *Eos, 67,* 1247, 1986.

Christiansen, R.L., and M.A. Clynne, [Chaos Crags map], in prep.

Clynne, M. A., Stratigraphy and major-element geochemistry of the Lassen Volcanic Center, California, *U.S. Geol. Sur. Open-File Rpt. 84-224,* 168 pp., 1984.

Clynne, M.A., [Redbook conference paper on volcanic centers of southernmost Cascades], in prep.

Crandell, D.S., Glaciation near Lassen Peak, northern California, *U.S. Geol. Sur. Prof. Pap. 800-C,* C179-C188, 1972.

Crandell, D.S., D.R. Mullineaux, R.S. Sigafoos, and M. Rubin, Chaos Crags eruptions and rockfall-avalanches, Lassen Volcanic National Park, California, *U.S. Geol. Sur. J. Res., 2,* 1, 49-59, 1974.

Eppler, D.B., J. Fink, and R. Fletcher, Rheological properties and kinematics of emplacement of the Chaos Jumbles rockfall avalanche, Lassen Volcanic National Park, California, *J. Geophys. Res., 92,* B5, 3623-3633, 1987.

Guffanti, M., M.A. Clynne, and L.J.P. Muffler, [Redbook Conference paper on volcanic-vent distribution and impingement of the Basin and Range on the Cascade Range in the Lassen region, northeastern California], in prep.

Guffanti, M., and C.S. Weaver, Distribution of late Cenozoic volcanic vents in the Cascade Range (USA): volcanic arc segmentation and regional tectonic considerations, *J. Geophys. Res., 93,* B6, 6513-6529, 1988.

Ingebritsen, S.E., and M.L. Sorey, A quantitative analysis of the Lassen hydrothermal system, north central California, *Water Res. Res., 21,* 6, 853-868, 1985.

Kane, P.S., Pleistocene glaciation, Lassen Volcanic National Park, California, *Geol., 35,* 5, 95-105, 1982.

Muffler, L.J.P., R. Jordan, and A.L. Cook, Thermal features and topography of Bumpass Hell and Devils Kitchen, Lassen Volcanic National Park, California, *U.S. Geol. Sur. Misc. Field Studies Map 1484,* 1:2,000, 1983.

Muffler, L. J. P., N.L. Nehring, A.H. Truesdell, C.J. Janik, M.A. Clynne, and J.M. Thompson, The Lassen geothermal system, *Proc. Pacific Geotherm. Conf.,* Auckland, New Zealand, 349-356, 1982.

Walter, S. R., V. Rojas, and A. Kollman, Seismicity of the Lassen Peak area, California, 1981-1983, *Geotherm. Res. Council, Trans., 8,* 523-527, 1984.

STRATIGRAPHY AND STRUCTURE OF THE NORTHERN SIERRA TERRANE, CALIFORNIA

David S. Harwood
U. S. Geological Survey, Menlo Park, California

INTRODUCTION

Metavolcanic rocks in the northeastern Sierra Nevada were deposited during three separate episodes of island-arc volcanism in Late Devonian to Early Mississippian, mid-Permian, and Early to Middle Jurassic time. Regional unconformities separate the volcanic sequences from each other and a locally profound angular unconformity separates the volcanic rocks from their basement of the lower Paleozoic Shoo Fly Complex. The volcanic rocks, together with the Shoo Fly Complex, comprise a multiply deformed but stratigraphically coherent package of rocks referred to as the northern Sierra terrane (Coney et al., 1980).

The northern Sierra terrane is bounded on the west by the Melones fault zone, which is marked along much of its length by intensely deformed serpentinite, peridotite, metagabbro, amphibolite, and local areas of blueschist that make up the Feather River peridotite belt (Figure 1) (Day et al., 1985). Tectonic slivers of Permian and Triassic rocks occur locally between the Feather River peridotite belt and the Shoo Fly Complex (Harwood et al., 1988). This diverse collection of rocks, as well as Paleozoic and Mesozoic oceanic and volcanic rocks farther to the west, were accreted to the northern Sierra terrane when it formed part of the Mesozoic continental margin of North America. The northern Sierra terrane is overlapped to the north by Tertiary and Quaternary volcanic rocks which prevent direct tracing of the Paleozoic and Mesozoic rocks into coeval rocks of the eastern Klamath terrane.

This field trip, following closely on a trip to the eastern Klamath terrane, provides an excellent opportunity to compare two Cordilleran volcanic assemblages whose tectonic evolution with respect to each other and to North America is the subject of on-going debate. One side of the debate views the volcanic rocks of the northern Sierra and eastern Klamath terranes as part of an extensive, west-facing island-arc system that formed, migrated, and reformed repeatedly off the active continental margin of North America from the early Paleozoic to the late Mesozoic (Burchfiel and Davis, 1972, 1975; Miller, 1987). This interpretation implies not only a paleogeographic link between the northern Sierra and eastern Klamath terranes, but also temporal and lithologic ties between these volcanic arc terranes and obducted back-arc sediments that now make up the Roberts Mountains and Golconda allochthons of northwestern Nevada. Alternatively, the northern Sierra and eastern Klamath terranes may be viewed as a far-travelled microplate of questionable paleogeographic affinity to North America (Speed, 1979; Speed and

Sleep, 1982; Coney et al., 1980). In this interpretation, the northern Sierra and eastern Klamath volcanic rocks formed part of a large exotic arc that collided with North America in Late Permian to Early Triassic time thrusting accretionary sediments of the Golconda allochthon over the Roberts Mountains allochthon and the miogeosyncline. This interpretation implies no pre-Late Permian stratigraphic ties between the arc terranes and rocks of the Golconda or Roberts Mountains allochthon.

This trip is a north-to-south traverse through the northern Sierra terrane that emphasizes major stratigraphic and structural changes in the Paleozoic and Mesozoic volcanic rocks along a strike distance of 120 km. Because the rocks dip steeply east and face in the same direction, we will be looking at a cross-sectional view of three successive volcanic-arc sequences that span a period of about 200 million years. Points of interest include; 1) general stratigraphic comparisons between the northern Sierra and eastern Klamath terranes, 2) lithologic changes in the Upper Devonian and Lower Mississippian volcanic sequence that may reflect a transition from arc to back-arc facies, 3) distribution and evolution of mid-Permian volcanic and epiclastic facies and 4) variations in major Late Jurassic structures through the terrane.

STRATIGRAPHIC REVIEW

The oldest rocks in the northern Sierra terrane belong to the lower Paleozoic Shoo Fly Complex. North of the 39th parallel the Shoo Fly is composed of four regionally extensive thrust blocks informally referred to as the (ascending): 1) Lang sequence, 2) Duncan Peak allochthon, 3) Culbertson Lake allochthon, and 4) Sierra City melange (Schweickert et al., 1984; Harwood, 1988). The Lang sequence is composed primarily of quartz arenite-pelite turbidites with interbedded lenses of chert and limestone. Conodonts from one lens of limestone are indicative of a late Middle to Late Ordovician age (Harwood and others, 1988), which is the oldest depositional age yet recorded in the Shoo Fly Complex. The Duncan Peak allochthon is composed of radiolarian chert and variable amounts of siliceous argillite that tectonically overlie and are locally interlayered with teconic slivers of the Lang sequence. The Duncan Peak allochthon is tectonically overlapped toward the north by the Culbertson Lake allochthon, which consists of pillow basalt, limestone, chert, quartz arenite, and volcanic lithic sandstone (Girty, 1983; Girty and Wardlaw, 1985). The Culbertson Lake allochthon is tectonically overlapped by the Sierra City melange,

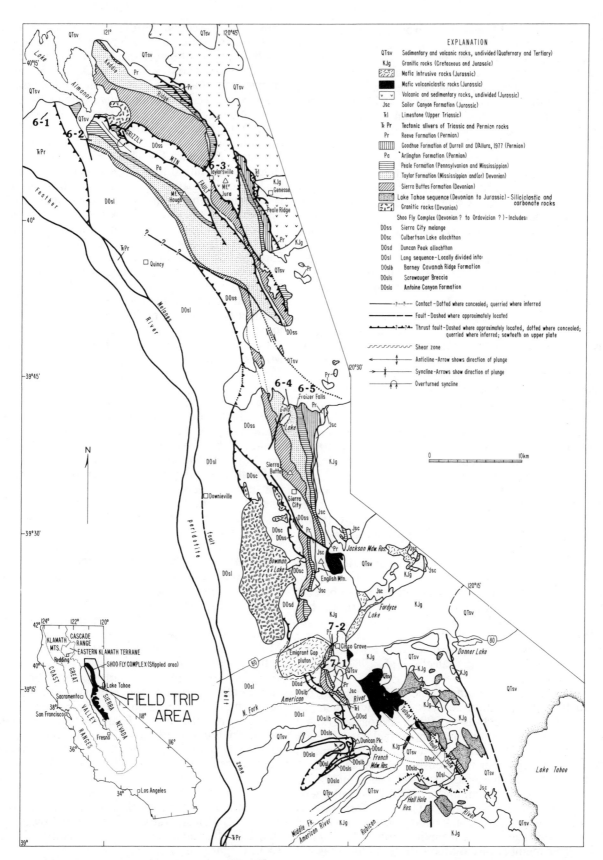

FIGURE 1. Geologic map of northern Sierra terrane showing field trip stops. Sources of data: D'Allura et al. (1977), Durrell and D'Allura (1977), Harwood (1983, unpublished data), Schweickert et al. (1984).

which contains blocks of serpentinite, gabbro, greenstone, chert, and limestone set in a matrix of disrupted sandstone and pelite. Conodonts and megafossils from scattered limestone blocks, are indicative of a Late Ordovician age and North American faunal affinities (Boucot and Potter, 1977; A. Harris, 1986, written communication; Hannah and Moores, 1986). Recent Pb/U studies on zircon from a felsic tuff interlayered with the matrix of the melange yielded a late Early Silurian (423±Ma) age (Saleeby et al., 1987). This is the youngest age obtained, thus far, from the Shoo Fly Complex and it indicates that the major fault blocks of the Shoo Fly Complex were thrust together between the late Early Silurian and the Late Devonian, the age of the oldest rocks that unconformably overlie the Shoo Fly.

The Shoo Fly Complex is unconformably overlain by discontinuous lenses of Upper Devonian epiclastic rocks assigned to the Grizzly Formation and regionally extensive submarine pyroclastic rocks of the Sierra Buttes Formation (Figure 2). The Grizzly Formation contains beds of conglomerate and breccia composed of chert, quartzite, and carbonate clasts derived from the Shoo Fly. The conglomerate and breccia occur as channel-fill debris-flow deposits cut into submarine fan deposits of quartz arenite and pelite. Chert lenses in the Grizzly have yielded Late Devonian (Famennian) conodonts and a submarine slump block of interbedded shallow-water limestone and quartz arenite has yielded Frasnian megafossils (Hanson and Schweickert, 1986). The Sierra Buttes Formation is composed of felsic tuff and tuff breccia, andesitic tuff breccia, and black phosphate-streaked chert. Megafossils and conodonts from the chert are indicative of a Famennian age (Anderson et al., 1974; Hanson and Schweickert, 1986).

In the northern and central parts of the northern Sierra terrane, the Sierra Buttes Formation is about 2 km thick and contains near-vent deposits of hyaloclastic breccia and abundant hypabyssal intrusive rocks. A rhyolite sill, interpreted to be contemporaneous with felsic tuff breccia of the Sierra Buttes Formation which it intrudes, has yielded a concordant Pb/U age of 368 Ma (Hanson et al., 1988). Granitoid plutons that represent deeper magmatic parts of the Devonian arc give discordant Pb/U ages that range from 364 to 355 Ma (Saleeby et al., 1987; Hanson et al., 1988).

In the southern part of the northern Sierra terrane, the Sierra Buttes Formation consists of a lower member composed of as much as to 300 m of coarse-grained felsic tuff breccia with abundant intraclasts of black phosphate-streaked chert. The lower member grades into an upper member that consists of 200 m of fine-grained, thinly laminated and cross-laminated tuffaceous siltstone and slate containing scattered lenses of chert-rich granule conglomerate. The conglomerate is composed of gray and black chert clasts and rounded clasts of quartzite derived from the Shoo Fly Complex. This upper member of the Sierra Buttes, which has not been identified in the near-vent deposits to the north, is interpreted as a distal facies deposited on the southern fringe of a submarine volcanic apron (Harwood, 1983).

The Sierra Buttes Formation is overlain by andesite flows, tuff breccia, lapilli tuff, and volcaniclastic sedimentary rocks of the Taylor Formation. In the northern part of the northern Sierra terrane, the Taylor Formation is about 2.7 km thick and is composed primarily of pillowed andesite flows and coarse-grained tuff breccia. The formation thins significantly toward the south and pinches out locally north of Jackson Meadow Reservoir (Figure 1) (Schweickert et al., 1984; Hanson and Schweickert, 1986). The Taylor reappears south of Interstate Highway 80 where it consists of 300 m of andesitic tuff turbidites. No fossils have been found in the Taylor Formation, but it can be no older than Late Devonian (Famennian) and no younger than Early Mississippian (late Kinderhookian), the age of the fossils in the overlying lower member of the Peale Formation; the Taylor is considered to be Late Devonian and (or) Early Mississippian in age.

The Peale Formation consists of a lower member, characterized in the north by alkali feldspar phyric keratophyre flows, tuff breccia, and tuff, and an upper member composed of radiolarian chert. In the north, the lower member also contains scattered lenses of shallow-water bioclastic limestone that have yielded Early Mississippian brachiopods and trilobites (McMath, 1966) and late Kinderhookian conodonts (A. Harris, 1984, written communication). The lower member attains a maximum thickness of 0.5 km in the northern part of the northern Sierra terrane and, like the underlying Taylor Formation, pinches out near the Middle Yuba River where the (upper) chert member rests disconformably on the Sierra Buttes Formation.

South of Interstate 80 the lower member of the Peale consists of finely laminated tuff, tuffaceous siltstone, and slate containing scattered lenses of chert- and quartzite-pebble conglomerate and volcaniclastic debris-flow units that locally contain blocks of jasper and volcanic breccia. Quartzite clasts in the conglomerate are well rounded and lithologically identical to quartzite in the Shoo Fly Complex. Intra-formational slump breccia, showing down-to-the-south layer-parallel extension, occurs locally in the tuffaceous siltstone.

The lower member of the Peale represents waning deposition within the Late Devonian to Early Mississippian volcanic arc (Harwood, 1988). Volcanic activity ceased in the Early Mississippian and, by Osagean time, the arc had cooled and subsided sufficiently to collect and preserve siliceous pelagic sediment. Based on radiolarian faunas recovered from the chert member of the Peale, pelagic sedimentation persisted without obvious interruption or influx of coarser detritus until the Middle(?) Pennsylvanian (Desmoinesian?); a period of at least 45 million years. Pelagic sedimentation may have continued for an additional 25 million years, but the depositional record is missing along an erosional unconformity that separates Middle(?) Pennsylvanian chert from Lower Permian (upper Wolfcampian(?) to lower Leonardian(?)) volcanic rocks (Figure 2).

The Permian volcanic sequence was deposited during late Wolfcampian(?) to early Guadalupian time and consists of four major lithofacies. The Arlington Formation consists of volcaniclastic sandstone, siltstone,

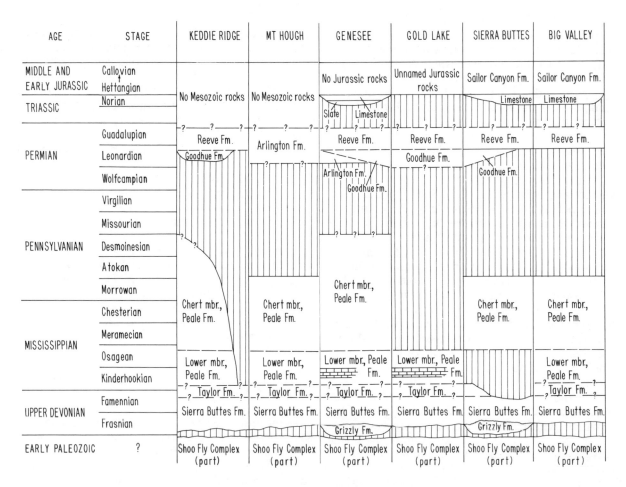

AGE	STAGE	KEDDIE RIDGE	MT HOUGH	GENESEE	GOLD LAKE	SIERRA BUTTES	BIG VALLEY
MIDDLE AND EARLY JURASSIC	Callovian ↑ Hettangian	No Mesozoic rocks	No Mesozoic rocks	No Jurassic rocks	Unnamed Jurassic rocks	Sailor Canyon Fm.	Sailor Canyon Fm.
TRIASSIC	Norian			Slate / Limestone		Limestone	Limestone
PERMIAN	Guadalupian	Reeve Fm.	Arlington Fm.	Reeve Fm.	Reeve Fm.	Reeve Fm.	Reeve Fm.
	Leonardian	Goodhue Fm.		Arlington Fm. / Goodhue Fm.	Goodhue Fm.	Goodhue Fm.	
	Wolfcampian						
PENNSYLVANIAN	Virgilian						
	Missourian						
	Desmoinesian						
	Atokan						
	Morrowan	Chert mbr, Peale Fm.	Chert mbr, Peale Fm.	Chert mbr, Peale Fm.		Chert mbr, Peale Fm.	Chert mbr, Peale Fm.
MISSISSIPPIAN	Chesterian						
	Meramecian						
	Osagean	Lower mbr, Peale Fm.	Lower mbr, Peale Fm.	Lower mbr, Peale Fm.	Lower mbr, Peale Fm.		Lower mbr, Peale Fm.
	Kinderhookian	Taylor Fm.	Taylor Fm.	Taylor Fm.	Taylor Fm.		Taylor Fm.
UPPER DEVONIAN	Famennian	Sierra Buttes Fm.	Sierra Buttes Fm.	Sierra Buttes Fm.	Sierra Buttes Fm.	Sierra Buttes Fm.	Sierra Buttes Fm.
	Frasnian			Grizzly Fm.		Grizzly Fm.	
EARLY PALEOZOIC	?	Shoo Fly Complex (part)	Shoo Fly Complex (part)	Shoo Fly Complex (part)	Shoo Fly Complex (part)	Shoo Fly Complex (part)	Shoo Fly Complex (part)

FIGURE 2. Stratigraphic columns from the northern Sierra terrane.

and slate with variable amounts of interbedded pebbly mudstone, breccia, and conglomerate. Coarse breccia deposits, which are most abundant in the lower 200 m of the principal reference section of the Arlington, contain blocks of volcanic rocks and chert, some as much as 50 m in length, and widely scattered blocks of bioclastic limestone, up to 2 m in size. Conglomerate lenses are composed primarily of chert clasts but also contain scattered rounded quartzite pebbles and cobbles derived from the Shoo Fly Complex. Conodonts from a limestone block near Mount Hough (Figure 1) are early Leonardian (A. Harris, 1985, written communication) and a poorly preserved shallow-water assemblage of brachiopods, gastropods and bivalves in the matrix of the breccia from the same area are indicative of a late Wolfcampian to early Leonardian age (J.T. Dutro, 1986, written communication).

On Peale Ridge (Figure 1), the Arlington is overlain gradationally by clinopyroxene-phyric pillow basalt of the Goodhue Formation. The basalt forms a lens-shaped deposit in the north-central part of the northern Sierra terrane where it attains a maximum thickness of about 2.7 km. Curiously, the Goodhue pinches out both to the north and south near the inferred vent areas of the older volcanic sequence suggesting that the distribution of the Goodhue was influenced in some way by the earlier volcanic substrate. The Goodhue is overlain by plagioclase-phyric andesite flows, tuff breccia, tuff, and

volcaniclastic sedimentary rocks of the Reeve Formation and volcaniclastic sedimentary rocks of the Robinson Formation of McMath (1958, 1966). Local accumulations of chert-rich conglomerate, typical of the Arlington Formation, occur in the lower part of the Reeve Formation. Possible vent areas for the Reeve have been identified in the northern and central parts of the northern Sierra terrane (D'Allura and others, 1977; Hannah and Moores, 1986). Calcareous tuff and tuffaceous siltstone in the Reeve Formation in the northern part of the terrane contain late Early Permian (early Leonardian) fusulinids equivalent to zone H of the McCloud Limestone (C.H. Stevens, 1988, oral communication). Early Guadalupian conodonts occur higher in the section in a lens of limestone in the Robinson Formation of McMath (1958, 1966). A shallow-water assemblage of brachiopods and gastropods, also of early Guadalupian age, occurs in the Reeve south of Taylorsville (D'Allura and others, 1977; J.T. Dutro, 1986, written communication). South of Interstate 80, the Reeve is composed of fine-grained tuffaceous siltstone that contains locally abundant *Nereites* trace fossils indicative of relatively deep-water environment. A tuffaceous debris-flow deposit near the base of the Reeve in this area contains abundant shelly debris indicative of a late Early Permian (Wordian or Roadian) age (Harwood, 1983, in press).

Like the underlying Upper Devonian to Lower

Mississippian arc assemblage, the Permian volcanic rocks had a generally northern source area or areas that gave way southward to a deeper water back-arc basin. However, Permian paleogeography and tectonism differed significantly from that of the older volcanic arc. For example, the Permian rocks rest disconformably on different stratigraphic units of the older arc sequence (Figure 2) indicating differential uplift and erosion of the substrate. Pillow basalt of the Goodhue Formation was deposited in an extensional tectonic setting that probably formed during Early Permian thermal uplift of the older arc sequence and its basement of the Shoo Fly Complex. Breccia and pebbly mudstone in the Arlington Formation were derived from uplifted fault blocks, which also must have been the source of limited Shoo Fly detritus. Available stratigraphic data indicate that the onset of Permian deposition was diachronous, beginning in the late Wolfcampian or early Leonardian in the north and in the latest Leonardian or early Guadalupian in the south.

Following deposition of the Permian arc rocks, the northern Sierra terrane was uplifted and locally deformed during latest Permian or Early Triassic time. Marine transgression began in the northern part of the terrane in the Middle Triassic, but did not reach the latitude of the North Fork of the American River until Late Triassic time. The oldest Triassic deposits are fossiliferous middle Ladinian tuffaceous siltstone and sandstone of the Pit Formation, which underlie the Upper Triassic (upper Karnian to lower Norian) Hosselkus Limestone in the northern part of the terrane. The Hosselkus underlies upper Norian black slate and thin-bedded limestone of the Swearinger Slate. At the North Fork of the American River, upper Karnian to lower Norian limestone, lithologically equivalent to the Hosselkus, lies gradationally above chert-rich conglomerate and sandstone that prograde southward unconformably overlapping the Paleozoic volcanic rocks and the underlying Shoo Fly Complex.

In the south, Upper Triassic rocks grade up into black slate, tuff, and volcaniclastic sandstone of the Lower and Middle Jurassic Sailor Canyon Formation. Sandstones in the lower part of the Sailor Canyon form thin, fine-grained, parallel-laminated beds locally associated with shallow-water sandy limestone and calcarenite. Coarse-grained, graded volcaniclastic turbidites associated with local slump deposits of chaotic sandstone and pelite, suggestive of inner-fan deposits, occur in the middle and upper parts of the formation. Middle Jurassic (Bajocian) rocks in the Sailor Canyon Formation are disconformably overlain by andesitic tuff breccia and tuff of the Tuttle Lake Formation. Scattered thin andesitic turbidite beds indicate submarine deposition for the Tuttle Lake Formation. No fossils have been found in the Tuttle Lake Formation but it probably is correlative with the upper part of the Mt. Jura section and with the Kettle Formation of McMath (1958, 1966) which contains fossils as young as late Middle Jurassic (Callovian) (Imlay, 1961).

In the northern part of the northern Sierra terrane, Jurassic rocks occur on Mt. Jura and to the east, but stratigraphic relations between these areas are complicated by Late Jurassic thrust faulting. At Mt. Jura, andesitic and dacitic pyroclastic rocks are interbedded with calcareous volcaniclastic sandstone and conglomerate, tuffaceous siltstone, and silty limestone. The pyroclastic rocks include fine-grained, sparsely porphyritic flows, tuff breccia, and tuff inferred to be distal submarine deposits. Interbedded volcaniclastic sedimentary rocks contain locally abundant shallow-water fossils (Diller, 1908) that range in age from Sinemurian to Callovian (Imlay, 1961). The Mt. Jura section is about 2.5 km thick (Diller, 1908).

East of Mt. Jura, Callovian ammonities (McMath, 1958; Imlay, 1961) occur near the top of an otherwise unfossiliferous section of volcanic and volcaniclastic rocks that may be as much as 12 km thick (Christe and Hannah, 1987). Christe (1987) identified four predominantly volcanic intervals, composed of andesitic to dacitic intrusive breccia, flows, tuff breccia, and ash flow tuff, separated by volcaniclastic conglomerate, sandstone and mudstone, which he considered to represent fluvial deposits. Christe (1987) and Christe and Hannah (1987) interpreted the eastern volcanic sequence to be a continental arc.

NORTHERN SIERRA-EASTERN KLAMATH CORRELATIONS

Determining when the northern Sierra and eastern Klamath terranes were juxtaposed is a major problem in the tectonic evolution of northern California with broader implications for terrane accretion and structural development in the western Cordillera. Paleomagnetic data, in this case, are not definitive because the northern Sierra terrane was metamorphosed and remagnetized in situ in the Late Jurassic (Hannah and Verosub, 1982) whereas Devonian and Permian paleolatitudes from the eastern Klamath terrane coincide with those of North America. Paleogeographic reconstructions, therefore, must be based largely on stratigraphic information. Correlation of arc-volcanic rocks is inherently difficult, but some conclusions about the paleogeographic association of the northern Sierra and eastern Klamath terranes can be drawn from the stratigraphic data shown in Figure 3.

Based on lithologic associations and shared faunas, it seems certain that the eastern Klamath and northern Sierra terranes were reasonably close, perhaps separated by a few hundred kilometers, during the Early Permian. Late Wolfcampian to early Leonardian fusulinids, equivalent to zone H of the McCloud Limestone, occur in the Reeve Formation (C.H. Stevens, 1988, oral communication). Furthermore, the gastropod, *Omphlatrocus*, which also is indicative of a late Wolfcampian to early Leonardian age and was first described from the McCloud Limestone, occurs in limestone blocks and the matrix of the Reeve and Arlington Formations (J.T. Dutro, 1986, written communication). It seems unlikely that the limestone blocks in the Reeve and Arlington Formations were derived directly from the McCloud Limestone. Instead, they were probably derived locally and indicate that shallow-water platform carbonate rocks, which are

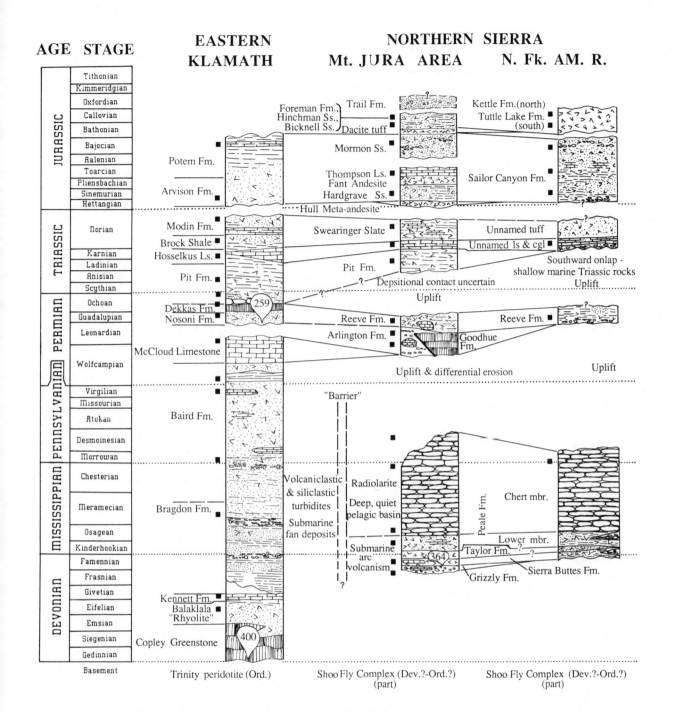

FIGURE 3. Stratigraphic columns showing possible correlations between the eastern Klamath and northern Sierra terranes. Eastern Klamath data from Miller (1987); Mt. Jura data from McMath (1958) and Imlay (1961, 1968). Trinity peridotite, Copley Greenstone, Balaklala "Rhyolite", Kennett Formation, Bragdon Formation, Baird Formation, McCloud Limestone, Nosoni Formation, Dekkas Formation, Pit Formation, Hosselkus Limestone, Brock Shale, Modin Formation (all) of Miller (1987); Arvison Formation of Sanborn (1960); Goodhue Formation; Kettle Formation of McMath (1958, 1966).

relatively widespread in the eastern Klamath terrane, formed on and were eroded from local shallow-water areas in the northern Sierra terrane during late Wolfcampian to early Leonardian time. During the early Wolfcampian, when the lower platform carbonate sequence of the McCloud Limestone was being deposited in the eastern Klamath terrane, the northern Sierra terrane was being differentially uplifted and extended apparently in response to the initial phases of southeast-directed subduction. Uplifted horsts were eroded producing the upper Wolfcampian(?) to lower Guadalupian (Wordian) epiclastic breccia, volcaniclastic pebbly mudstone, and conglomerate of the Arlington Formation. Pillow basalt of the Goodhue Formation was erupted into extensional troughs. By the mid-Permian (late Leonardian to early Guadalupian) andesitic volcanic rocks of the Reeve and Nosoni Formations and the Dekkas Formation of Miller (1987) erupted in both terranes.

Lithologic associations and faunal ties, established between the terranes in the mid-Permian, became stronger in the early Mesozoic. The middle Ladinian ammonite, *Meganoceras megani*, recently discovered in tuffaceous slate below the Hosselkus Limestone in the northern Sierra terrane (N.J. Silbering, 1987, written communication), establishes firm lithologic and paleontologic connections with the Pit Formation of the eastern Klamath terrane during the Middle Triassic. By the Late Triassic (late Karnian and early Norian) discontinuous shallow-water reefs, represented by the Hosselkus Limestone, extended from the northern part of the eastern Klamath terrane to the North Fork of the American River. Following deposition of the Hosselkus Limestone and extending into the Middle Jurassic, both terranes received shallow marine volcanic and volcaniclastic deposits that were interrupted by local hiatuses and brief periods of shallow-water carbonate deposition. Locally thick deposits of andesitic and dacitic submarine debris accumulated in the northern Sierra terrane during the Middle Jurassic (Bajocian to late Callovian).

The pre-Permian stratigraphy (Figure 3) of the northern Sierra terrane differs significantly from that in the eastern Klamath terrane, therefore, paleogeographic relations between the terranes are more tenuous and open to very different interpretations. In the northern Sierra terrane, Upper Devonian and Lower Mississippian submarine arc-volcanic rocks were deposited on amalgamated thrust blocks of the lower Paleozoic Shoo Fly Complex. Arc volcanism, which began in the Famennian and lasted for a period of about 30 m.y., waned in the late Kinderhookian and had ceased by the Osagean. During much of the Carboniferous, beginning in the Osagean and extending at least to the Desmoinesian(?) stage of the Middle(?) Pennsylvanian, the northern Sierra terrane was part of a deep-water basin that collected siliceous pelagic sediment represented by chert member of the Peale Formation. Although siliceous argillite comprises a minor part of the chert member, no interbeds of relatively coarse-grained volcaniclastic or siliciclastic debris have been found and sedimentation rates in the basin, uncorrected for compaction, were on the order of 1m/m.y. Correlative chert deposits in the Golconda allochthon (Murchey et al., 1986) indicate depositional and paleogeographic ties between the Peale basin in the northern Sierra terrane and obducted basinal sedimentary rocks now exposed in northwestern Nevada (Miller et al., 1984; Stewart et al., 1986).

During the Late Devonian and Early Mississippian the eastern Klamath terrane accumulated mixed pyroclastic and epiclastic deposits of the Bragdon Formation followed by predominantly pyroclastic rocks of the Carboniferous Baird Formation. Miller (1987; in press) concluded that the pyroclastic and epiclastic rocks of the Bragdon represent submarine-fan deposits derived from

a linear source area and deposited in a restricted extensional or transtensional intra-arc or arc-flank basin. The Bragdon basin could have developed adjacent to the Late Devonian and Early Mississippian arc in the northern Sierra terrane. Chert-rich conglomerate in the southern part of the Sierra Buttes and lower member of the Peale Formations is very similar to epiclastic deposits in the Bragdon and both contain quartz arenite and micaceous quartzite clasts probably derived from the Shoo Fly Complex. Both juvenile crystal-rich pyroclastic rocks and reworked volcanic-lithic deposits in the Bragdon could have been derived from the middle Paleozoic arc in the northern Sierra terrane. If the Bragdon correlates with the Sierra Buttes, Taylor, and lower member of the Peale Formations, and their shared lithologies are accepted as evidence of a paleogeographic affinity, significant changes occurred between the terranes in the Carboniferous. In the northern part of the eastern Klamath terrane, the Bragdon grades up into shallow-water volcanic deposits of the Baird Formation (Miller, in press) which is equal in age to, and possibly somewhat younger, than the chert member of the Peale Formation in the northern Sierra terrane. The lack of coarse volcanic debris in the chert member of the Peale suggests that the eastern Klamath and northern Sierra terranes were separated by either distance or by an unknown physical barrier, possibly a submarine barrier, during the middle and late Carboniferous.

ROAD LOG--DAY 6, July 3, 1989--CHESTER TO RENO

The road log begins at the junction of State Highways 36 and 89 four miles west of Chester, Calif.

Mileage

0.0   Proceed southeast on route 89 along the west side of Lake Almanor. Dyer Mountain and the north end of Keddie Ridge lie east of the lake and form one of the major Late Jurassic thrusts blocks (Keddie Ridge block) in the northern part of the northern Sierra terrane. The Keddie Ridge block contains an inverted east-facing, southwest-dipping section of Paleozoic volcanic rocks thrust eastward over autochothonous(?) Jurassic rocks. The prominent outcrops along the upper west slope of Keddie Ridge are blocky andesite tuff breccia and flows of the Upper Devonian and (or) Lower Mississippian Taylor Formation.

6.4   Turn right on Butt Lake Road.

8.5   STOP 6-1. Permian and Triassic tectonic sliver: Outcrops of tuffaceous slate and pebbly mudstone on the east side of the road are part of a large tectonic sliver of Permian and Triassic rocks that occurs between the Shoo Fly Complex to the east and the Feather River peridotite belt to the west (Figure 1). The tuffaceous slate contains pelecypods and ammonites of late Middle Triassic to Early Jurassic age (N.J.

Silberling, 1988, oral commun.). Thin parallel-laminated tuffaceous siltstone beds can be recognized locally in the slate. The southwestern part of this large outcrop is composed primarily of intensely cleaved pebbly tuffaceous mudstone that contains severely flattened clasts of felsic tuff and augen-shaped clasts of chert, quartz-mica schist and micaceous quartzite derived from the Shoo Fly Complex. The relatively coarse-grained muscovite in the Shoo Fly clasts contrasts sharply with the extremely low-grade quartz-sericite assemblage in the tuffaceous slate. Deformation of the more resistant chert and quartzite clasts in the C-S fabric indicate a west-over-east kinematic sense. Return to route 89.

10.6   Turn right on route 89 and proceed southeast.
16.0   Spillway of Lake Almanor - great view to the east of Keddie Ridge. The rocks at the road level are upper Cenozoic basalt flows from the Cascade province.
16.4   Turn right on Seneca Road. The bus will stop and turn around at end of the pavement. We will walk south on Seneca Road about 0.25 miles.
17.6   STOP 6-2. Arlington Formation: These outcrops of green and minor purple volcaniclastic siltstone, sandstone, and conglomerate form the northernmost extent of the Arlington Formation. On strike to the south, the size of the chert and volcanic fragments increase and the lower part of the unit is composed locally of blocky breccia and pebbly mudstone. D'Allura and others (1977) correlated these rocks with those at STOP 1 and, on the basis of that correlation, inferred that a major east-vergent anticline separated them. We now know that the Arlington Formation is mid-Permian (late Wolfcampian(?) and early Guadalupian) in age, and is not correlative with the Permian and Triassic rocks at Stop 1. Turn around and return to route 89.
18.8   Route 89, turn right, proceed toward Greenville following Wolf Creek.
21.1   Wolf Creek underpass.
22.1   Outcrops of deeply weathered granite left of road. This is the Wolf Creek stock from which Saleeby et al. (1987) obtained a Pb/U age on zircon of 378 + 5/-10 Ma. It presumably is part of the Devonian magmatic arc.
23.4   Rusty-weathering, hydrothermally altered felsic volcanic rocks of the Upper Devonian Sierra Buttes Formation.
27.1   Greenville. The town of Greenville lies at the northwest end of Indian Valley. Keddie Ridge to the east.
31.1   Turn left on Stampfel Lane.
       VIEW STOP - Panoramic view to the east of Indian Valley and the major Late Jurassic thrust blocks in the Taylorsville area.
       Continue east on Stampfel Lane. Small rounded hills in Indian Valley are underlain by the Sierra Buttes Formation.
34.1   Junction with Greenville Road, turn right.
34.8   Junction with Diamond Mine Road, turn right.

37.6   Taylorsville Rodeo ground to the right.
       STOP 6-3. Hardgrave Sandstone: Red tuff and tuffaceous siltstone of the Lower Jurassic (Pliensbachian) Hardgrave Sandstone, which occurs near the base of the Mt. Jura section. Imprints of pelecypods occur in some of the thin-bedded tuffaceous rocks which dip 60 degree west and are overturned.
       At the Rodeo grounds turn right on the Taylorville Road (County road A22) and proceed west through Taylorsville.
37.9   Taylorsville.
42.6   Junction of Taylorsville Road and Rt 89, turn left and proceed SW through Indian Creek gorge.
48.9   Junction Routes 89 and 70. Interbedded slate and quartz arenite are part of the Lang sequence, the lowest major thrust block in the Shoo Fly Complex in this area.
58.9   Quincy - follow routes 89 & 70 through town.
66.3   Large road cuts of serpentinite-block melange, which is part of the Sierra City melange, the highest major thrust block of the Shoo Fly Complex.
70.3   Lee Summit - large road cuts in this area expose Miocene lahar.
76.8   Two Rivers...Canyon walls of the Middle Fork of the Feather River form a narrow gorge here. During the Pleistocene, the gorge was blocked by landslide or glacial debris forming an extensive lake in the Mohawk Valley to the south.
79.6   Junction routes 70 and 89 turn right following route 89.
81.2   Graeagle - continue on route 89.
82.7   Junction route 89 and road to Lakes Basin - turn right.
88.6   STOP 6-4. Taylor Formation: Parking area to the right opposite large road cut of andesitic tuff breccia and tuff of the Upper Devonian and (or) Lower Mississippian Taylor Formation. We are south of the Late Jurassic overthrust blocks and in the east-facing homoclinal section of Paleozoic and Mesozoic volcanic rocks. Note the vertical northwest-trending cleavage in the fine-grained tuffaceous rocks which contrasts sharply with the gentle to moderate southwest-dipping thrust fabric that was dominant in the rocks at stops 1 and 2.
90.4   Plumas-Sierra County line.
91.3   Frazier Falls Road. Turn left and proceed to Frazier Falls trail head. Exposures adjacent to the road are andesitic tuff breccia of the Taylor Formation.
92.9   STOP 6-5. Lower member of Peale Formation: Frazier Falls trail head. This stop involves a 0.5 mile walk across the lower member of the Peale Formation. At this locality, the lower member is composed primarily of green tuff and lapilli tuff with scattered lenses of tuff breccia and tuffaceous debris flow deposits. Pink and white alkali feldspar and rare quartz phenocrysts occur in the lapilli tuff and tuff breccia fragments. The debris-flow deposits are composed primarily of

volcanic fragments but also contain keratophyre clasts (composed of alkali feldspar-quartz intergrowths) and sparse vein quartz and quartzite clasts similar to those in the Bragdon Formation in the eastern Klamath terrane. Minor purple tuff, locally associated with lenses of jasper, occur in the green tuff. Lenses of limestone at this locality have yielded Early Mississippian (late Kinderhookian) conodonts. Return to the bus and retrace the route to the Sierra City road.

94.5 Sierra City road - turn left.

96.6 Excellent view to the southwest of Sierra Buttes; type section of the Upper Devonian Sierra Buttes Formation. The upper slopes and ragged peaks of Sierra Buttes are underlain by felsic and andesitic hypabyssal rocks intrusive into felsic tuff and tuff breccia.

101.6 Junction with highway 49 - turn left and proceed to Reno. There are no more stops on this day of the field trip. As we proceed east and south, we leave the exposures of metamorphic rocks and traverse plutons of the Sierra Nevada batholith and great expanses of Tertiary volcanic rocks.

ROAD LOG--Day 7, July 4, 1989--RENO TO CISCO GROVE

The road log for this day's trip begins heading west on Interstate Highway 80 (I-80) at the California state line. Between Reno and the California state line, I-80 follows the canyon of the Truckee River, which flows east and north from Lake Tahoe to Pyramid Lake where the water ponds and evaporates. Cliffs and road cuts in the Truckee River canyon expose a variety of andesitic mudflows and lahars of Miocene and Pliocene age that were deposited on Cretaceous(?) granitic rocks of the Sierra Nevada batholith.

Mileage

0.0 California state line.

2.2 Road cut showing andesite dikes and lavas with local propylitic alteration. The granitic basement rocks are locally exposed beneath Miocene andesitic flows and lahars for the next 4.0 miles.

13.7 Cross Truckee River and the SPRR tracks. Note columnar jointing in basaltic lava in river cut to the south (left). This basalt (1.2 Ma K/Ar whole rock, Dalrymple, 1964) is part of the mafic suite erupted in the Tahoe-Truckee basin during the Pleistocene.

13.8 At this point we enter a broad flat alluvial plain underlain by outwash deposits derived from several episodes of Pleistocene glaciation in the Sierra Nevada (Birkeland, 1966).

19.1 Highway 89 exit south to Lake Tahoe. Extensive low outcrops of coarse bouldery outwash of the late Wisconsin Tioga glaciation occur south of the highway. From this outwash plain, the highway passes north of Donner Lake and rises toward Donner Pass on the Sierran crest.

23.0 View stop at rest area on north side of I-80. Peaks along the skyline to the west lie along the Sierran crest and are composed of remnants of andesitic flows (3 to 4 Ma, K/Ar whole rock ages; D. S. Harwood, unpublished data). The flows rest on Miocene(?) lahars that rest unconformably on Cretaceous granodiorite of the Sierra Nevada batholith. Oligocene and lower Miocene rhyolite ash flow tuffs occur locally below the lahars. Quaternary faults of the range-front fault system trend north and lie just east of the Sierran crest. There is about 600 m of stratigraphic separation on this fault system in this area. After crossing Donner Pass, we will travel through granitic plutons of the Sierra Nevada batholith for the next 35 km.

39.0 Cisco Grove exit.

43.6 Yuba Gap exit. Exit to the right from I-80 and cross over the highway.

44.1 Junction of secondary roads, bear right.

45.3 Junction of secondary roads, bear left and continue on the north side of Lake Valley reservoir.

51.1 STOP 7-1. Sierra Buttes and Taylor Formations: The north slope of Monumental Ridge contains a nearly continuous section of distal submarine deposits of the Sierra Buttes and Taylor Formations intruded on the west by late Middle Jurassic rocks of the (informal) Emigrant Gap mafic complex of James (1971). The lower member of the Sierra Buttes Formation is composed of coarse-grained felsic tuff breccia that contains abundant intraclasts of black phosphate-streaked chert. The upper member of the Sierra Buttes Formation consists of thin bedded, parallel and cross-laminated tuff and tuffaceous slate. Andesitic turbidites of the Taylor Formation conformably overlie the upper member of the Sierra Buttes Formation. To the south, the upper member of the Sierra Buttes Formation contains scattered lenses of chert-rich granule conglomerate that contains rounded clasts of quartzite derived from the lower Paleozoic Shoo Fly Complex. This stop emphasizes the southward thinning and fining of volcanic rocks in the Upper Devonian and Lower Mississippian arc sequence. Return to the bus and retrace the route back to I-80.

58.2 North ramp of I-80. Turn right and proceed east on I-80 to the Cisco Grove exit.

62.6 Cisco Grove exit. Exit right.

62.8 Cisco Grove. Turn left and cross over I-80.

62.9 On the north side of I-80 turn left on the entrance road to Thousand Trails campground and park in Thousand Trails parking lot. Walk 0.2 miles north on Fordyce Lake road to powerline crossing.

63.2 STOP 7-2. Triassic conglomerate and Sailor Canyon Formation: In this part of the northern Sierra terrane, Upper Triassic (upper Karnian to lower Norian) conglomerate, sandstone, and chert-limestone breccia rest unconformably on Lower Permian tuffaceous rocks of the Reeve Formation. Although the basal Triassic deposits are commonly composed of conglomerate reworked from the underlying rocks, at this locality the dominant lithology is a chert-limestone clast slump breccia that was derived from the south. The breccia contains scattered blocks of Upper Triassic limestone which outcrops as a mappable unit at the North Fork of the American River. Broad wavelength, cross-laminated chert-rich sandstone occurs locally above the breccia and locally grades into volcanic tuff. The Sailor Canyon Formation overlies the Upper Triassic rocks with apparent conformity but the presence of a hiatus here cannot be ruled out because ammonites in the lower part of the Sailor Canyon Formation are indicative of a late Sinemurian to early Pliensbachian age. Retrace the walk to the bus.

This is the last scheduled stop on this day of the field trip. Return to Reno via I-80.

REFERENCES CITED

Anderson, T.B., G.D. Woodard, S.M. Strathouse, M.K. Twichill, Geology of a Late Devonian fossil locality in the Sierra Buttes Formation, Dugan Pond, Sierra City quadrangle, California, *Geol. Soc. Am. Abstr. Programs, 6, 139,* 1974.

Birkeland, P.W., Tertiary and Quaternary geology along the Truckee River with emphasis on the correlation of Sierra Nevada glaceations with fluctuations of Lake Lahontan, *Geol. Soc. Am. Guidebook for field excurs. Cord. Sect.* pp. D1-D24, 1966.

Boucot, A.J., and A.W. Potter, Middle Devonian orogeny and biogeographical relations in areas along the North American Pacific rim, in *Western North America: Devonian:* edited by M.A. Murphy, W.B.N. Berry, and C.A. Sandberg, pp. 210-219, Univ. Calif. *Riverside Mus. Contrib., 4,* 1977.

Burchfiel, B.C., and G.A. Davis, Structural framework and evolution of the southern part of the Cordilleran orogen, western United States, *Am. J. Sci., 272,* 97-118, 1972.

Burchfiel, B.C., and G.A. Davis, Nature and controls of Cordilleran orogenesis, western United States,: extensions of an earlier synthesis, *Am. J. Sci., 275A,* 363-396, 1975.

Christe, Geoff, Geology of the Taylor Lake region, northern Sierra Nevada, California, M.S. thesis, Univ. Vermont, 1987

Christe, Geoff and J.C. Hannah, Continental arc volcanism in the eastern Mesozoic belt, northern Sierra Nevada, California: Implications for a revision of Jurassic paleogeography, *Geol. Soc. Am. Abstr. Programs, 19,* 366, 1987.

Coney, P.J., D.L. Jones, and J.W.H. Monger, Cordilleran suspect terranes, *Nature, 288,* 329-333, 1980.

D'Allura, J.A., E.M. Moores, and L. Robinson, Paleozoic rocks of the northern Sierra Nevada: their structural and paleogeographic implications, in *Paleozoic paleogeography of the western United States,* edited by J.H. Stewart, et al., *Soc. Econ. Paleo. Min.* Pacific Section, Pacific Coast Paleogeography Symposium 1, 395-408, 1977.

Dalrymple, G.B., Cenozoic chronology of the Sierra Nevada California, *Univ. Calif. Pubs. Geol. Soc. Bull., 47,* 41 pp., 1964.

Day, H.W., W.M. Moores, and A.C. Tuminas, Structure and tectonics of the northern Sierra Nevada, *Geol. Soc. Am. Bull., 96,* 436-450, 1985.

Diller, J.S., Geology of the Taylorsville region, California, *U.S. Geol. Surv. Bull.* 353, 128 pp., 1908.

Durrell, C. and J.A. D'Allura, Upper Paleozoic section in eastern Plumas and Sierra Counties, Sierra Nevada, California, *Geol. Soc. Am. Bull.,* 88, p.844-852.

Girty, G.H., The Culbertson Lake allochthon - a newly identified structureal unit in the Shoo Fly Complex: sedimentological, stratigraphic, and structural evidence for extension of the Antler orogenic belt to the northern Sierra Nevada, California, Ph. D. dissertation, Columbia Univ., New York, 155 pp., 1983

Girty, G.H., and M.S. Wardlaw, Petrology and provenance of pre-Late Devonian sandstones, Shoo Fly Complex, northern Sierra Nevada, California, *Geol. Soc. Am. Bull., 96,* 516-521, 1985.

Hannah, J.L., and E.M. Moores, Age relations and depositional environments of Paleozoic strata, northern Sierra Nevada, California, *Geol. Soc. Am. Bull., 97,* 787-797, 1986.

Hannah, J.L. and K.L. Verosub, Tectonic implications of remagnetized upper Palezoic strata of the northern Sierra Nevada, *Geology, 8,* 520-524, 1980.

Hanson, R.E., and R.A. Schweickert, Stratigraphy of mid-Paleozoic island-arc rocks in part of the northern Sierra Nevada, Sierra and Nevada Counties, California, *Geol. Soc. Am. Bull., 97,* 986-998, 1986.

Hanson, R.E., J.B. Saleeby, and R.A. Schweickert, Composite Devonian island-arc batholith in the northern Sierra Nevada, California, *Geol. Soc. Am. Bull., 100,* 446-457, 1988.

Harwood, D.S., Stratigraphy of upper Paleozoic volcanic rocks and regional unconformities in part of the northern Sierra terrane, *Geol. Soc. Am. Bull., 94,* 413-422, 1983.

Harwood, D.S., Tectonism and metamorphism in the northern Sierra terrane, northern California, in *Metamorphism and crustal evolution, western conterminous United States*, edited by W.G. Ernst, Rubey Vol. VII, Prentice-Hall, Englewood NJ., 764-788, 1988.

Harwood, D.S., A.S. Jayko, A.G. Harris, N.J. Silberling, and C.H. Stevens, Permian-Triassic rocks slivered between the Shoo Fly Complex and the Feather River peridotite belt, norther Sierra Nevada, California, *Geol. Soc. Am. Abstr. Programs, 20,* 167-168, 1988.

Imlay, R.W., Late Jurassic ammonites from the western Sierra Nevada, California, *U.S. Geol. Survey, Prof.*

Paper 374-D, D1-D30, 1961.

Imlay, R.W., Lower Jurassic (Pleinsbachian and Toarcian) ammonites from eastern Oregon and California, *U.S. Geol. Survey Prof. Paper 593-C*, C1-C51, 1968.

James, O.B., Origin and emplacement of the ultramafic rocks of the Emigrant Gap area, California, *J. Petr., 12*, 532-560, 1971.

McMath, V.E., The geology of the Taylorsville area, Plumas County, California, Ph.D. *diss.*, Univ. California Los Angeles, 199 pp., 1958.

McMath, V.E., Geology of the Taylorsville area, northern Sierra, Nevada, in *Geology of Northern California*, edited by E.H. Bailey, Calif. Div. Mines and Geol., Bull. 190, 173-183, 1966.

Miller, E.L., B.K. Holdsworth, W.B. Whiteford, D. Rodgers, Stratigraphy and structure of the Schoonover sequence, northeastern Nevada: Implications for Paleozoic plate-margin tectonics, *Geol. Soc. Am. Bull., 95,* 1063-1076, 1984.

Miller, M.M., Dispersed remnants of a northeast Pacific fringing arc: Upper Paleozoic terranes of Permian McCloud faunal affininty, western U.S., *Tectonics, 6,* 807-830, 1987.

Miller, M.M., Intra-arc sedimentation and tectonism: Late Paleozoic evolution of the eastern Klamath terrane, Callifornia, *Geol. Soc. Am. Bull.*, in press.

Murchey, B., D.S. Harwood, and D.L. Jones, Correlative chert sequences from the northern Sierra Nevada and Havallah sequence, Nevada, *Geol. Soc. Am. Abst. Programs, 18*, 162, 1986.

Saleeby, J.B., J.L. Hannah, and R.J. Varga, Isotopic age contraints on Middle Paleozoic deformation in the northern Sierra Nevada, California, *Geology, 15*, 757-760, 1987

Sanborn, A.F., Geology and paleontology of the southwest quarter of the Big Bend quadrangle, Shasta County, California, *Calif. Div. Mines Geol. Special Report 63,* 26 pp. 1960.

Schweickert, R.A., D.S. Harwood, G.H. Girty, and R.E. Hanson, Tectonic development of the northern Sierra terrane: An accreted late Paleozoic island-arc and its basement, in *Western Geological Excursions, 4,* edited by J. Lintz, Jr., pp. 1-65, Dept. Geol. Sciences, Mackay School of Mines, Reno, NV, 1984.

Speed, R.C., Collided Paleozoic microplate in the western United States, *J. Geol., 87,* 279-292, 1979.

Speed, R.C., and N.H. Sleep, Antler orogeny and foreland basin: a model, *Geol. Soc. Am. Bull., 93,* 815-828, 1982.

Stewart, J. H., Benita Murchey, D.L. Jones, and B.R. Wardlaw, Paleontologic evidence for complex tectonic interlayering of Mississippian to Permian deep-water rocks of the Golconda allochthon in Tobin Range, north-central Nevada, *Geol. Soc. Am. Bull., 97,* 1122-1132, 1986.

GRANITOIDS OF THE CENTRAL SIERRA NEVADA, CALIFORNIA

Franklin C. W. Dodge
U.S. Geological Survey, Menlo Park, California

INTRODUCTION

A glance at the geologic map of California shows that much of the Sierra Nevada is underlain by plutonic rocks (Figure 1). These rocks occur in numerous nearly vertical-walled plutons generally in sharp contact with one another or separated by thin remnants of metamorphic rocks, and make up the composite Sierra Nevada batholith, a geologic entity distinct from the range itself. The batholith dominates the southern and northeastern parts of the range, and extends eastward into the Great Basin. Although scattered granitoid masses of Paleozoic age have been discovered recently to intrude the older metamorphic rocks of the northwestern Sierra Nevada, these bodies are minor, and most of the plutonic rocks of the range are Mesozoic.

A broad transect across the central Sierra Nevada between 37° and 38°N. latitude has been intensively studied by geologists and geophysicists of the U.S. Geological Survey and by other workers for the last 40 years. Much of the present knowledge of the batholith is drawn from this region and is summarized in a recently completed synthesis by P.C. Bateman (in press). Currently, studies of the Geological Survey are being conducted both to the north and south.

Individual plutons within the transect range in outcrop area from less than 1 km² to over 1000 km². All of the large, and many of the small plutons are elongate in a northwest direction, parallel with the long axis of the batholith. Internal foliations, manifested both by oriented lens-shaped mafic inclusions and tabular and prismatic minerals, mimic the plutons external contacts. Many plutons vary internally in composition and texture, but abrupt changes usually occur only at contacts between plutons. Compositionally, the plutonic rocks range from gabbro to leucogranite, with tonalite and trondjhemite most common in plutons of the western part of the belt, and granodiorite and granite most common in those of the eastern part. Biotite, although rarely in amounts greater than 10 volume percent, and minor magnetite are ubiquitous. Hornblende is present in significant amounts in granodiorite and more mafic rocks. Mafic inclusions commonly occur in the hornblende-bearing granitoids, but metasedimentary inclusions are rare, and then only adjacent to similar wall rocks. Garnet is present only in rocks of a few late felsic or highly reduced plutons. Aluminum silicates and muscovite are rarely present and cordierite is absent. Highly felsic rocks are corundum normative, whereas more mafic rocks are diopside normative, but neither normative constitute ever exceeds a few weight percent. Sierran granitoids are metaluminous or only weakly peraluminous and all are considered I-types as defined by Chappell and White (1974) and White et al. (1986). The granitoids generally are medium-grained with hypidiomorphic granular textures. Most are equigranular or seriate, but some spanning the boundary between granite and granodiorite contain enhedral megacrysts of potassium feldspar.

INTRUSIVE SUITES AND AGE PATTERNS

Granitoids of a specific area that exhibit structural continuity and similar composition and textural characteristics, have been assigned to lithodemes (intrusive units equivalent in rank to formations). Contacts between lithodemes generally coincide with pluton boundaries, although some zoned plutons consist of two or more lithodemes. Assemblages of generally contiguous lithodemes with radiometric ages that permit assignment to individual intrusive epochs of limited duration have been grouped into intrusive suites (equivalent in rank to groups) believed to have been produced during the same magmatic episode as the result of a single fusion event. Relative ages of lithodemes of individual suites, determined by means of inclusions of one rock within another, by dikes of one rock cutting another, or by one rock truncating the structures of another, commonly, though not invariably, decrease sequentially from mafic to felsic. Lithodemes of concentrically zoned plutons generally vary inward from mafic to felsic.

Geographic age patterns are defined by distribution of the intrusive suites. Late Triassic granitoid plutons make up an extensive suite that crops out in scattered areas along and east of the eastern escarpment of the Sierra Nevada. Jurassic suites flank both sides of the batholith, and their locations indicate they originally followed a N.40°W. trend in the central Sierra Nevada, but were displaced in the batholith's interior by later Cretaceous plutons. Cretaceous suites trend more northerly, generally about N.20°W., parallel to the present-day Sierran crest. Older Cretaceous suites occur west of younger ones, which crop out along or near the crest; the locus of Cretaceous magmatism has been estimated to have migrated eastward at about 2.7 mm/yr (Chen and Moore, 1982). Enormous, individual zoned plutons, such as the Tuolumne Intrusive Suite (Figure 2) (Bateman and Chappell, 1979), make up the youngest of the suites and form a comagmatic chain extending far to both the north and south of the central transect.

COMPOSITIONAL VARATIONS ACROSS THE BATHOLITH

Several systematic compositional changes, seemingly independent of age of the granitoids, take place across the central part of the batholith and have been documented in numerous papers (for example, Bateman and Dodge, 1970; Dodge et al., 1982; Kistler and Peterman, 1973). The most noteworthy chemical change is that of K_2O, which increases progressively from about 1 percent in the western Sierra to an average of more than 4 percent east of the crest. A striking change is also shown by the iron oxidation ratio, which also increases eastward, and by H_2O+, which decreases; these changes are reflected by a decrease in hornblende and biotite

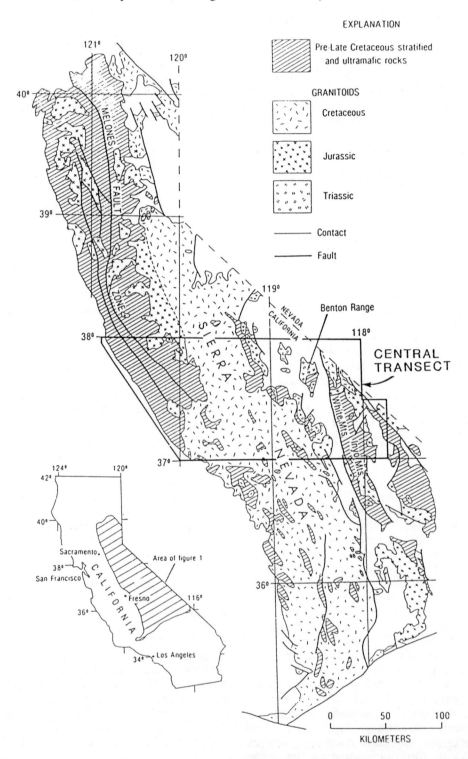

FIGURE 1 Sierra Nevada and adjacent areas in eastern California showing approximate age distribution of granitoids. Map is modified from maps by Evernden and Kistler (1970) and Stern et al. (1981).

concomitant with increase of magnetite eastward. Other major element changes include those of total Fe (as FeO), CaO, and MgO, all of which decrease eastward by small amounts. Na_2O is nearly constant, averaging about 3.4 percent across the batholith.

Minor element changes include those of U, Th, Rb, Be, and total rare earth elements all of which clearly increase eastward. Suprisingly, even though the light rare earth elements La and Ce dominate the total rare earth assemblage, the La/Yb ratio shows considerable scatter and no clearly discernible trend crossing the batholith.

In addition to these chemical changes, the initial $^{87}Sr/^{86}Sr$ increases eastward and a line of initial $^{87}Sr/^{86}Sr$ = 0.7060, trends N.30°W. about a third of the way eastward across the transect. Both initial $^{143}Nd/^{144}Nd$ and δD decrease from west to east; δ¹⁸O has a complex pattern, with no apparent systematic geographic variation (Masi et al., 1981).

COMPOSITIONAL VARIATIONS WITHIN INTRUSIVE SUITES

In addition to the chemical changes which take place across the batholith, considerable variation occurs within and between units of the granitoids. As lithodemes of individual intrusive suites generally become more felsic in time, silica variation provides a means of understanding evolutionary history of the granitoids. Silica contents range from less than 55 to nearly 80 weight percent, but average about 68 percent (Dodge, 1972). Hornblende is absent in rocks with more than about 72 weight percent SiO_2, and potassium feldspar occurs in most rocks with more than about 60 percent SiO_2. Commonly all the major element oxides, except those of the alkalis, but including H_2O, show pronounced tendencies to decrease with increasing SiO_2. Clearly K_2O increases with increasing SiO_2 content. The major element trends of individual intrusive suites show considerable overlap, but are somewhat displaced from one another according to geographic position; for example, eastern suites have higher K_2O but lower CaO values at specific SiO_2 contents than do western sequences. Iron oxidation shows no clear correlation with SiO_2.

Some trace elements also show systematic changes within intrusive suites that can be related to SiO_2 content. As with K_2O, Rb increases with SiO_2, however Cs and Ba, elements which also substitute for K, are both dispersed with no discernible trend relative to SiO_2, although some of the highest SiO_2 rocks have extremely low Ba contents. Magnitude of the Eu anomaly in rare earth element patterns and Sr show limited variation and no discernible trends at SiO_2 values below 72 percent, but they decrease systematically at higher SiO_2 values. On the other hand, Sc shows a decreasing trend with increasing SiO_2 to 70 to 72 percent, and flattens at higher SiO_2 values. The La/Yb ratio shows a rather ill-defined

increasing trend with SiO_2 to about 70 percent, then a general decrease.

Although overshadowed by large regional variations, small but significant differences in initial $^{87}Sr/^{86}Sr$ and initial $^{143}Nd/^{144}Nd$ values have been determined within the compositionally zoned Tuolumne Intrusive Suite (Kistler and others, 1986). The largest isotopic variations are found in the mafic outermost units of the suite and initial Sr values increase and initial Nd values decrease with increasing SiO_2.

MAGMA SOURCES AND DIFFERENTION PROCESSES

Intrusion of mantle-derived basaltic magmas into the base of a thickened Sierra Nevada lower crust caused anatectic melting of crustal materials leaving garnet-bearing residua. The primary basaltic magmas mixed to varying degrees with these melts generating parent granitic magmas. As these mixed magmas buoyantly rose into the upper crust during periods of crustal extension, they underwent incomplete homogenization and probably assimilated some wall rocks. Modification of the magmas began at deep levels by crystal-liquid fractionation, a process that continued in some cases after final magma emplacement.

The compositional changes that take place across the batholith may reflect a transition from oceanic to continental materials in the lower crust. As these changes are generally independent of age, it is unlikely that they are process related; trace element data, particularly La/Yb ratios, do not support laterally varying depths of magma generation as a cause of the changes.

Mixing processes were important during initial generation of the parental magmas, but the entire spectrum of granitoid rocks could not have been produced by simple intermingling between end-member mafic and felsic magmas. Although linear variaion diagrams of the major elements have been cited as evidence of the role of mixing in the evolution of the final variety of granitoid rocks of the batholith (Reid et al., 1983), trace element behavior indicates crystal fractionation processes were of upmost importance. The succession in time from mafic to felsic lithodemes in individual intrusive suites and the progressive decrease of Sc with increasing SiO_2 in rocks with less than 72 percent SiO_2 and the decrease of Sr accompanied by an increasingly negative Eu anomaly in rocks with greater than 72 percent SiO_2, indicates hornblende dominated fractionation of magmas at lower SiO_2 contents, whereas plagioclase feldspar became the controlling fractionate of higher SiO_2 magmas. The abnormally low Ba content of the most silicic leucocratic rocks indicates they have crystallized from true eutectic melts. The transition from hornblende to plagioclase fractionation may indeed be a consequence of crystallization depths with hornblende fractionation taking place at a range of crustal levels, but with plagioclase fractionation occurring at or near the final location of pluton emplacement.

ROAD LOG-- DAY 8, July 5, 1989-- RENO TO YOSEMITE

For several hours we will be traveling along the western margin of the Basin and Range province, paralleling the steep, faulted eastern escarpment of the Sierra Nevada. There are spectacular views of the range crest to the west. Most of the rocks seen from the road are either Mesozoic granitic and metavolcanic rocks or late Cenozoic volcanic flows and tuffs erupted prior to Basin and Range faulting. Tufa deposits are locally exposed in a few roadcuts.

STOP

8-1. Conway Summit Vista Point: This overlook provides a spectacular view of the entire Mono Basin. The water of Mono Lake, a small remnant of fresh-water Pleistocene Lake Russell,

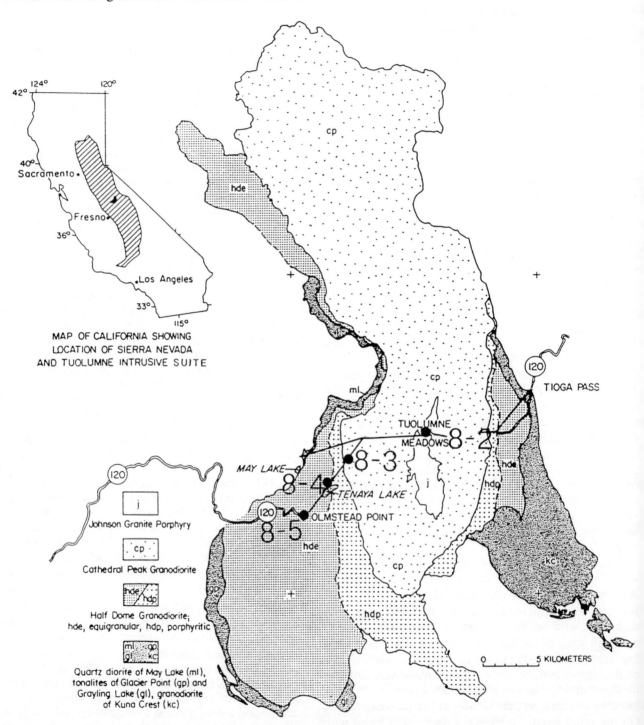

FIGURE 2 Generalized geological map of the Tuolumne Intrusive Suite showing location of field trip stops.

is so highly alkaline that only small brine shrimp now live in it. A terraced sublacustrine basaltic cinder cone forms Black Point, the low flat-topped buff mound on the near shore of the lake. The darker islands beyond Black Point are very young dacitic and rhyolitic domes and flows. Paoho, the largest, lighter-colored island, is a mass of lake sediment apparently uplifted by a shallow volcanic intrusion. Mono Craters, a 17 km long arcuate chain of rhyolite domes ranging in age from 550 to 35,000 years, lie beyond the south shore of the lake. Granitic peaks at the northern end of the White Mountains can be seen in the far distance.

In the cut on opposite side of the road from the overlook, till in a moraine of the Sherwin glaciation, one of the oldest glaciations recognized on the east side of the Sierra Nevada, rest on granitic rocks of Rattlesnake Gulch, dated at 92 m.y.

Yosemite National Park

Geologists of the U.S. Geological Survey began studies in Yosemite as early as the late 19th century, and their research has continued to the present. N. K. Huber has summarized the current geological knowledge of Yosemite National Park in a semi-technical book which is a supplement to the field trip guidebook.

STOP

8-2. Tuolumne Meadows (Lunch stop): The Late Cretaceous Tuolumne Intrusive Suite is an intensively studied (Bateman and Chappell, 1979; Bateman et al., 1983; Frey et al., 1978; Kistler et al., 1986) concentric texturally and compositionally zoned plutonic body that consists of a group of nested units progressively younger and more leucocratic inward (Figure 2). Low outcrops of the Johnson Granite Porphyry, the youngest, core-forming rock of the suite, can be seen in the meadows along the Tuolumne River, across the road from the store. The rock is very light-colored, with sparse, euhedral potassium feldspar phenocrysts set in a fine-grained matrix.

8-3 Pywiack Dome: The Cathedral Peak Granodiorite of the Tuolumne Intrusive Suite is well-exposed here and clearly displays its characteristic porphyritic texture, with potassium feldspar megacrysts commonly as long as 4 cm. The granodiorite has become progressively darker-colored and the megacrysts have become larger and more abundant than at Tuolumne Meadows. The contact between the Cathedral Peak Granodiorite and the Half Dome Granodiorite, also of the Tuolumne Intrusive Suite, is only a few hundred feet to the west. At most places, the contact is sharp and clearly intrusive, but it is gradational along the Tioga Road, with the large, blocky megacrysts of the Cathedral Peak Granodiorite giving way to smaller, tabular feldspar phenocrysts of the Half

Dome Granodiorite through a mixed zone 50 m wide.

8-4 Tenaya Lake: The potassium feldspar phenocrysts of the porphyritic facies of the Half Dome Granodiorite decrease in abundance and finally disappear in a km span along the north shore of Tenaya Lake. Many outcrops in this span are layered, grading from fine-grained and dark on one side to medium- to coarse-grained and light-colored on the other. The boundary surfaces between layers mark sharp discontinuities in the rate of magma flow.

8-5. Olmstead Point: The general appearance of the rock has changed little from Tenaya Lake, however some of the best exposures of the equigranular facies of the Half Dome Granodiorite surround this turnout. In addition to its equigranular texture, euhedral hornblende and biotite crystals typify the rock. A dark-colored rock of variable composition makes up the oldest, marginal western unit of the Tuolumne Intrusive Suite. It is not exposed along the road.

Glacial erratics of the last ice age are precariously balanced on the slopes above the overlook. The views, including the domes of the east end of Yosemite Valley to the south and the high peaks to the north, are spectacular from Olmstead Point.

ROAD LOG-- DAY 9, July 6, 1989-- YOSEMITE NATIONAL PARK

Locations of stops for this day's trip are plotted on generalized geologic map of the Yosemite Valley area (Figure 3).

STOP

9-1 Church Bowl: View of Glacier Point across Yosemite Valley. In the cliff face west of Glacier Point, note the contrast between unjointed Late Cretaceous Half Dome Granodiorite below and jointed granodiorite above. In the center of the valley south of here, as much as 600 m of glacio-lacustrine debris overlies the bedrock. In talus and in cliff faces on the north side of the valley at this stop, dark-colored Sentinel Granodiorite is cut by gently dipping light-colored dikes of coarse pegmatite and Half Dome Granodiorite at the margin of the large body of Half Dome Granodiorite. In some of the dikes, unequal concentration of dark minerals produces a nearly horizontal layering.

9-2 Bridalveil Fall and Bridalveil Moraine: From parking lot, walk along trail (about 300 meters) to viewpoint at base of Bridalveil Fall. Near viewpoint are fallen blocks of Cretaceous Leaning Tower Granite, Bridalveil Granodiorite, diorite, and El Capitan Granite. On the cliff face at the lip of Bridalveil Fall can be seen a thick horizontal sheet of smooth-weathering Bridalveil

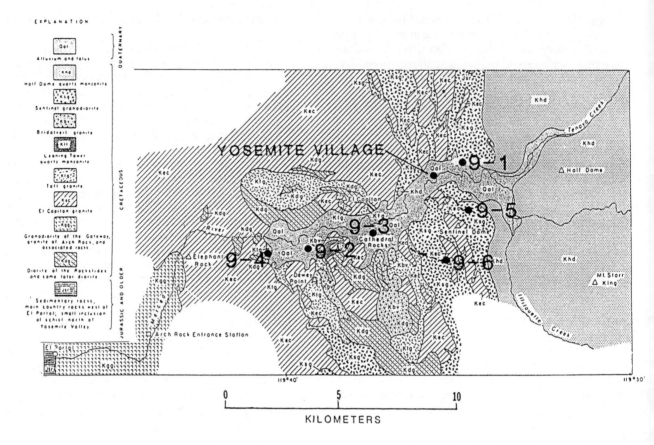

Within the image (map labels): EXPLANATION, QUATERNARY, Qal, Alluvium and talus, Khd, Half Dome quartz monzonite, Ksg, Sentinel granodiorite, Kbv, Bridalveil granite, Klt, Leaning Tower quartz monzonite, Kt, Taft granite, Kec, El Capitan granite, Kdg, Granodiorite of the Gateway, granite of Arch Rock, and associated rocks, Kdd, Diorite of the Rockslides and some later diorite, Jsr, Sedimentary rocks, main country rocks west of El Portal; small inclusion of schist north of Yosemite Valley, CRETACEOUS, JURASSIC AND OLDER, El Portal, YOSEMITE VILLAGE, Tenaya Creek, Half Dome, Merced River, Elephant Rock, Cathedral Rocks, Sentinel Dome, Dewey Point, Mt. Starr King, Illilouette Creek, Arch Rock Entrance Station, 9-1, 9-5, 9-3, 9-2, 9-4, 9-6, 119°40', 119°30'

0   5   10

KILOMETERS

FIGURE 3 Generalized geologic map of the Yosemite Valley area showing field trip stops. Map is modified from Calkins and Peck (1962).

Granitodiorite. Underneath is reddish rough-weathering Leaning Tower Granite, and to the east is dark diorite. Note that the lower part of Bridalveil Fall is in a slight recess flanked by buttresses. The borders of the recess are marked by the edges of slabs parallel to the surface; apparently the original continuations of these slabs across the face of the fall have dropped from the cliff and are represented by the cone of talus extending up the cliff to an apex just west of the base of the fall.

Return to trail junction and walk northwest (about 500 meters) to Bridalveil Moraine. The low morainal ridge is the westernmost of a series of early Wisconsin terminal moraines which held in the former Lake Yosemite. The lake was filled with as much as 100 m of silt and sand deposited in advancing deltas by the Merced River and Tenya Creek; this debris was probably supplied by glaciers of late Wisconsin, Tioga age, which did not reach as far downstream as Yosemite Valley. There are weathered boulders of Cathedral Peak Granodiorite, Half Dome Granodiorite, Sentinel Granodiorite (the dominant rock type), El Capitan Granite, and Bridalveil Granodiorite exposed in the moraine. The glacial erratics of Cathedral Peak Granodiorite, marked by distinctive large megacrysts of microcline, were transported at

9-3

9-4

least 20 km from the nearest outcrops to the east. View of diorite dikes on the face of El Capitan The irregular intrusive dikes in the southeast wall of El Capitan crudely resemble a map of North America. Recent workers (Reid et al., 1983, p. 244) have suggested that evidence of mixing of mafic and felsic magmas with some wall rock assimilation preserved in this wall indicate magma mixing produced the entire compositional variety of the granitoids of the Sierra Nevada batholith.

Parking lot at the east portal of Wawona Tunnel View to the east of El Capitan, Sentinel Rock Cathedral Rocks, the hanging valley of Bridalveil Creek, and Bridalveil Fall. Nearby, exposures of various granitic rocks and of diorite.

The abundance of joints in the diorite in the opposite valley wall (directly north of here) contrasts strongly with their scarcity in the massive cliffs of El Capitan and the Cathedral Rocks. The constriction in the valley between El Capitan and the Cathedral Rocks may be due to the massive nature of the granitic rocks at this point. The great abundance of talus in the cliffs directly north of here, in contrast to the paucity of talus farther up the valley, is due to the close jointing of the diorite of the cliffs.

The U-shape of Yosemite Valley, in contrast to the V-shape of the gorge of the Merced River below El Portal, is well displayed here. The bottom of the U, however, is much flatter than in typical glaciated valleys. According to seismic studies, the bedrock surface lies almost 300 m beneath the floor of the valley between El Capitan and Cathedral Rocks, and what we see is essentially a plain floored by lake sediments.

The top of the highest glacier in Yosemite Valley, according to Francois Matthes, reached about to the brow of El Capitan, and was about 100 m above the top of the Cathedral Rocks. The glacier swept around the flank of Sentinel Dome, but did not cover the dome. The upper 200 m of Half Dome, likewise, was unglaciated. These domes owe their form to concentric spalling of massive unjointed rock, not to glacial erosion. The steep lower course of Bridalveil Creek above Bridalveil Fall is graded to the level established by the Merced River during the most recent of three distinct stages of preglacial erosion (from oldest to youngest, the Broad Valley, Mountain Valley, and Canyon stages of Matthes); hence it helps to define the amount of glacial erosion in Yosemite Valley.

The V-shaped form of the gorge of the creek, although typical of steam erosion, is preserved because the sloping walls of the gorge coincide with throughgoing joints in the otherwise nearly unjointed rock. The upper part of Fireplace Creek, a little downstream from us on the opposite wall, is graded to the Mountain Valley stage of the Merced Canyon according to Matthes. Ribbon Creek, above the head of Ribbon Fall (which cannot be seen from here, but can be seen on the north wall of the canyon from places a kilometer or two down the road), is graded to the Broad Valley stage of the Merced according to Matthes.

The blasted rock face at the west end of the parking lot exposes a complicated mixture of diorite and El Capitan Granite. The porphyritic phase of the Taft Granite is well exposed on the slope just to the west. El Capitan Granite along the south side of the road contains blocks of partially assimilated diorite, and has a steeply dipping foliation. The El Capitan Granite has been dated at 103 m.y. by Pb-U techniques on zircon just to the west of here.

-5 Glacier Point: Walk from parking lot to the point.

View of upper Yosemite Valley, Tenya Canyon, Little Yosemite Valley, Vernal and Nevada Falls, and "the high country" to the east. The Half Dome Granodiorite-Sentinel Granodiorite contact can be seen across the valley, above Church Bowl, where we stopped this morning. Further down Yosemite Valley on the north wall in the vicinity of Yosemite Falls, large blocks of massive El Capitan Granite are included in jointed Sentinel Granodiorite. Up the valley, the form of the Royal Arches, Half Dome, and North Dome is controlled by

9-6

exfoliation of the Half Dome Granodiorite, resulting from expansion due to unloading brought about by denudation.

The rock at Glacier Point is Sentinel Granodiorite. We are east of any of the large inclusions of El Capitan Granite that can be seen across the valley. Exhaustive K-Ar dating of this eastern Sentinel Granodiorite and some fission track dating, led us to believe the rock was about 90 m.y. old. However, recent Rb-Sr work, although not definitive, suggests it may be considerably older.

This is our lunch stop. After lunch, those who wish, may leave the group and return to the valley floor by way of the Glacier Point Four Mile Trail. A Sentinel Granodiorite-El Capitan Granite contact is crossed near Union Point. Upon returning to the valley floor, you may wish to visit the visitor center and museum at Yosemite Village.

Junction of Taft Point Trail with Glacier Point Road: From here, we will walk to Taft Point, a distance of 1.8 km, and return.

In the cut on the south side of the road, a contact between the El Capitan Granite and the Sentinel Granodiorite is well exposed. It can be clearly demonstrated that the foliate granodiorite is the younger of the two rocks. This western Sentinel Granodiorite is undoubtedly Late Cretaceous. Age relations have not been seen between the El Capitan Granite and the eastern Sentinel Granodiorite.

A short distance along the trail, a small pegmatite pod, consisting mainly of striated quartz, crops out. A large interpenetrant twin K-feldspar crystal is present on the north side of the pod.

From here, the trail passes through roughly north-south trending, vertical dipping, foliate Sentinel Granodiorite. The foliation is particularly manifest in the orientation of elongate mafic inclusions. The Taft Granite is crossed into at the brink of the dropoff to Taft Point. Although the actual contact between the Taft Granite and Sentinel Granodiorite cannot be directly observed here, it can be located to within a few feet. The Taft Granite is probably simply a leucocratic, and here a fine-grained, phase of the El Capitan Granite. Taft Granite from near here has been dated at 96 m.y. by Pb-U techniques on zircon. The immediate area around Taft Point is underlain by a complex assemblage of Taft and El Capitan Granites and diorite. The fissures, which are developed in both granites, follow a system of northeasterly-trending joints which have been opened by solutions. This joint system controls the form of Sentinel Rock, Glacier Point, and the north face of Half Dome.

Taft Point affords one of the most spectacular views of the middle portion of Yosemite Valley. Across the valley, on the face of El Capitan, the irregular boundaries of the intrusive diorite of North America are clearly recognizable. Further east, the form of the Three Brothers can be seen to be controlled by joints dipping obliquely westward.

ROAD LOG-- DAY 10, July 7, 1989-- YOSEMITE TO SAN FRANCISCO

STOP

10-1   Rush Creek: The tonalite of Granite Creek has been dated at 163 m.y. at this locality. The presence of granitoids of this age on the west side of the Sierra Nevada and of contemporaneous granitoids on the east side indicates that Jurassic intrusions followed a general N. 40° W. trend in this region, whereas Cretaceous intrusions, which probably displaced the previously more widespread Jurassic granitoids in the central part of the batholith, have a general N.20°W. trend, parallel to the present-day Sierran crest. The tonalite is invariably strongly foliated and in some samples original igneous crystal shapes are lost, grain size is reduced, compositional zoning of mineral grains is obliterated, and larger crystals are bordered by granoblastic mortar, indicating the tonalite has been subjected to intense post-magmatic deformation. This deformation is attributed to the widely recognized Nevadan orogeny.

10-2   Stevens Bar: Harriman mine below water level. A large allochthonous limestone block in phyllite is exposed in the roadcut, north of the Stent Bridge. These highly deformed rocks have been interpreted as part of a melange occurring sporadically along the Melones fault zone. Elsewhere, similar limestone blocks have yielded poorly preserved Permian fossils.

10-3   Jamestown Mine: Although almost totally dormant for the last four decades, lode gold mining has been revived in the Sierra Nevada, one of the principal historic gold mining regions of the world. Sonora Mining Corporation began mining here on the site of the old Harvard mine in 1986. This deposit along with three others the Corporation plans to develop form the core of the Jamestown mining district, and account for about one-half of the $24 million recorded past gold production of the district or probably over 500,000 troy ounces; present production is about 60,000 ounces a year. The ore bodies are located along the Melones fault zone, which is locally characterized by the presence of mariposite-bearing carbonate rock and talc schist cut by numerous stringers and veins of gold-bearing quartz. The vein systems were extensively mined prior to World War II; however, economic interest is now focused on finely disseminated gold and electrum in mineralized country rocks of the hanging wall zone of the quartz vein system. A joint program of alteration and mineral paragenesis study is currently being carried on by members of the U.S. Geological Survey and staff of the Sonora Mining Corporation.

REFERENCES CITED

Bateman, P.C., Constitution and genesis of the central part of the Sierra Nevada batholith, California, *U.S Geol. Surv. Prof. Pap., 1483,* in press.

Bateman, P.C., and Chappell, B.W., Crystallization fractionation, and solidification of the Tuolumne Intrusive Series, Yosemite National Park, California *Geol. Soc. Am. Bull., pt., 1, 90,* 465-482, 1979.

Bateman, P.C., and Dodge, F.C.W., Variations of major chemical constituents across the central Sierra Nevada batholith, *Geol. Soc. Am. Bull., 81,* 409-420 1970.

Bateman, P.C., Kistler, R.W., Peck, D.L., and Busacca, A.J., Geologic map of the Tuolumne Meadows quadrangle, Yosemite National Park California, *U.S. Geol. Surv. Geol. Quad. map, GQ 1570,* scale 1:62,500, 1983.

Calkins, F.C., and Peck, D.L., Granitic rocks of the Yosemite Valley area, California, *Calif. Div. Mine. and Geol. Bull., 182,* 17-24, 1962.

Chappell, B.W., and White, A.J.R., Two contrasting granite types, *Pacific Geol., 8,* 173-174, 1974.

Chen, J.H., and Moore, J.G., Uranium-lead isotopic ages from the Sierra Nevada batholith, California, *J Geo. Res., 87,* B6, 4761-4784, 1982.

Dodge, F.C.W., Trace-element contents of some plutonic rocks of the Sierra Nevada batholith, *U.S Geol. Surv. Bull., 1314-F,* F1-F13, 1972.

Dodge, F.C.W., Millard, H.T., Jr., and Elsheimer H.N., Compositional variations and abundances of selected elements in granitoid rocks and constituer minerals, central Sierra Nevada batholith, California *U.S. Geol. Surv. Prof. Pap., 1248,* 24 pp., 1982.

Evernden, J.F., and Kistler, R.W., Chronology of emplacement of Mesozoic batholithic complexes in California and western Nevada, *U.S. Geol. Surv Prof. Pap., 623,* 42 pp., 1970.

Frey, F.A., Chappell, B.W., and Roy, S.D Fractionation of rare-earth elements in the Tuolumn Intrusive Series, Sierra Nevada batholith, California *Geol., 6,* 239-242, 1978.

Kistler, R.W., Chappell, B.W., Peck, D.L., an Bateman, P.C., Isotopic variations in the Tuolumn Intrusive Suite, central Sierra Nevada, Californi *Contrib. to Mineral. Petrol., 94,* 205-220, 1986.

Kistler, R.W., and Peterman, Z.E., Variations in S Rb, K, Na, and initial $^{87}Sr/^{86}Sr$ in Mesozoic graniti rocks and intruded wall rocks in central Californi *Geol. Soc. Am. Bull., 84,* 3489-3512, 1973.

Masi, Umberto, O'Neil, J.R., and Kistler, R.W., Stab isotope systematics in Mesozoic granites of central ar northern California and southwestern Oregon, *Contri to Mineral. Petrol., 76,* 116-126, 1981.

Reid, J.B., Jr., Evans, O.C., and Fates, D.G., Magm mixing in granitic rocks of the central Sierra Nevad California, *Earth and Planet. Sci. Lett., 66,* 243-26 1983.

Stern, T.W., Bateman, P.C., Morgan, B.A., Newe

M.F., and Peck, D.L., Isotopic U-Pb ages of zircon from the granitoids of the central Sierra Nevada, *U.S. Geol. Surv. Prof. Pap., 1185,* 17 pp., 1981.

White, A.J.R., Clemens, J.D., Holloway, J.R., Silver, L.T., Chappell, B.W., and Wall, V.J., S-type granites and their probable absence in southwestern North America, *Geol., 14,* 115-118, 1986.

Early Mesozoic Tectonics of the Western Great Basin, Nevada

Battle Mountain to Yerington District, Nevada
July 1–7, 1989

Field Trip Guidebook T122

Leaders:
R. C. Speed and N. J. Silberling

American Geophysical Union, Washington, D.C.

COVER Pleasant Valley Fault Scarp along western flank of the Tobin Range, Nevada.

IGC FIELD TRIP T122:
EARLY MESOZOIC TECTONICS OF THE WESTERN GREAT BASIN, NEVADA

R. C. Speed[1], N. J. Silberling[2], M. W Elison[1,3], K. M. Nichols[2,3], and W. S. Snyder[3,4]

INTRODUCTION

This excursion will present an overview of the stratigraphy, structure, and tectonic evolution of the margin of early Mesozoic North America in the Great Basin and of terranes tectonically accreted above and against this margin. Exposures to be visited will show the following features: 1) the Golconda allochthon, a tract of oceanic Paleozoic rocks that was thrust some 100 km across the edge of Triassic North America; 2) shelfal and platformal lower Mesozoic strata that unconformably cover the Golconda allochthon; 3) basinal lower Mesozoic strata that accumulated seaward of the inherited continental edge; 4) Mesozoic and Paleozoic arc volcanics; and 5) Jurassic and Cretaceous foreland structures developed in all the preceding rocks after accretion of terranes had left the early Mesozoic sialic margin well inboard of the active convergent margin.

This guide begins with introductions to the neotectonic and Phanerozoic palcotectonic evolutions of the Great Basin. Thereafter, its contents are organized in sequence with field trip stops which are as follows [route map in Figs. 5 and 6]:

Day 1. Battle Mt. Golconda allochthon.

Day 2. Tobin, Stillwater, and southern Humboldt Ranges. Stratigraphy of platform to basin transition of Triassic cover; Jurassic foreland thrusting; 1915 Pleasant Valley earthquake scarp.

Day 3. Sand Springs and Paradise Ranges. Terranes of volcanogenic Triassic and Paleozoic rocks structurally outboard of parautochthonous lower Mesozoic basinal and shelf edge rocks.

Days 4 and 5. Mina and Candelaria. North-south traverse across structurally condensed Early Triassic collision zone of exotic Paleozoic arc, Golconda allochthon plus serpentinite melange, Golconda foreland basin strata on Early Triassic North America; these units were covered by Middle and Upper Triassic strata and imbricated with the cover in late Mesozoic foreland thrusting.

Day 6. Yerington. Stratigraphy and structure of possibly oldest continental magmatic arc rocks in western Great Basin.

[1]Department of Geological Sciences, Northwestern University, Evanston, Illinois 60208.
[2]U.S. Geological Survey, Denver Federal Center, Đenver, Colorado 80225.
[3]Associate Leader.
[4]Department of Geology, Boise State University, Boise, Idaho 83725.

NEOTECTONICS

Our field trip is in the Great Basin which is a region of closed drainage equivalent to the Basin-Range morphotectonic province in the state of Nevada (Figs. 1A, 1B). The Basin-Range province is identified by normal faulting and block fault topography (Hamilton and Myers, 1966; Stewart, 1971, 1978; Smith, 1978; Zoback and others, 1981; Anderson and others, 1983). The northern Basin-Range is active tectonically, as indicated by high seismicity (Fig. 1C) and abundant fault scarps that are historic or Quaternary (Fig. 1D). The southern half, approximately south of Las Vegas (Fig. 1C) on the other hand, is generally inactive. The province may be viewed as an intercontinental rift zone of extraordinary width. Extension began widely in the Oligocene and continues to the present in the northern half.

Relative to Cenozoic tectonics at the western edge of the North American plate, the development of the Basin-Range can be traced in three stages: 1) 40-25 ma, the province was intra-arc and above a flat slab of the subducting Farallon plate (Lipman and others, 1972; Engebretson and others, 1985); 2) 25-10 ma, it was in back of a narrow, partly extinct southern Cascades magmatic arc, following the detachment or sinking of the Farallon slab (Christiansen and Lipman, 1972; Snyder and others, 1976; Eaton, 1982); and 3) 10-0 mybp, it has been landward of the San Andreas transform system in California and may have taken up part of the strike slip displacement between the North American and Pacific plates (Atwater, 1970; Minster and Jordan, 1984). Basin-Range extension in the first two stages may have been inspired by the rise and fall of a flat slab of the Farallon plate (Coney and Reynolds, 1977). The dynamics of extension in the last 10 my are not certain.

The extensional style of the western Basin-Range is almost entirely thick-skinned brittle; thin-skinned brittle style occurs at only a couple of places, and the ductile style seems not to exist at the surface. Within northern Nevada in the western half, faults that are range-bounding and those internal to ranges are high angle, and those still active apparently continue steeply to maximum seismogenic depths (15 km) (Ryall and Preistly, 1975; Okaya and Thompson, 1985; Anderson and others, 1983). These are mainly N to NE-striking and record EW extension that may have been supplanted by NW-SE extension in the last 10 my (Zoback and others, 1981). A local zone, the Walker Lane (Fig. 1D), which lies along and parallel to the California-Nevada border, also includes NW-striking right slip and E-striking left slip faults. It is unclear whether the Walker Lane takes up irrotational deformation with EW extension on conjugate shears or whether there is net simple shear on one of the strike slip sets.

Thin-skinned brittle extension of Cenozoic age is

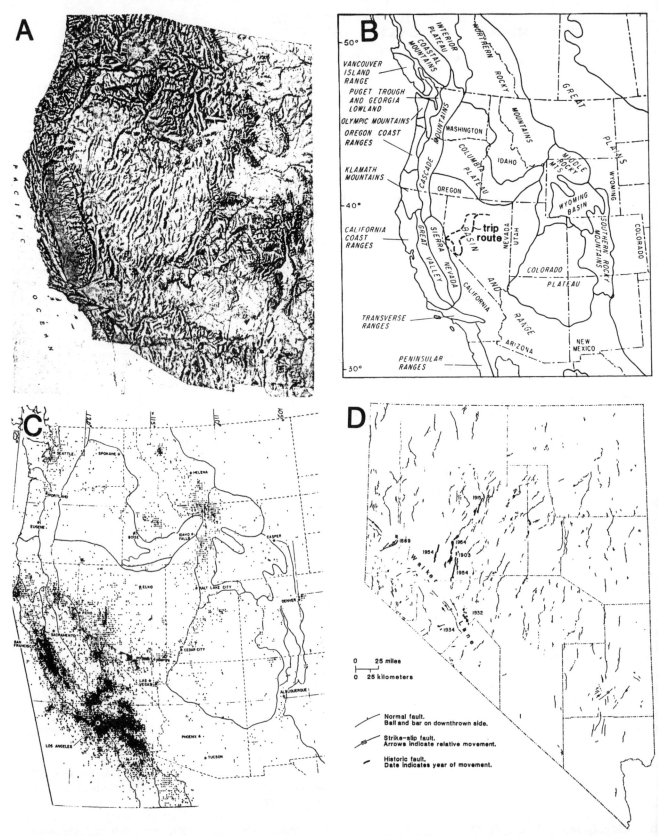

FIGURE 1 Neotectonic provinces and seismicity of the western United States. A) shaded relief map; B) mor-photectonic provinces showing trip route in northern Basin and Range province; C) epicenters 1950-1977, magnitudes ≥3 (Smith, 1978); dashed line in southern Nevada marks easternmost extent of pure strike slip events; and D) active and Quaternary fault traces in Nevada and epicenters of historic large shocks (M≥6) (Stewart, 1980).

demonstrated at Yerington (Proffett, 1977) and north of Austin (Smith, 1984). At other places in the western half, detachment faults that penetrate the basement to Tertiary cover have not been recognized at the surface or in seismic sections (Fig. 2) (Anderson and others, 1983; Hauge and others, 1987).

The western Basin-Range has undergone meager Cenozoic unroofing compared to the eastern, implying substantially less extension than the eastern. A maximum of 20% elongation is estimated for the western half (Speed and others, 1988), a value that corresponds to extension only by the thick-skinned style. Thus, the Basin-Range as whole appears to have undergone heterogeneous extension with a maximum in eastern Nevada along the zone of ductile metamorphic core complexes.

SEISMICITY

Fault block structure and widespread fault scarps (Fig. 1D) in the northern Basin-Range province indicate the entire province has probably been seismically active over Quaternary time. Large historic shocks, however, are concentrated in a NS belt in western Nevada, generally along the field trip route (Figs. 1B, 5). The implication is that displacements are concentrated in transient zones that skip around the northern Basin-Range, perhaps over periods of $>10^4$ yr.

The 1915 Pleasant Valley and 1954 Dixie Valley set of earthquakes (Fig. 1D) are the largest historic shocks, with magnitudes 7-7.5. All created large fault scarps with throw up to 7 m. The Dixie Valley shocks gave oblique-normal mechanisms and depths of 12-15 km.

CRUSTAL STRUCTURE

Deep seismic reflection profiling in the western Great Basin (Fig. 2) indicates a crustal thickness of 29-32 km above a smoothly undulating base of reflections taken to be the Moho (Klemperer and others, 1986; Hauge and others, 1987). The reflection Moho generally but not identically corresponds to the Moho determined by refraction above the layer with P velocities near 8 km/sec. The crust of the western Great Basin is noteworthy because 1) there are no thickness variations that correspond to pre-Cenozoic structures—the passive margin, sialic edge, and accreted terranes, and 2) the Moho has no offsets that relate to Basin-Range structures, whose throws are as great as 10 km. An explanation for such phenomena is that the Moho, upper mantle, and perhaps much of the lower crust are of Neogene age (Klemperer and others, 1986). The Moho is plausibly a differentiation surface across which basalt passes up into the lower crust and ultramafic residues stay below. The upper mantle of the western Great Basin is independently known to be anomalously hot by regional heat flow (80 m W/m^2) (Lachenbruch and Sass, 1978), thin lithosphere (45-65 km) (Priestly and Brune, 1978), and low refractor speeds (7.6-8.0 km/sec) (Eaton, 1963).

Within the crust, a zone of upper crustal reflectors from the surface to 5-10 km depth can easily be related to Basin-Range horsts and graben with low velocity fill (Fig.

2). Faults of the upper crust of the western Great Basin mainly extend planarly to seismogenic depths, unlike those of the eastern Great Basin which bottom in regional flat detachments. It is not clear whether any pre-Basin-Range structures are reflectors, although Figure 2 suggests that the Mesozoic Fencemaker and Golconda thrusts and the Paleozoic Roberts Mountains thrust may be resolved. The lower crust has numerous segmented dipping or flat reflector sets below an incoherent zone at mid-crustal levels. The lower crustal reflectors may be due to sills, magma bodies, and/or mylonite bodies.

PHANEROZOIC TECTONIC EVOLUTION

The Phanerozoic evolution of the North American continent and adjacent oceans in what is now Nevada was governed by three sequential tectonic regimes (Fig. 3A). The first, beginning in Late Proterozoic and continuing to Middle Cambrian time, created a passive margin to western sialic North America (Stewart, 1976). It caused the rifting and drifting away of an unknown portion of the continent and the growth of oceanic lithosphere against the new sialic edge. The second regime maintained a passive continental margin of western North America from Middle Cambrian to Middle or Late Triassic time but permitted collisions of outboard terranes with the sialic margin in Mississippi and Permian and Triassic times (Speed, 1983). Since late in Triassic time, western North America has existed in a regime of active margin tectonics (Hamilton, 1969).

Although similarly eventful histories probably occurred along the entire western margin of North America, a record of pre-Jurassic events is best preserved in Nevada, and in fact, many elements of the record are known only in Nevada. The pre-Jurassic margin of North America in Nevada evidently escaped strong tectonic erosion which elsewhere removed and rafted away sizeable fragments of the sialic continent and early accreted terraces (Speed, 1983). Thus, Nevada provides an almost unique glimpse into the past of marginal western North America.

Sialic Edge

The present edge of contiguous sialic Precambrian North America in Nevada is a basement feature and is cryptic due to burial by younger rocks and nappes or transformation by magmatism and metamorphism. The margin's surface trace (Fig 3B) is estimated by outermost outcrops of autochthonous continental platform or shelf facies, by ratios of initial Sr and Pb isotopes and mineralogy of autochthonous Phanerozoic magmatic rocks (Kistler and Peterman, 1973; Doe, 1973; Armstrong and others, 1977; Zartman, 1974; Miller and Bradfish, 1980). The edge of the continental basement in Nevada was probably formed at a Late Proterozoic passive margin. This was interpreted from autochthonous and parautochthonous upper Precambrian and lower Paleozoic shelf facies and basaltic rocks that crop out between the platform-shelf hinge (Fig. 3B) in Utah and southern Nevada and the sialic edge by Stewart (1972, 1976). West of the hinge, Paleozoic North America was a subsiding shelf, probably above

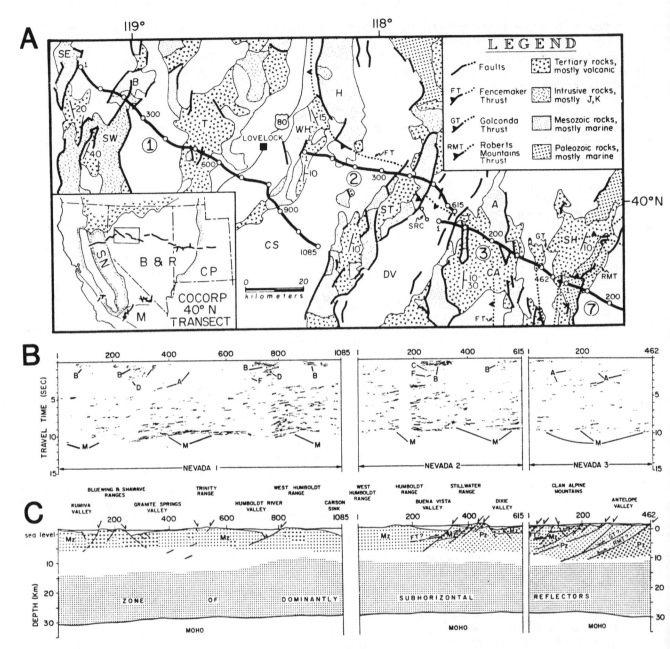

FIGURE 2 Crustal structure of northwestern Basin-Range by deep seismic profiling (Hauge and others, 1987). A) shotpoint (Vibroseis) tracks; SE, Selenite Range; A, Augusta Mountains; CA, Clan Alpine Mountains; SH, Shoshone Mountains; B) line drawings of immigrated stacks; M interpreted to be Moho; other lettered reflections are shallow crustal; geologic interpretation of B; fine stipple = Cenozoic basins, Pz = Paleozoic, Mz = Mesozoic, + = intrusions, small dots = eugeosynclinal strata; and large dots = miogeoclinal strata; Cenozoic intrusions are not shown. Shading in the lower crust denotes source of largely subhorizontal reflectors.

stretched diked continental crust.

Passive Margin Collisions

The Antler and Sonoma orogenies (Fig. 3A) (Silberling and Roberts, 1962) that occurred at the North American passive margin caused the emplacement of two regional terranes above the continental slope and outer shelf. These are the Roberts Mountains and Golconda allochthons (Figs. 3C-E). West of these allochthons, other terranes attached to North America in Mesozoic time, perhaps to

the toe of slope. The field trip leaders interpret these differently, as shown in Figures 3C and 3D, and discussed below.

The Roberts Mountains allochthon consists of a tectonic assemblage of pelagic, hemipelagic, turbiditic, and volcanic rocks of early Paleozoic age and probable oceanic derivation. It laps tectonically over lower Paleozoic strata of the North American continental shelf at least 130 km from the sialic edge and was almost certainly emplaced from the west early in Mississippian time (Roberts and others, 1958; Smith and Ketner, 1968; Stewart and Poole,

1974; Poole, 1974; Speed and Sleep, 1982; Dickinson and others, 1983). The Golconda allochthon possesses similar rocks and architecture to the Roberts Mountains allochthon except that the rocks are of Mississippian to Permian age (Silberling, 1973, 1975; Stewart and others, 1977; Speed, 1977, 1979; Miller and others, 1984; Stewart and others, 1986; Brueckner and Snyder, 1985). It was transported in Late Permian and/or in Early Triassic time at least 100 km inboard from the sialic edge and above the earlier Roberts Mountains allochthon and its upper Paleozoic and Lower Triassic cover (Fig. 3E) (Speed, 1979).

The outboard terranes are interpreted differently as follows: Silberling argues (Silberling and others, 1987) that the post-Sonoman (post-Early Triassic) cover in western Nevada varies discretely and allows identification of a number of terranes (Fig. 3C) that may have arrived by large and varied displacements to their present positions in mid-Mesozoic time. The Sonoma-age basements in such terranes, all poorly exposed and poorly understood, may have arisen at many different sites without relation to one another. In contrast, Speed argues (Speed, 1977, 1979) that a single terrane, Sonomia (Fig. 3D), underlies lower Mesozoic strata of western Nevada; it collided with North America in Early Triassic time and has since been parautochthonous.

Speed argues further that Sonomia is a lithospheric fragment of sequential Paleozoic arc-related tectonostratigraphic units, surmounted by a Permian magmatic arc. It collided with the edge of sialic North America early in the Triassic. Its main exposures are at the microplate margins where late Mesozoic deformation has transported Sonomia's rocks to the surface by imbricate thrusting. The central regions of Sonomia are deeply buried below thick Triassic flysch of the basinal sequence and continental arc volcanics that succeeded Sonomia (Speed, 1978a).

The emplacements of both the Golconda and Roberts Mountains allochthons had similar manifestations: transport from an oceanic region as a predeformed tectonic mass, absence of related magmatism and metamorphism within the continent, and lack of pervasive crustal shortening or mountain-building within the continental crust. Deformation within the overridden continental shelf strata consists only of local shear strain and/or thrust imbrication in a thin zone below the allochthon. These accretionary terranes resulted from collisions of the passive continental margin with migrating continent-facing arc systems (Speed, 1977, 1979, 1983; Speed and Sleep, 1982). Figure 3F shows a possible sequence of events according to Speed's scheme. Both major allochthons arrived as predeformed accretionary prisms to magmatic arcs, the Golconda to Sonomia, and the Roberts Mountains to the now-cryptic Antleria. The arcs surmounted subduction zones in which the downgoing slab was noncontinental lithosphere attached to the passive margin of North America. Closure ceased when continental lithosphere started down below the magmatic arc by which time the forearc, the allochthon, was almost fully emplaced on the outer continental shelf. Subsidence of the magmatic arcs upon welding to North America is explained by the development of new oceanfacing convergent zones to the west of the arc and thermal contraction due to loss of subduction-related heating. Deepwater sedimentary basins were successors to the subsided arcs.

A major effect of the Mississippian collision was the generation of an asymmetric foreland basin with amplitude of about 3.5 km that rimmed the continentward edge of the Roberts Mountain allochthon (Poole, 1974). In contrast to the extensive foreland basin developed during the Mississippian continental margin event, foreland basin deposits associated with the Permian and Triassic Golconda allochthon are evident only adjacent to the southern third of that allochthon (Fig. 3D). This may be because the northern Golconda allochthon was too small to cause significant flexure, because it remained submarine and provided no orogenic sediment to such as basin, or because it was later thrust over or laterally translated from related foreland basin strata.

Active Margin Events

An active margin developed on western North America (Hamilton, 1969) in Middle or Late Triassic time. The subduction trace between an east-dipping oceanic slab and the morphologic continent then existed at or west of the Foothills suture (Saleeby and Sharp, 1980; Schweickert, 1981) in California (Figs. 3C,D) such that at least part of Sonomia and perhaps other early accreted terranes were incorporated into Triassic North America.

During the passive-to-active margin transition in Nevada in the Triassic, a marine basin developed above the earlier accreted terranes and the outer 100-200 km of Precambrian North America (Fig. 4A). Lower Mesozoic strata are of four main facies (Fig. 4A): 1) shelfal carbonate and siliciclastic, 2) basinal carbonate and siliciclastic, 3) basinal volcanogenic sediments and carbonate, and 4) arc-edifice volcanogenic and associated rocks. The shelfal sequences were deposited along a subsiding platform margin that is approximately coincident with the edge of Precambrian North America (Fig. 3E), indicating perpetuation of the continent's freeboard in spite of passive margin collisions. The basinal sequences accumulated in mainly deeper water troughs between the shelf edge and the continental arc which lay to the west (Fig. 4A). The basin(s) are thought to have subsided by backarc spreading (Silberling) or by thermal contraction of Sonomia (Fig. 3F) (Speed). The arc-edifice facies represents varied environments above and on the continentward flank of the magmatic arc.

Jurassic and Cretaceous Deformation

Two deformational regimes greatly affected the western Great Basin in mid and late Mesozoic time: 1) foreland thrusting and folding, and 2) the Mina deflection and related intra-arc deformation (Fig. 4B).

Foreland Deformation

The region east of the continental arc to and including the Sevier thrust belt in Utah and southern Nevada (Fig. 4B) was the foreland of western North America that underwent mainly contractile deformation from Jurassic to Paleocene time. Within the Great Basin, foreland deformation was heterogeneous in surface distribution and

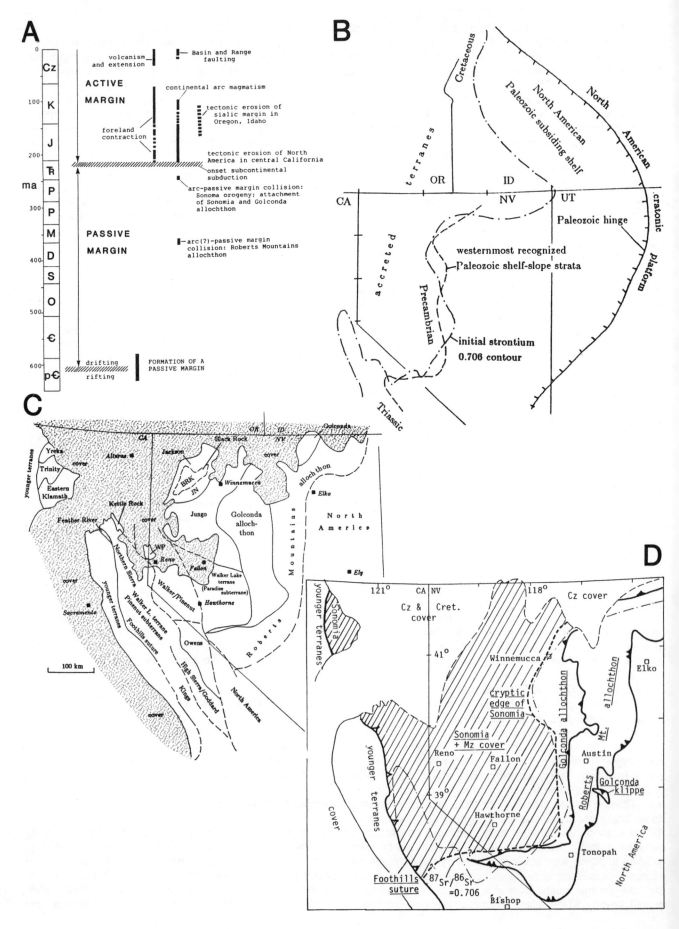

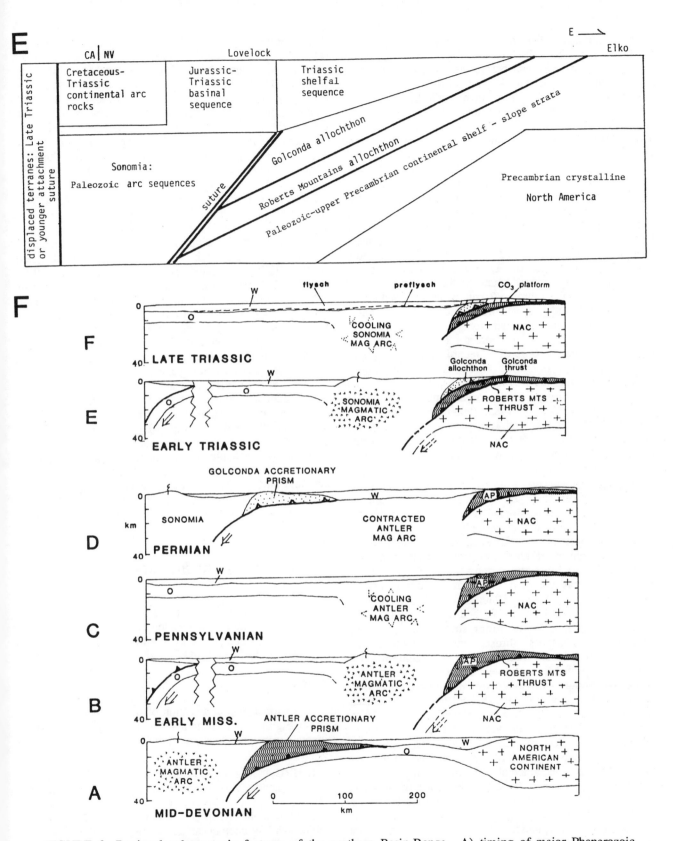

FIGURE 3 Regional paleotectonic features of the northern Basin-Range. A) timing of major Phanerozoic events; B) positions of late Precambrian-Paleozoic passive margin features: craton-shelf hinge, approximate position and age of edge of sialic continent (dash-dot line), and region of accreted terranes; C) division of accreted terranes by N. J. Silberling; D) division of accreted terranes by R. C. Speed; E) tectonostratigraphic stacking in a diagrammatic EW section related to scheme of Fig. 3D; F) model of passive margin collisions to explain emplacement of Roberts Mts and Golconda allochthons (Speed, 1983).

depth, apparently unlike that in its northerly prolongation, the Idaho-Wyoming belt (Fig. 4B). The Paleozoic-Triassic strata that were laid down before the onset of active margin tectonism record thin-skinned tectonics in a braided pattern (Speed, 1983). The three main thrust belts in which horizontal displacements ramped to the surface are the Sevier belt (Armstrong, 1968, 1972; Royse and others, 1975; Allmendinger and Jordan, 1981; Allmendinger and others, 1985), the Eureka belt (Speed, 1983), and the Winnemucca belt (Speed, 1978b, 1983; Oldow, 1981a, 1984a; Speed and others, 1988; Elison and Speed, 1988, 1989). Between the Sevier and Eureka belts, there is a tectonic enclave of little shallow deformation, and between the Eureka and Winnemucca belts (Fig. 4B) occurs a possible belt of moderate unroofing and exposure of formerly deep-seated, metamorphosed rocks, the Toiyabe uplift zone. Both the Sevier and Eureka thrust belts (Fig. 4B) are zones of displacement climb to the free surface with maximum recognized throw ≤7 km, syntectonic sedimentation, eastward vergence, and little evident involvement of crystalline basement or metamorphism reflecting emplacement of nappes from deep sources. The Sevier belt and regions immediately west of it record cratonward piggyback propagation of thrust sheets from Late Jurassic through Paleocene times with a total displacement of about 100 km (Armstrong, 1968; Royse and others, 1975; Allmendinger and others, 1985; Heller and others, 1986). The Eureka belt is less well known as to timing, propagation, and total displacement but includes Lower and Upper Cretaceous(?) sediments (Newark Canyon Group) that are probably syntectonic.

The tectonic enclave (Fig. 4B) is defined by an upper interval of cover strata of variable thickness from the Triassic down to Mississippian or deeper that underwent little or no deformation or stripping in Mesozoic time. Below this upper interval, however, Mesozoic deformation and metamorphism are known in deeper cover strata (Fig. 1) (mainly upper Precambrian and Cambrian) that are now exposed in areas unroofed in Cenozoic core complexes (Armstrong, 1972; Snoke, 1980; Miller, 1982; Allmendinger and others, 1985; Snoke and Miller, 1988). The Mesozoic structures in the lower stratal interval include recumbent fold nappes, thrusts, and low-angle normal faults. Such relations indicate that Mesozoic displacements were probably continuous between the Sevier and Eureka belts on an extensive detachment through the lower interval of the tectonic enclave (Speed, 1983). The detachment was a thrust flat whereas the two thrust belts were ramp zones. The upper interval in the enclave was thus a coherent thrust sheet of unusual dimensions. The Winnemucca deformation belt (Fig. 4B) is flanked on the west by the Mesozoic continental arc and on the east with poorly defined boundary by a possible zone of Mesozoic thermal doming and unroofing that we describe later. The Winnemucca belt probably lies astride the edge of Proterozoic sialic North America. Further description of the Winnemucca belt occurs later.

Mina Deflection

The predominant strike of most pre-Tertiary structures in western Nevada undergoes a 90° change from northerly east of about 117.5°W to westerly, west of that longitude (Fig. 4B). Strikes resume a general northerly trend farther west in the Sierra Nevada and White Mountains near Bishop. Thus, strike patterns form a regional Z-shaped kink with axial traces approximately N30W. The intermediate leg of the megakink which includes the Mina-Candelaria region has been called the Mina deflection (Wetterauer, 1977; Speed, 1979). The axial traces extend southeast and perhaps, northwest, for several hundred kilometers (Albers, 1967; Wright and Troxel, 1967; Stewart and others, 1968).

Origins conceived for the Mina deflection (Fig. 4C) are 1) an ancient (Precambrian?) bight in the edge of continental North America (Ferguson and Muller, 1949; Oldow, 1984b) that formed a sloping buttress against which accreting and covering rock masses subsequently flattened; Kistler (1978) postulated the bight was an early rift-rift transform; 2) an oroclinal bend created by right-slip faulting of unspecific kinematics but large displacement (Albers, 1967; Stewart and others, 1968) or by a right-slip intra-arc shear zone (Speed, 1978b); 3) a regional upright fold set imposed on initially west-dipping tectonostratigraphic sequences (Wetterauer, 1977); and 4) offset by EW-striking dextral faulting ± extension (Stewart, 1985).

Each idea is currently viable. Whatever its origin, the Mina deflection attained its present configuration by Late Cretaceous time because the eastern edge of the Late Cretaceous continental arc (Fig. 4B) is unaffected by the deflection and because preliminary paleomagnetic data from several such plutons and their wallrocks indicate no significant rotation with respect to North America (Geissman and others, 1982). If the deflection is a shear zone or megafold, it began during or after Early Jurassic time, the age of youngest rocks whose deformation strongly follows the deflection trends.

THE WINNEMUCCA REGION: STRATIGRAPHY AND STRUCTURE

This section presents the regional geology of the area between Winnemucca and Fallon (Figs. 5 and 6) for context on Days 1 and 2. Following this, information specific to each stop on Days 1 and 2 is given.

General

A regional map and section of the Winnemucca region are in Figure 7. The main features are: 1) edge of continental crust and suprajacent Paleozoic (Pz) shelf strata

FIGURE 4 Regional Mesozoic features of the northern Basin-Range. A) approximate distribution of lower Mesozoic facies (Speed, 1983). B) late Mesozoic deformation belts (Speed, Elison, and Heck, 1988) R, Reno; W, Winnemucca; A, Austin; El, Elko; E, Ely; and T, Tonopah. C) alternative origins of the Mina deflection in west-central Basin-Range (Stewart, 1985); axial trace of Mina deflection shown in Figure 4B; line with crosshatch is trend of sedimentary facies change; heavy lines are faults, arrows show relative movement; open arrows show motion of large blocks.

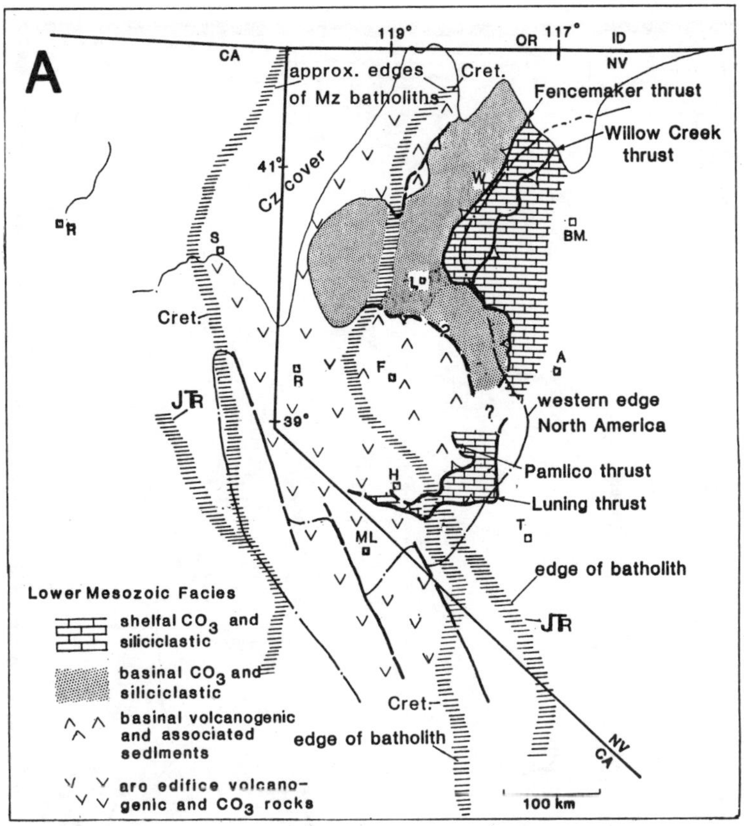

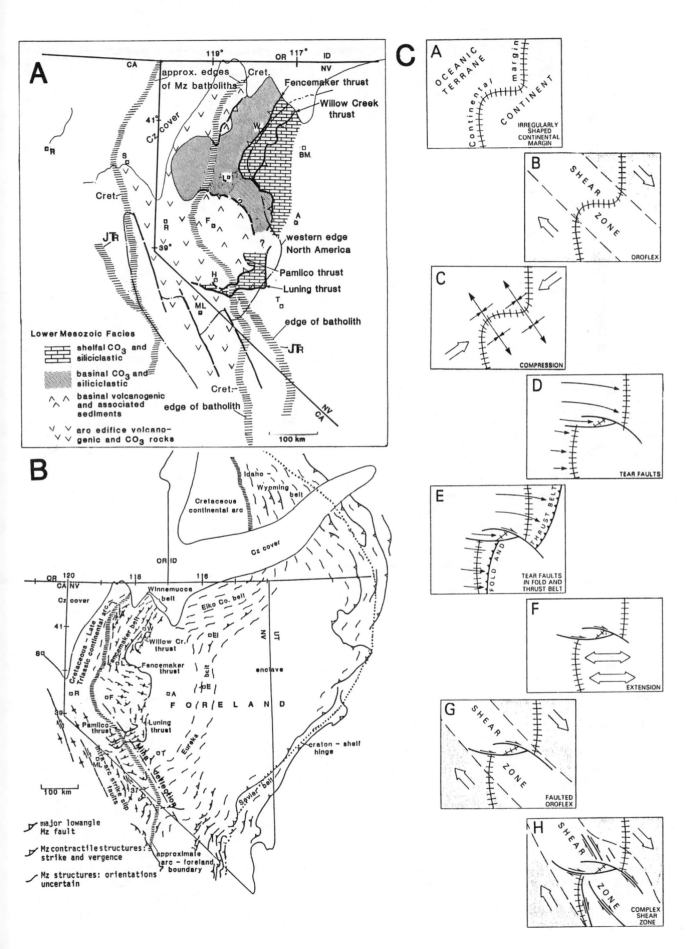

A
CA
119°
OR 117°
ID
NV
approx. edges of Mz batholiths
Cret.
Fencemaker thrust
Willow Creek thrust
41°
Cz cover
W
BM.
R
S
L
Cret.
A
R
F
JTR
39°
?
western edge North America
Pamlico thrust
H
Luning thrust
ML
T
edge of batholith
Lower Mesozoic Facies
shelfal CO_3 and siliciclastic
basinal CO_3 and siliciclastic
basinal volcanogenic and associated sediments
arc edifice volcanogenic and CO_3 rocks
JTR
Cret.
edge of batholith
NV
CA
100 km

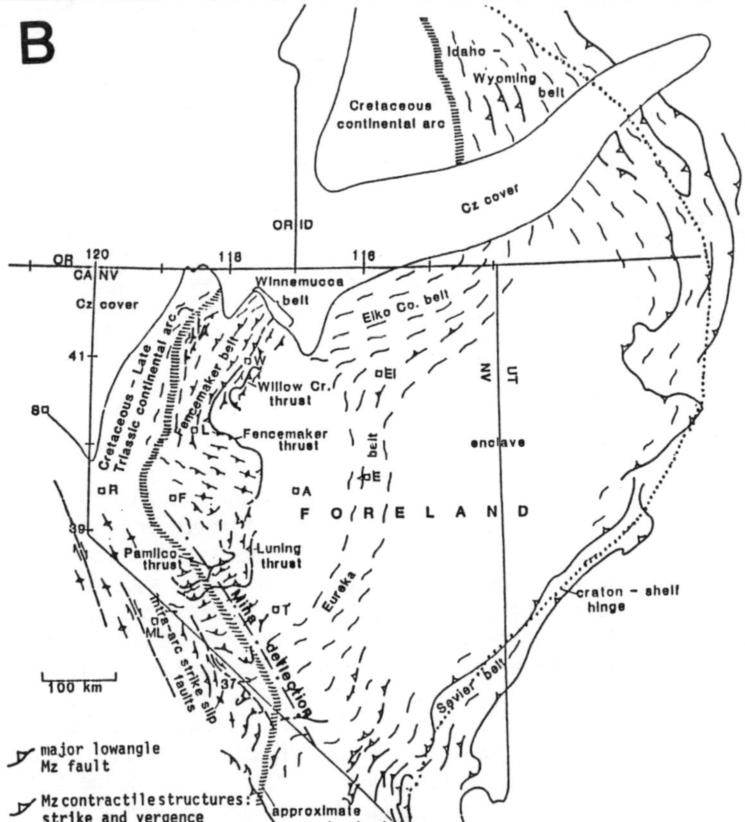

B
Idaho - Wyoming belt
Cretaceous continental arc
Cz cover
OR ID
OR 120
CA NV
118
116
Winnemucca belt
Elko Co. belt
Cz cover
41
Cretaceous - Late Triassic continental arc
Fencemaker belt
Willow Cr. thrust
El
NV
UT
Fencemaker thrust
belt
enclave
E
R
F
A
F O R E L A N D
39
Pamlico thrust
Luning thrust
Eureka
T
craton - shelf hinge
ML
Mina deflection
intra-arc strike slip faults
37
Sevier belt
100 km
major lowangle Mz fault
Mz contractile structures: strike and vergence
Mz structures: orientations uncertain
approximate arc - foreland boundary

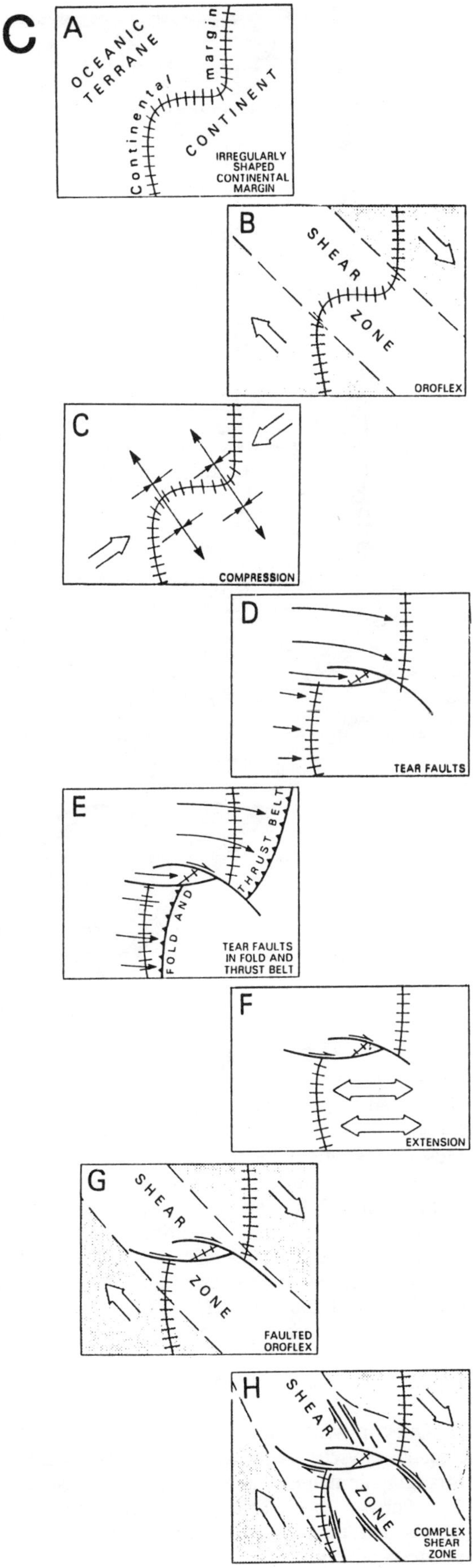

C
A
OCEANIC TERRANE
Continental margin
CONTINENT
IRREGULARLY SHAPED CONTINENTAL MARGIN
B
SHEAR ZONE
OROFLEX
C
COMPRESSION
D
TEAR FAULTS
E
FOLD AND THRUST BELT
TEAR FAULTS IN FOLD AND THRUST BELT
F
EXTENSION
G
SHEAR ZONE
FAULTED OROFLEX
H
SHEAR ZONE
COMPLEX SHEAR ZONE

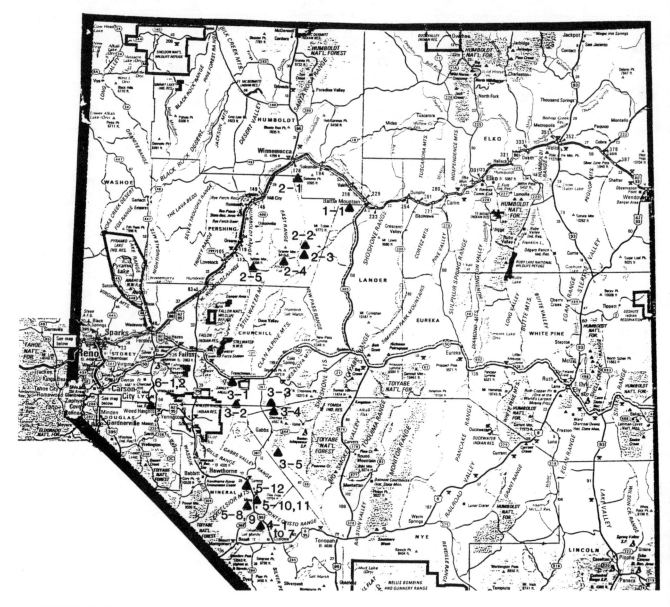

FIGURE 5 Route map with cultural features.

and oceanic allochthons and outboard accreted terrane(s) (Sonomia) of pre-mid-Spathian emplacement; 2) lower Mesozoic stratigraphic cover, continental magmatic arc, and foreland plutons that developed across and within the earlier collided fragments; 3) facies of lower Mesozoic strata: shelfal, basinal (comprising preflysch and terrigenous flysch), and volcanic arc edifice; and 4) major mid-Mesozoic structures, especially the east-verging Fencemaker thrust and west-verging Willow Creek thrust, both of the Winnemucca thrust belt (Fig. 4B).

Lower Mesozoic Stratigraphy

Our trip will view rocks of the shelfal and basinal (terrigenous flysch) sequences in the Winnemucca region. Stratigraphic nomenclature and ages for these are in Figure 8A. For part of the shelfal sequence, more detailed correlations and thickness distribution are in Figures 8B and 8C, respectively.

Shelfal Sequence

The shelfal sequence (Silberling and Roberts, 1962; Silberling and Wallace, 1969; Nichols and Silberling, 1977) consists of mainly carbonate and siliciclastic rocks of varied environments of accumulation: subtidal marine basins, platformal carbonate complexes and a littoral-deltaic siliciclastic complex. Preserved and probably original thicknesses increase from east to west from a few hundred meters to some 2 km. They represent deposition from late Spathian (late late Early Triassic) to at least middle Norian (middle late Late Triassic) time. The youngest preserved age may reflect the cessation of marine deposition in the shelf region. The section includes at least one pervasive unconformity and many others of local extent. Deformation of the shelf sequence varies from slight to intense in a systematic spatial pattern that is related to the proximity to the Fencemaker and Willow Creek thrust. Major open folds with NS axial traces pervade the

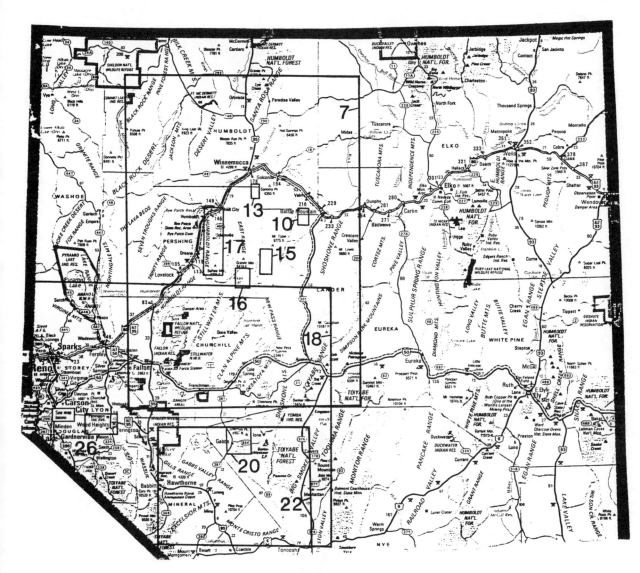

FIGURE 6 Index map to local area maps of this guide.

sequence.

Sediments near the top of the Koipato Formation (Fig. 8A) in the Humboldt Range yielded ammonites of the *subcolumbites* zone of late Early Triassic age (Silberling, 1973). Members and units of the Star Peak Group (Fig. 8A) are dated by ammonite zones in Figure 8B. The suprajacent Auld Lang Syne Group of the shelf sequence is poorly dated by ammonites or other faunas. Whereas correlatives in the basinal sequence are well dated in the Clan Alpine Range (Fig. 7), the assignments in Figure 8A to the Winnemucca and Dun Glen Formations of the Auld Lang Syne Group are not well controlled. Thickness gradients are between west and south (Fig. 8C). The Willow Creek thrust and the Tobin thrusts (Fig. 8C) cut the shelfal sequence, and these may have caused substantial transport of facies and thicknesses.

The following paleogeographic evolution of the shelfal sequence is interpreted within the uncertainties of structural juxtaposition by Silberling and Wallace (1969), Nichols and Silberling (1977), Speed (1978a), Lupe and Silberling (1985), and Elison and Speed (1988). The evolution is given as a sequence of events.

1. **Koipato volcanism and tectonics.** Koipato rhyolitic volcanism of Early Triassic age occurred with concurrent development of large local relief (volcanic troughs, upland sources of lithic debris, probable fault-bounded basins) and local marine conditions on the surface of the Golconda allochthon; Koipato vent systems are known from the Tobin Range west in the shelfal sequence outcrop area, and local trough-filling volcanogenic sediment occurs mainly east of there; cessation of volcanism seems to have been concomitant throughout the area.

2. **late Spathian transgression.** Subtidal organic muddy strata (Tobin, lower part of Prida) were laid down upon the Koipato surface; original distribution uncertain; inherited topography on the Koipato surface may have influenced late Spathian facies and thickness distributions; no contemporaneous CO_3 platform is recognized; trends in preserved thicknesses suggest increasing thicknesses of Tobin and lower part of Prida Formations SW below the Fencemaker allochthon near the Augusta Mountains (Fig. 7).

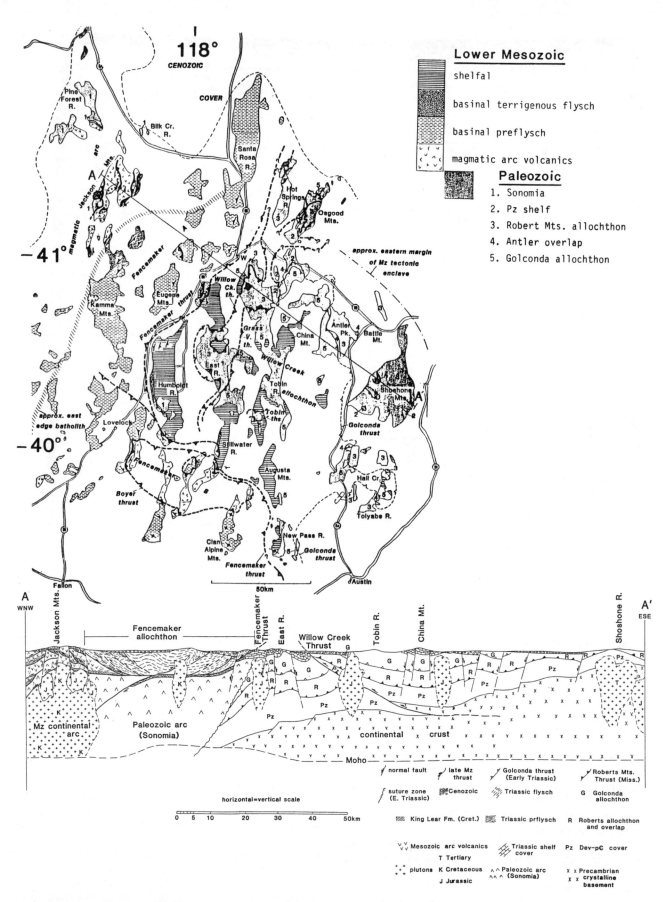

FIGURE 7 Generalized geologic map and cross section of the Winnemucca region. W is Winnemucca.

3. **early Anisian regression.** The unconformity above Tobin Formation, the succeeding conglomeratic Dixie Valley Formation, and the lacuna within the Prida Formation imply erosion during early Anisian time; southerly clastic transport in the Dixie Valley implies sources were in the northern part of the shelf outcrop area or north of there; it is uncertain whether the southward regression was eustatic or caused by tectonic uplift of the northern area.

4. **Fossil Hill transgression.** In middle and late Anisian, marine transgression progressed with a NE or N vector over the entire region and caused deposition of the organic muddy basinal carbonate of the Fossil Hill Members of the Favret and Prida Formations (Fig. 8F); as with the Tobin Formation, the Fossil Hill member includes no CO_3 platform facies, although the local lower member of the Favret Formation records the passage of an energetic environment before Fossil Hill deposition; the Fossil Hill transgression may have been eustatic or have been due to tectonic subsidence of the same northern region that may have been uplifted in early Anisian time. The eastward extent of this transgression between the Augusta Mountains and the Toiyabe Range is unknown.

5. **early Ladinian regression.** Partial erosion of the Fossil Hill Member occurred in the central northern outcrop area (China Mt., East Range); this is the same region to receive minimum thickness of the Fossil Hill during preceding transgression and to have a lacuna in Tobin Prida time; thus, there was either a persistent locus of cyclic uplift or a persistent static topographic feature against which eustatic onlap and offlap were substantial; the first suggestion of the existence of a carbonate platform comes at this time from slump masses and blocks of platform margin facies within the basinal upper beds of the Prida Formation in the westernmost outcrops of shelf sequence.

6. **late Ladinian to mid-Karnian carbonate platform** (Fig. 8F). Ladinian erosion of the central northern part of the outcrop area was followed by the establishment of a vast supratidal flat (Panther Canyon Member of Augusta Mountain Formation) with platform margin facies on its west edge in the East and Stillwater Ranges and basinal facies (upper part of Prida Formation) west of that; supratidal dolomite was laid down and subjacent carbonates underwent diagenetic dolomitization on the inner carbonate platform; a tongue of clastics was shed from the north on the northern half of inner platform (upper part of Panther Canyon Member of Augusta Mountain Formation), indicating the persistence or reactivation of elevated basement (Paleozoic) terrain in Early and Middle Triassic time north of China Mountain; during early Karnian time, the inner platform was transgressed, generating mainly intertidal (Smelser Pass Member) deposition; the platform margin prograded west above the diachronous upper part of Prida Formation; the transgression also spread east and extended intertidal deposition to and past the central Toiyabe Range; deeper water to the south yielded basinal facies in the New Pass Range, but there is no recognized platform-margin facies between the Augusta Mountains and New Pass Range; minor mafic volcanism occurred widely in the early

Karnian, suggesting that tectonic extension of unknown direction affected the shelf region.

7. **middle Karnian regression.** All but the southern part of the outcrop area was exposed, slightly eroded, and overrun by siliciclastic debris (lower part of Cane Spring Formation) of northerly provenance; once more, it is not clear whether eustatics or tectonics were at play.

8. **late Karnian transgression.** Outcrop area covered by intertidal inner platform carbonate (upper part of Cane Spring Formation); platform margin facies prograde west to edge of outcrop area.

9. **early Norian deltaic complex.** The entire carbonate platform was covered by a flood of terrigenous clastics of easterly provenance; the source has been considered alternatively to be the ancestral Rocky Mountains or an uplifted region including continental volcanoes in a belt from Death Valley through Arizona; the terrigenous complex included from east to west, littoral facies (Trgvo3 = Osobb facies (Fig. 8A)), possible delta top facies of the Osobb Formation, delta front facies of Grass Valley Formation, and turbidities of the basinal sequence west of shelf edge (Figs. 8D, 8E, 8G).

10. **middle Norian carbonate deposition.** The terrigenous influx stopped abruptly and an intertidal carbonate complex (Dun Glen Formation) covered the shelf; the terrigenous supply was either cut off at its source or bypassed to the north or south of the outcrop area of the shelf sequence; resumption of terrigenous influx in late middle Norian is indicated by the Winnemucca Formation, but its paleoenvironments are poorly known.

11. **final regression.** The sea withdrew to the west and south (?) from the outcrop area of the shelfal sequence late in Norian time, apparently never to return.

Major points in this history are the following: 1) a west-prograding carbonate platform developed from late Ladinian to late Karnian time; 2) cyclic transgression and regression occurred before the carbonate platform came into existence; they oscillated about a point at the center of the northern edge of the shelf outcrop area; more persistent subtidal or basinal deposition environments lay on western and southern fringes of the outcrop area; 3) the carbonate platform was succeeded by a deltaic-littoral complex of terrigenous sediment. It is important to note that facies of the carbonate complex and the deltaic complex parallel the probable depositional and tectonic boundary (northerly leg of the Fencemaker thrust) between shelf and basinal Triassic sequences. The most easterly reach of the Fencemaker thrust, however, cuts across the shelf facies at a high angle. This implies that the southeastern Fencemaker allochthon (B) has been transported a substantial distance across a now buried southerly prolongation of the shelf sequence.

Basinal Sequence

The lower Mesozoic basinal sequence (Figs. 4,7,8A)

Stratigraphic nomenclature for Triassic strata (late Spathian and younger) of northwestern Nevada.

A

East →

basinal sequence	shelfal sequence	
upper hemipelagite – CO₂ platform incls: Mud Sprs, Hoy + Fms.	lacuna	late Norian + Jur.?
turbidite = flysch incls: Raspberry, Mullinix, Andorno, Singas, O'Neil, Bernice, Dyer, Byers Fm.	Auld Lang Syne Group — Winnemucca Fm — Dun Glen Fm — Grass Valley Fm. Osobb Fm	early + middle Norian
lower hemipelagite = preflysch incls: Quinn River Fm.	Star Peak Group — Cane Springs Fm	late Karnian
Prida Fm upper Mmb.	Augusta Mt. Fm Smelser Pass Mmb. Panther C. Mmb. Home Sta. Mmb.	early Karnian and Ladinian
Fossil Hill Mmb.	Favret Fm Fossil Hill Mmb. Lower Mmb.	Anisian
Lower Mmb.	Dixie Valley Fm	late
	Tobin Fm	Spathian
	Koipato Fm	early Spath. and Older?

B

C

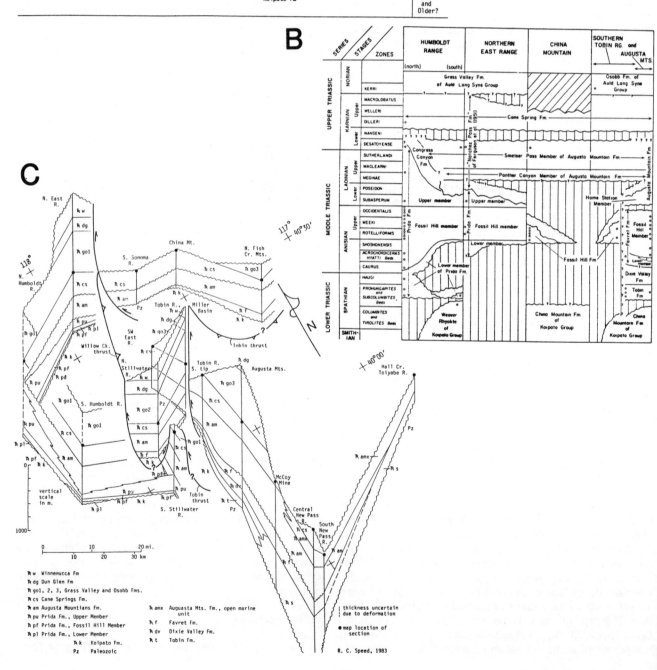

Ħ w Winnemucca Fm
Ħ dg Dun Glen Fm
Ħ go1, 2, 3, Grass Valley and Osobb Fms.
Ħ cs Cane Springs Fm.
Ħ am Augusta Mountians Fm.
Ħ pu Prida Fm., Upper Member
Ħ pf Prida Fm., Fossil Hill Member
Ħ pl Prida Fm., Lower Member
Ħ k Koipato Fm.
Pz Paleozoic

Ħ amx Auguasta Mts. Fm., open marine unit
Ħ f Favret Fm.
Ħ dv Dixie Valley Fm.
Ħ t Tobin Fm.

thickness uncertain due to deformation
● map location of section

R. C. Speed, 1983

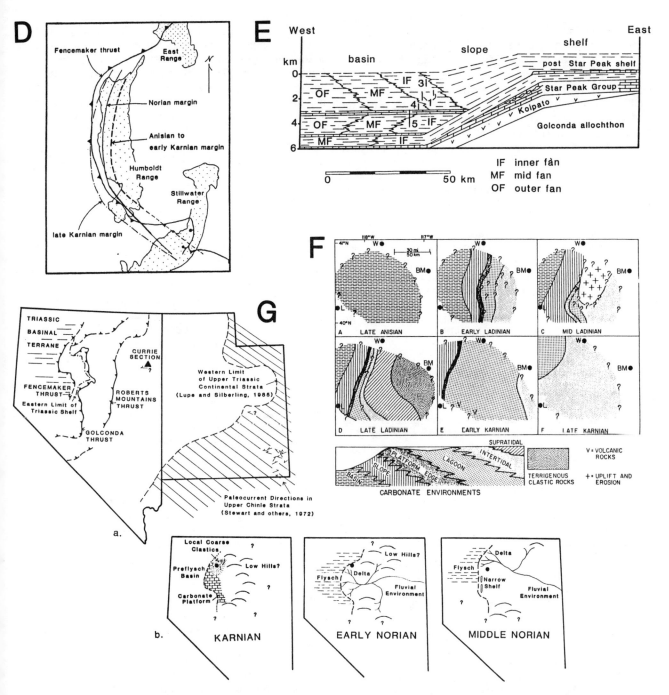

FIGURE 8 Stratigraphy and paleogeography of Triassic and ? Lower Jurassic marine strata of northcentral Nevada. A) division of basinal and shelfal sequences (mapped on Fig. 4A); B) time-position distribution of shelfal sequence (Nichols and Silberling, 1977); geographic positions on Fig. 6; C) thickness distribution of shelfal sequence and major structures (after Nichols and Silberling, 1977); D) Triassic shelf-basin transitions vs. age (Heck and Speed, 1987); E) reconstructed Triassic shelf-basin transition in East Range (Elison and Speed, 1988); F) Triassic carbonate platform progradation model at shelf-basin transition (Nichols and Silber-ling, 1977); G) model of Late Triassic terrigenous sediment transport across shelfal to basinal sequence (Elison and Speed, 1988).

occupies a belt seaward of the shelfal sequence. The basi-nal sequence is everywhere allochthonous. It is thought not to be far-traveled, however, because intermediate (slope) facies exist in the tectonic zone (the Fencemaker thrust zone) between basinal and shelfal sequences (Heck and Speed, 1987; Elison and Speed, 1988). The eastern part of the basinal sequence comprises the Fencemaker allochthons (Fig. 7). These allochthons contain deformed terrigenous and carbonate turbidite and hemipelagite of Norian age. Restoration of Norian stratigraphy in the Clan Alpine Range (Fig. 7)(Speed, 1978a) indicates a minimum thickness of 6 km of strata. To the west in the basinal sequence, probably west of the Fencemaker allochthons, turbidites regarded by Speed (1978a) as similar to those of the allochthons are as old as Middle Triassic.

The basinal sequence is interpreted to have filled a deep water basin that lay west of the outer continental shelf, probably from Early or Middle Triassic time on. The positions of the Triassic shelf-slope break in the Humboldt Range (Fig. 7) is shown in Fig. 8D. Initial strata of the basin were a condensed marly sequence or preflysch. In Late Triassic time, regional tectonics that affected the continental interior caused a flood of terrigenous debris to be delivered across the continental shelf, there depositing the deltaic formations of the Auld Lang Syne Group (Figs. 8A, 8B, 8G) above the earlier carbonate platform (Star Peak Group) (Fig. 8A). Seaward of the delta, the terrigenous debris together with shelf margin-derived carbonate debris spilled over the shelf edge to deliver the turbidites and thick hemipelagic mudstones (flysch) of the basinal sequence. By the end of the Triassic, the basin was filled or nearly filled with sediment. Figure 8E shows a reconstruction of the Triassic shelf-basin transition in or near the East Range (Fig. 7) (Elison and Speed, 1988).

Intermediate Facies

Triassic slope facies are recognized in the Fencemaker thrust zone in the East, Humboldt, and Stillwater Ranges (Fig. 7). These record the existence and position of the transition between the shelfal and basinal sequences (Fig. 8D, 8F). Slope facies are chiefly terrigenous mudstone and micrite, channel-filling graded sandstones, and carbonate olistostromes.

Structure

The Winnemucca thrust and fold belt of the Winnemucca Region (Figs. 4B, 7) contains four regional domains: from west to east: Fencemaker allochthons A and B, autochthon, and the Willow Creek allochthon. The Fencemaker allochthons are composed of mainly basinal lower Mesozoic sediments that were shortened and thrust shelfward on the Fencemaker thrusts that approximately follow the basin to shelf transition at the edge of Triassic North America (Speed and others, 1988; Heck and Speed, 1989; Elison and Speed, 1989). The two allochthons lie discretely on the NW- and SW-facing flanks of a regional bend in the Fencemaker thrust, contain partly different stratigraphic successions, and probably moved in different directions at different times. Fencemaker allochthon A (Fig. 7) moved ESE relative to the shelf in today's coordinates. Transport occurred after about 215 mybp, the age of youngest strata in the allochthon, and before the undated final emplacement of the Willow Creek allochthon (Fig. 7) whose movements imposed structures that overprint structures of Fencemaker A. Structures within the eastern part of Fencemaker A are uniformly ESE-vergent and suggest minimum horizontal contraction of 70%. The western flank of Fencemaker allochthon A is at the eastern flank of the Mesozoic continental arc (Fig. 7) (Russell, 1984). There, Fencemaker structures verge west, and are probably Middle Jurassic in part. Thus, Fencemaker allochthon A is bivergent, and its deep structure may be as

shown in the section of Figure 7. The eastern two-thirds of the allochthon contain only Upper Triassic flysch. Such strata probably detached from older basinal beds (preflysch, Fig. 7) and were imbricated against the apparently less deformable shelf and arc margins as those margins closed toward one another. Strata of the basinal sequence and Sonomia basement below the flysch detachment underrode the continental arc to the west in the Jurassic and/or Cretaceous. The choice of west vs. east underriding by the Sonomia basement stems from the constraint that the continental crust and its Phanerozoic tectonostratigraphic cover appear tied together in the Winnemucca autochthon.

Fencemaker allochthon B (Fig. 7) was transported NNE relative to the shelf edge. Structures generated within the autochthon by Fencemaker B overprint those caused by Fencemaker allochthon A. The age of final emplacement of Fencemaker allochthon B is uncertain, although contraction in Fencemaker B was ongoing between 150 and 165 ma, as indicated by syntectonic sedimentation and magmatism (Speed and Jones, 1969).

The Winnemucca autochthon to the Fencemaker and Willow Creek thrusts (Fig. 7) is capped almost completely by Triassic shelfal and shelf edge strata (Nichols and Silberling, 1977). Deformation in Triassic rocks of the autochthon, estimated by cleavage spacing and orientation with respect to bedding, fold tightness, and width of and distance between shear zones is proportional to proximity to the major Mesozoic thrusts.

The Willow Creek allochthon (Elison, 1987) (Fig. 7) contains a moderately disrupted tectonostratigraphic succession of rocks that lay above North America before thrusting. The allochthon's initial tectonostratigraphic thickness includes the Roberts Mountains and Golconda allochthons (Fig. 3E) which are known only to have exceeded the 2-3 km present exposed thicknesses. The Willow Creek allochthon was emplaced above the Triassic shelf cover and locally, lower rocks of the autochthon. Transport in the East and Sonoma Ranges (Fig. 7) was WNW, and the displacement is at least 35 km. Folds in the East Range related to motions of the Willow Creek allochthon are cut by an undeformed pluton that is dated at 151 ma (K-Ar, hornblende with slightly discordant biotite). Thus, at least some transport of the Willow Creek allochthon was before 151 ma, but the age of final emplacement is unknown. Fencemaker B as well as A may have preceded Willow Creek emplacement because open folds with northerly axial traces are widespread in the autochthon and deform the Fencemaker B thrust. Such folds are appropriately oriented to have been generated by motions related to Willow Creek transport.

As indicated in Figure 7, a southern terminus, if any, of the Willow Creek allochthon and the discrimination of it from autochthon in the Toiyabe and Shoshone Ranges are uncertain. South of the Winnemucca area, near localities to be visited in the Lodi Hills and Paradise Range (Figs. 5, 6, 18), the Quartz Mountain allochthon (Fig. 18) may be homologous with the Willow Creek allochthon (Silberling and John, 1989). Like the Willow Creek, this allochthon probably comprises Paleozoic rocks, including those of the older Roberts Mountains allochthon, and it is thrust westward over lower Mesozoic rocks that had already undergone SE- and NE-vergent deformations.

In the cross section of Figure 7, the sole of the Willow Creek allochthon is shown as a listric thrust that flattens east into sialic crust. This interpretation is based on the idea that the Willow Creek allochthon was caused by spreading away from a chain of thermal domes that lay immediately east, to be discussed presently.

The region of the western Great Basin between the Winnemucca and Eureka thrust belts (Fig. 4B) exposes moderately deep seated tectonostratigraphic units and moderate metamorphic grades with unroofing ages that are at least partly, if not entirely, Mesozoic. This is the Toiyabe uplift zone with distribution shown in Figure 9A (Speed and others, 1988). The zone may be a continuous belt of moderate unroofing, or more likely, it is a chain of domes, at least some of which are thermal and diapiric. Figure 9B-E shows a model of the Jurassic development of the Toiyabe uplift zone (TUZ) and concomitant bivergent thrusting of the Willow Creek thrust on the west of the TUZ and Eureka thrust belt (Fig. 4B) on the east. The model assumes that the Toiyabe uplift zone is a response to regional contraction because Mesozoic movements in the thrust belts on either side of the uplift zone indicate compression across the foreland in the Great Basin in that era. Crustal contraction takes place in a laterally narrow zone (±200 km wide) by volume transfer from a lower to an upper layer and doming of the free surface. The dome spreads laterally on surface thrusts, and extension occurs in the central region due to the lateral flow and stretching by progressive uparching. Magmas generated at the base of the sialic crust rise diapirically through the contracting part of the dome, increasing its ductility and the rate of local horizontal shortening. The plutons also provide a mechanism of mass transfer to higher levels. Because the unroofing is greater on the west side of the dome, uplift and diapirism presumably went on for a longer time there than on the east side.

Given an average of 10 km unroofing across the model dome and assuming no depression of the sialic crust below the dome for compensation of mountains, the mechanism shown in Figure 9E could accommodate 33% horizontal shortening relative to an initial 30 km thick crust. For a 200 km initial width of the domed region, about 66 km of contraction could have been taken up.

DAY 1: GOLCONDA ALLOCHTHON AT BATTLE MOUNTAIN

Battle Mountain (Fig. 7) provides an extensive exposure of the Golconda allochthon, its sole fault the Golconda thrust (Fig. 10A), and a relative autochthon comprising the Roberts Mountains allochthon (ЄDsv, Fig. 10A) and its local cover of shallow marine strata, the Antler overlap sequence (PPos, Fig. 10A; not shown on the section of Fig. 7) (Roberts, 1964; Miller and others, 1982; Brueckner and Snyder, 1985; Snyder and Brueckner, 1983). Our objective is to examine the rocks and structures of the allochthon in Willow Creek (Figs. 10A,B,C; 11; 12) with a view towards its origin as an accretionary complex emplaced across the continental edge.

The Golconda allochthon at Battle Mountain is divided into major lithotectonic units (numbered units, Fig. 10A) that were initially thought to be stratigraphic members by Roberts (1964). Each of these major units possesses a degree of lithic unity, but structural analyses and dating, mainly radiolarian, show each unit is a composite of fault-bound packets (Stewart and others, 1986; Brueckner and

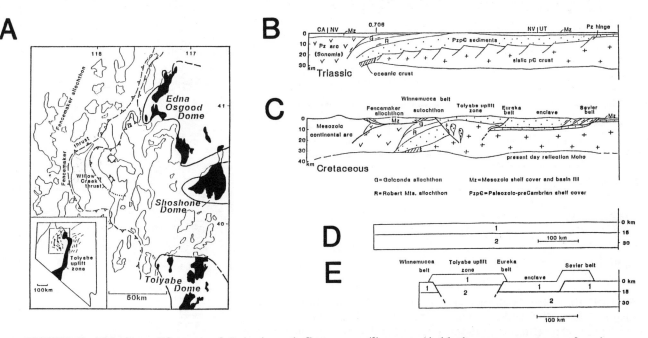

FIGURE 9 Toiyabe uplift zone of Jurassic and Cretaceous (?) age. A) black areas are zones of major unroofing and magmatism; B) model of crustal structure before TUZ; C) model of crustal structure after development of TUZ and cratonward foreland thrust belts; D) and E) simplified model showing balanced section.

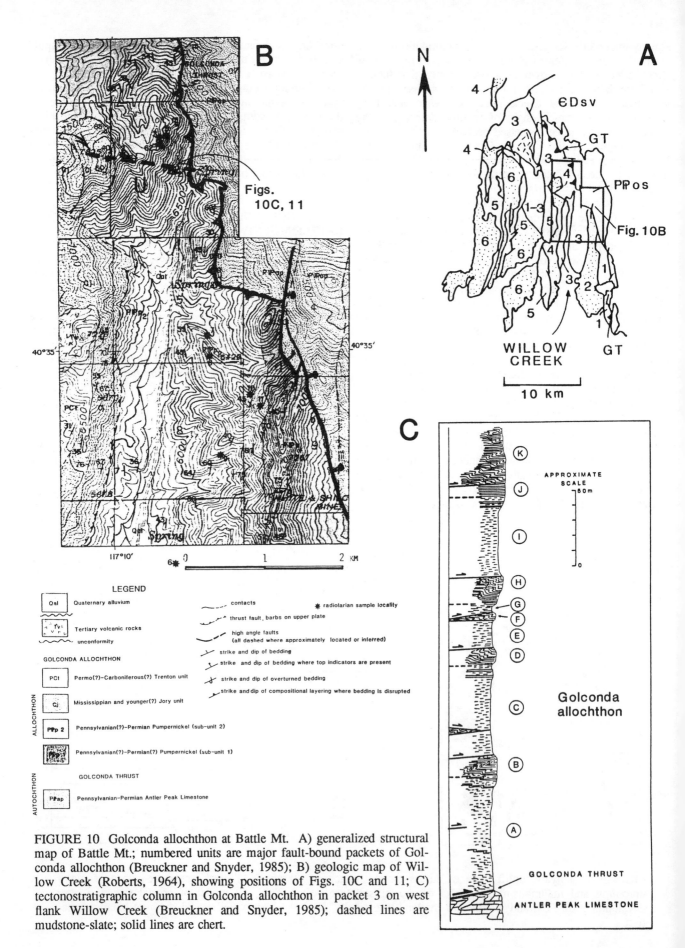

FIGURE 10 Golconda allochthon at Battle Mt. A) generalized structural map of Battle Mt.; numbered units are major fault-bound packets of Golconda allochthon (Breuckner and Snyder, 1985); B) geologic map of Willow Creek (Roberts, 1964), showing positions of Figs. 10C and 11; C) tectonostratigraphic column in Golconda allochthon in packet 3 on west flank Willow Creek (Breuckner and Snyder, 1985); dashed lines are mudstone-slate; solid lines are chert.

LEGEND

Qal Quaternary alluvium

Tv Tertiary volcanic rocks

unconformity

GOLCONDA ALLOCHTHON

PCt Permo(?)-Carboniferous(?) Trenton unit

Cj Mississippian and younger(?) Jory unit

ₚₚ2 Pennsylvanian(?)-Permian Pumpernickel (sub-unit 2)

ₚₚₚ1 Pennsylvanian(?)-Permian(?) Pumpernickel (sub-unit 1)

ALLOCHTHON

GOLCONDA THRUST

ₚₚap Pennsylvanian-Permian Antler Peak Limestone

AUTOCHTHON

contacts

thrust fault, barbs on upper plate

high angle faults
(all dashed where approximately located or inferred)

strike and dip of bedding

strike and dip of bedding where top indicators are present

strike and dip of overturned bedding

strike and dip of compositional layering where bedding is disrupted

radiolarian sample locality

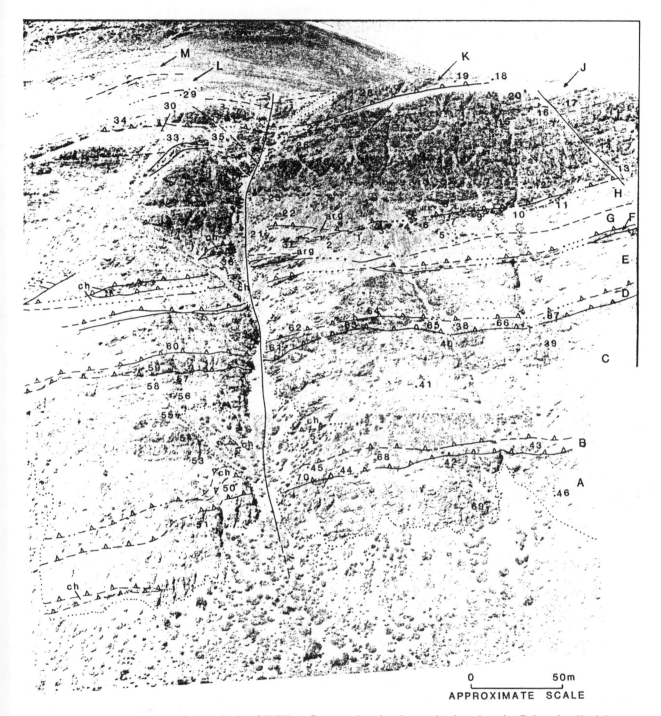

FIGURE 11 Photograph of west flank of Willow Canyon showing lettered subpackets in Golconda allochthon, packet 3 (Breuckner and Snyder, 1985).

Snyder, 1985; Miller and others, 1982; Stewart and others, 1987). We shall traverse unit 3 at the base of the allochthon. The packets within unit 3 are lettered and their boundaries shown in Figures 10C and 11. The packets are composed chiefly of either chert or argillite, and ages of cherts in different packets range from Late Mississippian to Early Permian, possibly to Guadalupian.

Figure 12A outlines the general sequence of structures of the Golconda allochthon at and near Battle Mountain. Figure 12B shows orientation data for structures in the allochthon of unit 3, in the Willow Creek traverse.

The packets are between a few tens of centimeters to tens of meters thick and up to 100's of meters long. Chert packets contain different intensities and styles of deformation and in fact some contain homoclinal beds (e.g. packets J, K). Some are characterized by boudinlike structures, and others have from one (K) to three generations (H) of homoaxial folds. The argillite packets are less folded.

The structures of the packets will be discussed in terms of initial conditions of porosity, strength by early diagenesis, defluidization, and detachment.

A

STRUCTURAL EVENTS OF THE HAVALLAH SEQUENCE, NORTH-CENTRAL NEVADA

Event	*Fabric elements*	*Orientation*	*Comments*
D_0	Fractures and qtz-veins, microstylolites (C_0), mound structures, dilation breccias? Slumps?	Fractures and veins variable, at high angles to beds. C_0 parallels bedding.	Sedimentary loading. Fractures and veins from volume changes during diagenesis.
D_1	Solution boudins, step planes, further formation of microstylolites (C_1), isoclinal folds with C_1 axial cleavage?	C_1 parallel to bedding. Step planes strike north-south. Boudins form north-plunging maximum.	Early tectonics with bedding-normal loading; some east-west extension. Extreme thinning by solution.
	Thrusting, cataclasis, tight to open folds, east-verging folds (F_2), solution cleavage (C_2), breccias and fractures by high pore pressure.	Folds parallel boudin line. C_2 at high angles to bedding. Thrusts at very low angles to bedding.	East-directed thrusting in accretionary prism. Chert packets juxtaposed. Episodes of high and low pore pressures.
	Asymmetric folds (F_{2A}) predate thrusts or refolded. C_2 in hinges.	Generally north-plunging hinges. Axial surfaces refolded.	Folds usually tight or chevron, local ductile behavior.
	Thrusts, thrust zones, clastic intrusions.	Very low angles to bedding.	Several generations of thrusts.
D_2	Asymmetric folds (F_{2B}), concentric, post-thrust.	Generally north-plunging hinges.	Folding more disharmonic than F_{2A}.
D_3	Golconda thrust and major internal thrusts. Large asymmetric concentric folds (F_3), vergence east.	Folds parallel D_2 fold fabric. Major thrusts cut off D_2 thrusts.	Permian-Triassic obduction onto North American craton (Sonoma orogeny).
D_4	Asymmetric, gentle to open concentric folds (F_4), vergence west.	Scattered, generally north-plunging.	Possibly related to Mesozoic events.

B

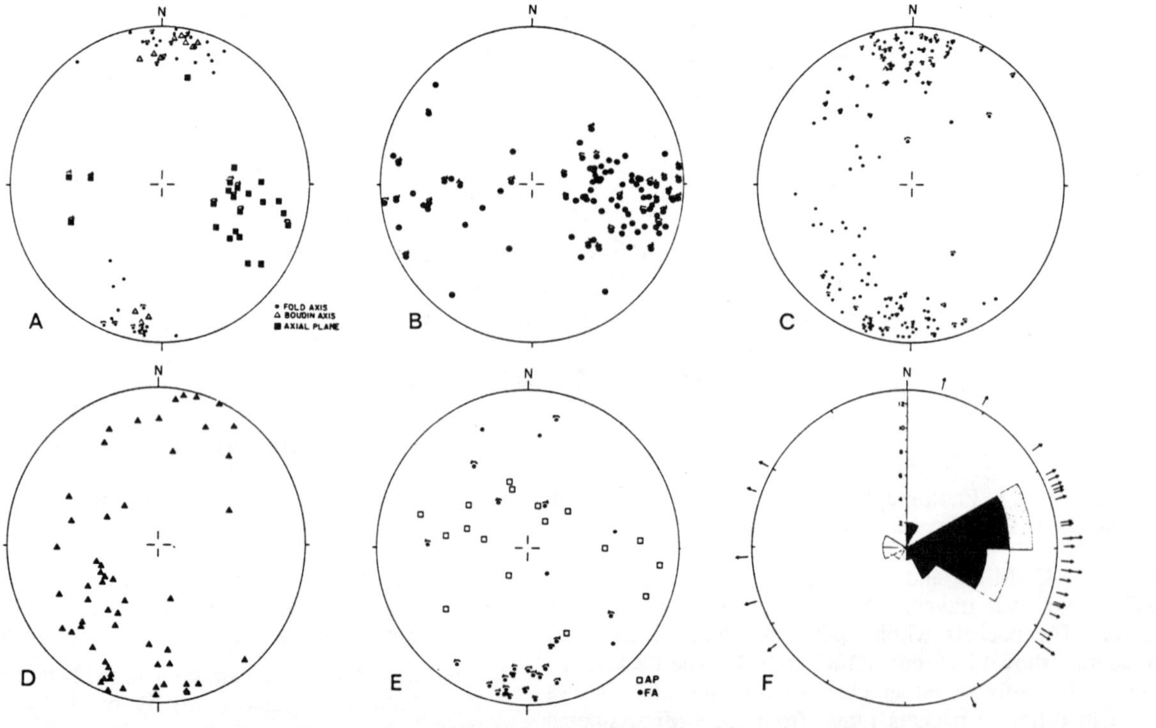

Structural data: (*a*) Fold axes, boudin axes, and poles to axial planes of subunit 1. (*b*) Poles to axial planes of subunit 2. (*c*) Fold axes of subunit 2. (*d*) Boudin axes of subunit 2. (*e*) F_2 kink fold axes and axial planes. (*f*) Rose diagram of slip line determinations. The actual trend of slip lines is shown by arrows on the outside of the diagram. Faulted fold determinations are shown in solid black and by solid arrows. Fault zone striae determinations are shown in stipple pattern and by open arrows.

FIGURE 12 Structural data for Golconda allochthon in north-central Nevada. A) structures in sequence (Breuckner and Snyder, 1985); B) orientation data in area of Figure 11 (Miller and others, 1982).

DAY 2: STRATIGRAPHY AND STRUCTURE OF LOWER MESOZOIC STRATA: SONOMA, TOBIN, STILLWATER, AND HUMBOLDT RANGES

Stop 1. Sonoma Range: Willow Creek Allochthon

The western flank of the Sonoma Range (Fig. 7) provides an excellent view of the Willow Creek allochthon and its footwall of deformed Triassic shelfal strata and Paleozoic rocks (Fig. 13) (Gilluly, 1967; Silberling, 1975; Stahl, 1987; Elison, unpubl. data). The Jurassic Willow Creek allochthon of the Sonoma Range includes the following tectonostratigraphic units (Figs. 7,13): Paleozoic shelf strata (Cambrian Preble Formation); Roberts Mountains allochthon (ePzr) (Valmy and Harmony Formations); late Paleozoic(?) overlap sequence consisting of a megabreccia (uPzo); and the Golconda allochthon. The footwall consists of the following, from top down: Triassic shelfal strata from the Winnemucca down to the Prida Formations (Figs. 8A, 13); Tallman Fanglomerate, an undated sedimentary breccia; Golconda allochthon; late Paleozoic overlap sequence, here an undated limestone; and Roberts Mountains allochthon.

In the Triassic strata of the footwall, Stahl (1987) recognized five generations of structures. First-phase folds (F1) are minor recumbent isoclines in the Grass Valley and Winnemucca Formations and west-verging inclined open folds in the Star Peak Group and Dun Glen Formation. Foliation is seen only in the Grass Valley Formation, where it is bedding parallel. Elison believes the folds in the Grass Valley and Winnemucca Formation are syndepositional; the foliation, compactional; and the folds in the carbonate units are of later generation. Second-phase folds (F2) are major folding of the entire Triassic sequence into a tight recumbent synform and tight inclined antiform (Fig. 13A, section); this generation is the result of blind thrusting within the Triassic rocks. Third-phase folds (F3) are solely in the Star Peak proximal to the overthrusted lower Paleozoic rocks in the central window (Fig. 13). Fourth-phase folds (F4) are minor inclined to upright west-verging tight folds of beds and earlier structural surfaces with near-vertical axial-plane foliation. Unlike earlier generations (F1 to F3), which are only in the Triassic rocks, a correlative to this phase is also in the lower Paleozoic rocks of the Willow Creek allochthon. Fifth-phase folds (F5) are open upright major folds with vertical axial-planar foliation. Correlatives to this phase are in both the Mesozoic autochthonous rocks and the Willow Creek allochthon.

Figure 13B shows a model of development of the F1 and F2 structures in the footwall, assuming the footwall is a duplex between an early sole fault in the footwall and out of sequence emplacement of the allochthon.

Stop 2. Tobin Range: Scarps of 1915 Pleasant Valley Earthquake (from Wallace, 1984a and b)

The fault scarps that formed during the earthquakes of October 2, 1915, are spectacular examples of surface faulting along a seismogenic fault. Block faulting characteristic of the Basin and Range province is demonstrated. Displacement is predominantly normal and dip slip. The upthrown Tobin Range block is a horst, and the downthrown Pleasant Valley block is a graben (Fig. 14A). The Tobin Range, China Mountain, and Sou Hills blocks are all tilted to the east. At Golconda Canyon and elsewhere along the scarps (Figs. 14B, 14E), the relation of active tectonism and geomorphic processes is well displayed.

The main earthquake of Richter magnitude 7-3/4 struck at 2252 PST, October 2, 1915. Two strong foreshocks occurred at 1541 and 1750 PST, and aftershocks continued for months. Four separate scarps formed during the earthquakes; from north to south they are the China Mountain, Tobin, Pearce, and Sou Hills scarps (Fig. 14).

The Golconda Canyon site lies midway on the Pearce scarp (Figs. 14B, 14E), and in the section a few kilometers long to the north are found the largest displacements (up to 5.8 m). Scarp heights are commonly greater than this because of slope and fault geometry and postearthquake erosion.

The 1915 fault scarps are generally characterized by a sharp crest, below which is a free face, a steep face from which sand and gravel spall and fall to a debris slope below. The sand and gravel of the debris slope stand at the angle of repose of about 30° to 35°. Commonly, coarser clasts of cobbles and boulders accumulate at the base of the debris slope. Where the original colluvium or alluvium is poorly indurated, the free face may have been covered by debris slope. Note the steep, almost unchanged free face at the end of the fence a few tens of meters south of the Golconda Canyon road (Fig. 14B,C). Apparently, the gravels underlying the swale are slightly more cemented than the colluvium to the north and south. Assuming that all of the original scarp face shown in Figure 14B stood at a position equivalent to that of the steep face at the fence line, one can estimate the amount (several feet; meters) of free face retreat of the rest of the scarp. Wallace (1984a and b) estimates that a free face will survive somewhere along the 60 km length of the 1915 scarps for at least several hundred years and possibly for as long as 2,000 years.

About 0.5 km north of Golconda Canyon is an old fence that was displaced right laterally about 2 m (Fig. 14E). Given an axis of regional extension of about N65°W, those segments that strike northwest, such as here, have a significant right-lateral component of displacement. Those segments that strike N.25°E, perpendicular to regional extension, have simple dip-slip displacement. On the hillside above the site can be seen arcuate scarps that represent slope-failure or landslide-like features.

From many points along the main Pleasant Valley road 5 to 7 km south of Golconda Canyon Road are excellent views of faceted spurs and wine glass-shaped valleys characteristic of fault-generated range fronts. During each displacement event, a steep scarp a few meters high is added to the base of the faceted spur. The planar nature of the entire facet persists for several million years, but erosion has caused the upper part of each triangular facet to decline to a lower slope angle than the basal part of the facet.

In many places the 1915 scarps formed along the line of older scarps, and remnants of the older scarps are commonly left as bevels at the crest of the 1915 scarp. Analysis of the profiles of these older scarps, uplift rates

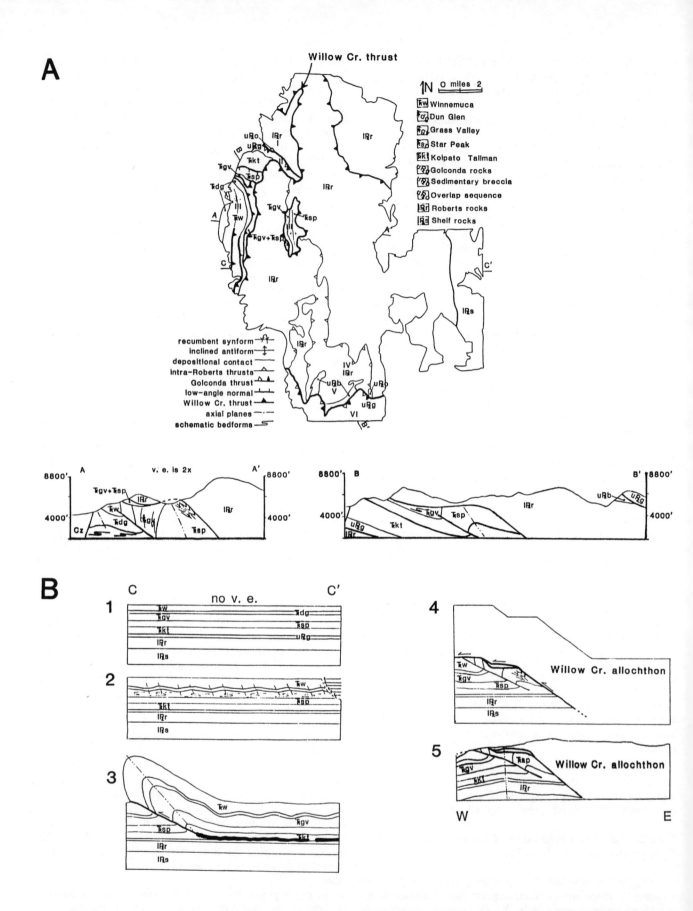

FIGURE 13 Mesozoic structures related to the Willow Creek thrust in the northern Sonoma Range (Fig. 7). A) map and sections; B 1-5) balanced reconstruction (Stahl, 1987).

of the range blocks, and dating by radiocarbon and other methods indicate that larger earthquakes (M>7) accompanied by surface faulting occur at intervals of several thousand years. An average displacement rate of a few tenths of a millimeter per year is indicated along the range front faults in the region.

Figure 14D shows a regional section that projects the Tobin Range frontal fault to seismogenic depths and a postulated zone of decoupling at about 15 km (Wallace, 1984b).

Stop 3. Tobin Range: Stratigraphy of the Star Peak Group

The southern Tobin Range (Figs. 7, 15A,B) exposes the full range of formations of the Triassic shelfal sequence, although there is no continuous section. Our objective at Stop 3 is to traverse the lower shelfal sequence from its depositional base with the Golconda allochthon. Younger strata outcrop mainly in the northern half and older in the southern half of the area, due in part to northward tectonic overriding in the Tobin thrust zone. An abrupt decrease in thickness from S to N or SW to NE occurs in the Lower Triassic section (Tobin Formation to Panther Canyon member of Augusta Mountain Formation) (Figs. 8A,B,C) in the hangingwall of the upper Tobin thrust (Fig 15A). Corresponding thickness and facies changes (Burke, 1973) indicate the thinning is depositional, not mainly tectonic. The thickness gradient indicates the existence of a S or SW-facing basin flank of 2-6 km horizontal width, about 1 km amplitude, and Spathian to mid-Ladinian duration that crosses the southern Tobin Range about midway NS in Figure 15A. The original position of the basin flank was south of its present one because of displacement on the Tobin thrusts. Although the tectonic transport is unknown, the displacement was more likely a few km than tens of km.

The succeeding carbonate platform section (Smelser Pass member of Augusta Mountain Formation and Cane Springs Formation) also thickens north to south in the southern Tobin Range (Fig. 8C). This is part of a regional doubling of thickness of these formations between China Mountain in the northern Tobin Range and the Augusta Mountains (Fig. 7), some 30 km south of this area. The position or amplitude of a flank of this Ladinian-Karnian basin is not known and may not have been related to the earlier Triassic basin flank.

The Norian Osobb Formation here contains 700 m of quartz arenite and minor carbonate. Its environment was probably a beach-barrier bar-lagoon, perhaps at the landward margin of the Osobb-Grass Valley delta (Fig. 8G). No control exists on regional thickness changes of the Auld Lang Syne Group (Osobb-Winnemucca Formations). Tertiary units consist of older tuff, widespread andesite that is 30-35 m.y. old, and younger tuff. It is not certain whether the older tuff is in mainly depositional or completely in tectonic contact with Triassic rocks because the older tuff is much covered and engulfed in andesite. Cross sections are drawn assuming depositional continuity of older tuff and Triassic beds (Fig. 15). Much structural interpretation hinges on this relationship.

Major structures of the southern Tobin Range (Figs.

15A, 15B) are 1) NS folds, 2) Tobin thrust zone, 3) omissional fault nappes, 4) Miller Basin fault of Tertiary age, 5) E-dipping homocline, and 6) high angle Cenozoic normal faults. Their sequence of formation is probably in the order given although the timing of structures 3 and 5 is particularly uncertain.

1) **NS folds.** Gentle to close folds with northerly striking axial planes affect Triassic beds nearly everywhere in the southern Tobin Range. Such folds are juxtaposed with a widespread E-dipping homocline (structure 5) that may be younger than the NS folds.

NS folding appears to have been the first deformation of Triassic rocks in the southern Tobin Range because the faults of other phases planarly cut Triassic beds whose dips are cylindrical with N or S-plunging axes. The geologic age of NS folding is known only as pre-Oligocene.

2) **Tobin thrust zone.** This is a local deformation zone that strikes EW across the central part of the southern Tobin Range. Structurally upward or from north to south (sections A-C, Fig. 15), the zone comprises: a) footwall strata deformed in a train of N-verging folds and abundant bedding-parallel faults in a 700-1000 m structural thickness, b) the lower Tobin thrust whose S-dipping hanging wall is 300 to 1000 m thick, and c) the upper Tobin thrust that presumably underlies the southern half of the area (Fig. 15A).

The two Tobin thrusts are nonplanar, south-dipping ramps; the thrust surfaces are in fact south-plunging monoclines with steeper W to SW-dipping western limb at the western range front. The dip of the eastern limb is to the south and is uncertain between about 10-50°. The nonplane faults are primary movement surfaces; they have undergone little or no folding. The eastern limb of the lower Tobin thrust surface cuts up across the footwall section to the north such that progressively younger strata exist below the eroded hangingwall north of the thrust trace. The western limb of the lower Tobin thrust cuts down across the full section of pre-Cane Springs Triassic strata in the footwall. Both Tobin thrusts cut across section in the hanging walls whose structure consists chiefly of open N-S folds (structure 1) juxtaposed on an E-dipping homocline. Because the Tobin thrusts are nonplane and cut folded strata, some segments of the thrusts are repetitional (older over younger) whereas others are omissional (younger over older).

Deformation in the footwall of the Tobin thrust zone is a train of north-verging asymmetric major folds (Fig. 15C) and abundant layer-parallel faults. Axial planes of the folds dip S to E and axes plunge 30-50° east. The dip of the axial planes is 20-30° at the lower thrust and appears to steepen structurally downward toward the base of the deformed interval of footwall strata. There is a concomitant decrease in limb appression with increasing dip of axial plane. Bedding-parallel slip in footwall strata occurred on faults spaced 1-20 m in the lower part of the deformed zone. Such slip evidently represents simple shear in the region of gradational displacement between the tightly folded footwall layers and the unaffected strata below the deformed interval.

Displacements on the Tobin thrusts were northerly as

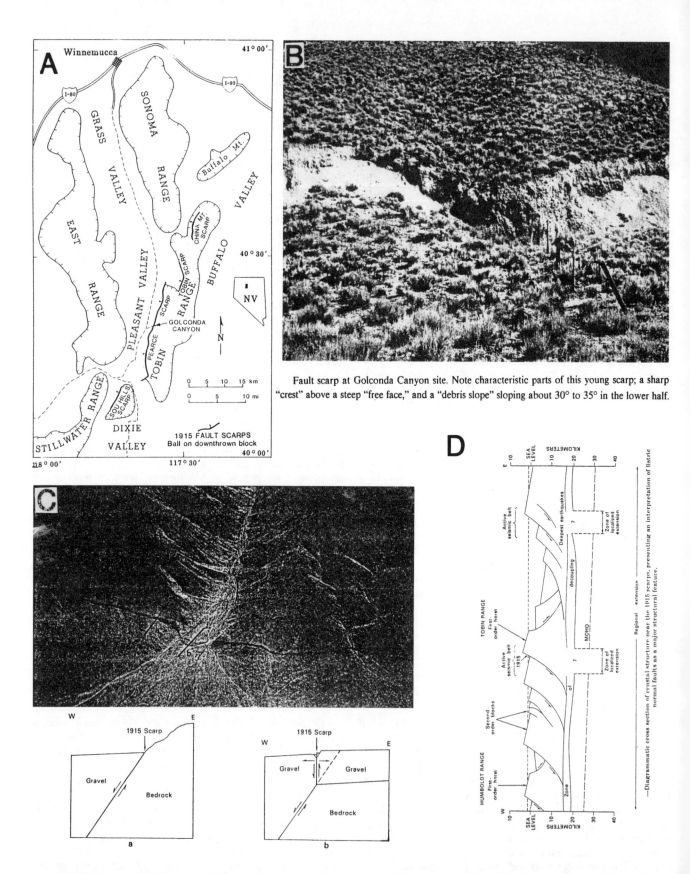

Fault scarp at Golconda Canyon site. Note characteristic parts of this young scarp; a sharp "crest" above a steep "free face," and a "debris slope" sloping about 30° to 35° in the lower half.

FIGURE 14   1915 Pleasant Valley normal fault scarp on western flank of Tobin Range (Wallace, 1984a). A) distribution of 1915 fault scarp; B) closeup of scarp at Golconda Canyon; C) oblique aerial view of 1915 scarp at stop 4 and sections at sites a and b; D) model of crustal structure showing horizontal detachment at depth and listric Tobin frontal fault.

E

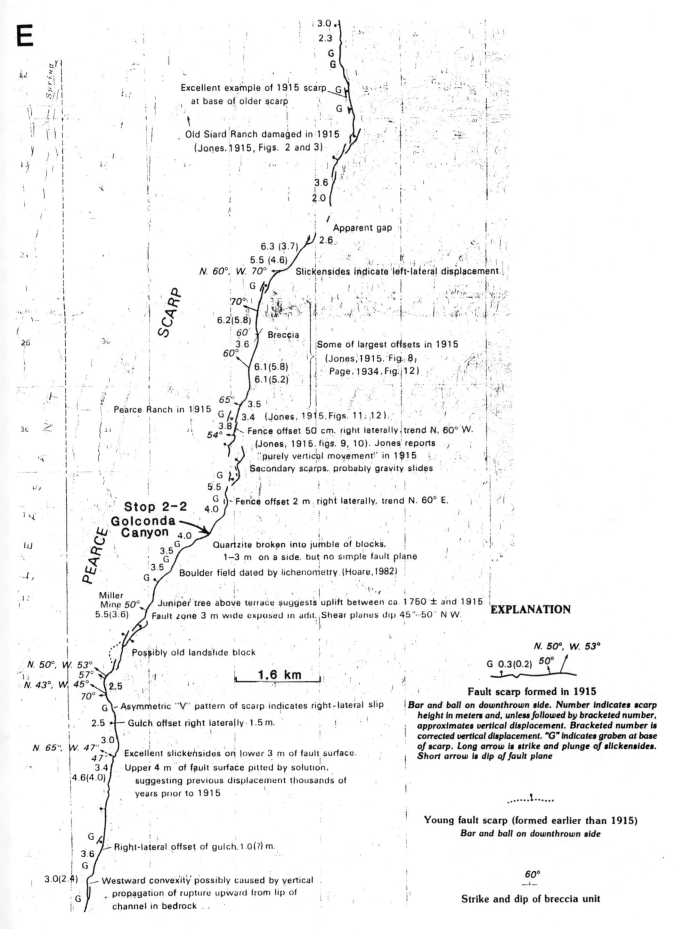

Spring

SCARP

PEARCE

i 3.0
2.3
G
G

Excellent example of 1915 scarp G
at base of older scarp
G

G

Old Siard Ranch damaged in 1915
(Jones, 1915, Figs. 2 and 3)

3.6
2.0

Apparent gap

6.3 (3.7)   2.6
5.5 (4.6)
N. 60°, W. 70°   Slickensides indicate left-lateral displacement
G

70°

6.2(5.8)
60°   Breccia
3.6
60°   Some of largest offsets in 1915
6.1(5.8)   (Jones, 1915, Fig. 8;
6.1(5.2)   Page, 1934, Fig. 12)

65°   3.5
Pearce Ranch in 1915   G   3.4   (Jones, 1915, Figs. 11, 12)
3.8
54°   Fence offset 50 cm. right laterally, trend N. 60° W.
(Jones, 1915, figs. 9, 10). Jones reports
"purely vertical movement" in 1915

G   Secondary scarps, probably gravity slides
5.5
G   Fence offset 2 m. right laterally, trend N. 60° E.
4.0

**Stop 2-2
Golconda
Canyon**   4.0
G   Quartzite broken into jumble of blocks,
3.5   1–3 m on a side, but no simple fault plane
G
3.5   Boulder field dated by lichenometry (Hoare, 1982)
G

Miller
Mine   50°   Juniper tree above terrace suggests uplift between ca. 1750 ± and 1915
5.5(3.6)   Fault zone 3 m wide exposed in adit. Shear planes dip 45°–50° N.W.

Possibly old landslide block

N. 50°, W. 53°   1.6 km
57°
N. 43°, W. 45°   2.5
70°
G   Asymmetric "V" pattern of scarp indicates right-lateral slip
2.5   Gulch offset right laterally 1.5 m.
3.0
N. 65°, W. 47°
47°   Excellent slickensides on lower 3 m of fault surface.
3.4   Upper 4 m of fault surface pitted by solution,
4.6(4.0)   suggesting previous displacement thousands of
years prior to 1915

G
3.6   Right-lateral offset of gulch, 1.0(?) m.
G

3.0(2.4)   Westward convexity possibly caused by vertical
G   propagation of rupture upward from lip of
channel in bedrock

EXPLANATION

N. 50°, W. 53°
G 0.3(0.2)   50°

Fault scarp formed in 1915

*Bar and ball on downthrown side. Number indicates scarp
height in meters and, unless followed by bracketed number,
approximates vertical displacement. Bracketed number is
corrected vertical displacement. "G" indicates graben at base
of scarp. Long arrow is strike and plunge of slickensides.
Short arrow is dip of fault plane*

........1......

Young fault scarp (formed earlier than 1915)
Bar and ball on downthrown side

60°

Strike and dip of breccia unit

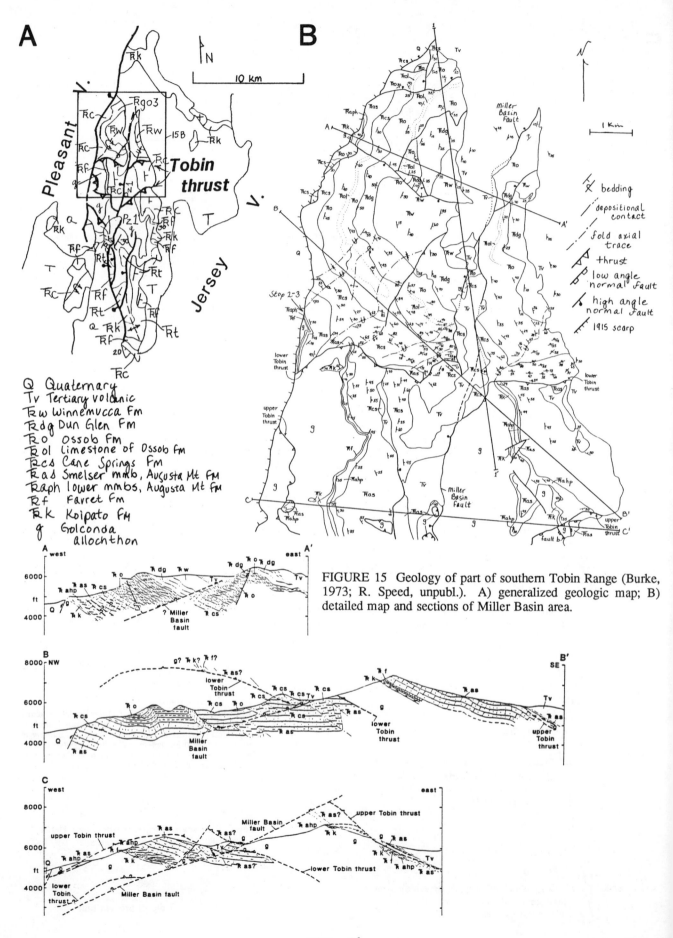

A

Q Quaternary
Ṛw Winnemucca Fm
Ṛdg Dun Glen Fm
Ṛo Ossob Fm
Ṛol Limestone of Ossob Fm
Ṛcs Cane Springs Fm
Ṛad Smelser mmb, Augusta Mt Fm
Ṛaph lower mmbs, Augusta Mt Fm
Ṛf Farret Fm
Ṛk Koipato Fm
g Golconda allochthon

B

bedding
depositional contact
fold axial trace
thrust
low angle normal fault
high angle normal fault
1915 scarp

FIGURE 15 Geology of part of southern Tobin Range (Burke, 1973; R. Speed, unpubl.). A) generalized geologic map; B) detailed map and sections of Miller Basin area.

indicated by fault plane striations, the footwall folds, and the shape of the thrust surfaces. Triassic facies of both walls of the fault zone are similar and represent strata deposited near to or north of the hinge of the Middle Triassic basin. Thus, displacement magnitude is probably less than 10 km. Because the footwall contains contractional structures and the Tobin thrusts bring Golconda rocks above the Triassic section the Tobin thrust zone can be regarded as a zone of local horizontal shortening and northward overriding. The eastern limb of the lower Tobin thrust evidently overrides a frontal ramp, and the western limb is probably a lateral or oblique ramp that transforms the exposed frontal ramp down to the west to a deeper structural level in Pleasant Valley (Fig. 15A). The proximity of the Tobin thrust zone and the region of abrupt thickening of Triassic strata in the hangingwall 2-6 km south of trace of upper Tobin thrust) implies that thrust zone is a climb of displacement due to the buttress effect of the flank of the Triassic basin in the southern Tobin Range. The distance of the basin hinge/flank in the footwall south of the trace of the Tobin thrust is unknown and depends on the displacement magnitude in the thrust zone.

It is inferred that the Tobin thrusts extend subhorizontally south and are footwall imbricates to the Fencemaker B thrust (Fig. 7).

If the east-dipping homocline of the area is of Tertiary age, attitudes of structures in the Tobin zone were rotated 30-50° east about a horizontal NS axis. Initial orientations in this case were: axes of footwall folds: subhorizontal, axial planes: SSW dipping, west limb (lateral ramp) of Tobin thrusts: subvertical. Moreover, tectonic transport indicators, when unrotated, would imply more northeasterly rather than northerly overriding.

3) **Omissional fault nappes.** These consist of fault-bound sheets of Triassic rocks that generally dip west less than 45° and are northerly striking. They mainly place younger over older strata. The orientation of folds in these nappes indicates they are not related to the Tobin thrust zone. The direction of motion of the omissional nappes is uncertain. The omissional nappes are younger than NS folding and are cut by Neogene high-angle faults (structure 6). The timing of omissional nappe motion with respect to the Tobin thrust zone (structure 2) or the Miller Basin fault (structure 4), however, is uncertain. It is reasonable to suppose that by virtue of similar attitudes, the omissional nappes and the hangingwall of the Miller Basin fault are contemporaneous west-moving normal fault nappes of Tertiary age. If correct, the idea implies that the west flank of the southern Tobin Range and the eastern Sou Hills contain a zone of distributed low-angle mid-Tertiary normal faults.

4) **Miller Basin fault.** This is a west-dipping low-angle normal fault (Fig. 15B) of known Tertiary age that forms a topographic defile in the northern half of the southern Tobin Range and a broader alluviated valley in the southern half. The alluviated valley separates the hanging wall exposed in the eastern Sou Hills from the footwall in the Tobin Range. The Miller Basin fault cuts older Tertiary tuff but is overlapped and invaded by

andesite, placing the age of slip at about 30-36 m.y.

5) **East-dipping homocline.** Triassic and Tertiary layering throughout much of the southern Tobin Range dips mainly east between 20 and 60°, averaging about 30°. Among Triassic beds, this represents a sheetdip upon which the NS folds are juxtaposed. The homoclinal structure of the Tertiary rocks almost certainly is due to rotation during normal faulting. The main question, yet unresolved, is whether a) detachment occurred within or at the base of the Tertiary section or b) whether faults causing rotation penetrate pre-Tertiary rocks.

6) **High-angle normal faults.** The youngest structures are steeply west-dipping normal faults of Neogene? and Holocene age. Scarps indicate steep surface dips, and aftershocks suggest that the Pleasant Valley fault continues planarly to depths of at least 8 km (Fig. 14). Regional evidence suggests that the main high angle normal faults have been active in the last 8-10 ma, after cessation of Tertiary volcanism.

Stop 4. Stillwater Range: Upper Strata of the Triassic Shelfal Sequence

Our objective in the Stillwater Range (Fig. 7) is to examine the Grass Valley and Dun Glen Formations, with views toward 1) facies differences of the deltaic Grass Valley and Osobb Formations and 2) the Dun Glen as a source bed for carbonate fragments in Triassic slope deposits to be seen at Stop 5.

The Grass Valley-Osobb Formation in the Stillwater Range (Fig. 16) is more pelitic and far more carbonate-rich than the Osobb formation in the Tobin Range. Various partial sections of the Grass Valley-Osobb appear to contain different proportions of sandstone, mudstone, and limestone within the limits of poor outcrop. This may represent local deltaic facies variations or may be due to tectonic juxtaposition.

The Dun Glen Formation of the Stillwater Range is not in a continuous section as previously thought and as in the East and Tobin Ranges where it is about 200 m thick. Partial sections give apparent thicknesses as great as 700 m in the Stillwater Range. This very high value could represent intraformational repetition by bedding-parallel thrusts or simply an exceptionally thick carbonate accumulation within the Norian Grass Valley-Osobb to Winnemucca deltaic system. Layer detachment is widespread in the Dun Glen, supporting the idea of tectonic thickening, but because of lithologic uniformity and meager conodont content, it will be difficult to prove which factor is more important.

The main structural features of the Stillwater Range, in their possible order of formation are: 1) Fencemaker allochthon and related footwall deformation, 2) NS folds, 3) structural domes, 4) nappes and related folds, 5) E-dipping homocline, 6) Tertiary nappe, and 7) high angle normal faults. The relative timing of 3 with respect to 1-6 and among 4-6 is poorly understood. Only 1 and 4 are discussed further.

Fencemaker allochthon and related structures. Fencemaker allochthon B underlies the entire Stillwater Range south of 40°N (Fig. 7). North of that latitude, the Fencemaker allochthon occupies a downthrown block of large Neogene throw on the west flank of the Stillwater Range with exposure as far north as 4 km N of Fencemaker Pass (Fig. 16). Thus, the allochthon probably extended originally at least as far north as Fencemaker Pass in the main body of the Stillwater Range (Figs. 7,16). Transport of the allochthon was probably between N and NE according to stretching lineations in its footwall. The transport was post 165 m.y.

Triassic shelf strata of the footwall to the Fencemaker thrust are penetratively (or nearly so) deformed in a zone from the thrust trace north to about 1 km north of Fencemaker Pass (Fig. 16). The proximity of such deformation to the Fencemaker allochthon allies the zone of penetrative structures to the emplacement of the allochthon. The penetrative structure is mainly bedding-parallel foliation that is axial planar to sporadic recumbent isoclinal minor folds. Stretching lineations within the foliation are well developed in the southern half of this zone. The northern or lower boundary of the deformation zone is a transition from penetrative foliation to widely spaced foliations where the latter probably define narrow regions of layer-parallel slip. The strain of ammonites and development of lineation diminish south to north and downward in the deformation zone.

Nappes and related folds. Many low angle faults cut the Triassic shelfal section of the Stillwater Range. These are mainly bedding-parallel or cut across bedding and formation boundaries without repetition. Such faults separate extensive nappes (Fig. 16) which collectively maintain a correct stratigraphy. Such characteristics imply the faults are omissional. Yet, within nappes, there are sets of recumbent folds that thicken the section (Fig. 16) and smaller fault-bounded packets that are repetitional, mainly within a single formation. The latter phenomena indicate at least local contraction. The nappes formed after NS folds and presumably, after Fencemaker thrusting. Their kinematic significance is not yet clear.

Stop 5. Humboldt Range: Norian Shelf-Basin Transition and Fencemaker Allochthon

The southern tip of the Humboldt Range (Figs. 7, 17A) provides views of the Fencemaker B allochthon (Figs. 17A,C); a major olistostrome, indicative of the Norian shelf-basin transition, in both hanging and footwalls of the Fencemaker thrust; and a major oblique shear zone in the footwall below the Fencemaker thrust due to Fencemaker A thrusting (Wallace and others, 1969; Heck and Speed, 1987, 1989).

General. The Humboldt Range exposes the autochthon of the late Mesozoic Winnemucca thrust belt (Fig. 4B) except at the southern tip where the Fencemaker B allochthon crops out (Fig. 17A). The Triassic shelfal strata of the northwestern and southern rims of the Humboldt Range are zones of strong deformation related to the Fencemaker A and B allochthons, respectively (Fig. 17A). The central (eastern) part of the Range is far less deformed and was apparently beyond (eastward of) the region covered by the Fencemaker allochthons. The younger

Willow Creek allochthon (Fig. 17A) backfolded Fencemaker-footwall structures in the northern Humboldt Range. Its effect in the southern Humboldt Range was weaker, an open S-plunging major anticline (Fig. 17A), due apparently to lesser proximity.

Stratigraphy. In the shelfal terrane of the Humboldt Range, the Prida Formation is long-time-ranging (Fig. 17B) and mainly composed of dark silty and cherty limestone, representing a basinal slope deposits. During Prida time east of the Humboldt Range, uplift occurred in early Ladinian time, producing a regional unconformity and succeeding platform carbonates. Westward progradation of this platform did not reach the southern Humboldt Range until late in the Ladinian. The migration is recorded by platform edge clastic carbonates (Congress Canyon Formation) which are surmounted by the Augusta Mountain and Cane Spring Formations.

The carbonate platform is overlain by delta-top and delta-front siliciclastic rocks of the Grass Valley Formation, perhaps the most distal facies of the Norian deltaic complex. In the autochthon of the southern Humboldt Range, the Grass Valley is succeeded by an olistostrome and turbidites and hemipelagites indicative of a slope environment.

Fencemaker allochthon. Detail of the Fencemaker B allochthon in the southern Humboldt Range is shown in Figure 17C. Its formational content comprises lower turbidite and hemipelagite probably contemporaneous with the Grass Valley Formation, a Norian olistostrome at intermediate levels, and upper Norian turbidite and hemipelagite. Lower Jurassic rocks crop out at the southern tip of the area (Wallace and others, 1969) but their continuity with the Norian strata is not clear. Muddy rocks of the allochthon contain first-phase cleavage that is subparallel to bedding and occasional isoclinal folds (Fig. 17G). The entire sequence of strata is in a train of major box folds and harmonics, which is second phase (section, Fig. 17A,G). The first phase structures are SE vergent and thought to have developed during Fencemaker A transport. The second phase structures underwent NE-SW contraction and are products of Fencemaker B transport.

Olistostrome. "Olistostrome" is used here as a general name for a mappable unit of submarine gravity-flow deposits composed mainly of coarse, angular, lithified rock fragments (Abbate and others, 1981). The term has typically been used to describe deposits that are probably debris flows, whereas Heck and Speed (1987) include several types of mass transport, including debris flows, turbidity flows, and density-modified grain flows, all of which are recognized in this body. Hemipelagic strata occur between the gravity flow deposits and are included as an olistostrome lithotype.

Figure 17F shows the unfolded geometry and extent of

FIGURE 16 Geology of northern Stillwater Range (R. Speed, unpubl.). A) geologic map-symbology same as Fig. 14A; B) diagrammatic NS section showing major structural features; C) detailed sections 1, 2, 3 and 4 located on Fig. 16A; D) model of possible duplication of Triassic section due to northward emplacement of Fencemaker allochthon.

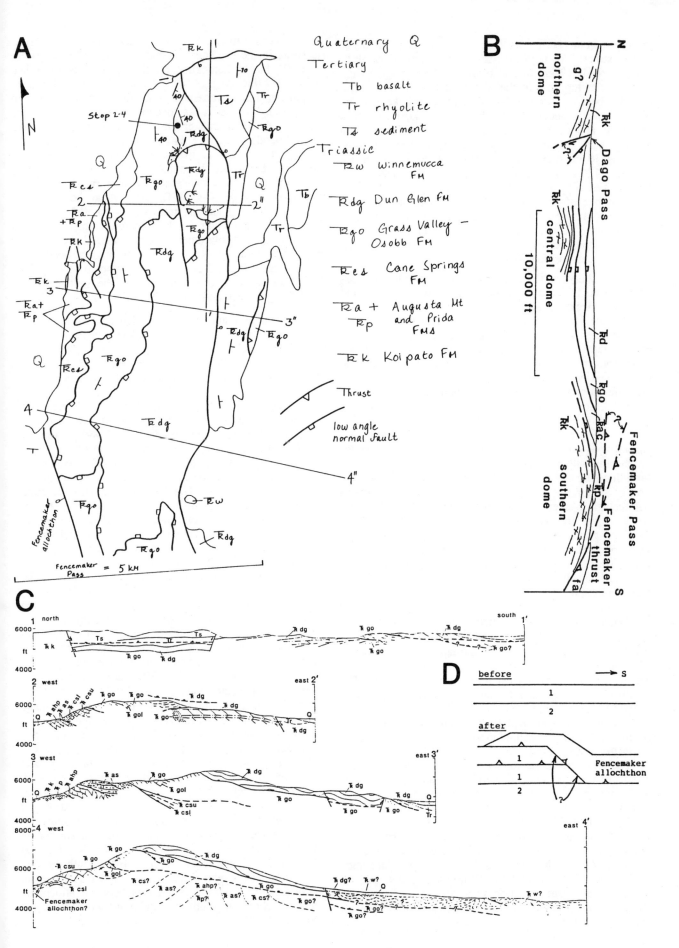

A

Quaternary Q
Tertiary
- Tb basalt
- Tr rhyolite
- Ts sediment

Triassic
- ℞w Winnemucca FM
- ℞dg Dun Glen FM
- ℞go Grass Valley – Osobb FM
- ℞cs Cane Springs FM
- ℞a + ℞p Augusta Mt and Prida FMS
- ℞k Koipato FM

⏗ Thrust

⏗ low angle normal fault

Stop 2-4

Fencemaker Pass = 5 KM

B

N
g?
northern dome
℞k
Dago Pass
℞k
central dome
10,000 ft
℞d
℞go
℞k
southern dome
℞ac
℞p
Fencemaker Pass
Fencemaker thrust
fa
S

C

1 north ... south 1'
6000 ft ... 4000

2 west ... east 2'
6000 ft ... 4000

3 west ... east 3'
6000 ft ... 4000

4 west ... east 4'
8000 6000 ft 4000

D

before → S
1
2

after
1
1
2
Fencemaker allochthon

Fencemaker allochthon?

T122: 29

the olistostrome body. Its unfolded width is about 6 km across strike. Its northwest-southeast length parallel to strike exceeds 17 km in and beyond the Pershing district. It has a flat top and, in the northwest half of the central outcrop belt, a northwest-southeast-trending keel. The olistostrome is about 720 m thick at the keel and thins rapidly to about 200 m in the southeast part of the central outcrop belt. In the northeast outcrop belt, the olistostrome is 50 m thick and locally absent; in the southwest belt, it is 120 m thick and locally absent. The olistostrome thickness at any location correlates directly with clast size and number of depositional layers; thickness variations within and between outcrop belts are therefore primarily depositional. Tectonic thickening and thinning by faulting and cleavage development occur locally but are small compared to the depositional variations.

Five lithotypes (Fig. 17F) occur within the olistostrome. Four of these (floatstone, grainstone, rudstone, and skeletal limestone) are submarine gravity-flow deposits, whereas the fifth (calcareous mudstone) is hemipelagic.

The megaclasts (up to 0.5 km long) of the olistostrome contain large folds of beds. Such folds are cut by the boundaries of the megalith, do not exist in the olistostrome matrix, and are discordant to tectonic structures. They are folds developed by downslope transport and are used with other data to determine the Norian slope vector in Figure 17D. This vector was between S56°W and S82°W.

To interpret, the olistostrome is a triangular-prism-shaped body with flat top and keel-like base that lies stratigraphically within dominantly fine-grained hemipelagic and turbiditic siliciclastic strata. It formed by episodic deposition of submarine carbonate gravity flows that punctuated hemipelagic background sedimentation of calcareous mudstone. Mass transport occurred by debris flow, density-modified grain flow transitional to debris flow, and turbidity flow. Strata that occur as large clasts in the olistostrome were folded during downslope transport, and these indicate that the debris flows were deposited down a west-southwest-facing slope that existed just northeast of the southern Humboldt Range. Early deposition occurred as frequent small flows, whereas later deposition transported huge volumes of shelfal material in two discrete events. The gravity flows filled and lapped over the flanks of a pre-existing trough at lower slope or base-of-slope. The trough geometry suggests that it is a half-graben. The olistostrome is late early Norian and formed by penecontemporaneous resedimentation of the upper lower Norian Dun Glen Formation, a platformal limestone that formed on the surface of a shallow-marine delta northeast of the olistostrome (Fig. 17E).

The olistostrome constrains the position and orientation of part of the Late Triassic shelf-basin transition and leads to the interpretation of an arcuate Triassic margin convex to the southwest (Fig. 17L) whose change in trend at the southern Humboldt Range was inherited from pre-Triassic time and not due to post-Triassic rotation. The olistostrome occurs in both walls of the Fencemaker thrust and indicates that displacements at the toe of the allochthon were small. Proximity of the Fencemaker thrust and the Triassic shelf-basin transition indicates that the transition strongly influenced the style, location, and orientation of Jurassic structures.

Southern autochthon. Shelfal and shelf-margin Triassic strata in the southern autochthon (Figs. 17A, 17H) are in a SW-facing homocline on the west flank of a late major anticline (Fig. 17H). Formations from Prida to Norian strata are represented (refer to tier nomenclature in Fig. 17B). These rocks occupy a major shear zone between the Fencemaker thrust and gradational lower strain boundary some 5 km northeast (Fig. 17H). The shear zone dips SW at 40° and is about 3 km thick (Figs. 17H, I). Stretching lineations and cleavage-shear zone boundary relations indicate the shear direction (tectonic transport) is ESE (Fig. 17I). The shear zone is thus mainly strike slip and its transport is parallel to that of Fencemaker allochthon A. It is therefore interpreted that Fencemaker A displacements carried around the southern end of the Humboldt Range, where an original bend in the shelf margin (Fig. 17K) caused the ramp to change from frontal southeast to the lateral one in the southern Humboldt Range (Heck and Speed, 1989).

The southern autochthon experienced post-shear zone deformation (Fig. 17G). The second phase was NE-verging kinks and minor fold trains. These developed during Fencemaker B emplacement. Figure 17J contrasts phase 2 structures in the allochthon and autochthon.

FIGURE 17 Geology of the Humboldt Range and vicinity (Silberling and Wallace, 1967; Wallace and others, 1969; Heck and Speed, 1987, 1989). A) major structural features and tectonostratigraphy of the Humboldt Range; B) stratigraphy of southern Humboldt Range (Pershing District); C) geology of Fencemaker allochthon in southern Humboldt Range; D) orientation data for structures in olistostrome in Fencemaker allochthon: (1) tectonic folds and slump folds for whole unit; (2) folds within two megaliths; (3) paleoslope direction interpreted from slump folds after removal of tectonic rotations; direction between S56W and S82W; E) reconstruction of shelf-basin slope and original olistostrome geometry in southern Humboldt Range; F) unfolded geometry of olistostrome; G) deformation phases in the Humboldt Range (areas identified in Fig. 17A); H) structure of southern Fencemaker autochthon; (1) map showing megashear zone between Fencemaker thrust and inner strain boundary and position of Relief Mine mesoshear zone (Fig. 17I); stratigraphic tiers: t$_1$: Prida and Koipato Fms., t$_2$: Augusta Mt. and Cane Springs Fms., t$_3$: Grass Valley Fm.; (2) diagrammatic section across megashear zone with p3 fold removed; (3) orientation data for phase 1 and 2 structures divided between domains (dI and dII, inset map) and tiers; sln: stretching lineations; distribution of bedding and cleavage in dII due to p3 folding (Heck and Speed, 1989); I) Relief Mine shear zone geometry and orientations; position on Fig. 17H; J) contrast in p2 structural style between Fencemaker allochthon and autochthon in southern Humboldt Range, due to constant shortening but different initial layer orientation; K) regional extents of Fencemaker A and B allochthons and footwall deformation zones; L) two models to explain contrast in p1 deformation facies between allochthon and autochthon in southern Humboldt Range (Heck and Speed, 1989).

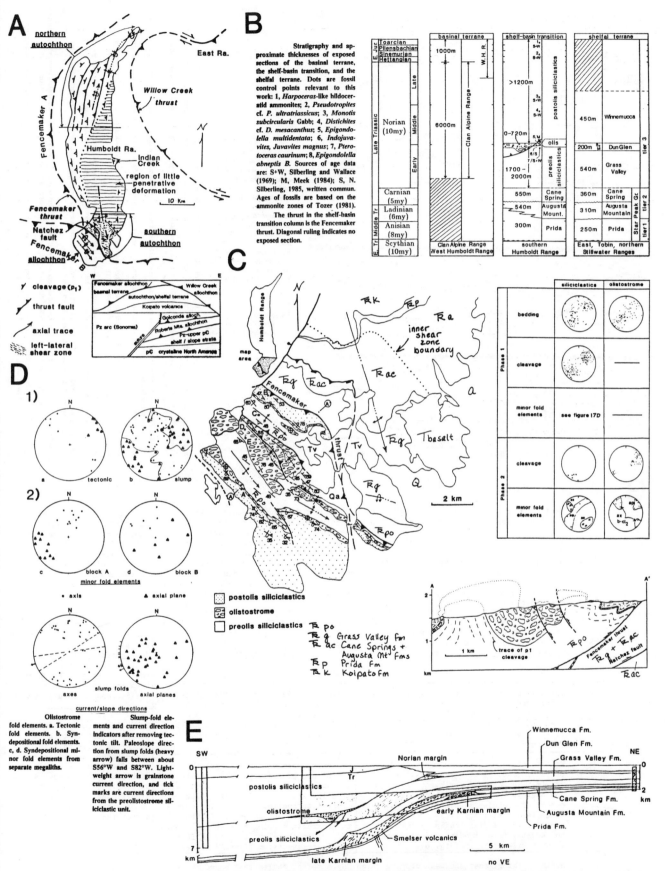

B Stratigraphy and approximate thicknesses of exposed sections of the basinal terrane, the shelf-basin transition, and the shelfal terrane. Dots are fossil control points relevant to this work: 1, *Harpoceras*-like hildoceratid ammonites; 2, *Pseudotropites* cf. *P. ultratriassicus*; 3, *Monotis subcircularis* Gabb; 4, *Distichites* cf. *D. mesacanthus*; 5, *Epigondolella multidentata*; 6, *Indojuvavites, Juvavites magnus*; 7, *Pterotoceras caurinum*; 8, *Epigondolella abneptis B*. Sources of age data are: S+W, Silberling and Wallace (1969); M, Meek (1984); S, N. Silberling, 1985, written commun. Ages of fossils are based on the ammonite zones of Tozer (1981).

The thrust in the shelf-basin transition column is the Fencemaker thrust. Diagonal ruling indicates no exposed section.

D Olistostrome fold elements. a. Tectonic fold elements. b. Syndepositional fold elements. c, d. Syndepositional minor fold elements from separate megaliths.

Slump-fold elements and current direction indicators after removing tectonic tilt. Paleoslope direction from slump folds (heavy arrow) falls between about S56°W and S82°W. Light-weight arrow is grainstone current direction, and tick marks are current directions from the preolistostrome siliciclastic unit.

minor fold elements
• axis ▲ axial plane

current/slope directions

postolis siliciclastics
olistostrome
preolis siliciclastics

ꓥpo
ꓥg Grass Valley Fm
ꓥac Cane Springs + Augusta Mt Fms
ꓥp Prida Fm
ꓥk Koipato fm

E Northeast-southwest cross section at end of about Early Jurassic showing inferred shelf/basin relations at the southern Humboldt Range before Middle Jurassic contractile deformation. Areas providing stratigraphic control are boxed; other relations are extrapolated or are based on interpretations in text. We have assumed the trough containing the olistostrome keel is a half-graben.

F

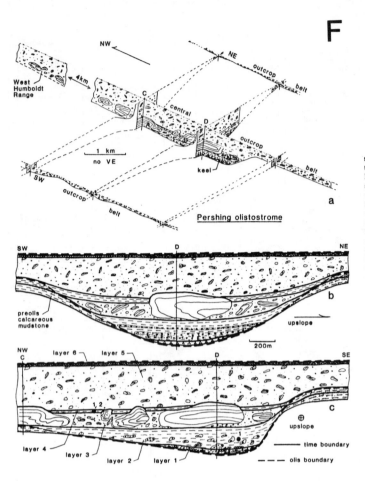

SW outcrop belt

West Humboldt Range

←4km

NW

NE

C

central outcrop belt

1 km
no VE

D

keel

outcrop belt

SW outcrop belt

Pershing olistostrome

a

SW   D   NE

preolis calcareous mudstone

upslope

200m

b

NW   layer 6   layer 5   SE
C   D

layer 4
layer 3   layer 2   layer 1

1, 2

upslope

⊕

c

time boundary
olis boundary

LITHOTYPES OF THE PERSHING OLISTOSTROME

Floatstone: includes all particle sizes; matrix supported except for some megaliths and large mesoliths; no internal stratification; deposited by debris flow; constitutes 90%-95% of olistostrome.

Megalith floatstone: holds all megaliths; mesoliths and smaller clasts float in matrix of dominantly carbonate with subordinate siliciclastic mudstone; occurs in olistostrome keel as a discrete layer.

Mesolith floatstone: holds only mesolith and smaller size clasts; matrix dominantly carbonate with subordinate siliciclastic mudstone; occurs in all outcrop belts as a discrete layer.

Pebbly floatstone: holds mostly pebbles and smaller size clasts; matrix dominantly carbonate mudstone; occurs in olistostrome keel as part of a discrete layer, and supports megaliths and mesoliths in other floatstone layers.

Grainstone: limestone, dolomite sand; caps olistostrome; gradational or in sharp contact above rudstone; well sorted; commonly shows normal grading, cross-lamination, plane lamination; occurs in all outer 2 belts; turbidity current deposition; constitutes 2%-4% of olistostrome.

Rudstone: limestone, dolomite pebbles, clast supported; well sorted; above and gradational with mesolith floatstone; commonly shows normal grading; occurs in all outcrop belts; turbidity current deposition; 2%-4% of olistostrome.

Skeletal limestone: mostly supported by pebbles of cemented biogenic fragments; matrix of coarse and skeletal debris and lime mud; occurs in lenses exhibiting faint internal layering 1- 5 m thick; occurs at base of olistostrome keel; probably deposited by density-modified grain flow; <1% of olistostrome.

Calcareous mudstone: siliciclastic mudstone with some calcite and containing thin planar beds of dark impure limestone and laminae to thin beds of calcareous very fine grained quartz sandstone; occurs between floatstone layers; hemipelagic deposition.

Cross sections showing olistostrome stratigraphy and geometry. Layer 1, skeletal limestone (triangle pattern) with pebbly floatstone; layer 2, pebbly floatstone with hemipelagic calcareous mudstone; layer 3, megalith floatstone; layer 4, hemipelagic calcareous mudstone; layer 5, mesolith floatstone; layer 6, rudstone/grainstone. Intralayer distribution of particles and lithotypes is mostly schematic.

Large dots are fossil positions: 1, *Indojuvavites*; 2, *Juvavites magnus*.

G

	NORTHERN AUTOCHTHON	SOUTHERN AUTOCHTHON	ALLOCHTHON
PHASE 1	Northwest to west dipping, bedding parallel, penetrative cleavage axial planar to southeast to east verging minor folds of bedding and containing dip parallel stretching lineations; also in small east verging shear zones; becomes spaced and oblique to bedding on east side of range. Late stage east verging folds and thrusts deform the penetrative cleavage.	Southwest dipping, southeast verging, bedding parallel, left-lateral megashear zone with penetrative, bedding parallel cleavage containing a strike parallel stretching lineation in the shear zone center and spaced cleavage oblique to bedding in the shear zone boundary region. Olistostrome is penetrativley deformed.	Southeast verging minor folds and thrusts and southeast verging cleavage all confined to the fine-grained siliciclastic strata. Olistostrome is mostly undeformed.
PHASE 2	Northwest trending, subvertical crenulation cleavage that is widely spaced and spottily developed in fine-grained siliciclastics.	West to northwest trending, subvertical crenulation cleavage that is locally well developed and axial planar to north to northeast verging minor folds of p1 cleavage. Natchez thrust fault formed.	Northwest trending major upright box folds of beds and p1 structures with development of subvertical axial planar cleavage in the olistostrome. Northwest trending reverse faults and the Fencemaker thrust formed.
PHASE 3	North to northeast trending, west overturned, subrecumbent syncline with local development of west verging cleavage and minor folds. Overturns p1 structures to the west at the north end of the range.	North trending upright major anticline that reorients part of p1 major shear zone and p2 minor structues from southwest to southeast dips on southeast side of southern autochthon.	The allochthon contains numerous northwest to northeast striking high angle normal and strike-slip faults whose origin with respect to p1-p4 is uncertain. They do not significantly reorient demonstrable p1 and p2 structures.
PHASE 4	West to northwest verging minor folds of bedding and p1 cleavage localized proximal to west and northwest dipping low angle normal faults. Local development of crenulation cleavage in fine-grained siliciclstics. Partly responsible for distribution about northeast axis of bedding and cleavage on west side of range (Fig. 6).	Southwest verging minor folds of beds, and p1 and p2 cleavage localized beneath the Natchez fault which was reactivated as a low angle normal fault. Subhorizontal to shallow west dipping normal faults offset and occur at high angles to southeast dipping beds on southeast side of southern autochthon. Effects mostly removed from Figure 7 stereonets.	
PHASE 5	North striking, east and west dipping, high angle normal faults that bound the range and offset older structures in the range. Tertiary volcanics dip from 0 to 25 degrees east due to rotation on these faults about subhorizontal northerly axes. Older structural trends change by less than 10 degrees with the maximum 25 degree rotation which is considered insignificant.		

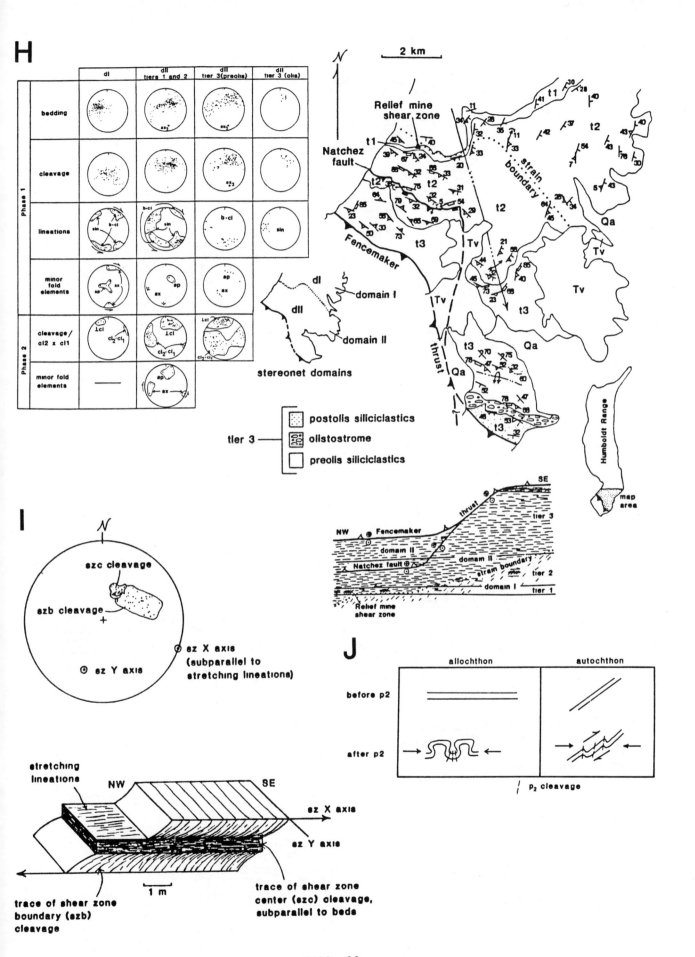

K

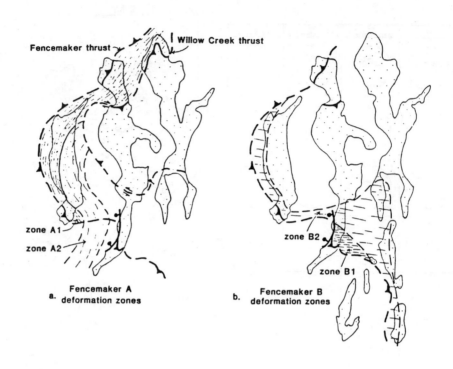

Fencemaker thrust

Willow Creek thrust

zone A1

zone A2

a. Fencemaker A deformation zones

zone B2

zone B1

b. Fencemaker B deformation zones

L

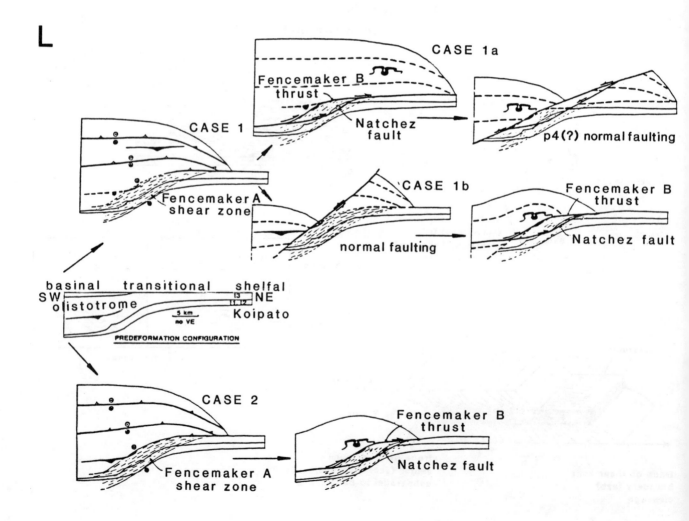

CASE 1a

Fencemaker B thrust

Natchez fault

p4(?) normal faulting

CASE 1

Fencemaker A shear zone

CASE 1b

normal faulting

Fencemaker B thrust

Natchez fault

basinal   transitional   shelfal

SW   olistotrome   NE

Koipato

5 km
no VE

PREDEFORMATION CONFIGURATION

CASE 2

Fencemaker A shear zone

Fencemaker B thrust

Natchez fault

Triassic platform-basin slope. The major expression of slope deposition in the southern Humboldt Range is the olistostrome, described above, that contains huge carbonate blocks, debris flows, and turbidity current flows, and is contemporaneous with the Dun Glen Formation (Heck and Speed, 1987). Syndepositional folds in olistostrome clasts indicate deposition down a northwest to north-northwest striking shelf-basin transition. This orientation of the Late Triassic margin at the southern Humboldt Range, together with the orientation of the Middle Triassic margin (Congress Canyon Formation) which trends north-northeast from the southern to the northern Humboldt Range (Nichols and Silberling, 1977) indicates a long standing Triassic margin with an arcuate trace convex to the southwest, approximately along the Fencemaker thrust trace (Fig. 7). North of the Humboldt Range the margin is apparently beneath the Fencemaker allochthon because only shelf interior Upper Triassic strata are exposed in the Fencemaker autochthon in the East Range (Fig. 7) (Elison, 1987). To the southeast, the margin turns south beneath the Fencemaker allochthon because the New Pass Range (Fig. 7) contains Upper Triassic shelf interior strata. The arcuate shape of the Triassic platform-basin margin (Fig. 17K) approximates that of the western sialic edge of North America according to isotopic and other data (Kistler and Peterman, 1973) and may therefore have been inherited from late Precambrian rifting and passive margin formation.

Fencemaker allochthon development. The Fencemaker thrust system is a consequence of two Jurassic deformation events. First was Fencemaker A during which basinal sediments were transported east to southeast. Second was Fencemaker B which transported basinal rocks north to northeast. Deformation of the Fencemaker A autochthon in the Humboldt Range was heterogeneous in orientation and style as a consequence of an arcuate Triassic shelf margin relative to rectilinear Fencemaker A motion (Fig. 17Ka). Frontal thrust ramp structures formed at the northern Humboldt Range where the motion direction was about normal to slope contours (Triassic shelf-basin transition and uplifted basement at the core of the shelf) and oblique thrust ramp structures formed in the southern Humboldt Range where the motion was subparallel to slope contours. East and southeast of the southern Humboldt Range, Fencemaker A deformation did not occur, (Fig. 17Ka) indicating a deformation gradient existed in rocks south of the frontal ramp. Penetrative fabrics formed early in the Fencemaker A autochthon as thrust slices of the basinal terrane piled up at the shelf margin. Footwall deformation diminished rapidly shelfward as the allochthon thinned across the shelf margin. In the northern Humboldt Range, brittle failure of the footwall with east-southeast transport deformed ductile Fencemaker A fabrics implying footwall unroofing late in Fencemaker A. In the southern Humboldt Range, late Fencemaker A brittle deformation occurred in the allochthon due either to denudation of the allochthon or earlier southeastward transport to shallow levels.

The Fencemaker B allochthon converged north to northeast against the southwest facing segment of the shelf-basin transition (Fig. 17Kb). Fencemaker B structures overprint Fencemaker A structures in the Humboldt Range and in the allochthon west and north of the Humboldt Range. Fencemaker B imposed the first deformation on rocks of the basinal sequence east and southeast of the Humboldt Range. The Fencemaker thrust at the southeast of the southern Humboldt Range is either a structure of Fencemaker B time or, perhaps in the Humboldt Range, reactivated as a southwest dipping normal fault(Fig. 17L). The position and orientation of the Fencemaker thrust, the orientation of structures in the Fencemaker allochthon, and variations in the style of footwall deformation in the Humboldt Range were mostly controlled by the position and orientation of the Triassic shelf-basin transition. The regional pattern of Fencemaker deformation in the autochthon was apparently controlled more by the shape of the uplifted basement ridge on which the Triassic shelf was built (Heck and Speed, 1987).

DAY 3: STRATIGRAPHY AND STRUCTURE OF LOWER MESOZOIC ROCKS OF THE WALKER LANE TERRANE: SAND SPRINGS RANGE, LODI HILLS, AND PARADISE RANGE

Introduction

In travelling late on Day 2 from the Humboldt Range to Fallon, by way of Lovelock, the field trip route crosses the entire belt of lower Mesozoic, continentally derived, basinal siliciclastic and carbonate rocks (Figs. 4B, 7) that form much of the Fencemaker allochthons and, in the terminology of Silberling and others (1987), the Jungo terrane. The dual tectonostratigraphic terminology used by the trip leaders reflects emphasis by Speed, on the one hand, on Sonomia and the pervasive basement thought to have been created by its Permian-Triassic collision with North America, and emphasis by Silberling, on the other hand, on large displacements of lower Mesozoic rocks which may have greatly disrupted any original basement formed of Sonomia.

At Fallon, beneath the thick Cenozoic fill of the Carson Sink, (Fig. 6), the pre-Tertiary basement would presumably include lower Mesozoic rocks originally formed more Pacificward than those of the Jungo terrane. To the south of Fallon in west-central Nevada, such rocks are part of the Walker Lake terrane (Figs. 3C, 18) which includes important amounts of volcanogenic rocks but which received no clastic sediment of obvious continental derivation until about the beginning of Jurassic time.

On Day 3 the field trip route heads east from Fallon to the indefinitely located outboard edge of the Jungo terrane and then courses southward through rocks of the Walker Lake terrane, eventually arriving in the town of Hawthorne.

Walker Lake Terrane

The Walker Lake terrane (Fig. 3C, 18) (Silberling and others, 1987) is a tectonically and stratigraphically complex assemblage of upper Paleozoic and lower Mesozoic rocks, commonly volcanogenic, whose structural and depositional histories are somewhat alike. The terrane contacts

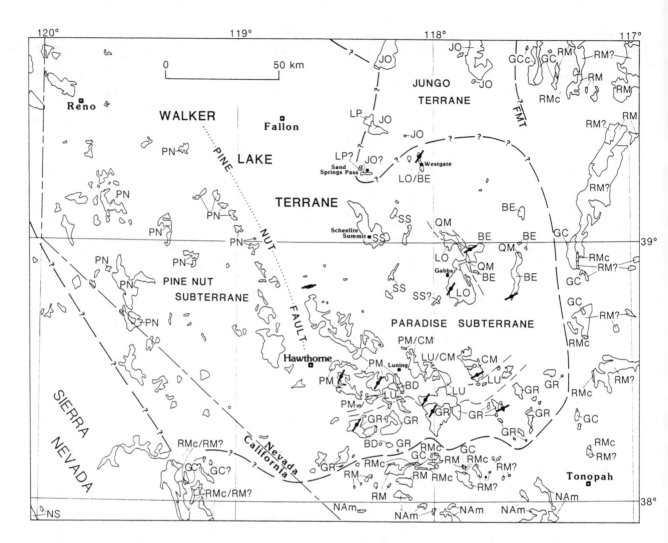

FIGURE 18 Outcrop map showing distribution of pre-Tertiary stratified rocks in the Walker Lake terrane and surrounding parts of Nevada and California, and their terrane and surrounding parts of Nevada and California, and their terrane, subterrane, and allochthon assignments (from Oldow, 1984; Schweickert and Lahren, 1987; Silberling and others, 1987). Bold lines with central dots are reconstructed strike lines of axial planes of initial deformation within Paradise subterrane (modified from Oldow, 1984); BD, Black Dyke allochthon; CM, Cedar Mountain allochthon; GC, Golconda terrane; Gcc, lower Mesozoic cover of Golconda terrane; GR, Gold Range allochthon; JO, Jungo terrane; LO, Lodi allochthon; LP, La Plata allochthon; LU, Luning allochthon; NAm, nonaccretionary North America; NS, Northern Sierra terrane; PM, Pamlico allochthon; PN, Pine Nut subterrane; RM, Roberts terrane; RMc, pre-Golconda cover of Roberts terrane; QM, Quartz Mountain allochthon; SS, Sand Springs allochthon.

the Golconda allochthon at its southern and eastern boundaries and the Jungo terrane at its northern (Figs. 3C, 18). The Jungo terrane is the basinal terrigenous Triassic sequence of north-central Nevada (Figs. 4A, 7). The Paleozoic rocks of the Walker Lake terrane, all volcanogenic, are included in Sonomia by Speed (1979) (Fig. 3D).

Tectonostratigraphic divisions of the Walker Lake terrane consist of subterranes (Pinenut and Paradise) and allochthons within the Paradise (Fig. 18). The two subterranes consist of somewhat similar stratigraphic sequences but markedly different structural histories (Oldow, 1984a); they are divided by an unexposed discontinuity in structural style called the Pinenut Fault by Oldow (1984a) (Fig. 18).

Paradise Subterrane

Several different allochthons of stratigraphically interrelated lower Mesozoic and upper Paleozoic rocks are grouped together into the Paradise subterrane. Among these are the Lodi, Pamlico, Berlin, Cedar, Luning, and Pilot allochthons (Fig. 18). All of these share some of the same general Upper Triassic-Lower Jurassic stratigraphic section (Fig. 19), but they are distinguished by marked stratigraphic variations that are juxtaposed across the allochthon-bounding faults. Traces of these allochthon boundaries are shown on Figure 18 only where they are relatively well constrained. Because they are the end result of Cenozoic extensional tectonics superimposed on

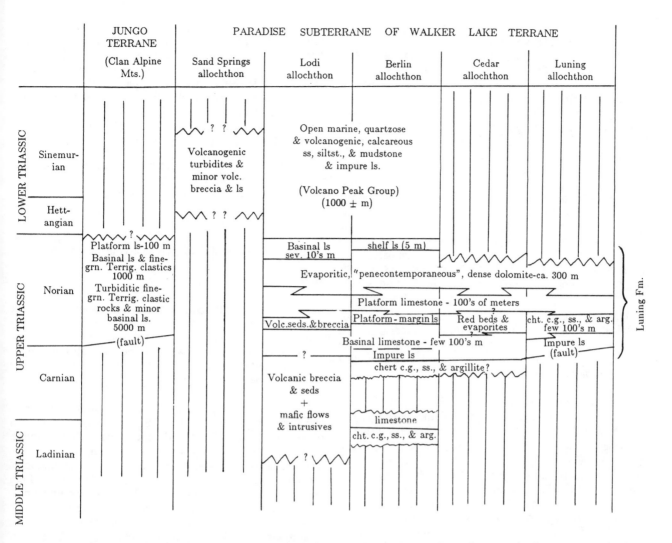

	JUNGO TERRANE	PARADISE SUBTERRANE OF WALKER LAKE TERRANE				
	(Clan Alpine Mts.)	Sand Springs allochthon	Lodi allochthon	Berlin allochthon	Cedar allochthon	Luning allochthon

FIGURE 19 Correlation chart of upper Middle Triassic to lower Lower Jurassic strata in Jungo terrane and Paradise subterrane of Walker Lake terrane.

geometrically different generations of Mesozoic deformations, their configuration beneath the extensive Cenozoic cover is speculative.

In general, these allochthons are formed of volcanogenic rocks regarded as part of Sonomia as well as thick, mainly marine, sections of lower Mesozoic rocks. The older parts of the Triassic are dominated by volcanogenic and locally derived clastic rocks, which give way upwards into the marine carbonate rocks of the Luning Formation (and its equivalents) that characterize the Upper Triassic. Beginning abruptly in very latest Triassic time, deposition changed to largely quartzose, fine-grained, open-marine, calcareous clastic sediments which are the first that were clearly derived from continental sources. These deposits, of the Volcano Peak Group (Taylor and others, 1983), range up through most of the Lower Jurassic and then grade upwards in the Pliensbachian or Toarcian into quartzose, partly eolian, sandstones, evaporitic supratidal dolomites, and nonmarine volcanogenic and coarse clastic rocks of the Dunlap Formation.

In the interpretation that views the Mina Deflection as a bight in the Paleozoic continental margin formed prior to Jurassic time (Fig. 4C), rocks of the Paradise subterrane were imbricated and thrust over the southern margin of the deflection during pronounced Middle Jurassic or younger, northwest-southeast tectonic contraction. This is the earliest, and most intense, of Jurassic-Cretaceous deformations to affect these rocks; the aggregate tectonic shortening is estimated to be on the order of hundreds of kilometers (Oldow, 1984a). The degree to which older structures that bounded the rocks of the Golconda and Roberts Mountain allochthons now found along the southern margin of the Mina Deflection have been disrupted and displaced southeastward by this Mesozoic deformation is uncertain but could be substantial. The northeast-trending structures formed by the first Jurassic-Cretaceous deformation of rocks of the Paradise subterrane are cut and reoriented by northwest-trending folds and thrusts formed during several subsequent phases of northeast-southwest late Mesozoic shortening. It is tempting to relate these two main phases of deformation to Fencemaker A and B events to the north.

Lower Mesozoic rocks in the more southern and eastern allochthons of the Paradise subterrane are depositionally related to the Mina Formation (Speed, 1977), an upper Paleozoic chert-volcaniclastic complex regarded as part of Sonomia, if not a part of the Golconda allochthon itself. Lower Jurassic orogenic clastic rocks of the Dunlap Formation and younger volcanogenic rocks rest depositionally on the Mina Formation in the Pilot allochthon, and coarse chert conglomerates in the Upper Triassic Luning Formation of the Luning and Berlin allochthons are evidently derived from the Mina Formation. Thus, the shelfal Upper Triassic strata of at least the Luning and Berlin allochthons are paleotectonically analogous to the Triassic shelf deposits, such as those of the Star Peak Group, that depositionally overlie the Golconda allochthon farther north in northcentral Nevada. The more northern and western, and inferred structurally higher, allochthons of the Paradise subterrane, such as the Pamlico and Lodi allochthons, are markedly more volcanogenic compared with rocks of the Luning and Berlin allochthons, as are the partly contemporaneous rocks of the Sand Springs allochthon.

The Sand Springs allochthon, which cannot be stratigraphically related to other parts of the Paradise subterrane, is regarded as the structurally highest part of the accretionary complex resulting from the SE-vergent, first phase of Jurassic-Cretaceous deformation.

Sand Springs allochthon. In the Sand Springs Range and the hills surrounding Gabbs Valley, the rocks assigned to the Sand Springs allochthon are mainly turbiditic volcanogenic sandstones and some volcanic breccia and very minor amounts of limestone. These rocks have been complexly deformed by intrusion of extensive granitic plutons, such as the Sand Springs pluton of Cretaceous age, and by earlier polyphase Mesozoic deformations. The best evidence of the age of these rocks is a poorly preserved ammonite of probable Early Jurassic age collected by A.W. Gelber from volcanogenic turbidites in the Black Hills at the edge of Gabbs Valley.

Lodi allochthon. In the Paradise Range and Lodi Hills, lower Mesozoic strata of the Lodi allochthon are nearly juxtaposed with those of the Berlin allochthon, being separated from them only by a 1-2 kilometer wide strip of disrupted intrusive rocks and presumed lower Paleozoic rocks of the Quartz Mountain allochthon. The Quartz Mountain allochthon is thought to have been carried westward over the Berlin allochthon and imbricated within the Paradise subterrane by a second-phase Jurassic-Cretaceous deformation. The stratigraphically lowest exposed part of the Lodi allochthon is composed of volcaniclastic rocks. These contrast strongly with correlative epiclastic chert conglomerate, sandstone, and argillite in the nearby Berlin allochthon, although intercalated and overlying Upper Triassic carbonate units are similar in both allochthons (Fig. 19). The volcaniclastic lower part of the Lodi allochthon is exposed in the overturned section forming the partly sheared out lower limb of the Gabbs nappe (LOu, Figs. 20, 21), which crops out continuously along the west side of the Lodi Hills and Paradise Range. The Gabbs nappe formed during the southeast-vergent first phase of Jurassic-Cretaceous deformation and has been refolded and faulted against its lower plate by major

second phase Jurassic-Cretaceous structures. In the southwestern part of the Paradise Range, second phase refolding is minimal, and the effects of the first deformation are well displayed in rocks representing structurally thinned, but successive overturned parts of the stratigraphic section. Strong metamorphic fabrics and stretching lineation developed here are among the most intense and deep-crustal late Mesozoic deformational effects observed among Mesozoic rocks in western Nevada.

As compared with the Berlin allochthon to the east, rocks of the Lodi allochthon manifest much more shortening resulting from the first-deformation southeast tectonic transport. This suggests that the original tectonic boundary between the Lodi and Berlin allochthons may have been a left-lateral fault prior to northeast-southwest contraction during later phases of Jurassic-Cretaceous deformation.

Stop 1. Fairview Valley

Three very different, tectonically juxtaposed, lower Mesozoic sequences crop out in the ranges surrounding Fairview Valley. To the west of our vantage point at the intersection of U.S. Highway 50 and the Scheelite Highway, metamorphic rocks that crop out along Highway 50 where it crosses the Sand Springs Range may belong to the Jungo terrane. About 15 km north, at the south end of the Stillwater Range (Figs. 7, 18), the Jungo terrane is unquestionably represented by less metamorphosed basinal limestones and fine-grained terrigenous clastic rocks. About 15 km to the east of our vantage point are carbonate and calcareous clastic rocks having the stratigraphic sequence characteristic of the upper part of either the Lodi or the Berlin allochthons which have similar uppermost Triassic and lower Jurassic sequences. To the south in the Sand Springs Range are volcaniclastic rocks and minor limestones of the Sand Springs allochthon. The Lodi, Berlin, and Sand Springs allochthons are the northernmost rocks included in the Paradise subterrane of the Walker Lake terrane.

The spotted quartz-mica phyllite, which may be Jungo terrane in cuts along Highway 50 on the west side of Sand Springs Pass, is capped by klippen of marble in the low hills south of the highway from the summit eastward. The significance of this thrust and terrane assignment of the marble is unknown.

About 3 km northwest of Sand Springs Pass, terrigenous clastics like those of the Jungo terrane are thrust over a sequence of marble; fossiliferous, early Mesozoic, calc-silicate, quartzitic hornfels; and felsic volcanic conglomerate and sedimentary breccia having quartzitic matrixes. This peculiar metamorphosed sequence resembles that of the La Plata allochthon which is in fault contact with the Jungo terrane farther north in the Stillwater Range (Fig. 18) (Silberling, unpubl. data). Rocks of the La Plata allochthon are more like Upper Triassic and Jurassic strata in various allochthons of the Walker Lake terrane than those known to occur in the Jungo terrane, and with considerable uncertainty they constrain the Walker Lake-Jungo terrane boundary as drawn on Figure 18.

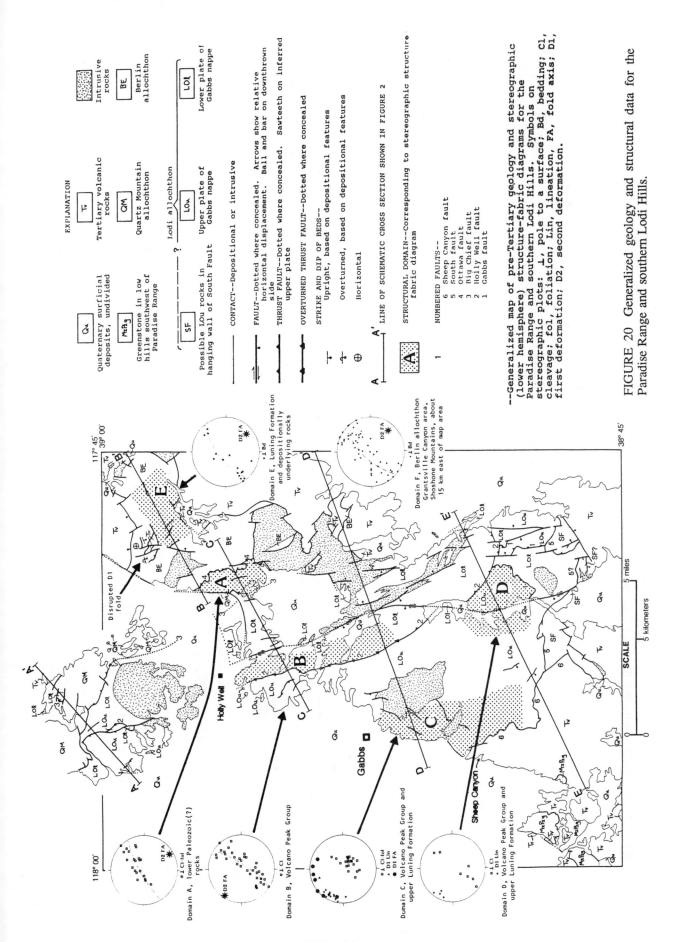

FIGURE 20 Generalized geology and structural data for the Paradise Range and southern Lodi Hills.

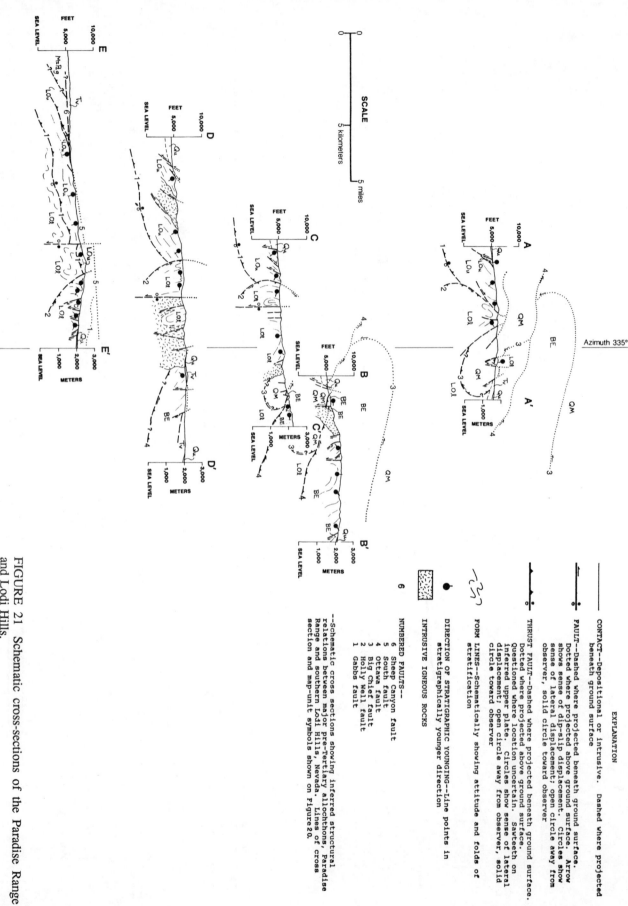

FIGURE 21 Schematic cross-sections of the Paradise Range and Lodi Hills.

EXPLANATION

CONTACT--Depositional or intrusive. Dashed where projected beneath ground surface

FAULT--Dashed where projected above ground surface. Arrow shows sense of dip-slip displacement. Circles show sense of lateral displacement; open circle away from observer, solid circle away from observer

THRUST FAULT--Dashed where projected beneath ground surface. Dotted where projected above ground surface. Questioned where location uncertain. Sawteeth on inferred upper plate. Circles show sense of lateral displacement; open circle away from observer, solid circle toward observer

FORM LINES--Schematically showing attitude and folds of stratification

DIRECTION OF STRATIGRAPHIC YOUNGING--Line points in stratigraphically younger direction

INTRUSIVE IGNEOUS ROCKS

NUMBERED FAULTS--
6 Sheep Canyon fault
5 South fault
4 Ottawa fault
3 Big Chief fault
2 Holly Well fault
1 Gabbs fault

--Schematic cross sections showing inferred structural relations between major pre-Tertiary allochthons, Paradise Range and southern Lodi Hills, Nevada. Lines of cross section and map-unit symbols shown on Figure 20.

Stop 2. Scheelite Summit

Penetratively cleaved, volcanogenic turbidites and associated rocks of the Sand Springs allochthon are well exposed in roadcuts near the summit and in the wash along the east side of the highway. Several attempts to obtain conodonts from the marbles here have been fruitless.

Stop 3. Westgate

Returning to Highway 50, Upper Triassic and Lower Jurassic strata belonging to either the Lodi or Berlin allochthon can be viewed north of the highway from a vantage point in Stingaree Valley just west of Westgate. On the east side of the valley, Lower Jurassic strata of the Volcano Peak Group (Fig. 19) describe a large, north-trending, west-verging, isoclinal syncline on the west flank of the southern Clan Alpine Range. Because of their areal association with rocks characteristic of the Volcano Peak Group, the carbonate rocks that form most of Chalk Mountain, the isolated hill on the west side of the valley, can be confidently assigned to the Luning Formation and are its northernmost outcrop (Willden and Speed, 1974, p. 10). The dark part of Chalk mountain is underlain by dense dolomite like that which forms the upper part of the Luning Formation regionally (Fig. 19) and which is referred to as the dolomite of Milton Canyon. This dolomite, to be examined at Stop 4, is an algally laminated, supratidal deposit that serves as a key unit in the tectonic analysis. It correlates with thick, deep basinal, terrigenous-clastic rocks of the Jungo terrane that crop out just to the north and elsewhere throughout the Jungo terrane. Because of the large areal extent in the Jungo and Walker Lake terranes of contemporaneous rocks representing very different depositional environments, juxtaposition of these terranes requires major structural displacement.

Stop 4. Holly Well in Lodi Valley

Looking south from Holly Well (Fig. 20), strata of the Lodi allochthon and their NW-trending, second-deformation structures can be viewed in cross-section (c-c', Fig. 21) along the northwest edge of the Paradise Range. The main objectives here are to see the extent of the Gabbs nappe (LOu on Fig. 20) and to examine the dolomite of Milton Canyon where its stratigraphic relations are clear and it is relatively unrecrystallized.

In the Gabbs nappe the stratigraphic sequence is continuously overturned and ENE facing throughout, from the west side of the Lodi Hills all the way along the west slope of the Paradise Range to the south end of the range. The Holly Well fault (number 2 on Fig. 20), a folded, second-deformation structure, displaces the Gabbs nappe against its lower plate (LO1 on Figs. 20 and 21) to the east. Cleavage, where not destroyed by late Mesozoic thermal metamorphism, in both plates of the Gabbs thrust is folded around NW-trending second-deformation axes, and it may have formed originally during emplacement of the Gabbs nappe during the Middle Jurassic initial deformation.

Exposures of the dolomite of Milton Canyon and of

the Volcano Peak Group to be visited south of Lodi Well are in the lower plate of the Gabbs thrust.

Between Stops 4 and 5. Gabbs

The town of Gabbs was established to support the Gabbs magnesite-brucite mine, the large quarry and mill to the east of the town at the edge of the Paradise Range. More than 80,000 tons of magnesium metal was produced during World War II, and as of 20 years ago, an additional 5,000,000 tons of magnesite and 1,000,000 tons of brucite had been produced for industrial chemicals (Schilling, 1968). The magnesite was formed by hydrothermal replacement of carbonate rocks of the Luning Formation in the Gabbs nappe. It is localized in the west-dipping, overturned platform-limestone unit where it is sandwiched between the stratigraphically overlying dense, penecontemporaneous dolomite unit in the uppermost Luning Formation and a thrust fault carrying this same dolomite unit over the limestone (Silberling and John, 1989). The brucite is the result of contact-metasomatic replacement of the earlier formed magnesite at the margins of the Late(?) Cretaceous Gabbs pluton that crops out immediately south of the mine area (Schilling, 1968).

Stop 5. Sheep Canyon, Southwest Paradise Range

A traverse up the first east tributary to Sheep Canyon crosses the very shallowly dipping Sheep Canyon detachment fault that separates the Tertiary volcanic rocks along the range front from the overturned pre-Tertiary rocks of the Gabbs nappe. In the overturned section, rocks of the dolomite of Milton Canyon and stratigraphically subjacent units are strongly flattened, lineated, and recrystallized. If the well developed NW-SE mineral-stretching lineation in these rocks parallels the structural-transport direction and is related to overturning of the Gabbs nappe, then the 25 km dimension of the nappe in this direction demonstrates major NW-SE tectonic shortening during the regional initial deformation.

Stop 6. Near Luning

AT the Junction of the Gabbs-Luning highway and U.S. Highway 95, the afternoon light provides a good view to the east of the west side of the Gabbs Valley Range. Folded and thrust faulted rocks of the Luning and Volcano Peak, and the suprajacent Dunlap Formation, are intruded to the north by large late Mesozoic granitic bodies. The dark stratified rocks (including those forming "Volcano Peak") are the dolomite of Milton Canyon.

DAYS 4 AND 5: MINA-CANDELARIA TRAVERSE OF THE EARLY TRIASSIC SONOMIAN ARC-CONTINENT COLLISION ZONE

Synopsis

The Mina-Candelaria region (Fig. 6) straddles the boundary between the passive margin of Paleozoic North America and accreted terranes which is here east-striking

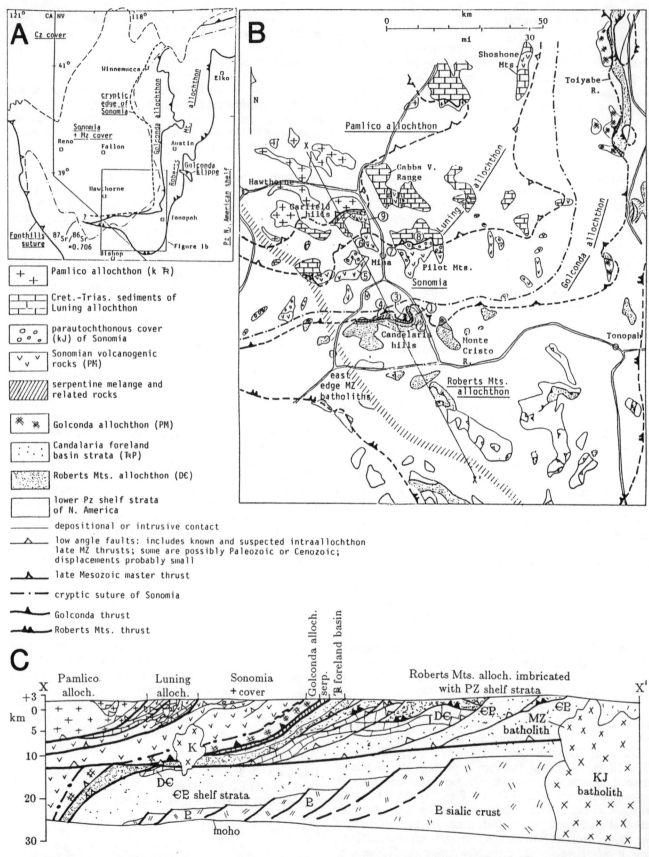

Legend (Figure 22B):

- Pamlico allochthon (k Ꝯ)
- Cret.-Trias. sediments of Luning allochthon
- parautochthonous cover (kJ) of Sonomia
- Sonomian volcanogenic rocks (PM)
- serpentine melange and related rocks
- Golconda allochthon (PM)
- Candalaria foreland basin strata (ꝮP)
- Roberts Mts. allochthon (DꞒ)
- lower Pz shelf strata of N. America
- depositional or intrusive contact
- low angle faults: includes known and suspected intraallochthon late MZ thrusts; some are possibly Paleozoic or Cenozoic; displacements probably small
- late Mesozoic master thrust
- cryptic suture of Sonomia
- Golconda thrust
- Roberts Mts. thrust

FIGURE 22 Geology of Mina region (Speed, 1984). A) regional map showing major faults and location of 22B; terrane nomenclature of Fig. 3D used; B) tectonic map of Mina region; C) diagrammatic NS crustal section through Mina region.

owing to its position in the Mina deflection (Figs. 4B, 4C, 22A, 22B). The traverse is a south to north route across a north-dipping tectonic stack (Fig. 22C) that includes the Antler and Sonoman arc-continent collision zones and superposed late Mesozoic foreland (or back of the arc) imbrication (Speed, 1977; 1984). As shown on the section (Fig. 22C) from south to north, the pre-Middle Triassic tectonic units are: 1) parautochthonous North American shelf strata of early Paleozoic and late Precambrian ages; 2) Roberts Mountains allochthon together with upper Paleozoic cover and succeeding Lower Triassic strata of the Candelaria Formation; according to Speed (1977, 1984) the Candelaria is a foreland basin deposit related to Sonomia; 3) serpentinite melange; 4) Golconda allochthon; and 5) Sonomia. The region almost certainly includes the Proterozoic edge of North American crust at depth (Fig. 22C).

Sonomia of the Mina-Candelaria region is included in the Walker Lake terrane by Silberling and others (1987), and called the Gold Range allochthon (Fig. 18), by Oldow (1984a).

The Roberts Mountains allochthon was probably emplaced against North America here as in northern Nevada in the Mississippian although the oldest local stitching is the Cretaceous batholith (Fig. 3F). According to Speed (1979) Golconda allochthon and Sonomia were probably paired as forearc and arc platform components of an arc system that overran the passive margin in Early Triassic time (Fig. 3F). The Candelaria foreland basin was due to loading of the continental edge by the encroaching Sonomian arc, and the bathymetry, ages, and provenance of the Candelaria are the principal record of the Sonomian collision. The serpentinite melange underlies the Golconda thrust, which is the shallow arc-continent suture. The melange is interpreted to have been of late synemplacement diapiric origin. The source of the serpentinite may have been buried oceanic terranes attached to Paleozoic North America or associated with the Roberts Mountains allochthon (Antleria, Fig. 3F) (Speed, 1984).

Sonomia in the Mina-Candelaria region, as elsewhere, was covered by marine Mesozoic strata after emplacement. Such strata do not exist south of Sonomia above structurally lower units 1-4, as given above and on Figure 22. It is a question whether the postcollisional Mesozoic cover once extended south across the collision zone onto Paleozoic North America and was since completely eroded, whether the region south of Sonomia has been elevated since the Sonomian collision, or whether Jurassic-Cretaceous thrusting modified the collision zone and displaced Sonomian rocks and their cover far southward beyond the extent of lower Mesozoic strata on the continental margin. Above Sonomia, the maximum age of cover is Middle Triassic. The age generally decreases southward, and facies are concomitantly more paralic. The southernmost preserved Mesozoic cover above Sonomia (Camp Douglas, Stop 6) is probably Late Cretaceous (Speed and Kistler, 1980).

Jurassic and Cretaceous deformation strongly affected the Mina-Candelaria region by S to SE-vergent foreland thrusting and folding and probably, by the development of the Mina deflection (Fig. 4B, 4C). The foreland contraction caused the imbrication and piggyback transport of two

major allochthons, Pamlico, the higher, and Luning below it (Fig. 22B, 22C). Both contain rocks of Sonomia and its Mesozoic cover. South of the Luning allochthon, flat faults of known and suspected late Mesozoic age repeat the tectonostratigraphy in the Sonomian collision and probably, well south of there (Fig. 22C). Such thrusting was active in Jurassic through Late Cretaceous times (Speed and Kistler, 1980). A basal detachment probably exists at depth (Fig. 22C) above which all the late Mesozoic allochthons and subsidiary thrust slices compose an extensive south-vergent imbricate stack that propagated cratonward, north to south.

The late Mesozoic imbricate stack of the Mina-Candelaria region in fact extends north and includes the Lodi, Berlin, and Sand Springs allochthons as structurally higher units (Fig. 18). These higher allochthons expose rocks from greater structural depth than those of the Mina-Candelaria region. Individual and integral tectonic transport probably increases northward through the imbricate stack, as does also the depth of unroofing.

The Mina-Candelaria region contains late Cenozoic structures that are characteristic of the Walker Lane of the Basin-Range Province (Fig. 4A). Its main features include high seismicity (Ryall, 1977), Holocene and(or) late Neogene volcanism, and three sets of active high angle faults: N-striking mainly dipslip, NW-striking mainly right slip, and NE to E-striking mainly left slip. The mainly strike-slip faults at least have been active since 25 ma (Speed and Cogbill, 1979a,b; Ekren and Byers, 1984). The relative slip magnitudes among the three sets are unknown, and it is not certain whether the deformation they permit is irrotational with westerly maximum extension or takes up net simple shear on the NW-trending strike-slip faults.

Roberts Mountains allochthon. The lithology of this unit (Stop 1, Fig. 22) is similar in the Mina-Candelaria region to that elsewhere in Nevada (Roberts and others, 1958): radiolarian chert, pelitic slate, thin bedded organic limestone, quartzose and calcareous sandstone, and basaltic pillow lava and breccia. Such rocks give ages of Late Cambrian, Ordovician, and Devonian (R.C. Speed, unpubl. data). They occur in packets bounded by mainly low angle faults. First folds and foliation in such rocks formed before the fault packet did whereas second and locally, third phase major folds deform the packets (Oldow, 1984b). The first and perhaps, second folds are apparently earlier than the emplacement of the allochthon as judged by overlap of Mississippian limestone and the Diablo Formation. The third phase is clearly younger as it deforms harmonically all pre-Tertiary rocks in the Candelaria Hills.

Candelaria Formation and subjacent strata. The Lower Triassic Candelaria Formation and the local subjacent thin Permian beds (Diablo Formation) (stops 2, 3; Fig. 22) occupy a narrow belt (Fig. 22B, 23A) between the Golconda allochthon and serpentinite melange above and a depositional base above the Roberts Mountains allochthon (Speed, 1977). Scarce remnants of Mississippian limestone (Stop 2) above the Roberts Mountains allochthon are also included in this unit.

The Candelaria Formation is a vital element in Speed's reconstruction of the Sonoma orogeny. In the Candelaria

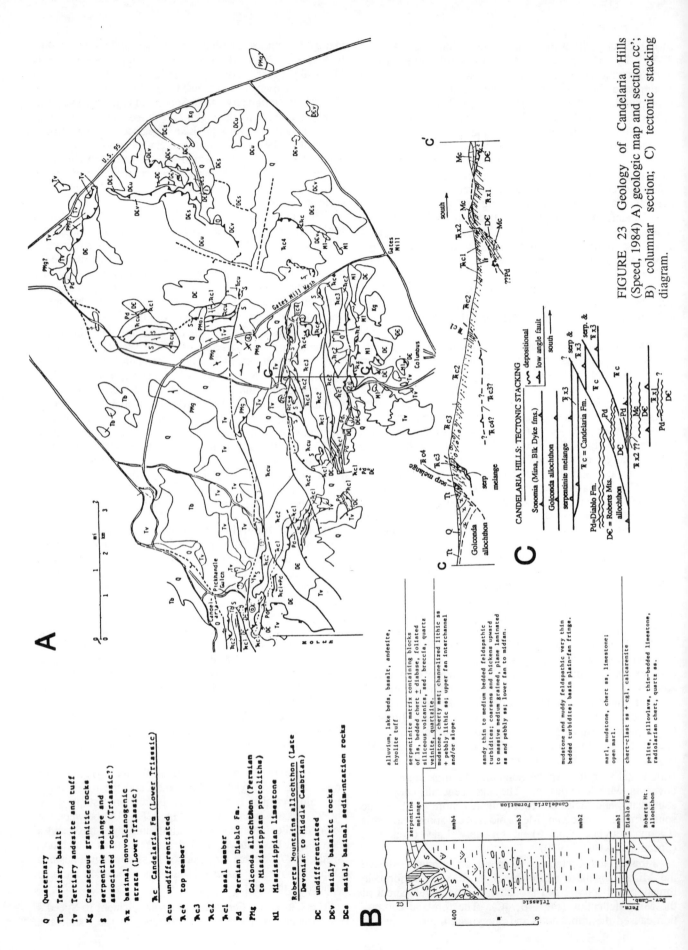

FIGURE 23 Geology of Candelaria Hills (Speed, 1984) A) geologic map and section cc'; B) columnar section; C) tectonic stacking diagram.

Hills, it is at least 1 km thick and consists of a deformed succession (Fig. 23) of basal open marine marl and limestone surmounted by turbiditic volcanogenic sediments. The vertical succession of turbidites indicates increasing proximality of the volcanic source during Early Triassic time. Fan facies from bottom to top are basin plain, outer fan, midfan, and braided channel fills together with interchannel or slope and crevasse splay deposits. Sediment transport features imply provenance from the N to E, in present coordinates. Basal strata of the Candelaria are earliest Triassic (Griesbachian) (Silberling and Tozer, 1968) and the highest are late Early Triassic (early Spathian, R.C. Speed and B. Wardlaw, 1982, unpubl. data).

The thin Diablo Formation (≤35 m) is mainly terrigenous and calcareous sandstone and conglomerate, containing megafossils whose age is between late Wolfcampian and Guadalupian. It rests on the Roberts Mountains allochthon. Patches of Mississippian limestone in the Candelaria Hills (Stop 2) were also deposited on the Roberts, but these occupy different thrust slices of Mesozoic(?) age from those of the Diablo and Candelaria Formations. The relationship of the Diablo and Mississippian limestone is uncertain.

The sediments of this unit record erosion at or near the continental shelf-slope break in late Paleozoic time, strandline accumulation in the Permian (Diablo Formation), and progressive subsidence in the Early Triassic (Candelaria Formation). Volcanogenic debris in the Triassic succession came from an encroaching northerly source, for which the Sonoma arc volcanics are the only recognized candidate (Speed, 1977). Thus, the Candelaria Formation is interpreted by Speed as the fill of a foreland basin that was generated by the emplacement of the paired Golconda allochthon and Sonomia arc above the edge of Early Triassic North America.

Serpentinite melange. Outcrops of serpentinite melange (Stops 2, 3; Fig. 22) form a poorly exposed and narrow belt in the Candelaria Hills, Monte Cristo, Toquima, and Toiyabe Ranges of western Nevada (Fig. 22A) (Speed, 1977). The belt intervenes between the Golconda allochthon above and Lower Triassic foreland basin strata below. Serpentinite melange varies in thickness from a few meters to greater than 500m. It comprises blocks of diverse protoliths in a locally foliated serpentinite matrix. The blocks are of diabase, pillow lava, flowfoliated volcanic rocks, radiolarian chert some of which was intruded in soft state by basalt, quartz wacke, quartz veinite, crystalline limestone, and pebbly mudstone. Block ages range from Mississippian to Early Triassic (R.C. Speed, unpubl. data). At places in the Candelaria Hills (Fig. 23A), serpentinite melange is underlain by thin fault slivers of fossiliferous well bedded Lower Triassic (Spathian) marls of starved basin aspect. The base of the melange is everywhere tectonic and lies mainly on Lower Triassic foreland basin strata. An Early Triassic age of the melange is suggested by the concurrent youngest ages of blocks it contains and of strata it cuts. The melange appears to be of tectonic assembly rather than olistostromal. It apparently forms a sole to the Golconda allochthon. The origin of the serpentinite is problematic and could be diapiric.

Golconda allochthon. This allochthon (Stop 4) consists of fault-bounded packets of Mississippian to Permian biogenic chert, hemipelagic chert and pelite, terrigenous and calcareous turbidite, and mafic volcanic rocks that represent deep marine sites of accumulation. Such rocks are multiply deformed within packets, and muddy rocks contain pervasive foliation that is axial planar to tight folds of varied wavelengths. Such structures are probably pre-emplacement. All deformations occurred in shallow, nonmetamorphic environments. Orientations of early structures generally indicate contraction approximately normal to the North American margin, but these are locally markedly reoriented by late Mesozoic and Cenozoic tectonism.

The Mina-Candelaria region includes the westernmost outcrops of rocks that can confidently be correlated with the Golconda allochthon. Here, the allochthon occupies a narrower width of outcrop than to the northeast (Fig. 3C, 3D). This can be explained by 1) an inherently narrower allochthon (originally narrower forearc to Sonomia) in its southern reaches, 2) greater imbrication by Mesozoic thrusting in the south, 3) greater overrunning of the forearc by Sonomian magmatic arc rocks during collision in the south.

The upper surface of the Golconda allochthon is nowhere exposed. Sonomia outcrops are north and west of the Golconda belt and presumably contact the Golconda allochthon at depth (Fig. 22C). Figure 22C shows a postulated north dipping suture between the Sonomia and Golconda belts.

The lower or continentward boundary of the allochthon is the Golconda thrust. The most prevalent tectonostratigraphic sequence below the thrust is serpentinite melange, foreland basin strata of the Candelaria Formation and local older parautochthonous strata, and the Roberts Mountains allochthon.

Sonomia. This unit contains Paleozoic lava, breccia, intrusions, and copious sedimentary rocks of basalt and basaltic andesite composition. The sedimentary rocks (Mina Formation) (stops 8, 9) are commonly turbidites and grain flows in individual beds as thick as 15m and are intercalated with thick chert of mainly nonbiogenic origin (Speed, 1977, 1979). Radiometric and radiolarian ages indicate Late Mississippian to Late Permian age of accumulation. Chemical and strontium isotopic compositions indicate probable island arc affinities of the Sonomian rocks.

Sonomian volcanic rocks occur in two tectonic settings in the Mina-Candelaria traverse (Figs. 22, 23). First, they occupy a discrete belt (Sonomia) that is bounded on the south with unexposed contact by the Early Triassic Golconda allochthon and on the north, by the younger Mesozoic Luning allochthon; the Sonomian volcanics of this belt are the parautochthon to the foreland thrust belt, including the Luning allochthon. The Sonomian volcanics are locally overlain in this belt by synorogenic Mesozoic volcanic and sedimentary cover that is dated as Cretaceous and latest Jurassic at two places (Speed and Kistler, 1980).

The second tectonic setting is as nappes intercalated with slices of shelfal Triassic and synorogenic younger rocks in the Luning allochthon (Speed, 1978b; Oldow, 1981a and b). The Sonomian volcanics are thought to have extended far to the north of the Sonomia belt (Fig. 3D) and to have composed the depositional basement to successor Triassic deposits as represented in the Luning and Pamlico allochthons. Depositional continuity between such rocks in the Luning allochthon is known in the Shoshone Mountains (Fig. 23) where Silberling calls it the Berlin allochthon (Fig. 13). Jurassic and/or Cretaceous foreland thrusting caused detachment of the Mesozoic cover to form the many allochthons of the Walker Lake terrane and incorporated slices of the basement in the process (Speed, 1977).

Parautochthonous Mesozoic cover of Sonomia. This (Figs. 22, 24) includes sparsely dated volcanic and sedimentary rocks (stop 8,10) that are depositional on Sonomian volcanics of the belt south of the Luning allochthon. Dated rocks include Early Jurassic, Late Jurassic (142 ma), and mid-Cretaceous (100 ma) (Speed and Kistler, 1980) ages. The rocks are mainly conglomeratic, terrigenous, and volcanogenic sediments and silicic extrusive rocks that accumulated in varied paralic and possibly subaerial environments with local relief. These beds were originally mapped as Excelsior, Luning, and Dunlap Formations by Muller and Ferguson (1939) and as Gold Range Formation by Speed (1977); formational nomenclature for these rocks is probably unwise. The Upper Jurassic and Cretaceous extrusive cover rocks indicate a protracted history of continental arc volcanism following the more widely recognized Triassic volcanism in this region. They are inferred to have accumulated on the fringe of the backarc because no Mesozoic volcanics are known farther east in Nevada. They also demonstrate thrust imbrication in later Mesozoic and in particular, Late Cretaceous time within the Sonomia belt because upper Paleozoic Sonomian volcanics are locally thrust above the cover (stop 8).

Luning allochthon. The Luning allochthon (stops 10, 11) (Speed, 1978b) consists of nappes of Mesozoic strata and of Sonomian volcanic rocks of late Paleozoic age. North of Luning (Fig. 24), the Luning allochthon is called the Berlin allochthon by Silberling and others (1987). The Luning allochthon was emplaced during foreland thrusting in a southeasterly direction (Oldow, 1981a and b) above an autochthon of Sonomia and its parautochthonous Mesozoic cover. The probable maximum age of final emplacement of the Luning allochthon is mid-Cretaceous. A small cross-cutting pluton with a biotite K-Ar age of 67 ma gives the minimum age of emplacement. Oldow (1981a) thought that nappes of the Luning allochthon probably accreted structurally downward.

Nappes of the Luning allochthon are related by their content of Upper Triassic carbonate and siliciclastic rocks (Luning Formation of Muller and Ferguson, 1939) that represent paralic environments of shelf deposition. Oldow (1981a) traced facies from conglomeratic marginal marine to open shallow marine in approximately contemporaneous horizons from bottom to top of the structural stack and estimated 50 km of contraction. It is noteworthy that volcanogenic debris is sparse in these Triassic beds in con-

trast to partly contemporaneous strata of the suprajacent Pamlico allochthon. The Triassic rocks of the Luning allochthon are surmounted by Lower Jurassic open marine strata (Volcano Peak Group, Fig. 19) which are in turn succeeded by undated but probably Jurassic and/or Cretaceous synorogenic sediments (Speed, 1978b). The latter strata are volcanogenic, terrigenous, and conglomeratic. Nappes of Sonomian volcanics (stop 12) in the Luning allochthon are considered slices of the depositional basement to the Triassic shelf rocks of the allochthon.

Pamlico allochthon. In present coordinates, transport of the Pamlico allochthon was southeasterly (Oldow, 1978, 1984a) above the Luning allochthon. The Pamlico is constituted by thrust slices of deformed Mesozoic volcanic and sedimentary rocks. The oldest rocks are Lower and Upper Triassic silicic and intermediate lava, breccia, and protrusion, volcanogenic sediments, and carbonate rocks assigned to the Pamlico Formation (Oldow, 1978, 1984a) These are probably surmounted by open marine Lower Jurassic deposits and succeeding the synorogenic quartzose and volcanogenic strata and arc-related volcanic rocks (Speed, 1978a). The latter, originally called Dunlap Formation by Muller and Ferguson (1939), are undated but probably Jurassic and/or Cretaceous. Rocks of the Pamlico allochthon are thought to represent mainly marine eruptives, derivative sediment, and fringing carbonate deposits that accumulated on the crest and/or back wall of the Triassic continental arc (Speed, 1978a).

Stop 1. Rock Hill

Thrust-bound slices of well preserved successions of pillow lava, breccia, and intercalated sediments and others of thin bedded chert, slate, quartzose and calcareous sandstone; nappes are folded and here rotated in the core of a major third fold. (Fig. 22).

Stop 2. North of Columbus

Foot traverse across southern Candelaria Hills to examine: a) Lower Triassic strata overthrust by a nappe of Roberts Mountains allochthon and depositionally overlying Mississippian limestone; b) thrust contact of the Candelaria Formation above Mississippian limestone; c) stratigraphy and structure of Lower Triassic Candelaria Formation in its most complete section (Fig. 22B); traverse starts in open marine marls and limestone of Griesbachian age (Trc1) and proceeds up through 1 km of upward coarsening and thickening strata of mainly volcanogenic origin; Trc2: basin plain and outer fan tabular turbidites; Trc3: tabular sandy turbidites and thick plane laminated sandstone and pebbly sandstone, locally channelized, of the lower and upper midfan, respectively; Trc4: innerfan or slope mudstone deposits and small sandstone channel fills d) serpentinite melange, here with a sole of Spathian calcareous sediments at its base and very few included blocks.

Stop 3. Gates Mills Wash

More extensive outcrops of slaty rocks of unit Trc4 of the Candelaria Formation, overthrust by sedimentary brec-

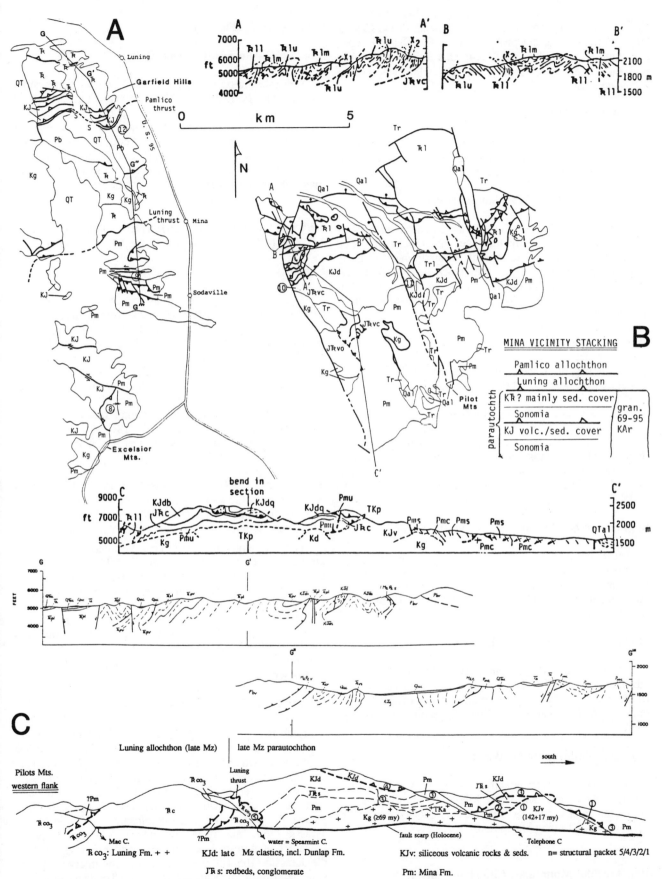

FIGURE 24 Geology of Pilot Mts., Garfield Hills, and eastern Excelsior Mts. A) geologic map after Oldow (1981a) and Speed (1984); B) tectonic stacking diagram; C) perspective geologic section of western flank of Pilot Mts.

cia and other lensy coarse-grained sediments of the Candelaria Formation that are probably inner fan channel deposits equivalent to Trc3 or Trc4; this thrust apparently underlies the entire section of Candelaria Formation traversed at Stop 2.

Stop 4. Gates Mill Wash

Golconda allochthon; slaty terrigenous and calcareous turbidites of Permian age; fault slice of Lower Pennsylvanian radiolarian chert and pelite.

Stop 5. 1.5 km east of Pickhandle Gulch

Serpentinite melange; sharp thrust contact with Trc2 of the Candelaria Formation; melange with blocks of varied protoliths: crystalline limestone of Mississippian and Early Triassic ages, quartz veinite grit, black bedded chert and sedimentary breccia of diverse sediments of probable volcanogenic provenance; serpentinite matrix is variably foliated and nonfoliated.

Stop 6. Candelaria Mine

Main pit of Candelaria Mine operated by Nerco Minerals Co., ores related to structures.

Stop 7. East wall of Pickhandle Gulch

Serpentinite melange; examination of varied compositions of blocks: dark chert intruded by diabase in soft state, various mafic igneous rocks, limestone and sedimentary breccia and of serpentinite fabrics; foot traverse to Gulch bottom and pickup at canyon mouth.

Stop 8. Excelsior Mountains, Silver Dyke Canyon

Mina Formation of Sonomia, Permian volcanogenic sediments thrust above Cretaceous cover of Sonomia; late phase major fold in overthrust packet; Cretaceous rocks are volcanogenic and terrigenous beds dated by Rb-Sr whole rock isochron at 100 ma.

Stop 9. Garfield Hills, Douglas Canyon

Mina Formation of Sonomia (Fig. 23A); examine various facies and lithologies of volcanogenic turbidites: fan fringe and outer fan; massive margin flows of upper midfan or innerfan channel; massive diagenetic nonbiogenic chert.

Stop 10. Pilot Mountains, Water Canyon

View of tectonostratigraphy of western front Pilot Mts; Luning allochthon and late Mesozoic imbrication of Sonomian rocks and its Jurassic or Cretaceous cover; examine basal Luning allochthon.

Stop 11. Pilot Mountains, Cinnabar Canyon

Structure of Luning allochthon as worked out by Oldow (1981a) (Fig. 23).

Stop 12. Garfield Hills, Black Dyke

Volcanic breccia, intrusions, and volcanogenic strata (Black Dyke Formation) of Permian age (Fig. 23A); these occur in a nappe within the Luning allochthon; they were thrust south in the Luning allochthon and are thought to represent the depositional basement of the Triassic carbonate-deltaic complex of the Luning allochthon that was initially as much as 50 km north; these volcanic rocks may represent the main magmatic arc of Sonomia.

DAY 6: STRATIGRAPHY AND STRUCTURE OF THE LOWER MESOZOIC ARC EDIFICE NEAR YERINGTON

Synopsis

The Yerington district (Figs. 5, 25A) provides good exposures of Lower Mesozoic rocks of the magmatic arc edifice (Fig. 4A) west of sites of the basinal sequences seen in the Fencemaker, Sand Springs, Lodi, and Pamlico allochthons. This district has been studied extensively for its copper ore deposits (e.g., Einaudi, 1977; Dilles, 1987); its geology is comprehensively dealt with by Dilles and Wright (1988) from which most of the following is gained.

The Triassic and Jurassic arc edifice rocks at Yerington occupy the Pinenut subterrane (Fig. 18) and are thought to have stratigraphic links to sequences in the Paradise subterrane east of the Pinenut fault (Silberling and others, 1987). They are presumed to be built on Sonomia (Dilles and Wright, 1988) and outboard of Precambrian North America by virtue of low values of initial Sr isotopic ratios in plutonic equivalents (≤ 0.704). No Paleozoic rocks have been recognized, however, in the Yerington region.

The Mesozoic succession near Yerington is of particular significance because it contains the oldest dated volcanics of the Mesozoic batholith (Fig. 4A) and it shows the episodicity in arc magmatism, at least in the Yerington area.

The Yerington district is also widely known for local high extensional Basin-Range strain (100%), demonstrated by Proffett (1977). The extension was taken up by 50-90° domino rotation of blocks and bounding planar faults, west side down.

Mesozoic Arc Edifice Rocks

Figures 25A and 25B indicate the lithologic units and their ages and succession (Dilles and Wright, 1988). There were two discrete periods of local magmatism, each causing thick accumulation of calcalkaline high K andesitic and silicic breccia, protrusion, and lava, and abundant intrusion of up to batholithic dimension. The first phase yielded the McConnell Canyon volcanics which are Middle Triassic (232+ ma by zircon dating). The second phase, volcanics of Artesia Lake and of Fulstone Spring are Middle Jurassic (165-169 ma by zircon). The intervening time saw deposition of Karnian to lower Toarcian sediments (Fig. 25B). These include shelfal and basinal carbonate, organic mudstone, and volcanogenic pyro- and epiclastic

A

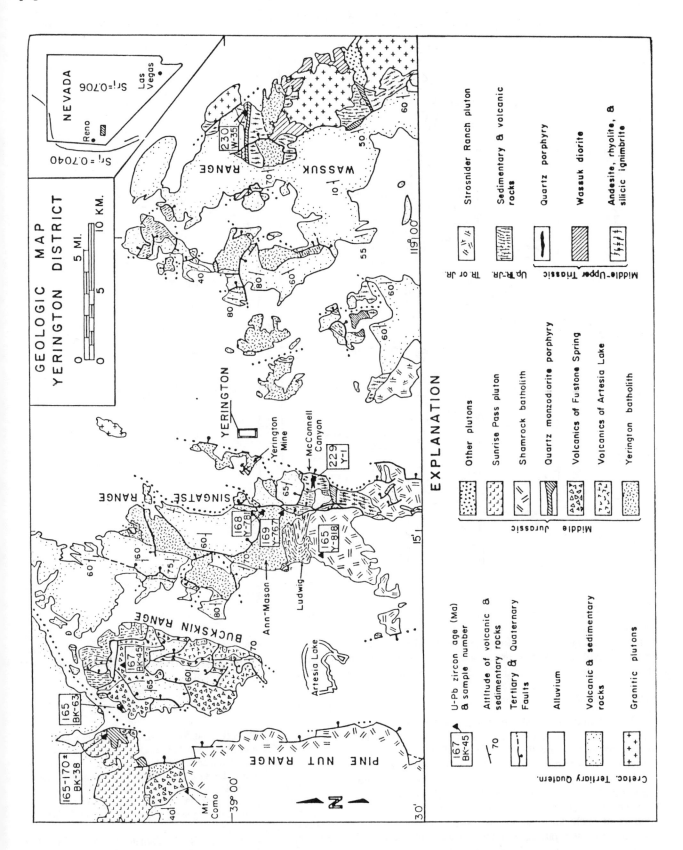

FIGURE 25 Geology of Yerington area after Dilles and Wright (1988). A) map; B) stratigraphic column.

B

COMPOSITE MESOZOIC COLUMNAR SECTION

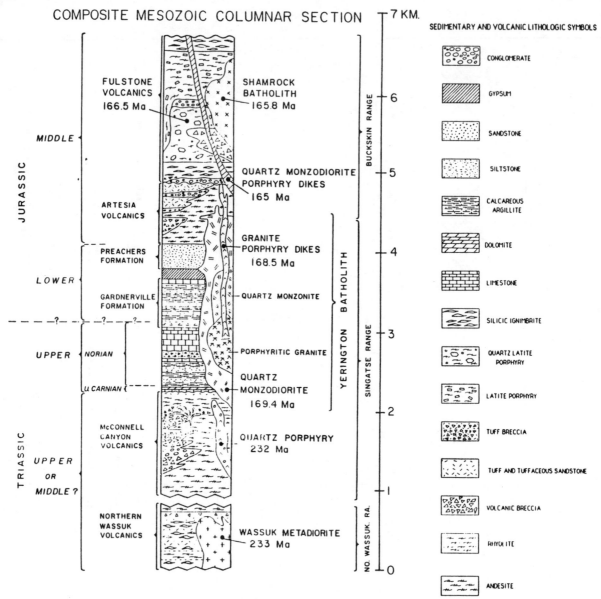

SEDIMENTARY AND VOLCANIC LITHOLOGIC SYMBOLS

CONGLOMERATE

GYPSUM

SANDSTONE

SILTSTONE

CALCAREOUS ARGILLITE

DOLOMITE

LIMESTONE

SILICIC IGNIMBRITE

QUARTZ LATITE PORPHYRY

LATITE PORPHYRY

TUFF BRECCIA

TUFF AND TUFFACEOUS SANDSTONE

VOLCANIC BRECCIA

RHYOLITE

ANDESITE

rocks. The lowest carbonate (Fig. 25B) was dated late Karnian by Silberling (1984). Near the top of the sedimentary succession is the Preachers Formation, a quartz arenite of probable early Toarcian age. This arenite is part of a widespread eolian and fluvial sand blanket at the western fringes of the extensive Navajo, Nuggett, and Aztec sandstones of the Colorado Plateau.

Deformation of the McConnell Canyon volcanics, synchronous intrusive rocks, and succeeding sedimentary rocks occurred before the deposition of the Artesia Lake volcanics. Therefore, deformation was apparently short-lived and concurrent with intrusion of the Yerington batholith (Fig. 25A). The very large reclined west-facing fold west of McConnell Canyon in the Singatse Range is a product of such deformation. Its current altitude is partly due to Tertiary rotation, and upon restoration, the fold's Mesozoic axial plane is NW-striking and its axis plunges shallowly NW.

Stops 1 and 2

These stops provide a traverse from McConnell Canyon to Ludwig (Fig. 25A) through the older volcanics and succeeding sedimentary section

REFERENCES

Abbate, E., V. Bortolli, and M. Sagri, Olistostromes in the Oligocene Macigno Fm. (Florence area), in *Excursion guidebook with contributions on sedimentology of some Italian basins*, edited by F. Ricci Lucchi, pp. 163–203, International Association of Sedimentologists, Oxford London, 1981.

Albers, J. P., Belt of sigmoidal bending and right-lateral faulting in the western Great Basin, *Geol. Soc. Am. Bull., 78,* 143–156, 1967.

Allemdinger, R. W., and T. E. Jordan, Mesozoic evolution, hinterland of the Sevier orogenic belt., *Geology, 9,* 308–314, 1981.

Allmendinger, R. W., D. M. Miller, and T. E. Jordan, Known and inferred Mesozoic deformation in the hinterland of the Sevier Belt, northwestern Utah, in *Geology of northwestern Utah, southern Idaho, and northeastern Nevada*, edited by G. J. Kerns, pp. 21–34, Utah Geological Association Publication no. 13, 1985.

Anderson, R. E., M. L. Zoback, and G. A. Thompson, Implications of selected subsurface data on the structural form and evolution of some basins in the northern Basin and Range province, Nevada and Utah, *Geol. Soc. Am. Bull., 94*, 1055–1072, 1983.

Armstrong, R. L., Sevier orogenic belt in Nevada and Utah., *Geol. Soc. Am. Bull., 79*, 439–458, 1968.

Armstrong, R. L., Low-angle (denudation) faults, hinterland of the Sevier orogenic belt, eastern Nevada and western Utah, *Geol. Soc. Am. Bull., 83*, 1729–1754, 1972.

Armstrong, R. L., W. H. Taubeneck, and P. O. Hales, Rb-Sr and K-Ar geochronometry of Mesozoic granitic rocks and their Sr isotopic composition, Oregon, Washington, and Idaho, *Geol. Soc. Am. Bull., 88*, 321–331, 1977.

Atwater, T., Implications of plate tectonics for the Cenozoic tectonic evolution of western North America, *Geol. Soc. Am. Bull., 81*, 3513–3536, 1970.

Brueckner, H. K., and W. S. Snyder, Structure of the Havallah sequence, Golconda allochthon, Nevada: Evidence for prolonged evolution in an accretionary prism., *Geol. Soc. Am. Bull., 96*, 1113–1130, 1985.

Burke, D. B., *Reinterpretation of the "Tobin thrust": Pre-Tertiary geology of the southern Tobin Range, Pershing County, Nevada*, 82 pp., [Ph.D. thesis] Stanford University, Stanford California, 1973.

Christiansen, R. L., and P. W. Lipman, Cenozoic volcanism and plate-tectonic evolution of the western United States, Part II. Late Cenozoic, *Philosophical Transactions of the Royal Society of London, A271*, 249–284, 1972.

Coney, P. J., and S. J. Reynolds, Cordilleran benioff zones, *Nature, 270*, 403–406, 1977.

Dickinson, W. D., D. W. Harbaugh, A. H. Saller, P. L. Heller, and W. S. Snyder, Detrital modes of upper Paleozoic sandstone derived from Antler orogen in Nevada: implications for nature of Antler orogeny, *Am. J. Sci., 283*, 481–509, 1983.

Dilles, J. H., The petrology of the Yerington batholith: Evidence for the evolution of porphery copper ore fluids, *Economic Geology, 82*, 1750–1789, 1987.

Dilles, J. H., and J. E. Wright, The chronology of early Mesozoic arc magmatism in the Yerington district of western Nevada and its regional implications, *Geol. Soc. Am. Bull., 100*, 644–652, 1988.

Doe, B. R., Variations in lead-isotopic compositions in Mesozoic granitic rocks of California - A preliminary investigation, *Geol. Soc. Am. Bull., 84*, 3513–3526, 1973.

Eaton, G. P., The Basin and Range province - Origin and tectonic significance, *Annual Review of Earth and Planetary Sciences, 10*, 409–440, 1982.

Eaton, J. P., Crustal structure from San Francisco, California, to Eureka, Nevada from seismic refraction measurements, *J. Geophys. Res., 68*, 5789–5806, 1963.

Einaudi, M. T., Petrogenesis of the copper-bearing skarn at the Mason Valley mine, Yerington district, Nevada, *Economic Geology, 72*, 769–795, 1977.

Ekren, E. B., and F. M. Byers, Jr., The Gabbs Valley Range - A well-exposed segment of the Walker Lane in west-central Nevada, in *Western Geological Excursions v.4*, edited by J. Lintz, Jr., pp. 203–215, Geological Society of America Annual Meeting, Mackay School of Mines, Reno, Nevada, 1984.

Elison, M. W, Structural geology and tectonic implications of the East Range, Nevada, Ph.D. Thesis, Northwestern Univ., 306 pp, 1987.

Elison, M. W., and R. C. Speed, Triassic flysch of the Fencemaker allochthon, East Range, Nevada: Fan facies and provenance, *Geol. Soc. Am. Bull., 100*, 185–199, 1988.

Elison, M. W., and R. C. Speed, Structural development during flysch basin collapse: The Fencemaker allochthon, East Range, Nevada, *Journal of Structural Geology (in press)*, 1989.

Engebretson, D. C., A. Cox, and R. G. Gordon, *Relative motions between oceanic and continental plates in the Pacific basin*, 59 pp., Geological Society of America Special Paper 206, 1985.

Ferguson, H. G., and S. W. Muller, *Structural geology of the Hawthorne and Tonopah quadrangles, Nevada*, 55 pp., U. S. Geological Survey Professional Paper 216, 1949.

Garside, L. J., and N. J. Silberling, New K-Ar ages of volcanic and plutonic rocks from the Camp Douglas Quadrangle, Mineral County, Nevada, *Isochron/West, 22*, 29–32, 1978.

Geissman, J. W., J. T. Callian, J. S. Oldow, and S. E. Humphries, Paleomagnetic assessment of oroflexural deformation in west-central Nevada and significance for emplacement of allochthonous assemblages, *Tectonics, 3*, 179–200, 1984.

Gilluly, J., *Geologic Map of the Winnemucca quadrangle, Pershing and Humboldt Counties Nevada*, U. S. Geological Survey Map GQ-15, 1967.

Hamilton, W., Mesozoic California and the underflow of the Pacific mantle, *Geol. Soc. Am. Bull., 80*, 2409–2430, 1969.

Hamilton, W., and W. B. Myers, Cenozoic tectonics of the western United States, *Reviews of Geophysics, 4*, 509–549, 1966.

Hauge, T. A., R. W. Allmendinger, C. Caruso, E. C. Hauser, S. L. Klemperer, S. Opdyke, C. J. Potter, W. Sanford, L. Brown, S. Kaufman, and J. Oliver, Crustal structure of western Nevada from COCORP deep seismic-reflection data, *Geol. Soc. Am. Bull., 98*, 320–329, 1987.

Heck, F. R., *Mesozoic foreland deformation and paleogeography of the western Great Basin, Humboldt Range, Nevada*, 214 pp., [Ph.D. thesis] Northwestern University, Evanston Illinois, 1987.

Heck, F. R., and R. C. Speed, Triassic olistostome and shelf-basin transition in the western Great Basin: Paleogeographic implications, *Geol. Soc. Am. Bull., 99*, 539–551, 1987.

Heck, F. R., and R. C. Speed, Normal and oblique thrust ramping related to arcuate shelf margin, Mesozoic Winnemucca Thrust Belt, Nevada, *Geological Society of America Bulletin (in press)*, 1989.

Heller, P. L., S. S. Bowdler, H. P. Chambers, J. C. Coogan, E. S. Hogen, M. W. Shuster, N. S. Winslow, and T. F. Lawton, Time of initial thrusting in the Sevier orogenic belt, Idaho-Wyoming and Utah., *Geology, 14*, 388–391, 1986.

Kistler, R. W., Mesozoic paleogeography of California: A viewpoint from isotope geology, in *Mesozoic paleogeography of the western United States, Pacific Coast Paleogeography Symposium 2*, edited by D. G. Howell and K. A. McDougall, pp. 75–84, Society of Economic Paleontologists and Mineralogists, Pacific Section, 1978.

Kistler, R. W., and Z. E. Peterman, Variations in Sr, Rb, K, Na, and initial Sr87/Sr86 in Mesozoic granitic rocks and intruded wall rocks in central California, *Geol. Soc. Am. Bull., 84*, 3489–3512, 1973.

Klemperer, S. L., T. A. Hauge, E. C. Hauser, J. E. Oliver, and C. J. Potter, The Moho in the northern Basin and Range province, Nevada, along the COCORP 40° N seismic reflection transect, *Geol. Soc. Am. Bull., 97*, 603–618, 1986.

Lachenbruch, A. H., and J. H. Sass, Models of an extending lithosphere and heat flow in the Basin and Range province, in *Cenozoic tectonics and regional geophysics of the western Cordillera*, edited by R. B. Smith and G. P. Eaton, pp.

209–251, Geological Society of America Memoir 152, 1978.

Lipman, P. W., H. J. Prostka, and R. L. Christiansen, Cenozoic volcanism and plate-tectonic evolution of the western United States, Part I. Early and middle Cenozoic, *Philosophical Transactions of the Royal Society of London, A271,* 217–248, 1972.

Lupe, R., and N. J. Silberling, Genetic relationship between lower Mesozoic continental strata of the Colorado Plateau and marine strata of the western Great Basin: Significance for accretionary history of Cordilleran lithotectonic terranes., in *Tectonostratigraphic Terranes of the Circum-Pacific Region.,* edited by D. G. Howell, pp. 263–271, Circum-Pacific Council for Energy and Mineral Resources, Earth Science Series, No. 1., 1985.

Miller, C. F., and L. J. Bradfish, An inner Cordilleran belt of muscovite-bearing plutons, *Geology, 8,* 412–416, 1980.

Miller, D. M., Relations between younger-on-older and older-on-younger low-angle faults, Pilot Range, Nevada and Utah, *Geol. Soc. Am. Abstr. Programs, 14,* 216, 1982.

Miller, E. L., L. R. Kanter, D. K. Larue, R. J. Turner, B. Murchey, and D. L. Jones, Structural fabric of the Paleozoic Golconda allochthon, Antler Peak Quadrangle, Nevada: Progressive deformation of an oceanic sedimentary assemblage., *J. Geophys. Res., 87,* 3795–3804, 1982.

Miller, E. L., B. K. Holdsworth, W. B. Whiteford, and D. Rogers, Stratigraphy and structure of the Schoonover sequence, northeastern Nevada: Implications for Paleozoic plate-margin tectonics., *Geol. Soc. Am. Bull., 95,* 1063–1076, 1984.

Minster, J. B., and T. H. Jordan, Vector constraints on Quaternary deformation of the western United States east and west of the San Andreas fault, in *Tectonics and sedimentation along the California margin,* edited by J. K. Crouch and S. B. Bachman, pp. 1–16, Society of Economic Paleontologists and Mineralogists, Pacific Section, n. 38, 1984.

Muller, S. W., and H. G. Ferguson, Mesozoic stratigraphy of the Hawthorne and Tonopah quadrangles, Nevada, *Geol. Soc. Am. Bull., 50,* 1573–1624, 1939.

Nichols, K. M., and N. J. Silberling, *Stratigraphy and depositional history of the Star Peak Group (Triassic), northwestern Nevada.,* 73 pp., Geological Society of America Special Paper 178, 1977.

Noble, D. C., *Mesozoic geology of the southern Pine Nut Range, Douglas County, Nevada,* 200 pp., [Ph.D. thesis] Stanford University, Stanford California, 1962.

Okaya, D. A., and G. A. Thompson, Geometry of Cenozoic extensional faulting: Dixie Valley, Nevada, *Tectonics, 4,* 107–125, 1985.

Oldow, J. S., Triassic Pamlico Formation: an allochthonous sequence of volcanogenic-carbonate rocks in west-central Nevada, in *Mesozoic paleogeography of the western United States, Pacific Coast Paleogeography Symposium 2,* edited by D. G. Howell and K. A. McDougall, pp. 233–236, Society of Economic Paleontologists and Mineralogists, Pacific Section, 1978.

Oldow, J. S., Structure and stratigraphy of the Luning allochthon and kinematics of allochthon emplacement, Pilot Mountains, west-central Nevada, *Geol. Soc. Am. Bull., 92,* 889–991, 1647-1669, 1981a.

Oldow, J. S., Kinematics of late Mesozoic thrusting, Pilot Mountains, Nevada, *J. Struct. Geol., 3,* 39–51, 1981b.

Oldow, J. S., Evolution of a late Mesozoic back-arc fold and thrust belt in northwestern Nevada, *Tectonophysics, 102,* 245–274, 1984a.

Oldow, J. S., Spatial variability in the structure of the Roberts Mountains allochthon, *Geol. Soc. Am. Bull., 95,* 174–185, 1984b.

Poole, F. G., Flysch deposits of the Antler foreland basin, in *Tectonics and sedimentation,* edited by W. R. Dickinson, pp. 58–82, Society of Economic Paleontologists and Mineralogists Special Publication 22, 1974.

Preistly, K. F., and J. N. Bruhn, Surface waves and the structure of the Great Basin of Nevada and Utah, *J. Geophys. Res., 83,* 2265–2272, 1978.

Proffett, J. M., Cenozoic geology of the Yerington district, Nevada, and implications for origin of basin-range faulting, *Geol. Soc. Am. Bull., 88,* 247–266, 1977.

Roberts, R. J., *Stratigraphy and structure of the Antler Peak quadrangle, Humboldt and Lander counties, Nevada,* 93 pp., U. S. Geological Survey Professional Paper 459A, 1964.

Roberts, R. J., P. E. Hotz, J. Gilluly, and H. G. Ferguson, Paleozoic rocks of north-central Nevada, *Am. Assoc. Petrol. Geol. Bull., 42,* 2813–2857, 1958.

Royse, F., M. A. Warner, and D. L. Reese, Thrust belt structural geometry and related stratigraphic problems, Wyoming - Idaho - northern Utah, in *Symposium on Deep Drilling Frontiers in the Central Rocky Mountains,* edited by D. W. Bolyard, pp. 41–54., Rocky Mountains Association of Geologists, Denver, Colorado, 1975.

Russell, B. J., Mesozoic geology of the Jackson Mountains, northwest Nevada., *Geol. Soc. Am. Bull., 95,* 313–323, 1984.

Ryall, A., Seismic hazard in the Nevada region, *Seismological Society of America Bulletin, 67,* 517–532, 1977.

Ryall, A., and K. Preistly, Seismicity, secular strain, and maximum magnitude in the Excelsior Mountains area, western Nevada and eastern California, *Geol. Soc. Am. Bull., 86,* 1585–1592, 1975.

Saleeby, J. B., and W. Sharp, Chronology of structural and petrologic development of the SW Sierra Nevada foothills, *Geol. Soc. Am. Bull., 91,* 317–320, 1980.

Schilling, J. H., The Gabbs magnesite-brucite deposit, Nye County, Nevada, *Ore deposits of the United States (Graton-Sales Volume) v. 2,* pp. 1607–1622, American Institute of Mining, Metallurgy, and Petroleum Engineers, NeW York, 1968.

Schweickert, R. A., Tectonic evolution of the Sierra Nevada Range, in *Rubey V. 1, Geotectonic development of California,* edited by G. W. Ernst, pp. 87–132, Prentice Hall, New Jersey, 1981.

Schweickert, R. A., and M. M. Lahren, Continuation of the Antler and Sonoma orogenic belts to the eastern Sierra Nevada, California, and Late Triassic thrusting in a compressional arc, *Geology, 15,* 270–273, 1987.

Silberling, N. J., Geologic events during Permian-Triassic time along the Pacific margin of the United States., in *The Permian and Triassic Systems and their mutual boundary.,* edited by A. Logan and L. V. Hills, pp. 345–362, Canadian Society of Petroleum Geologists Memoir 2, 1973.

Silberling, N. J., *Age relationships of the Golconda thrust fault, Sonoma Range, north-central Nevada,* 28 pp., Geological Society of America Special Paper 163, 1975.

Silberling, N. J., *Map showing localities and correlation of age-diagnostic lower Mesozoic megafossils, Walker Lake 1° X 2° Quadrangle, Nevada and California,* U. S. Geological Survey Miscellaneous Field Studies Map MF-1382-0, scale 1:250,000, 1984.

Silberling, N. J., and D. A. John, *Geologic map of the pre-Tertiary rocks of the Paradise Range and southern Lodi Hills, west-central Nevada,* U. S. Geological Survey Miscellaneous Field Studies Map MF-2062, 1989.

Silberling, N. J., and R. J. Roberts, Pre-Tertiary stratigraphy and structure of northwestern Nevada., *Geological Society of America Special Paper 72,* 58, 1962.

Silberling, N. J., and E. T. Tozer, *Biostratigraphic classification of the marine Triassic in North America,* 63 pp., Geological Society of America Special Paper 110, 1968.

Silberling, N. J., and R. E. Wallace, *Geology of the Imlay quadrangle, Pershing County, Nevada*, U.S. GeologicalSurvey Quadrangle Map GQ-666, 1967.

Silberling, N. J., and R. E. Wallace, *Geology of the Imlay quadrangle, Pershing County, Nevada*, U. S. Geological Survey Geological Quadrangle Map GQ-666, 1967.

Silberling, N. J., and R. E. Wallace, *Stratigraphy of the Star Peak Group (Triassic) and overlying lower Mesozoic rocks, Humboldt Range, Nevada.*, 50 pp., U. S. Geological Survey Professional Paper 592, 1969.

Silberling, N. J., D. L. Jones, M. C. Blake, Jr., and D. G. Howell, *Lithotectonic terrane map of the western coterminous United States*, U. S. Geological Survey Miscellaneous Field Studies Map MF-1874-C, scale 1:2,500,000, 1987.

Smith, D. L., Effects of unrecognized Oligocene extension in central Nevada on the interpretation of older structures., *Geol. Soc. Am. Abstr. Programs, 16*, 660, 1984.

Smith, J. F., and K. B. Ketner, Devonian and Mississippian rocks and the date of the Roberts Mountains thrust in the Carlin-Pinon Range area, Nevada, *U. S. Geological Survey Bulletin, 1251-I*, 11–118, 1968.

Smith, R. B., Seismicity, crustal structure and intraplate tectonics of the interior of the western Cordillera, in *Cenozoic tectonics and regional geophysics of the western Cordillera*, edited by R. B. Smith and G. P. Eaton, pp. 111–144, Geological Society of America Memoir 152, 1978.

Snoke, A. W., Transition from infrastructure to suprastructure in the northern Ruby Mountains, Nevada, in *Cordilleran metamorphic core complexes*, edited by M. D. Crittenden, P. J. Coney and G. H. Davis, pp. 287–333, Geological Society of America Memoir 153, 1980.

Snoke, A. W., and D. M. Miller, Metamorphic and tectonic history of the northeastern Great Basin, in *Rubey V. 7, Metamorphism and crustal evolution in the western United States*, edited by G. Ernst, pp. 606–648, Prentice-Hall, New Jersey, 1988.

Snyder, W. S., and H. K. Brueckner, Tectonic evolution of the Golconda allochthon, Nevada: Problems and perspective., in *Pre-Jurassic rocks in western North American suspect terranes.*, edited by C.H. Steven, pp. 103–123, Society of Economic Paleontologists and Mineralogists, Pacific Section, 1983.

Snyder, W. S., W. R. Dickinson, and M. L. Silberman, Tectonic implications of space-time patterns of Cenozoic magmatism in the western United States, *Earth Planet. Sci. Lett., 32*, 91–106, 1976.

Speed, R. C., Island arc and other paleogeographic terranes of late Paleozoic age in the western Great Basin, in *Paleozoic paleogeography of the western United States: Pacific Coast Paleogeography Symposium 1*, edited by J. H. Stewart, C. H. Stevens and A. E. Fritsche, pp. 349–362, Society of Economic Paleontologists and Mineralogists, Pacific Section, 1977.

Speed, R. C., Paleogeography and plate tectonic evolution of the early Mesozoic marine province of the western Great Basin., in *Mesozoic Paleogeography of the Western United States: Pacific Coast Paleogeography Symposium 2*, edited by D. G. Howell and K. A. McDougall, pp. 253–270, Society of Economic Paleontologists and Mineralogists, Pacific Section, 1978a.

Speed, R.C., Basinal Terrane of the Early Mesozoic Marine Province of the Western Great Basin., in *Mesozoic Paleogeography of the Western United States: Pacific Coast Paleogeography Symposium 2*, edited by D. G. Howell and K. A. Mcdougall, pp. 237–252, Society of Economic Paleontologists and Mineralogists, Pacific Section, 1978b.

Speed, R. C., Collided Paleozoic microplate in the western United States, *Journal of Geology, 87*, 279–292, 1979.

Speed, R. C., Evolution of the sialic continental margin in the central-western United States., in *Studies in continental margin geology*, edited by J. S. Watkins and C. L. Drake, pp. 457–468, American Association of Petroleum Geologists Memoir 34 (Hedberg Volume), 1983.

Speed, R. C., Paleozoic and Mesozoic continental margin collision zone features: Mina to Candelaria, NV, traverse, in *Western Geological Excursions v.4*, edited by J. Lintz, Jr., pp. 66–80, Geological Society of America Annual Meeting, Mackay School of Mines, Reno, Nevada, 1984.

Speed, R. C., and A. H. Cogbill, Candelaria and other left-oblique slip faults of the Candelaria region, Nevada, *Geol. Soc. Am. Bull., 90*, 149–163, 1979a.

Speed, R. C., and A. C. Cogbill, Deep fault trough of Oligocene age, Candelaria, Nevada, *Geol. Soc. Am. Bull., 90*, 494–527, 1979b.

Speed, R. C., and T. A. Jones, Synorogenic quartz sandstone in the Jurassic mobile belt of western Nevada, *Geol. Soc. Am. Bull., 80*, 2551–2584, 1969.

Speed, R. C., and R. W. Kistler, Cretaceous volcanism, Excelsior Mountains, Nevada, *Geol. Soc. Am. Bull., 91*, 292–298, 1980.

Speed, R. C., and N. H. Sleep, Antler orogeny and foreland basin: A model, *Geol. Soc. Am. Bull., 93*, 815–828, 1982.

Speed, R. C., M. W. Elison, and F. R. Heck, Phanerozoic tectonic evolution of the Great Basin., in *Rubey V. 7, Metamorphism and Crustal Evolution of the Western United States*, edited by G. Ernst, pp. 572–605, Prentice Hall, New Jersey, 1988.

Stahl, S. D., *Pre-Cenozoic structural geology and tectonic history of the Sonoma Range, north-central Nevada*, 287 pp., [Ph.D. Thesis] Northwestern University, Evanston Illinois, 1987.

Stewart, J. H., Basin and Range structure: A system of horsts and grabens produced by deep-seated extension, *Geol. Soc. Am. Bull., 82*, 1019–1044, 1971.

Stewart, J. H., Initial deposits of the Cordilleran geosyncline: evidence of a late Precambrian (850my) continental separation, *Geol. Soc. Am. Bull., 83*, 1345–1360, 1972.

Stewart, J. H., Late Precambrian evolution of North America: plate tectonic implications, *Geology, 4*, 11–15, 1976.

Stewart, J. H., Basin and Range structure in western North America - A review, in *Cenozoic tectonics and regional geophysics of the western Cordillera*, edited by R. B. Smith and G. P. Eaton, pp. 1–32, Geological Society of America Memoir 152, 1978.

Stewart, J. H., *Geology of Nevada*, 136 pp., Nevada Bureau of Mines and Geology Special Publication 4, 1980.

Stewart, J. H., East-trending dextral faults in the western Great Basin: An explanation for anomalous trends of pre-Cenozoic strata and Cenozoic faults, *Tectonics, 4*, 547–564, 1985.

Stewart, J. H., and F. G. Poole, Lower Paleozoic and uppermost Precambrian Cordilleran miogeocline, Great Basin, in *Tectonics and sedimentation*, edited by W. R. Dickinson, pp. 27–57, Society of Economic Paleontologists and Mineralogists Special Publication 22, 1974.

Stewart, J. H., J. P. Albers, and F. G. Poole, Summary of regional evidence for right-lateral displacement in the western Great Basin, *Geol. Soc. Am. Bull., 79*, 1407–1413, 1968.

Stewart, J. H., J. R. MacMillan, K. M. Nichols, and Stevens C. H., Deep-water upper Paleozoic rocks in north-central Nevada - A study of the type area of the Havallah Formation, in *Paleozoic paleogeography of the western United States: Pacific Coast Paleogeography Symposium 1*, edited by J. H. Stewart, C. H. Stevens and A. E. Fritsche, pp. 337–347, Society of Economic Paleontologists and Mineralogists, Pacific Section, 1977.

Stewart, J. H., B. Murchey, D. L. Jones, and B.R. Wardlaw,

Paleontologic evidence for complex tectonic interlayering of Mississippian to Permian deep-water rocks of the Golconda allochthon in Tobin Range, northcentral Nevada., *Geol. Soc. Am. Bull., 97,* 1122–1132, 1986.

Taylor, D. G., P. L. Smith, R. A. Laws, and J. Guex, The stratigraphy and biofacies trends of the lower Mesozoic Gabbs and Sunrise Formations, west-central Nevada, *Canadian Journal of Earth Science, 20,* 1598–1608, 1983.

Wallace, R. E., *Fault scarps formed during the earthquakes of October 2, 1915 in Pleasant Valley, Nevada and some tectonic implications,* 33 pp., U. S. Geological Survey Professional Paper 1274-A, 1984a.

Wallace, R. E., Patterns and timing of Late Quaternary faulting in the Great Basin province and relation to some regional tectonic features, *J. Geophys. Res., 88,* 5763–5769, 1984b.

Wallace, R. E., D. B. Tatlock, N. J. Silberling, and W. P. Irwin, *Geologic map of the Unionville quadrangle, Pershing County, Nevada,* U. S. Geological Survey Geological Quadrangle Map GQ-820, 1969.

Wetterauer, R. H., *The Mina deflection - a new interpretation based on the history of the Lower Jurassic Dunlap Formation, western Nevada,* 155 pp., [Ph.D. Thesis] Northwestern University, Evanston Ilinois, 1977.

Willden, R., and Speed. R. C., Geology and mineral deposits of Churchill County, Nevada, *Nevada Bureau of Mines and Geology Bulletin, 83,* 95, 1974.

Wright, L. A., and B. W. Troxel, Limitations on right-lateral strike-slip displacement, Death Valley and Furnace Creek fault zones, California, *Geol. Soc. Am. Bull., 78,* 933–950, 1967.

Zartman, R. E., Lead isotopic provinces in the Cordillera of the western United States and their geologic significance, *Economic Geology, 69,* 7992–805, 1974.

Zoback, M. L., R. E. Anderson, and G. A. Thompson, Cenozoic evolution of state of stress and stlye of tectonism of the Basin and Range province of western U. S., *Philosophical Transactions of the Royal Society of London, A300,* 407–434, 1981.

Tectonics of the Eastern Part of the Cordilleran Orogenic Belt, Chihuahua, New Mexico and Arizona

El Paso, Texas to Tucson, Arizona
June 29–July 7, 1989

Field Trip Guidebook T121

Leader: *Harald Drewes*

Associate Leader: *Russ Dyer*

American Geophysical Union, Washington, D.C.

Leader:

Harald Drewes
U.S. Geological Survey
Federal Center
Bldg. 25, MS 905
Denver, CO   80225

Associate Leader:

Russ Dyer
Department of Geological Sciences
University of Texas at El Paso
El Paso, TX   79968

IGC FIELD TRIP T121:
TECTONICS OF THE EASTERN PART OF THE CORDILLERAN OROGENIC BELT
CHIHUAHUA, NEW MEXICO, AND ARIZONA

Harald Drewes[1], J.S. Pallister[1], Russ Dyer[2], D.V. LeMone[3], W.R.Seager[4], R.E. Clemons[4], F.E. Kottlowski[5], Sam Thompson III[5], Robert Munn[6], Andy Alpha[6], D. J. Brennan[7], and J.M. Guilbert[8]

[1]U.S. Geological Survey, Denver, Colorado
[2]Dept. of Energy, Las Vegas, Nevada
[3]Univ. of Texas at El Paso, Texas
[4]New Mexico State Univ., Las Cruces, New Mexico
[5]New Mexico Bureau of Mines and Mineral Resources, Socorro, New Mexico
[6]Denver, Colorado
[7]Arizona Oil and Gas Conservation Commission, Phoenix, Arizona
[8]Univ. of Arizona, Tucson, Arizona

COVER ILLUSTRATION: View northeast, of the northern part of the Helvetia mining district. Quarry on the distant ridge is in the Mississippian Escabrosa Limestone, Me, which is thrust faulted on Upper Cambrian Abrigo Formation, Ca, Devonian Martin Formation, Dm, and a northwest-trending sinistral fault that is out of view but shown on map by Drewes (1972a, plate 4). These units are, in turn, truncated by another northwest-trending sinistral fault seen on the bold hill, at mid-distance, of Proterozoic granodiorite, Yg, and Middle Cambrian Bolsa Quartzite, Cb, that are intruded, in the foreground, by Paleocene granite, Tg

TABLE OF CONTENTS

TABLE 1 Summary of Daily Events of IGC Trip T-121

DAY	LOCAL EXPERT	INSTITUTE	STOP Key stop*	GEOLOGY	EVENING Town; Hotel	Program
0	–	–	–	(Arrival throughout day)	–	–
1	R. Dyer	University of Texas at El Paso	Sierra Juárez*	Cretaceous stratigraphy Fold and thrust zone style of deformation	El Paso, TX; Airport Hilton	Possible meeting with El Paso Geological Society
1a	D. LeMone	University of Texas at El Paso	Sierra Cristo Rey, Franklin Mts. Intermountain Road	Cretaceous stratigraphy Paleozoic stratigraphy Proterozoic stratigraphy		
2	D. LeMone R. Dyer	University of Texas at ElPaso	West side of Franklin Mts.	Paleozoic stratigraphy Basin-and-Range faults	Deming, NM; Grand Motor Inn	Meeting with people from NM State University, NM Technical Institute, and NM Bureau of Mines and Mineral Resources
	R. Clemons	NM State University	Florida Mts.*	Paleozoic stratigraphy Foreland style of def. NW-trending fault		
3	S. Thompson	NM Bureau of Mines and Mineral Resources	Hachita area	General review	Portal, AZ; Cave Creek Lodge	Facilities of SW Research Station, tectonic review
	G. Mack	NM State University	Little Hatchet Mts.	General review		
			Animas Mts.*	Overturned plate, gypsum plug; East intermediate style of deformation		
4			Chiricahua Mts.* Ft. Bowie area	Pz-Mz rocks pɛ, Olig. granitic rocks Major thrusts, two movement phases; West Intermediate style of deformation	Willcox, AZ; Royal Western Lodge	Elks Club for dinner and music
5			Dragoon Mts.*	pɛPz-Mz rocks, three major thrust plates, West Intermediate style	Tombstone, AZ; Adobe Lodge, Larian Motel	Tombstone ore deposits; discuss options for next day
5a	S. Reynolds	Arizona Geological Survey		Pearce mining camp		
5b	Personnel at AMERIND			Archeology		
6			(Rest day - 4th of July holiday - western lore)			
7			Whetstone Mts. Box Canyon Helvetia mining area* Florida Canyon	Pz, R Jr rocks Thrust plates; pɛ-T rocks 2 phases of movement, dating of movement Ore controls R rocks	Madera Canyon; Santa Rita Lodge	Barbecue
8	J. Guilbert	University of Arizona at Tucson	Sierrita Mts. Rincon Mts. (west side)*	Porphyry Cu pit Gneiss-cored dome; 2 phases of movement and str environments	Tucson, AZ; Howard Johnson Hotel at Airport	Meeting with people from AZ Geological Survey and University of AZ
9			Rincon Mts. (north and northeast sides)	2-stage dome history, Hinterland and West Intermediate styles	END OF TRIP	Discuss options for seeing Tucson and travelling to Washington

IGC FIELD TRIP T121:
TECTONICS OF THE EASTERN PART OF THE CORDILLERAN OROGENIC BELT CHIHUAHUA, NEW MEXICO, AND ARIZONA: ROADLOG

Harald Drewes

U.S. Geological Survey, Denver Federal Center, Colorado

SUMMARY

Trip 121 begins in El Paso, Texas, on mid-morning of Thursday, June 29, 1989, and ends in Tucson, Arizona, at mid-afternoon of Friday, July 7. Daily excursions include driving across the region, short stops alongside the road, and walking tours at key stops. With exception of a rest day in Tombstone, Arizona, each night will be spent in a different hotel or lodge. While late arrivals in El Paso may be arranged for early morning of June 29, and early departures from Tucson may also be arranged for the evening of July 7, trip members are encouraged to plan a more leisurely itinerary, perhaps through using the facilities of the first and last nights more than once.

Although reviewing the geology of the El Paso-Tucson region is our main objective, the trip has been planned to show also some of the cultural and historical features of our western heritage. Overtones of a frontier environment still are noticeable in our towns, our manners, and our politics; doubtless they have also left a mark on our scientific endeavors.

The leaders, associate leaders, and other occasional participants on this trip will present some diverse geologic ideas. This has been arranged intentionally to illustrate that our science is healthy and growing. We need each other at least in order to test our own ideas and to keep our own biases in check. We trust you will participate in the discussions, share with us your own knowledge, and laugh with us when we overextend ourselves.

Geological Objectives

Trip 121 is designed to illustrate the systematic changes in style of deformation across parts of four tectonic zones of the Cordilleran orogenic belt; the foreland, fold and thrust, intermediate (two subdivisions) and a deeper level or hinterland zone. Not all of these zones are throughgoing features, a situation that reflects the semi-cratonic characteristic of the Colorado Plateau block, oblique convergence of plates along the southwestern border of the block, and the influence of other tectonic features and events.

Tectonic overprinting is widespread in the region of this excursion, and is largely responsible for the many geologic uncertainties and controversies you will hear about on this trip. Resolution of problems arising from tectonic overprinting requires both studies of extensive regions and of many topics. Consequently, the synthesis that is a focus of this excursion has been slow in its development and doubtless remains imperfect.

In the El Paso-Tucson region structural features of Proterozoic age exerted a pervasive and long-lived influence on later structural development. Cordilleran compressive stress was superimposed on this already faulted terrane, and the new deformation varied in intensity, age, and duration across the region. Subsequently, extensional stress led to the development of Basin-and-Range block faulting, to other extensional movement, and to widespread volcanism.

This complex tectonic development of the region is intertwined with other geologic processes. Therefore, during the excursion we will discuss or illustrate features of stratigraphy of sedimentary rocks, petrography of igneous rocks, ore deposits and oil and gas potential, and geomorphology. Regrettably, these geologic features cannot be presented in as systematic a manner as the tectonic features will be, owing to constraints of both time and publication space. Where this guidebook is lacking, we encourage the participants to develop further contacts with the various topical experts they will meet during this excursion, or with the geologic institutes of the region.

Itinerary

A summary of our daily plans is given in table 1. The daily excursions include a mix of driving, roadside stops, and a key stop at which we will walk about 2-4 hours. Typically, the walks will be on trails or ranch roads but some will be cross country. Participants will thereby get a close-up view of geology that is both typical of a zone and as nearly self-evident as possible.

Travel will be in small vans. At the few places where the roads are too steep or rough for vans we will utilize one or more 4-wheel-drive vehicles. Steep climbs are few and are either short or on ranch roads. A day of rest in Tombstone, Arizona, is planned for the 4th of July national holiday.

Overnight accomodations vary from first class to utilitarian. Most are in towns with basic shopping and other facilities. Two are in rural settings in delightful

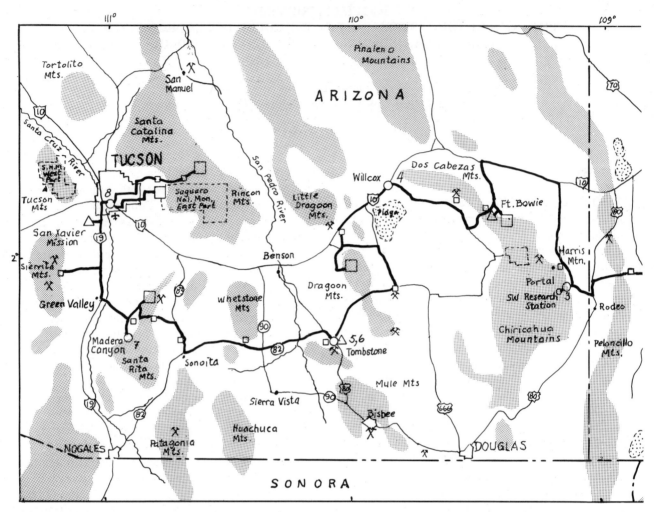

FIGURE 1 Index map showing trip route and location of stops. Large squares are sites of key stops; small squares are sites of other stops; triangles are historic sites; and circles are overnight stops.

wooded canyons. First and last night stops are near the airports of major cities for the convenience of our travelers. Rooms are to be shared, normally with another trip member, and in two places possibly with two members. Meals will be provided at our quarters or at a place within walking distance or a short drive from our quarters.

The general route and evening stopping places are shown on figure 1. Complimentary highway maps may be available to show the route in greater detail. Other supplementary material should be available from the various leaders.

The uncaptioned sketches marking the ends of the separately authored articles within the body of the roadlog are the products of Andy Alpha, and were made by him in the field trip region.

Historical Background

The region to be visited by excursion 121 has had a long and lively history by North American standards.

American Indians were for a long time the only inhabitants but many groups were highly mobile and none left written records. Spanish explorers travelled in or near the region during the 16th century and the first Spanish settlers arrived during the 17th century and were well established by the early 18th century. Other European settlers arrived during the 19th century. These various newcomers brought with them traditions of herdsmen, miners, and more varied agricultural practices than the original inhabitants had. The blending of these disparate societies has not always been peaceful but it was always colorful. Oftentimes the colorful lore of the old west is "bigger than life," yet it is always instructive.

During the excursion we will see some signs of the past and learn about others. We will see an old mining town at its 20th century wildest moment. Stops will be made at the ruins of an old cavalry fort, Ft. Bowie National Monument, and at a Spanish Mission, San Xavier del Bac, which is still in use. We will see traces of the old Butterfield Stage route and learn about the

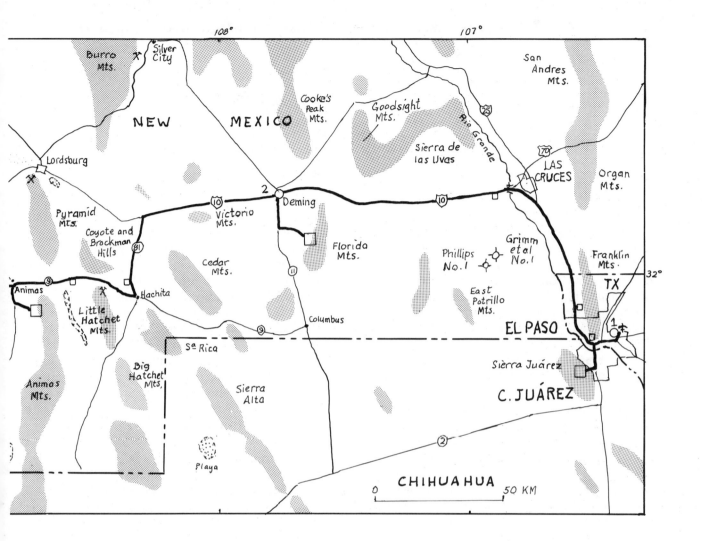

heliograph signal network. Possibly other arrangements may also be made.

GEOLOGIC DESCRIPTION

Day 1

During the afternoon of the first day, beginning by mid-morning, we will visit the Cerro Bola area of the Sierra Juárez, not far from the southwest side of Ciudad Juárez (fig. 2). In the Cerro Bola area we will see the style of deformation characteristic of the fold and thrust zone of the Cordilleran orogenic belt. These structures are seen in profile along the road up Cerro Bola, where the sequence consists of alternating massive or thick-bedded limestone and shale or thin-bedded limestone and marlstone, all of Aptian and Albian age. Dikes of Eocene age (47.8-48.2 Ma) are undeformed.

15 mi Stop 1. Base of steep grade up Cerro Bola. Prepare for short steep walk up the road, and return.
(All distances driven are shown in miles to better fit the local highway maps and odometers; they are incremental--stop to stop--and some are only approximate.)

GEOLOGY OF THE CERRO BOLA AREA, SIERRA JUAREZ, CHIHUAHUA, MEXICO

Russ Dyer
University of Texas at El Paso;
Department of Energy, Las Vegas, Nevada

Harald Drewes
U.S. Geological Survey, Denver Federal Center, Colorado

Cerro Bola is one of the highest peaks in the Sierra Juárez and is conspicuous for its various communications towers. The area is underlain by a sequence of five formations of Aptian and Albian age, and by Cenozoic gravel deposits. The lower three formations are each divided into three informal members, a medial thick-bedded or massive limestone and overlying and underlying thin-bedded limestone, shale, and sandstone. The next overlying formation is entirely thick-bedded limestone, and the uppermost unit is largely shale. Their place in the Sierra Juárez sequence is given on figure 3.

The lowest formation, the Cuchillo Formation, is typically brownish gray and includes coarse-grained, crossbedded, quartz sandstone and conglomerate. A pelecypod fauna, including Exogyra- and Unio-like shells (oysters and clams), characterize these rocks. Conformably overlying the Cuchillo is the thick light-gray Benigno Formation, made up of micritic limestone and biohermal reef limestones. Corals, algae, and rudistid pelecypods characterize the reef fauna, and rudistid beds and Orbitolina packstones occur in the thin-bedded upper member. The conformably overlying Lágrima Formation has a thin medial limestone resembling that of the Benigno, and thick lower and upper members of very pale orange marl and shale. The upper two formations have not been subdivided. Finlay Limestone is thick bedded and cliffy, resembling the middle member of the Benigno Formation, except that it contains dictyoconid foraminifera. The Del Norte Formation contains dark-gray shale, and some thin limestone beds. In nearby areas, these formations are the upper parts of a conformable sequence, but here they occur only in a structurally low major thrust plate that has been overridden by the older formations. The Lágrima Formation occurs in both plates.

Folds of great length and amplitude characterize the structures of the Sierra Juárez. These folds are generally tight to isoclinal and overturned to the northeast along the northeast flank of the mountains, but they are more open and upright to slightly inclined on the southwest flank. Fold axes strike northwest in the Sierra Juárez and throughout much of the El Paso-Tucson region. In the Cerro Bola area (fig. 2) the folds in the lower major plate strike west-northwest, about 30° differently from those in the adjacent parts of the upper major plate. Throughout the sierra, plunges are gentle and dominantly southeast; at the Cerro Bola area, however, most folds plunge gently northwest.

The folds and thrust faults are closely associated, as shown in the structure sections of several workers (fig. 4). Folds are commonly truncated upward or downward against thrust faults, thereby forming disharmonic structures. Locally, thrust plates are cut by strike-slip faults that do not affect overlying or underlying thrust plates; these are also disharmonic structures. In general, deformation is more intense to the northeast in structurally deeper rocks, and less intense to the southwest, at higher levels.

The thrust faults of the Cerro Bola area generally place older rocks over younger ones, but the reverse relations also occur. Most of these faults dip gently or moderately southwest, more or less parallel to adjacent beds. A major thrust fault, shown on figure 2 by a heavier line than the other faults, is marked by a repetition of the entire Lower Cretaceous sequence (the Arroyo Colorado thrust fault of fig. 4). In a few places thrust faults steepen or merge with reverse faults that separate groups of fault platelets from one another, as happens half a kilometer east of Cerro Bola. Such structures may be small ramps.

Elsewhere, as along the road above the STOP 1 site, both major plates are offset along a transverse fault along which there is right-slip offset. This fault trends about north, oblique to trends of most other features that it cuts, but about normal to the fold axes of the lower plate. Along strike the transverse fault appears to die out along bedding of shaley units. Possibly this fault is an oblique-slip ramp-like structure that formed as an adjustment to local intra-plate stress late during the compressive deformation.

Structural features found in other parts of the Sierra Juárez but not at Cerro Bola include: (1) disharmonic strike-slip faults that formed concurrently with thrust faults and folds, (2) back thrusts and southwest-inclined folds related to a backthrust, and (3) folds that vanish through the merger of axial planes of an anticline-syncline pair.

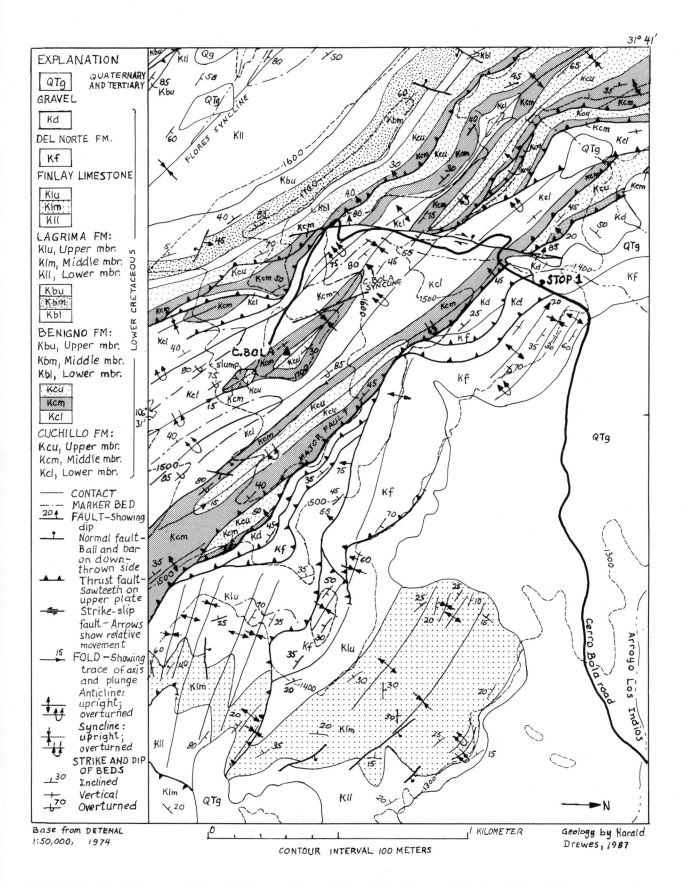

FIGURE 2 Geologic map of Cerro Bola area, Sierra Juárez, Chihuahua, and showing fold and thrust zone style of deformation.

AGE		STAGE	FORMATION	DESCRIPTION	THICKNESS(M)
QUATERNARY	Holocene			Gravel and sand	Thin
	Pleistocene		Camp Rice Fm	Gravel and sand, high terrace	Thin
TERTIARY	Pliocene		Fort Hancock Fm	Gravel, sand, clay; bolson fill	Thick
	Miocene / Oligocene		Local names *	Mainly rhyolitic rocks; some andesitic lava	Thick
	Eocene		Campus Andesite	Hornblende andesite plugs, dikes	—
			Love Ranch Fm, upper part *	Conglomerate, sandstone (to north)	300+
	Eocene and Paleocene ?		Love Ranch Fm, lower part *	Sandstone, shale, conglomerate (to north)	300+
CRETACEOUS	Late Cretaceous	Maastrichtian	El Picacho Fm *	Shale and limestone (to SE)	200-300?
		Campanian to Coniacian	San Carlos Fm *	Shale and limestone (to southeast)	500-1000?
		Turonian and Cenomanian	Ojinaga Fm, * ≅ Boquillas Fm	Shale and limestone (to SE) / Shale, thin-bedded limestone	200-300? / 10-265
		Cenomanian?	Buda Limestone	Thick-bedded, nodular	10-12
			Del Rio Fm	Shale, thin nodular limestone	24-30
			Anapra Fm	Sandstone and siltstone	67
			Mesilla Valley Fm	Shale, siltstone, thin limestone	42-55
			Muleros Fm	Nodular limestone, shale	21-38
	Early Cretaceous	Albian	Smeltertown Fm	Shale, siltstone, thin limestone	44-140
			Del Norte Fm	Shale and thin limestone	15-27
			Finlay Limestone	Massive reef limestone	130-186
			Lagrima Fm	Shale, marlstone, reef limestone	208-239
			Benigno Fm	Massive reef limestone	206
			Cuchillo Fm	Limestone, sandstone, shale	278
		Aptian	Las Vigas Fm *	Limestone and siltstone	?
		Neocomian	—	Siltstone, sandstone, gypsum(s)	Thick
JURASSIC ?				Sandstone, conglomerate, phyllite	Thick
PERMIAN	Early Permian		Hueco Limestone *	Light-gray, cherty (to east)	300-1000
			Colina Limestone	Dark-gray (to west)	300-500
			Epitaph Dolomite	Dark- and light-gray (to west)	300-500
PRE-PERMIAN			Older rocks *	Limestone, dolomite, quartzite, shale, and granite	—

FIGURE 3 Generalized stratigraphic column for the Sierra Juárez, Chihuahua. *, units present in nearby area which may have been present at or near the Sierra Juárez. Application of stage names follow local usage. Unconformities shown by vertical-ruled intervals. Rocks of Cerra Bola are in box. Modified from Drewes and others (1982, fig. 4, in part)

Eocene rhyolitic to andesitic dikes, not found in the Cerro Bola area, occur in many parts of the Sierra Juárez, where they commonly intrude northeast-trending strike-slip faults and are found to cut folds and thrust faults and are themselves undeformed. They are radiometrically dated by the K-Ar method on amphibole and biotite at 47.8 + 2.2 and 48.2 ± 1.7 Ma, respectively, ages similar to that of the Campus Andesite of the El Paso area.

In the Sierra Juárez the age of compressive deformation is constrained by the lower Upper Cretaceous Boquillas Formation, youngest to be deformed, and the intrusive rocks, oldest to be undeformed, as late Late Cretaceous to early Eocene. A young age within this range is more likely because the style of folding indicates a considerable thickness of cover, perhaps as much as 3-4 km, presumably of upper Upper Cretaceous, Paleocene and some lower Eocene

rocks, which have subsequently been eroded. Remnants of such rocks are found about 300 km to the southeast, and recently have also been discovered in a drill hole about 70 km to the northwest.

In summary, the region was compressively deformed probably in early Eocene time, when the rocks now exposed in the Sierra Juárez were 3-4 km beneath the surface. Stress orientation was generally northeast-southwest, possibly shifting 30° more northerly during the orogenic period.

Some additional inferences may be made from studies in other nearby ranges. The basal fault shown in most sections of figure 4, perhaps a décollement, followed levels either at the base of Mesozoic rocks or high in the Permian sequence. The fold and thrust zone extends about 150 km to the west, where the allochthon may have been thicker, and where the exposed rocks are deformed in the same style as those in the Sierra Juárez

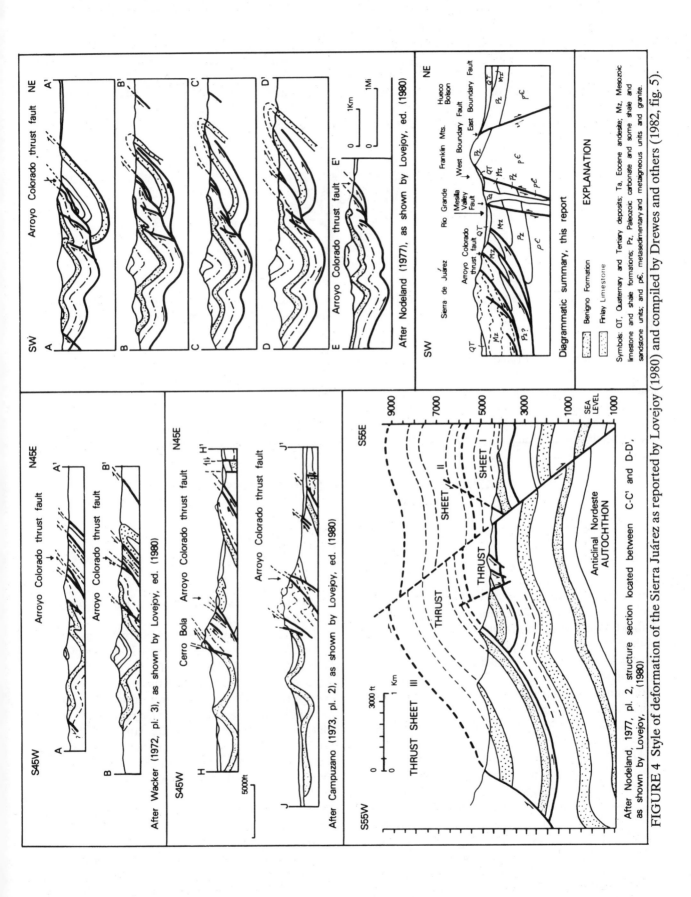

FIGURE 4 Style of deformation of the Sierra Juárez as reported by Lovejoy (1980) and compiled by Drewes and others (1982, fig. 5).

but less intensely. The relations between the fold and thrust zone and the foreland zone are concealed and will be reviewed later. Likewise, relations between the fold and thrust zone and the terrane to the northwest will be discussed after we have seen more.

Generalized structure sections of the Sierra Juárez have been presented by Lovejoy (1980), following his students, Wacker (1972), Campuzano (1973), and Nodeland (1977), and by Drewes and others (1982, fig. 4). The relations between the more-deformed northeastern part of the fold and thrust zone and the less deformed southwestern part of this zone, as well as the relations between this zone and the eastern intermediate zone are shown on figure 9. Relations along the northwest end of the fold and thrust zone are explained in a later section.

15 mi Day 1. Return to the airport Hilton Hotel.

Day 2

Depart early for the Franklin Mountains, Las Cruces, the Florida Mountains, and Deming (fig. 1) to see the style of deformation of the foreland zone and an area believed to have been exhumed from beneath the edge of the fold and thrust zone. The entire area has undergone Basin-and-Range or Rio Grande Rift style of block faulting, mainly during late Tertiary time. The deformation of the exhumed area is also strongly influenced by a major northwest-trending fault, one of a system of such faults that are common in the El Paso-Tucson region, and generally in the mobile belt of the Cordilleran orogenic belt southwest of the Colorado Plateau (and craton) region.

10 mi Stop 1 (optional). Tom Lea Park at the south end of the Franklin Mountains.

GEOLOGY OF THE FRANKLIN MOUNTAINS, TEXAS

D.V. LeMone
University of Texas at El Paso

The Franklin Mountains are very different geologically from most mountains to the east and west in that they are an uplifted block in which rocks as old as Middle Proterozoic are exposed. Key stratigraphic and structural features of the Franklin Mountains are conveniently seen at Tom Lea Park near the south end of the mountains and at Tom Mays Park on the west flank of the mountains.

From Tom Lea Park (Stop 1) the contrast in rock types and in structural style between the Sierra Juárez and Franklin Mountains is most apparent. Some conspicuous gravel terraces occur in the gap between these mountains through which the Rio Grande flows. The park itself lies on a remnant of the Pleistocene Kern Place terrace, cut on gravels of the Pleistocene (Irvingtonian) Camp Rice Formation, inset into the Pliocene and (late Blancan) Fort Hancock Formation (fig. 3). Two lower and two higher terraces are recognized; the highest is the La Mesa surface and forms the prominent level above the valley scarp west of the Rio Grande.

The Franklin Mountains form a north-trending fault-bounded linear raised block underlain mainly by Paleozoic rocks 1,600 m thick (fig. 5). Near the southern end of the mountains these rocks are raised enough to expose underlying Proterozoic metasedimentary and metavolcanic rocks 2,050 m thick, which are intruded by red granite dated as 995-953 Ma. Lower Cretaceous rocks 1,250 m thick occur low on the southwest flank of the mountains and extend to Cerro Cristo Rey, on the International Boundary just west of the Rio Grande. These rocks are intruded by a laccolith at Cristo Rey and by plugs of 47.1 Ma Campus Andesite between Cristo Rey and Tom Lea Park.

Major faults separate the Franklin Mountains from adjacent terranes. The east and west flanks of the mountains are bounded by normal faults of the Basin-and-Range type. To the south, along the Rio Grande is

AGE	STAGE		FORMATION	DESCRIPTION	THICKNESS, M
CRETACEOUS				(see figure 3)	
PERMIAN	Early Permian	Leonardian and Wolfcampian	Hueco Limestone	Gray, cherty; some yellowish gray siltstone	413
PENNSYLVANIAN	Late Penn.	Virgilian+Missourian	Magdalena Group — Panther Seep Fm.	Silty limestone, shale, gypsum	418
	Middle Penn.	Desmoinesian and Atokan	Bishop Cap Fm.	Yellowish-brown shale; limestone	180-194
			Berino Formation	Gray limestone, brown shale	160
	Early Pennsylvanian	Morrowan	La Tuna Formation	Medium-gray, cliffy limestone	85-107
MISSISSIPPIAN	Late Miss.	Chesterian	Helms Formation	Gray shale; olive green limestone	30
	Early Mississippian	Meramecian	Rancheria Fm.	Brown to light-gray limestone; sls	117
		Osagean	Las Cruces Fm.	Light-gray, cliffy limestone; shale	217
DEVONIAN	Late Devonian		Percha Shale	Black fissile shale	12-21
	Middle Devonian		Canutillo Formation	Siltstone and shale	15-30
SILURIAN	Early + Middle Silurian	Niagran Alexandrian	Fusselman Dolomite	Light-gray, cherty, cliffy dolomite	150-186
ORDOVICIAN	Late Ordovician	Cincinnatian	Montoya Group — Cutter Dolomite	Fine-grained, dark-gray; weathers lt.	52-60
			Aleman Dolomite	Fine-grained, dark-gray, cherty	30-44
	Middle Ordovician	Champlainian	Upham Dolomite	Medium-grained, gray /Cable Canyon Mbr	15-27
	Early Ordovician	Canadian	El Paso Group (locally Formation) — Florida Mountains Fm.	Dolomite, limestone	0-10
			Scenic Drive Fm.	Dolomite reef	87
			McKelligon Canyon Fm.	Dolomite	207
			José Formation	Dolomite, oolitic	22
			Victorio Hills Fm.	Dolomite, limestone	88
			Cooks Formation	Dolomite, limestone	33
			Sierrite Formation	Dolomite	37
CAMBRIAN	Late Cambrian		Bliss Sandstone	Quartz, glauconite grains; pebbles	0-300
PROTEROZOIC	Middle Proterozoic		Red Bluff Granite	Coarse-grained, some porphyritic	pluton
		Thunderbird Gp. — Tom Mays Park Fm.	Welded rhyolitic tuff	480	
		Smuggler Pass Fm	Trachyte	46	
		Coronado Hills Cgl	Conglomerate	11-27	
			Lanoria Quartzite	Metaquartzite, hornfels	215-260
			Mundy Breccia	Basalt breccia; marble chips	0-58
			Castner Marble	Marble /Cut by basalt sill	496

FIGURE 5 Stratigraphy of the Franklin Mountains, Texas. Most stage names not widely used in the El Paso-Tucson region. Unconformities shown by vertical-ruled intervals. Modified from Drewes and others (1982).

the Clint fault, known from a pair of oil wells to have a stratigraphic separation of 3,050 m, down to the southwest. This fault has been considered as related to, or even to be continuous with, the Texas lineament farther southeast (Hill, 1902; Muehlberger, 1980), and a branch of this fault may be the feature identified from geophysical evidence as the Mesilla Valley fault 5-10 km west of the Franklin Mountains. This entire system of faults may be part of a zone of anastomosing, generally northwest-trending basement flaws originally developed along the southwest margin of the North American craton and subsequently diversely reactivated (Drewes, 1981a).

About 150 km southeast of El Paso, the Rio Grande thrust fault is recognized to involve Paleozoic and Mesozoic rocks in outcrop. Closer to El Paso the thrust fault is concealed by extensive bolson and terrace deposits. However, its presence in the subsurface toward El Paso is shown by the repetition of part of the Cretaceous sequence in two drill holes. Everywhere the fault separates compressively deformed terrane to the southwest from undeformed terrane to the northeast, and likely it is the inferred basal thrust fault underlying the folded terrane of the Sierra Juárez (fig. 4). It thus corresponds with the frontal fault of the southeastern lobe of the Cordilleran orogenic belt of Drewes (1981a).

The junction of the three major fault systems, the northwest-trending Clint fault, Rio Grande thrust fault, and Basin-and-Range block faults, in the El Paso area has resulted in the strong contrast in deformational style between the Sierra Juárez and the Franklin Mountains. Due to lack of subsurface structural control, relations among these faults are unclear. One hypothesis by Drewes (in press), based on broad regional study, proposed that early movement on the Clint fault, perhaps left slip of large amount, emplaced some

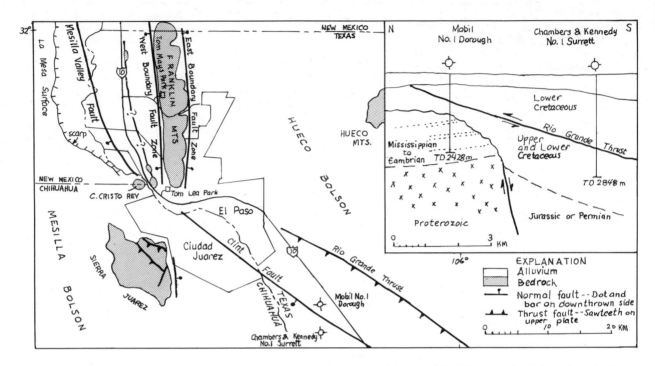

FIGURE 6 General structural setting of the El Paso border region, modified from Uphoff (1978) and after LeMone (1982). Faults largely or entirely concealed.

crystalline basement rock into a barrier-like position across the movement plane or basal décollement of the younger Rio Grande thrust fault; this age relation was also suggested by Lovejoy (1980). This barrier became the site of a ramp zone beneath the northeastwardly overriding thrust plates. Some compressional stress was absorbed in the development of tight folds and abundant thrust faults, such as occur on the northeast side of the Sierra Juárez. Other stress may have been absorbed in the generation of ramp and overthrust platelets, such as those found 150 km southeast of El Paso in the Quitman Mountains; the leading edges of these platelets have been eroded back from their initial extent in post-orogenic time. The Basin-and-Range block faults may abut the concealed Clint fault (modified by thrust faulting). Other normal faults developed along the east flank of the Sierra Juárez and along the ranges in Mexico. The theme of interaction of these major fault systems will recur throughout the excursion.

Our route from Tom Lea Park to Interstate Highway 10 takes us through the University of Texas at El Paso (UTEP) campus, home of our speakers yesterday and this morning.

17 mi Stop 2. Tom Mays Park provides another view of the geology of the Franklin Mountains. The high peak to the northeast (arbitrarily taken as 12:00 by the watch compass) is Anthony's Nose. The hill at 2:00 is underlain by Proterozoic, Ordovician, Silurian, Devonian, Mississippian, and Pennsylvanian rocks, summarized on figure 5. Local structural complexities in these rocks reflect in part their proximity to the Western Boundary fault zone (fig. 6), and in part younger landslide movement.

At 4:30 is North Mount Franklin, and in the distance at 6:00 are the Sierra Juárez. Between 8:00 and 10:00 is the north flank of an anticline in the Lower Permian Hueco Limestone. Near us the stratigraphic separation along the West Boundary fault zone is 2,200 m. The low retaining wall at 9:00 is composed of ignimbrite of the Middle Proterozoic Tom Mays Park Formation of the Thunderbird Group (fig. 5), described by Thomann (1981)

The eastern boundary of the Basin-and-Range province near El Paso, is placed along the western margin of the Hueco Mountains (fig. 6). The Rio Grande graben, which may extend south to the El Paso region, either becomes a broader more diffuse feature, extending from the west side of the Franklin Mountains to the east flank of the Florida Mountains or may be concealed beneath the basin east of the Florida Mountains. Many workers view the graben as being formed mainly or entirely in response to tensional stress of late Tertiary age, whereas others emphasize an earlier stage of compressional stress.

Basaltic volcanic rocks occur in the Rio Grande graben and in some of the adjacent areas. The Aden volcanic field, some cinder cones and several maars (phreato-magmatic depressions) are found west and northwest of the area of Stop 2 beyond the scarp marking the edge of the La Mesa surface. Kilbourne hole, one of the best known of the maars, is associated with basalt containing mantle-derived xenoliths of peridotite, granulite, pyroxenite, charnockite.

27 mi No stop. Commentary en route to Stop 3. Several exploratory oil wells were drilled a few tens of kilometers west of the Rio Grande at about the latitude of the Texas-New Mexico State line (fig. 1). One of the deeper of these wells, the Grimm et al., No. 1 Mobil 32, bottomed at about 6,580 m in the Middle and Upper Ordovician Montoya Group. It penetrated a sedimentary sequence that included beds of possible Jurassic age. Another of these wells, the Phillips No. 1 Sunland Park unit, bottomed at 5,620 m in Maastrichtian rocks. It encountered no Jurassic beds but penetrated about 1,000 m of Upper Cretaceous, Paleocene, and Eocene marine rocks (marine as determined through nannofossils) not known in nearby mountains. The lowermost Upper Cretaceous rocks occur in the Sierra Juárez; Paleocene rocks are found only many hundred kilometers to the southeast and Eocene rocks are even farther to the southeast.

The occurrence of these Mesozoic rocks, apparently preserved locally in concealed basins of Tertiary age, has several structural implications. First, the Jurassic beds of Chihuahua contain evaporites and are commonly the horizon of major thrust faulting above which the fold and thrust zone style of deformation is well developed. This style of deformation occurs in the Sierra Juárez and probably also in the East Potrillo Mountains (fig. 1) where, however, the basal décollement horizon is unexposed. It is possible, then, to have had the same zone of structural weakness extend this far north. Second, the development of the folds of the fold and thrust zone requires a substantial cover to provide adequate confining pressure. This cover could only have been provided by Upper Cretaceous and lower Tertiary rocks, which were largely eroded during post-orogenic time (Drewes, 1988a, in press). The presence of Upper Cretaceous to Eocene marine beds in the subsurface provides evidence that such a cover did extend across this region, as well as evidence for the large amount of erosion that subsequently took place along the northern and northeastern edge of the mobile zone of the Cordilleran orogenic belt.

17 mi No stop. New Mexico State University is in Las Cruces just north of the junction of Interstate Highways 10 and 25. This institute is the home of our speakers at the next stops.

5 mi Stop 3. An overlook on the bluff west of the Rio Grande provides views of more mountains in the foreland zone of the Cordilleran orogenic belt. First option for lunch.

GEOLOGY OF THE LAS CRUCES AREA, NEW MEXICO

W.R. Seager and R.E. Clemons
New Mexico State Univ., Las Cruces

The area is underlain by a basement of Proterozoic granite, and local belts of metasedimentary rocks. Paleozoic sedimentary rocks, mostly carbonates, overlie the basement and are about 1,900 m thick. Mesozoic strata, about 850 m thick, include Lower Cretaceous marine strata to the southwest of Las Cruces and Upper Cretaceous marine and nonmarine, dominantly clastic, strata to the north of Las Cruces. Tertiary sedimentary and volcanic rocks include lower Tertiary syn- and post-orogenic Laramide clastic wedges, middle Tertiary silicic to intermediate-composition volcanic rocks, and

upper Tertiary basaltic rocks and thick basin-fill deposits of the Rio Grande rift.

Rocks as young as the basal half of the Paleocene or Eocene Love Ranch Formation were deformed by northeast-directed compressional stress during the Laramide orogeny. Structural features, uplifts, and basins, created by this event in south-central New Mexico have been compared to those of the central Rockies of Wyoming (Seager, 1983; Seager and Mack, 1986; Seager and others, 1986). The structural style in both areas is distinguished by basement-cored block uplifts separated by broad basins filled with synorogenic to post-orogenic clastics of the Love Ranch Formation. Thrust or reverse faults border uplift-basin margins, at least on one side, and these, as well associated tight, overturned folds indicate substantial crustal shortening.

In south-central New Mexico, uplifts and basins trend northwest to nearly east-west, oblique to the northerly trend of late Tertiary fault blocks of the Rio Grande rift. Thus, the Laramide uplifts are truncated and segmented

by the younger faults so that fragments of the Laramide structures and uplifts are often revealed in cross section in the modern ranges. Such fragments, exposed near the Rio Grande from the Caballo Mountains and Black Range south to the Robledo Mountains, constrain the trend and geometry of the Rio Grande uplift (fig. 7). Drill-hole data, as well as outcrops of thick lower Tertiary basin-fill deposits and preserved Cretaceous strata, help identify complementary basins to the

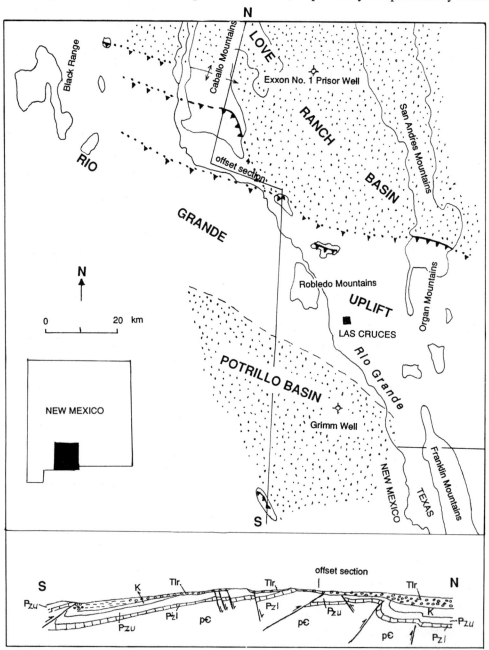

FIGURE 7 Tectonic map showing location of Laramide Rio Grande uplift and complementary basins in south-central New Mexico. Outlined areas in modern mountains show location of outcrop control for reconstruction of map and cross section. Heavy lines with barbs are uplift-boundary thrust faults. Tlr, Tertiary Love Ranch Formation; K, Cretaceous rocks; P u, upper Paleozoic rocks; P l, lower Paleozoic rocks; pC, Precambrian rocks.

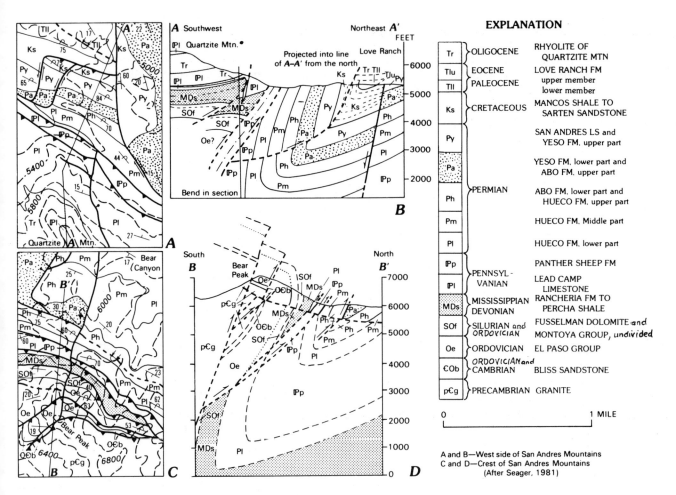

EXPLANATION

Tr	OLIGOCENE	RHYOLITE OF QUARTZITE MTN
Tlu	EOCENE	LOVE RANCH FM upper member
Tll	PALEOCENE	lower member
Ks	CRETACEOUS	MANCOS SHALE TO SARTEN SANDSTONE
Py		SAN ANDRES LS and YESO FM, upper part
Pa		YESO FM, lower part and ABO FM, upper part
Ph	PERMIAN	ABO FM, lower part and HUECO FM, upper part
Pm		HUECO FM, Middle part
Pl		HUECO FM, lower part
ℙp		PANTHER SHEEP FM
ℙl	PENNSYL-VANIAN	LEAD CAMP LIMESTONE
MDs	MISSISSIPPIAN DEVONIAN	RANCHERIA FM TO PERCHA SHALE
SOf	SILURIAN and ORDOVICIAN	FUSSELMAN DOLOMITE and MONTOYA GROUP, undivided
Oe	ORDOVICIAN	EL PASO GROUP
€Ob	ORDOVICIAN and CAMBRIAN	BLISS SANDSTONE
pЄg	PRECAMBRIAN	GRANITE

0   1 MILE

A and B—West side of San Andres Mountains
C and D—Crest of San Andres Mountains
(After Seager, 1981)

FIGURE 8   Geologic maps and structure sections of parts of the Bear Spring fold and thrust zone, southern San Andres Mountains, northeast of Las Cruces, New Mexico, after Seager (1981), compiled by Drewes and others, (1982, fig. 11).

northeast and southwest of the Rio Grande uplift--the Love Ranch and Potrillo basins, respectively (fig. 7).

The Organ Mountains to the east, with the strongly serate skyline, are underlain by Proterozoic granite, Paleozoic strata (thinned by Laramide erosion), thin lower Tertiary Love Ranch conglomeratic strata, and a thick pile of Oligocene (33 Ma) rhyolitic ash-flow tuff and lavas intruded by comagmatic quartz syenite. The quartz granite body is of batholith size, is 32.8 Ma, and is interpreted to be the remains of a magma chamber whose silicic cap erupted as pyroclastic flows and lavas to form the Organ cauldron (Seager and McCurry, 1988). Subsequent uplift of the modern range may have involved an early stage of closely spaced faulting and moderate west tilting or east downwarping. A later stage (post 9 Ma) of uplift and west tilting on a single range-boundary fault, located along the eastern margin of the range, probably continues to the present as indicated by very young Holocene (1,000-yr) scarps (Gile, 1986).

The Sierra de las Uvas to the northwest (fig. 1) are centered in the Sierra de las Uvas ash-flow field, of

Oligocene age. The field covers about 3,700 sq km and structurally coincides with the Goodsight-Cedar Hills volcano-tectonic depression. The Cedar Hills vent zone, which includes about 27 rhyolite domes, a diatreme, collapsed area, andesite dikes and buried cone, and two Uvas Andesite vents marks the asymmetric eastern margin. Its western margin is the Goodsight Mountains vent zone, which includes one rhyolite plug-flow complex, and a series of dacite-latite-andesite vents, tuff breccias, and flows. The age of these rocks is about 36 to 37 Ma. The ash-flow field contains up to 500 m of six ash-flow tuff units and interbedded volcaniclastic units which formed about 36 to 35 Ma (Clemons, 1976).

In the southern part of the San Andres Mountains rocks as young as the lower part of the Eocene or Paleocene Love Ranch Formation were deformed by northeast-directed compressional stress. Structural features formed by this event are widely scattered asymmetric anticline and syncline pairs, such as the Bear Springs folds of the southern part of the San Andres Mountains (fig. 8), and reverse faults that dip southwest (Seager, 1981). In several places such folds and faults occur together, with the fault formed along the axial plane of the anticline. Basement rocks occur in the cores of some of these folds, for the associated reverse faults continue their moderate to steep southwest dip into the basement rock. Neither surface mapping nor subsurface geophysical evidence indicates that the faults flatten downward, ultimately to merge with the structures of the mobile belt. Apparently, the faults continue down into the ductile terrane where they are lost. These conditions--scattered structures rather than a field of folds, and reverse faults not known to be tied directly to faults of the mobile belt--are used to distinguish the foreland from the fold and thrust zones.

50 mi Stop 4 (optional). The east side of Deming provides an alternate site for lunch.

20 mi Stop 5. The Florida Mountains (fig. 1) present a variation of the foreland style of deformation of the Cordilleran orogenic belt in which thrust faults and folds are present along a major northwest-trending fault zone that was diversely and recurrently active. Because the structural complexities of the area are greater than those of the foreland zone to the northeast, and because the interaction with the northwest-trending fault is so characteristic of the mobile belt, I propose that this area once underlay the fold and thrust belt near its northern edge and has been exhumed through uplift and deep erosion. Therefore, this area is labelled the sub-fold and thrust area on figure 9, on which an attempt is made to illustrate the distribution of the tectonic terranes of eastern part of our traverse.

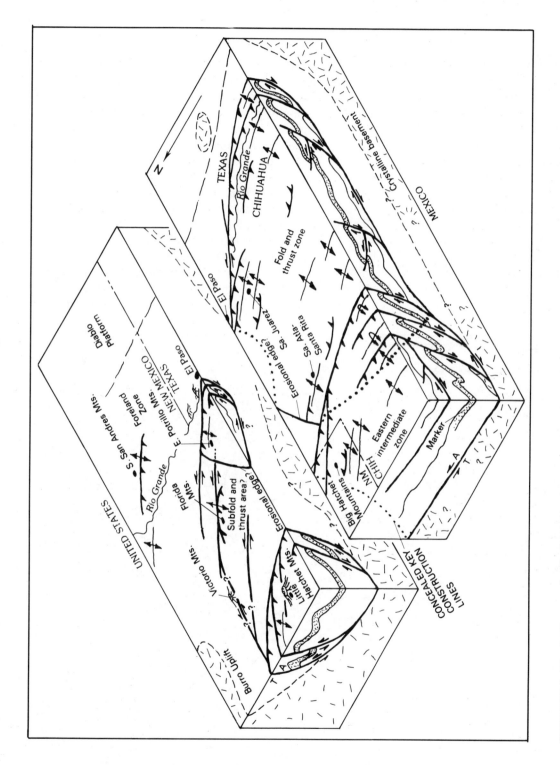

FIGURE 9 Block diagram (pulled apart), showing variations of tectonic styles between the fold and thrust zone and adjacent terranes. Not drawn to scale. Erosional edge subsequently concealed by Tertiary and Quaternary deposits.

GEOLOGY OF THE FLORIDA MOUNTAINS, NEW MEXICO

R.E. Clemons
New Mexico State Univ., Las Cruces

The Florida Mountains, southeast of Deming, New Mexico, are underlain by a wide variety of igneous and sedimentary rocks, most of which are only slightly deformed but some of which, in the central part of the mountains, are moderately folded and faulted (fig. 10).

The central part of the mountains is underlain mainly by Proterozoic granite and syenite and an overlying sequence of Cambrian to Mississippian sedimentary formations. Tertiary andesite and basalt dikes cut all rocks and most structures of the core of the range. A major northwest-trending fault, the South Florida fault zone, cuts obliquely through the central part of the Floridas. The rocks near this fault are folded and abundantly faulted as shown by Clemons (1985) because the clast assemblage is mixed, but is here so labelled for cartographic simplicity.

Along the South Florida fault zone are two slices of Upper Devonian Percha Shale and one of the Upper Mississippian Rancheria Formation. The Percha is an olive-gray to black shale 75 m thick that is poorly exposed. In a separate fault slice, the Rancheria Formation is a medium- to dark-gray cherty, coarse-grained limestone 68 m thick, this unit has also been correlated with Lake Valley Limestone and Escabrosa Limestone (in part) in other studies.

North of the South Florida fault zone, on the eastern flank of the Florida Mountains, this Paleozoic sequence is overlain by 130 m of Lower Permian limestone belonging to the Hueco Formation, and the Cretaceous or lower Tertiary Lobo Formation. The Lobo rests unconformably on the Permian rocks 2 km southeast of the area of figure 10A, and on the syenite 2 km north of

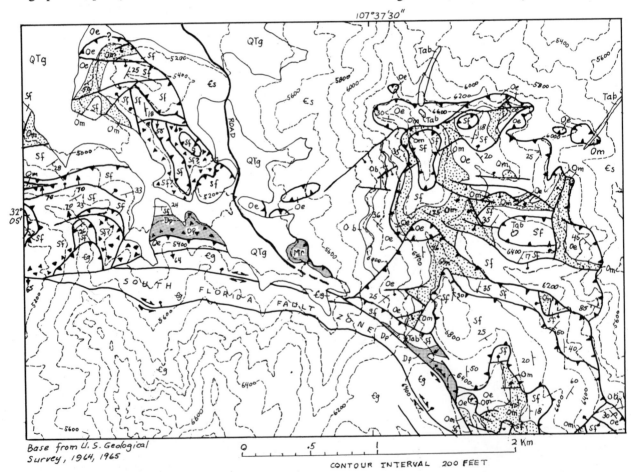

FIGURE 10 Geologic map (10A) and structure sections (10B) of the central part of the Florida Mountains, New Mexico, after Clemons and Brown (1983) and Clemons (1985). In sections, SFMF is the South Florida fault zone.

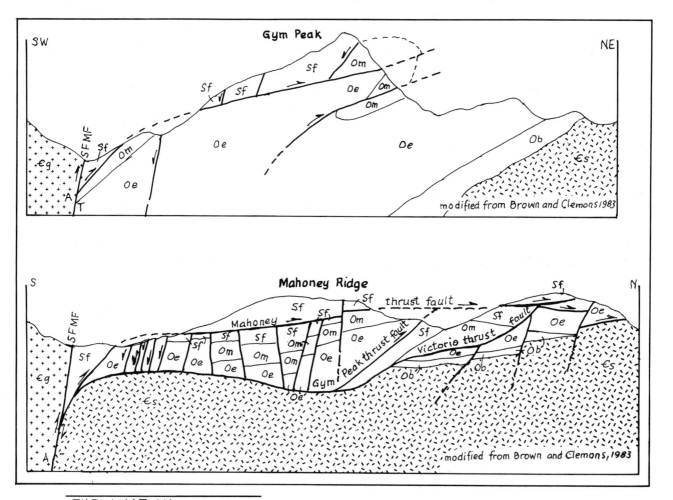

Gym Peak

SW ... NE

modified from Brown and Clemons 1983

Mahoney Ridge

S ... N

thrust fault

Mahoney

Gym Peak thrust fault

Victorio thrust

Sf fault

modified from Brown and Clemons, 1983

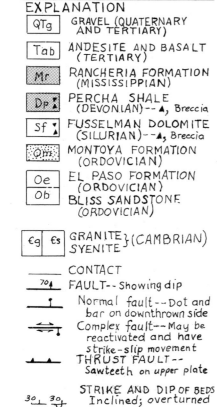

EXPLANATION

QTg	GRAVEL (QUATERNARY AND TERTIARY)
Tab	ANDESITE AND BASALT (TERTIARY)
Mr	RANCHERIA FORMATION (MISSISSIPPIAN)
Dp ▲	PERCHA SHALE (DEVONIAN)-- ▲, Breccia
Sf ●	FUSSELMAN DOLOMITE (SILURIAN)-- ▲, Breccia
Om	MONTOYA FORMATION (ORDOVICIAN)
Oe	EL PASO FORMATION (ORDOVICIAN)
Ob	BLISS SANDSTONE (ORDOVICIAN)
€g €s	GRANITE } (CAMBRIAN) SYENITE

_____ CONTACT

_70_ FAULT-- Showing dip

Normal fault--Dot and bar on downthrown side

Complex fault--May be reactivated and have strike-slip movement

THRUST FAULT-- Sawteeth on upper plate

30⊥ 30⊤ STRIKE AND DIP OF BEDS Inclined; overturned

the area of figure 10A. Within the area small irregular intrusions of finely crystalline monzonite and dikes of andesite and basalt cut all structures except the youngest movement on the South Florida fault zone, and outside the area a rhyolite dike cuts two branches of this fault zone and ends against, or is cut by the third branch.

The South Florida fault zone dips steeply southwest and has the offset of a reverse fault. The systematic change in strike of bedding along the east flank of the range (Clemons and Brown, 1983) suggests that there was a substantial right-slip component of movement on the fault zone. Movement on the fault zone is believed to have generated the imbricate thrust faulting of the Paleozoic rocks. Along a lower thrust fault, younger rocks have been placed upon older ones; commonly the Bliss Sandstone and at least part of the El Paso Formation are faulted out (fig. 10). Along a middle-level thrust fault are imbricate slices of two members of the Montoya Dolomite, and truncated recumbent folds. At least 600 m of movement is required along part of this thrust fault. Along a high-level thrust dips increase from subhorizontal, a kilometer northwest of the South Florida fault zone to nearly vertical along that fault zone (fig, 10), suggesting a genetic tie between the movement (or an early phase of movement?) on the fault zone and the thrust fault. This high-level fault also places younger rocks over older ones, and it truncates structures in the underlying rocks, suggesting that it is younger than the middle-level faults.

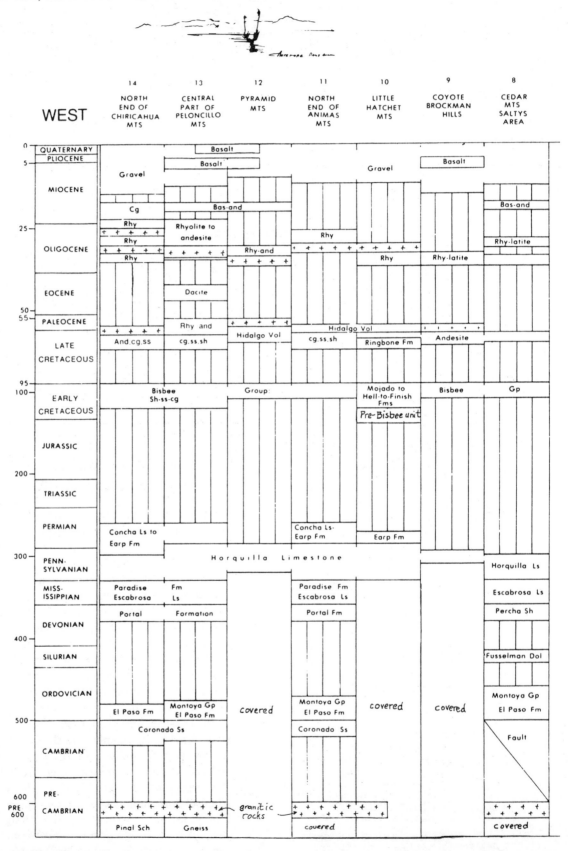

FIGURE 11 Stratigraphic correlation diagram of rocks in the region between Sierra Juárez,

20 mi Deming. Return to Deming for our motel, night 2.

Day 3

Departure on Interstate Highway 10 for Hachita, Animas Mountains, and Portal. On Day 3 we cross more of the region of foreland style of deformation (or subfold and thrust zone area) and enter the eastern intermediate zone of the Cordilleran orogenic belt. Because this tectonic zone was thicker than the fold and thrust zone, its lower part is still preserved in such

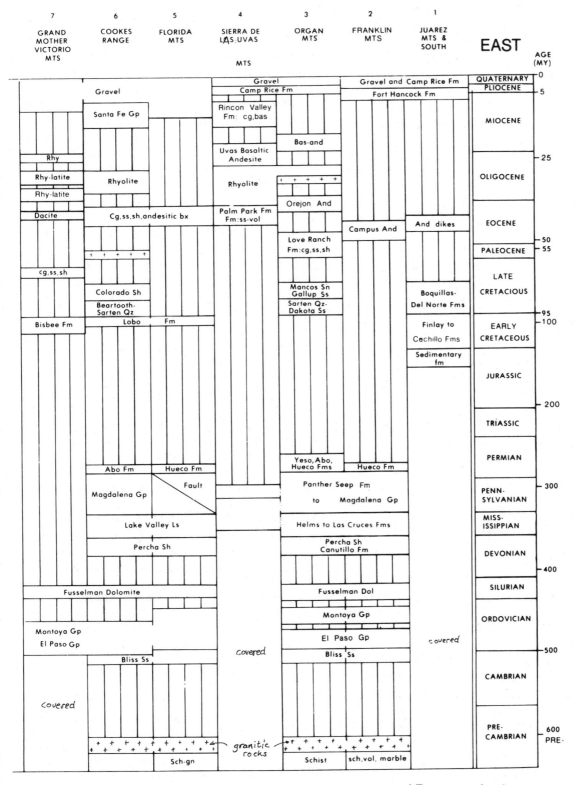

Chihuahua, and the Chiricahua Mountains, southeastern Arizona (Drewes and others, 1982).

ranges as the Sierra Rica, Little Hatchet Mountains, and the Animas Mountains. Unfortunately, the contact between the zones is covered, or has not yet been recognized because detailed mapping is incomplete.

17 mi No stop. The low hills to the south are the Victorio Mountains. They are underlain by Ordovician and Silurian dolomites, Lower and Upper(?) Cretaceous shale and conglomerate and, along the range crest, Eocene (41.7 Ma) dacitic volcanic rocks. An east-trending fault through the southern part of the mountains may be a thrust fault (Kottlowski, 1963; Corbitt and Woodward, 1970) or a strike-slip fault (Thorman and Drewes, 1980). Argentiferous galena was mined in the area south of the fault; production, in the 1880's, was $1.5 million.

The low hills north of the highway are underlain by Oligocene rhyolite and latite volcanic rocks and intercalated sedimentary rocks (Thorman and Drewes, 1979a).

10 mi No stop. Leave Interstate Highway 10. Low hills 2 km to the north are underlain by Ordovician El Paso Formation and Montoya Dolomite. The higher hills beyond them are the Burro Mountains, mainly underlain by Proterozoic granite (Drewes and others, 1985).

15 mi Stop 1. Overview of the geology of several ranges from rise in road 5 mi north of Hachita village, taken at the 12:00 position. At 12:30 to 1:00, in the distance, are the Big Hatchet Mountains, which are underlain by a Paleozoic section almost 3,500 m thick and, to the south, by a thick sequence of Lower Cretaceous sedimentary rocks and Oligocene volcanic rocks (fig. 11). Stratigraphic studies on some of the units were made by Armstrong (1958), Zeller (1965), and Thompson and Jacka (1981), and geologic maps by Zeller (1975) and Drewes (in preparation). The area was studied for inclusion in the wilderness system (Drewes and others, 1988).

At 1:00-1:30 in Hatchet Gap is Proterozoic granodiorite. The gap is the site of a zone of northwest-trending strike-slip faults, some of which probably were reactivated as range-front normal faults.

At 1:30-2:00 are the Little Hatchet Mountains, which are underlain by fragmentary sections of Paleozoic and Mesozoic rocks, a stock and andesitic rocks of Late Cretaceous or Paleocene age, and a stock and rhyolitic rocks of Oligocene age. At the north end of the mountains, the Lower Cretaceous Hell-to-Finish Formation (equivalent to the Morita Formation farther west) is thrust upon Upper Cretaceous to lower Tertiary Hidalgo Volcanics, and the Lower Cretaceous U-Bar Formation (partly equivalent to the Mural Limestone

farther west) is thrust upon Upper Cretaceous sedimentary rocks of the Ringbone Formation (Lasky, 1947; Zeller, 1970). This shingled style of thrust faulting recurs in the Sierra Rica.

At 2:30-4:30 are, respectively, the Coyote and Brockman Hills. The Coyote Hills are mainly underlain by Oligocene and Miocene rhyolite (Thorman, 1977). The Brockman Hills are an anomalously east-west-trending ridge of the Lower Cretaceous Cintura Formation, here anomalous in that it is made up of a sandstone facies rather than the more characteristic shale-rich facies, and also anomalous in that the beds are folded along east-west axes. To the northwest these rocks are faulted against or upon the Hidalgo Volcanics, and some klippen of Paleozoic rocks occur in structural position that seems to be between the two terranes; the cover in the area is extensive (Thorman, 1977). Exploratory oil wells northeast of the Brockman Hills encountered thrust-faulted Paleozoic rocks.

At 7:00-8:00 are the Cedar Mountains, which are underlain mainly by Oligocene rhyolitic rocks but, along a northwest-trending fault zone, also by Proterozoic, Paleozoic, and Mesozoic rock in very fragmentary sequence (Thorman and Drewes, 1981a). The style of thrust faulting of these Paleozoic rocks resembles that of the Florida Mountains.

At 10:30-11:00 are the Sierra Rica Mountains and, not separately identified on figure 1, the Apache Hills. The rocks in this area are thrust-faulted Paleozoic and Mesozoic rocks that are intruded by Cretaceous or Tertiary granite(?) and rhyolite and are capped by Oligocene(?) rhyolite (Zeller, 1975; Vanderspuy, 1975; Drewes, in press). The deformation style is that of shingled thrust platelets like that of the northern part of the Little Hatchet Mountains.

The regional structural relations are summarized on figure 12. Other structural features germane to this interpretation that occur in the Big Hatchet and Animas Mountains and in the Sierra Alta will be discussed during an evening session. An alternative structural interpretation based on local unspecified uplifts and resultant reverse faults and glide faults will be presented during the trip.

27 mi Stop 3. Northern part of Animas Mountains: Lunch stop and walking tour. This tour is to show the subtle expression of some major thrust faults, ones readily overlooked because the fault surface is not conspicuous and juxtaposition of units does not seem to

FIGURE 12 Block Diagram (pull apart) showing relations among structural features in northwestern Chihuahua and southwestern New Mexico. Mountains (shaded) are identified in key; EZ, eastern intermediate zone. Incipient faults, shown with minimal offset, develop slightly later than the others.

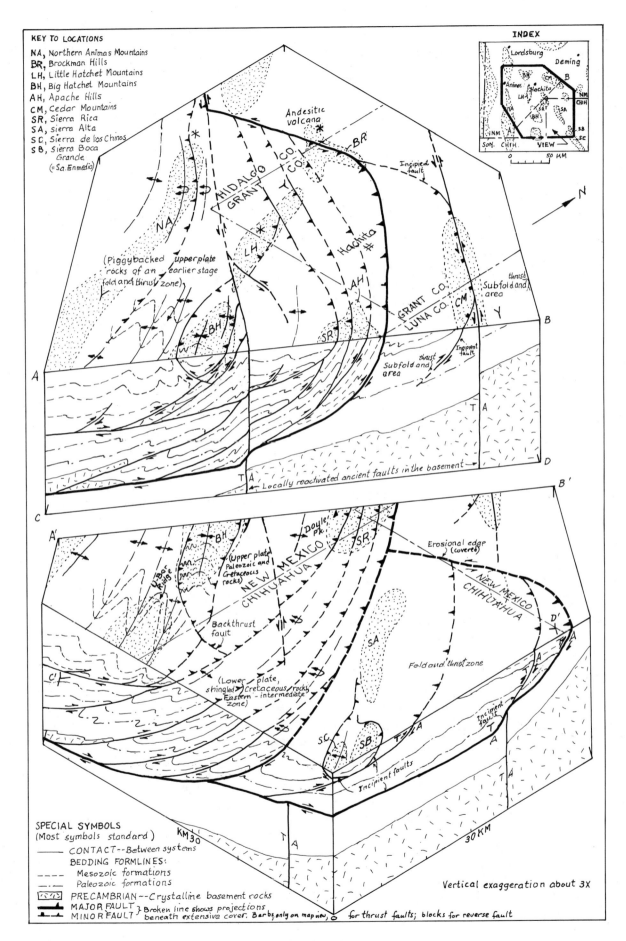

KEY TO LOCATIONS
NA, Northern Animas Mountains
BR, Brockman Hills
LH, Little Hatchet Mountains
BH, Big Hatchet Mountains
AH, Apache Hills
CM, Cedar Mountains
SR, Sierra Rica
SA, Sierra Alta
SC, Sierra de los Chinos
SB, Sierra Boca
   Grande
   (= Sa.Enmedio)

INDEX

Andesitic
volcana

N

(Piggybacked upperplate
rocks of an earlier stage
fold and thrust zone)

Incipient
fault

thrust
Subfold and
area

Subfold and
area

Incipient
fault

T A

T A

Locally reactivated ancient faults in the basement

(Upper plate
Paleozoic and
Cretaceous
rocks)

Erosional edge
(covered)

Backthrust
fault

Fold and thrust zone

(Lower plate,
shingled Cretaceous rocks)
Eastern - intermediate
zone

Incipient
fault

T A

Incipient faults

SPECIAL SYMBOLS
(Most symbols standard)

KM 30

30 KM

Vertical exaggeration about 3X

——— CONTACT -- Between systems
BEDDING FORMLINES:
- - - - Mesozoic formations
-·-·- Paleozoic formations
PRECAMBRIAN -- Crystalline basement rocks
MAJOR FAULT } Broken line shows projections
MINOR FAULT   beneath extensive cover. Barbs only on map view,
              for thrust faults; blocks for reverse fault

require much movement. This area also illustrates some stratigraphic units and rock types not seen elsewhere along our tour.

The area of figure 13 is taken from Drewes (1986); the formations in the normal Paleozoic sequence not present in the area of this figure occur nearby. The walking route takes us counter clockwise around a 2-km loop, cross country.

The hill northeast of the stop is capped by a klippe of the lower, or Pennsylvanian part, of the Horquilla Limestone, which dips gently to moderately southwest. The contact beneath the Horquilla is faulted, inasmuch as it overlies rocks as young as the Permian Epitaph Dolomite. Where we climb down the hill, a low-angle fault is exposed near the base of the limestone and a thrust fault is seen to separate the two formations. Fossils in the Horquilla provide evidence that the klippe is overturned. For one, a series of fusulines collected with Sam Thompson indicate the older fauna to overlie the younger one. Additionally, the Chaetetes that occur in place in a select bed fan out downward rather than upward. Apparently then, the klippe of Horquilla is overturned but the underlying rocks are not. The klippe thus represents a remnant of a major recumbent fold of a plate all but eroded off the Animas Mountains.

Continuing the tour, down the southwest side of hill, a tectonic plug of gypsum occurs between the Epitaph Dolomite and an elongate body of Oligocene granite. Such pods occur in several ranges of southwest New Mexico and southeast Arizona, and they also occur in southwestern Arizona, far west of our excursion route. Typically, they occur in areas of abundant and complex faulting. That they were derived from the adjacent Permian rocks seems unlikely in this region because those rocks provide no evidence, such as salt casts or solution breccias (rauwacke) of ever having contained gypsum or anhydrite. An alternative source here is from Jurassic formations that are known to contain evaporites in Chihuahua. Such rocks may underlie major thrust faults; hints of their presence or possible presence came from the Grimm et al. well southwest of Las Cruces, New Mexico, and from one group of undated gypsiferous beds beneath the Bisbee Group in the Little Hatchet Mountains. Although nongypsiferous, the Walnut Gap Formation of southeast Arizona indicates that rocks of Triassic or Jurassic age were once more widespread than they are now.

The third site to be visited is a klippe of upended Concha Limestone, the highest formation but one of the Paleozoic sequence north of the Mexican border. Here it, too, may be a part of the major plate inferred to have overlain the Animas Mountains. Two other klippen of this limestone are found in the range, one south of the area of figure 12. Fold axes in the range strike northwest to west-northwest; compression thus was oriented northeast-southwest. Some Upper Cretaceous volcaniclastic rocks are also involved in the deformation, which must have been of latest Cretaceous to early Tertiary age.

20 mi Stop 4 (optional). Low road cuts are in olivine basalt flows, here dated as 140,000 years old, that issued from near a cinder cone about 8 km away at 10:00. This basalt is probably related to the more extensive San Bernardino volcanic field, south of Rodeo in the valley west of the Peloncillo Mountains. Young fault scarps are found along segments of both flanks of the Peloncillos; one of these scarps in Mexico indicates fault movement in historic time.

The central part of the Peloncillo Mountains, at 2:00, are underlain by Proterozoic granodiorite, by Paleozoic and Mesozoic sedimentary sequences, by Upper Cretaceous or Paleocene andesitic volcanic rocks, and by Eocene (41.7 Ma) dacitic volcanic rocks. Both north and south of the central part of the mountains, Oligocene rhyolitic ash-flow deposits cover the older rocks. Cauldron structures, beyond our field of view both at 2:30 and at 10:00, have been proposed as sources of the extensive ash-flow deposits. The bold peak at 2:00 is underlain by a stock of Oligocene granite.

The rocks of the central part of the Peloncillo Mountains are cut by one of the northwest-trending strike-slip fault zones, shown by Gillerman (1958), Armstrong and others (1978), and Drewes and Thorman (1980a, 1980b). Rocks in and southwest of the fault zone are thrust faulted and folded along northwest-trending axes; those northeast of the fault zone are less deformed and typically dip gently northeast. The northeast-moving thrust plates probably were forced against more resistant rocks, such as some crystalline basement rocks, northeast of the fault zone and responded in part through left slip movement along the zone and in part as an oblique slip movement on a ramp structure.

At 4:00 to 5:00 are the Pyramid Mountains, which are underlain mainly by Oligocene rhyolitic to andesitic rocks related to another cauldron (Elston and Deal, unpub. map). In the northern part of the Pyramid Mountains these rocks are underlain by a 56 Ma granodiorite stock, 67 Ma andesite volcanic host rocks to the stock, and some Bisbee Group (Lasky, 1938; Flege, 1959; and Thorman and Drewes, 1978). The Lordsburg mining district is centered on a vein system related to the stock; silver, lead, and copper were produced as recently as about 10 years ago. The Lightning Dock KGRA (Known Geothermal Resource Area) geothermal prospect lies in the valley west of the highest part of these mountains.

34 mi Portal area. Cave Creek Lodge will be our quarters tonight and meals will be at the Southwestern

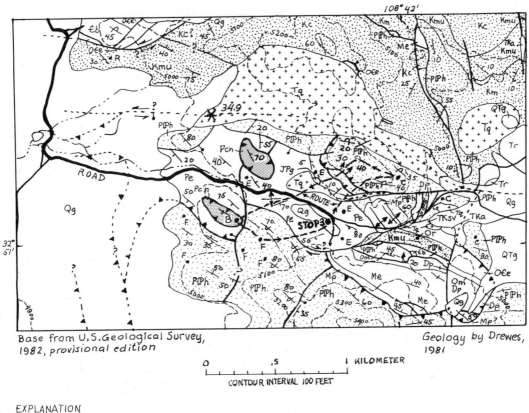

Base from U.S.Geological Survey,
1982, provisional edition

Geology by Drewes,
1981

```
0          .5          1 KILOMETER
|___|___|___|___|___|___|
    CONTOUR INTERVAL 100 FEET
```

EXPLANATION

Qg	} QUATERNARY	GRAVEL -- Deposits on low terraces
QTg		GRAVEL -- Deposits on high terraces
Tr	} TERTIARY	RHYOLITE -- Tuff
Tq		QUARTZ MONZONITE -- Fine-grained, porphyritic
TKsv	TERTIARY OR	SEDIMENTARY AND VOLCANIC ROCKS -- Clasts include andesite
TKa	CRETACEOUS	ANDESITE
Kc		BISBEE GROUP
		Cintura Formation -- Dark gray shale, and sandstone
Kmu	CRETACEOUS	Mural Limestone -- R, rudistid; O, orbitolina fossils
Km		Morita Formation -- Dark gray shale
JPg	} JURASSIC(?) OR PERMIAN	GYPSUM AND ANHYDRITE (?)
Pcn	PERMIAN	CONCHA LIMESTONE -- Cherty; B, dictyoclostid brachiopod
Pe		EPITAPH DOLOMITE -- Dark-gray; E, Echinoid spine
PIPh	} PENNSYLVANIAN	HORQUILLA LIMESTONE -- C, Chaetetes; F, large fuseline
Mp		PARADISE FORMATION -- Olive-gray shale
Me	} MISSISSIPPIAN	ESCABROSA LIMESTONE -- Massive, crinoidal
Dp	} DEVONIAN	PERCHA SHALE -- Gray shale and bioclastic limestone
Om	} ORDOVICIAN	MONTOYA DOLOMITE -- Brownish-gray dolomite
OЄe		EL PASO FORMATION -- Dolomitic limestone
Єb	} CAMBRIAN	BLISS SANDSTONE -- Arkosic and quartzitic sandstone

CONTACT
FAULT -- Showing dip. Dotted where concealed or intruded
   Normal fault -- Dot and bar on downthrown side
   Strike-slip fault -- Arrows show relative movement
   Thrust fault -- Sawteeth on upper plate
MARKER HORIZON
ANTICLINE
STRIKE AND DIP OF BEDS
   Inclined, Vertical; overturned
SAMPLE SITE -- Radiometrically dated rock, and age in million years
ROUTE OF TRAVERSE

FIGURE 13 Geology of the northern part of the Animas Mountains, New Mexico, showing route of traverse on foot, after Drewes (1986).

Research Station, a short drive up the valley.

Day 4

Departure for the northern part of the Chiricahua and Dos Cabezas Mountains, Ft. Bowie National Monument, and Apache Pass. On this day we are at the transition between the eastern and western divisions of the intermediate zone of the Cordilleran orogenic belt, marked by the occurrence of plates of Proterozoic basement rocks among plates of younger rocks and, to the west, also marked by the presence of several major thrust plates. We will also see evidence for left slip movement on the northwest-trending faults and for multiple phases of movement on them, all before the emplacement of an Oligocene stock. Metamorphic conditions occur along the lower part of a suspected ramp structure. A regional tectonic map of southeast Arizona shows the setting of this area (Drewes, 1980).

0.0 Stop 1. Before leaving Portal, a view back into the central part of the Chiricahua Mountains in the morning light helps to review the history of development of nested cauldrons of Oligocene and Miocene ages, presently being studied by John Pallister and Ed du Bray.

MID-TERTIARY RHYOLITIC MAGMATISM IN THE CHIRICAHUA MOUNTAINS, ARIZONA

J.S. Pallister

U.S. Geological Survey, Denver Federal Center, Colorado

A fine overview of Silver Peak and its volcanic rocks (fig. 14) is seen from near the lodge. Tuff of Horseshoe Canyon is down-dropped in a graben along the northwest side of Cave Creek. The northern boundary faults of the graben project through the low saddle on the southeast flank of Silver Peak. Rhyolite lava is overlain by tuff of Horseshoe Canyon in the graben block; equivalent rhyolite lava forms the skyline of Silver Peak, and an erosional remnant of tuff of Horseshoe Canyon is exposed near the top of Silver Peak (not visible from here). Similarly, rhyolite lava is overlain by tuff of Horseshoe Canyon near the crest of Portal Peak, to the south (Bryan, in press).

A relatively intact section through the pre-caldera volcanic rocks is exposed in the northeast face of Silver Peak, northwest of the Cave Creek graben; in contrast, the southwestern flank of Silver Peak is complicated by numerous normal faults. The lowest exposed rocks are monolithologic ash-matrix rhyolite breccias (units Trl/Tpf of fig. 14) of carapace-breccia and/or pyroclastic-flow origin. These breccias extend as an apparent stratiform unit around the northern flank of Silver Peak where they overlie graywacke and andesite (Bisbee Group or possibly another volcanic unit).

The rhyolite breccia is overlain by a section of maroon-weathering dacite lavas, which, in turn, are overlain by a variegated tan and white unit composed of pyroclastic flow deposits, ash-rich lahars, and ash-matrix volcanic sandstones. The pyroclastic flow deposits interfinger with rhyolite lavas (Trl of fig. 14), which probably represent distal parts of lava domes that were source rocks for the flow deposits. The sequence is capped by red-weathering rhyolite lava flows (Trl). A rhyolite lava feeder is exposed at the head of the main canyon south of "The Fingers" (note arrowheads on fig. 14). View of the vent is obscured by the foreground ridge, but vertically jointed lava in the vent throat is visible as the proiminent greenish-white (lichen-covered) cliffs near the head of the valley (labeled Trl, fig. 14). The vent area is eroded back into an amphitheatre at the head of the valley, such that throat-lava can no longer be traced continuously into the overlying lava flow that caps the skyline. A similar, but more strikingly eroded, vent plug is exposed at Cathedral Rock, southeast of Cave Creek (above Stewart Campground).

Summary of Volcanism of the Turkey Creek Caldera

At least two mid-Tertiary (circa 25 Ma) caldera fragments are exposed in the Chiricahua Mountains (Pallister and du Bray, in press). The 20-km-diameter Turkey Creek caldera (Marjaniemi, 1969) underlies the high terrane west of the Southwest Research Station; this is the source caldera for outflow welded-tuff sheets of the Rhyolite Canyon Formation, which are well-exposed in Chiricahua National Monument. The slightly older Portal caldera is centered near Horseshoe Canyon, about 6 km south of the Research Station. Welded tuffs from both calderas (Rhyolite Canyon Formation from the Turkey Creek caldera, and tuff of Horseshoe Canyon from the Portal caldera) are petrographically similar high-silica quartz-sanidine rhyolites. Bryan (in press) notes that the ring fault of the Turkey Creek caldera intersects and truncates the boundary of the Portal structure; trap-door collapse of

Rhyolite lava (Trl)

Silver Peak

The Fingers

Tuff of Horseshoe Canyon (Tth)

Trl

Tth

Trl

Trl

Tpf

Tdl

Tth

Trl

Pyroclastic flow deposits (Tpf)

Trl

Tdl

Dacite lava (Tdl)

Tdl

Tdl

Trl/Tpf

Tdl

FIGURE 14 View of the eastern flank of Silver Peak, near Portal, Arizona, showing stratigraphic relations of mid-Tertiary pre-caldera volcanic rocks. Unit Trl, rhyolite lava; Tth, tuff of Horseshoe Canyon (from Portal caldera); Tpf, pyroclastic flow deposits (locally derived); Tdl, dacite lava; Trl/Tpf, rhyolite lava carapace breccia and/or pyroclastic flow breccia.

the Portal caldera predates collapse at Turkey Creek. However, evolution of the two calderas is closely linked in time; resurgence of the Portal caldera took place after collapse of Turkey Creek (outflow tuff from the Turkey Creek caldera fills the Portal caldera and is resurgently domed), but prior to eruption of Turkey Creek moat lavas, which overlie volcanic and intrusive rocks of both calderas along an erosional surface.

Deep levels of erosion into the Turkey Creek caldera allow observation of hypabyssal features of this rhyolitic magmatic system, a situation that is analogous in many respects to the deeply eroded Questa caldera of northern New Mexico (Lipman, in press). The structural boundary of the Turkey Creek caldera is intruded by a thick ring dike of quartz monzonite porphyry; the dike appears to represent margins of a saucer-shaped (upper surface) intrusion that is also exposed centrally within the caldera. Stratigraphic and structural data indicate that intrusion of monzonite closely followed ash-flow eruption of the ~500-km^3 Rhyolite Canyon Formation welded tuff.

Gradations in petrography and geochemistry are observed between the high-silica rhyolite welded tuff and the monzonite porphyry. The ring intrusion is the

feeder for basal dacite lava flows within the caldera moat; textures grade from porphyritic granophyre to glass-matrix lava and compositions are intermediate between monzonite porphyry and rhyolite tuff. Rocks that are mineralogically and texturally transitional between ring-dike porphyry and ash-flow tuff also occur along the interior margin of the dike, as well as along fault-controlled feeder zones within the caldera floor. The latter occurrence suggests brittle failure of the magma chamber roof during ash-flow eruption, resulting in central, as well as ring-fracture, eruption of ash-flow tuff. The roof of the Turkey Creek magma chamber was brecciated, either by mechanical failure during inflation-deflation cycles (related to caldera formation and resurgence) or possibly by shock metamorphism accompanying magmatic explosions.

Pre-caldera floor rocks are not observed in the Turkey Creek caldera, despite deep levels of exposure. Either the underlying magma chamber (now represented by monzonite porphyry) was very close to the surface prior to ash-flow tuff eruption, or monzonitic magma intruded as a thick sill within the intracaldera tuff during or just following eruption, a relationship observed in the Portal caldera.

Following intrusion of monzonite porphyry and eruption of dacite lavas in the caldera moat, high-silica rhyolite volcanism resumed, and the caldera moat was filled with lavas and local pyroclastic flows. The stratigraphic and temporal progression from high-silica ash-flow tuff to dacite moat lava and monzonite porphyry to aphyric rhyolite moat lavas and dikes, defines a chemical trend from >75% SiO_2 to <65% SiO_2, then back to >75% SiO_2. However, thick sedimentary breccias and volcanic sandstones in the southern part of the caldera unconformably overlie, and contain clasts of, Rhyolite Canyon Formation tuff and monzonite porphyry. The sedimentary rocks are overlain, in turn, by aphyric high-silica rhyolite moat lavas. These relations indicate a hiatus in magmatism following ash-flow tuff eruption and emplacement of the monzonite porphyry before the return to high-silica rhyolite magmatism.

The features described above are consistent with a magma chamber and eruptive episode model in which the less-evolved lower parts of a stratified magmatic reservoir were drawn up into, and locally erupted effusively from, the same conduits and vents that initially fed explosive eruptions of high-silica rhyolite from the upper part of the reservoir. These initial caldera-forming events were followed by a period of resurgence and erosion, which in turn was followed by a return to high-silica rhyolite moat-lava magmatism. In contrast to the earlier high-silica eruptions, which were explosive, the later rhyolites were erupted effusively to form mainly aphyric lava flows. The magmatic hiatus, and the shift to aphyric lavas may record arrival of a new batch of magma in the Turkey Creek system.

Exposed basement rocks to the mid-Tertiary magmatic rocks of the Chiricahua Mountains range in age from Early Proterozoic (Pinal Schist) to Early Cretaceous Bisbee Group. These rocks are exposed as fault-bounded blocks throughout southeast Arizona (Drewes, 1981a). In addition, poorly understood, possible lower to middle Tertiary sedimentary rocks, and basaltic to andesitic lava flows locally overlie the Bisbee Group (Nipper Formation of Sabins, 1957a).

Overlying these basement rocks, and stratigraphically below the tuff of Horseshoe Canyon (outflow from the Portal caldera), is a thick sequence of dacite and 26 Ma rhyolite lavas and locally derived pyroclastic flow deposits. These rocks are well exposed in the prominent cliffs bordering lower Cave Creek near Portal and in the castellated flanks of Silver Peak. Exposures in this area suggest a composite sequence composed dominantly of dacite lavas and rhyolitic breccias. A wedge of rhyolite lavas and pyroclastic flow deposits overlies the dacite lavas. This wedge thickens to the south and near the high peak south of Portal is about 1 km thick.

The abundance of rhyolite lava flows and pyroclastic flow deposits, without major ash-flow tuffs and evidence of caldera collapse, records an early phase of dominantly effusive rhyolite volcanism in the central Chiricahua Mountains. Vulcanian to Strombolian eruption of rhyolitic magma implies relatively low volatile contents, a feature that may have characterized several mid-Tertiday rhyolite magma systems in the southern Basin-and-Range. Strombolian eruption, analogous to Hawaiian fire-fountaining, and low volatile contents are inferred for rhyolite lavas of the Black Range in southwestern New Mexico, on the basis of field evidence and mineral chemistry (W.A. Duffield, written commun., 1988). Caldera formation in the Chiricahua Mountains was preceeded by development of a high-silica rhyolite lava field; it would appear that volatile contents only episodically built up to the point necessary to drive large-volume pyroclastic eruptions and caldera collapse.

Portal Area
Chiricahua Mtns

10.0 Stop 2. The Harris Mountain area provides exposures of a thick Paleozoic sequence that is cut by many thrust faults and normal or strike-slip faults (Drewes, 1982). The sequence at Blue Mountain (fig. 15) a few kilometers to the northwest, is much like that at Harris Mountain. The area is instructive in showing that seemingly undisturbed contacts may be the locus of major thrust faults (fig. 16). To put it another way, contacts that seem to be depositional in some locations may prove to be faults in nearby locations. Local studies alone may thus generate erroneous structural conclusions.

The plan here is to compare the rocks across the basal contact of the Cambrian Coronado Sandstone in two places. Pebble beds occur low in the local Coronado and commonly are taken to mark the base of the formation; however, they are found at scattered levels throughout its lower half in nearby sections. Where we

AGE		STAGE	FORMATION		DESCRIPTION	THICKNESS (m)
QUATERNARY					Gravel and sand	
	Pliocene					
TERTIARY	Miocene		Rhyolite Canyon and Cave Creek Fms		Rhyolite ash-flow tuffs, and dikes	0·700 +
	Oligocene		Volcanics group of Cochise Head		Rhyolite to andesite; tuff, welded tuff, flows, and conglomerate	0·2340 +
			unnamed		Granodiorite and quartz monzonite stock	
			Nipper Fm and dacite of Davis Mtn		Lava flows and conglomerate	0·200
	Eocene Paleocene					
CRETACEOUS	Late Cretaceous		unnamed		Conglomerate to shale; andesite flows and rhyolite welded tuff	450 +
	Early Cretaceous	Albian	Bisbee Group	Cintura Fm	Shale and sandstone	750
				Mural Ls	Limestone and shale	
		Aptian		Morita Fm	Shale, sandstone, and conglomerate	
				Glance Cg	Conglomerate	0·300
JURASSIC						
TRIASSIC						
PERMIAN	Late Permian					
	Early Permian (Wolfcampian)		Concha Ls		Cherty limestone, fossiliferous	75
			Scherrer Fm		Quartzite and red shale	120·150
			Epitaph Dolomite		Dark-gray dolomite	0·30
			Colina Ls		Dark-gray limestone	473·505
			Earp Fm		Marlstone and limestone	300 +
PENNSYLVANIAN	Virgilian to Morrowan		Horquilla Ls		Medium-bedded limestone and siltstone	900 ±
MISSISSIPPIAN			Paradise Fm		Shale and limestone	0·47
			Escabrosa Ls		Massive crinoidal limestone	190
DEVONIAN			Portal Fm		Shale and limestone	120
SILURIAN					Light gray, fine-grained limestone	
ORDOVICIAN			El Paso Fm		Dolomite and sandy dolomite	137
CAMBRIAN			Coronado Ss		Sandstone and quartzite	115
PRECAMBRIAN					Granodiorite porphyry	
			Pinal Schist		Phyllite, schist, quartzite, and metavolcanics	very thick

FIGURE 15 Stratigraphic column of the Blue Mountain area (near Harris Mountain) (Sabins, 1957a) and vicinity, about 10 km north of Portal, Arizona.

first cross the contact, the sandstone appears to lie unconformably upon Proterozoic granodiorite porphyry, although the contact itself is concealed by talus (fig. 16).

Where we next cross the contact, the underlying rock is no longer granodiorite but is Pennsylvanian Horquilla Limestone. This limestone is juxtaposed against the granodiorite along a fault that probably is a strike-slip structure and is truncated upward by the Coronado Sandstone. The basal contact of the Coronado here must therefore be a bedding-plane fault, and the pebble beds in the lowest part of the Coronado need not constitute a basal conglomerate. Whether any of the formation is faulted out and, if so, how much is missing, remains unknown. In any case, the bedding-plane fault implies substantial stratigraphic repetition and thus must have a large amount of displacement, likely parallel to the trace of the strike-slip fault.

A further point may be made about the likely situation downdip on the strike-slip fault. Such structures are usually found to be disharmonic with respect to the underlying rocks, just as they are here with respect to the overlying ones. In other words, the combined mass of Horquilla Limestone and granodiorite (as well as some Devonian rocks along the

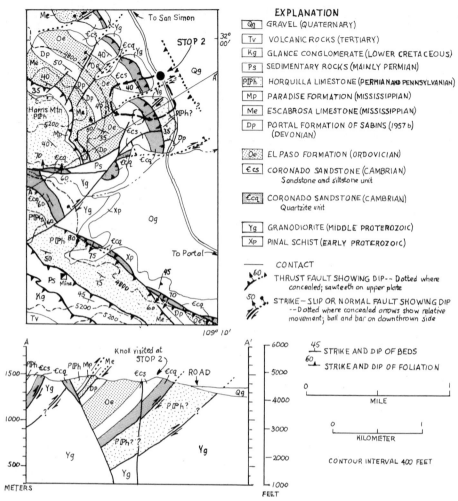

EXPLANATION

Qg	GRAVEL (QUATERNARY)
Tv	VOLCANIC ROCKS (TERTIARY)
Kg	GLANCE CONGLOMERATE (LOWER CRETACEOUS)
Ps	SEDIMENTARY ROCKS (MAINLY PERMIAN)
PlPh	HORQUILLA LIMESTONE (PERMIAN AND PENNSYLVANIAN)
Mp	PARADISE FORMATION (MISSISSIPPIAN)
Me	ESCABROSA LIMESTONE (MISSISSIPPIAN)
Dp	PORTAL FORMATION OF SABINS (1957b) (DEVONIAN)
Oe	EL PASO FORMATION (ORDOVICIAN)
€cs	CORONADO SANDSTONE (CAMBRIAN) Sandstone and siltstone unit
€cq	CORONADO SANDSTONE (CAMBRIAN) Quartzite unit
Yg	GRANODIORITE (MIDDLE PROTEROZOIC)
Xp	PINAL SCHIST (EARLY PROTEROZOIC)

——— CONTACT

THRUST FAULT SHOWING DIP -- Dotted where concealed; sawteeth on upper plate

STRIKE-SLIP OR NORMAL FAULT SHOWING DIP --Dotted where concealed arrows show relative movement; ball and bar on downthrown side

45 STRIKE AND DIP OF BEDS

60 STRIKE AND DIP OF FOLIATION

CONTOUR INTERVAL 400 FEET

FIGURE 16 Geologic map of *Stop 2* area about 10 km north of Portal, Arizona, after Drewes and others (1982).

road south of the Horquilla Limestone) are probably a thrust plate that overlies a concealed thrust fault whose position as shown on figure 16 is conjectural.

5.0 No stop. After passing the Paradise road junction we head north past Blue Mountain, at 9:00. The regional stratigraphic relations between the Blue Mountain section, described by Sabins (1957b) and other sections of the region are summarized on figure 17. In the large reentrant in the mountain front another thick pile of Oligocene volcanic rocks overlies the Paleozoic and Mesozoic formations, perhaps filling a fault bounded basin. Ten km north of Blue Mountain is another spur of Paleozoic and Mesozoic rocks that is much thrust faulted. The next stop will be at the west end of this spur.

46 mi Stop 3. Emigrant Canyon, beyond an unoccupied ranch. The objective of this stop is to see the relations between early major-phase movement on Cordilleran thrust faults and late-phase movement on a branch of the Apache Pass fault zone. Upper

Cretaceous volcanic rocks are cut by early thrust faults and Oligocene granite intrudes the branch of the Apache Pass fault zone. At the end of our walk, we will see, above us and at a distance, the first fully exposed plate of Proterozoic granodiorite, here thrust in among plates of Cretaceous sedimentary rocks (Sabins, 1957a; Drewes, 1981b, 1982), shown on figure 18.

Note the rib of upended Cambrian Coronado Sandstone near Stop 3; it is faulted out (offset outside of the field of view behind a hill on the Emigrant Canyon fault) at a point 0.4 km west of the canyon bottom. To the east the sandstone is overlain by more formations, and across the hill this sequence dips moderately south and comprises thrust platelets of alternating Paleozoic and Mesozoic units. On our walk we will see what happens to the rocks west of the Emigrant Canyon fault. These features were mapped by Sabins (1957a).

3a--Walk or use 4-wheel-drive vehicle shuttle to end of rough road. Outcrops are all of Proterozoic porphyritic granodiorite.

3b--Granodiorite in canyon bottom is cut by a minor

fault that dips 50° N. Along trail the position of the Emigrant Canyon fault is marked by a sliver of Mississippian Escabrosa(?) Limestone.

3c--Oligocene granite stock, dated as 32.7 Ma (K-Ar, hornblende) and 34.2 Ma (K-Ar, biotite), cuts the Emigrant Canyon fault.

3d--South of the stock the fault reappears at a sliver of Pennsylvanian and Permian limestone that lies between the granodiorite to the west and shale of the Lower Cretaceous Bisbee Group to the east. Faults bounding the sliver dip 45-70° E. The zone of shingled upended Paleozoic rocks at Stop 3 is offset to a site a few hundred meters south of 3d.

3e--The Emigrant Canyon fault is marked by a sliver of Mural Limestone of the Bisbee Group. Far to the south (100 km) the Mural is a thick biohermal limestone but here it is a sequence of thinner pelecypod-bearing limestone units and interbedded shale. East of this fault sliver are Upper Cretaceous volcaniclastic rocks correlative with the Ringbone Formation of New Mexico or the Fort Crittenden Formation, and others, of Arizona. West of the sliver are Escabrosa and Horquilla Limestones.

3f--The Paleozoic rocks west of Emigrant Canyon fault are not simply offset from site 3; their internal sequence differs slightly, perhaps reflecting additional movement among the thrust plates as well as offset along the Emigrant Canyon fault. Despite considerable recrystallization of the limestone, some silicified Chaetetes milliporaceous sponge fossils are found in beds of Late Pennsylvanian age, in this area usually from a stratigraphic level about 300 m above the base of the Horquilla Limestone.

Note that high up the hill 2 km to the southwest is a cliffy zone. The cliff is underlain by a thrust plate of the Proterozoic granodiorite faulted between two plates of Bisbee rocks. The Glance Conglomerate, basal formation of the Bisbee Group, is hundreds of meters thick in the plate overlying this granodiorite but is only tens of meters thick in the plate beneath the granodiorite. A large amount of tectonic transport is required both by the change of facies and by the intercalation of a basement-rock plate among Cretaceous beds. Elsewhere in the mountains facies changes in Ordovician and Devonian formations are also seen tied to adjacent thrust plates.

3g--Several pods of Oligocene granite cut rocks between sites 3f and 3g. Locally, block faulting brings together various formations, and a granite mass in the canyon bottom cuts all faults. Downstream from 3g note the diverse attitudes of two blocks of Coronado Sandstone.

Rejoin trail near site 3c, return to vehicles, and drive back toward the Bowie road but turn west at the National Monument.

6.0 mi Stop 4. The geology of the service entrance to Fort Bowie National Monument is noteworthy for its evidence of amphibolite-grade metamorphism. The geology of the area suggests that major thrust faults and the Apache Pass fault zone (one of the major thrust faults of the northwest-trending system) have merged (Drewes, 1981a, 1984). This sequence is overlain with subtle angular unconformity by the pelecypod-bearing Mural Limestone and the Cintura Formation, both of the Bisbee Group. Along the unconformity older parts of the Bisbee seem to lap upon older units of the Paleozoic sequence, suggesting the past existence of gentle relief.

A major thrust plate of Paleozoic rocks overlies the Bisbee and trends northwest to Apache Pass, where the plate is broken into smaller platelets that include some Permian formations, and is warped into an overturned fold. North of the pass the Bisbee is also overlain by Upper Cretaceous rocks, upon which the plate of Permian rocks is faulted.

Slaty cleavage and large porphyroblasts occur in the argillaceous beds of the Lower Cretaceous Cintura Formation exposed along the road up to the Monument. The porphyroblasts were identified by B.F. Leonard, U.S. Geological Survey, as andalusite, biotite pseudomorphs after staurolite, andalusite after kyanite, chloritoid, tourmaline, and some flakes of graphite. A thicker seam of graphite occurs in the brushy gully below the trailers in the Monument. These metamorphic features occur in a belt of high pressure and possibly moderate temperature along that segment of the Apache Pass fault zone which is obliquely crossed by the Hidalgo thrust fault system. A remnant of that thrust fault system occurs in the northern part of the Monument where there is a plate of Permian Concha Limestone overlying tightly folded Permian and Lower Cretaceous rocks, which in turn are thrust over Upper Cretaceous rocks. Such a metamorphic environment and distribution suggests that the rocks were at the base of a ramp structure and that the metamorphism occurred on the stoss side (or upstream side) of the ramp where the confining pressure due to load was slightly augmented by local conditions of tectonic stress.

The bold peaks southwest of the Monument are underlain by Proterozoic metaquartzite which was intricately folded before intrusion by Proterozoic granodiorite (Drewes, 1981b). The likely continuation of this quartzite northeast of the Apache Pass fault zone is tens of kilometers to the northwest in the Dos Cabezas quadrangle (Drewes, 1984). Such an offset supports the inference of major left-slip movement on the fault zone in Proterozoic time.

Return to vehicles, drive back to the county road, and turn left to Apache Pass. The same sequences of rocks are crossed that were seen near Fort Bowie.

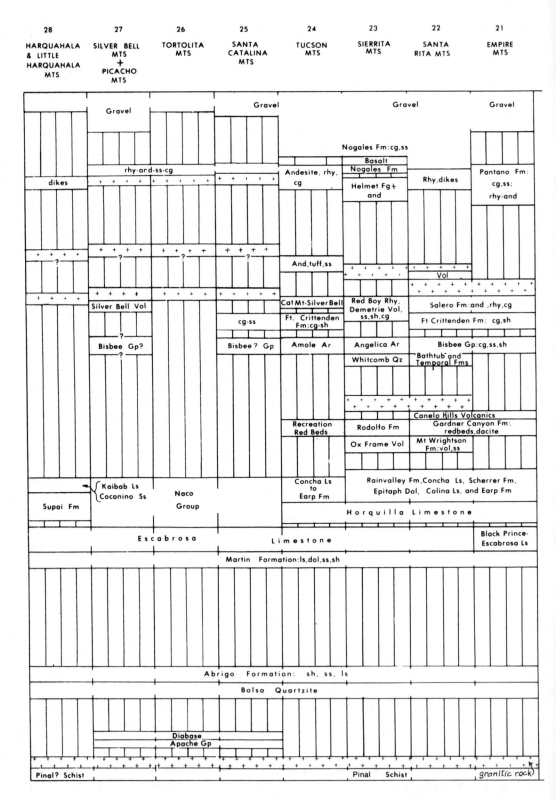

FIGURE 17 Stratigraphic correlation diagram for the region between the Chiricahua Mountains, southeastern Arizona, and the Wickenburg area, after Drewes and others (1982).

5.0 No stop. Between the parking place at the trailhead for the Fort and the Apache Pass, the tracks of the Butterfield Stage line are still visible, and marked by small posts.

11.0 Stop 5 (optional stop). Review geology near the Dos Cabezas village area. Signs of Cordilleran compressional deformation are minimal here, and consequently the effects of the deformation along the

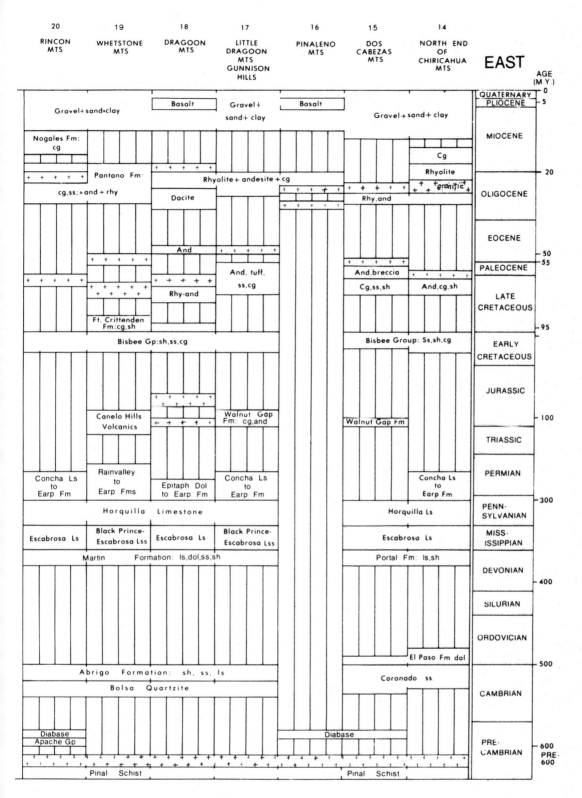

FIGURE 17 Continued

reactivated northwest-trending basement flaws are more apparent. The thrust faults that crossed the Apache Pass fault zone near Apache Pass were offset to the south near the boldest of the quartzite-capped peaks, and may be projected to an area of some structural complications at the east end of the rib of Paleozoic rocks south of Dos Cabezas village, thence to head west beneath the gravels of Sulphur Springs Valley to the

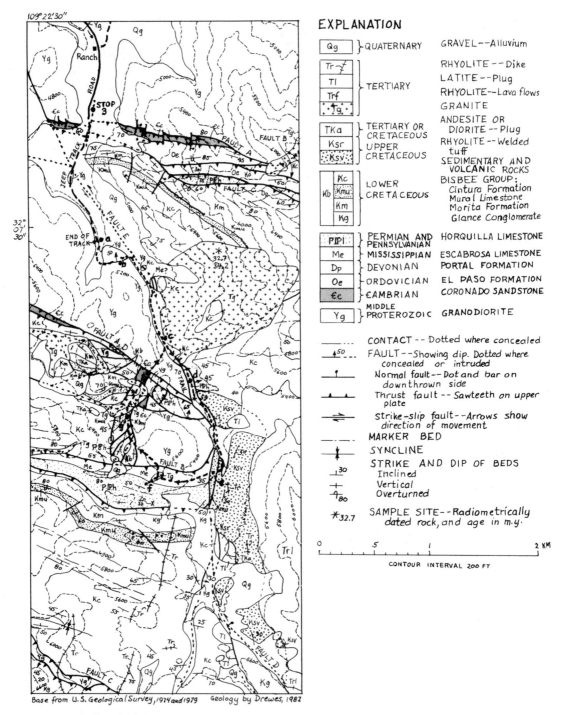

EXPLANATION

Qg	QUATERNARY	GRAVEL--Alluvium
Tr		RHYOLITE--Dike
Tl	TERTIARY	LATITE--Plug
Trf		RHYOLITE--Lava flows
Tg		GRANITE
TKa	TERTIARY OR CRETACEOUS	ANDESITE OR DIORITE--Plug
Ksr	UPPER CRETACEOUS	RHYOLITE--Welded tuff
Ksv		SEDIMENTARY AND VOLCANIC ROCKS
Kc / Kmu / Km / Kg	Kb LOWER CRETACEOUS	BISBEE GROUP: Cintura Formation, Mural Limestone, Morita Formation, Glance Conglomerate
PIPl	PERMIAN AND PENNSYLVANIAN	HORQUILLA LIMESTONE
Me	MISSISSIPPIAN	ESCABROSA LIMESTONE
Dp	DEVONIAN	PORTAL FORMATION
Oe	ORDOVICIAN	EL PASO FORMATION
Єc	CAMBRIAN	CORONADO SANDSTONE
Yg	MIDDLE PROTEROZOIC	GRANODIORITE

CONTACT -- Dotted where concealed

FAULT--Showing dip. Dotted where concealed or intruded

Normal fault--Dot and bar on downthrown side

Thrust fault -- Sawteeth on upper plate

Strike-slip fault--Arrows show direction of movement

MARKER BED

SYNCLINE

STRIKE AND DIP OF BEDS
Inclined
Vertical
Overturned

SAMPLE SITE--Radiometrically dated rock, and age in m.y.

0 .5 1 2 KM

CONTOUR INTERVAL 200 FT

Base from U.S. Geological Survey, 1974 and 1979   Geology by Drewes, 1982

FIGURE 18 Geologic map of the Emigrant Canyon area, Chiricahua Mountains, Arizona, showing route of traverse on foot and stops 3a-3g. Faults *A*, *B*, and *C* are the lower, middle, and upper branches, respectively, of the Hidalgo thrust fault system. Fault *D* is the northeast strand of the Apache Pass fault zone (in places merged with part of the Hidalgo thrust fault), and fault *E* is the Emigrant Canyon fault, a splay off fault *D*.

northern part of the Dragoon Mountains (fig. 1).

The geology of the Dos Cabezas Mountains is shown on maps of the Dos Cabezas and Simmons Peak quadrangles (Drewes, 1985a, 1985b) and the volcanic deposits and mineral potential of the core of the range are described by Drewes and others (1988). In essence,

there are two major structural blocks separated by a third wedge-shaped terrane of the Apache Pass fault zone that lies roughly at mid-slope on the range flank north of the stop.

The structural block southwest of the fault zone, on which we are standing, is largely underlain by

T121: 32

FIGURE 19 Texas Canyon stock, an Eocene 2-mica granitic rock that weathers to spectacular knobs and boulders, near the Dragoon exit of I-10, Arizona. Photo by Jack Rathbone.

Proterozoic granodiorite or granite porphyry that intrudes Pinal Schist in the foothills 1-2 km north of the village. A sequence of little-disturbed Paleozoic rocks overlies these Proterozoic rocks, and to the west there are also some outcrops of the Triassic or Jurassic Walnut Gap Formation (volcaniclastic rocks and red shale) and Lower Cretaceous Glance Conglomerate of the Bisbee Group. A high-angle fault separating the Paleozoic from the Mesozoic sequence can be projected beneath the gravel deposits of the valley on the basis of the position of an anomalous gravity gradient. Two small magnetic highs along the ridge of Paleozoic rocks coincide with swarms of Oligocene rhyolite dikes, and thus are inferred to mark concealed plugs.

The Apache Pass fault zone of the Dos Cabezas Mountains is only a few hundred meters wide to the west but increases in width to about 1 km to the east, and it continues to widen in the Chiricahua Mountains. North of the village it is made up of slices of

metamorphosed Bisbee Group and Paleozoic rocks; metamorphism decreases and Upper Cretaceous rocks are also present east of the village. A conspicuous magnetic trough that is believed to reflect the presence of these sedimentary rocks is offset southwest of the axis of the fault zone, and so suggests that at depth the sedimentary rocks flatten to the southwest, and thus diverge from the Apache Pass fault. Some copper, lead, and silver mines and prospects lie along the fault zone.

The northeastern block also is underlain mainly by Proterozoic granitic rock, but there it is overlain by a thick pile of Upper Cretaceous or Paleocene volcanic rocks. These rocks form a composite andesitic volcanic cone, modified by volcano-tectonic collapse along the Apache Pass fault zone. Massive rhyolitic ash-flow sheets initiated and ended the volcanic event, and voluminous breccia deposits marked the intervening volcanic episode during which breccia pipes and some hypabyssal rocks were intruded. Several Paleocene granite stocks also were emplaced in the volcanic pile. Locally, exotic-block breccias are present and some partly detached slabs of wall or floor rocks have been dragged into the volcanic pile. Locally also, these rocks are mineralized, and most are propylitized. Several magnetic highs occur at sites of exposed stocks or of inferred concealed stocks. A conspicuous magnetic trough runs transverse to the volcanic pile and Apache Pass fault zone, apparently separating a thicker part of the volcanic pile to the southeast from a thinner part to the northwest. Several attractive sites for additional ore deposits are proposed through the combined results of geologic, geophysical, and geochemical study (Drewes and others, 1988).

Continue to Willcox. Note the dune deposits on the east side of Willcox playa. Ground water occurs at shallow depth in several parts of Sulphur Springs Valley, and so agricultural centers have developed. Also, the available water provides coolant for the power plant on the west side of the playa. During wet years the playa contains a large body of water. This body was more permanent during the ice ages and reached a level marked by conspicuous beach lines along the north and west shores. Several Cochise Culture (paleo-indian) occupation sites occur on these beach bars.

15 mi Willcox. This will be our overnight stop.

Day 5

We should make an early departure for the northern part of the Dragoon Mountains, where we will see the full range of structural complexities characteristic of the northern part of the western intermediate zone style of Cordilleran deformation. These complexities include small tight folds associated with thrust platelets; three major thrust plates, two with upended metamorphosed rocks facing in opposite directions and the third with gently inclined unmetamorphosed rocks; evidence of telescoping not only of metamorphic facies but also of sedimentary facies; and massive involvement of Proterozoic basement rocks in one of the major plates. Apparently, these structural complexities account for the reported thickness of 15,000 m of rock beneath Willcox playa that have slow-velocity seismic refraction(?) signals characteristic of sedimentary rocks, a thickness three times that known in local stratigraphic sequences.

10.0 mi Stop 1. An overview of geology of several ranges is available from the vantage point of a tourist stop. The geology of the area near the stop, in the Little Dragoon Mountains quadrangle, is given by Cooper and Silver (1964). The area to the west with the bouldery and knobby landforms is underlain by the Texas Canyon stock (fig. 19). This stock is a massive 2-mica- and

garnet-bearing granite, and resembles the peraluminous 2-mica granites commonly associated with gneiss-cored domes (metamorphic core complexes). This rock is dated as early Eocene, 50-52 Ma, a date that in older literature was considered late Paleocene. South of the stock the Paleozoic and Mesozoic metasedimentary rocks are strongly deformed. North of the stock, and including the highest part of the mountains, the Paleozoic rocks are very little deformed.

The low hills to the east are underlain mainly by Paleozoic rocks, as well as by some Triassic and Jurassic and Lower Cretaceous rocks, almost all nearly undeformed, aside from being tilted. However, at the extreme southern end of the Gunnison Hills (fig. 20) an east-trending strike-slip fault brings a slice of Bisbee Group rocks among locally folded Pennsylvanian and Permian rocks.

The high mountains to the south are the Dragoon Mountains in which the rocks are highly varied and strongly deformed (fig. 21). The rocks include thick sequences of Proterozoic metasedimentary rocks and granitic rocks, Paleozoic sedimentary and metasedimentary rocks, and Mesozoic sedimentary and metasedimentary rocks, and stocks, plugs, and dikes of various Tertiary ages. Farther south in the mountains,

EXPLANATION

CORRELATION OF MAP UNITS

		DESCRIPTION OF MAP UNITS
Qg	QUATERNARY	Gravel
Tg	TERTIARY	Granitic rocks
TKr	TERTIARY OR CRETACEOUS	Rhyolite porphyry
Kb	CRETACEOUS	Bisbee Formation: r. Conglomerate-rich facies of plate 1 p. Conglomerate-poor facies of plate 3
PIPs	PERMIAN AND PENNSYLVANIAN	Scherrer to Earp Formations
IPh	PENNSYLVANIAN	Horquilla Limestone
MDs	MISSISSIPPIAN AND DEVONIAN	Escabrosa Limestone and Martin Formation
€s	CAMBRIAN	Abrigo Formation and Bolsa Quartzite
Yg	PROTEROZOIC	Granodiorite Pinal Schist
Xp Xpq		Pinal Schist, quartzite unit

——— CONTACT

———··· FAULT ·· Dotted where concealed

——— Normal fault ·· Ball and bar on downthrown side

⇒ Strike-slip fault ·· Arrow couple shows direction of movement. Heavier line indicates major fault

▲▲▲ Thrust fault ·· Sawteeth on upper plate. Heavier line indicates major fault

—┼— FOLD ·· Anticline showing strike of axial plane and direction of plunge of axis

STRIKE AND DIP OF BEDS

30 Inclined

Vertical

80 Overturned

FIGURE 20 Generalized geologic map of the northern part of the Dragoon Mountains, Arizona (Drewes and others, 1982; Drewes, 1987).

Oligocene volcanic rocks are the youngest rock sequence (Drewes, 1987).

In general, the highly deformed terrane is separated by a northeasterly trending strike-slip fault zone from the little-deformed terrane; this fault must be a major boundary at the northern edge of this part of the mobile terrane of the Cordilleran orogenic belt. One branch of this fault cuts the southern end of the Gunnison Hills; another is exposed low on the spurs southeast of Dragoon Village, near some marble quarries. The main

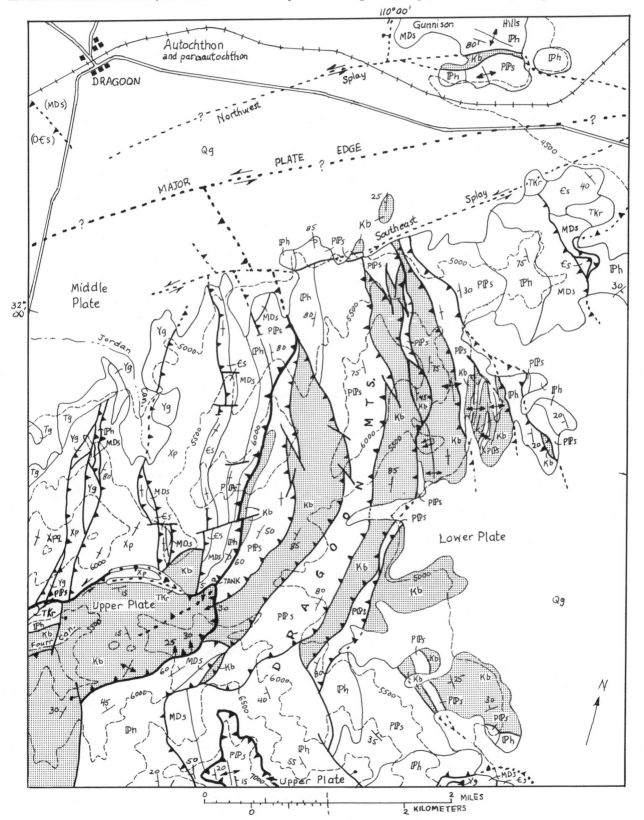

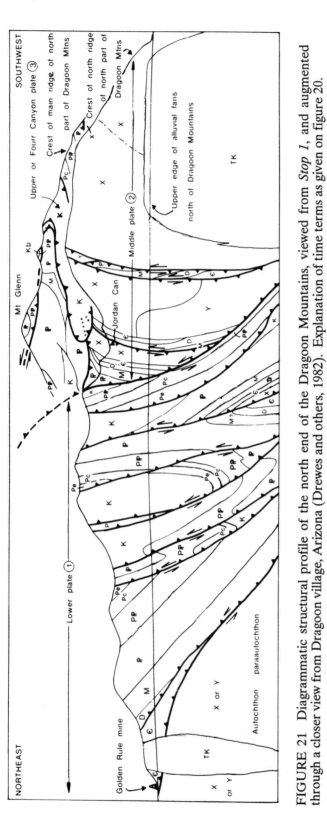

FIGURE 21 Diagrammatic structural profile of the north end of the Dragoon Mountains, viewed from *Stop 1*, and augmented through a closer view from Dragoon village, Arizona (Drewes and others, 1982). Explanation of time terms as given on figure 20.

part of the strike-slip fault is inferred to underlie the alluvium south of the railroad line, and it is a left-slip fault to judge from the evidence of northeast tectonic transport in the mobile terrane plus the offset indicated across sites like the gap between the Gunnison Hills and the Dragoon Mountains.

10 mi Stop 2. Fourr Canyon, above ranch. Prepare for a walk of about 5 hours, mainly on an old ranch road. The geology and route are shown on figure 22, as adapted from Drewes (1987); the geology of the area was also mapped by Gilluly (1956). The actual site where the vans will be parked will depend on the condition of the road in Fourr Canyon; probably it will be 1 km west of site a (2a).

2a--At first major fork in the canyon and the track bear left (east), and at the second, 0.15 km beyond, bear right (east).

2b--Beyond the gate the track in the canyon may be obliterated for a few hundred meters, past outcrops of the Lower Cretaceous Morita Formation(?) of the Bisbee Group.

2c--The track leaves the canyon bottom and follows a local remnant of gravel terrace. Cobbles derived from this gravel deposit include phyllite that has large kyanite(?) porphyroblasts. These cobbles are probably derived from the northwestern major thrust plate (the middle plate of fig. 20), on the high ridge north of Fourr Canyon.

2d--At the windmill, pause to review the geology of the middle and upper major thrust plates. Under foot is the Fourr Canyon, or upper, plate, which extends across the saddle to the northeast, down the canyon to the southwest, and caps the high peak to the southeast. It is made up mainly of Permian and Cretaceous rocks. The basal Glance Conglomerate of the Bisbee Group is a pebble and cobble deposit about 2 m thick. The northwest edge of the Fourr Canyon plate about 0.2 km from the Windmill, is a strike-slip fault that locally is intruded by Tertiary rhyolite.

The high terrain northwest of the windmill is underlain by the middle major thrust plate, which extends north and west to the end of the mountains, and east to a mid-slope position on the east flank of Jordan Canyon. The middle major plate is made up of platelets of lower Paleozoic rocks and of Early and Middle Proterozoic rocks, all upended and metamorphosed. The western platelet of lower Paleozoic rocks is cut out (out of our sight) northwest of the high ridge along an upended thrust fault that runs up the gully northwest of the windmill. The other slices of Paleozoic rock extend to the edge of the Fourr Canyon plate farther east along our traverse.

2e--Along the steep part of the track between sites 2d and 2e, are some thin limestone beds concentrated in a

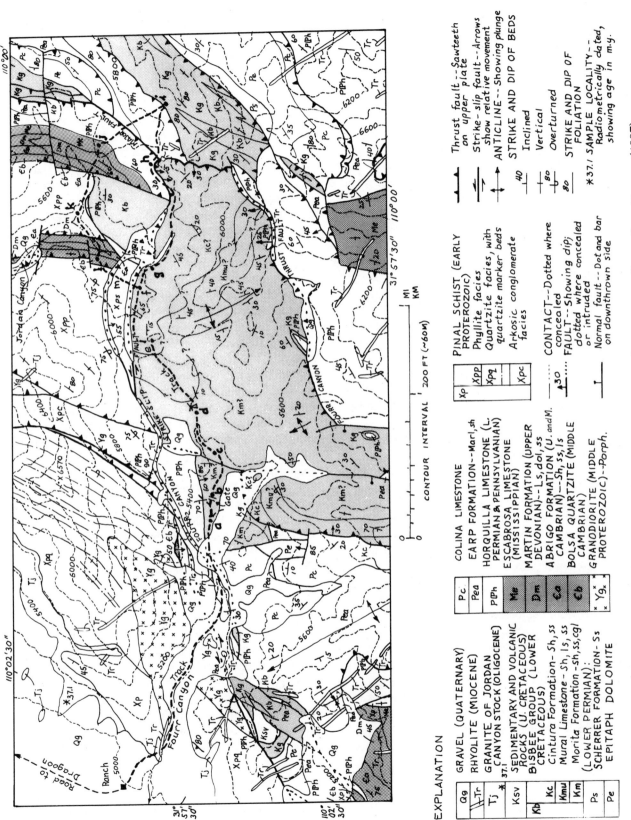

EXPLANATION

Qg	GRAVEL (QUATERNARY)
Tr	RHYOLITE (MIOCENE)
Tj *37.1	GRANITE OF JORDAN CANYON STOCK (OLIGOCENE)
Ksv	SEDIMENTARY AND VOLCANIC ROCKS (U. CRETACEOUS)

BISBEE GROUP (LOWER CRETACEOUS)

Kb	
Kc	Cintura Formation–Sh, ss
Kmu	Mural Limestone–Sh, ls, ss
Km	Morita Formation–Sh, ss, cgl

(LOWER PERMIAN):

Ps	SCHERRER FORMATION–Ss
Pe	EPITAPH DOLOMITE

Pc	COLINA LIMESTONE
Pea	EARP FORMATION–Marl, sh
PlPh	HORQUILLA LIMESTONE (L. PERMIAN & PENNSYLVANIAN)
Me	ESCABROSA LIMESTONE (MISSISSIPPIAN)
Dm	MARTIN FORMATION (UPPER DEVONIAN)–Ls, dol, ss
Ɛa	ABRIGO FORMATION (U. and M. CAMBRIAN)–Sh, ss, ls
Ɛb	BOLSA QUARTZITE (MIDDLE CAMBRIAN)
Yg	GRANODIORITE (MIDDLE PROTEROZOIC)–Porph.

PINAL SCHIST (EARLY PROTEROZOIC)

Xp	
Xpp	Phyllite facies
Xpq	Quartzite facies, with quartzite marker beds
Xpc	Arkosic conglomerate facies

CONTACT--Dotted where concealed

⎯⎯30⎯⎯ FAULT--Showing dip; dotted where concealed or intruded

⊥ Normal fault--Dot and bar on downthrown side

Thrust fault--Sawteeth on upper plate

Strike-slip fault--Arrows show relative movement

ANTICLINE--Showing plunge

STRIKE AND DIP OF BEDS

40 Inclined

Vertical

Overturned

STRIKE AND DIP OF FOLIATION

*37.1 SAMPLE LOCALITY-- Radiometrically dated, showing age in m.y.

FIGURE 22   Geologic map of the Four Canyon area, showing the route of traverse on foot, after Drewes (1987).

CONTOUR INTERVAL 200 FT (~60M)

zone that may be correlative with the Mural Limestone of the Bisbee Group. Normally, the Mural is a thick-bedded biohermal formation; these beds may represent a lagoonal facies of the Mural. Some shale along the track has a slaty cleavage, a feature common in scattered localities of known or suspected high pressure conditions as far east as the fold and thrust tectonic zone in southern New Mexico and northern Chihuahua.

2f--A rhyolite dike follows the Fourr Canyon strike-slip fault. The moderate dip on the intrusive contact probably follows the fault where it is beginning to flatten and merge with the Fourr Canyon thrust fault.

2g--A small mass of Horquilla Limestone occurs here between two branches of the strike-slip fault zone, each followed by the arm of a dike. Where the arms merge, in the gully east of site 2g, the rhyolite mass thickens.

2h--Lunch stop near tank.

The Fourr Canyon fault dips about 30° SE a few hundred meters west of the tank but only 5° SE at the tank, near which the trace of the fault trends south. About 300 m south of the tank the fault dips 30° SW. Away from the traverse and 1-2 km southwest of the tank the fault beneath the Fourr Canyon thrust plate dips 25-30° NW.

The trace of another major thrust fault trends north from a site along the Fourr Canyon fault near the tank. This poorly exposed thrust fault separates the middle from the lower major thrust plates (fig. 20). The presence and importance of this fault are indicated by the opposite-facing direction of the component thrust platelets of each major plate. In other words, the rocks of each platelet of the middle major plate get younger to the east (the sequence is repeated in each platelet), whereas the rocks in each platelet of the lower major plate get younger to the west. Rocks of both of these plates are essentially upended here, but to the south those of the lower plate dip more moderately. Also, the rocks of both of these major plates are metamorphosed to low amphibolite grade, whereas those of the upper plate, aside from the local slaty cleavage, are unmetamorphosed.

From site 2h we go across country, along the edge of the flat area, to the east and descend a spur about 40 m.

2i--The rocks of the lower main plate are made up of platelets of upper Paleozoic formations and of the Bisbee Group. The Glance Conglomerate of this major plate is largely a cobble and boulder deposit hundreds of meters thick. It also contains landslide breccia deposits (Drewes, 1987). At the station 2i the Glance is a stretched-pebble conglomerate.

Climb back up to the flat area and continue the traverse, to the northwest. Near a jeep track note a zone of slaty shale, probably the Permian Earp Formation. West of this shale zone is a coarsely crystalline limestone or marble, identified as Horquilla

Limestone of the middle main plate, based on mapping to the north. A zone of this coarse carbonate rock has relict crinoid stems, commonly a feature diagnostic of the Mississippian Escabrosa Limestone, but also locally found in the Horquilla. The choice here of an Horquilla correlation provides a simpler structural picture than the alternate option.

2j--The spur of the flat area that lies south of the upended rib of Middle Cambrian Bolsa Quartzite is underlain by Escabrosa Limestone. At a dead mountain mahogany tree, 50 m east of this spur, is a bed of Horquilla with fusulines. A 3-m-thick shale or slate zone between the Horquilla and Escabrosa is probably the Upper Mississippian Paradise Formation, common in mountains to the east but thinning out in easternmost Arizona. Elsewhere in the Dragoons (in a higher structural plate?) this unit may be marked by one or two thin clastic marker beds.

The traverse continues down the spur to the west, across the Escabrosa Limestone, the Upper Devonian Martin Formation, and the Middle and Upper Cambrian Abrigo Formation. The Bolsa is faulted out. The traverse continues down the east branch of Jordan Canyon.

2k--Near the junction of the two branches of Jordan Canyon examine the phyllite unit of the Pinal Schist. The traverse continues up the west fork of the canyon, past an old cabin and up the spur to the west. Note the mine dump and pyrite enrichment. Cross the second platelet of upended lower Paleozoic rocks of the middle major thrust plate.

2l--On the knoll is Bolsa Quartzite. Traverse south across two dikes of rhyolite porphyry, the northern of which truncates the thrust platelets. The local fault slice between the faults intruded by the two dikes is made up of Pinal. The Fourr Canyon, or upper thrust plate, is to the south of this slice. Rejoin the traverse at site 2f, and return to the vans on the jeep track.

The continuation of the trip of Day 5 may be run along one of two routes, to be selected according to the time and wishes of the group. The quick way to Tombstone is via Benson and the San Pedro River valley, with options of one brief stop to review the geology of the mountains that come into view. An archeology stop can also be made at the Amerind Foundation museum. The slower way, partly on unpaved road and via Pearce in Sulfur Springs Valley, offers a closer look at geology, described as follows.

22 mi Stop 3. Pearce. This near-ghost town was a silver camp about 70 years ago, with nearly the entire production coming from the Commonwealth mine, near the large mine dump to the southeast. Heap-leaching of these dumps is the only recent mining activity. The mineralization mainly consists of veins and replacement deposits in Tertiary andesite and rhyolite breccia; some gold placers were also worked (Elsing and Heineman, 1936). Production was of moderate size ($10-100 million).

9 mi Stop 4. Middlemarch Pass area provides a view of the southward extension of the structural complexities of the northern part of the mountains. The Miocene Stronghold stock, a granite body that has primary fluorite, separates the two terranes. The widespread metamorphism displayed by the lower and middle major thrust plates to the north is not present near this pass; such metamorphism as does occur is typically Precambrian, or of Miocene contact type.

The core of the range is underlain by a fault block or thrust slice of Proterozoic and Cambrian rocks that to the northwest form a keel-like prong between two lobes of the Stronghold stock. Most of the rocks to the northeast and southwest of this core are Paleozoic and Cretaceous rocks in thrust platelets. Both of the terranes flanking the core also contain some Proterozoic crystalline rocks. The Glance Conglomerate of the Bisbee Group is hundreds of meters thick to the northeast but only tens of meters thick to the southwest.

Agreement exists among all workers in this area that thrust faults exist, but the major faults of steep inclination bounding the core terrane have been variously interpreted in local studies. Indeed, it is likely that these faults have had a polyphase history of multiple movement including some early strike-slip movement, some reverse movement, and some late thrust-fault or ramping movement. The evidence for early strike-slip movement comes from the fact that the faults extend to the southeast, where they are cut by a Jurassic stock, and from the many slivers of rock stratigraphically out of place with respect to the adjacent blocks. The evidence of reverse fault movement comes from the present attitudes of the faults and the offsets across them. Furthermore, the contrast in lithologic facies of the Glance Conglomerate suggests the presence of local relief during Glance time. Such reverse faults may have been paleo-foreland zone structures, analogs to the Bear Spring fault zone in the San Andres Mountains of Day 2. The evidence for thrust faults is ample; low angle structures emplacing younger rocks over older ones (or the reverse) abound.

13 mi Tombstone. Our lodgings for the next two nights are in this town.

Day 6

Rest day in Tombstone, with a chance to absorb some local history and lore. Possible brief tour of the mining camp.

Day 7

Early departure for Sonoita, Box Canyon, the Helvetia mining district, and Madera Canyon. This traverse is entirely across the western intermediate zone of the Cordilleran orogenic belt. This zone is characterized by strong deformation and involvement of crystalline basement rocks in the thrust plates, but the first part of the traverse is across an area of mild compressive deformation (Drewes, 1980, 1981a). This area, centered on the Whetstone Mountains, is surrounded by mountains in which deformation is far more intense, and in which thrust faults typically dip inward toward the Whetstones. These relations suggest that major thrust faults extend beneath the rocks exposed in the Whetstones and that the Whetstones are part of a major thrust plate that has escaped strong deformation because a greater thickness gave it strength to better resist the compressional stress. The second part of the traverse is in the Santa Rita Mountains in which the more typical complex structures are not only well exposed but are closely datable.

20 mi Stop 1. The Whetstone Mountains, north of Highway 82, form a west-dipping block of Proterozoic granite or granodiorite and capping homoclinal Paleozoic and Mesozoic rocks (Creasey, 1967). An Upper Cretaceous stock cuts these rocks a few kilometers from the highway. Sedimentary sequences of the central part of the mountains are essentially unfaulted but to the north and south small thrust faults are reported by Drewes (1981a, fig. 11) and Creasey (1967), respectively.

The low Mustang Mountains to the west and southwest of the stop are underlain by Permian rocks overlain by red siltstone and conglomerate and capping rhyolite lava flows and welded tuff, all of Triassic or Jurassic age (Hayes and Raup, 1968). These Mesozoic rocks are not found northeast of the Mescal Spring fault zone, a northwest-trending high-angle fault that probably had both strike-slip and dip-slip movements. Along the river south of the Mustang Mountains is another major fault, the Babocomari fault, mapped as a thrust fault by Hayes and Raup (1968) and as a left-slip fault by Drewes (1981a, pl. 7).

An east-west-trending graben underlies the lowlands between the nearby hills and the distant Huachuca Mountains. The bounding faults of the graben have had not only dip-slip offset but probably also have had much left-slip movement. A thick sequence of Oligocene and Miocene Pantano Formation occurs in the graben, and overlies Upper Cretaceous and older Mesozoic rocks. The Pantano is made up mainly of moderately consolidated conglomerate and sandstone but also has some basaltic andesite lava flows and rhyolite tuff that have provided ages of 24-39 Ma. The graben wedges out eastward and its bounding faults merge westward with the Sawmill Canyon fault zone, another of the northwest-trending reactivated basement flaws.

21 mi Stop 2. The high grassy valley north of Sonoita provides a vantage point from which to review the geology of surrounding mountains. The small mountain range to the north is underlain by rocks much like those of the Whetstone Mountains east of the stop. Thrust faults are recorded mainly along the west flank of the range (Finnell, 1971). The hills to the south are underlain mainly by Permian formations and the Triassic or Jurassic volcanic and sedimentary rocks; to the south there are also Upper Cretaceous sedimentary rocks and granite and Cretaceous or Tertiary volcanic and sedimentary rocks that mark a major igneous center.

To the west, in the Santa Rita Mountains, Mesozoic volcanic and sedimentary sequences that have a combined thickness of more than 10 km (Drewes, 1971a, 1971b, 1971c) overlie the usual Paleozoic sequence and Proterozoic basement rocks. The northern low part of the mountains is underlain mainly by Proterozoic, Paleozoic, and Cretaceous rocks, cut by small stocks of Paleocene age. The southern high part of the mountains is underlain by some much-faulted Paleozoic rocks, a thick sequence of Triassic and Cretaceous

rocks, and large stocks of granite to diorite composition and of Triassic, Jurassic, Cretaceous, and Paleocene ages. Major fault zones occur along the west flank of the high mountains and another, the Sawmill Canyon Fault zone, cuts northwest across the mountains between the high and low parts. To the far southwest, Oligocene volcanic rocks associated with a rectilinear cauldron system overlap the older rocks (Drewes, 1972b). This volcanic site also has genetically related vitrophyre laccoliths and their feeder dikes, as well as a small granite stock. Our traverse next takes us through the low part of the mountains and to some structures in their north end.

8 mi Stop 3. In the upper part of Box Canyon (fig. 23) are several plates of Paleozoic and Mesozoic rocks thrust over the Proterozoic Continental Granodiorite. Of particular interest is the contrast in facies in the basal parts of two Bisbee sequences; the eastern plate has a limestone cobble conglomerate typical of the Glance Conglomerate, and the western (or lower) plate has an arkosic sedimentary breccia.

Several northwest-trending grabens of Cretaceous rock are probably related to a late phase of movement, in part left slip. Locally these structures conceal or modify earlier thrust faults (fig. 24). A swarm of Oligocene rhyolite dikes crosses the lower part of Box Canyon, beyond the area of Stop 3. The Continental

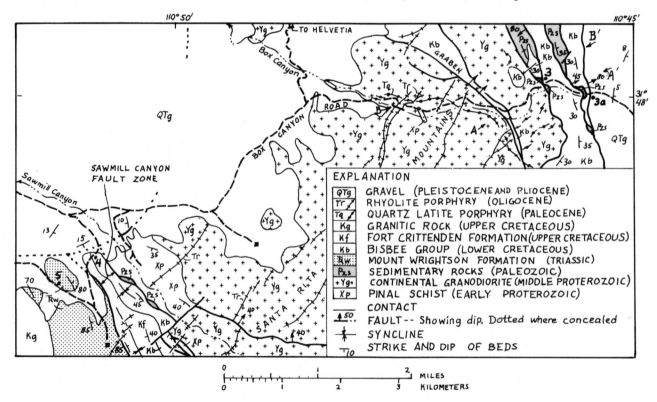

FIGURE 23 Generalized geologic map of the Box Canyon-Sawmill Canyon area, Santa Rita Mountains, southeastern Arizona, after Drewes (1971a).

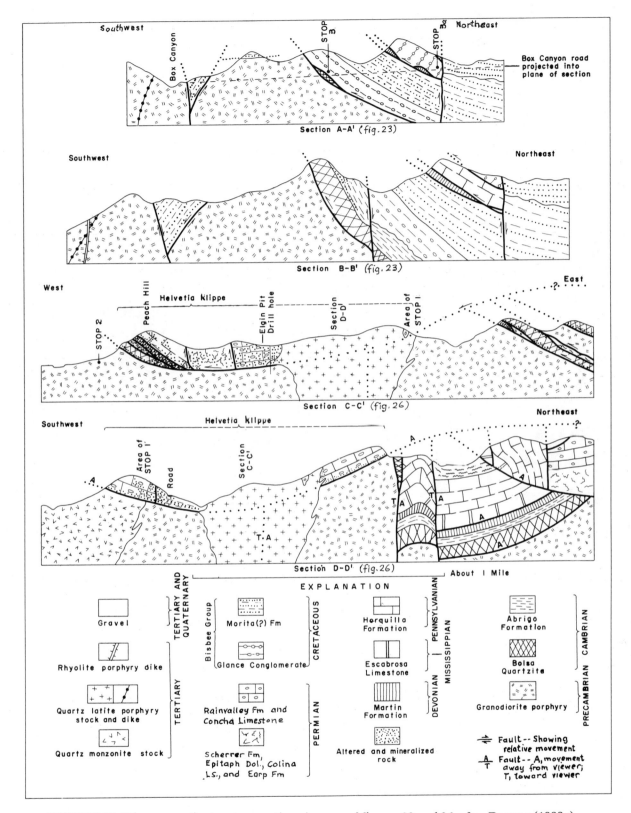

FIGURE 24 Diagrammatic structure sections in area of figures 23 and 26, after Drewes (1988a).

Granodiorite between Box Canyon and the north end of the mountains was recrystallized in Paleocene time.

9 mi No stop. Pass Helvetia ghost town. Rugged outcrops along crest of the range to the east are mainly of the Middle Cambrian Bolsa Quartzite, but Proterozoic Continental Granodiorite underlies much of the cliff to the right of Gunsight Notch, through which some major northwest-trending faults run (fig. 25).

Gunsight Notch

FIGURE 25   View of crest of the Santa Rita Mountains, capped by Bolsa Quartzite, from Helvetia ghost town. Photo by Jack Rathbone.

10 mi Stop 4. Helvetia area lunch stop and walking tour of about 3 km, first in a loop northeast of the vans, and then past the vans and down the road, with the vans driven to the end of the tour near Stop 4i. The geology of the Helvetia area is given by Drewes (1971a; 1972a, pl. 4), and the route we follow is shown on figures 26 and 24.

4a--Drill site at lower member of the Permian Epitaph Dolomite, which is mainly composed of marlstone, dolomitic siltstone, and interbedded argillite and limestone. Note the tectonic pod of gypsum. Slope to east is underlain by the Permian Scherrer Formation (composed mainly of fine-grained, very light pinkish gray quartzite or sandstone), by the overlying Permian Concha Limestone, and by the capping Permian Rainvalley Formation (limestone, dolomite, and sandstone). The local stratigraphic column is shown on figure 27.

4b--Hilltop, underlain by upended, metamorphosed Pennsylvanian Horquilla Limestone, provides an

overview of the Paleozoic rocks and of the thrust faults and tear faults of Paleocene age (Helvetian phase of the Cordilleran orogeny). Geology reviewed on panorama (fig. 28).

4c--Thrust-faulted slivers of Upper Devonian Martin Formation (brown dolomite, limestone, and some sandstone), here strongly altered to calc-silicate minerals, and of Middle and Upper Cambrian Abrigo Formation (thin-bedded shale, siltstone, and some limestone), also altered to calc-silicates, and of Middle Cambrian Bolsa Quartzite (coarse-grained, light-gray to brownish- or purplish-gray, arkosic rock). To the west is the northeast corner of the Helvetia klippe (fig. 29); prospect pits are located along its base.

4d--Continental Granodiorite thrust slice. Rock is coarse grained and very coarsely porphyritic. It is dated at 1,450 Ma but may be as old as 1,600-1,700 Ma. Locally, however, its age reflects a 55 Ma thermal event, which was probably the emplacement age of the late orogenic stocks (Drewes, 1976, p. 9-17).

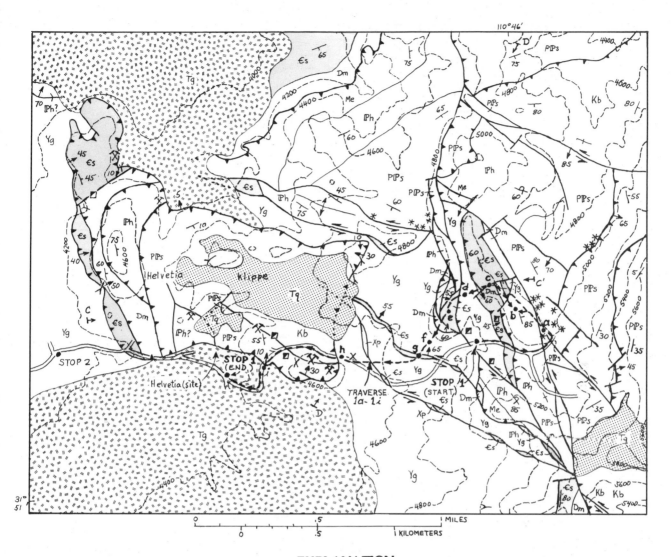

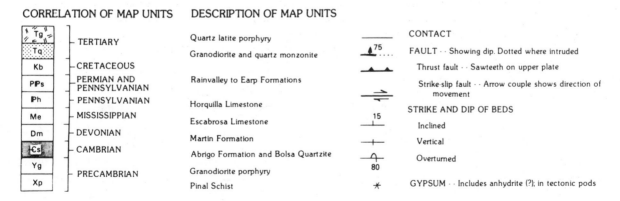

EXPLANATION

CORRELATION OF MAP UNITS

Tg	— TERTIARY
Tq	
Kb	— CRETACEOUS
PPs	— PERMIAN AND PENNSYLVANIAN
Ph	— PENNSYLVANIAN
Me	— MISSISSIPPIAN
Dm	— DEVONIAN
€s	— CAMBRIAN
Yg	— PRECAMBRIAN
Xp	

DESCRIPTION OF MAP UNITS

Quartz latite porphyry

Granodiorite and quartz monzonite

Rainvalley to Earp Formations

Horquilla Limestone

Escabrosa Limestone

Martin Formation

Abrigo Formation and Bolsa Quartzite

Granodiorite porphyry

Pinal Schist

————— CONTACT

FAULT · · Showing dip. Dotted where intruded

Thrust fault · · Sawteeth on upper plate

Strike-slip fault · · Arrow couple shows direction of movement

STRIKE AND DIP OF BEDS

Inclined

Vertical

Overturned

GYPSUM · · Includes anhydrite (?); in tectonic pods

FIGURE 26 Geologic map of the central part of the Helvetia area, Santa Rita Mountains, Arizona, showing route of the foot traverse, after Drewes (1972a).

4e--Second thrust plate of lower Paleozoic formations.

4f--Another slice of Continental Granodiorite.

4g--Recrystallized Bolsa Quartzite along the Gunsight Notch fault zone, which is a tear fault that bounds the southwest sides of some of the local thrust plates. These thrust and tear faults merge in such a way as to show that they were active concurrently. The largest offset on a tear fault here is about 2 km.

4h--The Paleocene granodiorite stock (54 Ma) of the

AGE		FORMATION	DESCRIPTION	THICKNESS (m)
QUATERNARY	Holocene Pleistocene	unnamed	Gravel and sand	
TERTIARY	Pliocene			
	Miocene			
	Oligocene	unnamed	Rhyolite dikes	
	Paleocene	unnamed	Quartz latite dikes and plugs of the Greaterville intrusives Quartz monzonite of the Helvetia stocks	
CRETACEOUS	Late Cretaceous	Elephant Head Quartz Monzonite Madera Canyon Granodiorite	Stocks	
		Salero Formation	Rhyolite, andesite, and sedimentary rocks	2000 ±
		Fort Crittenden Formation	Conglomerate, sandstone, and shale; tuff in upper part	3000 ±
	Early Cretaceous	Bisbee Group	Arkose, sandstone, shale, and siltstone	2700 ±
JURASSIC				
JURASSIC OR TRIASSIC		Canelo Hills Volcanics	Arkose, conglomerate, shale, and tuff	180 +
		Gardner Canyon Formation	Red mudstone, dacite, and conglomerate	600 +
TRIASSIC		Mount Wrightson Formation	Rhyolite, andesite, and sandstone	2600 ±
PERMIAN	Early Permian	Rainvalley Formation	Limestone, dolomite, and sandstone	0-90
		Concha Limestone	Cherty limestone	120-175
		Scherrer Formation	Quartzite, dolomite, and sandstone	220 ±
		Epitaph Dolomite	Limestone and dolomite	300 ±
		Colina Limestone	Limestone	110 ±
PENNSYLVANIAN	Late Pennsylvanian	Earp Formation	Siltstone and marlstone	250 ±
	Late to Early Pennsylvanian	Horquilla Limestone	Fine-grained limestone	300 ±
MISSISSIPPIAN		Escabrosa Limestone	Coarse-grained limestone	170 ±
DEVONIAN	Upper Devonian	Martin Formation	Dolomite, limestone, and siltstone	120 ±
SILURIAN				
ORDOVICIAN				
CAMBRIAN	Late Cambrian	Abrigo Formation	Shale, sandstone, and limestone	225-275
	Middle Cambrian	Bolsa Quartzite	Quartzite	140 ±
	Early Cambrian			
PRECAMBRIAN		Continental Granodiorite	Granodiorite porphyry	
		Pinal Schist	Gneiss and schist	

FIGURE 27 Stratigraphic section of the Helvetia area and the Sahuarita quadrangle, Arizona (Drewes, 1971a).

flats south of Helvetia is cut by the thrust fault beneath the klippe, and that thrust fault is cut by the quartz latite porphyry stock (56 Ma), thereby closely dating this phase of faulting. The dates of the two stocks are the reverse of their relative ages based on field relations but overlap within the analytical errors.

4i--Road cut of thrust fault at base of klippe located 0.1 mi down the road from a sharp curve. Lower quartzite member of the Scherrer Formation is thrust faulted on the Paleocene granodiorite.

1 mi Stop 5. Southwest side of Helvetia klippe. The brown ledge is underlain by the Bolsa Quartzite and the gentle slope above the ledge by the Abrigo and Martin Formations. The Horquilla Limestone caps the hill, because the Escabrosa Limestone, which normally overlies the Martin, has been cut out by a low-angle fault. The thrust fault beneath the klippe is well exposed in adits all around the klippe, and it has also been penetrated by many drill holes that establish the fault to be saucer shaped. Copper-bearing hydrothermal solutions related to the so-called "ore porphyry" spread from the stock along the faults and replaced such units as the Abrigo and Martin

Formations (Schrader and Hill, 1915, Creasey and Quick, 1955; Drewes, 1972a).

11 mi Stop 6 (optional). Triassic rocks at the mouth of Florida Canyon. The low hill is underlain by the Triassic Mount Wrightson Formation, which in the main part of the mountains is at least 3 km thick. At the stop, and typical of the middle member of the formation, is rhyodacite welded tuff with some thin intercalated eolian quartzitic sandstone. The formation may have been deposited in a block-fault basin bounded to the northeast by the Sawmill Canyon fault zone.

Like the Apache Pass fault zone, the Sawmill Canyon fault zone is narrow to the northwest and widens to the southeast, where the included rock sequences are more complete and less deformed. Near Stop 6, the southwestern branch of the fault zone separates Triassic rocks from volcaniclastic rocks of the Upper Cretaceous Fort Crittenden Formation that are exposed in the ditches along the road northeast of the hill. These volcaniclastic rocks conformably overlie the Lower Cretaceous Turney Ranch Formation, highest unit of the Bisbee Group of this area. The northeastern branch of the fault zone separates the Turney Ranch from Pinal Schist, and along this branch are tectonic lenses of Paleozoic formations. Recurrent movements along this fault are discussed by Drewes (1972a) and its regional significance is reviewed by Drewes (1981a).

7 mi Madera Canyon. Our lodging for the night is in this canyon.

Day 8

During the morning a trip to one of the porphyry copper open-pit mines of the Sierrita Mountains may be arranged. Then, on our way to Tucson we will visit the San Xavier Mission. In the afternoon we will visit the west side of the Rincon Mountains gneiss-cored dome (metamorphic core complex) in the Saguaro National Monument. With the descent to the low valley of the Santa Cruz River drainage the tour enters the Sonoran Desert vegetation assemblage. Pecan orchards along the Santa Cruz River indicate the availability of ground water and the warmer climate of this area.

15 mi. Green Valley may be a stop to meet the associate leader of the morning and to check on final arrangements for the mine tour.

8 mi(?) Stop 1. At mine, to be selected by associate leader according to conditions at the time of the trip. The geology of the Sierrita Mountains was mapped by Cooper (1973), who showed the core of the mountains to be underlain by assorted Triassic to Cretaceous sedimentary, volcanic, and plutonic rocks. Much of the

northeast flank of the mountains is underlain by the Ruby Star granite stock of Paleocene age. Low on the northeast flank are Proterozoic granodiorite and thrust-faulted and folded Paleozoic and Mesozoic rocks, also host to the large stock. Additionally, an Oligocene or Miocene gravity-impelled San Xavier glide plate of Helmet Fanglomerate and some interbedded volcanic rocks, plus some piggy-backed previously deformed Paleozoic and Mesozoic formations underlie much of the low foothills on the northeast flank of the mountains.

Some of the porphyry copper deposits occur among small stocks and plugs of diorite and quartz latite porphyry along the southern border of the Ruby Star stock. Other such deposits occur beneath pediment gravel, either in a part of the San Xavier glide plate, or beneath it.

The existence and geometric configuration of the San Xavier glide plate is known not only from surface mapping but from many drill holes that have penetrated the plate in the course of copper exploration. The plate is inferred to have moved north-northwest from a topographic high no longer present. Presumably this high area was subsequently faulted down as part of the graben of the Santa Cruz River valley, whose nearby part contains young fault scarps and geophysically demarked concealed normal faults.

18 mi Stop 2. San Xavier Del Bac Mission on the San Xavier Indian Reservation. This mission, built by the early Spanish settlers, still serves the local community.

23 mi Stop 3. Saguaro National Monument, for lunch stop and visit to nature museum.

8 mi Stop 4. The Loop Drive brings us to a key site on the southwest flank of the Rincon Mountains gneiss-cored dome (also called core complex). This area was the site of a G.S.A. Penrose conference in 1977 (see G.S.A. Memoir 153 on the origin of these features). These domes have received considerable attention in recent years, and their genesis still is incompletely understood. Some work focused on their Miocene or post-Miocene development, and their association with an extensional event (Keith and others, 1980), but the study based on extensive geologic mapping indicates that they have an older origin and an association with early compressional tectonism as well as with late extensional tectonism. The rocks and field relations to be seen at this stop, along with those to be seen tomorrow, will illustrate some key features of the core and carapace of the dome and of the fault between them.

The gneiss-cored domes are recognized in much of the Cordilleran orogenic belt, both in regions of

FIGURE 28 Panorama from east to southeast of Stop 4b. Slope to east is underlain mainly by the Permian Scherrer Formation, Ps, and Concha Limestone, Pcn, which are separated by a bedding plane thrust fault. The range crest to the east-southeast is underlain by a Paleocene quartz latite porphyry plug, Tql. The bold peak to the southeast is mostly Middle Proterozoic Continental Granodiorite, Yg, except for the rugged left flank that is mostly underlain by Middle Cambrian Bolsa Quartzite, Cb. For the other units, see the stratigraphic section of figure 27. Photos by Jack Rathbone.

FIGURE 28 (Continued)

FIGURE 28 (Continued)

FIGURE 28 (Continued)

FIGURE 29. View, looking west, of the Helvetia klippe (left and central masses of Pennsylvanian Horquilla Limestone, Ph, and other rocks not visible from here), and a left-slip fault between Middle Proterozoic Continental Granodiorite (Yg) and more Horquilla (right mass), seen from stop 4c. Units Ca, Cambrian Abrigo Formation; C, Cambrian Bolsa Quartzite; Tql, Paleocene quartz latite porphyry. Photo by Jack Rathbone.

subsequent Basin-and-Range style faulting or by vast extensional movements, as well as regions where these late movements are not known. Thus their initial formation may be of Cordilleran orogeny age--Late Jurassic(?) to Eocene. These tectonized terranes are also characteristic of the interior part, or hinterland zone of the orogenic belt (more common west of Phoenix than east of Tucson). The cores of the domes of southeastern Arizona resemble outliers of the deep-seated and hinterland zone terrane. This relationship between tectonic zones of the orogenic belt implies, of course, that the zone boundaries are not vertical but are inclined outward from the orogenic core so that a specific terrane may overlap another either with or without subhorizontal fault movement between them.

The Rincon Mountains dome is one of three such features that form a major high elongated in a northwest direction. The Santa Catalina and Tortolita domes form the central and northwestern culminations, respectively. Smaller scale irregularities occur on the surfaces of the southwest flank of the domes.

The Rincon Mountains gneiss-cored dome consists of several distinct terranes, two of which are separated by a fault that has had a history of diverse and recurrent movement (Drewes, 1974, 1977, in press). The core terrane is made up of various Proterozoic and Phanerozoic crystalline rocks, among them the Continental Granodiorite. Large masses of a 2-mica granite, whose oldest phase is probably Late Cretaceous or Paleocene (and was once erroneously believed by me to be Precambrian?), have core-terrane affinity but their late-phase counterparts are found in the cover rocks also. Locally, some anomalous rock types occur in the core, too.

Rocks of the core terrane are typically tectonized. A foliation is developed in all rocks and is the feature most responsible for the shape of the dome. Foliation is most strongly developed near the overlying (surrounding) fault and along internal zones of strong shear; it is weakly developed in the middle-phase 2-mica granite bodies and is even absent on the northeast side of the youngest of these middle-phase granite bodies near Youtcy Ranch, as we will see tomorrow. The tectonized rocks also have a widespread lineation that is mostly aligned N. 60-70° E. However, over an area of many tens of square kilometers on the north side of the dome the lineation swings gradually north and even to N. 10° W. Lineation is absent in places on the east side of the dome and in the middle- and late-phase 2-mica granites.

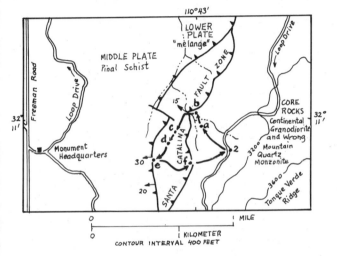

FIGURE 30   Sketch map of the Saguaro National Monument Headquarters area, Tucson, Arizona, showing route of the traverse on foot.

Rocks of the cover terrane of the Rincon Mountains dome include assorted sedimentary rocks of Proterozoic to Miocene age and some untectonized crystalline basement rocks. Three major thrust plates are recognized; these form extensive sheets on the northeast flank of the dome but are disrupted or lenticular masses on the southwest flank. The lower plate is made up largely of metamorphosed Paleozoic and Mesozoic formations. The middle plate is made up of crystalline basement rocks: Rincon Valley Granodiorite to the south, a septum of Pinal Schist in the center, and Oracle Granite to the north. The upper plate contains mainly unmetamorphosed Paleozoic and Mesozoic formations, including some of the same units that occur in the lower plate. This repetition of formations in stacked thrust plates, viewed in the direction of tectonic transport, indicates a minimum of 16-32 km of tectonic transport.

The cover rocks are not foliated or lineated like the core rocks, but they are internally folded and faulted in the tectonic style of the rocks of the western

intermediate zone, a style that implies development under a thick cover, and probably over a protracted time span. Latest phase undeformed 2-mica granite intrudes the cover rocks.

A third terrane, made up of a thick sequence of Oligocene and Miocene Pantano Formation, has some affinities with the cover terrane, as well as some characteristics of its own. These rocks are mainly semi-indurated clastics, but they include some volcanic rocks, a few of which are dated at 24-29 Ma. The rocks low in the Pantano include clay and gypsiferous silt of a valley-center facies; overlying rocks include conglomerate, monomictic fanglomerate, and megabreccia landslide deposits. Clast assemblages of the coarse-grained deposits change upward in a way that reflects, in reverse, the stratigraphy of the cover and core rocks of their source. Like the rocks of the cover terrane, the Pantano is faulted upon the core rocks but unlike the cover rocks, these rocks display brittle deformation that implies very little cover during deformation. The Pantano rocks (and some associated older plate rocks) are inferred to have moved, as glide plates, outward from the core of the dome as that dome rose rapidly in a near-surface, tectonically brittle environment, in Miocene time.

The Santa Catalina complex fault separates the cover rocks and Pantano Formation from the core rocks. The fault dips outward from the dome, essentially lying parallel to the foliation of the core rocks and to the shear zones within those rocks. It also wraps around the subordinate arches (turtlebacks?) and troughs on the southwest flank of the dome. Along this southwest flank, the fault typically is marked by a mylonitic sheet 0.5-1 m thick that commonly is greenish-gray and strongly indurated. However, along the northeast flank the strong mylonitic characteristic of the fault is absent. Mullion structures and slickensides that trend N. 60-70° E. are well developed on the southwest flank; some slickensides are found to the northeast, but neither feature was noted along the southeast or northwest sides.

The contrast in deformation styles between the Paleozoic and Mesozoic cover rocks and the Pantano Formation indicates that movement on the Santa Catalina complex fault took place during two different times in diverse structural environments; the earlier movement was deep-seated and the later movement was shallow. So large a change in structural environment further implies a hiatus between the two tectonic events, during which the region was deeply eroded. The evidence of unidirectional transport given by the orientation of lineation and mullions, and of radial transport direction given by orientations of folds, studied by Davis (1975), also implies two deformation times and conditions. Many of these concepts are

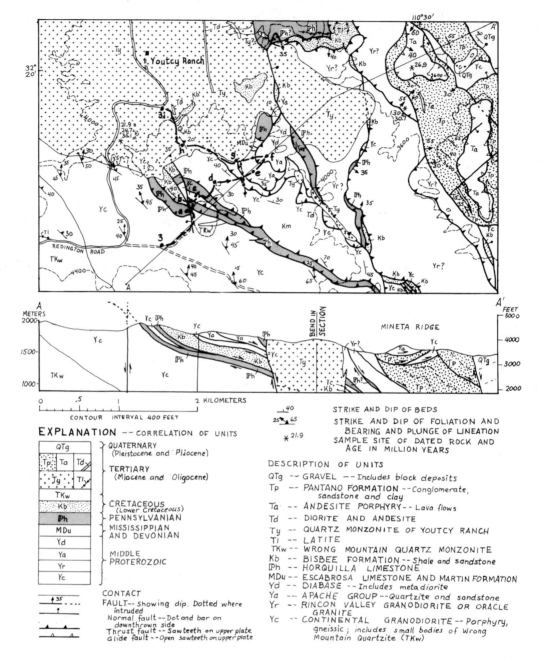

FIGURE 31 Geologic map of the Youtcy Ranch area, Rincon Mountains, Arizona (Thorman and Drewes, 1981b), showing route of the traverse on foot.

developed further in Drewes (in press).

Returning now to the area of Stop 2, the main tectonic units and the route to be followed are shown on figure 30. The core rocks of this part of Tanque Verde Ridge are mixed Continental Granodiorite and the Wrong Mountain Quartz Monzonite, one of the 2-mica granites. The rocks of the lower plate of cover rocks are a mixture, chiefly of blocks of Paleozoic rocks as much as 100 m across, in a matrix of red siltstone or marlstone of the Pennsylvanian and Permian Earp Formation. The rocks of the middle plate are of the phyllite and quartzite member of the Early Proterozoic Pinal Schist

of the septum between the two untectonized granitoid bodies of northern and southern parts of this plate. The upper plate is not present here.

4a--Overlook into gully. En route to a, note the lit-par-lit gneiss, the foliation and lineation, and the protomylonitic fabric. Both ductile and brittle fabric elements are present. On the opposite canyon wall, the ledge, which dips 15° W., is the indurated brecciated protomylonite, which separates the underlying crystalline core rocks from an overlying melange sheet. Continue across the canyon to the stripped surface on the Santa Catalina complex fault.

4b--Fault surface. The ledge is an indurated breccia of the protomylonitized core rock. The overlying rocks form a tectonic melange sheet, in which occur scattered blocks of assorted Paleozoic and Mesozoic formations as much as 100 m across. The matrix is red siltstone or marlstone of the Pennsylvanian and Permian Earp Formation. Some of the Paleozoic limestone blocks are internally folded, and these folds have been systematically studied by Davis (1975) and his students, who infer the melange to have moved northwest off the adjacent part of the Tanque Verde Ridge during or after Pantano time. This movement is clearly different from, and later than, that during which the ductile features of the lineation were formed.

4c--First knoll on the ridge. Nearby blocks in the melange are of red jasper-pebble conglomerate (Earp) and dark-gray dolomite (Permian Epitaph Dolomite). Outcrops on the hill below, to the west, are of quartzitic Pinal Schist, which overlies the melange sheet. Continue south along the ridge.

4d--Second knoll on the ridge. Nearby blocks are brownish-gray quartzite (Bolsa) and sandstone (Bisbee). In the next saddle to the south the melange matrix is exposed. Cross the third knoll and descent to the southwest.

4e--At small gully west of the knoll. Stripped fault surface between the underlying melange sheet and overlying Pinal Schist.

Return to top of third knoll and head east along ridge. Bold outcrop on the northeast side of the knoll is Glance Conglomerate Member at the base of the Bisbee Formation. Descend to the east, toward the vehicles.

4f--Base of melange sheet, here poorly exposed at midslope. Return to vehicle and continue around Loop Drive.

18 mi. Return to lodgings near Tucson Airport.

Day 9

Early departure for Redington Pass for a last traverse that will focus on the rocks and structural features of the north and northeast flanks of the Rincon Mountains gneiss-cored dome, with a stop at the Italian Trap klippe in the pass (from which an intermediate phase 2-mica granite mass is seen), and a foot traverse of some core rock and lower plate cover rock near where they are intruded by the Youtcy Ranch stock of 2-mica granite. Radiometric and fission-track dates show that (1) these masses of the 2-mica granites are earliest Oligocene or Eocene; (2) cooling took place over a protracted period, extending into the Miocene; and (3) some movement on the fault between cover and core rocks of this area predates the deposition of the Pantano Formation (29-24 Ma), thus indicating that a simple development of the dome in response only to late Tertiary gravity gliding is untenable. Also of key interest is the contrast in intensity of deformation across the dome. Explanation of this contrast provides a tie between Cordilleran compressive deformation and emplacement of the early phase of the 2-mica granite; the gneiss-cored domes are primarily features of the hinterland zone of the Cordilleran orogenic belt.

26 mi Stop 1. Road cut of Oligocene and Miocene Pantano Formation, which is here a tilted and faulted pinkish-gray pebble conglomerate. Clasts are subangular and comprise mid-Tertiary volcanic rocks, Middle Proterozoic Rincon Valley Granodiorite, and Paleozoic and Mesozoic sedimentary rocks. The formation is capped by a terrace deposit that contains coarser detritus derived from the core of the domes. From this vantage point the main geologic features of the Rincon gneiss-cored dome are reviewed. The low hills toward the base of the Rincons are underlain by rocks of the cover and Pantano terranes visited yesterday. The profile of the Tanque Verde Ridge left of the low hills is the southwest-plunging crest of an antiform on the flank of the dome that has been compared with turtleback structures of California and Utah.

18 mi Stop 2. From the knoll 2.4 mi beyond entrance to Bellota Ranch, review the geography of Redington Pass area and the geology of the Italian Trap klippe, shown on the geologic map of Thorman and Drewes, 1981b). The radiometric age of about 46 Ma, on the Wrong Mountain Quartz Monzonite (2-mica granite) probably is the closest available to its emplacement age; yet it also is likely to have been reset or to have cooled slowly. From regional considerations on tectonic development, plus the age of the oldest untectonized 2-mica granite in the region, I favor an emplacement age of the oldest phase of 2-mica granite to be late Paleocene or early Eocene (about 58-50 Ma).

2.6 mi Stop 3. From here we will walk about 4 hours and take along lunch. Van drivers should see leaders about arranging a shuttle and rejoining the group. The route and main stops are shown on figure 31, adapted from the geologic map of Thorman and Drewes (1981b).

3a--Saddle area and spur to west. Horquilla Limestone is faulted over Continental Granodiorite. Fault in and near copper prospect dips about 60° NE. Return to saddle and contour around the hill to the northeast, following cattle track.

3b--Spur. Horquilla Limestone faulted over the granodiorite here also; offset from site a by a north-trending high-angle fault. On ridge to the west the brown rock overlying the Horquilla is the Bisbee

Formation. Continue about 200 m to the jeep track in the small saddle.

3c--Small draw. Thrust platelet of Continental Granodiorite in the larger plate, which is mainly Paleozoic and Mesozoic sedimentary rocks. Note that the foliation and lineation in the platelet is oriented like that of the underlying crystalline core rocks. Continue on jeep track to second saddle; leave the track and climb the knoll to the north.

3d--Knoll underlain by metamorphosed quartzitic sandstone and siltstone of the Bisbee Formation. Review of geology of the Mineta Ridge area to the east. Also indicate site of next stop.

3e--Gully underlain by diabase (or metadiorite) associated with the Middle Proterozoic Apache Group. Plate of Continental Granodiorite to the southwest overlies the Bisbee. To the east and northeast are red argillite of the Pioneer Formation and quartzitic sandstone and round-clast conglomerate of the Barnes Conglomerate Member of the Dripping Spring Formation, all of the Apache Group. Continue to knoll to the east.

3f--Another quartzitic sheet has been mapped as Dripping Spring but may be Bolsa Quartzite. The spur to the northeast has Martin Formation, Escabrosa Limestone, and Horquilla Limestone, which will be visited if time permits. Note that all plates seen thus far are part of the lower main thrust plate. To the east, beyond our traverse, is the second major thrust plate of Precambrian rock, made up of the Middle Proterozoic Oracle Granite (or granodiorite) of Peterson (1938). The upper major plate, comprised of Paleozoic and Mesozoic formations, is present to the southeast of the area of figure 31. On Mineta Ridge is the glide plate of Tertiary Pantano Formation (locally and informally called Mineta beds by some and Mineta Formation by others) and associated volcanic rocks (a very coarsely porphyritic andesite, a nonporphyritic andesite, and a rhyolite welded tuff). Both the fossils in the sedimentary rocks and a 26 Ma radiometric age on the porphyry date the rocks as probable latest Oligocene. Continue west to the bouldery outcrops.

3g--Bold outcrops of the Tertiary Youtcy Ranch stock of quartz monzonite.

3h--On spur. The Youtcy Ranch rock is subtly foliated here but is not lineated. A single sample gives ages that range from 35 to 16 Ma, depending on mineral and method used. Perhaps these ages reflect a period of gradual (19 Ma) cooling through various annealing temperatures that spanned about 150 °C. Foliation in this pluton must be younger than that in the core rocks and in the thrust plates. The oldest ages available for the early phase of this 2-mica suite (the Wrong Mountain Quartz Monzonite of the Rincon Mountains or the granite of the Wilderness Area of the Santa Catalina Mountains) are 45 Ma. The age of the more distant 2-mica granite at Texas Canyon is about 52 Ma, which may most closely represent the emplacement age of this phase. Continue down the spur and across the gully, past the edge of a reddish outcrop of a pendant or inclusion of the Bisbee Formation, to the vans.

50 mi. Return to hotel near Tucson Airport; end of excursion. See leader about departure plans, and some options for late afternoon activities, if time permits.

REFERENCES

Armstrong, A.K., 1958, The Mississippian of west-central New Mexico: *New Mexico Bureau of Mines and Mineral Resources Memoir 5*, 32 p.

Armstrong, A.K., Silverman, M.L., Todd, V.R., Hoggatt, W.C., and Carton, R.B., 1978, Geology of the central Peloncillo Mountains, Hidalgo County, New Mexico: *New Mexico Bureau of Mines and Mineral Resources Circular 158*, 18 p.

Bryan, C.R., in press, Mid-Tertiary volcanism in the eastern Chiricahuas: The Portal Caldera, *in* Sawyer, D.A., and Pallister, J.S., eds, From silicic calderas to mantle nodules: Cretaceous to Quaternary volcanism , southern Basin and Range Province, Arizona and Mexico: *New Mexico Geological Society Guidebook for the 1989 IAVCEI meeting in Santa Fe, New Mexico.*

Campuzano, J., 1973, The structure of the Cretaceous rocks in the southeastern part of Sierra de Juárez, Chihuahua, Mexico: El Paso, *University of Texas M.S. thesis*, 85 p.

Clemons, R.E., 1976, Sierra de las Uvas ash-flow field, south-central New Mexico, *in* Tectonics and Mineral Resources of Southwestern North America: L.A. Woodward and S.A. Northrop, eds., *New Mexico Geological Society Special Publication 6*, p. 115-121.

_____1985, Geology of the South Peak quadrangle, Luna County, New Mexico: *New Mexico Bureau of Mines and Mineral Resources Map 59*, scale 1:24,000.

Clemons, R.E., and Brown, G.A., 1983, Geology of the Gym Peak quadrangle, Luna County, New Mexico: *New Mexico Bureau of Mines and Mineral Resources Geologic Map 58*, scale 1:24,000.

Cooper, J.C., 1973, Geologic map of the Twin Buttes quadrangle, southwest of Tucson, Pima County, Arizona: *U.S. Geological Survey Miscellaneous Geologic Investigations Map I-745*, scale 1:48,000.

Cooper, J.C., and Silver, L.T., 1964, Geology and ore deposits of the Dragoon quadrangle, Cochise County, Arizona: *U.S. Geological Survey Professional Paper 416*, 196 p., scale 1:31,680.

Corbitt, L.L., and Woodward, L.A., 1970, Thrust faults of the Florida Mountains, New Mexico, and their regional tectonic significance, *in* Tyrone-Big Hatchet Mountains-Florida Mountains region: *New Mexico Geological Society 21st Field Conference Guidebook*, p. 69-74.

Creasey, S.C. 1967, Geologic map of the Benson quadrangle, Cochise and Pima Counties, Arizona: *U.S. Geological*

Survey Miscellaneous Geologic Investigations Map I-470, scale 1:48,000.

Creasey, S.C., and Quick, G.L., 1955, Copper deposits of part of the Helvetia mining district, Pima County, Arizona: *U.S. Geological Survey Bulletin 1027-F,* p. 301-323.

Davis, G.H., 1975, Gravity-induced folding of a gneiss dome complex, Rincon Mountains, Arizona: *Geological Society of America Bulletin,* v. 86, no. 7, p. 979-990.

Drewes, Harald, 1971a, Geologic map of the Sahuarita quadrangle, southeast of Tucson, Pima County, Arizona: *U.S. Geological Survey Miscellaneous Geologic Investigations Map I-613,* scale 1:48,000.

_____1971b, Geologic map of the Mount Wrightson quadrangle, southeast of Tucson, Santa Cruz and Pima Counties, Arizona: *U.S. Geological Survey Miscellaneous Geologic Investigations Map I-614,* scale 1:48,000.

_____1971c, Mesozoic stratigraphy of the Santa Rita Mountains, southeast of Tucson, Arizona: *U.S. Geological Survey Professional Paper 658-C,* 81 p.

_____1972a, Structural geology of the Santa Rita Mountains, southeast of Tucson, Arizona: *U.S. Geological Survey Professional Paper 748,* 35 p., scale 1:12,000.

_____1972b, Cenozoic rocks of the Santa Rita Mountains, southeast of Tucson, Arizona: *U.S. Geological Survey Professional Paper 746,* 66 p.

_____1974, Geologic map and sections of the Happy Valley quadrangle, Cochise County, Arizona: *U.S. Geological Survey Miscellaneous Geologic Investigations Map I-832,* scale 1:48,000.

_____1976, Plutonic rocks of the Santa Rita Mountains, southeast of Tucson, Arizona: *U.S. Geological Survey Professional Paper 915,* 75 p.

_____1977, Geologic map of the Rincon Valley quadrangle, Pima and Cochise Counties, Arizona: *U.S. Geological Survey Miscellaneous Geologic Investigations Map I-997,* scale 1:48,000.

_____1980, Tectonic map of southeastern Arizona: *U.S. Geological Survey Miscellaneous Geologic Investigations Series Map I-1109,* scale 1:125,000.

_____1981a, Tectonics of southeastern Arizona: *U.S. Geological Survey Professional Paper 1144,* 96 p.

_____1981b, Geologic map and sections of the Bowie Mountain South quadrangle, Cochise County, Arizona: *U.S. Geological Survey Miscellaneous Geologic Investigations Map I-1363,* scale 1:24,000.

_____1982, Geologic map and sections of the Cochise Head quadrangle and adjacent areas, southeastern Arizona: *U.S. Geological Survey Miscellaneous Geologic Investigations Series Map I-1312,* scale 1:24,000.

_____1984, Geologic map and structure sections of the Bowie Mountain north quadrangle, Cochise County, Arizona: *U.S. Geological Survey Miscellaneous Geologic Investigations Series Map I-1492,* scale 1:24,000.

_____1985a, Geologic map and structure sections of the Simmons Peak quadrangle, Cochise County, Arizona: *U.S. Geological Survey Miscellaneous Geologic Investigations Series Map I-1569,* scale 1:24,000.

_____1985b, Geologic map and structure sections of the Dos Cabezas quadrangle, Cochise County, Arizona: *U.S. Geological Survey Miscellaneous Geologic Investigations Series Map I-1560,* scale 1:24,000.

_____1986, Geology and structure sections of part of the northern Animas Mountains, Hidalgo County, New Mexico: *U.S. Geological Survey Miscellaneous Investigations Series Map I-1686,* scale 1:24,000.

_____1987, Geologic map and cross sections of the Dragoon Mountains, southeastern Arizona: *U.S. Geological Survey Miscellaneous Geologic Investigations Series Map I-1662,* scale 1:24,000.

_____1988, Southeast Arizona Tectonics, *in* [G. Holden, ed., GSA Field Trips]: *Geological Society of America.*

_____1989 (in press), Description and development of the Cordilleran orogenic belt in the southwestern United States: *U.S. Geological Survey Professional Paper.*

Drewes, Harald, Keith, S.B., LeMone, D.V., Seager, W.R., Clemons, R.E., and Thompson, Sam, III, 1982, Styles of deformation in the southern Cordillera, U.S.A.--A transect of the Cordilleran orogenic belt between El Paso, Texas, and Wickenburg, Arizona, *in* Drewes, Harald, ed., *Cordilleran Overthrust belt,* Texas to Arizona, field conference: Denver, Colorado, Rocky Mountain Association of Geologists.

Drewes, Harald, Houser, B.B., Hedlund, D.C., Richter, D.H., Thorman, C.H., and Finnell, T.L., 1985, Geologic map of the Silver City 1°x2° quadrangle, New Mexico and Arizona: *U.S. Geological Survey Miscellaneous Investigations Series Map I-1310-C,* scale 1:250,000.

Drewes, Harald, Klein, D.P., and Birmingham, S.C., 1988, Volcanic and structural controls of mineralization in the Dos Cabezas Mountains of southeastern Arizona: *U.S. Geological Survey Bulletin 1676,* 45 p.

Drewes, Harald, and Thorman, C.H., 1980a, Geologic map of the Steins and part of the Vanar quadrangles, New Mexico and Arizona: *U.S. Geological Survey Miscellaneous Geologic Investigations Series Map I-1220,* scale 1:24,000.

_____1980b, Geologic map of the Cotton City and part of the Vanar quadrangles, New Mexico and Arizona: *U.S. Geological Survey Miscellaneous Geologic Investigations Series Map I-1221,* scale 1:24,000.

Elsing, H.J., and Heineman, R.E., 1936, Arizona metal production: *Arizona Bureau of Mines Bulletin 140,* 112 p.

Flege, R.F., 1959, Geology of the Lordsburg quadrangle, Hidalgo County, New Mexico: *U.S. Bureau of Mines and Mineral Resources Bulletin 62,* 36 p.

Finnell, T.L., 1971, Preliminary geologic map of the Empire Mountains, Arizona: *U.S. Geological Survey Open-File Map,* scale 1:48,000.

Gile, L.H., 1986, Late Holocene displacement along the Organ Mountains fault in southern New Mexico--A summary: *New Mexico Geology,* v. 8, p. 1-4.

Gillerman, Elliot, 1958, Geology of the central Peloncillo Mountains, Hidalgo County, New Mexico, and Cochise County, Arizona: *New Mexico Bureau of Mines and Mineral Resources Bulletin 57,* 152 p.

Gilluly, James, 1956, General geology of central Cochise County, Arizona, *with sections on* Age and correlation by A.R. Palmer, J.S. Williams, and J.E. Reeside, Jr.: *U.S. Geological Survey Professional Paper 182,* 129 p.

Hill, R.T., 1902, The geographic and geologic features and their relation to the mineral products of Mexico: *American Institute of Mining Engineers Transactions,* v. 32, p. 163-178.

Hayes, P.T., and Raup, R.B., 1968, Geologic map of the Huachuca and Mustang Mountains, southeastern Arizona:

U.S. Geological Survey Miscellaneous Geological Investigations Map I-509.

Kelley, Shari, and Matheny, J.P., 1983, Geology of Anthony quadrangle, Doña Ana County, New Mexico: *New Mexico Bureau of Mines and Mineral Resources Geologic Map 54,* scale 1:24,000.

Keith, S.B., Reynolds, S.J., Damon, P.E., Shafiqullah, M., Livingston, D.E., and Pushkar, P.D., 1980, Evidence for multiple intrusion and deformation within the Santa Catalina-Rincon-Tortolita crystalline complex, southeastern Arizona, *in* Crittenden, M.D., Jr., Coney, P.J., and Davis, C.H., eds., Metamorphic core complexes, *Geological Society of America Memoir 153,* p. 217-268.

Kottlowski, R.E., 1963, Paleozoic and Mesozoic strata of southwestern and south-central New Mexico: *New Mexico Institute of Mining and Technology, State Bureau of Mines and Mineral Resources Bulletin 791,* 100 p.

Lasky, S.C., 1938, Geology and ore deposits of the Lordsburg mining district, New Mexico: *U.S. Geological Survey Bulletin 885,* p. 56-60.

_____1947, Geology and ore deposits of the Little Hatchet Mountains, Hidalgo and Grant Counties, New Mexico: *U.S. Geological Survey Professional Paper 208,* 101 p.

LeMone, David, 1958, The Devonian stratigraphy of Cochise, Pima, and Santa Cruz Counties, Arizona, and Hidalgo County, New Mexico: *University of Arizona unpublished M.S. thesis.*

_____1982, Geology of the El Paso border region from the Tom Lea Park, El Paso, Texas, *in* Drewes, Harald, ed., *Cordilleran overthrust belt,* Texas to Arizona, field conference: Denver, Colorado, Rocky Mountain Association of Geologists, p. 96-100.

Lipman, P.W., in press, Evolution of silicic magma in the upper crust: The mid-Tertiary volcanic field and its cogenetic granitic batholith, northern New Mexico, U.S.A.: *Royal Society of Edinburgh, Proceedings Volume of a symposium on "The origin of granites".*

Lovejoy, E.M., ed., 1980, Sierra de Juárez, Chihuahua, Mexico, Structure and stratigraphy: *El Paso Geological Society Guidebook,* 59 p.

Marjaniemi, D.K., 1968, Tertiary volcanism in the northern Chiricahua Mountains, Cochise County, Arizona, *in Southern Arizona Guidebook 3:* Tucson, Ariz., Arizona Geological Society, p. 209-214.

Muehlberger, W.R., 1980, Texas lineament revisited, in trans-Pecos Region, 31st Field Conference: *New Mexico Geological Society.*

Nodeland, S.K., 1977, Cenozoic tectonics of Cretaceous rocks in the northeast Sierra de Juárez, Chihuahua, Mexico: El Paso, Texas, *University of Texas at El Paso unpublished M.S. thesis,* 79 p.

Pallister, J.S., and du Bray, E.A., 1989 (in press), A field guide to volcanic and plutonic features of the Turkey Creek caldera, Chiricahua Mountains, southeast Arizona, *in* Sawyer, D.A., and Pallister, J.S., eds., From silicic calderas to mantle nodules--Cretaceous to Quaternary volcanism, southern Basin and Range, Arizona and New Mexico: *New Mexico Geological Society Guidebook for the 1989 IAVCEI meeting in Santa Fe, New Mexico.*

Peterson, N.P., 1938, Geology and ore deposits of the Mammoth mining camp area, Pinal County, Arizona:

Arizona Bureau of Mines Bulletin 144, Geology Series 11, 63 p.

Sabins, F.F., Jr., 1957a, Geology of the Cochise Head and western part of the Vanar quadrangles, Arizona: *Geological Society of America Bulletin,* v. 65, no. 10.

_____1957b, Stratigraphic relations in the Chiricahua and Dos Cabezas Mountains, Arizona: *American Association of Petroleum Geologists Bulletin,* v. 41, no. 3.

Schrader, F.C., and Hill, J.M., 1915, Mineral deposits of the Santa Rita and Patagonia Mountains: *U.S. Geological Survey Bulletin 582.*

Seager, W.R., 1981, Geology of the Organ Mountains: *New Mexico Bureau of Mines and Mineral Resources Memoir 36.*

_____1983, Laramide wrench faults, basement-cored uplifts, ane complementary basins in southern New Mexico: *New Mexico Geology,* v. 5, p. 69-76.

Seager, W.R., Mack, G.H., Raimonde, M.S., and Ryan, R.G., 1986, Laramide basement-cored uplift and basins in south-central New Mexico, *in* Clemons, R.E., King, W.E., and Mack, G.H., eds., Truth or Consequences region: *New Mexico Geological Society, Guidebook to 37th Field Conference,* p. 123-130.

Seager, W.R., and Mack, G., 1986, Laramide paleotectonics of southern New Mexico, *in* J.A. Peterson, ed., Paleotectonics and sedimentation in the Rocky Mountain region: *American Association of Petroleum Geologists Memoir 41,* p. 669-685.

Seager, W.R., and McCurry, M., 1988, The Organ cauldron and cogenetic Organ batholith--Evolution of a large volume silicic magma chamber: *Journal of Geophysical Research,* May 10, 1988.

Thomann, W.F., 1981, Ignimbrites, trachytes, and sedimentary rocks of the Precambrian Thunderbird Group, Franklin Mountains, El Paso, Texas: *Geological Society of America Bulletin,* v. 92, no. 2, p. 94-100.

Thompson, Sam III, and Jacka, A.D., 1981, Pennsylvanian stratigraphy, petrography, and petroleum geology of the Big Hatchet Peak section, Hidalgo County, New Mexico: *New Mexico Bureau of Mines and Mineral Resources Circular 176.*

Thorman, C.H., 1977, Geologic map of the Coyote Peak and Brockman quadrangles, Hidalgo and Grant Counties, New Mexico: U.S. Geological Survey Miscellaneous Field Studies Map MF-924, scale 1:24,000.

Thorman, C.H., and Drewes, Harald, 1978, Geologic map of the Gary and Lordsburg quadrangles, Hidalgo County, New Mexico: *U.S. Geological Survey Miscellaneous Geologic Investigations Series Map I-1151,* scale 1:24,000.

_____1979a, Geologic map of the Grandmother Mountain East and Grandmother Mountains West quadrangles, New Mexico: *U.S. Geological Survey Miscellaneous Field Studies Map MF-1088,* scale 1:24,000.

_____1979b, Geologic map of the Saltys quadrangle, Grant County, New Mexico: *U.S. Geological Survey Miscellaneous Field Studies Map MF-1137,* scale 1:24,000.

_____1980, Geologic map of Victoria Mountains, Luna County, New Mexico: *U.S. Geological Survey Miscellaneous Field Studies Map 1175, scale 1:24,000.*

_____1981a, Geologic map of the Gage southwest quadrangle, Grant and Luna Counties, New Mexico: *U.S. Geological Survey Miscellaneous Geologic Investigations*

Map I-1231, scale 1:24,000.

_____1981b, Geology of the Rincon Wilderness Study Area: *U.S. Geological Survey Bulletin 1500-A*, 62 p.

Uphoff, T.L., 1978, Subsurface stratigraphy and structure of the Mesilla and Hueco bolsons, El Paso region, Texas and New Mexico: El Paso, Texas, *University of Texas at El Paso M.S. thesis*.

Vanderspuy, 1975, Geology and geochemical investigation of geophysical anomalies, Sierra Rica, Hidalgo County, New Mexico: *New Mexico Bureau of Mines and Mineral Resources Open-File Report 62*.

Wacker, H.J., 1972, The stratigraphy and structure of the Cretaceous rocks in north-central Sierra de Juárez,

Chihuahua, Mexico: El Paso, Texas, *University of Texas at El Paso M.S. thesis*.

Zeller, R.A., 1965, Stratigraphy of the Big Hatchet Mountains area, New Mexico: *New Mexico Bureau of Mines and Mineral Resources Memoir 16*.

_____1970, Geology of the Little Hatchet Mountains, Hidalgo and Grant Counties, New Mexico: *New Mexico Bureau of Mines and Mineral Resources Bulletin 96*, scale about 1:32,000.

_____1975, Structural geology of Big Hatchet Peak quadrangle, Hidalgo County, New Mexico: *New Mexico Bureau of Mines and Mineral Resources Circular 146*.

A BRIEF HISTORY OF THE EL PASO-TUCSON REGION, TEXAS-ARIZONA

Harald Drewes, Robert Munn, and Andy Alpha
U.S. Geological Survey, Denver Federal Center, Colorado

The recorded human events of the region of International Geologic Congress trip 121 are, by world standards, of brief duration and minor consequence. Nevertheless, this local history has been so colorful that it has captured the imagination of generations of story and song writers, and of movie and television producers. Included among these writers are the world-reknowned Zane Gray and Karl Mai, and among the producers is John Ford. Seeing the movie "Stagecoach," starring John Wayne, in Switzerland, with German subtitles is clear indication of the widespread appeal of the lore and history of western frontier time. Naturally, such sources distort history to some extent, but even the more traditional sources of history, such as eyewitnesses accounts, official documents, and professionally written books, have their own distortions, for the viewpoints of the common folk and the vanquished people are rarely heard. From such mixed sources, then, we offer some glimpses of the main actors on this "Wild West Stage" and some of the dramas presented on it.

THE INDIANS

The first people on the stage were the Indians, a people given this name by 15th Century Europeans who at first failed to recognize that they were on two new continents and not in India. The American Indians arrived from northeastern Asia probably over a span of a few tens of thousands of years. Typically, those Indians encountered by the European explorers were not the original migrants from Asia but more recent arrivals who pushed aside or absorbed their predecessors, some of whom had developed fairly complex cultures. Territorial conflict was thus an old drama in the New World, much as in the Old. This conflict continued, and took some new turns with the arrival in our region first of the Spaniards, and then of other Europeans (later identified as Americans).

The earliest occupation sites (9500-5600 B.C.) are known for their Clovis and Folsom type of projectile points, found at such sites as Murray Spring, Lehner, and Naco. These earliest people were accompanied by, and hunted, late Pleistocene fauna, now partly extinct. Archaic Hunters (5500 B.C.-1 A.D.) are known from Double Adobe site, and perhaps also from the shores of ancient lake Willcox (Cochise Culture). Hohokam people (A.D. 1-1450) arrived from Mexico, and settled along the western rivers, where they made pottery and pursued agricultural activities. Mogollon culture people (A.D. 1-1150) settled farther east where they are known at sites at San Simon and Cave Creek, among others.

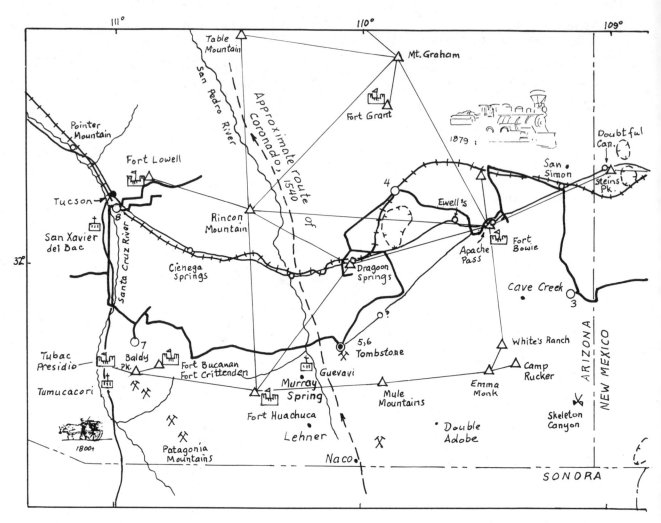

FIGURE 1 Historic map of the region of trip T-121, El Paso, Texas, to Tucson, Arizona. Prepared by Andy Alpha.

During the 13-15th centuries, the early people underwent a major disruption, in general suggesting that population pressures were too great for the available resources. Likely a decrease in rainfall and a change in the yearly distribution of precipitation contributed to the problem. By A.D. 1500-1650 the eastern river valleys were occupied by the Jumano and Suma tribes and the western valleys by the Pima tribe, all speaking languages of the Aztec-Tanoan family. The intervening area, as well as the area to the north and east, were occupied by relative newcomers, some of the Apache tribes, speaking a language of the Athapascan family. Other Apaches seem to have moved into the area as recently as 1700 A.D. from the Great Plains, from which they were expelled by more powerful tribes. Whatever mastery of agriculture these late arrivals had was abandoned and they returned to a hunting, gathering, and raiding style of life.

Serious problems existed between these two cultural groups of Indians--agriculturalists and nomads--even before the advent of the Spanish people. These problems took a new turn with the arrival of the Spaniards, for they brought with them a new mobility, the horse, and a new weapon, the firearm. Early Spanish settlements, such as El Paso (around 1590) and Tubac (around 1700) were attacked many times. A major Apache uprising occurred during the 1860's, coincident in part with the time of the American Civil War, during which early mining camps and ranches were decimated. The last major uprising took place under Cochise, an Apache chief, in the late 1870's, just over 100 years ago. These events made their mark on our relatively recent history, then provided numerous picturesque geographic place names, left relics of many forts, and led to an improved group of transportation and communication facilities (fig. 1).

SPANIARDS

The first Europeans known to have arrived in the El

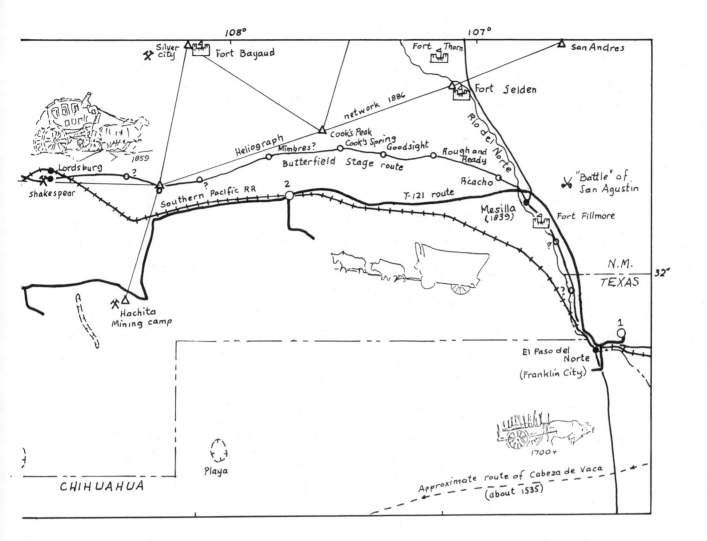

Paso-Tucson region were Spaniards. Following their initial conquest under Hernando Cortés, of the Aztec Indians of Mexico, who themselves were conquerors a few generations earlier of a more advanced society, the Spaniards set about exploring northern Mexico for mineral deposits, agricultural opportunities, and souls. Their explorations reached the El Paso-Tucson region early in the 16th century, and their settlements were started late in that century and early in the next one. While most of the exploratory trips were planned, a few of the more intriguing journeys developed accidentally.

Perhaps the earliest journey, one of the accidental ones, was that of Alvar Nuñez Cabeza de Vaca and his colored or moorish companion, Esteban, who during 1530-1536 wandered across southern and western Texas to near El Paso, across northern Chihuahua and Sonora, and then down southward in western Mexico. Their journey began with a large group in Florida, whose remnants coasted westward in small boats they were obliged to build. These boats were driven by a storm

upon the shore of southern Texas, where the few survivors were enslaved by the local Indians for 2 years. Cabeza de Vaca's medical skills enabled him gradually to move freely among neighboring villages, to locate Esteban, and to formulate plans to escape. With this mistique as a healer, and the particular customs of the natives, their journey became a sort of religious progress; they were accompanied in festive manner by an ever-changing large entourage of Indians. His place in history was probably played down by the Spaniards themselves because his accounts of mistreatment of Indians in western Mexico led to unpopular reforms among Spanish mine owners.

In 1539, Fra Marco de Niza explored in the region of present Arizona and New Mexico, at first accompanied by, and then gradually preceded by the experienced Esteban. Although Esteban tried to revive to the medical-religious style of travel of his earlier mentor, Cabeza de Vaca, he met with little success and ultimately was killed in northern New Mexico. The actual results of de Niza's trip remain unclear; in any event, he seems to have favored an easier lifestyle than that of explorer.

A more ambitious and successful exploration trip was led by Francisco Vásquez de Coronado in 1540-1541. With a large group of soldiers, he visited several of the southwestern and Great Plains states, returning with some of the earliest accounts of mineral wealth. Upon crossing Cabeza de Vaca's earlier route, he recorded that this procession was still talked about by the natives.

In 1582 Antonio de Espejo reached central Arizona and in 1604 Juan de Oñate crossed the region on his way to western Colorado.

Settlement of the region began around 1590, with mixed success. El Paso was well established in 1659, and the Santa Cruz River of Arizona about 50 years later. In 1687 an Austrian Jesuit, Eusebio Francisco Kino, established the first of his many Missions in Pimeria Alta, as northern Sonora and southern Arizona were then known. Around 1700, work was begun on the missions of Tumacacori and San Xavier del Bac (see trip itinerary, Day 8). Despite occasional setbacks from depredations of Apaches, both El Paso and Tucson were soon flourishing.

Mexico revolted in 1821 and gained its independence from Spain. For a short time the French attempted to colonize the new Mexican republic, but the Mexicans soon regained control. Most of this turmoil took place in southern and central part of the country; the El Paso-Tucson region remained a remote frontier. Doubtless, this situation fostered persistent Indian troubles, banditry, and the immigration of other Europeans. Part of the region was lost to the growing United States at the termination of the Mexican War in 1848. The greater part of the region, known as the Gadsden Purchase, was bought in 1853 with the intent to develope a low-level transcontinental railroad route.

THE MINERS

Mining activities developed slowly until transportation improved. The Indians needed few mined products for use as tools and ornaments. Some native copper was obtained by them near Silver City. The Spaniards were experienced prospectors and metal miners, and many of the rich near-surface deposits, particularly of precious metals, were known to them. Among these were the silver-lead deposits in the Santa Rita and Patagonia Mountains of Arizona and the copper deposits at Silver City. Mining activities

increased again during the American Civil War, when the Confederate States sought sources of lead. Union troops ended these efforts after minor battles northeast and northwest of the region, plus one bloodless skirmish in the mountains east of Mesilla. Following the Apache uprising during the Civil War period, mining activities were resumed and many of the fabulous mining camps of the region were developed during the 1870's, when railroad construction was begun. None of these was more colorful than Tombstone (see itinerary, Day 5 and 6).

Tombstone, the most famous of Arizona's old mining camps, was discovered in 1877 by prospector Ed Schieffelin. Schieffelin came to territorial Fort Huachuca in the early 1870's with a party of soldiers. When he left the safety of the fort to prospect, his comrades were alleged to have said, "that's Apache country, all you will find is your tombstone," thus the name he chose for his discovery. Rumors of the rich strikes made a boom town of Tombstone, and by 1888, $30 million in silver had been taken from 11 major mines. Unfortunately, ground water was abundant, and as early as 1883 the mines started to flood. In 1903 new pumps were installed, but by 1909 strikes and flooding had taken their toll and mining at Tombstone was finished.

"The Town too tough to die" was aptly named, and like all robust mining camps, Tombstone attracted droves of people looking for a piece of the action. You would find on one street the girls, gamblers, and countless saloons, and on the next street lawyers, doctors, and engineers. The backbone of the mining people were the Cornish hard rock miners, who spent most of their lives underground. They lived in small houses, raised families, and earned no more than $6 for a 10-hour shift. They and the riches they uncovered brought the Arizona Territory out of its economic woes and prompted the Southern Pacific to build a railroad across the territory, opening up the area and bringing in other settlers and merchants.

The Tombstone of today is primarily known for its infamous characters of the early 1900's--like Doc Holliday, Wyatt and Virgil Earp, and the Clantons-- whose colorful exploits helped make the erstwhile boom town into a major tourist attraction. Tombstone is now designated as a National Historic Site, and the Bird Cage Theatre, Boothill Cemetery, the OK Corral (site of the famous Earp-Clanton gun battle), and the offices of the Tombstone Epitaph (where the newspaper was and still is being printed), among others, are preserved for our enjoyment.

GENERAL GEOLOGY OF THE SIERRA DE JUÁREZ, CHIHUAHUA, MEXICO

Russ Dyer

University of Texas at El Paso; currently Dept. of Energy, Las Vegas

Sierra de Juárez (also shown on U.S. maps as Sierra Juárez, to be followed hereafter) is the northernmost range of a series of more-or-less continuous ranges that trend southeast along the international border from Juárez, Chihuahua, Mexico. This belt of ranges has been termed the Chihuahua Tectonic Belt. The Chihuahua Tectonic Belt is part of the more extensive Sierra Madre Oriental, the Mexican segment of the North American Cordillera. Because of its superb exposures and easy accessibility, Sierra Juárez has become a standard for stratigraphic and structural comparisons in northern Chihuahua, west Texas and southwestern New Mexico. The geologic contrast between the Sierra Juárez and the tilted fault block of Proterozoic and Paleozoic rocks in the Franklin Mountains to the northeast is striking.

Sierra Juárez is composed entirely of Cretaceous rocks which have been intruded by Eocene igneous rocks. The Cretaceous rocks have been folded and thrust faulted (fig. 1). Most fold axes strike northwest and are subhorizontal with fold geometries ranging from open to isoclinal and overturned. Imbricate thrusting transported thrust sheets from the southwest to the northeast. The deformed Cretaceous strata are intruded by postdeformational sills or northeast-trending dikes of Eocene andesites and rhyolites.

STRATIGRAPHY

All strata exposed in Sierra Juárez is Early Cretaceous in age, although a PEMEX drill hole in northwest Sierra Juárez encountered Late Cretaceous rocks in the subsurface, beneath Aptian strata. Most of the strata present in the range are assigned to four formations: Cuchillo, Benigno, Lágrima, and Finlay (see Road log, fig. 3). The lower three formations are each divided into three members: a medial thick-bedded or massive limestone overlain and underlain by thin-bedded limestone, shale, and sandstone. The next overlying formation is entirely thick-bedded limestone. These four formations are all Aptian-Albian in age. Locally present are shales and sandstones of various formations of the Cristo Rey Group (informal usage) of Albian-Cenomanian age. All formation contacts are conformable.

The lowest formation, Cuchillo, is typically brownish gray and is characterized by the 10- to 20-m-thick ridge or cliff of the massive middle member. Coarse-grained,

crossbedded, quartz sandstone and conglomerate is interbedded with limestone and shale in the lower and upper members. Silicified wood is locally abundant in the lower member, but the dominant fossils throughout the formation are a pelecypod fauna, including Exogyra- and Unio-like shells (oysters and clams).

The overlying Benigno Formation is easily recognized by the thick (120 m), massive, light-gray cliff-forming biohermal reef limestones of the middle member. Corals, algae, and caprinid-rudistid pelecypod characterize the reef fauna. Orbitolina sp. first becomes abundant in the Benigno. Orbitolina sp. packstones are found in the thinner bedded lower and upper members.

Lágrima Formation consists of interbedded marly shales and limestones, in about equal proportions, commonly giving the formation a stair-step appearance. A medial massive-bedded rudistid-reef limestone (20 m) is present in the western thrust plate, but is not found in the eastern plates.

Finlay Formation is a thick-bedded, cliffy limestone resembling the middle member of the Benigno. Distinction is based on the olive-gray to blue-gray color of the Finlay and the presence of Dictyoconus sp. wackestones and packstones in this formation.

At Cerro Cristo Rey, 15 km north of Sierra Juárez, the Finlay is conformably overlain by 340 m of calcareous shale, sandstone, and limestone. These strata have been divided into eight formations: Del Norte, Smeltertown, Muleros, Mesilla Valley, Anapra, Del Rio, and Buda. Representatives of the Cristo Rey group are found on the two northeastern allochthons, as well as the autochthon.

The Boquillas Formation is the youngest Cretaceous unit preserved in the area. At Cerro Cristo Rey the Boquillas unconformably overlies the Buda Formation. Boquillas is a dark-gray, sparsely fossiliferous, calcareous shale containing thin-bedded limestone laminae. Boquillas underlies the western thrust sheet in Sierra Juárez.

Eocene dikes and sills of rhyolitic to andesitic composition are locally abundant in Sierra Juárez. Dikes commonly intrude northeast-trending faults which cut folds and thrust faults, while the dikes themselves are undeformed. Radiometric dating by the K-Ar method on amphibole and biotite from dike rocks of Sierra Juárez gives dates of 47.8+2.2 and 48.2±1.7 Ma, comparable to a date obtained on the Campus Andesite of the El Paso area.

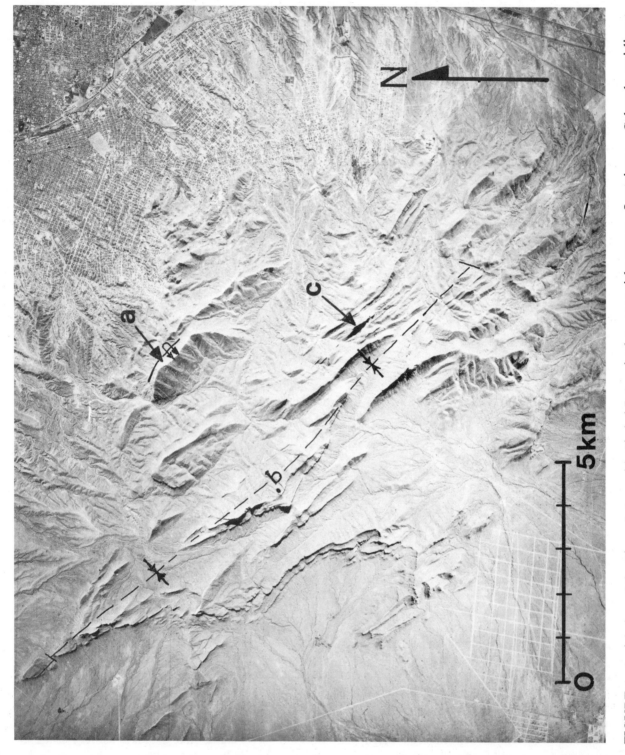

FIGURE 1   Aerial view of the Sierra Juárez with Ciudad Juárez in the upper right corner.   Locations, *a*, Colorado anticline; *b*, Flores Peak and Flores syncline; and *c*, Cerro Bola.   Strata are mainly Early Cretaceous limestones and shales

STRUCTURE

Large amplitude, subhorizontal, northwest-trending folds are the most visible structures in Sierra Juárez (fig. 1). Folding is commonly disharmonic, with thick, competent units forming broad, open folds while folds in thinner bedded, less competent units above and below may be tight to isoclinal. The style of folding is also a function of position within the range; open and upright to slightly inclined folds are common in the southwestern part of the range, while folding in the northeastern part is commonly tight to isoclinal and overturned to the northeast.

Three large thrust faults are recognized in the northeastern part of Sierra Juárez. Numerous minor thrust faults are also present. The major thrust faults define imbricate allochthonous plates which have moved to the northeast over presumed authchonous rocks. These folds and thrust faults are closely associated, as shown in the structures offered by several workers (see Road log, fig. 4). Folds are commonly floored or roofed by a thrust fault, giving rise to disharmonic structures.

Some of these strike-slip faults may be lateral ramps which accommodated different displacements within the allochthons.

In general, deformation is more intense to the northeast, in structurally deeper rocks, and less intense to the southwest, at higher structural levels. The shaly Boquillas Formation is the major décollement horizon.

TIMING OF DEFORMATION

The compressive deformation in Sierra Juárez must postdate the lower Upper Cretaceous Boquillas Formation, and predate the Eocene intrusive rocks. A young age within this range is preferred because the style of folding is indicative of considerable cover, perhaps 3-4 km thick, presumably of upper Upper Cretaceous and some Paleocene rocks, which have subsequently been removed by erosion. Remnants of such rocks are now found about 300 km to the southeast, but recently have also been discovered in a drill hole about 70 km to the northwest.

Apache Geronimo: woodcut
by Werner Drewes

OIL AND GAS EXPLORATION WELLS DRILLED TO PRECAMBRIAN BASEMENT IN SOUTHEASTERN ARIZONA AND SOUTHWESTERN NEW MEXICO

D.J. Brennan
Arizona Oil and Gas Conservation Commission, Phoenix

Sam Thompson III
New Mexico Bureau of Mines and Mineral Resources, Socorro

Abstract.

In the Tucson-El Paso area of southeastern Arizona, southwestern New Mexico, and northwestern Chihuahua, 18 oil and gas exploration wells have been drilled to Precambrian basement. Structural elevations of the Precambrian top in these wells range from + 1,626 ft (above sea level) to -11,710 ft (below sea level). However, the Precambrian rises to a surface elevation of over 7,000 ft in the Florida Mountains and is estimated to lie at a subsea elevation of about -24,600 ft beneath the recently drilled Phillips No. 1 Sunland Park well in Doña Ana County, New Mexico. A major part of the structural relief may be attributed to Basin and Range normal faulting in Miocene-Pliocene time. A significant part may be attributed to Laramide folding, thrusting, and wrench-faulting. The Phillips No. A1 Tombstone State well in Cochise County, Arizona is the only one that appears to have penetrated a basement-involved thrust; in that well, Precambrian granite overlies Cretaceous sedimentary rocks.

INTRODUCTION

Figure 1 is a map of the Tucson-El Paso area covering southeastern Arizona, southwestern New Mexico, and northwestern Chihuahua; it shows the 18 oil and gas exploration wells that have been drilled to Precambrian basement. These and other key wells provide the most important data used in evaluation of the petroleum potential of this frontier area. Although commercial production has not been established yet, several wells have encountered significant shows of oil and gas in the Paleozoic and Mesozoic sedimentary section.

or below sea level (shown in parentheses) is determined by subtracting the depth of the top from the elevation of the kelly bushing (or other datum). Indicated structural relief ranges from + 1,626 ft (above sea level) (Hachita Dome No. 1 Tidball Berry well in Hidalgo County, New Mexico) to -11,710 ft (below sea level) (Pemex No. 1 Moyotes well in Chihuahua, Mexico). However, in the Florida Mountains (southeast of Deming) the

Precambrian rises to a surface elevation of over 7,000 ft. If a pre-Tertiary section similar to that in the Grimm et al. No. 1 Mobil 32 well lies below the Phillips No. 1 Sunland Park Unit well, the top of the Precambrian may lie at a depth of about 29,000 ft, or a subsea elevation of about -24,600 ft. A structure contour map is being prepared at a larger scale to show both the surface and subsurface control in this complexly faulted Basin and Range area.

Thompson and others (1978) discussed all wells that had been drilled to Paleozoic or Precambrian rocks in this area up to that time. Because of space limitations in this paper, several of those wells are not discussed and are shown only with well-location symbols on Fig. 1. In addition to discussions of the Precambrian tests, the following sections include brief reviews of selected pre-1978 wells as well as more recently drilled key wells that bottomed in Cenozoic, Mesozoic, or Paleozoic rocks.

SOUTHEASTERN ARIZONA

Table 1 lists the 3 wells drilled to Precambrian basement and 4 other key wells drilled in southeastern Arizona. A total of 43 oil and gas exploration wells have been drilled in this part of the map area (fig. 1). For further information, see the latest report by the Arizona Oil and Gas Conservation Commission (1987).

The depths to the top of Precambrian (underlined on Fig. 1) generally indicate the depths to economic basement; however, Paleozoic-Mesozoic sedimentary rocks may underlie Precambrian granitic rocks as a result of Laramide thrusting in some areas. Total depths of other key wells (not underlined) provide some constraints on estimated depths to Precambrian where the bottom was in Paleozoic, Mesozoic, or Cenozoic rocks. Parts of the Paleozoic-Mesozoic section were duplicated locally by thrusting in a few cases.

Structural elevation of the top of Precambrian above

Pima County

The Humble (now Exxon) No. 1 State (32) well was drilled as a stratigraphic test. It penetrated Cenozoic

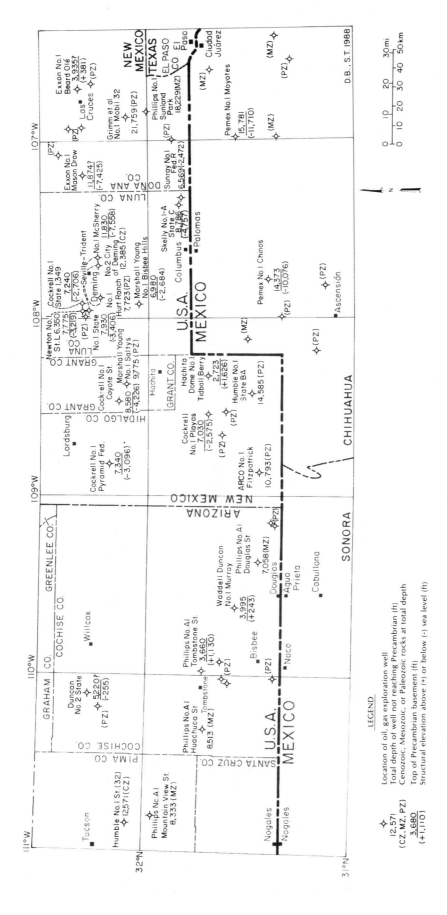

FIGURE 1 Map of oil and gas exploration wells drilled to Precambrian basement in southeastern Arizona, southwestern New Mexico, and northwestern Chihuahua. Names of wells and total depths are given for Precambrian tests and other key wells discussed in text. Wells without names and total depths are discussed in Thompson, Tovar, and Conley (1978).

TABLE 1 Oil and Gas Wells Drilled to Precambrian Basement in Southeastern Arizona. Other Key Wells Also Are Listed In This Table.

Section, Township, Range	Name of well	Elevation (ft)	Top Precamb. (ft)	Total depth (ft)	Date of Completion
PIMA COUNTY:					
5-16S-15E	Humble (Exxon) No. 1 State (32)	2,886KB		12,571	12-18-72
10-17S-15E	Phillips No. A1 Mtn View State	3,141GL		8,333	7-22-82
COCHISE COUNTY:					
33-13S-22E	Duncan No. 2 State	4,965DF	5,220?	5,310	12-20-60
16-20S-20E	Phillips No. A1 Huachuca State	4,261GL		8,513	6- 3-82
14-20S-23E	Phillips A1 Tombstone State	4,790KB	3,660	10,571	9-14-81
5-22S-27E	Waddell-Duncan No. 1 Murray	4,238KB		3,995	5-16-52
14-23S-29E	Phillips No. A1 Douglas State	4,295GL		7,058	6-23-82

(Elevation datum: GL = ground level, DF = derrick floor, KB = kelly bushing)

sedimentary and volcanic rocks down to 11,987 ft, and Tertiary quartz monzonite to 12,571 ft (total depth). Tertiary sedimentary rocks include evaporites and slide blocks of Paleozoic limestones. In the paper by Thompson and others (1978), the quartz-monzonite was interpreted to be Precambrian(?). However, after reconsideration of the evidence it is now interpreted to be Eocene, based mainly on the rock type and the potassium-argon date of 46 Ma.

The Phillips No. A1 Mountain View State well penetrated Lower Cretaceous (Barremian-Albian) sedimentary rocks down to 8,333 ft. This well was part of an extensive exploration program in this region conducted by Phillips and their partners during the late 1970's and early 1980's.

The main objective of this program was to test the possibility of an extension of the Western Overthrust Belt from the productive Utah-Wyoming segment, through southeastern Arizona and southwestern New Mexico, with continuation to the Chihuahua Tectonic Belt of eastern Chihuahua and western Texas. Drill sites were chosen on the basis of extensive seismic coverage and surface/subsurface geologic data. However, the complex Basin and Range faulting made difficult the interpretations of Laramide overthrusting.

Cochise County

The Waddell-Duncan No. 1 Murray well penetrated

Cenozoic sedimentary and volcanic rocks down to 1,755 ft (possible fault or major unconformity), Pennsylvanian to Cambrian sedimentary rocks to 3,995 ft, and Precambrian granite to 4,400 ft. This was the first well to reach Precambrian rocks in this region. The Duncan No. 2 State well penetrated Cenozoic sedimentary and volcanic rocks down to 5,220 ft (major unconformity), and Precambrian(?) basement to 5,310 ft. The generalized sample description indicates only igneous rock below 3,600 ft. Gamma ray and resistivity logs indicate that a Tertiary volcanic rock/Precambrian granite contact may be present at 5,220 ft, as seen in nearby outcrops.

The Phillips No. A1 Huachuca State well penetrated Lower Cretaceous sedimentary rocks down to 8,513 ft. A siltstone interval from 3,330 to 3,930 ft was dated as Aptian-Cenomanian.

The Phillips No. A1 Tombstone State well penetrated Cretaceous(?) sandstones and conglomerates down to 1,780 ft (major unconformity), Upper Paleozoic limestones to 3,660 ft (probable normal fault), Precambrian granite and quartzite to 6,000 ft (probable thrust fault), and probable Cretaceous sedimentary rocks to 10,571 ft (Betton, 1982). The Precambrian age was determined radiometrically. This well is the only one in the region that appears to have penetrated a basement-involved thrust surface. The indicated thrust probably is Laramide in age, but seismic or other information is needed to determine whether it is part of a regional low-angle overthrust, a local high-angle upthrust, or a flower structure associated with wrench faulting. Although less likely, the surface may be a result of gravity sliding off an adjacent uplift.

The Phillips No. A1 Douglas State well penetrated Cenozoic sedimentary and volcanic rocks down to 4,240 ft, and Lower Cretaceous sedimentary rocks to 7,058 ft. Coal was reported from 6,716 to 6,720 ft. The operator reported that the well was in Lower Cretaceous (Barrenian-Albian) at total depth.

SOUTHWESTERN NEW MEXICO

Table 2 lists the 13 wells drilled to Precambrian basement and 7 other key wells drilled in southwestern New Mexico. A total of 75 oil and gas exploration wells have been drilled in this part of the map area (Fig. 1). Wells drilled to Mesozoic, Paleozoic, or Precambrian rocks were discussed by Thompson (1982). For further information, contact the New Mexico Bureau of Mines and Mineral Resources.

Hidalgo County

In the northern part of the Animas Valley, the Cockrell No. 1 Pyramid Federal well penetrated Cenozoic sedimentary and volcanic rocks down to 5,795 ft, Mississippian to Cambrian sedimentary rocks (with several Tertiary igneous intrusions) to 7,340 ft, and Precambrian granite to 7,404 ft (total depth). In the southern part of the Animas Valley, the ARCO No. 1 Fitzpatrick well penetrated Cenozoic sedimentary and volcanic rocks down to 5,582 ft, Cretaceous limestones, mudstones, and sandstones to 7,209 ft (major unconformity), Permian limestones and dolostones to 8,718 ft, metamorphosed Paleozoic carbonates to 10,780 ft, and green hornfels to 10,793 ft.

In the Playas Valley, the Cockrell No. 1 Playas State well penetrated Cenozoic sedimentary and volcanic rocks down to 2,480 ft, lower Permian to Cambrian sedimentary rocks to 7,030 ft, and Precambrian granite to total depth. The Humble (now Exxon) No. 1 State BA well drilled Quaternary sediments down to 230 ft, Lower Cretaceous sedimentary rocks to 995 ft, and Permian to Ordovician sedimentary rocks to 14,585 ft. At 3,310 ft a reverse fault repeated part of the Permian section with about 970 ft of stratigraphic throw.

In the Hachita Valley, the Hachita Dome No. 1 Tidball-Berry Federal well penetrated Quaternary sediments down to 224 ft, Mississippian to Cambrian sedimentary rocks to 2,723 ft, and Precambrian granite to 2,726 ft. The structural elevation of the top of the Precambrian at +1,626 ft (above sea level) is the highest in all of the wells drilled in this region. This well was located on the upper plate of the Sierra Rica thrust and may have reached Cretaceous sedimentary rocks in the lower plate if it had been drilled a few thousand feet deeper.

Grant County

The Cockrell No. 1 Coyote State well penetrated Cenozoic sedimentary and volcanic rocks down to 1,790 ft, Lower Cretaceous sedimentary rocks to 7,240 ft (major unconformity), Ordovician to Cambrian sedimentary rocks to 8,580 ft, and Precambrian granite to 9,282 ft. The recent Marshall Young No. 1 Saltys well penetrated Tertiary volcanic and sedimentary rocks down to 770 ft, Permian to Cambrian sedimentary rocks to 4,494 ft (thrust fault), Lower Cretaceous sedimentary rocks to 8,344 ft (same major unconformity as in the Cockrell well), and Ordovician to Cambrian sedimentary rocks to 9,775 ft. The thrust fault has a vertical separation of 4,588 ft on top of the Ordovician. Horizontal separation may be only a few miles. The thicker Paleozoic section above the thrust may be part of a block that was downfaulted on the southern flank of the Burro Uplift in Triassic(?) time, buried by Lower Cretaceous sediments, and thrust in Laramide time over a more deeply eroded Paleozoic section nearer the crest of the uplift.

Luna County

The Cockrell No. 1 State 1,349 well penetrated Cenozoic sedimentary and volcanic rocks down to 1,310 ft, Cretaceous-Tertiary redbeds to 6,500 ft (major unconformity), Ordovician to Cambrian sedimentary rocks to 7,240 ft, and Precambrian granite to 7,375 ft. The Sycor Newton No. 1 State L6,350 well penetrated Cenozoic sedimentary and volcanic rocks down to about 2,000 ft, Cretaceous-Tertiary redbeds to 6,530 ft (major unconformity), Ordovician to Cambrian sedimentary rocks to 7,775 ft, and Precambrian granite and gneissic granite to 10,090 ft. The 2,315 ft of Precambrian section drilled here is nearly the same as the 2,320 ft drilled in the Phillips No. A1 Tombstone well, but this Sycor Newton well did not encounter a thrust fault with Cretaceous sedimentary rocks below. The thrust possibility may be tested further by drilling another few thousand feet or by obtaining seismic evidence.

In the Deming area, Seville-Trident drilled 4 key wells. Clemons (1986) made a petrographic study of the drill cuttings. The No. 1 State well penetrated probable Cenozoic sedimentary and volcanic rocks and definite Cretaceous-Tertiary redbeds (first sample at 6,000 ft) down to 6,450 ft (major unconformity), Ordovician to Cambrian sedimentary rocks down to 7,930 ft, and Precambrian granite to 8,240 ft. The No. 1 Hurt Ranch well penetrated Cenozoic sedimentary and volcanic

TABLE 2 Oil and Gas Wells Drilled to Precambrian Basement in Southwestern New Mexico. Other Key Wells Also Are Listed In This Table.

Section, Township, Range	Name of well	Elevation (ft)	Top Precamb. (ft)	Total depth (ft)	Date of Completion
HIDALGO COUNTY:					
31-24S-19W	Cockrell No. 1 Pyramid Fed.	4,244KB	7,340	7,404	9-30-69
12-30S-15W	Hachita Dome No. 1 Tidball Berry Fed.	4,349DF	2,723	2,726	5-23-57
14-30S-17W	Cockrell No. 1 Playas State	4,455KB	7,030	7,086	6-11-70
25-32S-16W	Humble No. 1 State BA	4,587KB		14,585	12-24-58
10-33S-20W	ARCO No. 1 Fitzpatrick	5,192KB		10,793	4- 5-85
GRANT COUNTY:					
33-25S-15W	Marshall Young No. 1 Saltys	4,406KB		9,775	12-21-85
14-25S-16W	Cockrell No. 1 Coyote St.	4,354KB	8,580	9,282	8-24-69
LUNA COUNTY:					
7-23S-10W	Cockrell No. 1 State 1,349	4,534KB	7,240	7,375	11- 6-69
10-23S-11W	Sycor Newton No. 1 State L6350	4,556KB	7,775	10,090	10-22-74
15-23S-11W	Seville-Trident No. 1 State	4,524GL	7,930	8,240	1-21-83
4-24S-8W	Seville-Trident No. 1 McSherry	4,272KB	11,830	12,495	11- 7-81
6-24S-8W	Seville-Trident No. 2 City of Deming	4,282GL		12,385	6- 1-83
8-24S-10W	Seville-Trident No. 1 Hurt Ranch	4,407GL		7,723	3-30-83
11-26S-11W	Marshall Young No. 1 Bisbee Hills	4,296KB	6,980	7,120	9-18-83
19-28S-5W	Skelly No. 1-A State C	4,209KB	8,786	9,437	1-13-64
27-28S-5W	Sunray No. 1 Federal R	4,097KB	6,569	6,626	3-24-62
DONA ANA COUNTY:					
13-23S-4W	Exxon No. 1 Mason Draw Fed.	4,449KB	11,874?	11,948	2-19-84
11-23S-2E	Exxon No. 1 Beard Olé Fed.	4,316KB	3,935?	4,001	12-31-83
32-25S-1E	Grimm et al. No. 1 Mobil 32	4,220GL		21,759	10-12-73
4-27S-1E	Phillips No. 1 Sunland Park	4,223KB		18,229	9-25-85

rocks (and possibly Cretaceous-Tertiary redbeds) down to 7,100 ft (major unconformity), and Ordovician carbonate rocks to 7,723 ft. The No. 2 City of Deming well penetrated Cenozoic sedimentary and volcanic rocks (with some monzonite intrusives) down to the total depth of 12,385 ft. (In a preliminary study of the small-size cuttings, Cretaceous-Tertiary redbeds were determined from 11,100 to 11,660 ft and Precambrian granite from 11,660 to 12,385 ft). The No. 1 McSherry well penetrated Cenozoic sedimentary and volcanic rocks (with some monzonite intrusives) down to 11,300 ft, Cretaceous-Tertiary redbeds to 11,600 ft, Tertiary monzonite intrusive to 11,830 ft, and Precambrian gneiss and schist to 12,495 ft. The monzonite intrusive may occupy the position of a major unconformity or a fault between the redbeds and the Precambrian. A fault downthrown to the west probably is present between the No. 1 McSherry and No. 2 City of Deming wells, especially if no Precambrian was encountered in the latter.

The Marshall Young No. 1 Bisbee Hills well penetrated Cenozoic sedimentary and volcanic rocks down to 1,800 ft, Cretaceous-Tertiary redbeds to 2,060 ft, reported Lower Cretaceous sedimentary rocks to 5,035 ft (major unconformity), Silurian to Cambrian sedimentary rocks to 6,980 ft, and Precambrian quartzite and chlorite schist to 7,120 ft.

The Skelly No. 1-A State C well penetrated Cenozoic sedimentary and volcanic rocks down to 3,900 ft, Cretaceous-Tertiary redbeds to 5,130 ft, Cretaceous volcanic rocks to 8,388 ft (major unconformity), Ordovician sedimentary rocks to 8,786 ft (normal fault), and Precambrian granite to 9,437 ft. The Sunray No. 1 Federal R well penetrated Cenozoic sedimentary and volcanic rocks down to 2,886 ft (major unconformity), Silurian to Cambrian sedimentary rocks (with several Tertiary intrusives) to 6,569 ft, and Precambrian gabbro to 6,626 ft.

Doña Ana County

The Exxon No. 1 Mason Draw and No. 1 Beard Olé wells apparently penetrated Paleozoic and Precambrian rocks. Samples of drill cuttings have not been released yet. Using only the wire-line logs, the questionable tops of the Precambrian are determined as 11,874 ft and 3,935 ft, respectively.

The Grimm et al. No. 1 Mobil 32 well penetrated Tertiary-Quaternary sediments down to 1,940 ft, Tertiary volcanic rocks to 5,880 ft, Tertiary redbeds to 12,100 ft, Lower Tertiary (Paleocene-Eocene) sedimentary rocks to 13,300 ft (unconformity), Lower Cretaceous sedimentary rocks to 14,100 ft, Jurassic (Oxfordian-Callovian) sedimentary rocks to 15,550 ft (major unconformity), and Permian to Ordovician

sedimentary rocks to 21,759 ft. The redbeds in this well are definitely Tertiary, whereas to the west similar redbeds are inferred to be uppermost Cretaceous to Tertiary. The Lower Tertiary and Jurassic sedimentary units were dated by palynology (Thompson, 1982). The marine Jurassic section is the first to be documented between Mexico and Wyoming. The total depth of 21,759 ft is the deepest in the region. If it had been about 2,000 ft deeper, this well probably would have reached Precambrian basement.

The Phillips No. 1 Sunland Park well penetrated Tertiary-Quaternary sediments down to 1,530 ft, Tertiary volcanic rocks to 11,020 ft, Tertiary redbeds(?) to 15,350 ft, Lower Tertiary (Paleocene to Middle Eocene) sedimentary rocks to 18,200 ft, and Upper Cretaceous (Maastrictian) sedimentary rocks to 18,229 ft. The Upper Cretaceous-Lower Tertiary unit was dated with nannofossils by Young in Bordine and others (1986). These marine deposits are the first to be documented in the Tertiary of this region. The Lower Tertiary section in the Grimm well may also be marine. Studies are underway to determine the lateral extent of this unit. Using the pre-Tertiary section in the Grimm well as a guide, the depth of the top of Precambrian basement below this Phillips well location is estimated to be about 29,000 ft. The subsea elevation of about -24,600 ft, would be the lowest in the region.

NORTHWESTERN CHIHUAHUA

All 10 of the wells drilled by Petroleos Mexicanos (Pemex) in the map area of northwestern Chihuahua (Fig. 1) were described by Tovar in Thompson and others (1978). The Pemex No. 1 Chinos penetrated Permian to Cambrian sedimentary rocks down to 4,381 meters (14,373 ft), and Precambrian granite gneiss to 4,411 meters (14,473 ft). The Pemex No. 1 Moyotes penetrated Cenozoic rocks down to 685 meters (2,247 ft) (unconformity), Lower Cretaceous sedimentary rocks to 2,365 meters (7,759 ft), Upper Jurassic sedimentary rocks to 3,395 meters (11,138 ft) (major unconformity), Permian sedimentary rocks to 4,810 meters (15,781 ft), and Precambrian granite gneiss to 4,943 meters (16,217 ft). Precambrian rocks in these two wells were dated by the rubidium-strontium method as 1,327 Ma and 890 Ma, respectively.

Acknowledgments

We gratefully acknowledge the well operators, service companies, geologists, geophysicists, and others who have provided us with their data and interpretations. They are credited individually in other papers. We also appreciate the reviews of the manuscript by Frank E. Kottlowski and Ronald F. Broadhead at the New

Mexico Bureau of Mines and Mineral Resources. We accept full responsiblity for the contents of this paper.

REFERENCES

Arizona Oil and Gas Conservation Commission, Well location map four--State of Arizona, Phoenix, Ariz., 37 p., revised Aug. 1987.

Betton, C.W., Phillips' Tombstone State A-1 exploration well, Cochise County, Arizona, *Geologic studies of the Cordilleran thrust belt*, Rocky Mountain Association of Geologists, Symposium, p. 675, 1982.

Bordine, B.W., E.B. Robertson, and C.R. Young, Biostratigraphy and petroleum source-rock potential, Phillips Petroleum Co. No. 1 Sunland Park Unit well, Doña Ana County, New Mexico, *Open-file Report No. OF-327*, 16 p., New Mexico Bureau of Mines and Mineral Resources, Socorro, New Mex., 1986.

Clemons, R.E., Petrography and stratigraphy of Seville-Trident exploration wells near Deming, New Mexico, *New Mexico Geology*, v. 8, no. 1, p. 5-11, 1986.

Thompson, Sam III, Oil and gas exploration wells in southwestern New Mexico, *Geologic studies of the Cordilleran thrust belt*, Rocky Mountain Association of Geologists, Symposium, p. 521-536, 1982.

Thompson, Sam III, J.C. Tovar, and J.N. Conley, Oil and gas exploration wells in the Pedregosa basin, *Land of Cochise*, New Mexico Geological Society, Guidebook 29th Field Conference, p. 331-342, 1978.

Cheyenne: woodcut by Werner Drewes

PALEOZOIC ROCKS BETWEEN EL PASO AND THE NEW MEXICO-ARIZONA BORDER

F.E. Kottlowski

New Mexico Bureau of Mines and Mineral Resources, Socorro

Paleozoic rocks in southwestern New Mexico rest unconformably on Proterozoic rocks. These Proterozoic rocks are mainly granites and granite gneiss, but include locally thick sequences of metasedimentary and metaigneous rocks of about 1650 and 1000 Ma.

Early Paleozoic sedimentation began with Cambrian to early Ordovician clastic deposits of coastal to shallow-marine origin, chiefly reddish quartzose sandstones containing glauconite, hematite, and feldspar. In the New Mexico area, this unit is the Bliss Sandstone, a lithologic equivalent and roughly a time equivalent to the Coronado Sandstone and Bolsa Quartzite of southeastern Arizona. The base of the Bliss is an evenly beveled surface in most places, but in the late Cambrian and early Ordovician seas there were some hills of Proterozoic rocks, such as those in the west-central Franklin Mountains. The Bliss Sandstone is 100-300 feet thick in the New Mexico area, thinning northward onto an early Paleozoic upland (Thompson and Potter, 1981).

Conformably overling the Bliss Sandstone are shallow marine to supratidal limestones and dolostones of the Lower Ordovician El Paso Formation. Where the Bliss is absent locally, the El Paso onlaps Proterozoic basement hills. Late Cambrian fossils occur in basal Bliss sandstones of the Silver City-Black Range-Hatch-Truth or Consequences region, but in other areas the lower Bliss appears to be of early Ordovician age. In the Florida Mountains, for example, the Bliss Sandstone overlies an irregular surface eroded onto syenite and alkali-feldspar granite dated by Evans and Clemons (1987), by U-Th-Pb method on zircon, as $495-507 \pm 10$ Ma, thus the strata are on the Cambrian-Ordovician borderline.

The El Paso is truncated by pre-Montoya erosion over much of the New Mexico region, with the northern and western limits being the results of pre-Devonian erosion. The sequence is 1,400 feet thick near El Paso, thinning to about 600 feet in the Lordsburg area (Hayes, 1975). Basal beds are arenaceous glauconitic carbonate rocks, and arenaceous beds occur at several higher stratigraphic positions. Clemons (1986) subdivided the El Paso Formation, in ascending order, as Hitt Canyon Member (silty limestone), José Member (oolitic arenaceous limestone), McKelligon Member (stromatolitic limestone), and Padre Member.

The Middle and Upper Ordovician Montoya Dolomite consists of a lower Cable Canyon Sandstone Member that has rounded quartz grains and dolomitic cement, a massive dark-gray, cliff-forming Upham Dolomite Member, a cherty dark-gray Aleman Dolomite Member, and a light-gray-weathering calcic dolomite, the Cutter Dolomite Member (locally also referred to as the Valmont Dolomite Member). The thickness of the Montoya decreases northward and westward owing chiefly to erosion before deposition of Upper Devonian strata. The Montoya is about 450 feet thick near El Paso and 300-400 feet thick between Deming and Lordsburg. Westward it is thinned drastically by pre-Devonian erosion.

Overlying the Montoya Dolomite on a gently undulatory disconformity is the Lower to Middle Silurian Fusselman Dolomite, a shallow-marine shelf deposit that has some intertidal and supratidal units. It is about 1,000 feet thick near El Paso, but its northern and western limits, the results of pre-Middle Devonian erosional truncation, occur abruptly about 100 miles north of El Paso and between Deming and Lordsburg.

In southwestern and south-central New Mexico, prior to deposition of Upper Devonian rocks, the Fusselman and all older Paleozoic formations were tilted towards the south and east, with consequent truncation by widespread erosion of the northern and western limits. Then an extensive blanket of black, shallow marine shale of the Upper Devonian Percha Formation progressively overlapped older Paleozoic beds from south to north.

In many places, the contact of the Percha with overlying Mississippian units is that of a low-relief unconformity but in southwesternmost New Mexico it is a transitional contact. In southwestern New Mexico, the Mississippian, mapped as the Escabrosa Group, is dark, argillaceous, cherty limestones and white crinoidal limestones. These form shallow-marine deposits and are thickest along a northwest-southeast trend. They could give the earliest evidence of subsidence along the axis of the Pedregosa Basin, which may have occurred in southwestern New Mexico and in adjoining southeastern Arizona, extending into northern Mexico. The uppermost part of the Escabrosa is transitional into the overlying Upper Mississippian Paradise Formation, which consists of shallow-marine to coastal deposits of oolitic, skeletal, and arenaceous limestones.

In south-central New Mexico, the basal Mississippian is the Caballero Formation, which consist of olive- to tan-weathering silty calcareous shales that have some lenses of calcareous siltstone and silty limestone. The Caballero is overlain unconformably by the Lake Valley Formation that consists of alternating and intertonguing members of crinoidal bioclastic cherty limestone and of slope-forming shaly fossiliferous marls. This unit is missing in the Franklin Mountains where deep marine black calcareous mud of the Las Cruces and Rancheria Formations buried an irregular karsted and channeled surface on the Lake Valley carbonate rocks or on Devonian strata. Withdrawal of the Rancheria seas southward in latest Mississippian time is recorded by the nearshore clastic rocks and oolitic limestones of the overlying Helms Formation. Mississippian rocks near El Paso are about 500 feet thick; they thin northward owing to erosion during early Pennsylvanian time. In the Pedregosa Basin area, the Mississippian Escabrosa and Paradise Formations are as much as 1,400 feet thick.

The basic sedimentary framework changed in Pennsylvanian time to south-central and southwestern New Mexico. Most of the Pennsylvanian structural trends are north-south in south-central New Mexico. Before the Pennsylvanian strata were deposited, pre-Pennsylvanian beds were tilted to the south and their northern parts were truncated by erosion. In places, this pre-Pennsylvanian surface has relief of more than 1,000 feet. In the general area about 30 miles north of Las Cruces, widespread channels were cut into Mississippian and older strata and were filled with detritus of the eroded units, particularly of Mississippian cherts, Morrowan (Early Pennsylvanian) carbonate rocks occur in the Franklin Mountains and onlap the pre-Pennsylvanian surface to the northwest. The Morrowan and Atokan (Middle Pennsylvanian) strata consist of deltaic to shallow marine clastic units and of dark-gray limestones. In the south-central area, the Desmoinesian-Missourian (Middle and Late Pennsylvanian) rocks are mainly shallow-marine limestones in which are local lenses of arkosic sandstone and shale derived from the early stages of the Pedernal Upland to the northeast of El Paso and east of Alamogordo (Pray, 1961). In the southwestern area, shallow-marine limestones covered most of the area, but along part of the axis of the southeastward-trending Pedregosa Basin, subsidence exceeded sedimentation and deep-marine limestones, shales, and sandstones were deposited; these basinal deposits are fringed to the northeast by the carbonate-rock reefs. Similar deep-marine sedimentation continued in the Pedregosa Basin area during Missourian, Virgilian (Late Pennsylvanian), and Early Wolfcampian (Early Permian) times.

Thick, porous dolostone units are exposed in the Big Hatchet Mountains along the northeast side of the basin (Thompson and Jacka, 1981). Transition of shelf-to-basin rocks is shown in outcrops of the central Big Hatchet Mountains (Zeller, 1965). Deep-marine deposits include carbonate turbidites and debris flows interstratified with basinal mudstones and limestones.

An ancient submarine canyon along the shelf margin indicates that proximal parts of the basin may contain significant quantities of sandstone in submarine-fan deposits. These Middle and Upper Pennsylvanian basinal facies extend southeastward into Chihuahua, as shown by the deep-marine mudstones and limestones in that area.

The entire carbonate rock sequence of Pennsylvanian and Early Wolfcampian age in the Big Hatchet Mountains (Zeller, 1965) and adjoining areas is known as the Horquilla Formation after its occurrence in southeastern Arizona. In south-central New Mexico, the area from Deming to Las Cruces and El Paso and northward, the Pennsylvanian (Kottlowski, 1960) is mainly combined into the Magdalena Formation or Group. Where known as group rank, it is subdivided into numerous formations in various mountain ranges. In shelf areas, such as near Las Cruces, the Pennsylvanian is only 500-800 feet thick, but the sequence is 2,000 to more than 3,000 feet thick in the Pedregosa and Orogrande Basins.

The Orogrande Basin had its beginnings in Mississipian time, although it is mainly outlined by thick deposits of Virgilian age; it extended at least 110 miles north from the latitude of El Paso and averaged 35 miles in width in the areas of the present day Hueco and Tularosa Basins. It was bounded on the east by the Pedernal Uplift. This north-south-trending upland ran from about the New Mexico-Texas border northward to and beyond the present day Pedernal Hills which are in central New Mexico, 50 miles southeast of Albuquerque. At its maximum extent, it was probably about 70 miles wide, east to west.

In Virgilian time, and locally in late Missourian time, the north-south complimentary downwarping west of this Pedernal Uplift, produced the main development of the Orogrande Basin. Thick, cyclic, mostly clastic deposits of the Panther Seep Formation accumulated in the basin, the infilling processes approximately keeping pace with subsidence (Wilson, 1967; Kottlowski and others, 1956). Intertidal and supratidal conditions were common, as indicated by mud cracks, ripple marks, stromatolites, local occurrences of caliche and gypsum beds, and dolomitized limestone. Fluctuating sea level seems to explain the cyclic repetition of rock types; the coarser clastics were formed during the periods of low water, and the fine-grained clastics or laminated carbonate mud were deposited during higher strands of sea level. Maximum thickness of these Virgilian basinal sediments is about 2,500 feet.

In early Permian time, subsidence of the Orogrande Basin continued, resulting in a thick lower Wolfcampian sequence; this sequence is somewhat similar to the lower Wolfcampian carbonate phase of the Horquilla Formation in the Pedregosa Basin area. However, in south-central New Mexico, depositional conditions changed as marine limestones and shales dominate the section with cyclic influxes of terrigenous clastics. In mid-Wolfcampian time, the Pedernal Uplift rose again, and southeast of Alamogordo, tight folds and vertical faults developed on its western flank. After widespread erosion, nonmarine redbed clastics of the Abo Formation were spread over the northwest part of the Basin area. To the southeast, in the general region of Las Cruces, these red beds intertongued with shallow marine limestones of the Hueco Formation. This Abo-Hueco transition can be seen in the Sacramento Mountains (Pray, 1961) and San Andres Mountains, taking place in an east-west belt that is 15-30 miles wide north-south. Bioherms occur locally near the shorelines. On the east side of the Orogrande Basin (towards the Pedernal Uplift), the shaly carbonate-rock Hueco facies grades northeastward within a few miles into a red-bed facies. In southwestern New Mexico, the Horquilla Formation is unconformably overlain by the Earp Formation of Upper Wolfcampian and Leonardian age. It consists of a red-bed facies in the Big Hatchet Mountains area and of a limestone facies to the southeast (Greenwood and others, 1977), similar to the Abo-Hueco transition in the Orogrande Basin area. The Earp is overlain conformably by the Colina-Epitaph Leonardian units, a shallow-marine limestone-dolostone transition that, in the upper beds, contains some gypsum and clastic deposits. Overlying a low-relief disconformity on top of the Epitaph are the Scherrer Sandstone and the overlying Concha carbonate unit of Leonardian and Guadalupian (Late Permian) age.

During Leonardian time, marine waters first covered south-central New Mexico with interbedded dolomitic limestones, gypsum beds, sandstones, and siltstones of the Yeso Formation, followed by the shallow marine

seas in which the San Andres Formation carbonate rocks were deposited (Kottlowski, 1963). These units are roughly equivalent to the Colina-Epitaph and Scherrer-Concha units of southwestern New Mexico.

Younger Guadalupian and Ochoan (Late Permian) rocks do not occur in southwestern and south-central New Mexico. However, Jurassic rocks are reported in a drill hole southwest of Las Cruces. The region apparently was uplifted and eroded until Early Cretaceous time when shallow-marine seas from the south spread Lower Cretaceous sandstones and overlying limestone units.

In the Franklin Mountains area, the Hueco Formation, with a few thin intertonguing red shales of Abo lithology, are about 2,200 feet thick. Wolfcampian beds of the upper Horquilla Limestone, Earp Formation, and Colina limestone in the Big Hatchet Mountains area are about 2,500 feet thick. In many places in south-central New Mexico, the Yeso and San Andres Formations have been removed by erosion prior to deposition of Cretaceous or younger rocks. But in the southern San Andres Mountains, they are about 600 feet thick and thicken northward to more than 1,200 feet in the northern San Andres Mountains (Kottlowski and others, 1956).

Along a northwest-trending axis that passes through the Deming area there were small uplifts of low relief referred to as the Florida Uplift, or to the southeast of Deming, the Moyotes Uplift. The Florida Uplift began in Late Pennsylvanian time and was buried by the Hueco limestone of Wolfcampian age with the Moyotes Uplift of similar age buried by Abo-Hueco intertongued units. These positive areas were probably connected and formed the northeast boundary of the Pedregosa Basin in Wolfcampian time. Eastward was a broad shelf bordering the west edge of the Orogrande Basin.

REFERENCES

Clemons, R.E., 1986, in Clemons, R.E., and Osburn, G.R., Geology of the Truth or Consequences area: New Mexico Geological Society Guidebook to Truth or Consequences region, p. 69-81.

Evans, K.V., and Clemons, R.E., 1987, U-Pb geochronology of the Florida Mountains, New Mexico--New evidence for Latest Cambrian-Earliest Ordovician alkalic plutonism (abstract): Geological Society of America, Abstract with Programs, 19, 7, 657-658.

Greenwood, E., Kottlowski, F.E., and Thompson, Sam III, 1977, Petroleum potential and stratigraphy of Pedregosa Basin--Comparison with Permian and Orogrande Basins: American Association of Petroleum Geologists Bulletin 61, p. 1448-1469.

Hayes, P.T., 1975, Cambrian and Ordovician rocks of southern Arizona and New Mexico and westernmost Texas: U.S. Geological Survey Professional Paper 873.

Kottlowski, F.E., 1960, Summary of Pennsylvanian sections in

southwestern New Mexico and southeastern Arizona: New Mexico Bureau of Mines and Mineral Resources Bulletin 66.
_____1963, Paleozoic and Mesozoic strata of southwestern and south-central New Mexico: New Mexico Bureau of Mines and Mineral Resources Bulletin 79.

Kottlowski, F.E., Flower, R.H., Thompson, M.L., and Foster, R.W., 1956, Stratigraphic studies of the San Andres Mountains: New Mexico Bureau of Mines and Mineral Resources Memoir 1.

Pray, L.C., 1961, Geology of the Sacramento Mountains escarpment, Otero County, New Mexico: New Mexico Bureau of Mines and Mineral Resources Bulletin 35.

Thompson, Sam III, and Jacka, A.D., 1981, Pennsylvanian stratigraphy, petrography, and petroleum geology of Big Hatchet Peak section, Hidalgo County, New Mexico: New Mexico Bureau of Mines and Mineral Resources Circular 176.

Thompson, Sam III, and Potter, P.E., 1981, Paleocurrents of

Bliss Sandstone, south-western New Mexico and western Texas: *New Mexico Bureau of Mines and Mineral Resources Annual Report*, 1979-1980, p. 36-51.

Wilson, J.L., 1967, Cyclic and reciprocal sedimentation in Virgilian strata of southern New Mexico: *Geological Society of America Bulletin 78*, p. 805-817.

Zeller, R.A., 1965, Stratigraphy of the Big Hatchet Mountains, New Mexico: *New Mexico Bureau of Mines and Mineral Resources Memoir 16.*

Courtesy of Arizona Pioneer's Historical Society

THE GEOLOGY OF SOUTHWESTERN NORTH AMERICAN PORPHYRY BASE METAL DEPOSITS

J. M. Guilbert
University of Arizona, Tucson, Arizona

INTRODUCTION

Any modern discussion of 'porphyry base metal deposits',or 'PBMDs', almost immediately involves a retinue of related deposit sub-types such as skarns, Cordilleran veins, and epithermal precious metals, so continuous and far-reaching are the bloodlines of this ore-forming environment. But the heart of these related deposits is the porphyry base metal deposit itself. 'Porphyries' in general were talked about in the 20s and 30s, when Parsons (1933) first attempted their definition, but it remained until the period 1965-1975 that they were collectively considered and clarified in 5 major publications. They were[1]: Titley and Hicks' (1966) compendium volume, the Graton-Sales volumes of AIME (Ridge,1968), Lowell and Guilbert (1970), Sillitoe (1973), and Gustafson and Hunt (1975), the latter two with momentum from the second Penrose Conference on porphyries in 1968. From these and many other contributors, most notably another edited volume by Titley (1982) and his entry with Richard E. Beane in the 75th Anniversary Volume of the Society of Economic Geologists, has come a clearer picture of the dynamics of porphyry ore deposit formation and their tectonic settings. Research by Meyer and Hemley (1967) on alteration processes, following upon the senior author's career studies at Butte, Montana, and stable isotope studies by Hugh P. Taylor, Jr. and his students were also crucial to advances of the period.

It should be noted that PBMDs do indeed carry the retinue of associated deposits listed above, externally and as functions of the nature of the wallrocks (skarns, limestone replacement deposits) and depth of exposure (Cordilleran veins, epithermal deposits). PBMDs as a group are generally taken to include porphyry copper deposits (PCDs) and porphyry molybdenum deposits ('porphyry molies'or PMoDs), the two being Cu- and Mo-rich endmembers of a calc-alkalic-pluton-affiliated series and quite distinct from Climax-type molybdenum occurrences. The important Climax-type ores can be reviewed in White, et al.(1981). PBMDs also include porphyry tin deposits such as that at Llallagua, Bolivia, but the terms "porphyry gold" and "porphyry uranium", etc., have not been accepted by those who know 'porphyries' best. Discussion is restricted here to the PCD-PMoD series as it occurs in the southwestern

United States and northern Mexico. General relationships and a more detailed modern summary of PBMDs can be found in Guilbert and Park (1986). The International Geological Congress field trip group will visit the Sierrita porphyry copper-moly deposit.

This overview comprises the above introduction followed by sections on the definition of porphyry copper-molies, alteration-mineralization zoning and dynamics, lithotectonic settings, and finally a brief description of Sierrita.

DEFINITION

Porphyry copper deposits are large, low- to medium-grade deposits primarily of copper as chalcopyrite and molybdenum as molybdenite in which hypogene sulfide and silicate zones span inner potassic to outer propylitic alkali metasomatic alteration assemblages, typically overprinted by phyllic and argillic hydrolitic alteration, and that are temporally and spatially related to epizonal calc-alkalic porphyritic intrusives. Co-product--by-product silver, gold, zinc, lead, and manganese are common in porphyry coppers, as is rhenium in some porphyry molies. The descriptor 'large' casts these deposits into the billion (109) tonne range; several, for example Bingham Canyon in Utah, Morenci and San Manuel-Kalamazoo in Arizona, and Cananea in Sonora, probably contain 2 to 3 billion tonnes each, although the average deposit size is more at 250 million tonnes. In outcrop area, the original surface expressions averaged 1 to 2 km^2, although the open pit mines are generally more than twice that size because of outward-dipping contacts, zones that expand laterally with depth, and the need for inward-sloping outer mine walls.

The PCDs represent mining engineering, mine geology, and economics in 'high gear'. Many mines produce more than 100,000 tonnes of ore a day, with equal or greater amounts of discarded waste rock moved daily. A vital geologic and economic aspect of each 'porphyry copper' deposit that is not included in the definition revolves around whether or not supergene enrichment has occurred. If it has not, the ores can be seen to be invariably epigenetic, later in textural age than the altered rocks that contain them, and they are said to be 'hypogene'. One's location in the lateral and vertical zoning determines the appearance of the mineralization, as will be discussed later with figure 1. If supergene enrichment has occurred, leached outcrops

---

[1] Copies will be available for field trip participant use.

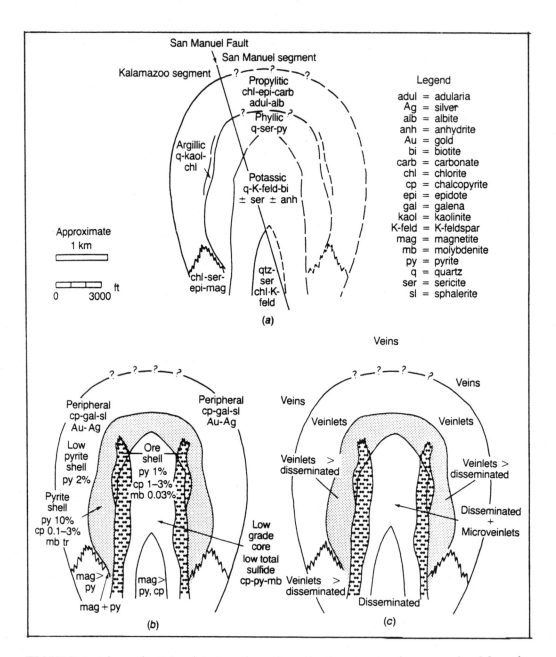

FIGURE 1 Schematic of coaxial alteration-mineralization-structural zones at San Manuel-Kalamazoo, Arizona. (a) Alteration zones. Broken lines indicate uncertain continuity or location of contacts. (b) Mineralization zones. (c) Structural zoning and sulfide occurrence modes. From Guilbert and Park (1986).

show textures and mineralogies characteristic of weathered sulfides, subjacent lithologies are leached and limonitic, and deeper hypogene sulfides that have been protected by depth and a stable water table are coated and replaced along fractures with thin films of chalcocite-djurleite, digenite, and covellite and enriched many fold in copper values. Such 'supergene blankets' are generally a few tens of meters beneath the modern surface and subparallel to it. Figures 17-13, -14, and-15 (p. 799-801) and the back cover of Guilbert and Park (1986) depict these relations; little more of supergene effects will be described.

Part of the definition refers to the association of PCDs to epizonal calc-alkalic rocks. Many PCDs appear to have formed virtually in the roots of stratovolcanoes, and all of the occurrences that would be considered 'porphyries' from other aspects of the definition also prove to be intimately associated with dikes, stocks, or plutons of porphyritic intermediate to felsic units, typically rhyodacite, latite, and rhyolite. It is certain that the same P-T environments spawn both the porphyritic rock textures and the separation-circulation of fluids that generate the alteration-mineralization.

ALTERATION-MINERALIZATION

Most of the geologic aspects of PCDs prove to be parts of a vertical and horizontally coaxial zoning pattern. The major aspects of that zoning were perceived and described in the 1960s and early 1970s, but it remained for Gustafson and Hunt (1975) to account for many of the aspects in a time-staged dynamic model involving early deuteric hypogene manifestations with later convection-related superimposed, or 'overprinted', effects. Figure 1 shows three vertical cross-sections of San Manuel, here taken to be representative of a porphyry system. The three parts overlain constitute the composite deposit, but the fact that the quartz-sericite-pyrite phyllic zone is overprinted across the potassic-propylitic boundary zone is not obvious from it. Figure 1 does show, however, that there is an outward progression from a potassic low-grade core across generally transitional contacts through a high-chalcopyrite potassically altered ore zone marked by disseminations and microveinlet fillings of sulfides, across the mineralized phyllic zone that is marked by more abundant and coarser veinlets and microveinlets and by a vastly higher pyrite content, still with enough chalcopyrite to constitute ore grades. In detail, this material is texturally younger than the potassic alteration on its inward flank and the propylitic alteration on its outer side. Other aspects of the schematic representation are clear from the figure. It has been discerned at many deposits that there exists an upper zonal element called the lithocap (Norton and Knight, 1977) that represents Advanced Argillic relatively acidic, low alkali, hot-spring alteration with sulfides and arsenides dominant. All of these alteration assemblage arrays depend to some extent on the presence of calc-alkalic intruded host rocks; nonreactive hosts, or for example a Paleozoic carbonate-rich section, create perturbations that can be interpreted by the 'porphyry' geologist but that need not be covered here.

The physical appearance of the ore zone is worth describing, as it may be seen in many districts in which hypogene assemblages have survived supergene overprint. Quite commonly, as at Sierrita, one finds disseminations and hair-line-wide microveinlets of minute grains of chalcopyrite-pyrite stippling or cutting potassic alteration assemblage minerals such as quartz, orthoclase, and biotite, commonly with minor anhydrite or apatite. Primary K-feldspar and biotite are commonly recrystallized to new deuterically controlled compositions. Volume percent chalcopyrite rarely exceeds 1, which would produce about a 0.7 weight percent copper material, nearly the average grade at many hypogene deposits, although the equivalent figure at Sierrita is only 0.30% Cu, with molybdenum at 0.03%. The highest hypogene copper values usually occur in an inverted-cup-shaped annular zone that surrounds a barren central core zone. The core is also potassically altered but it contains low total sulfide content, represented by the coaxial walls of figure 1, with the closure across the 'roof' attested to at several major districts. Ore in the phyllic zone is commonly far less obviously mineralized—a few percent chalcopyrite and minor bornite can be less conspicuous in the coarser, pyrite-choked stockworks of veins and veinlets a few millimeters wide on decimeter centers that characterize the phyllic assemblage. Once again, the reader is referred to pages 406-426 in Guilbert and Park (1986) for greater detail.

It can be said here that professional-level understanding of the interplay between lateral and vertical aspects of the geology, geochemistry, mineralogy, and structural geology of the PCDs, coupled with recognition of the perturbations induced by system size, wallrock variation, and timing of many aspects, permits confident interpretation of each occurrence. Particular interest has been shown lately in high-level and lateral concentrations of gold in skarns, in suprajacent lithocaps and epithermal settings, and in the more alkalic (shoshonitic) porphyry systems themselves.

LITHOTECTONICS

It was noted above that PCD-PMoDs are associated with calc-alkalic intrusive-extrusive systems, both in time and space. The occurrence of porphyry copper systems through time closely parallels the development of major belts of calc-alkaline rocks along continental margins. There are few Precambrian age porphyries, the Haib district in Namibia in 1.8 Ga andesites of the Richtersveld series being the oldest well-developed porphyry deposit. More appeared in the Paleozoic, and in the Mesozoic and Cenozoic they multiplied almost proportionally to the production of continental-scale calc-alkaline belts. They are forming today along subductive consuming margins around the rim of the Pacific Basin, so it is compelling to assert that porphyry-related deposits do indeed form above subduction zones and in magmatic arcs, although specific links to melting of the asthenosphere, the subducted slab, the upper mantle, or the lower continental crust have not yet been forged. Samarium-neodymium isotope studies are starting to reveal complex patterns requiring partial melting of continental crustal rocks in several porphyry systems in the western and southwestern US, including Sierrita (Anthony and Titley, 1988), and more studies are continuing. These problems are significant not only petrogenetically but also in that they may more meaningfully constrain the sources of metals in the related ore deposits.

As described above, virtually all porphyry deposits occur in recently or currently active subduction areas

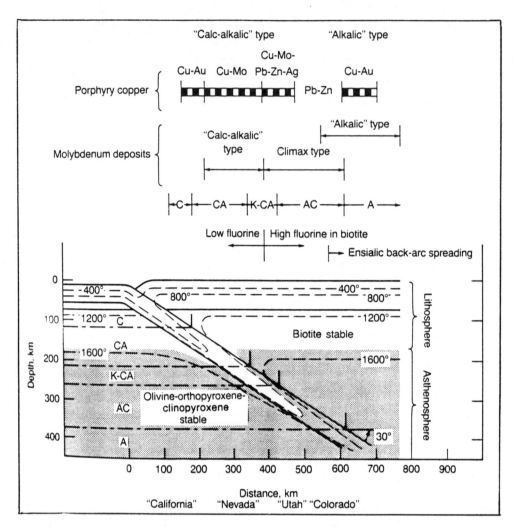

FIGURE 2 A schematic section of the interaction of subduction, igneous rock major-element chemistry, and PBMDs proposed by Westra and Keith (1981) as applicable to the western United States and, by inference, elsewhere. Lithotectonic settings can be inferred. From Guilbert and Park (1986).

with ages ranging from 200 My in British Columbia to 1.5 at Panguna, Bougainville, and 4.5 My at El Teniente in Chile. Jurassic occurrences are known in Arizona, especially the Bisbee district, but the majority of southwestern North American deposits are of Laramide age, from 75 to 54 My. Many are clearly related to major global structures such as medial transcurrent faults in the Phillipines and the remarkable West Fissure (Falla Oeste) in Chile. Many more are relatable to less-clearcut, at least district-scale faults, as at Bagdad and Silver Bell, Arizona, that may have regional significance. Many porphyry plutons show jointing strain patterns that are reactivations of existing or inherited stress directions in their containing rocks. But it remains difficult to assert that the economic metals have come preponderantly from the mantle or a slab associated with subduction, from the continental crust by partial melting and assimilation above a subduction system, by lateral secretion-mobilization of metals by fluids

convecting around a cooling pluton, or from some other source. Fluid inclusion, geochemical, and isotopic evidence is mounting that hypersaline, high-temperature, hydrothermal deuteric-magmatic plumes are normal early manifestations that inject sulfur and metals into the proto-porphyry environment.

Another line of evidence is the specificity of time-constrained, major-element covariable ore-host rock settings. Briefly, Keith and his colleagues have assembled whole-rock analyses of thousands of fresh, ore-related, radiometrically dated rocks and compiled trace and economic metal ratios and profiles of them as functions of calc-alkalinity of various types and oxidation-reduction character. When this data was related to subduction geometries as developed for the southwestern US in recent years, a pattern of porphyry metalization related to calc-alkaline systematics emerged (fig. 2). When this and the radiometric data was further mounted for the southwestern United States

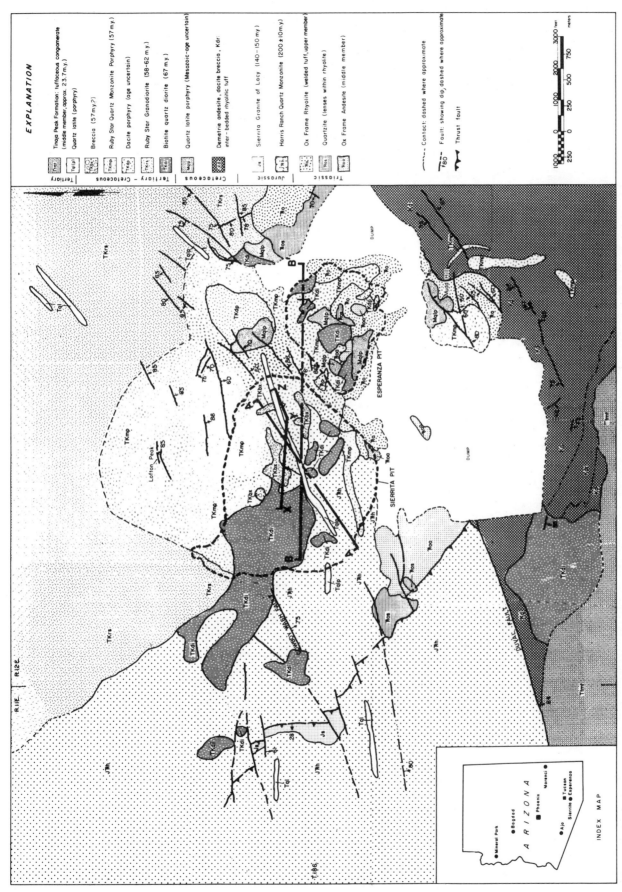

FIGURE 3 A generalized geologic map of the Sierrita-Esperanza mine area. The Ruby Star Granodiorite, generally in the northeast portion, and the Ruby Star quartz monzonite porphyry are thought to be the progenitors of the PCD-PMoD system. From West and Aiken (1982).

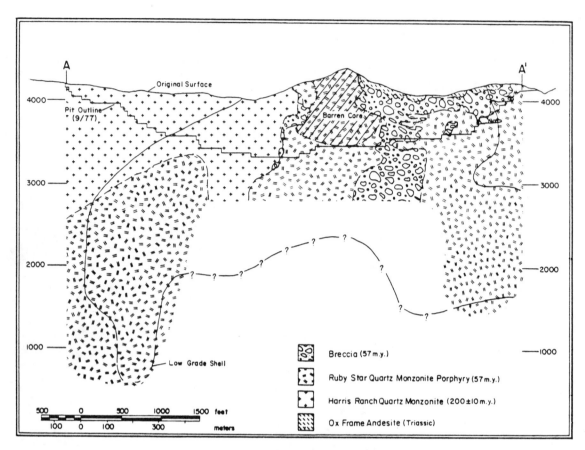

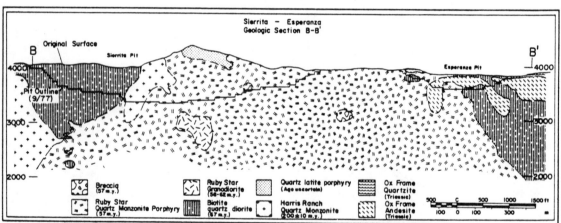

FIGURE 4 Cross-sections A-A' northeasterly through the Sierrita pit and B-B' east-west through both the Sierrita and Esperanza mines as shown on figure 3. Also refer to figure 1. From West and Aiken (1982).

and northern Mexico, several time-constrained lithotectonic PCD-related belts were discernible. Keith (1986) subdivided four assemblages, the Wilderness, Morenci, Tombstone, and Hillsboro units, each succeeding the other in time-progressive pulses at a given location. PCDs occur in the earliest Hillsboro metaluminous alkalic suite and in the 70-54 My Morenci silicic metaluminous calc-alkalic one, with initial

strontium ratios of 0.7065-0.7095. The Morenci deposit, San Manuel, Cananea, and Sierrita belong in this subset. The evidence is still preliminary and in part proprietary, but subduction-plate tectonic-lithotectonic control of PCD-PMoDs and related deposits is indicated by Keith's data, which also ascribes to them a specific lithotectonic setting.

SIERRITA-ESPERANZA MINE

This major PCD, now owned and operated by Cyprus Metals Company, has suffered a split identity, having been opened in 1959 with operations in the eastern, more supergene-enriched portion called the Esperanza deposit. The Sierrita portion, larger but lower in grade than the Esperanza, was opened in 1970. The Sierrita body is now the mainstay of the district and is one of the lowest-grade PCD-PMoDs being mined in the world.

The principal reference to Sierrita is that of West and Aiken (1982), from which figures 3 and 4 have been reproduced (see footnote, page 1). The general geology, as seen in both the district geologic map (fig. 3) and the mine cross-sections (fig. 4), is complex. The oldest rocks in the area are Triassic sediments, volcaniclastics, flows, and tuffs that were intruded by the Jurassic Harris Ranch quartz monzonite, one of the major host rocks of the deposit, as part of the southwest-facing Mesozoic magmatic arc that characterized the continental margin. A second phase of consuming-margin tectonics in the Laramide brought extrusion of late Cretaceous Demetrie Volcanics and then a 10-million-year Laramide series of biotite quartz diorite plugs (68 My), the Ruby Star Granodiorite batholith (63 My), the Ruby Star Quartz Monzonite Porphyry (the granite porphyry of Anthony and Titley (1988)) at 58 My, and finally a series of breccia pipes and quartz latite dikes (58 My). Hypogene mineralization is zoned, with a low-grade core of <0.2% Cu roughly 200 by 300 meters that lies within a crudely annular ore zone of pyrite-chalcopyrite-molybdenite, with lesser pyrrhotite and trace galena, sphalerite, and magnetite. The system occurs within three controlling regional ENE, WNW, and NNW trends, all of which have local significance keynoted by the NNW conduit of the several episodes of magmatism.

Space does not permit description of the development of alteration-mineralization zones in space and time. In general, 'overprinted' phyllic alteration is best developed at the Esperanza side, but sericitization overall is less well developed at these mines than at many. West and Aiken describe 'Potassic-Early Hydrothermal' with quartz-orthoclase veins and flooding, with the sulfides listed above, minor sericite, chlorite, epidote, and albite, and common anhydrite followed by 'Phyllic-Middle Hydrothermal' dominated by veinlet- and vein-controlled and pervasive sericitization still accompanied by quartz, chalcopyrite, pyrite and especially quartz-molybdenite veinlets. These were followed by later propylitic and Late Hydrothermal quartz-sphalerite-galena-gypsum veins. These alteration-mineralization events were neatly tracked through space and time by Haynes and Titley (1980), who showed nearly smooth decreases from the heart of the ore body outward of earliest quartz-K-feldspar veins (340-360°), most-abundant Middle Hydrothermal veins (250-320°), and least abundant later K-feldspar-epidote veins (170-230°).

Visitors to the mine may see any of the assemblages described above, most probably the Early-Potassic Hydrothermal ore-grade material. Epidote, presumably reflecting higher than normal calcium and ferric iron ion activities, is relatively common, and light purple anhydrite may be seen. Chalcopyrite to pyrite ratio is about 2. Further details will be provided by the mine staff.

Ongoing research at Sierrita and Esperanza includes the isotopic work of Anthony and Titley and studies of the Ruby Star batholith and its structural setting with regard to possible tilted caldera models developed within the U.S. Geological Survey by C. Fridrich. Other petrographic and structural studies by the U.S. Geological Survey, Cyprus Metals Company geologists, and researchers in other institutions continue to clarify the nature of the Jurassic and Laramide magmatic arc systems and the geology of PBMDs in general and Sierrita-Esperanza in particular.

REFERENCES

Anthony, E.Y. and S. R. Titley, 1988, Progressive mixing of isotopic reservoirs during magma genesis at the Sierrita porphyry copper deposit, Arizona: inverse solutions. *Geochimica et Cosmochim Acta*. v. 52, p. 2235-2249.

Guilbert, J.M. and Charles F. Park, 1986, *The Geology of Ore Deposits*. New York: W.H. Freeman, 985 pp.

Gustafson, L.B. and J. P. Hunt, 1975. The porphyry copper deposit at El Salvador, Chile. *Economic Geology*. v. 70, p. 857-912.

Keith, S.B., 1986, Petrochemical variations in Laramide magmatism and their relationship to Laramide tectonic and metallogenic evolution in Arizona and adjacent regions, p. 89-101 in Beatty, B. and P.A.K. Wilkinson, Eds., *Frontiers in Geology and Ore Deposits of Arizona and the Southwest*. Arizona Geological Digest v. 16, 554 pp.

Lowell, J.D. and J.M. Guilbert, 1970, Lateral and vertical alteration-mineralization zoning in porphyry ore deposits. *Economic Geology*. v. 65, p. 373-408.

Meyer, C., and J.J. Hemley, 1967, Wall rock alteration, pp. 166-235 in H.L. Barnes, Ed., *Geochemistry of Hydrothermal Ore Deposits*. New York: Holt Rinehart and Winston, 670 pp.

Norton, D. and J.E. Knight, 1977, Transport phenomena in hydrothermal systems: cooling plutons. *American Journal of Science*. v. 277, p. 581.

Parsons, A.B., 1933, *The Porphyry Coppers*. New York: American Institute of Mining Engineers, 581 pp.

Ridge, J.D., Ed., 1968, *Ore Deposits of the United States, 1933/1967*, Graton-Sales Vols. New York: American Institute of Mining Engineers, 1880 pp.

Sillitoe, R.H., 1973, The tops and bottoms of porphyry copper deposits. *Economic Geology.* v. 68 p. 799-815.

Titley, S.R. and R. E. Beane, 1981, Porphyry copper deposits. *Economic Geology 75th Anniversary Volume.* 964 pp.

Titley, S.R. Ed., 1982, *Advances in Geology of the Porphyry Copper Deposits, Southwestern North America.* Tucson: University of Arizona Press, 560 pp.

Titley, S.R. and C. Hicks, Eds., 1966, *Geology of the Porphyry Copper Deposits, Southwestern North America.* Tucson: University of Arizona Press, 287 pp.

West, R.J. and D.M. Aiken, 1982, Geology of the Sierrita-Esperanza deposit. pp. 433-465 in Titley, S.R. Ed., *Advances in Geology of the Porphyry Copper Deposits, Southwestern North America.* Tucson: University of Arizona Press, 560 pp.

Westra, G. and S.B. Keith, 1981, Classification and genesis of stockwork molybdenum deposits. *Economic Geology.* v. 76 p. 844-873.

White, W.H., A.A. Bookstrom, R.J. Kamilli, M.W. Ganster, R.P. Smith;, D.E. Ranta, and R.C. Steininger, 1981, Character and origin of Climax-type molybdenum deposits. *Economic Geology.* v. 72 p. 686-690.

ISBN 0-87590-676-1